Main groups

							18 8A
							2 **He** 4.00260
		13 3A	14 4A	15 5A	16 6A	17 7A	
		5 **B** 10.81	6 **C** 12.011	7 **N** 14.0067	8 **O** 15.9994	9 **F** 18.998403	10 **Ne** 20.1797

10	11 1B	12 2B	13 **Al** 26.98154	14 **Si** 28.0855	15 **P** 30.97376	16 **S** 32.066	17 **Cl** 35.453	18 **Ar** 39.948
28 **Ni** 58.69	29 **Cu** 63.546	30 **Zn** 65.39	31 **Ga** 69.72	32 **Ge** 72.61	33 **As** 74.9216	34 **Se** 78.96	35 **Br** 79.904	36 **Kr** 83.80
46 **Pd** 106.42	47 **Ag** 107.8682	48 **Cd** 112.41	49 **In** 114.82	50 **Sn** 118.710	51 **Sb** 121.757	52 **Te** 127.60	53 **I** 126.9045	54 **Xe** 131.29
78 **Pt** 195.08	79 **Au** 196.9665	80 **Hg** 200.59	81 **Tl** 204.383	82 **Pb** 207.2	83 **Bi** 208.9804	84 **Po** (209)	85 **At** (210)	86 **Rn** (222)
110 (269)	111 (272)	112 (277)						

63 **Eu** 151.96	64 **Gd** 157.25	65 **Tb** 158.9254	66 **Dy** 162.50	67 **Ho** 164.9304	68 **Er** 167.26	69 **Tm** 168.9342	70 **Yb** 173.04	71 **Lu** 174.967
95 **Am** (243)	96 **Cm** (247)	97 **Bk** (247)	98 **Cf** (251)	99 **Es** (252)	100 **Fm** (257)	101 **Md** (258)	102 **No** (259)	103 **Lr** (260)

Principles and Applications of Geochemistry

PRINCIPLES AND APPLICATIONS OF
GEOCHEMISTRY

A Comprehensive Textbook for Geology Students

Second Edition

GUNTER FAURE
THE OHIO STATE UNIVERSITY

Prentice Hall
Upper Saddle River, New Jersey 07458

Library of Congress Cataloging-in-Publication Data

Faure, Gunter.
  Principles and applications of geochemistry : a comprehensive
  textbook for geology students/Gunter Faure.—2nd ed.
      p.   cm.
  Rev. ed. of: Principles and applications of inorganic
  geochemistry. c1991.
  Includes bibliographical references and index.
  ISBN 0-02-336450-5
  1. Geochemistry   I. Faure, Gunter. Principles and applications
of inorganic geochemistry. II. Title.
QE515.F28 1998   97-44563
551.9—dc21   CIP

Executive Editor: Robert A. McConnin
Total Concept Coordinator: Kimberly P. Karpovich
Art Director: Jayne Conte
Cover Designer: Karen Salzbach
Manufacturing Manager: Trudy Pisciotti
Production Supervision/Composition: WestWords, Inc.

©1998, 1991 by Prentice-Hall, Inc.
Upper Saddle River, New Jersey 07458

Printed in the United States of America

ISBN 0-02-336450-5

Prentice-Hall International (UK) Limited, *London*
Prentice-Hall of Australia Pty. Limited, *Sydney*
Prentice-Hall of Canada, Inc., *Toronto*
Prentice-Hall Hispanoamericana, S. A., *Mexico*
Prentice-Hall of India Private Limited, *New Delhi*
Prentice-Hall of Japan, Inc., *Tokyo*
Prentice-Hall Asia Pte. Ltd., *Singapore*
Editora Prentice-Hall do Brasil, Ltda., *Rio de Janeiro*

For Terri

Contents

vii

Part V

APPLICATIONS OF GEOCHEMISTRY TO THE SOLUTION OF GLOBAL PROBLEMS

19
Consequences of Chemical Weathering   342

20
Chemical Composition of Surface Water   368

21
Chemical Weathering of Mineral Deposits   400

22
Geochemical Cycles   425

Preface

This textbook is intended as an *introduction to geochemistry* for geology students in their senior year or in their first year of graduate work. At that time in their education, students are ready to *relearn* those principles of chemistry that are especially applicable to the study of geological processes. Geochemistry can enhance their understanding of these processes and can teach them to apply principles of chemistry to the solution of geological problems. Geologists in virtually all branches of the science can benefit from an introductory course in geochemistry because it can help them to make quantitative predictions about the outcome of chemical reactions that occur in the context of many geological processes.

The subject matter of this book is presented in five parts. Part I: Planet Earth in the Solar System (Chapters 1 through 4) presents the big picture starting with the Big Bang, stellar evolution, nucleosynthesis, the solar system, and the geochemistry of the Earth. This part also contains a brief history of geochemistry leading up to a statement of its goals and stating the theme of this book: Studies of geochemistry convert idle speculation into understanding.

The second part, Principles of Inorganic Geochemistry (Chapters 5 through 8), starts with the electronic structure of atoms and demonstrates how wave mechanics and the Aufbau principle lead directly to the periodicity of the chemical properties of the elements. Once the periodic table has been constructed, systematic variations in bonding, ionic radii, and the structure of crystals become apparent, thus preparing the way for a discussion of ionic substitution based on the rules of Goldschmidt and Ringwood. Part II concludes with discussions of coupled substitution in feldspar, distribution coefficients, and the geochemical classification of the elements.

Part III is entitled Aqueous Geochemistry and the Stability of Minerals (Chapters 9 through 15).

These chapters explain the concept of chemical equilibrium and use the Law of Mass Action to study the dissociation of weak acids and bases and to calculate the solubility of amorphous silica as a function of the pH. Next, the dissociation of salts, the hydrolysis of their ions, and the necessity of using the activities of ions rather than their molal concentrations in Mass Action problems are presented. All of these concepts come into play in the carbonate equilibria involving calcite, its ions in aqueous solution, and CO_2 in the atmosphere. However, the incongruent dissolution of microcline to form koalinite plus ions leads to an impasse because the value of the equilibrium constant for that reaction has not been measured.

Therefore, Chapter 11 contains a guided tour of the principles of thermodynamics leading up to a derivation of the Law of Mass Action for reacting mixtures of ideal gases at equilibrium. In addition, this chapter demonstrates that equilibrium constants at the standard temperature and pressure can be calculated from the Gibbs free energy change of chemical reactions. This skill is immediately used to construct ion–activity diagrams for aluminosilicate minerals including zeolites (Chapter 12) and clay minerals (Chapter 13).

Finally, oxidation–reduction reactions are introduced in Chapter 14, and the previously acquired knowledge of thermodynamics is used to explain some basic principles of electrochemistry. These insights are then used to define the stability limits of water on the Earth and to construct Eh–pH diagrams for the oxides of iron and their ions. Chapter 15 deals with the kinetics of chemical reactions, and with the transport of ions by diffusion and advection. These concepts are used to model the growth of concretions during diagenesis and the formation of monomineralic layers at the sediment–water interface.

Part IV is entitled Isotope Geochemistry and Mixing (Chapters 16 through 18). Following the presentation of the different modes of decay and a statement of the Law of Radioactivity, Chapter 16 continues with the derivation of the geochronometry equation based on the accumulation of stable radiogenic daughter nuclides. This equation is then used to explain the principles and assumptions of dating by the K–Ar, Rb–Sr, Sm–Nd, Re–Os, and U, Th–Pb methods. In addition, the chapter contains descriptions of the $^{230}Th/^{238}U$ method of measuring sedimentation rates and of dating by the cosmogenic radionuclides ^{14}C, ^{10}Be, and ^{26}Al.

Chapter 17 is devoted to the fractionation of the stable isotopes of H, O, C, and S in nature. This chapter includes derivations of the basic mathematical relationships used to interpret variations of the isotope compositions of meteoric water, marine calcite, clay minerals, oilfield brines, hydrocarbons, and sulfide minerals. In addition, Chapter 17 presents a discussion of the time-dependent variation of the isotope compositions of Sr and S in marine carbonates of Phanerozoic age and the use of $^{87}Sr/^{86}Sr$ ratios for dating Cenozoic marine carbonates.

Part IV concludes with Chapter 18 on mixing and dilution of water and sedimentary rocks that are mixtures of two or three components having different chemical compositions. The chapter continues with two-component mixtures containing elements having distinctive isotope compositions (Sr, Pb, O, H, etc.). The systematics of such mixtures are briefly illustrated using oilfield brines formed by mixing two components having different $^{87}Sr/^{86}Sr$ ratios and Sr concentrations.

The fifth and last part of this book is entitled Applications of Geochemistry to the Solution of Global Problems (Chapters 19 through 25). This part treats the consequences of chemical weathering of rocks, such as the formation of soils and their agricultural productivity (Chapter 19); the chemical composition of surface water, its evolution by evaporative concentration, and its contamination by natural and anthropogenic causes (Chapter 20); and the weathering of metallic mineral deposits based on Eh–pH diagrams and including a section on the importance of bacteria in the oxidation of iron sulfides (Chapter 21). The

resulting release of metals into the environment is the basis for geochemical methods of prospecting. The chapter ends with a discussion of mineral economics including King Hubbert's prediction of the ultimate depletion of finite and nonrenewable resources such as petroleum.

The migration of ions released by weathering of rocks and metallic mineral deposits on the continents leads to a consideration of geochemical cycles in Chapter 22. After presenting the concept of mass balance, this chapter contains summaries of the geochemical cycles C-H-O-N, including water. The links between the reservoirs of these and other elements play an important role in stabilizing the chemical composition of the atmosphere and the oceans.

Chapter 23 on the chemistry of the atmosphere continues this train of thought by presenting the basic facts about the structure and chemical composition of the atmosphere. This information is then used to explain the formation of ozone in the stratosphere and its destruction by anthropogenic emissions of Cl-bearing gases, especially over Antarctica. The chapter also contains a presentation of the radiation balance of the Earth and the perturbation of this balance by emissions of gases that enhance the absorption of infrared radiation in the atmosphere. Chapter 23 ends with a long-range forecast of the effects of global warming on the time scale of 10^4 years based on the work of W. S. Broecker.

The dangers of environmental contamination are illustrated in Chapter 24 by consideration of the problems associated with high-level radioactive waste generated by nuclear reactors. The chapter explains how this waste is generated and how it is (or should be) processed prior to permanent storage in underground repositories. The geochemical properties of the transuranium elements, which play an important role in the long-term safety of such repositories, are briefly discussed using Eh–pH diagrams for neptunium and plutonium. Chapter 24 ends with a summary of the events and consequences of the explosion of a nuclear reactor in Chernobyl, Ukraine, on April 26, 1986.

The final chapter summarizes the geochemistry of Pb in the environment and its effect on human health. Lead is dispersed primarily through the atmosphere in the form of aerosol particles. Some civilizations have suffered the consequences of Pb toxicity because this metal was used as a food additive and in household utensils. The effects of Pb poisoning on humans are subtle and difficult to identify because all humans alive today are contaminated with Pb released primarily because of its use as an additive in gasoline. The impairment of mental functions as well as the toxic effects of Pb were tragically demonstrated by the members of Sir John Franklin's expedition to the Canadian Arctic in 1845–47.

The presentation of the subject matter in this book starts with *basic principles* and *emphasizes quantitative* methods of problem solving in order to gain better *understanding* of natural phenomena. Although this book is intended to be an *introduction* to geochemistry, the references at the end of each chapter will permit students to pursue topics of their choice in the library and thereby approach the *state of the art*. In addition, the end-of-chapter problems enable students to test their understanding of the principles. The book ends with a compilation of geochemical data in Appendix A and of *thermodynamic data* for a large number of elements and their compounds in Appendix B. In addition, most chapters contain data tables and diagrams that will make the book a *useful source of geochemical data*. The diagrams have lengthy captions that should *encourage browsing* and will help users to extract information from them without necessarily having to reread the entire chapter.

This geochemistry text demonstrates that the principles and applications of geochemistry are internally consistent, useful for understanding the world around us, and accessible to anyone willing to make the effort. In addition, this book makes the point that geochemists have an important task to perform in society by assisting in the prudent use of natural resources and in the safe disposal of the resulting waste products.

In conclusion, I thank the reviewers who read the manuscript of this book and made many suggestions which helped me to improve it: James H. Crocket, McMaster University; William J. Green,

Miami University (Ohio); Donald I. Siegel, Syracuse University, Robert D. Shuster, University of Nebraska, Omaha; and A. Russell Flegal, University of California, Santa Cruz. The second edition has benefited significantly from comments by colleagues who have pointed out errors of various kinds in the first edition and suggested improvements in the presentation of the subject matter. I thank these individuals by name in alphabetical order: Walther M. Barnard, Georges Beaudoin, Phillip Boger, Daniel Marcos Bonotto, Steven E. Bushnell, Christopher Conaway, Ethan L. Grossman, Giehyeon Lee, Peter C. Lichtner, James O'Neil, Risto Piispanen, Robert D. Shuster, and Terri L. Woods. I also thank Robert A. McConnin, Executive Editor of Prentice Hall, for his interest in this book, Elisabeth H. Belfer for her role in its production, and Susan M. McMullen for drafting the diagrams of the first edition. I am grateful to Betty Heath for typing part of the manuscript for the second edition.

The ultimate purpose of this book is to encourage students of the Earth to adopt the quantitative approach to problem solving that is exemplified by geochemistry, and to make them aware of the important role Earth scientists must play in the effort to preserve our habitat on the Earth.

GUNTER FAURE

Postscript: A manual with solutions to the end-of-chapter problems is available to instructors from the publisher by writing on school stationery to the Geology Editor, Prentice Hall, 1 Lake Street, Upper Saddle River, NJ 07458.

I

PLANET EARTH IN THE SOLAR SYSTEM

Future generations may remember the 20th century primarily for the advances in technology that have enabled us to begin the exploration of the solar system. Our resulting awareness of the celestial environment of the Earth has raised questions about the origin and evolution of the solar system, of the Milky Way Galaxy, and of the Universe at large. The start of an introductory geochemistry course is an appropriate time to satisfy our desire to know the origins of the world in which we live.

1

What Is Geochemistry?

Geology began when early man first picked up a stone, considered its
qualities, and decided that it was better than the stone he already had.
Good stones were useful and they were collected, mined, and traded.

HENRY FAUL AND CAROL FAUL (1983, p. 1)

Geochemistry is based on the urge to understand why some stones are "good" and how they formed. It involves applications of the principles of chemistry to the solution of geological problems and therefore could not develop until chemistry and geology had been established as scientific disciplines. These applications were practiced first during the 16th century in the mines of Europe where knowledge of minerals and their chemical compositions was used to recognize and follow veins of ore. For example, in 1574 Lazarus Ercker, the superintendent of mines for King Rudolf II of Austria, published a manual in which he described ores and outlined procedures for their analysis (Ercker, 1951). However, several centuries passed before geochemistry in the present sense of the word became established as an integral component of Earth and Planetary Science.

1.1   Early History

Two essential prerequisites for the growth of geochemistry were the discovery of the chemical elements and the development of sensitive and accurate methods for the analysis of rocks and minerals. Until the end of the 16th century matter was thought to consist of earth, water, fire, and air, whose basic qualities were described as warm, cold, dry, and wet. The metals gold, silver, copper, iron, tin, mercury, and lead together with sulfur and carbon had been known for thousands of years but were not recognized as elements in the present sense of the word. Antimony, arsenic, bismuth, and phosphorus were studied by the alchemists during the Middle Ages.

The development of analytical chemistry during the 18th century resulted in the discovery of 46 chemical elements between 1720 and 1850. After Bunsen and Kirchhoff invented the optical emission spectrograph, 30 additional elements were added from 1850 to 1925. The transuranium elements and certain other radioactive elements were discovered later in the 20th century (Correns, 1969).

As the number of known chemical elements increased during the 19th century, chemists became aware that they could be organized into groups based on similarities of their chemical properties. These tendencies became the foundation of the periodic table of the elements proposed independently by D. I. Mendeleev in 1869 and by J. L. Meyer in 1870 (Asimov, 1965). The cause for the periodicity of the chemical properties of the elements was not known then and was recognized only after the internal structure of atoms was worked out during the first 30 years of the 20th century.

The periodic table is a manifestation of the relationship between the internal structure of atoms and the chemical and physical properties of the elements. It facilitates a rational explanation of the chemical properties of the elements and of their observed distribution in nature.

The roots of geochemistry lie mainly in the 19th century, but it acquired its present multifaceted complexity during the 20th century. The word *geochemistry* was first used in 1838 by Christian Friedrich Schönbein, who was a chemistry professor at the University of Basel in Switzerland. The term was used again in 1908, 70 years later, when Frank W. Clarke of the U.S. Geological Survey published the first edition of his book, *The Data of Geochemistry.*

Clarke was the Chief Chemist of the U.S. Geological Survey from 1884 until 1925. He started the tradition for excellence in the analysis of rocks and minerals that has helped to make the Survey one of the largest geochemical research centers in the world. The first edition of Clarke's book was amazingly modern in its approach to geochemistry. He opened with chapters on the chemical elements, the atmosphere, lakes and rivers, and the ocean followed by chapters on saline lakes and springs and on evaporite deposits. Next he took up volcanic gases, magma, rock-forming minerals, and igneous rocks and then weathering, sedimentary rocks, metamorphic rocks, and ore deposits. The book ends with chapters on fossil fuels and their origin. That is pretty much how we would present the data of geochemistry today, but Clarke did it this way in 1908! More than 15 years passed before the great Soviet geochemist V. I. Vernadsky published his geochemistry book in 1924. Clarke was the first geochemist in the modern sense of the word.

Clarke's book went through five editions, the last of which appeared in 1924. It is now being updated and expanded by the U.S. Geological Survey under the editorship of Michael Fleischer. However, the *Handbook of Geochemistry,* edited by K. H. Wedepohl from 1969 to 1978, is the most complete source of geochemical data currently available.

1.2 Geochemistry in the U.S.S.R.

Geochemistry became an important subject in the U.S.S.R. because of the need to develop the mineral resources of the country. In 1932 Soviet geologists began to analyze systematically collected soil samples using optical spectroscopy (Fersman, 1939). In subsequent years the geologists of the Central Geological and Prospecting Institute in Moscow carried out extensive "metallometric" surveys based on soil samples in search of geochemical anomalies.

Geochemistry prospered in the U.S.S.R. because of the popular support it received partly in response to the remarkable work of Alexander Fersman (1883–1945). Fersman studied mineralogy at the University of Moscow under V. I. Vernadsky, who took his students on many field trips and encouraged them to regard minerals as the products of chemical reactions. These were radical departures from conventional mineralogy practiced at the turn of the century. Fersman graduated in 1907 and was elected to a professorship in mineralogy in 1912 at the age of 27. He embarked on an extraordinary career of teaching and research in geochemistry with emphasis on applications. He traveled widely in the U.S.S.R. and explored regions rarely visited before. His sustained interest in the exploration of the Kola Peninsula in the north ultimately led to the discovery of deposits of apatite and nickel ore. As a result, a remote wilderness area was transformed into an important center of mining and industry. Fersman also explored central Asia and discovered a native sulfur deposit at Kara Kum that was actively mined for decades.

Between 1934 and 1939 Fersman published a comprehensive four-volume work entitled *Geochemistry* in which he applied the principles of physical chemistry to the distribution of the chemical elements in nature. Fersman, above all, aroused public interest in geology and geochemistry by his inspiring lectures and popular books. Alexander K. Tolstoy called him a "poet of stones." Fersman earned this accolade by writing books intended for the layman: *Mineralogy for Everyone* (1928), *Recollections about a Stone* (1940), *My Travels, Stories about Precious Stones,* and *Geochemistry for Everyone* (1958). The last two were published posthumously. Fersman's conviction that geochemistry should serve his country is reflected in his words: "We cannot be merely idle admirers of our vast country; we must actively help reshape it and create a new life" (Shcherbakov, 1958).

The scientific legacy of Fersman's former teacher, Vladimir Ivanovich Vernadsky (1863–1945), is similarly impressive, but it has not been appreciated in the western world. Vernadsky emphasized the importance of the activities of living organisms in geological and geochemical processes. In fact, he regarded living matter as the "most powerful geological force" and wrote that the Earth's crust originated from the biosphere. These convictions about the importance of life on Earth were not widely accepted by geologists in North America, partly because Vernadsky wrote in many languages (Russian, German, French, Czech, Serbo-Croatian, and Japanese). His most important books (Vernadsky, 1924, 1929) were written in French and published in Paris. However, Vernadsky was also far ahead of his contemporaries by concluding that the biosphere may have controlled the environment on the surface of the Earth from the very beginning, long before the existence of life in Early Precambrian time was known (Lapo, 1986). Vernadsky's views on this subject are now reflected in the "Gaia Hypothesis" of Lovelock (1979), made popular on American television in the "Planet Earth" series by the Public Broadcasting System. Vernadsky also identified the biosphere as the principal transformer of solar energy into chemical energy. This idea is now widely accepted. For example, Cloud (1983) described the biosphere as "a huge metabolic device for the capture, storage and transfer of energy." Fossil fuels (coal, oil shale, petroleum, and natural gas) are now regarded as deposits of solar energy collected by the biosphere.

Vernadsky's views on the importance of the biosphere (Vernadsky, 1945) are becoming increasingly relevant as we attempt to predict the climatic consequences of the "greenhouse effect" caused by the discharge of carbon dioxide and other gases into the atmosphere. We are carrying out an experiment on a global scale the outcome of which could be detrimental to life. The response of the biosphere to this perturbation of the environment may be crucial to the ultimate outcome of this experiment.

V. I. Vernadsky is revered in the U.S.S.R. as one of the giants of science in the 20th century. His teachings still motivate the geochemists working at the Vernadsky Institute of Geochemistry and Analytical Chemistry in Moscow, which is among the foremost geochemical research institutes in the world.

1.3   V. M. Goldschmidt

The Institute of Geochemistry at the University of Göttingen in Germany was made famous by the work of Victor M. Goldschmidt (1888–1947). Goldschmidt earned his doctorate at the University of Oslo in 1911 with a study of contact metamorphism based on the thermodynamic phase rule. In the following year Max von Laue discovered diffraction of x-rays by crystals and thereby provided a method for determining the crystal structure of a mineral and the radii of the ions of which it is composed. From 1922 to 1926 Goldschmidt and his associates at the University of Oslo used x-ray diffraction to determine the crystal structures of many minerals (Goldschmidt, 1930). In 1930 Goldschmidt moved to the University of Göttingen, where he studied the distribution of the chemical elements using an optical spectrograph. From this body of data he deduced that the chemical composition of minerals is determined by the requirements of "closest packing" of ions. Moreover, the substitution of the ions of a major element by the ions of a trace element depends on the similarity of their radii and charges. These generalizations provided a rational explanation for the observed distribution of the elements in the minerals of the crust of the Earth.

Goldschmidt returned to Oslo in 1935 but was unable to continue his work after the

German invasion of Norway in 1940. He fled to Sweden in 1942 and ultimately made his way to England, where he worked at the Macaulay Institute for Soil Research. His health had seriously deteriorated as a consequence of imprisonment in concentration camps in Norway. Goldschmidt returned to Oslo in 1946 but died there prematurely in 1947 at the age of 59. He left behind the manuscript of a partially completed book entitled *Geochemistry,* which was completed by Alex Muir and published posthumously in 1954. A special commemorative issue of *Applied Geochemistry* (1988) was devoted to Goldschmidt on the 100th anniversary of his birthday.

Goldschmidt's principal contribution to geochemistry is the rational explanation of isomorphous substitution in crystals based on the compatibility of the radii and charges of the ions. He thereby achieved one of the major goals of geochemistry. Nevertheless, the urge to know and to understand the distribution of the chemical elements has continued to motivate the work of several outstanding geochemists, notably L. H. Ahrens in South Africa, S. R. Nockolds in England, S. R. Taylor in Australia, K. Rankama in Finland, K. H. Wedepohl in Germany, A. P. Vinogradov in the U.S.S.R., D. M. Shaw in Canada, and K. K. Turekian in the United States.

1.4   Modern Geochemistry

Beginning in the 1950s, geochemists turned increasingly to the study of chemical reactions and processes. The roots of this line of research can be found in the work of A. Fersman, who used the concepts of thermodynamics to study the stability of minerals in their natural environment. In addition, J. H. van't Hoff (1852–1911) studied the crystallization of salts by evaporating seawater in Berlin and, in 1904, the Geophysical Laboratory of the Carnegie Institution of Washington was founded in the United States. The principal objective of the Geophysical Laboratory has been to study the origin of igneous rocks and ore deposits by experimental methods. The results achieved by N. L. Bowen and his colleagues, and by their successors, at that laboratory have become the foundation of modern igneous petrology. Bowen's book, *The Evolution of the Igneous Rocks,* published in 1928, turned igneous petrology from a preoccupation with the description and classification of igneous rocks toward a concern for their origin and the geochemical differentiation of the Earth by magmatic activity.

In 1952 Brian Mason published the first edition of *Principles of Geochemistry,* which was widely used as a textbook at universities and helped to establish geochemistry as a legitimate component of Earth Science. In the next decade Robert M. Garrels and Konrad B. Krauskopf used thermodynamics and solution chemistry to determine the stability of minerals and the mobility of their ions at the surface of the Earth. They trained the modern generation of geochemists by their own research and through their popular textbooks (Garrels, 1960; Garrels and Christ, 1965; Krauskopf, 1967).

Geochemical prospecting, as we know it today, was initiated in 1947 by the U.S. Geological Survey (Hawkes and Lakin, 1949) and was based on colorimetry. The techniques developed in the United States were used in 1953 for soil surveys in the United Kingdom and in Africa (Webb, 1953). Subsequently, the Geochemical Prospecting Research Center at the Imperial College of Science and Technology in London became a center of activity in this aspect of geochemistry. Geochemical prospecting continues to be one of the most important practical applications of geochemistry and now relies on a wide range of sophisticated analytical techniques.

Environmental geochemistry has arisen recently because of the need to monitor the dispersion of metals and various organic compounds that are introduced into the environment as anthropogenic contaminants. This new application of geochemistry to the welfare of humankind is closely related to pollutant hydrogeology and medical geochemistry. Geochemical prospecting and environmental geochemistry are of paramount importance because they contribute to the continued well-being of the human species on Earth.

Since about the middle of the 20th century geochemistry has become highly diversified into many subdivisions, among them inorganic geochemistry, organic geochemistry, isotope geochemistry, geochemical prospecting, medical geochemistry, aqueous geochemistry, trace-element geochemistry, and cosmochemistry. Each of these subdivisions has its own rapidly growing body of literature. However, the geochemists of the world have organized scientific societies that transcend the boundaries of specialization and help to unify the field. These societies also publish journals that promote the free flow of information and ideas among their members. For example, the Geochemical Society and the Meteoritical Society both sponsor *Geochimica et Cosmochimica Acta*. The European Association for Geochemistry has adopted the journal *Chemical Geology,* and the International Association of Geochemistry and Cosmochemistry sponsors *Applied Geochemistry.* Among other journals that publish papers dealing with geochemistry are

Earth and Planetary Science Letters
Contributions to Mineralogy and Petrology
Journal of Geochemical Exploration
Economic Geology
American Journal of Science
Nature
Science

Some journals are closely associated with certain regions, for example,

Geokhimia
Doklady (Earth
   Science Section)
Lithology and   } Russia
   Mineral Resources
Geochemistry
   International
Geochemical Journal   Japan
Chinese Journal   People's Republic
   of Geochemistry   of China

In fact, most Earth Science journals now publish papers in which geochemical data or principles are used to explain a geological process or to solve a particular problem. The strength of geochemistry lies in the fact that most geological processes involve chemical reactions in some significant way. Geochemistry really is for everyone, as Fersman claimed in his popular book.

What then is geochemistry? There is no easy answer to this question because geochemists concern themselves with a wide range of natural phenomena and use many different techniques to study them. According to Fersman, "Geochemistry studies the history of chemical elements in the Earth's crust and their behavior under different thermodynamic and physicochemical natural conditions." Goldschmidt's definition of geochemistry was "Geochemistry is concerned with the laws governing the distribution of the chemical elements and their isotopes throughout the Earth" (Correns, 1969). We could go on and on. Geochemistry is a highly diversified and constantly evolving subject that can be described in many different ways.

For our purposes the major goals of geochemists are

1. To know the distribution of the chemical elements in the Earth and in the solar system.
2. To discover the causes for the observed chemical composition of terrestrial and extraterrestrial materials.
3. To study chemical reactions on the surface of the Earth, in its interior, and in the solar system around us.
4. To assemble this information into geochemical cycles and to learn how these cycles have operated in the geologic past and how they may be altered in the future.

These are the lofty goals of a scientific discipline. Their intrinsic merit is apparent. However, geochemists practicing their craft today also have an obligation to humankind to assist in the development and management of natural resources, to monitor the quality of the environment both locally and on a global scale, and to warn humanity against dangerous practices that may threaten the quality of life in the future.

With so much at stake it is not easy to express succinctly what geochemistry is all about. K. O. Emery and J. M. Hunt (1974, p. 586) put it this way: "Studies of geochemistry ... convert idle speculation into ... understanding ...". That will be the theme of this book. We can no longer afford idle speculation in the Earth Sciences. We must have understanding. Geochemistry can show us how to achieve it.

References

ASIMOV, I., 1965. *A Short History of Chemistry.* Doubleday, Garden City, NY, 263 pp.

BOWEN, N. L., 1928. *The Evolution of the Igneous Rocks.* Princeton University Press, Princeton, NJ, 34 pp.

CLARKE, F. W., 1908. *The Data of Geochemistry.* U.S. Geol. Surv. Bull. 330. (The fifth edition in 1924 is U.S.G.S. Bull. 770.)

CLOUD, P., 1983. The biosphere. *Sci. Amer.,* **249:**132–144.

CORRENS, C. W., 1969. The discovery of the chemical elements. The history of geochemistry. Definitions of geochemistry. In K. H. Wedepohl (Ed.), *Handbook of Geochemistry,* vol. 1, 1–11. Springer-Verlag, Berlin, 442 pp.

EMERY, K. O., and J. M. HUNT, 1974. Summary of Black Sea investigations. In E. T. Degens and D. A. Ross (Eds.), The Black Sea—geology, chemistry, and biology, 575–590. *Amer. Assoc. Petrol. Geol., Mem.* 20, 633 pp.

ERCKER, L., 1951. *Treatise on Ores and Assaying.* From the German Edition of 1580. A Grünhaldt Sisco and C. Stanley Smith, trans. and ann. University of Chicago Press.

FAUL, H., and C. FAUL, 1983. *It Began with a Stone.* Wiley, New York, 270 pp.

FERSMAN, A. YE., 1939. Geochemical and mineralogical methods of prospecting. Chapter IV in *Special Methods of Prospecting.* Akad. Nauk S.S.R., Moscow. Translated by L. Hartsock and A. P. Pierce, U.S. Geol. Surv. Circ. 127, 1952.

FERSMAN, A., 1958. *Geochemistry for Everyone.* Foreign Language Publishing House, Moscow, 454 pp.

GARRELS, R. M., 1960. *Mineral Equilibria.* Harper & Row, New York.

GARRELS, R. M., and C. L. CHRIST, 1965. *Solutions, Minerals and Equilibria.* Harper & Row, New York (later Freeman and Cooper, San Francisco), 450 pp.

GOLDSCHMIDT, V. M., 1930. Geochemische Verteilungsgesetze und kosmische Häufigkeit der Elemente. *Naturwissenschaften,* **18:**999–1013.

HAWKES, H. E., and H. W. LAKIN, 1949. Vestigial zinc in surface residuum associated with primary zinc ore in east Tennessee. *Econ. Geol.,* **44:**286–295.

KRAUSKOPF, K. B., 1967. *Introduction to Geochemistry.* McGraw-Hill, New York.

LAPO, A. V., 1986. V. I. Vernadsky's ideas on the leading role of life in the generation of the Earth's crust. *Earth Sci. Hist.,* **5:**124–127.

LOVELOCK, J. E., 1979. *Gaia. A New Look at Life on Earth.* Oxford University Press, Oxford, England, 158 pp.

MASON, B., 1952. *Principles of Geochemistry.* Wiley, New York.

SHCHERBAKOV, D., 1958. Foreword. In A. Fersman, *Geochemistry for Everyone.* Translated by D. A. Myshne, Foreign Languages Publishing House, Moscow, 453 pp.

VERNADSKY, V. I., 1924. *La Geochimie.* Alcan, Paris, 404 pp.

VERNADSKY, V. I., 1929. *La Biosphere.* Alcan, Paris, 232 pp.

Vernadsky, V. I., 1945. The biosphere and the noosphere. *Amer. Scientist,* **33:**1–12.

WEBB, J. S., 1953. A review of American progress in geochemical prospecting. *Inst. Mining Metallurgy Trans.,* **62:**321–348.

WEDEPOHL, K. H. (Ed.), 1969. *Handbook of Geochemistry,* vol. I. Springer-Verlag, Berlin. Vol. II (1970).

2

In the Beginning

The urge to trace the history of the universe back to its beginnings is irresistible.

STEVEN WEINBERG (1977, p. 1)

Certain questions about our existence on Earth are so fundamental that they have been incorporated into religious mythologies. These questions not only concern the origin of the Earth and the evolution of life but also extend to the origin of the universe and to the nature of space and time. Did the universe have a beginning and will it ever end? What existed before the universe formed? Does the universe have limits and what exists beyond those limits? It is proper to raise these questions at the beginning of a geochemistry course because they are within the scope of cosmochemistry.

2.1 The Big Bang

The universe started like a bubble in a stream. At first it was not there, and suddenly it formed and expanded rapidly as though it were exploding (Gott, 1982). Science has its share of practical jokers who immediately referred to the start of the expansion of the universe as the Big Bang (Gamow, 1952). From the very beginning the universe had all of the mass and energy it contains today. As a result, its pressure and temperature, say 10^{-32} sec after the Big Bang, were so high that matter existed in its most fundamental form as "quark soup." As the universe expanded and cooled, the quarks combined to form more familiar nuclear particles that ultimately became organized into nuclei of hydrogen and helium.

Formation of atomic nuclei began about 13.8 sec after the Big Bang when the temperature of the universe had decreased to 3×10^9 K. This process continued for about 30 min, but did not go beyond helium because the nuclear reactions could not bridge a gap in the stabilities of the nuclei of lithium, beryllium, and boron. At that time the universe was an intensely hot and rapidly expanding fireball.

Some 700,000 years later, when the temperature had decreased to about 3×10^3 K, electrons became attached to the nuclei of hydrogen and helium. Matter and radiation were thereby separated from each other, and the universe became transparent to light. Subsequently, matter began to be organized into stars, galaxies, and galactic clusters as the universe continued to expand to the present time (Weinberg, 1977).

But how do we know all this? The answer is that the expansion of the universe can be seen in the "red shift" of spectral lines of light emitted by distant galaxies, and it can be "heard" as the "cosmic microwave radiation," which is the remnant of the fireball, that still fills the universe. In addition, the properties of the universe immediately after the Big Bang were similar to those of atomic nuclei. Therefore, a very fruitful collaboration has developed among nuclear physicists and cosmologists that has enabled them to reconstruct the history of the universe back to about 10^{-32} sec after the Big Bang. These studies have shown that the forces we recognize at low temperature are, at least in part, unified at extremely high temperatures and densities. There is hope that a Grand Unified Theory (GUT) will eventually emerge that may permit us to approach even closer to understanding the start of the universe.

What about the future? Will the universe continue to expand forever? The answer is that the future of the universe can be predicted only if we know the total amount of matter it contains. The matter that is detectable at the present time is not sufficient to permit gravity to overcome the expansion. If expansion continues, the universe will become colder and emptier with no prospect of an end. However, a large fraction of the mass of the universe is hidden from view in the form of gas and dust in interstellar and intergalactic space, and in the bodies of stars that no longer emit light. In addition, we still cannot rule out the possibility that neutrinos have mass even when they are at rest. If the mass of the universe is sufficient to slow the expansion and ultimately to reverse it, then the universe will eventually contract until it disappears again in the stream of time.

Since the universe had a beginning and is still expanding, it cannot be infinite in size. However, the edge of the universe cannot be seen with telescopes because it takes too long for the light to reach us. As the universe expands, space expands with it. In other words, it seems to be impossible to exceed the physical limits of the universe. We are trapped in our expanding bubble. If other universes exist, we cannot communicate with them.

Now that we have seen the big picture, let us review certain events in the history of the standard model of cosmology to show that progress in Science is sometimes accidental.

In 1929 the American astronomer Edwin Hubble reported that eighteen galaxies in the Virgo cluster are receding from Earth at different rates that increase with their distances from Earth. He calculated the recessional velocities of these galaxies by means of the "Doppler effect" from observed increases of the wavelengths of

characteristic spectral lines of light they emit. This "red shift" is related to the recessional velocity by an equation derived in 1842 by Johann Christian Doppler in Prague:

$$\frac{\lambda'}{\lambda} = 1 + \frac{v}{c} \qquad (2.1)$$

where λ' is the wavelength of a spectral line of light emitted by a moving source, λ is the wavelength of the same line emitted by a stationary source, c is the velocity of light, and v is the recessional velocity. Hubble's estimates of the distances to the galaxies were based on the properties of the Cepheid Variables studied previously by H. S. Leavitt and H. Shapley at Harvard University. The Cepheid Variables are bright stars in the constellation Cepheus whose period of variation depends on their absolute luminosity, which is the total radiant energy emitted by an astronomical body. Hubble found such variable stars in the galaxies he was studying and determined their absolute luminosities from their periods. The intensity of light emitted by a star decreases as the square of the distance increases. Therefore, the distance to a star can be determined from a comparison of its absolute and its apparent luminosity, where the latter is defined as the radiant power received by the telescope per square centimeter. In this way, Hubble determined the recessional velocities and distances of the galaxies in the Virgo cluster and expressed their relationship as:

$$v = Hd \qquad (2.2)$$

where v is the recessional velocity in km/sec, d is the distance in 10^6 light years, and H is the Hubble constant (Hubble, 1936).

The Hubble constant can be used to place a limit on the age of the universe. If two objects are moving apart with velocity v, the time (t) required for them to become separated by a distance d is:

$$t = \frac{d}{v} = \frac{1}{H} \qquad (2.3)$$

The initial results indicated that the Hubble constant had a value of 170 km/sec/10^6 light years, which corresponds to an expansion time of less than 2×10^9 years. This result was very awkward because age determinations based on radioactivity had established that the Earth is older than this date. Eventually Walter Baade discovered an error in the calibration of the Cepheid Variables, and the value of the Hubble constant was revised (Baade, 1968). The presently accepted value is 15 km/sec/10^6 light years, which indicates an expansion time for the universe of less than 20×10^9 years. This date is compatible with independent estimates of its age based on consideration of nucleosynthesis and the evolution of stars. By combining all three methods Hainebach et al. (1978) refined the age of the universe to $(14.5 \pm 1.0) \times 10^9$ years.

The Big Bang theory of cosmology was not accepted for many years for a variety of reasons. The turning point came in 1964 when Arno A. Penzias and Robert W. Wilson discovered a microwave background radiation that corresponds to a blackbody temperature of about 3 K. The discovery of this radiation was accidental, even though its existence had been predicted twenty years earlier by George Gamow and his colleagues Ralph A. Alpher and Robert Herman. Because they were unaware of Gamow's work, Penzias and Wilson were skeptical about the phenomenon they had discovered and took great care to eliminate all extraneous sources of the background radiation. For example, they noticed that two pigeons had been nesting in the throat of the antenna they were using at Holmdel, New Jersey. The pigeons were caught and taken to a distant location, but promptly returned. Therefore, they were caught again and dealt with "by more decisive means." The pigeons had also coated the antenna with a "white dielectric material," which was carefully removed. However, the intensity of the background radiation remained constant and independent of time in the course of a year.

Word of this phenomenon reached a group of astrophysicists at nearby Princeton University who were working on models of the early history of the universe under the guidance of Robert H. Dicke. Eventually, Penzias called Dicke, and it

was agreed that they would publish two companion letters in the *Astrophysical Journal*. Penzias and Wilson announced the discovery, and Dicke and his colleagues explained the cosmological significance of the microwave background radiation (Penzias and Wilson, 1965). In 1978 Penzias and Wilson shared the Nobel Prize in physics for their discovery.

The radiation discovered by Penzias and Wilson is a remnant of the radiation that filled the universe for about 700,000 years when its temperature was greater than about 3000 K. During this early period, matter consisted of a mixture of nuclear particles and photons in thermal equilibrium with each other. Under these conditions the energy of radiation at a specific wavelength is inversely proportional to the absolute temperature. According to an equation derived by Max Planck at the start of the 20th century, the energy of blackbody radiation at a particular temperature increases rapidly with increasing wavelength to a maximum and then decreases at longer wavelengths. Radiation in thermal equilibrium with matter has the same properties as radiation inside a black box with opaque walls. Therefore, the energy distribution of radiation in the early universe is related to the wavelength and to the absolute temperature by Planck's equation. The wavelength near which most of the energy of blackbody radiation is concentrated (λ_{max}) is approximately equal to:

$$\lambda_{max} = \frac{0.29}{T} \qquad (2.4)$$

where λ_{max} is measured in centimeters and T is in Kelvins (Weinberg, 1977).

The original measurement of Penzias and Wilson was at a wavelength of 7.35 cm, which is much greater than the typical wavelength of radiation at 3 K. Since 1965, many additional measurements at different wavelengths have confirmed that the cosmic background radiation does fit Planck's formula for blackbody radiation. The characteristic temperature of this radiation is about 3 K, indicating that the typical wavelength of photons has increased by a factor of about 1000 because of the expansion of the universe since its temperature was 3000 K (Weinberg, 1977).

2.2 Stellar Evolution

Matter in the universe is organized into a "hierarchy of heavenly bodies" listed below in order of decreasing size.

clusters of galaxies	comets
galaxies	asteroids
stars, pulsars, and	meteoroids
black holes	dust particles
planets	molecules
satellites	atoms of H and He

On a subatomic scale, space between stars and galaxies is filled with cosmic rays (energetic nuclear particles) and photons (light).

Stars are the basic units in the hierarchy of heavenly bodies within which matter continues to evolve by nuclear reactions. Many billions of stars are grouped together to form a galaxy, and large numbers of such galaxies are associated into galactic clusters. Stars may have stellar companions or they may have orbiting planets, including ghostly comets that flare briefly when they approach the star on their eccentric orbits. The planets in our solar system have their own retenue of satellites. The space between Mars and Jupiter contains the asteroids, most of which are fragments of larger bodies that have been broken up by collisions and by the gravitational forces of Jupiter and Mars. Pieces of the asteroids have impacted as meteorites on the surfaces of the planets and their satellites and have left a record of these events in craters.

On an even smaller scale, space between stars contains clouds of gas and solid particles. The gas is composed primarily of hydrogen and of helium that were produced during the initial expansion of the universe. In addition, the interstellar medium contains elements of higher atomic number that were synthesized by nuclear reactions in the interiors of stars that have since exploded. A third component consists of compounds of hydrogen and carbon that are the precursors of life. These clouds of gas and dust may contract to form new stars whose evolution depends on their masses and on the H/He ratio of the gas cloud from which they formed.

The evolution of stars can be described by specifying their luminosities and surface temperatures. The luminosity of a star is proportional to its mass, and its surface temperature or color is an indicator of its volume. When a cloud of interstellar gas contracts, its temperature increases, and it begins to radiate energy in the infrared and visible parts of the spectrum. As the temperature in the core of the gas cloud approaches 20×10^6 K, energy production by hydrogen fusion becomes possible, and a star is born. Most of the stars of a typical galaxy derive energy from this process and therefore plot in a band, called the *main sequence,* on the Hertzsprung–Russell diagram shown in Figure 2.1. Massive stars, called *blue giants,* have high luminosities and high surface temperatures. The Sun is a star of intermediate mass and has a surface temperature of about 5800 K. Stars that are less massive than the Sun are called *red dwarfs* and plot at the lower end of the main sequence.

As a star five times more massive than the Sun converts hydrogen to helium while on the main sequence, the density of the core increases, causing the interior of the star to contract. The core temperature therefore rises slowly during the hydrogen-burning phase. This higher temperature accelerates the fusion reaction and causes the outer envelope of the star to expand. However, when the core becomes depleted in hydrogen, the rate of energy production declines and the star contracts, raising the core temperature still further. The site of energy production now shifts from the core to the surrounding shell. The resulting changes in luminosity and surface temperature cause the star to move off the main sequence toward the realm of the *red giants* (Figure 2.1).

The helium produced by hydrogen fusion in the shell accumulates in the core, which continues to contract and therefore gets still hotter. The resulting expansion of the envelope lowers the surface temperature and causes the color to turn red. At the same time, the shell in which hydrogen

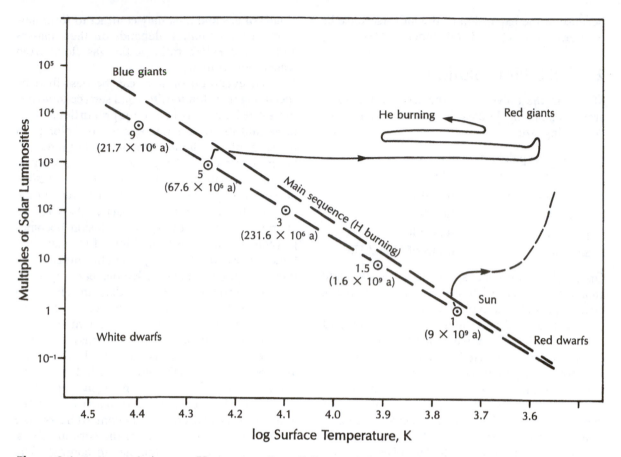

Figure 2.1 Stellar evolution on a Hertzsprung–Russell diagram for stars ranging from 1 to 9 solar masses. When a star has used up the hydrogen in its core, it contracts and then moves off the main sequence and enters the realm of the red giants, which generate energy by helium fusion. The evolutionary track and the life expectancy of stars are strongly dependent on their masses. Stars five times more massive than the sun are nearly 1000 times brighter, have surface temperatures of about 18,000 K—compared to 5800 K for the Sun—and remain on the main sequence only about 68 million years. Their evolution to the end of the major phase of helium burning takes only about 87 million years (Iben, 1967).

is reacting gradually thins as it moves toward the surface, and the luminosity of the star declines. These changes transform a main-sequence star into a bloated red giant. For example, the radius of a star five times more massive than the Sun increases about 30-fold just before helium burning in the core begins.

When the core temperature approaches 100×10^6 K, helium fusion by means of the "triple-alpha process" begins and converts three helium nuclei into the nucleus of carbon-12. At the same time, hydrogen fusion in the shell around the core continues. The luminosities and surface temperatures (color) of red giants become increasingly variable as they evolve, reflecting changes in the rates of energy production in the core and shell. The evolutionary tracks in Figure 2.1 illustrate the importance of the mass of a star to its evolution. A star five times as massive as the Sun is 1000 times brighter while on the main sequence and has a more eventful life as a red giant than stars below about two solar masses (Iben, 1967, 1974).

The length of time a star spends on the main sequence depends primarily on its mass and to a lesser extent on the H/He ratio of its ancestral gas cloud. In general, massive stars (blue giants) consume their fuel rapidly and may spend only 10×10^6 years on the main sequence. Small stars (red dwarfs) have much slower "metabolic" rates and remain on the main sequence for very long periods of time exceeding 10×10^9 years. The Sun, being a star of modest magnitude, has enough hydrogen in its core to last about 9×10^9 years at the present rate of consumption. Since it formed about 4.5×10^9 years ago, the Sun has achieved middle age and will provide energy to the planets of the solar system for a very long time to come. However, ultimately its luminosity will increase, and it will expand to become a red giant, as shown in Figure 2.1. The temperature on the surface of the Earth will then rise and become intolerable to life forms. The expansion of the Sun may engulf the terrestrial planets, including Earth, and vaporize them. When all of its nuclear fuel has been consumed, the Sun will assume the end stage of stellar evolution that is appropriate for a star of its mass and chemical composition.

Toward the end of the giant stage, stars become increasingly unstable. When the fuel for a particular energy-producing reaction is exhausted, the star contracts and its internal temperature rises. The increase in temperature may trigger a new set of nuclear reactions. In stars of sufficient mass, this activity culminates in a gigantic explosion (supernova) as a result of which a large fraction of the outer envelope of the star is blown away. The debris from such explosions mixes with hydrogen and helium in interstellar space to form clouds of gas and dust from which new stars may form.

As stars reach the end of their evolution they turn into white dwarfs, or neutron stars (pulsars), or black holes, depending on their masses (Wheeler, 1973). Stars whose mass is less than about 1.2 solar masses contract until their radius is only about 1×10^4 km and their density is between 10^4 and 10^8 g/cm^3. Stars in this configuration have low luminosities but high surface temperatures and are therefore called white dwarfs (Figure 2.1). They gradually cool and fade from view as their luminosities and surface temperatures diminish with time. Stars that are appreciably more massive than the Sun develop dense cores because of the synthesis of heavy chemical elements by nuclear reactions. Eventually, such stars become unstable and explode as supernovas. The core then collapses until its radius is reduced to about 10 km and its density is of the order of 10^{11} to 10^{15} g/cm^3. Such stars are composed of a "neutron gas" because electrons and protons are forced to combine under the enormous pressure and the abundance of neutrons greatly increases as a result. Neutron stars have very rapid rates of rotation and emit pulsed radio waves that were first observed in 1965 by Jocelyn Bell, a graduate student working with A. Hewish at the Cavendish Laboratory of Cambridge University in England (Hewish, 1975). The Crab Nebula contains such a "pulsar," which is the remnant of a supernova observed by Chinese astronomers in 1054 A.D. (Fowler, 1967). The cores of the most massive stars collapse to form black holes in accordance with Einstein's theory of general relativity. Black holes have radii of only a few kilometers and densities in excess of 10^{16} g/cm^3. Their gravitational field is so great that neither light nor matter can escape from them, hence the name "black hole." Observational evidence supporting the existence of black holes is growing and they are believed to be an important phenomenon in the evolution of galaxies.

Stars, it seems, have predictable evolutionary life cycles. They form, shine brightly for a while, and then die. Hans Bethe (1968, p. 547) put it this way:

> If all this is true, stars have a life cycle much like animals. They are born, they grow, they go through a definite internal development, and finally die, to give back the material of which they are made so that new stars may live.

2.3 Nucleosynthesis

The origin of the chemical elements is intimately linked to the evolution of stars because the elements are synthesized by the nuclear reactions from which stars derive the energy they radiate into space. Only helium and deuterium, the heavy isotope of hydrogen, were synthesized during the initial expansion of the universe. The entire theory of nucleosynthesis was presented in a detailed paper by Burbidge, Burbidge, Fowler and Hoyle (1957) referred to affectionately as B_2FH. Subsequent contributions (e.g., Schramm and Arnett, 1973) have been concerned with specific aspects of the theory and its application to the evolution of stars of different masses and initial compositions.

The theory presented by B_2FH (1957) evolved from the work of several other scientists, among whom George Gamow deserves special recognition. Gamow received his doctorate from the University of Leningrad and came to the University of Göttingen in the spring of 1928 for some postdoctoral studies. There he met Fritz Houtermans, with whom he spent many hours in a café making calculations (Gamow, 1963). Subsequently, Houtermans moved to the University of Berlin where he met the British astronomer Robert Atkinson. Houtermans and Atkinson used Gamow's theory of alpha decay to propose that stars generate energy in their interiors by the formation of helium nuclei from four protons captured by the nucleus of another light element. The initial title of their paper was "Wie kann man einen Heliumkern im Potentialtopf kochen." (How one can cook a helium nucleus in a pressure cooker.) Needless to say, the editors of the *Zeitschrift für Physik* insisted that the title be changed (Atkinson and Houtermans, 1929).

Ten years later, in April of 1938, Gamow organized a conference in Washington, D.C., to discuss the internal constitution of stars. The conference stimulated one attendee, Hans Bethe, to examine the possible nuclear reactions between protons and the nuclei of the light elements in order of increasing atomic number. Bethe (1939) found that the "pressure cooker" of Atkinson and Houtermans was the nucleus of $^{12}_{6}C$ and made it the basis of his famous CNO cycle for hydrogen fusion in stars (to be presented later in this section) (Bethe, 1968).

Gamow influenced the evolution of the theory of nucleosynthesis in many other ways. In 1935 he published a paper in the *Ohio Journal of Science* (an unlikely place for a nuclear astrophysics paper) on the buildup of heavy elements by neutron capture and subsequently championed the idea that the chemical elements were synthesized during the first 30 min after the Big Bang. In the mid-1940s Alpher and Gamow wrote a paper detailing the origin of the chemical elements based on that assumption. At Gamow's suggestion, Hans Bethe's name was added "in absentia," thus creating the famous triumvirate Alpher, Bethe, and Gamow (1948).

The abundance of the chemical elements and their naturally occurring isotopes is the blueprint for all theories of nucleosynthesis. For this reason, geochemists and stellar spectroscopists have devoted much time and effort to obtaining accurate analytical data on the concentrations of the elements in the Sun and other nearby stars from the wavelength spectra of the light they emit. Information on the abundances of nonvolatile elements has come also from chemical analyses of stony meteorites, especially the carbonaceous chondrites, which are the most undifferentiated samples of matter in the solar system available to us (Mason, 1962). Table 2.1 lists the abundances of the elements in the solar system compiled by Anders and Ebihara (1982). The abundances are expressed in terms of the number of atoms relative to 10^6 atoms of silicon. Figure 2.2 is a plot of these data and illustrates several important observations about the abundances of the elements.

1. Hydrogen and helium are by far the most abundant elements in the solar system, and the atomic H/He ratio is about 12.5.

2. The abundances of the first 50 elements decrease exponentially.

3. The abundances of the elements having atomic numbers greater than 50 are very low and do not vary appreciably with increasing atomic number.

4. Elements having even atomic numbers are more abundant than their immediate neighbors with odd atomic numbers (Oddo–Harkins rule).

Table 2.1 Abundances of the Elements in the Solar System in Units of Number of Atoms per 10^6 Atoms of Silicon

Atomic no.	Element	Symbol	Abundance[a]	Atomic no.	Element	Symbol	Abundance[a]
1	hydrogen	H	2.72×10^{10}	53	iodine	I	9.0×10^{-1}
2	helium	He	2.18×10^{9}	54	xenon	Xe	4.35×10^{0}
3	lithium	Li	5.97×10^{1}	55	cesium	Cs	3.72×10^{-1}
4	beryllium	Be	7.8×10^{-1}	56	barium	Ba	4.36×10^{0}
5	boron	B	2.4×10^{1}	57	lanthanum	La	4.48×10^{-1}
6	carbon	C	1.21×10^{7}	58	cerium	Ce	1.16×10^{0}
7	nitrogen	N	2.48×10^{6}	59	praseodymium	Pr	1.74×10^{-1}
8	oxygen	O	2.01×10^{7}	60	neodymium	Nd	8.36×10^{-1}
9	fluorine	F	8.43×10^{2}	61	promethium	Pm	0
10	neon	Ne	3.76×10^{6}	62	samarium	Sm	2.61×10^{-1}
11	sodium	Na	5.70×10^{4}	63	europium	Eu	9.72×10^{-2}
12	magnesium	Mg	1.075×10^{6}	64	gadolinium	Gd	3.31×10^{-1}
13	aluminum	Al	8.49×10^{4}	65	terbium	Tb	5.89×10^{-2}
14	silicon	Si	1.00×10^{6}	66	dysprosium	Dy	3.98×10^{-1}
15	phosphorus	P	1.04×10^{4}	67	holmium	Ho	8.75×10^{-2}
16	sulfur	S	5.15×10^{5}	68	erbium	Er	2.53×10^{-1}
17	chlorine	Cl	5.240×10^{3}	69	thullium	Tm	3.86×10^{-2}
18	argon	Ar	1.04×10^{5}	70	ytterbium	Yb	2.43×10^{-1}
19	potassium	K	3.770×10^{3}	71	lutetium	Lu	3.69×10^{-2}
20	calcium	Ca	6.11×10^{4}	72	hafnium	Hf	1.76×10^{-1}
21	scandium	Sc	3.38×10^{1}	73	tantalum	Ta	2.26×10^{-2}
22	titanium	Ti	2.400×10^{3}	74	tungsten	W	1.37×10^{-1}
23	vanadium	V	2.95×10^{2}	75	rhenium	Re	5.07×10^{-2}
24	chromium	Cr	1.34×10^{4}	76	osmium	Os	7.17×10^{-1}
25	manganese	Mn	9.510×10^{3}	77	iridium	Ir	6.60×10^{-1}
26	iron	Fe	9.00×10^{5}	78	platinum	Pt	1.37×10^{0}
27	cobalt	Co	2.250×10^{3}	79	gold	Au	1.86×10^{-1}
28	nickel	Ni	4.93×10^{4}	80	mercury	Hg	5.2×10^{-1}
29	copper	Cu	5.14×10^{2}	81	thallium	Tl	1.84×10^{-1}
30	zinc	Zn	1.260×10^{3}	82	lead	Pb	3.15×10^{0}
31	gallium	Ga	3.78×10^{2}	83	bismuth	Bi	1.44×10^{-1}
32	germanium	Ge	1.18×10^{2}	84	polonium	Po	~0
33	arsenic	As	6.79×10^{0}	85	astatine	At	~0
34	selenium	Se	6.21×10^{1}	86	radon	Rn	~0
35	bromine	Br	1.18×10^{1}	87	francium	Fr	~0
36	krypton	Kr	4.53×10^{1}	88	radium	Ra	~0
37	rubidium	Rb	7.09×10^{0}	89	actinium	Ac	~0
38	strontium	Sr	2.38×10^{1}	90	thorium	Th	3.35×10^{-2}
39	yttrium	Y	4.64×10^{0}	91	protactinium	Pa	~0
40	zirconium	Zr	1.07×10^{1}	92	uranium	U	9.00×10^{-3}
41	niobium	Nb	7.1×10^{-1}	93	neptunium	Np	~0
42	molybdenum	Mo	2.52×10^{0}	94	plutonium	Pu	~0
43	technetium	Tc	0	95	americium	Am	0
44	ruthenium	Ru	1.86×10^{0}	96	curium	Cm	0
45	rhodium	Rh	3.44×10^{-1}	97	berkelium	Bk	0
46	palladium	Pd	1.39×10^{0}	98	californium	Cf	0
47	silver	Ag	5.29×10^{-1}	99	einsteinium	Es	0
48	cadmium	Cd	1.69×10^{0}	100	fermium	Fm	0
49	indium	In	1.84×10^{-1}	101	mendelevium	Md	0
50	tin	Sn	3.82×10^{0}	102	nobelium	No	0
51	antimony	Sb	3.52×10^{-1}	103	lawrencium	Lr	0
52	tellurium	Te	4.91×10^{0}				

[a]The terrestrial abundances of the radioactive daughters of uranium and thorium are very low and are therefore indicated as "~0." In addition, neptunium and plutonium are produced in nuclear reactors and therefore occur on the Earth. However, the transuranium elements having atomic numbers of 95 or greater do not occur in the solar system. Their abundances are therefore stated as "0." SOURCE: Anders and Ebihara (1982).

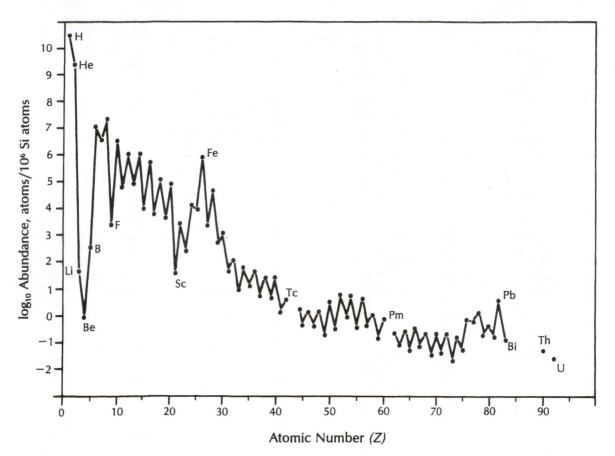

Figure 2.2 Abundances of the chemical elements in the solar system in terms of atoms per 10^6 atoms of Si. The data were derived primarily by analysis of carbonaceous chondrite meteorites and by optical spectroscopy of light from the Sun and nearby stars (Anders and Ebihara, 1982).

5. The abundances of lithium, beryllium, and boron are anomalously low compared to other elements of low atomic number.

6. The abundance of iron is notably higher than those of other elements with similar atomic numbers.

7. Two elements, technetium and promethium, do not occur in the solar system because all of their isotopes are unstable and decay rapidly.

8. The elements having atomic numbers greater than 83 (Bi) have no stable isotopes, but occur naturally at very low abundances because they are the daugh-

ters of long-lived radioactive isotopes of uranium and thorium.

The nucleosynthesis model of B_2FH (1957) includes eight different kinds of nuclear reactions that occur at specified temperatures in the course of the evolution of a star. Several of these reactions may take place simultaneously in the cores and outer shells of massive stars. As a result, the core of a star may have a different chemical composition than the shell surrounding it. Moreover, not all of the nuclear processes take place in all stars. Consequently, other stars in the Milky Way Galaxy do not necessarily have the same chemical composition as the Sun and her planets.

All stars on the main sequence generate energy by hydrogen fusion reactions. This process results in the synthesis of helium either by the direct proton–proton chain (equations 2.5–2.8) or by the CNO cycle (equations 2.9–2.14). The proton–proton chain works as follows. Two nuclei of hydrogen, consisting of one proton each, collide to form the nucleus of deuterium ($_1^2H$) plus a positron (β^+) and a neutrino (ν). (The designation of atomic species is presented in Chapter 6.) Each reaction of this kind liberates 0.422 million electron volts (MeV) of energy. The positron (positively charged electron) is annihilated by interacting with a negatively charged electron giving off additional energy of 1.02 MeV. The deuterium nucleus collides with another proton to form the nucleus of helium-3 ($_2^3He$) plus a gamma ray (γ) and 5.493 MeV of energy. Finally, two helium-3 nuclei must collide to form helium-4 ($_2^4He$), two protons, and 12.859 MeV. The end result is that four hydrogen nuclei fuse to form one nucleus of helium-4, a gamma ray, a neutrino, and 19.794 MeV of energy.

The entire process can be described by a series of equations in which the nuclei of hydrogen and helium are represented by the symbols of the appropriate isotopes (see Chapter 6), even though these isotopes do not actually exist in atomic form in stellar interiors where their electrons are removed from them because of the high temperature:

$$_1^1H + {}_1^1H \rightarrow {}_1^2H + \beta^+ + \nu + 0.422\ \text{MeV} \quad (2.5)$$

$$\beta^+ + \beta^- \rightarrow 1.02\ \text{MeV (annihilation)} \quad (2.6)$$

$$_1^2H + {}_1^1H \rightarrow {}_2^3He + \gamma + 5.493\ \text{MeV} \quad (2.7)$$

$$_2^3He + {}_2^3He \rightarrow {}_2^4He + {}_1^1H + {}_1^1H + 12.859\ \text{MeV} \quad (2.8)$$

The direct proton–proton fusion to form helium-4 can only take place at a temperature of about 10×10^6 K, and even then the probability of its occurrence (or its "reaction cross section") is very small. Nevertheless, this process was the only source of nuclear energy for first-generation stars that formed from the primordial mixture of hydrogen and helium after the Big Bang.

Once the first generation of stars had run through their evolutionary cycles and had exploded, the interstellar gas clouds contained elements of higher atomic number. The presence of carbon-12 ($_6^{12}C$) synthesized by the ancestral stars has made it easier for subsequent generations of stars to generate energy by hydrogen fusion. This alternative mode of hydrogen fusion was discovered by Hans Bethe and is known as the CNO cycle:

$$_6^{12}C + {}_1^1H \rightarrow {}_7^{13}N + \gamma \quad (2.9)$$

$$_7^{13}N \rightarrow {}_6^{13}C + \beta^+ + \nu \quad (2.10)$$

$$_6^{13}C + {}_1^1H \rightarrow {}_7^{14}N + \gamma \quad (2.11)$$

$$_7^{14}N + {}_1^1H \rightarrow {}_8^{15}O + \gamma \quad (2.12)$$

$$_8^{15}O \rightarrow {}_7^{15}N + \beta^+ + \nu \quad (2.13)$$

$$_7^{15}N + {}_1^1H \rightarrow {}_6^{12}C + {}_2^4He \quad (2.14)$$

The end result is that four protons are fused to form one nucleus of $_2^4He$, as in the direct proton–proton chain. The nucleus of $_6^{12}C$ acts as a sort of catalyst and is released at the end. It can then be reused for another revolution of the CNO cycle.

The Sun contains elements of higher atomic number than helium including $_6^{12}C$ and therefore carries on hydrogen fusion by the CNO cycle. In fact, most stars in our Milky Way Galaxy are second-generation stars because our Galaxy is so old that only the very smallest first-generation stars could have survived to the present time.

The low reaction cross section of the proton–proton chain by which the ancestral stars generated energy has been a source of concern to nuclear astrophysicists. When this difficulty was pointed out to Sir Arthur Eddington, who proposed hydrogen fusion in stars in 1920, he replied (Fowler, 1967):

> We do not argue with the critic who urges that the stars are not hot enough for this process; we tell him to go and find a hotter place.

After the hydrogen in the core has been converted to helium "ash," hydrogen fusion ends, and the core contracts under the influence of gravity. The core temperature rises toward 100×10^6 K, and the helium "ash" becomes the fuel for the next set of energy-producing nuclear reactions. The critical reaction for helium burning is the fusion of three alpha particles (triple-alpha process) to form a nucleus of $^{12}_{6}C$:

$$^{4}_{2}He + ^{4}_{2}He \rightarrow ^{8}_{4}Be \qquad (2.15)$$

$$^{8}_{4}Be + ^{4}_{2}He \rightarrow ^{12}_{6}C + \gamma \qquad (2.16)$$

This is the critical link in the chain of nucleosynthesis because it bridges the gap in nuclear stability of the isotopes of lithium, beryllium, and boron. The crux of the problem is that the nucleus of $^{8}_{4}Be$ is very unstable and decays quickly with a "half-life" of about 10^{-16} sec. Therefore, $^{8}_{4}Be$ must absorb a third helium nucleus very soon after its formation to make it safely to stable $^{12}_{6}C$ (Fowler, 1967). An alternative reaction involving the addition of a proton to the nucleus of $^{4}_{2}He$ has an even smaller chance to succeed because the product, $^{5}_{3}Li$, has a half-life of only 10^{-21} sec and decomposes to form helium and hydrogen:

$$^{4}_{2}He + ^{1}_{1}H \rightarrow ^{5}_{3}Li \qquad (2.17)$$

$$^{5}_{3}Li \rightarrow ^{4}_{2}He + ^{1}_{1}H \qquad (2.18)$$

The triple-alpha process is indeed the key to the synthesis of all elements beyond helium. Without it, stellar evolution would be short-circuited, and the universe would be composed only of hydrogen and helium.

Helium burning sustains red giants only for a few tens of millions of years or less. With increasing temperature in the core, alpha particles fuse with the nuclei of $^{12}_{6}C$ to produce nuclei of still higher atomic number:

$$^{12}_{6}C + ^{4}_{2}He \rightarrow ^{16}_{8}O \qquad (2.19)$$

$$^{16}_{8}O + ^{4}_{2}He \rightarrow ^{20}_{10}Ne \qquad (2.20)$$
$$\text{etc.}$$

However, electrostatic repulsion between positively charged nuclei and alpha particles limits the size of the atoms that can form in this way. The heaviest atom produced by the addition of alpha particles is $^{56}_{28}Ni$, which decays to $^{56}_{27}Co$ and then to stable $^{56}_{26}Fe$. These nuclear reactions therefore cause the enhanced abundance of the elements in the iron group illustrated in Figure 2.2.

During the final stages of the evolution of red giants, several other kinds of nuclear reactions occur (B_2FH, 1957). The most important of these

are the neutron-capture reactions, which produce a large number of the atoms having atomic numbers greater than 26 (Fe). These reactions involve the addition of a neutron to the nucleus of an atom to produce an isotope having the same atomic number but a large mass number. For example, the equation:

$$\underset{\text{neutron}}{^{62}_{28}Ni + ^{1}_{0}n} \rightarrow ^{63}_{28}Ni + \gamma \qquad (2.21)$$

indicates that the nucleus of $^{62}_{28}Ni$ absorbs a neutron, which changes it to an excited state of $^{63}_{28}Ni$, which then deexcites by emitting a gamma ray. Nickel-63 is radioactive and decays to stable $^{63}_{29}Cu$ by emitting a β^{-} particle:

$$^{63}_{28}Ni \rightarrow ^{63}_{29}Cu + \beta^{-} + \bar{\nu} + 0.0659 \text{ MeV} \quad (2.22)$$

where $\bar{\nu}$ is an antineutrino. $^{63}_{29}Cu$ is a stable isotope of copper and can absorb another neutron to form $^{64}_{29}Cu$:

$$^{63}_{29}Cu + ^{1}_{0}n \rightarrow ^{64}_{29}Cu + \gamma \qquad (2.23)$$

Copper-64 is radioactive and undergoes branched decay to form $^{64}_{30}Zn$ and $^{64}_{28}Ni$, both of which are stable:

$$^{64}_{29}Cu \rightarrow ^{64}_{30}Zn + \beta^{-} + \bar{\nu} + 0.575 \text{ MeV} \quad (2.24)$$

$$^{64}_{29}Cu \rightarrow ^{64}_{28}Ni + \beta^{+} + \nu + 1.678 \text{ MeV} \quad (2.25)$$

This process of successive additions of neutrons is illustrated in Figure 2.3. It takes place during the red giant stage of stellar evolution when the neutron flux is low enough to permit the product nucleus to decay before the next neutron is added. This process therefore is characteristically *slow* and is therefore referred to as the *s-process*.

By examining Figure 2.3 closely we see that the track of the s-process bypasses $^{70}_{30}Zn$, which is one of the stable isotopes of zinc. In order to make this isotope by neutron capture reactions, the pace must be speeded up so that unstable $^{69}_{30}Zn$ can pick up a neutron to form $^{70}_{30}Zn$ before it decays to stable $^{70}_{31}Ga$:

$$^{69}_{30}Zn + ^{1}_{0}n \rightarrow ^{70}_{30}Zn + \gamma \qquad (2.26)$$

An even more rapid rate of neutron capture is required to make $^{70}_{30}Zn$ from $^{65}_{29}Cu$ by addition of five neutrons in succession to form $^{70}_{29}Cu$, which then decays by β^{-} emission to stable $^{70}_{30}Zn$:

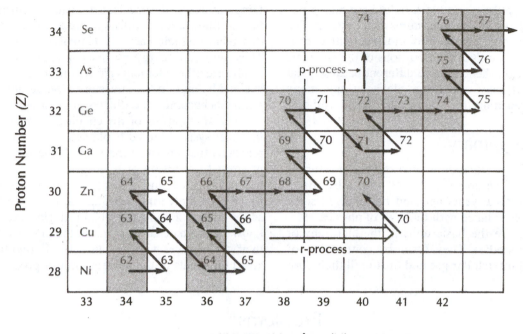

Figure 2.3 Nucleosynthesis in red giants by neutron capture on a slow time scale (s-process) followed by beta decay. The dark squares are stable isotopes, whereas the unshaded ones are radioactive. The process starts with stable $^{62}_{28}$Ni, which absorbs a neutron to form unstable $^{63}_{28}$Ni, which decays to stable $^{63}_{29}$Cu by emitting a β^- particle. The main line of the s-process, as indicated by arrows, proceeds from $^{62}_{28}$Ni to $^{77}_{34}$Se and beyond but bypasses $^{70}_{30}$Zn and $^{74}_{34}$Se. $^{70}_{30}$Zn is produced by neutron capture on a rapid time scale (r-process) from $^{68}_{30}$Zn via unstable $^{69}_{30}$Zn and from $^{65}_{29}$Cu, which captures five neutrons in rapid succession to form $^{70}_{30}$Cu followed by decay to stable $^{70}_{30}$Zn. $^{74}_{34}$Se is a proton-rich nuclide that cannot form by either the s-process or the r-process and requires the addition of two protons (p-process) to stable $^{72}_{32}$Ge.

$$^{65}_{29}\text{Cu} + 5\,^{1}_{0}\text{n} \rightarrow\, ^{70}_{29}\text{Cu} + 5\gamma \qquad (2.27)$$

$$^{70}_{29}\text{Cu} \rightarrow\, ^{70}_{30}\text{Zn} + \beta^- + \bar{\nu} +\ \sim7.2\,\text{MeV} \qquad (2.28)$$

Capture of neutrons at such *rapid* rates is characteristic of the *r-process,* which requires a much greater neutron flux than the s-process and therefore takes place only during the last few minutes in the life of a red giant when it explodes as a supernova.

However, no neutron capture on any time scale can account for the formation of some atoms such as stable $^{74}_{34}$Se shown also in Figure 2.3. This nuclide is synthesized by the addition of two protons to stable $^{72}_{32}$Ge in the so-called *p-process:*

$$^{72}_{32}\text{Ge} + 2\,^{1}_{1}\text{H} \rightarrow\, ^{74}_{34}\text{Se} \qquad (2.29)$$

This process also takes place at the very end of the giant stage of stellar evolution.

The system of nuclear reactions originally proposed by B$_2$FH (1957) can account for the observed abundances of the chemical elements in the solar system and in nearby stars. Nucleosynthesis is taking place at the present time in the stars of our galaxy and in the stars of other galaxies throughout the universe. We have good evidence in the wavelength spectra of light from distant galaxies that the chemical elements we find on Earth also occur everywhere else in the universe.

However, pulsars and black holes have high internal pressures and temperatures that cause atomic nuclei to disintegrate into more primitive constituents. The relative proportions of the chemical elements in other stars are different because local conditions may affect the yields of the many nuclear reactions that contribute to their synthesis.

2.4   Summary

We live in an expanding universe whose future is uncertain. The universe started with a Big Bang about 15×10^9 years ago and has evolved since then in accordance with the laws of physics.

Stars are the basic units in the hierarchy of heavenly bodies. They form by contraction of clouds of interstellar gas and dust until their core temperatures are sufficient to cause hydrogen fusion. Stars evolve through predictable stages depending on their masses and initial compositions. They generate energy by nuclear reactions that synthesize other elements from primordial hydrogen and helium. In the end, stars explode, and the remnants become solid objects of great density.

The abundances of the chemical elements in the solar system can be explained by the nuclear reactions that energize the stars. These reactions progress from fusion of hydrogen and helium to neutron capture and to other reactions, most of which occur only for a short time at the end of the active life of a star. The chemical elements we know on Earth occur throughout the universe, but their abundances vary because local conditions affect the yields of the nucleosynthesis processes.

Problems*

1. How has the abundance of hydrogen (H) in the universe changed since the Big Bang?

2. Why are elements with even atomic numbers more abundant than their neighbors with odd atomic numbers?

3. How do we know that the Sun is at least a "second-generation" star?

4. Why do technetium (Tc) and promethium (Pm) lack stable isotopes?

5. What other elements also lack stable isotopes?

6. Why do the elements of Problem 5 exist on the Earth, whereas Tc and Pm do not?

*Not all of these questions have unequivocal answers.

7. Why is lead (Pb) more abundant than we might have expected?

8. Check the abundance of argon (Ar) and determine whether it is greater than expected. If so, suggest an explanation.

9. How did lithium (Li), beryllium (Be), and boron (B) form?

10. Compare the abundances of the "rare earths" (lanthanum to lutetium) to such well-known metals as Ta, W, Pt, Au, Hg, and Pb. Are the rare earths really all that "rare"?

References

ALPHER, R. A., H. BETHE, and G. GAMOW, 1948. The origin of the chemical elements. *Phys. Rev.,* **73**:803–804.

ANDERS, E., and M. EBIHARA, 1982. Solar-system abundances of the elements. *Geochim. Cosmochim. Acta,* **46**:2363–2380.

ATKINSON, R. d'E., and F. G. HOUTERMANS, 1929. Zur Frage der Aufbaumöglichkeit der Elemente in Sternen. *Zeit. Phys.,* **54**:656–665.

BAADE, W., 1968. *Evolution of Stars and Galaxies.* Harvard University Press, Cambridge, MA.

BETHE, H. A., 1939. Energy production in stars. *Phys. Rev.,* **55**:434–456.

BETHE, H. A., 1968. Energy production in stars. *Science,* **161**:541–547.

BURBRIDGE, E. M., G. R. BURBRIDGE, W. A. FOWLER, and F. HOYLE, 1957. Synthesis of the elements in stars. *Rev. Mod. Phys.,* **29**:547–650.

FOWLER, W. A., 1967. Nuclear astrophysics. *Amer. Phil. Soc. Mem.* **67,** 109 pp.

GAMOW, G., 1935. Nuclear transformations and the origin of the chemical elements. *Ohio J. Sci.,* **35**:406–413.

GAMOW, G., 1952. *The Creation of the Universe.* Viking Press, New York, 144 pp.

GAMOW, G., 1963. My early memories of Fritz Houtermans. In J. Geiss and E. D. Goldberg (Eds.), *Earth Science and Meteoritics.* North-Holland Publ. Co., Amsterdam, 312 pp.

GOTT, J. R., III, 1982. Creation of open universes from de Sitter space. *Nature,* **295**:304–307.

HAINEBACH, K., D. KAZANAS, and D. N. SCHRAMM, 1978. A consistent age for the universe. Short papers of the Fourth International Conference, Geochronology, Cosmochronology and Isotope Geology. *U.S. Geol. Sur. Open-File Rept.,* 78–701, 159–161.

HEWISH, A., 1975. Pulsars and high density physics. *Science,* **188**:1079–1083.

HUBBLE, E., 1936. *The Realm of the Nebulae.* Yale University Press, New Haven, CT. Reprinted 1958 by Dover Pub., Inc., New York.

IBEN, I., Jr., 1967. Stellar evolution: Comparison of theory with observation. *Science,* **155**:785–796.

IBEN, I., Jr., 1974. Post main sequence evolution of single stars. *Ann. Rev. Astron. and Astrophys.,* **12**:215–256.

MASON, B., 1962. *Meteorites.* Wiley, New York, 274 pp.

PENZIAS, A. A., and R. W. WILSON, 1965. A measurement of excess antenna temperature at 4,080 Mc/s. *Astrophys. J.,* **142**:419.

SCHRAMM, D. N., and W. D. ARNETT, 1973. *Explosive Nucleosynthesis.* University of Texas Press, Austin and London, 301 pp.

WEINBERG, S., 1977. *The First Three Minutes.* Bantam Books, New York, 177 pp.

WHEELER, J. C., 1973. After the supernova, what? *Amer. Scientist,* **61**:42–51.

3

The Solar System

The Sun formed from a cloud of gas and dust particles, as did all other stars in the Milky Way Galaxy and elsewhere in the universe. Sometimes the process leads to the formation of two companion stars. However, in the case of the Sun a small fraction of the original cloud accreted to form a set of nine planets, including the Earth. Compared to stars, these planets are insignificant objects, but to us they are the very basis of existence. The human race appeared on the Earth only two or three million years ago and, after a slow start, learned to fly and to explore the solar system within a span of a mere century.

Geochemistry today not only encompasses the study of the composition and chemical processes occurring on the Earth, but is also concerned with all of the planets and their satellites. Information for geochemical studies of the solar system is derived by analysis of meteorites and rock samples from the Moon and by remote sensing of planetary surfaces. The exploration of the solar system has expanded our horizon and has provided the basis for comparative planetary geochemistry. The satellites of the large gaseous planets are of special interest in this new field of study because some of them are larger than our Moon and have very different chemical compositions and surface features than the Earth.

We therefore need to become acquainted with the new worlds we must explore before we concentrate our attention on the conventional geochemistry of the Earth.

3.1 Origin of the Solar System

The origin of the planets of the solar system is intimately linked to the formation of the Sun. In the beginning there was a diffuse mass of interstellar gas and dust known as the *solar nebula*. It had formed about six billion years ago as a result of the terminal explosions of ancestral stars, which added the elements they had synthesized to the primordial hydrogen and helium that originated from the Big Bang. The chemical composition of the solar nebula was given in Chapter 2 (Table 2.1). The dust cloud was rotating in the same sense of the Milky Way Galaxy and was acted upon by gravitational, magnetic, and electrical forces.

As soon as the main mass of the solar nebula began to contract, order began to be imposed on it by the physical and chemical conditions that existed during this phase of star formation. These included the development of pressure and temperature gradients and an increase in the rate of rotation. Certain kinds of solid particles that had formed in the nebula evaporated as the temperature increased in order to maintain equilibrium between solids and gases. As a result, only the most refractory particles (Fe–Ni alloys, Al_2O_3, CaO, etc.) survived in the hottest part of the contracting nebula, whereas in the cooler outer regions a larger variety of compounds remained in the solid state. The increase in the rate of rotation caused part of the nebula outside of the protosun to form a central disk. The solid particles congregated in this disk and made it sufficiently opaque to absorb infrared radiation. The temperature in the central disk therefore increased until it ranged from about 2000 K at the center to about 40 K at approximately 7.5×10^9 km from the protosun. The pressure ranged from less than 0.1 atm to about 10^{-7} atm near the edge of the disk (Cameron, 1978; Cameron and Pine, 1973).

The development of pressure and temperature gradients within the disk caused the first major chemical differentiation of the solar nebula. Compounds with low vapor pressures persisted throughout the nebula and formed "dust" particles, whereas compounds with high vapor pressures could exist only in the cooler outer regions. The condensation temperatures of various compounds that existed in the solar nebula are listed in Table 3.1. The condensates accreted to form larger bodies as a result of selective adhesion caused by electrostatic and magnetic forces. The resulting solid bodies, called *planetesimals*, had diameters ranging from about 10 m to more than 1000 km, and their chemical compositions varied with distance from the center of the planetary disk. The planetesimals close to the protosun were composed of refractory compounds dominated by oxides and metallic iron and nickel; farther out were Mg and Fe silicates and farthest out, ices composed of water, ammonia, methane, and other volatiles.

The initial rate of evolution of the solar system was remarkably fast. The time required for the Sun to reach the ignition temperature for hydrogen fusion was less than 100,000 years. The Sun's initial luminosity was two to three times greater than is consistent with the main sequence because the Sun contained excess thermal energy generated during the initial contraction. This "superluminous" phase of the Sun lasted about 10 million years and resulted in the expulsion of about 25% of its original mass in the form of "solar wind" composed of a proton/electron plasma. This is called the "T-Tauri stage" of stellar evolution after the star that is the prototype for this process. All gaseous matter in the vicinity of the Sun was blown away during this period, and only the solid planetesimals having diameters larger than about 10 m remained.

The planetesimals in the inner region of the planetary disk subsequently accreted to form the so-called earthlike planets—Mercury, Venus, Earth, and Mars—and the parent bodies of the meteorites now represented by asteroids. Instabilities in the outer part of the disk resulted in the formation of the gaseous outer planets: Jupiter, Saturn, Uranus, Neptune, and Pluto.

Table 3.1 Condensates from the Solar Nebula at Different Temperatures

Temperature, °C	Condensates
1325	refractory oxides: CaO, Al_2O_3, TiO_2, REE oxides
1025	metallic Fe and Ni
925	enstatite ($MgSiO_3$)
925–220	Fe forms FeO, which reacts with enstatite to form olivine [$(Fe,Mg)_2SiO_4$]
725	Na reacts with Al_2O_3 and silicates to form feldspar and related minerals. Condensation of K and other alkali metals
400	H_2S reacts with metallic Fe to form troilite (FeS)
280	H_2O vapor reacts with Ca-bearing minerals to make tremolite
150	H_2O vapor reacts with olivine to form serpentine
−100	H_2O vapor condenses to form ice
−125	NH_3 gas reacts with water ice to form solid $NH_3 \cdot H_2O$
−150	CH_4 gas reacts with water ice to form solid $CH_4 \cdot 7H_2O$
−210[a]	Ar and excess CH_4 condense to form Ar and CH_4
−250[a]	Ne, H, and He condense

[a]These reactions probably did not occur because the temperature in the planetary disk did not decrease to such low values.
SOURCE: Glass (1982), based on Lewis (1974).

The origin and chemical composition of Pluto are still not well known because this planet is difficult to observe from Earth. Pluto also does not conform to the so-called *Titius–Bode law,* which appears to govern the distances of the planets from the Sun when these are expressed in *astronomical units* (A. U.), defined as the average distance between the Earth and Sun. The Titius–Bode law was published in 1772 by J. E. Bode, director of the Astronomical Observatory in Berlin, and is based on a series of numbers discovered by J. D. Titius of Wittenberg in 1766. The series is composed of the numbers 0.4, 0.7, 1.0, 1.6, 2.8, . . . , which are obtained by writing 0, 3, 6, 12, 24, . . . , adding 4 to each number, and dividing by 10. The resulting numbers match the distances of the planets from the Sun remarkably well up to and including Uranus (Mehlin, 1968). However, the radius of the orbit of Neptune is only 30.1 A.U., whereas the Titius–Bode value is 38.8 and the discrepancy in the case of Pluto is even greater (Table 3.2). The Titius–Bode law predicts a value of 77.2, but the actual orbital radius of Pluto is only 39.4 A.U. The discrepancy may suggest that Pluto did not form in the orbit it now occupies.

The descriptive physical properties of the solar system listed in Table 3.2 indicate that 99.87% of the total mass of the solar system $(2.052 \times 10^{33}$ g) is concentrated in the Sun. The remaining 0.13% is distributed among the nine major planets, among which Jupiter is by far the largest with about 71% of the planetary masses. The densities and sizes of the planets, shown in Figure 3.1, vary widely and imply the existence of large differences in their chemical compositions. The inner planets (Mercury, Venus, Earth, and Mars), as well as the Moon and the asteroids, are solid objects composed primarily of elements and compounds having low vapor pressures. The outer planets (Jupiter, Saturn, Uranus, and Neptune) have low densities and are essentially gaseous, although all of them probably have condensed cores. The inner planets resemble the Earth in chemical composition and are therefore referred to as the *terrestrial* or *earthlike* planets. The outer planets consist primarily of hydrogen and helium with small amounts of the other elements and resemble the Sun in chemical composition.

The earthlike planets (plus the Moon and asteroids), taken together, make up only 0.0006% of the total mass of the solar system and only

Table 3.2 Properties of the Sun and Her Planets

Object	Mean solar distance		Equatorial diameter, km	Mass, g	Density g/cm³	Number of satellites
	10^6 km	A.U.				
Sun	—		1,391,400	1.987×10^{33}	1.4	—
Mercury	57.9	0.387	4,878	3.30×10^{26}	5.44	0
Venus	108.2	0.723	12,100	4.87×10^{27}	5.25	0
Earth	149.6	1.00	12,756	5.98×10^{27}	5.52	1
Moon	—		3,476	7.35×10^{25}	3.34	—
Mars	227.9	1.52	6,786	6.44×10^{26}	3.94	2
Ceres (asteroid)	414.0	2.77	1,020	1.17×10^{24}	2.2	—
Jupiter	778.3	5.20	142,984	1.90×10^{30}	1.33	16
Saturn	1,427.0	9.54	120,536	5.69×10^{29}	0.70	21
Uranus	2,869.6	19.2	51,118	8.66×10^{28}	1.30	15
Neptune	4,504	30.1	49,562	1.0123×10^{29}	1.76	8
Pluto	5,900	39.4	3,000	1.5×10^{25}	1.1	1

SOURCE: Carr et al. (1984), Glass (1982), and fact sheet, Jet Propulsion Laboratory.

0.44% of the planetary masses. All of these objects are so close to the Sun that the orbital radius of the outermost asteroid is only about 7% of the total radius of the solar system. Evidently, the terrestrial planets are not typical of the solar system and owe their existence to the special conditions in the planetary disk close to the Sun. Earth is the largest of the inner planets with about 50.3% of the mass, followed by Venus (40.9%), Mars (5.4%), and Mercury (2.8%).

The earthlike planets, viewed in Figure 3.1 in the perspective of the entire solar system, are a physical and chemical anomaly. The Earth is unique among its earthlike neighbors in having about 71% of its surface covered by liquid water within which life developed early in its history and evolved into the multitude of species of the plant and animal kingdoms. Insofar as we know, life forms do not exist at the present time anywhere else in the solar system.

3.2 Origin of the Earthlike Planets

Our current understanding of the origin of the solar system indicates that the earthlike planets were hot when they formed and that their internal geochemical differentiation may have begun with the *sequential accretion* of planetesimals of differing compositions (Murray et. al., 1981). Planetesimals composed of metallic iron and oxides accreted first to form a core that was subsequently buried by the planetesimals composed of silicates. The earthlike planets were initially molten because of the heat generated by the rapid capture of the planetesimals and because of radioactive heating.

The last phase of formation of Earth, Venus, and Mars involved the capture of planetesimals composed of volatile compounds that had formed in the outer reaches of the planetary disk, perhaps

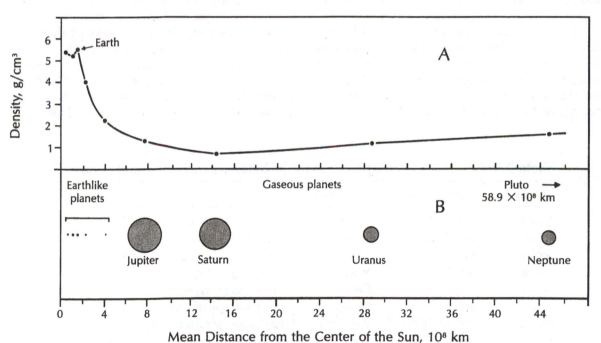

Figure 3.1 A: Variation of density of the planets with mean distance from the Sun. Note that the Earth has the highest density among the earthlike planets, which, as a group, are more dense than the outer gaseous planets. **B:** The planets of the solar system magnified 2000 times relative to the distance scale. The earthlike planets are very small in relation to the Sun and the gaseous planets of the solar system.

beyond the orbit of Jupiter. These volatile-rich planetesimals, also known as *cometesimals*, deposited solids composed of water, ammonia, methane, and other volatiles on the surfaces of the planets. The water and other volatiles deposited on the Earth promptly evaporated to form a dense atmosphere from which water ultimately condensed as the surface of the Earth cooled. Mercury and the Moon do not have atmospheres partly because they are too small to retain gaseous elements of low atomic number and their compounds.

According to this scenario, the earthlike planets have been cooling since the time of their formation. Mercury and the Moon have cooled sufficiently to become geologically "inactive" in the sense that their interiors no longer interact with their surfaces. Venus and Earth, being the largest of the earthlike planets, have retained more of the initial heat than their neighbors in the solar system and are still active. Mars is intermediate in size and has had volcanic eruptions in the not-too-distance geologic past. However, the age of the last martian volcanism is not known.

In spite of the similarity in size and overall composition of Venus and Earth, their surfaces have evolved very differently. Venus has a dense atmosphere composed of CO_2 that has caused its surface to become extremely hot and dry. The surface of Earth cooled rapidly, allowing oceans

to form more than 4×10^9 years ago by condensation of water vapor in the atmosphere. The presence of a large volume of water on the surface of Earth permitted geological processes to operate and created conditions conducive to the development and evolution of life. Neither the Earth nor any of the earthlike planets ever had atmospheres composed of the hydrogen and helium of the solar nebula because these gases were expelled from the inner region of the solar system during the T-Tauri stage of the Sun.

3.3 Satellites of the Outer Planets

All of the outer planets have satellites, some of which are larger than the Moon and the planet Mercury. In addition, these satellites have a wide range of chemical compositions and have responded in very different ways to the forces acting on them. The large satellites of Jupiter were seen by Galileo Galilei on January 7, 1610, with his newly built telescope. Actually, they may have been observed even earlier by Simon Marius. However, detailed images of their surfaces were obtained only recently during "flybys" of the American Pioneer and Voyager space probes. A chronology of these events in Table 3.3 reveals the sudden burst of activity in the exploration of the solar system between 1973 and 1989.

Table 3.3 Chronology of Exploration of the Satellites of Jupiter and Saturn

Planet	Encounter date	Spacecraft	Comments
Jupiter	Dec. 3, 1973	Pioneer 10	left solar system on June 14, 1983
Jupiter	Dec. 2, 1974	Pioneer 11	continued to Saturn
Jupiter	March 5, 1979	Voyager 1	1408 images of jovian satellites
Jupiter	July 9, 1979	Voyager 2	1364 images of jovian satellites
Jupiter	Dec. 7, 1995	Galileo	mission in progress in 1997
Saturn	Sept. 1, 1979	Pioneer 11	low-resolution images
Saturn	Nov. 13, 1980	Voyager 1	900 images of saturnian satellites
Saturn	Aug. 26, 1981	Voyager 2	1150 images of saturnian satellites
Uranus	Jan. 25, 1986	Voyager 2	images of all satellites as well as the rings
Neptune	Aug. 25, 1989	Voyager 2	images of the planet and its satellite Triton

After Greeley (1985).

The planet Jupiter has 16 satellites and a ring guarded by two small satellites. The so-called Galilean satellites Io, Europa, Ganymede, and Callisto were actually named by Marius after mythological lovers of the Greek god Zeus whose Latin name was Jupiter. A fifth satellite was discovered 282 years later by Barnard who named it Amalthea after a nymph who once nursed Jupiter. These satellites, together with three small ones found in 1979, move in nearly circular orbits close to the equatorial plane of Jupiter. Four small satellites (Leda, Himalia, Lysithea, and Elara) are located about 11 million km from Jupiter and have eccentric orbits that are inclined nearly 30° to its equatorial plane. A third group (Ananke, Carme, Pasiphaë, and Sinope) at a distance of about 22 million km from Jupiter have retrograde orbits whose inclination is between 150° and 160°. The Galilean satellites form in effect a small-scale planetary system of their own with Jupiter as their central "star." Their densities, listed in Table 3.4 and displayed in Figure 3.2, decrease with increasing orbital radius from 3.53 (Io) to 1.79 (Callisto) g/cm³; therefore they are believed to have significantly different chemical compositions.

Io is composed primarily of silicate material and may have an iron sulfide core. Before the encounters with the Voyager spacecraft, Peale et al. (1979) calculated the amount of heat generated within Io by tidal friction caused by the gravitational pull of Jupiter and Europa and predicted that active volcanoes may be present. This prediction was supported by images sent back by Voyager 1 on March 9, 1979, which indicated that the surface of this planet was *not* cratered as had been expected. Subsequently, Morabito et. al. (1979) discovered a huge volcanic eruption on Io that was obliterating impact craters and all other topographic features in a large area surrounding the volcano. Eventually, nine active volcanoes were identified and named after mythological deities associated with fire (Amirani, Loki, Marduk, Masubi, Maui, Pele, Prometheus, Surt, and Voland). The volcanic plumes contain sulfur dioxide and the lava flows may be composed of liquid sulfur, consistent with the fact that the surface of Io is yellow to red in color. The surface heat flow of Io is about 48 microcalories/cm²/sec and is therefore about 30 times greater than that of the Earth. Io is clearly the most volcanically active object in the solar system.

The other Galilean satellites of Jupiter (Europa, Ganymede, and Callisto) have lower densities than Io and are composed of silicate material with crusts of water ice and mantles of liquid water. They are not volcanically active at the present time, and their surfaces are cratered. Europa appears to be completely covered by a frozen ocean 75–100 km deep. The icy crust may be underlain by liquid water that does not freeze because of heat generated by tidal friction. The surface of Europa is crisscrossed by a multitude of curving bands, some of which have been traced for more than 1000 km. The bands appear to be fractures in the crust caused by internal tectonic activity and by meteorite impacts. The fractures were subsequently filled with subcrustal water that froze to form ice dikes.

Ganymede is larger than the planet Mercury and appears to be composed of water and silicate material in about equal proportions. Its surface is composed of water ice mixed with impurities that cause it to darken in color. The dark terrains are fragmented and more heavily cratered than the light-colored terrain in which they are embedded. The ice crust is about 100 km thick and is underlain by a mantle of liquid water between 400 and 800 km deep. The water may have intruded the crust locally in the form of ice slush to produce intrusive bodies of water ice.

Callisto is the outermost of the Galilean satellites. It is darker in color than the others and has a heavily cratered icy crust about 200 km thick. A prominent feature on its surface is a very large multiringed basin called *Valhalla* whose diameter is nearly 2000 km. A second ringed basin near the north pole is called *Asgard*. The icy crust may be underlain by a liquid mantle about 1000 km thick composed of water. Callisto apparently became inactive very early in its history partly because the amount of heat generated by tidal friction is less than that of the other Galilean satellites. In

Table 3.4 Physical Properties of the Major Satellites of Jupiter, Saturn, Uranus, and Neptune

Name	Mean distance from planet, km	Diameter, km	Mass, g	Density, g/cm^3
Jupiter				
Io	421,600	3,640	8.91×10^{25}	3.53
Europa	670,900	3,130	4.87×10^{25}	3.03
Ganymede	1,070,000	5,280	1.49×10^{26}	1.93
Callisto	1,883,000	4,840	1.06×10^{26}	1.79
Saturn				
Mimas	188,000	390	$(3.7 \pm 0.1) \times 10^{22}$	1.2
Enceladus	240,000	500	$(7.4 \pm 3.4) \times 10^{22}$	1.1
Tethys	297,000	1,050	$(6.26 \pm 0.17) \times 10^{23}$	1.0
Dione	379,000	1,120	$(1.05 \pm 0.03) \times 10^{24}$	1.42
Rhea	528,000	1,530	$(2.16 \pm 0.7) \times 10^{24}$	1.3
Titan	1,221,000	5,140	$(1.348 \pm 0.017) \times 10^{26}$	1.9(+)
Hyperion	1,502,000	~ 290		
Iapetus	3,559,000	1,440	$(2.8 \pm 0.7) \times 10^{24}$	1.2
Phoebe	10,583,000	160		
Uranus				
Miranda	130,000	470	7.34×10^{22}	1.35
Ariel	192,000	1,150	1.32×10^{24}	1.66
Umbriel	267,000	1,170	1.26×10^{24}	1.51
Titania	438,000	1,580	3.47×10^{24}	1.68
Oberon	586,000	1,520	2.90×10^{24}	1.58
Neptune				
Triton	354,000	3,000	9.273×10^{25}	
Nereid	5,510,000	940	1.3×10^{18}	

Data from Glass (1982), Greeley (1985), Hunt and Moore (1981), Hamblin and Christiansen (1990), and fact sheet, Jet Propulsion Laboratory.

addition, less heat is generated by radioactivity because its rocky core is relatively small.

The satellites of Saturn, Uranus, and Neptune are identified in Table 3.4. All of them were "seen" during flybys of Voyager 2. We will mention only Titan, the largest of the satellites of Saturn. It has a dense atmosphere composed principally of methane, nitrogen, and smaller amounts of other gases including ethane, acetylene, ethylene, and hydrogen cyanide. Like the Gallilean satellites of Jupiter, each of the satellites of Saturn, Uranus, and Neptune is a unique body in the solar system that has recorded on its surface a history of events caused by internal and external processes. Images and remote-sensing data of these satellites were received only recently during

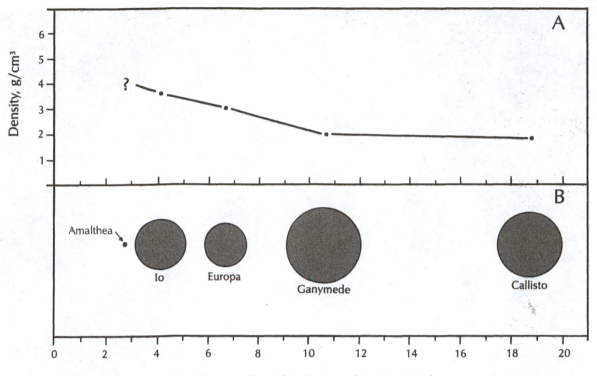

Figure 3.2 A: Variation of the density of the Galilean satellites of Jupiter with increasing distance from the planet. The decrease in density is caused by increases in the proportion of water relative to silicate material. **B:** The Galilean satellites magnified 50 times relative to the distance scale. Amalthea is much smaller than Io but appears to be a silicate object.

the flyby of the Voyager space probes in 1981 (Saturn), 1986 (Uranus), and 1989 (Neptune).

3.4   Pictures of Our Solar System

The exploration of the solar system relies primarily on unmanned space probes that have the capability of taking pictures of planetary surfaces and of returning the images to Earth. The two Viking landers and the Pathfinder mission on Mars, as well as the two Voyager spacecraft and the spacecraft Galileo in the Jupiter system are spectacular examples of this technique, whereas the Moon has been explored primarily by American astronauts who actually walked on its surface.

The exploration of the new worlds, which are suddenly within our reach, will become an important task of the community of Earth Scientists. Although an understanding of these new worlds must ultimately be based on studies of the chemical compositions of matter and of reactions and processes that take place on them, images of the surfaces of planets help to identify the problems that need to be solved.

For this reason, we now examine some of the images of our solar system to confront the challenge that lies ahead. Excellent photographs of landforms on planetary surfaces appear in textbooks by Hamblin and Christiansen (1990) and by Greeley (1985).

The Moon is a familiar image in the sky. Its surface is pockmarked with craters formed by impacts of meteoroids, which continue to fall, though at a greatly reduced rate. The lunar landscape consists of dark plains, called mare (singular) and maria or mares (plural), and bright mountainous highlands. The mare basins, formed by impacts of large objects during the early history of the Moon, are filled with sheetlike basalt flows. The highlands are older than the mare basins and are composed of anorthositic gabbro. The surface of the Moon is covered by a layer of regolith (colloquially called "soil") that consists of rock and mineral particles, beads of impact glass, and chunks of regolith breccia.

   This view of the Moon was taken from space by the astronauts of Apollo 17 in December 1972. It shows the roughly circular, black Mare Crisium in the left upper quadrant. Southwest of Mare Crisium is the Mare Fecunditatis, and directly west of Crisium is the Mare Tranquilitatis. The Mare Serenitatis is located northwest of Tranquilitatis and extends northwesterly beyond the horizon. (Photo by NASA.)

In the late evening of July 20, 1969 (EDT on Earth), Neil A. Armstrong and Edwin E. Aldrin descended from their spacecraft and set foot on the surface of the Moon. They had landed near the southwest margin of the Sea of Tranquility (Mare Tranquilitatis). The picture shows Edwin Aldrin on his way down just before he stepped onto the lunar surface. The dusty plain in the background contains scattered boulders ejected from craters excavated by impact of meteoroids. (Photo by NASA.)

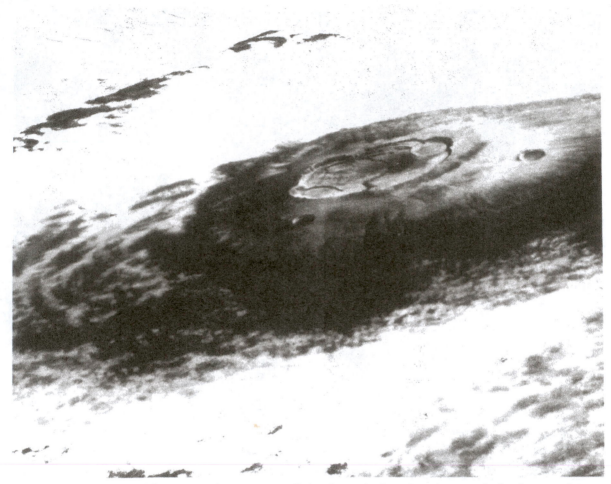

The surface of the planet Mars comes closest among all the planets in the solar system to the landscapes of Earth. Mars has an atmosphere composed of N_2 and CO_2 with a small amount of water. It also has roughly circular plains called planitia and highly cratered highlands resembling those of the Moon. Mars has been an active planet, as indicated by large shield volcanoes and rift valleys. In some places on Mars the surface is dissected by valleys in dendritic patterns similar to stream valleys on Earth. Therefore, there is reason to believe that liquid water has existed on the surface of Mars and that ice, in the form of permafrost, may still occur on Mars at the present time.

The picture shows the summit of the volcano Olympus Mons protruding through clouds on a frosty morning on Mars much like Mauna Loa on the island of Hawaii. The summit contains several overlapping calderas whose presence suggests a long history of volcanic activity. The volcano is 550 km in diameter at its base, and it rises 25 km above the surrounding plain—far higher than any mountain on Earth. (Photo by NASA.)

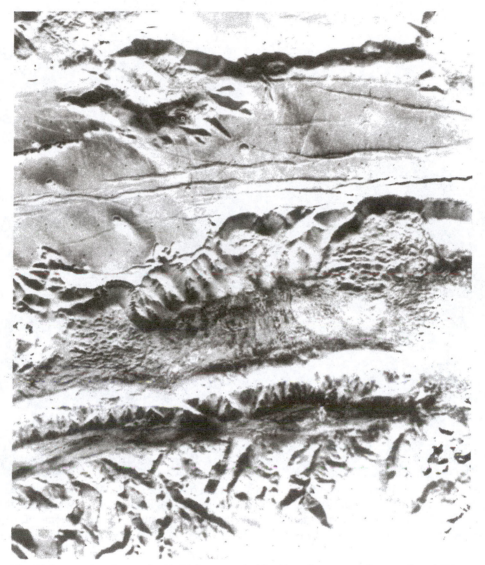

The Valles Marineris on Mars, which are probably rift valleys, extend more than 2400 km in an east–west direction near the martian equator. Some of the valleys are up to 200 km wide and 7 km deep. The walls of the valleys have been extensively modified by slides and by erosional channels. The rocks into which the valleys are cut are layered and may be sheetlike flows of basalt. (Photo by NASA.)

The first Viking lander touched down on Mars in Chryse Planitia on July 20, 1976. As shown, this area is a stony desert containing small sand dunes and angular boulders. The view at the landing site of Viking 2 in the Utopia Planitia is quite similar, indicating the importance of wind in shaping the surface of Mars. Some of the boulders are pitted or vesicular, perhaps because they originated from underlying lava flows. (Photo by NASA.)

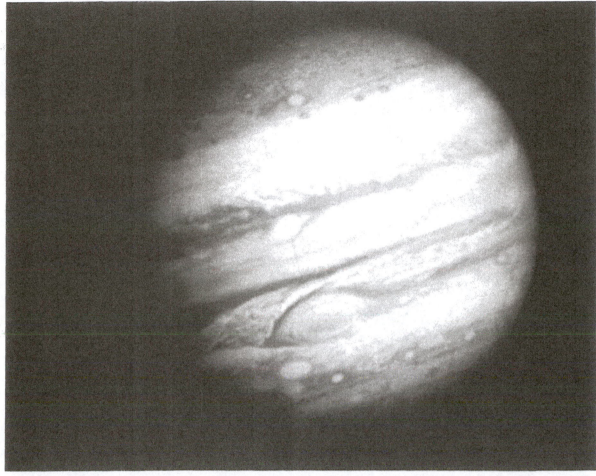

Jupiter is by far the largest and most massive of the planets in the solar system. It has a turbulent atmosphere that contains several cyclonic storm centers, including the "Great Red Spot." The atmosphere is composed primarily of hydrogen and helium with small amounts of other elements of higher atomic number. Jupiter is not massive enough to initiate hydrogen fusion in its core. Nevertheless, it has several sets of satellites, and some of the satellites are comparable in size to the Moon of the Earth. (Photo by NASA.)

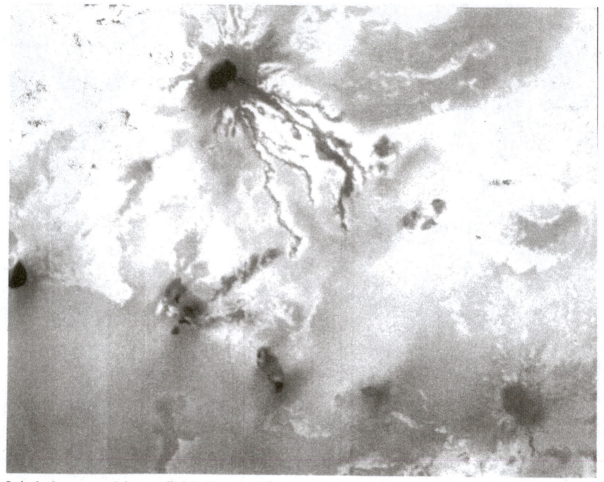

Io is the innermost of the so-called Galilean satellites of Jupiter. It is probably composed of silicate rocks like the Moon and Mars, but its many active volcanoes emit sulfur-rich lavas and gases. The surface of Io is not cratered because impact craters are quickly buried by lava flows and volcanic ash. The heat that causes the volcanic activity is generated by tidal forces caused by the gravitational fields of Jupiter and Europa, the neighbor of Io. (Photo by NASA.)

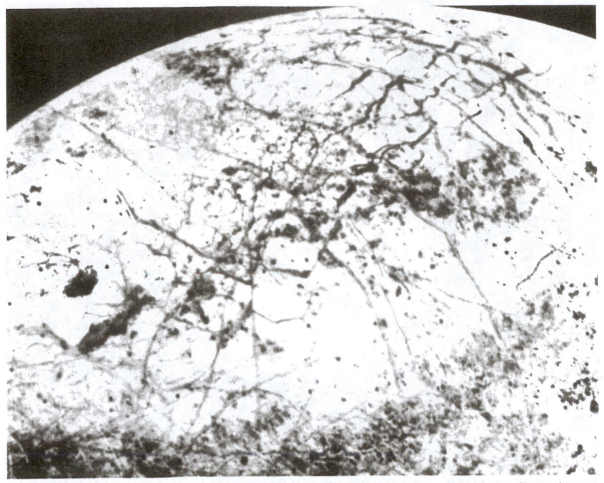

Europa is the second of the Galilean satellites of Jupiter. Unlike Io, Europa is encased in a layer of water ice beneath which liquid water may be present. The icy surface is cut by sets of lines that may be fractures filled with "ice-dikes" from below. Impact craters are present but are not common, presumably because of periodic renewal of the surface of extrusion of water through the system of fractures. The interior of Europa consists of silicate rocks and a small dense core. Ganymede and Callisto, the other two Galilean satellites, are also ice-covered, but each is distinctly different from its neighbors. (Photo by NASA.)

Saturn is justly famous for its intricate rings, although all of the gaseous planets are now known to have rings. Saturn, like Jupiter, has a retinue of satellites, the largest of which is Titan. This satellite is only slightly smaller than Ganymede and has an atmosphere composed mainly of nitrogen with some methane and other hydrocarbon gases. Because of the low surface temperature on Titan, most of the methane (or ethane, C_2H_6) may form oceans of liquid hydrocarbons. (Photo by NASA.)

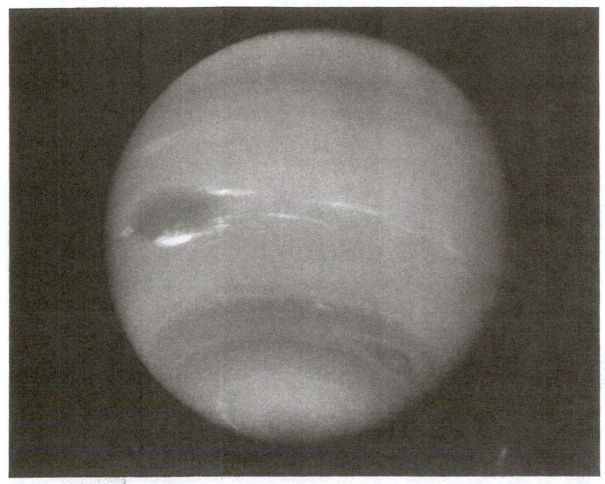

Neptune, the last of the great gaseous planets, was photographed in August 1989 by Voyager 2. It is blue in color and has a turbulent atmosphere like that of Jupiter in spite of its great distance from the Sun (more than 4×10^9 km) and its much smaller size compared to Jupiter (about 6%). Neptune has at least eight satellites, Triton being the most massive. The surface temperature of Triton is only 37 K or $-236\,°C$. Nevertheless, cryovolcanic activity is occurring on Triton involving the eruption of liquid N_2 and perhaps methane. Consequently, the surface of Triton contains few impact craters and is composed of landforms never seen before on any satellite. (Photo by NASA.)

Planet Earth is our home in the solar system. It is the only planet or satellite with liquid water on its surface and with an atmosphere containing molecular oxygen. Earth is also the only place in the solar system that can sustain life as we know it.

This view of the southern hemisphere showing Africa, the Indian Ocean with Madagascar, and the Atlantic Ocean was taken from space by the astronauts of Apollo 17, December 1972. (Photo by NASA.)

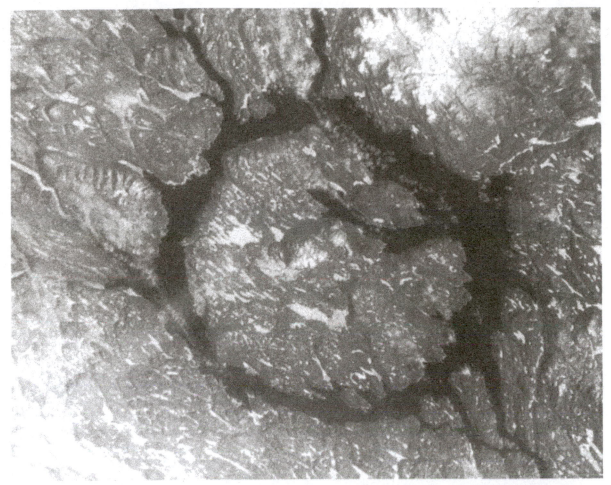

Manicouagen-Mushalagen Lakes, north shore, St. Lawrence River, Quebec, Canada. These two lakes form a circular structure because they are the remnants of a deeply eroded impact crater. This and hundreds of other such impact craters testify to the fact that the Earth has been bombarded by meteoroids and comets just as have the Moon, Mars, and all other bodies in the solar system. In fact, such impacts are now recognized as an important geological process that has disrupted the surface environment by causing short-term catastrophic climate changes and biological extinctions. (Photo by NASA.)

3.5 Summary

The sequence of events leading to the formation of the solar system can be reconstructed as a direct extension of stellar evolution by applying the laws of physics and chemistry to a diffuse cloud of gas and dust particles in interstellar space.

The earthlike planets constitute a very small fraction of the total mass of the solar system and are dwarfed even by the outer gaseous planets. Nevertheless, Earth is the only planet in the entire solar system on which the surface environment is conducive to the development and evolution of life forms.

The satellites of Jupiter form a miniature planetary system of their own. The four largest satellites are similar in size to Mercury and the Moon but differ significantly in their chemical compositions and surface features.

The satellites of Saturn, Uranus, and Neptune are likewise of great interest in the study of the solar system but are less well known than the satellites of Jupiter.

References

ANONYMOUS, 1989. *Voyager Fact Sheet; Neptune Encounter.* Jet Propulsion Laboratory, Pasadena, CA.

CAMERON, A. G. W., 1978. Physics of the primitive solar accretion disk. *Moon Planets,* **18**:5–40.

CAMERON, A. G. W., and M. R. PINE, 1973. Numerical models of the primitive solar nebula. *Icarus,* **18**:377–406.

CARR, M. H., R. S. SAUNDERS, R. G. STROM, and D. E. WILHELMS, 1984. *The Geology of the Terrestrial Planets.* NASA SP-469, 317 pp.

GLASS, B. P., 1982. *Introduction to Planetary Geology.* Cambridge University Press, Cambridge, England, 469 pp.

GREELEY, R., 1985. *Planetary Landscapes.* Allen and Unwin, London, 265 pp.

HAMBLIN, W. K., and E. H. CHRISTIANSEN, 1990. *Exploring the Planets.* Macmillan, New York, 449 pp.

HUNT, G., and P. MOORE, 1981. *Jupiter.* Rand McNally, New York, 96 pp.

LEWIS, J. S., 1974. The chemistry of the solar system. *Sci. Amer.,* **230**(3):51–65.

MEHLIN, T. G., 1968. *Astronomy and the Origin of the Earth,* Brown Co., Dubuque, IA, 131 pp.

MORABITO, L. A., S. P. SYNNOTT, P. N. KUPFERMAN, and S. A. COLLINS, 1979. Discovery of currently active extraterrestrial volcanism. *Science.* **204**:972.

MURRAY, B., M. C. MALIN, and R. GREELEY, 1981. *Earthlike Planets.* W. H. Freeman, San Francisco, 387 pp.

PEALE, S. J., P. CASSEN, and R. T. REYNOLDS, 1979. Melting of Io by tidal dissipation. *Science,* **203**:892–894.

4

Chemical Differentiation
of the Earth

The Earth is a highly differentiated planet, and its geological activity continues to produce diversified suites of igneous and sedimentary rocks. Although some of the other terrestrial planets may also have remained active in certain limited and specific ways, the Earth still has "fire in her belly" and is cloaked in envelopes of corrosive water and atmospheric gases.

4.1   Internal Structure of the Earth

The Earth was probably initially molten and differentiated very early in its history into several layers or shells having different chemical compositions. The resulting internal structure of the Earth has been determined by seismologists based on the reflection and refraction of compressional (P) and shear (S) seismic waves. On the basis of these results, the interior of the Earth has been divided into the *core,* the *mantle,* and the *crust,* as shown in Figure 4.1. The densities of these major interior units of the Earth range from 2.8 g/cm^3 for rocks in the crust to more than 12 g/cm^3 for the core (Bullen, 1963). The increase of the density and of the velocities of seismic waves with depth is caused both by changes in the chemical composition and by the recrystallization of minerals into more closely packed structures. Figure 4.1 shows that the seismic velocities and densities have two major discontinuities marking the boundaries between the crust and mantle and between the mantle and core. The crust–mantle boundary, located at a depth of

about 40 km under the continents, is known as the Mohorovicic discontinuity or "Moho," whereas the mantle–core boundary, at a depth of 2883 km, is referred to as the Wiechert–Gutenberg discontinuity, and both are named after eminent seismologists. There are additional, but more subtle, transition regions within the Earth caused by other important changes in its physical properties. Table 4.1 contains a summary of the physical properties of the Earth.

The mantle of the Earth has been subdivided into three parts based on the presence of boundaries at depths of 413 and 984 km. The *upper mantle,* between the Moho and the boundary at about 400 km, is probably heterogeneous because of the formation of magma by partial melting. Xenoliths of ultramafic rocks derived from this region indicate that the upper mantle is composed of olivine, pyroxene, and garnet with lesser amounts of spinel, amphibole, and phlogopite. The *transition zone,* located between depths of about 400 and 1000 km, is a region of pressure-induced phase transformations of the orthosilicates of Fe and Mg from the structure of olivine to that of spinel. The *lower mantle,* below a depth of about 1000 km, is probably homogeneous and may be richer in Fe than the upper mantle (Henderson, 1982).

The chemical composition of the mantle is difficult to determine with certainty because it is inaccessible and heterogeneous. Nevertheless, ultramafic xenoliths, as well as the stony meteorites, have provided useful clues. In addition, large masses of alpine-type peridotite may have originated from the upper mantle at convergent

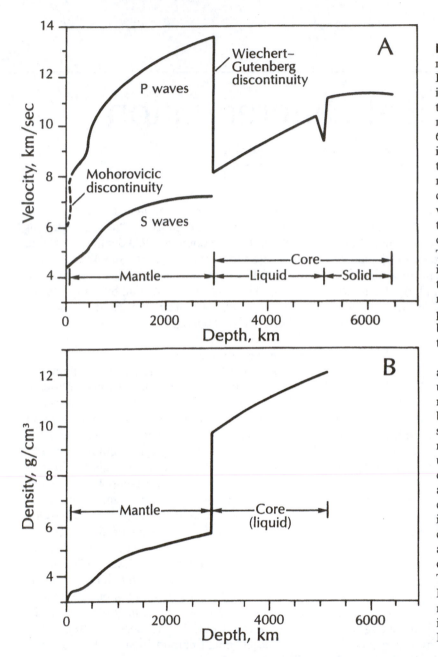

Figure 4.1 **A:** Variation of seismic velocities with depth in the Earth. The velocity of P waves increases abruptly at the base of the crust (Mohorovicic discontinuity or Moho) from about 6 km/sec to nearly 8 km/sec. It increases with depth in the mantle but drops sharply again at the mantle–core boundary. A third discontinuity in the *P*-wave velocities within the core marks the transition from the liquid outer core to the solid inner core. The velocity of *S* waves also increases at the Moho and continues to rise with depth in the mantle. However, *S* waves do not penetrate the mantle–core boundary, which indicates that the outer core is liquid.

B: The density of rocks rises abruptly at the Moho and continues to increase with depth in the mantle. At the mantle–core boundary the density rises sharply from about 6 g/cm³ to nearly 10 g/cm³, and then continues to increase to values in excess of 12 g/cm³ at the boundary between the outer and inner core. The variation of these physical properties leads to the conclusion that the mantle is solid and is composed primarily of silicates and oxides of Mg and Fe. The core consists of an alloy of Fe, Ni, and other siderophile elements and is liquid except for the inner part which is solid (after Bullen, 1963).

plate boundaries. Estimates of the chemical composition of the mantle must meet certain geophysical requirements regarding density, seismic velocities, and heat production by the decay of U, Th, and K. In addition, the mantle must be capable of forming the major types of basaltic magmas by partial melting under the physical conditions likely to prevail in that region of the Earth.

These requirements are generally satisfied by a rock called *pyrolite* invented by Ringwood (1966) by combining peridotite and basalt in the ratio of 3

Table 4.1 Physical Dimensions of the Earth Including the Continental Crust

	Thickness, km	Volume, 10^{27} cm³	Average density, g/cm³	Mass, 10^{27} g	Mass, %
Whole Earth	6371[a]	1.083	5.52	5.976	100
Core	3471[a]	0.175	10.7	1.876	31.5
Mantle	2883	0.899	4.5	4.075	68.1
Crust[b]	40	0.008 42	2.8	0.023 6	0.4
Hydrosphere	3.8 (av.)	0.001 37	1.03	0.001 41	0.024
Atmosphere	—	—	—	0.000 005	0.000 09

[a]Radius.
[b]Continental crust.
SOURCE: Parker (1967) and Taylor and McLennan (1985).

Table 4.2 Estimates of the Chemical Composition of the Mantle of the Earth

	Sample 1[a]	Sample 2[b]	Sample 3[c]
SiO_2	45.2	48.1	45.0
MgO	37.5	31.1	39.0
FeO	8.0	12.7	8.0
Al_2O_3	3.5	3.1	3.5
CaO	3.1	2.3	3.25
Na_2O	0.57	1.1	0.28
Cr_2O_3	0.43	0.55	0.41
MnO	0.14	0.42	0.11
P_2O_5	0.06	0.34	—
K_2O	0.13	0.12	0.04
TiO_2	0.17	0.12	0.09
NiO	—	—	0.25
Sum	98.8	99.95	99.93

[a]Pyrolite (Ringwood, 1966).
[b]Mantle plus crust based on meteorites (Mason, 1966).
[c]Undepleted mantle based on lherzolites (Hutchison, 1974).

to 1. The resulting chemical composition of the mantle is similar to one proposed by Mason (1966), which was based on meteorites, and also resembles the estimate made by Hutchison (1974), which was derived from ultramafic lherzolite inclusions in igneous rocks. The estimates of the chemical composition of the mantle are compiled in Table 4.2.

The great interest by the scientific community in the chemical composition of the mantle and in the Moho caused one of the more colorful episodes in the history of science. In the course of a gathering of some members of the American Miscellaneous Society (AMSOC) in Washington, D.C., chaired by Professor Harry Hess of Princeton University, someone casually suggested that a hole should be drilled through the oceanic crust to find out what causes the Moho. The idea ignited the enthusiastic support of the group, and a proposal was promptly submitted to the National Science Foundation to determine the feasibility of this idea. Thus began the so-called *Mohole Project,* which received much attention in the popular press but never achieved its objective. Nevertheless, engineering studies were made with support from the National Science Foundation, a drilling ship was outfitted, and some test holes were drilled. These tests demonstrated that it was indeed possible to drill in deep water from a ship positioned over the selected site. Eventually AMSOC disbanded, but the equipment was taken over by the Deep-Sea Drilling Program, which has been exploring the history of the ocean basins by drilling first from the *Glomar Challenger* and later from the *Glomar Explorer.* The AMSOC gave up the quest to drill to the Moho partly because it became apparent that this important boundary is actually exposed in some deep-sea trenches and that rocks from the upper mantle could be, and in fact had been, recovered by dredging.

4.2   The Continental Crust: Major Elements

The crust of the Earth is important because it contains all of the natural resources that sustain us. It includes the *atmosphere, hydrosphere,* and *biosphere,* as well as part of the *lithosphere.* The term *lithosphere* is now defined as the rigid outer portion of the Earth consisting of the crust and upper mantle in contrast to the underlying *asthenosphere,* which deforms plastically in response to tectonic stress. The study of the chemical composition of the *continental* crust, and of the different types of rocks of which it is composed, has been a major objective of geochemistry (Wedepohl, 1969; Taylor and McLennan, 1985).

The crust is the outer shell of the Earth, which lies above the mantle. It not only makes up the continents but also occurs under the oceans. The *oceanic* crust is only 5–8 km thick and differs from the continental crust in chemical composition and origin. The continental crust is composed primarily of igneous and metamorphic rocks. Metamorphic rocks are concentrated in orogenic belts, but the distinction between igneous and metamorphic rocks is not always obvious because most of the igneous rocks in the continental crust have been metamorphosed. Sedimentary and volcanic rocks form at the surface of the Earth and therefore tend to cover the underlying igneous and metamorphic rocks. Sedimentary and volcanic rocks may also become folded into elongated masses by compression of tectonic basins in which they were originally deposited. Clarke and Washington (1924) estimated that the continental crust consists of 95% igneous and metamorphic rocks and 5% sedimentary rocks. The latter include 4% shale, 0.75% sandstone, and 0.25% limestone. Evidently, sedimentary rocks make up only a minor portion of the continental crust, but their occurrence at the surface of the continents enhances their importance.

The chemical composition of the continental crust has been estimated by (1) averaging large numbers of chemical analyses, (2) combining chemical analyses of different rock types weighted in terms of their abundances, (3) analyzing sediment derived from the continent, (4) combining the compositions of acidic and mafic rocks in varying proportions, and (5) by modeling. The results of these different methods are surprisingly similar as indicated in Table 4.3.

Clarke and Washington (1924) estimated the chemical composition of the crust not only by averaging a large number of chemical analyses of igneous rocks from all continents and from the ocean basins, but also by combining the average chemical composition of igneous rocks with those of shale, sandstone, and limestone in accordance with their observed abundances. The results in columns 1 and 2 of Table 4.3 are very similar. An enormous amount of work was required to collect and analyze the well over 5000 rock samples on which the estimate in column 1 is based. Therefore, V. M. Goldschmidt decided in the early 1930s to analyze the clay-size fraction of till in southern Norway because, he argued, glacial sediment of Pleistocene age was produced by mechanical erosion of bedrock in Scandinavia with little chemical weathering. In addition, transport at the base of the ice sheet and subsequent deposition by meltwater caused glacial clay to be well mixed, making it a representative sample of the rocks in the path of the ice sheet. The average chemical composition of 77 samples of glacial clay from southern Norway (column 3, Table 4.3) is, in fact, quite similar to the estimates of Clarke and Washington. We note, however, that the concentrations of CaO and Na_2O of the glacial clay are somewhat *lower* than their values in columns 1 and 2, presumably because these elements were partially lost from the clay by leaching. In addition, the glacial clays have a lower TiO_2 content than the crustal averages of Clarke and Washington, which may be a regional characteristic of the rocks of Scandinavia.

Table 4.3 Estimates of the Average Chemical Composition of the Continental Crust in Weight Percents

	1[a]	2[b]	3[c]	4[d]	5[e]	6[f]	7[g]	8[h]
SiO_2	59.12	59.07	59.19	60.06	59.4	59.3	57.3	68.4
TiO_2	1.05	1.03	0.79	0.90	1.2	0.7	0.9	0.4
Al_2O_3	15.34	15.22	15.82	15.52	15.6	15.0	15.9	14.8
Fe_2O_3	3.08	3.10	3.41	3.55	2.3	2.4	—	1.3
FeO	3.80	3.71	3.58	4.06	5.0	5.6	9.1	3.2
MnO	0.12	0.11	0.11	0.21	0.1	0.1	—	—
MgO	3.49	3.45	3.30	3.56	4.2	4.9	5.3	1.7
CaO	5.08	5.10	3.07	5.62	6.6	7.2	7.4	3.4
Na_2O	3.84	3.71	2.05	3.28	3.1	2.5	3.1	3.1
K_2O	3.13	3.11	3.93	2.88	2.3	2.1	1.1	3.6
P_2O_5	0.30	0.30	0.22	0.36	0.2	0.2	—	0.1
CO_2	—	0.35	0.54	—	—	—	—	—
H_2O	1.15	1.30	3.02	—	—	—	—	—
Sum	99.50	99.56	99.03	100.00	100.0	100.0	100.1	100.0

[a]Clarke and Washington (1924) based on the average of 5159 analyses of igneous rocks from all continents and the oceans.
[b]Clarke and Washington (1924) based on 95% igneous rocks, 4% shale, 0.75% sandstone, 0.25% limestone.
[c]Goldschmidt (1954) based on 77 analyses of glacial clay from southern Norway.
[d]Daly (1914) based on 1 : 1 mixture of average granite and basalt.
[e]Poldervaart (1955), based on the average of young folded belts and continental shield region shown in Table 4.4.
[f]Ronov and Yaroshevsky (1976) based on a detailed model like that of Poldervaart (1955).
[g]Taylor and McLennan (1985, Table 3.4, No. 10) based on 75% Archean crust and 25% *andesite* crust.
[h]Sederholm (1925) based on the Precambrian rocks of Finland.
SOURCE: Parker (1967); Taylor and McLennan (1985).

Columns 4–8 of Table 4.3 contain estimates of the chemical composition of the continental crust derived by various methods. Daly (1914) estimated its composition by combining analyses of granite and basalt in equal proportions (column 4, Table 4.3) because these rock types were thought to originate from fundamental magmas of granitic and basaltic compositions. Poldervaart (1955) constructed a detailed model by dividing the crust into the *continental shields, young folded belts,* the *suboceanic region,* and the *deep oceanic region.* He assigned each region certain lithologic compositions (indicated in Table 4.4) and combined the chemical compositions of the different rock types according to their abundances. Column 5 of Table 4.3 contains his estimate of the chemical composition of the continental crust as a whole. Poldervaart's estimates of the chemical composition of each of the four structural units are shown separately in Table 4.4. These results illustrate the differences in the chemical composition of the *continental crust* (continental shield + young folded belts) and the *oceanic crust* (deep oceanic region). It is evident that the oceanic crust has higher average concentrations of Fe_2O_3 + FeO, MgO, CaO, MnO, P_2O_5, and TiO_2 and lower concentrations of SiO_2, Al_2O_3, Na_2O, and K_2O than the continental crust (column 5, Table 4.4). These differences arise because the continental crust is composed primarily of granitic rocks (granodiorite?), whereas the oceanic crust was assumed by Poldervaart (1955) to consist largely of olivine basalt. However, the study of *ophiolite complexes,*

Table 4.4 Estimates of the Chemical Composition of Four Major Structural Units of the Lithic Crust of the Earth in Weight Percent

	Poldervaart				Ronov and Yaroshevsky
	1 Continental shields[a]	2 Young folded belts[b]	3 Suboceanic regions[c]	4 Deep oceanic[d]	5 Oceanic crust
SiO_2	59.8	58.4	49.4	46.6	49.4
TiO_2	1.2	1.1	1.9	2.9	1.4
Al_2O_3	15.5	15.6	15.1	15.0	15.4
Fe_2O_3	2.1	2.8	3.4	3.8	2.7
FeO	5.1	4.8	6.4	8.0	7.6
MnO	0.1	0.2	0.2	0.2	0.3
MgO	4.1	4.3	6.2	7.8	7.6
CaO	6.4	7.2	13.2	11.9	12.5
Na_2O	3.1	3.1	2.5	2.5	2.6
K_2O	2.4	2.2	1.3	1.0	0.3
P_2O_5	0.2	0.3	0.3	0.3	0.2
Sum	100.0	100.0	100.0	100.1	100.0

[a]Sedimentary cover ~0.5 km: 41% shale, 43% sandstone, 16% limestone. Igneous rocks vary in composition with depth such that there are 22 km of granodiorite, 3 km of diorite, and 10.5 km of basalt.

[b]Sedimentary cover ~5 km: 52% shale, 13% sandstone, 22% limestone, 5% greywacke, 6% andesite, 2% rhyolite. Igneous rocks: 22.5 km of granodiorite, 2.7 km of diorite, and 7.3 km of basalt. In addition, 40% of the sediment layer is assumed to be granodiorite based on the outcrop area of granitic batholiths.

[c]Shelf sediment ~4 km, similar in composition to that of folded mountain belts. In addition: 74% terrigenous mud, 22% coral mud, 4% volcanic mud. Igneous rocks: 4 km of diorite, 2 km of tholeiite, and 5 km of olivine basalt.

[d]Sediment thickness ~0.3 km: 72% calcareous sands and oozes, 19% red clay, 9% siliceous ooze. Igneous rocks: 5.57 km of olivine basalt.

SOURCE: Poldervaart (1955); Ronov and Yaroshevsky (1976).

summarized by Moores (1982), indicates that the oceanic crust is actually composed of mafic and ultramafic rocks, whose average composition may approach that of basalt, although trace element concentrations have been altered by interaction with heated seawater. Ronov and Yaroshevsky (1976) also modeled the continental and oceanic crust and arrived at compositions that are similar to those of Poldervaart (1955). Their estimate of the chemical composition of the continental crust is listed in column 6 of Table 4.3, whereas column 5 of Table 4.4 contains their estimate for the oceanic crust.

Taylor (1964) revived Daly's method of estimating the chemical composition of the continental crust by demonstrating that the average concentrations of rare earth elements (REEs) in sedimentary

rocks can be duplicated by combining chemical analyses of mafic and felsic igneous rocks in the proportions 1:5. The resulting chemical composition resembles that of diorite or andesite, which is why this crustal model is generally referred to as the *andesite* model. Subsequently, Taylor and McLennan (1985) proposed that the continental crust consists of 75% average Archean rocks and 25% andesite crust. The resulting composition is shown in column 7 of Table 4.3. It differs from all other estimates listed in Table 4.3 by having somewhat *lower* concentrations of SiO_2, TiO_2, Na_2O, and K_2O and *higher* concentrations of Al_2O_3, Fe as FeO, MgO, and CaO. Accordingly, the chemical composition of the continental crust proposed by Taylor and McLennan (1985) is somewhat more mafic than those of their predecessors. However, it does meet the following important constraints.

1. It satisfies the observed rate of heat flow based on the decay of U, Th, and K.
2. It is capable of generating the more highly differentiated granodiorites of the upper crust by partial melting.
3. It restricts crustal growth by andesitic island–arc volcanism to post-Archean time.
4. It recognizes that 75% of the continental crust is composed of rocks that are more than 2500 million years old, which did not necessarily form by the same tectonic-magmatic processes that became important in post-Archean time.

In conclusion, we note that the chemical composition of the continental crust of the Earth is very different from that of the solar nebula and stony meteorites. The nine most abundant elements and their average concentrations in the continental crust are:

	Wt.%		Wt.%
O	45.5	Mg	3.2
Si	26.8	Na	2.3
Al	8.40	K	0.90
Fe	7.06	Ti	0.5
Ca	5.3		

All of the remaining elements make up only a very small fraction of the crust of the Earth.

4.3   Differentiation of Igneous and Sedimentary Rocks

The chemical compositions of igneous rocks vary widely because of geochemical differentiation that takes place during their formation. The differentiation starts during magma formation by partial melting of rocks in the lower crust or upper mantle. The chemical composition of the resulting magma depends on the composition of the source rocks and on the extent of melting, which is controlled by the pressure and temperature at the source. The composition of magmas may be modified subsequently by assimilation of country rocks or by mixing with magmas derived from different sources. The crystallization of magmas depends on their chemical compositions and on the physical conditions in the magma chamber. The rocks produced by fractional crystallization of magma may vary widely in their mineral composition from ultramafic cumulates of early-formed ferromagnesian minerals (olivine or pyroxene) to quartz-rich late-stage differentiates.

The most comprehensive compilation of the chemical compositions of the highly diverse igneous rocks can be found in the *Handbook of Geochemistry,* edited by K. H. Wedepohl from 1969 to 1978. The compilation of data in Table 4.5 is based on earlier compilations by Turekian and Wedepohl (1961) and Vinogradov (1962). The data are grouped into ultramafic rocks, basalt, high-Ca granite, and low-Ca granite and therefore represent most of the compositional spectrum of igneous rocks. The category "basalt" includes both volcanic and plutonic rocks of basaltic composition, whereas the "high-Ca granites" represent igneous rocks of intermediate composition, which may not actually be *granites* in the strict sense of that term.

The sedimentary rocks are just as highly diversified in their chemical compositions as igneous rocks. However, the processes that cause this diversification operate on the *surface* of the Earth rather than at depth and include *weathering, transport, deposition,* and *lithification.* The study of these processes encompasses a major portion of the subject of geochemistry and will occupy the remainder of this book.

Sedimentary rocks can be divided into two groups depending on whether they are composed of mineral and rock particles of preexisting rocks or whether they precipitated from aqueous solution. This division is not completely justified because clastic sedimentary rocks may contain chemically precipitated mineral cements, whereas the chemical precipitates may contain a component of detrital sediment. Moreover, the chemical compositions of rocks in both groups are modified

Table 4.5 Chemical Composition of Igneous and Sedimentary Rocks (in parts per million unless otherwise indicated)

Element, Z	Ultramafic[a]	Basalt[a]	High-Ca granites[b]	Low-Ca granites[b]	Shale[b]	Sandstone[b]	Carbonate rocks[b]	Deep-sea clay[b]
3 Li	0.5	16	24	40	66	15	5	57
4 Be	0.2	0.7	2	3	3	—	—	2.6
5 B	2	5	9	10	100	35	20	230
9 F	100	385	520	850	740	270	330	1300
11 Na (%)	0.49	1.87	2.84	2.58	0.96	0.33	0.04	4
12 Mg (%)	23.2	4.55	0.94	0.16	1.50	0.70	4.70	2.10
13 Al (%)	1.2	8.28	8.20	7.20	8	2.50	0.42	8.40
14 Si (%)	19.8	23.5	31.40	34.70	7.30[e]	36.80	2.40	25
15 P	195	1130	920	600	700	170	400	1500
16 S	200	300	300	300	2400	240	1200	1300
17 Cl	45	55	130	200	180	10	150	21000
19 K (%)	0.017	0.83	2.52	4.20	2.66	1.07	0.27	2.50
20 Ca (%)	1.6	7.2	2.53	0.51	2.21	3.91	30.23	2.90
21 Sc	10	27	14	7	13	1	1	19
22 Ti	300	11400	3400	1200	4600	1500	400	4600
23 V	40	225	88	44	130	20	20	120
24 Cr	1800	185	22	4.1	90	35	11	90
25 Mn	1560	1750	540	390	850	—	1100	6700
26 Fe (%)	9.64	8.60	2.96	1.42	4.72	0.98	0.33	6.50
27 Co	175	47	7	1	19	0.3	0.1	74
28 Ni	2000	145	15	4.5	68	2	20	225
29 Cu	15	94	30	10	45	—	4	250
30 Zn	40	118	60	39	95	16	20	165
31 Ga	1.8	18	17	17	19	12	4	20
32 Ge	1.3	1.4	1.3	1.3	1.6	0.8	0.2	2
33 As	0.8	2.2	1.9	1.5	13	1	1	13
34 Se	0.05	0.05	0.05	0.05	0.6	0.05	0.08	0.17
35 Br	0.8	3.3	4.5	1.3	4	1	6.2	70
37 Rb	1.1	38	110	170	140	60	3	100
38 Sr	5.5	452	440	100	300	20	610	180
39 Y	—	21	35	40	26	40	30	90
40 Zr	38	120	140	175	160	220	19	150
41 Nb	9	20	20	21	11	—	0.3	14
42 Mo	0.3	1.5	1	1.3	2.6	0.2	0.4	27
47 Ag	0.05	0.11	0.051	0.037	0.07	0.01	0.01	0.11
48 Cd	0.05	0.21	0.13	0.13	0.3	—	0.035	0.42
49 In	0.012	0.22	—	0.26	0.1	—	—	0.08
50 Sn	0.5	1.5	1.5	3	6	0.1	0.1	1.5
51 Sb	0.1	0.6	0.2	0.2	1.5	0.01	0.2	1
53 I	0.3	0.5	0.5	0.5	2.2	1.7	1.2	0.05
55 Cs	0.1	1.1	2	4	5	0.1	0.1	6
56 Ba	0.7	315	420	840	580	10	10	2300
57 La	1.3[c]	6.1[d]	45	55	92	30	1	115
58 Ce	3.5[c]	16[d]	81	92	59	92	11.5	345
59 Pr	0.49[c]	2.7[d]	7.7	8.8	5.6	8.8	1.1	33

Table 4.5 (continued)

Element, Z	Ultramafic[a]	Basalt[a]	High-Ca granites[b]	Low-Ca granites[b]	Shale[b]	Sandstone[b]	Carbonate rocks[b]	Deep-sea clay[b]
60 Nd	1.9[c]	14[d]	33	37	24	37	4.7	140
62 Sm	0.42[c]	4.3[d]	8.8	10	6.4	10	1.3	38
63 Eu	0.14[c]	1.5[d]	1.4	1.6	1	1.6	0.2	6
64 Gd	0.54[c]	6.2[d]	8.8	10	6.4	10	1.3	38
65 Tb	0.12[c]	1.1[d]	1.4	1.6	1	1.6	0.2	6
66 Dy	0.77[c]	5.9[d]	6.3	7.2	4.6	7.2	0.9	27
67 Ho	0.12[c]	1.4[d]	1.8	2	1.2	2	0.3	7.5
68 Er	0.30[c]	3.6[d]	3.5	4	2.5	4	0.5	15
69 Tm	0.041[c]	0.60[d]	0.3	0.3	0.2	0.3	0.04	1.2
70 Yb	0.38[c]	3.2[d]	3.5	4	2.6	4	0.5	15
71 Lu	0.036[c]	0.55[d]	1.1	1.2	0.7	1.2	0.2	4.5
72 Hf	0.4	1.5	2.3	3.9	2.8	3.9	0.3	4.1
73 Ta	0.5	0.8	3.6	4.2	0.8	0.01	0.01	0.1
74 W	0.5	0.9	1.3	2.2	1.8	1.5	0.6	1
79 Au	0.006	0.004	0.004	0.004	—	—	—	—
80 Hg	0.01	0.09	0.08	0.08	0.4	0.03	0.04	0.1
81 Tl	0.04	0.21	0.72	2.3	1.4	0.82	0.01	0.8
82 Pb	0.5	7	15	19	20	7	9	80
90 Th	0.0045	3.5	8.5	17	12	1.7	1.7	7
92 U	0.002	0.75	3	3	3.7	0.45	2.2	1.3

[a]Average of Turekian and Wedepohl (1961) and Vinogradov (1962).
[b]Turekian and Wedepohl (1961).
[c]Calculated from data listed by Herrmann (1970).
[d]Average values calculated by Hermann (1970).
[e]Krauskopf (1979) listed a value of 23.8%.

by ion exchange processes associated with electrical surface charges of mineral grains regardless of whether they are detrital or formed as chemical precipitates.

The compilation of data in Table 4.5 includes two rock types that are predominantly detrital (shale and sandstone), one that is predominantly precipitated (carbonate rocks), and deep-sea clay, which is a mixture of detrital grains and chemical precipitates. The selection of these rock types has an additional virtue because it includes rocks that form in shallow marine basins in the vicinity of continents as well as sediment deposited in deep ocean basins far from land.

The apparent geochemical differentiation indicated by these data reflects the way in which geological processes operating on the surface of the Earth "sort" the elements on the basis of their chemical properties. We can think of the surface environment of the Earth as a giant *machine* that processes igneous and metamorphic rocks into a wide variety of sedimentary rocks. The products of this machine are highly diversified because the chemical elements, and the minerals they form, respond differently to the treatment they experience as they pass through the machine.

The representation of geological processes operating on the surface of the Earth as a machine is helpful because it emphasizes the *coherence* of these processes and because it promotes a *global* view of the geochemistry of the Earth's surface. We return to this theme in Part V of this book, where we consider the geochemical cycles of selected elements (Chapter 22).

4.4   Differentiation of the Hydrosphere

The hydrosphere consists of a number of reservoirs of water that are connected by means of the hydrologic cycle. Table 4.6 indicates that the oceans are by far the largest reservoirs of water with 97.25% of the total volume, followed by ice sheets and glaciers (2.05%) and deep groundwater (0.38%). That leaves a very small fraction of the water in the hydrosphere in shallow groundwater (0.30%), lakes (0.01%), and rivers (0.0001%) where it is accessible to humans for municipal and industrial use (see also Chapter 20).

When igneous, metamorphic, or sedimentary rocks come in contact with water at or near the surface of the Earth, different kinds of chemical reactions occur, which collectively constitute *chemical weathering* or *water–rock interaction.* The outcome of these reactions depends not only on the chemical and physical conditions at the site but also on the properties of the minerals. Minerals vary widely in their susceptibility to chemical weathering although most are quite insoluble in water. Therefore, the chemical elements do *not* enter the aqueous phase in the same proportions in which they occur in the rocks of the continental crust. As a result, the chemical composition of surface water or groundwater differs from that of the rocks of the upper continental crust in ways that reflect the geochemical properties of the elements and of the minerals they form. Whitfield and Turner (1979) actually expressed these differences in terms of partition coefficients (water/upper crust) and showed that they are related to the ionic character of the metal–oxygen bonds in the minerals.

Once an element has entered the hydrosphere, it is subjected to other kinds of processes that may affect its concentration. For example, some ions are selectively adsorbed on the charged surfaces of clay minerals or particles of oxides and hydroxides of Fe, Mn, and Al. Other elements may enter the biosphere as nutrients and become associated with organisms and biogenic carbon compounds. The noble gases are released into the atmosphere, whereas oxygen, carbon dioxide, and

Table 4.6 Inventory of Water in the Hydrosphere

Reservoir	Volume, $10^6 \ km^3$	Percent, of total
Oceans	1370	97.25
Ice sheets and glaciers	29	2.05
Deep groundwater (750–4000 m)	5.3	0.38
Shallow groundwater (<750 m)	4.2	0.30
Lakes	0.125	0.01
Soil moisture	0.065	0.005
Atmosphere[a]	0.013	0.001
Rivers	0.001 7	0.000 1
Biosphere	0.000 6	0.000 04
Total	1408.7	100

[a]Liquid equivalent of water vapor.
SOURCE: Berner and Berner (1987).

nitrogen of the atmosphere dissolve in the water. Therefore, the chemical composition of water varies widely and reflects not only the mineral composition of the rocks it has interacted with, but is also affected by the geochemical environment, which is determined by climate, topography, vegetation, and by the possible discharge of industrial or municipal waste. These matters are considered in much greater detail in Part V.

Streams ultimately transport the elements that dissolve in the water to the oceans where they reside for varying periods of time. The geochemical processes operating in the oceans *selectively* remove elements from solution in such a way that seawater is *not* simply average river water that has been concentrated by evaporation. This statement is supported by inspection of the average concentrations of the elements in streams and in the oceans (Table 4.7). The fifth column in Table 4.7 contains enrichment factors calculated by dividing the concentration of each element in the oceans by its concentration in average river water. The resulting enrichment factors vary widely from 3350 for Br to about 0.0006 for Th. Evidently, geochemical processes in the oceans are changing its

Table 4.7 Average Composition of Water in Streams and in the Oceans in micrograms per gram

Element	Classification[a]	Stream water	Seawater	Seawater enrichment	MORT[b]
Li	I	3×10^{-3}	1.7×10^{-1}	56.7	2.5×10^6
Be	IV	1×10^{-5}	2×10^{-7}	0.02	6.3×10^1
B	I	1×10^{-2}	4.5	450	1.6×10^7
F	I	1×10^{-3}	1.3	1300	7.9×10^5
Na	I	6.3	1.08×10^4	1714	2.0×10^8
Mg	I	4.1	1.29×10^3	315	5.0×10^7
Al	IV	5×10^{-2}	8×10^{-4}	0.016	7.0
Si	II	6.5	2.8	0.43	7.9×10^3
P	II	2×10^{-2}	7.1×10^{-2}	3.6	4.0×10^4
S	I	3.7	9.0×10^2	243	5.0×10^8
Cl	I	7.8	1.95×10^4	2500	6.3×10^8
K	I	2.3	3.99×10^2	173	1.3×10^7
Ca	I	15	4.13×10^2	27.5	1.3×10^6
Sc	IV	4×10^{-6}	6.7×10^{-7}	0.17	2.5×10^1
Ti	IV	3×10^{-3}	$<9.6 \times 10^{-4}$	0.32	$<1.6 \times 10^2$
V	IV	9×10^{-4}	1.2×10^{-3}	1.3	7.9×10^3
Cr	IV	1×10^{-3}	2×10^{-4}	0.2	1.6×10^3
Mn	III	7×10^{-3}	3×10^{-4}	0.04	3.2×10^1
Fe	IV	4×10^{-2}	6×10^{-5}	0.0015	6.9×10^{-1}
Co	III	1×10^{-4}	2×10^{-6}	0.02	2.0×10^1
Ni	III	3×10^{-4}	5×10^{-4}	1.7	1.6×10^3
Cu	III	7×10^{-3}	3×10^{-4}	0.04	1.0×10^3
Zn	III	2×10^{-2}	4×10^{-4}	0.02	1.3×10^3
Ga	IV	9×10^{-5}	2×10^{-5}	0.2	7.9×10^2
Ge	IV	5×10^{-6}	5×10^{-6}	1	2.0×10^3
As	IV	2×10^{-3}	1.7×10^{-3}	0.85	1.0×10^5
Se	III	6×10^{-5}	1.3×10^{-4}	2.2	6.3×10^5
Br	I	2×10^{-2}	6.7×10^1	3350	7.9×10^8
Rb	I	1×10^{-3}	1.2×10^{-1}	120	7.9×10^5
Sr	I	7×10^{-2}	7.6	109	5.0×10^6
Y	III	4×10^{-5}	7×10^{-6}	0.18	1.3×10^2
Zr	IV	—	3×10^{-5}	—	1.6×10^2
Nb	IV	—	$<5 \times 10^{-6}$	—	$<2.5 \times 10^2$
Mo	I	6×10^{-4}	1.1×10^{-2}	18.3	3.2×10^5
Ag	III	3×10^{-4}	2.7×10^{-6}	0.009	2.0×10^4
Cd	IV	1×10^{-5}	8×10^{-5}	8	7.9×10^4
In	IV	—	1×10^{-7}	—	1.0×10^3
Sn	IV	4×10^{-5}	5×10^{-7}	0.013	1.3×10^2
Sb	III	7×10^{-5}	1.5×10^{-4}	2.1	1.3×10^5
I	I	7×10^{-3}	5.6×10^{-2}	8	$<4.0 \times 10^6$
Cs	I	2×10^{-5}	2.9×10^{-4}	14.5	4.0×10^4
Ba	III	2×10^{-2}	1.4×10^{-2}	0.7	5.0×10^3
La	III	4.8×10^{-5}	4.5×10^{-6}	0.094	7.9×10^1
Ce	III	7.9×10^{-5}	3.5×10^{-6}	0.044	3.2×10^1

Table 4.7 (continued)

Element	Classification[a]	Stream water	Seawater	Seawater enrichment	MORT[b]
Pr	III	7.3×10^{-6}	1.0×10^{-6}	0.14	7.9×10^1
Nd	III	3.8×10^{-5}	4.2×10^{-6}	0.11	7.9×10^1
Sm	III	7.8×10^{-6}	8.0×10^{-7}	0.10	7.9×10^1
Eu	III	1.5×10^{-6}	1.5×10^{-7}	0.10	6.3×10^1
Gd	III	8.5×10^{-6}	1.0×10^{-6}	0.11	1.0×10^2
Tb	III	1.2×10^{-6}	1.7×10^{-7}	0.14	1.0×10^2
Dy	III	7.2×10^{-6}	1.1×10^{-6}	0.15	1.0×10^2
Ho	III	1.4×10^{-6}	2.8×10^{-7}	0.20	1.3×10^2
Er	III	4.2×10^{-6}	9.2×10^{-7}	0.22	1.6×10^2
Tm	III	6.1×10^{-7}	1.3×10^{-7}	0.21	1.6×10^2
Yb	IV	3.6×10^{-6}	9.0×10^{-7}	0.25	2.0×10^2
Lu	IV	6.4×10^{-7}	1.4×10^{-7}	0.22	2.0×10^2
Hf	IV	—	$<7 \times 10^{-6}$	—	$<1.3 \times 10^3$
Ta	IV	—	$<2.5 \times 10^{-6}$	—	$<2.0 \times 10^3$
W	IV	3×10^{-5}	1×10^{-4}	3.3	7.9×10^4
Re	IV	—	4×10^{-6}	—	3.2×10^6
Au	III	2×10^{-6}	4.9×10^{-6}	2.5	1.6×10^6
Hg	III	7×10^{-5}	1×10^{-6}	0.14	7.9×10^3
Tl	IV	—	1×10^{-5}	—	6.3×10^3
Pb	III	1×10^{-3}	2×10^{-6}	0.002	5.0×10^1
Bi	IV	—	2×10^{-5}	—	2.5×10^4
Th	IV	$<1 \times 10^{-4}$	6×10^{-8}	~ 0.0006	3.4
U	I	4×10^{-5}	3.1×10^{-3}	2.7	1×10^6

[a]I. Conservative element whose concentration is directly related to the salinity of seawater.

II. Nonconservative element whose concentration varies with depth or regionally within the oceans, or both, generally because of involvement in biological activity.

III. Nonconservative element whose concentration varies irregularly and is not related to salinity, depth, or geographic factors.

IV. Unclassified, but probably nonconservative.

[b]Mean oceanic residence time in years.

SOURCE: Taylor and McLennan (1985).

chemical composition in comparison with river water and thereby contribute to the geochemical differentiation of the hydrosphere.

The eventual fate of all elements and compounds dissolved in the ocean is to be removed from it. Most elements are incorporated into the sediment that is accumulating at the bottom of the oceans, but a few elements escape from the ocean into the atmosphere, whose composition is summarized in Table 4.8. These changes in residence are part of the *migration* of elements on the surface of the Earth and contribute to the operation of *geochemical cycles* (see Chapter 22). Every

element is moved from one reservoir to the next by geological, geochemical, or biological processes operating on the surface of the Earth. This is the geochemical machine we spoke of earlier.

The dissolved constituents in the oceans can be divided into the *conservative* (I in Table 4.7) and *nonconservative* groups (II and III in Table 4.7). The conservative elements occur in *constant proportions* throughout the oceans, although their *concentrations* may vary because of dilution or evaporative concentration. This group includes the major elements and complex anions—Na, K, Mg, Ca, Cl, sulfate, and borate. The concentrations

Table 4.8 Chemical Composition of Dry Air

| Constituent | Concentration by Volume | |
	%	ppm
N_2	78.084	—
O_2	20.946	—
CO_2	0.033	—
Ar	0.934	—
Ne	—	18.18
He	—	5.24
Kr	—	1.14
Xe	—	0.087
H_2	—	0.5
CH_4	—	2
N_2O	—	0.5

Data from the *CRC Handbook of Chemistry and Physics* (Weast et al., 1986).

of the nonconservative elements vary with depth as well as regionally within the oceans, and their concentration ratios with chloride are *not* constant. The nonconservative elements in seawater include most of the trace elements along with dissolved nitrate, bicarbonate, silicic acid, phosphate, and dissolved oxygen. The concentrations of many nonconservative constituents vary with depth because they are involved in the biological activity of the oceans. As a result, they have low concentrations in the surface layer of the oceans where most of the biological activity is concentrated. When the organisms die, their bodies sink and decompose, thereby returning nutrients to the water. Consequently, deep ocean water is enriched in nutrients compared to surface water.

The conservative constituents have large enrichment factors, indicating that they are *unreactive* and therefore become concentrated in seawater compared to river water. The average enrichment factor of the major (conservative) constituents of seawater is 775. These elements are generally joined by their congeners, many of which are trace elements. For example, the alkali metals, alkaline earths (except Be and Ba), and halogens all have higher concentrations in seawater than in

river water. Their average enrichment factors in the oceans are

Alkali metals	416	(14.5–1714)
Alkaline earths (except Be and Ba)	151	(27.5–315)
Halogens	1790	(8–3350)

In addition, B and S, both of which are major conservative elements, have large enrichment factors indicating that they too tend to accumulate in the oceans. They are followed by P, V, Ni, Ge, Se, Mo, Sb, W, Au, and U, whose enrichment factors are all <20 but >1. Only Mo and U are conservative among the elements in this group.

The enrichment factors of the remaining nonconservative elements are all <1, an indication that they are rapidly removed from seawater by geochemical processes. The concentrations of these elements in seawater are controlled primarily by ion exchange reactions with particulate matter, by ionic substitution in calcium carbonate, calcium phosphate, and amorphous silica, which are in large part biogenic in origin, by direct precipitation caused by evaporation of seawater, and by exchange reactions with volcanic rocks along spreading ridges (Hart and Staudigel, 1982). The concentrations of most elements in seawater are *not* controlled by the solubilities of their compounds.

In contrast to the conservative elements (and a few nonconservative ones), the majority of the nonconservative elements do *not* accumulate in the ocean but are rapidly removed from it by one or several of the geochemical processes mentioned earlier. The adsorption of ions on the charged surfaces of both inorganic and organic particles plays a very important role in this regard and was described by Broecker and Peng (1982) as "the great particulate sweep." Other important books on the geochemistry of seawater are by Berner and Berner (1987), Holland (1978), Goldberg (1972), and Riley and Chester (1971).

The wide variation of the enrichment factors we have been discussing implies that the unreactive conservative elements remain in the oceanic reservoir much longer than the reactive nonconservative elements. This phenomenon is reflected by the *residence time* defined as:

$$t = \frac{A_x}{dX/dt} \qquad (4.1)$$

where t is the residence time in years, A_x is the total amount of an element (x) in solution in the oceans, and dX/dt is the average annual input of that element into the oceans. The amount A_x of an element in the oceans is assumed to be constant, which is probably valid for the open ocean but may not apply to the continental shelves and estuaries. The annual input into the oceans can be estimated from the discharge and chemical composition of the major rivers of the Earth. Such estimates are uncertain, however, because both discharge and elemental concentrations of streams vary seasonally and are not known for all rivers with equal certainty (see also Chapter 20). In addition, the runout of groundwater into the oceans is disregarded because of lack of information. Furthermore, the discharge of industrial and municipal waste water has significantly altered the chemical composition of many rivers. For this reason Taylor and McLennan (1985) estimated the average annual input to the oceans from the rate of deposition and the chemical composition of deep-sea sediment as originally suggested by Li (1982). Accordingly, the mean oceanic residence times (MORT) listed in Table 4.7 were calculated by this method based on sedimentation rates provided by Lisitsyn et al. (1982). The MORTs calculated from the output to deep-sea sediment differ from those obtained from river-input data, but the two sets of results are well correlated.

Actually, the residence time of elements in the oceans was first used to estimate the age of the Earth. The idea was originally suggested in 1715 by Edmund Halley, but nothing was done about it for nearly 200 years. Finally, in 1899 John Joly took up Halley's suggestion and calculated the residence time of Na in the oceans. His result was 90 million years, which is less than half the 200 million years calculated by Taylor and McLennan (1985). Nevertheless, Joly's result was an important milestone in the history of geology because it contradicted Lord Kelvin's calculations based on the cooling of the Earth, which indicated that its age was less than 40 million years.

The MORTs of the elements in Table 4.7 vary over nine orders of magnitude from 0.69 years for Fe to 7.9×10^8 years for Br. As expected, the conservative elements have long residence times, whereas many nonconservative elements have significantly shorter residence times. The average residence times in the oceans of the *conservative* elements are

	Years
Alkali metals	43.3×10^6
Alkaline earths (except Be and Ba)	18.8×10^6
Halogens	356.2×10^6

The other two major elements (B and S) have residence times of 16×10^6 and 500×10^6 years, respectively.

The lithophile elements have widely different enrichment factors and residence times, whereas a majority of the chalcophile and siderophile elements are depleted in seawater and have short residence times.

Evidently, the oceans play a very important role in the geochemical differentiation of the crust of the Earth by storing some elements and letting others pass rapidly to the sediment accumulating at the bottom of the oceans. The elements that are removed from the oceans in the form of sediment may ultimately reenter the rock cycle and become the input for another pass through the geochemical machine.

4.5 Summary

The Earth differentiated very early in its history into a core, mantle, and crust with characteristic chemical compositions. The core consists of metallic Fe, Ni, and related elements. The mantle is composed primarily of silicates and oxides of Mg and Fe. The chemical composition of the continental crust has been estimated in different ways with generally concordant results. The nine most abundant chemical elements in the continental crust of the Earth are O, Si, Al, Fe, Ca, Mg, Na, K, and Ti. All other elements together make up only a very small fraction of the continental crust.

The distribution of trace elements among different kinds of igneous rocks depends on their ability to replace major elements in the ionic crystals that form during crystallization of magma. As a result, different kinds of igneous rocks are enriched in specific trace elements, some of which are important industrial metals. Certain kinds of igneous rocks may therefore host ore deposits of these metals.

The chemical compositions of sedimentary rocks are just as diversified as those of the igneous rocks. They form by chemical weathering of igneous, metamorphic, and sedimentary rocks and by the transport, deposition, and lithification of the resulting products. Many elements become concentrated in deep-sea clay and may reenter the rock cycle when deep-sea sediment melts in subduction zones to form magma.

The oceans play an important role in the chemical differentiation of the crust because some elements accumulate in seawater, whereas others pass through it rapidly. The geological and geochemical processes associated with chemical weathering of minerals, transport of weathering products, and their ultimate deposition in the oceans are the subjects of Part III of this book.

Problems

1. Recalculate the chemical composition of pyrolite (Table 4.2, column 1) in terms of the weight percent concentrations of the elements and rank them in terms of their abundance.

2. Recalculate the concentrations of Si, Al, Fe, Ca, Mg, Na, K, Ti, Mn, and P of shale, sandstone, and carbonate rocks (Table 4.5) in terms of the appropriate oxides.

3. Combine the concentrations calculated in Problem 2 in proportions of 4 parts shale, 0.75 parts sandstone, and 0.25 parts limestone.

4. Compare the results of Problem 3 to the estimates of the average chemical composition of the continental crust (Table 4.3). Take note of discrepancies and similarities and relate them to the geochemical properties of the individual elements.

5. Assess the validity of the hypothesis that the sedimentary rocks (shale, sandstone, and limestone) collectively represent the "weathering crust" on the continents. What additional sources or sinks are needed to adequately describe the redistribution of elements from the continental crust into different kinds of marine sediment?

References

BERNER, E. K., and R. A. BERNER, 1987. *The Global Water Cycle.* Prentice-Hall, Upper Saddle River, NJ, 397 pp.

BROECKER, W. S., and T.-H. PENG, 1982. *Tracers in the Sea.* Eldigio Press, Palisades, NY, 690 pp.

BULLEN, K. E., 1963. *An Introduction to the Theory of Seismology,* 3rd ed. Cambridge University Press, Cambridge, England, 381 pp.

CLARKE, F. W., and H. S. WASHINGTON, 1924. *The Composition of the Earth's Crust.* U.S. Geol. Surv. Prof. Paper 127, 117 pp.

DALY, R. A., 1914. *Igneous Rocks and Their Origin.* McGraw-Hill, New York, 563 pp.

GOLDBERG, E. D. (Ed.), 1972. *The Sea.* Wiley, New York, 5 vols.

GOLDSCHMIDT, V. M., 1954. *Geochemistry.* Oxford University Press, 730 pp.

HART, S. R., and H. STAUDIGEL, 1982. The control of alkalies and uranium in seawater by ocean crust alteration. *Earth Planet. Sci. Lett.,* **58**:202–212.

HENDERSON, P., 1982. *Inorganic Geochemistry.* Pergamon Press, Oxford, England, 353 pp.

HERRMANN, A. G., 1970. Yttrium and the lanthanides. In K. H. Wedepohl (Ed.), *Handbook of Geochemistry,* vol. 2, part 5, ch. 39, sec. E, 57–71. Springer-Verlag, Berlin.

HOLLAND, H. D., 1978. *The Chemistry of the Atmosphere and Oceans.* Wiley, New York, 351 pp.

HUTCHISON, R., 1974. The formation of the Earth. *Nature,* **250**:556–568.

KRAUSKOPF, K. B., 1979. *Introduction to Geochemistry,* 2nd ed. McGraw-Hill, New York, 617 pp.

LI, Y.-H., 1982. A brief discussion on the mean oceanic residence time of elements. *Geochim. Cosmochim. Acta,* **46**:2671–2675.

LISITSYN, A. P., et al., 1982. The relation between element influx from rivers and accumulation in ocean sediments. *Geochem. Int.,* **19**:102–110.

MASON, B., 1966. *Principles of Geochemistry,* 3rd ed. Wiley, New York, 329 pp.

MOORES, E. M., 1982. Origin and emplacement of ophiolites. *Rev. Geophys. Space Phys.,* **20**:735.

PARKER, R. L., 1967. Composition of the Earth's crust. In M. Fleischer (Ed.), *Data of Geochemistry.* U.S. Geol. Surv. Prof. Paper 440D, 1–19.

POLDERVAART, A., 1955. Chemistry of the earth's crust. In A. Poldervaart (Ed.), *Crust of the Earth—A Symposium.* Geol. Soc. Amer. Spec. Paper 62, 119–144.

RILEY, J. P., and R. CHESTER, 1971. *Introduction to Marine Chemistry.* Academic Press, New York, 465 pp.

RINGWOOD, A. E., 1966. The chemical composition and origin of the Earth. In P. M. Hurley (Ed.), *Advances in Earth Sciences,* 287–356. MIT Press, Cambridge, MA.

RONOV, A. B., and A. A. YAROSHEVSKY, 1976. A new model for the chemical structure of the Earth's crust. *Geochem. Int.,* **13**(6):89–121.

SEDERHOLM, J. J., 1925. The average composition of the earth's crust in Finland. *Comm. géol. Finlande Bull.,* **12**(70), 20 pp.

TAYLOR, S. R., 1964. The abundance of chemical elements in the continental crust—a new table. *Geochim. Cosmochim. Acta,* **28**:1273–1285.

TAYLOR, S. R., and S. M. MCLENNAN, 1985. *The Continental Crust: Its Composition and Evolution.* Blackwell Scientific Publ., Oxford, England, 312 pp.

TUREKIAN, K. K., and K. H. WEDEPOHL, 1961. Distribution of the elements in some major units of the Earth's crust. *Geol. Soc. Amer. Bull.,* **72**:175–192.

VINOGRADOV, A. P., 1962. Average contents of chemical elements in the principal types of igneous rocks of the Earth's crust. *Geochemistry,* **1962**(7):641–664.

WEAST, R. C., M. J. ASTLE, and W. H. BEYER (Eds.), 1986. *CRC Handbook of Chemistry and Physics.* CRC Press, Boca Raton, FL.

WEDEPOHL, K. H., 1969. *Handbook of Geochemistry,* vol. 1. Springer-Verlag, Berlin, 442 pp.

WHITFIELD, M., and D. R. TURNER, 1979. Water–rock partition coefficients and the composition of seawater and river water. *Nature,* **278**:132–137.

II

PRINCIPLES OF INORGANIC CHEMISTRY

An important task of geochemists is to understand the chemical compositions of minerals and rocks based on the physical and chemical properties of the elements and their atoms. The regular pattern of variation of the electronic structures of atoms determines the chemical properties of the elements and hence their positions in the periodic table. Consequently, our understanding of the distribution of chemical elements among the minerals and rocks of the Earth rests securely on the principles of physics and chemistry embodied in the periodic table.

5

The Electronic Structure
of Atoms

We now end our tour of the stars and the solar system and turn to some basic principles of chemistry that will help us to understand the chemical processes taking place on the earthlike planets. The intellectual excitement generated by the exploration of the solar system cannot hide the fact that only on the Earth can we live without elaborate protection against the harsh environments of the other planets. This lesson of planetary exploration should increase our interest in our own planet and in the natural processes that take place on its surface and in its interior. We need to understand these processes well in order to assure adequate living conditions for future generations. For these reasons we now move to a description of atoms whose orbiting electrons resemble the orbiting planets and their satellites of the solar system.

5.1   The Atom of Thomson and Rutherford

Throughout the 19th century physicists accepted Sir Isaac Newton's opinion that atoms are "solid, massy, hard, impenetrable, moveable particles" (Jastrow, 1967, p. 9). However, in 1897 J. J. Thomson (1856–1940) discovered the electron and used it to develop a new model of the atom in which negatively charged electrons were thought to be embedded in a positively charged matrix like raisins in a plum pudding. In 1911 Niels Bohr (1885–1962) came to the Cavendish Laboratory at Cambridge as a postdoctoral fellow to work with Thomson on his electron theory. He argued

that the plum-pudding model of the atom could not be correct because it did not provide for the quantization of electromagnetic radiation that had been discovered by Max Planck (1858–1947) in Germany (Planck, 1900). However, Thomson was in no mood to argue with the 26-year-old Dane who had little knowledge of the English language at the time. Sir John, who was the director of the Cavendish Laboratory, had won the Nobel Prize in 1906, and had been knighted in 1908, was not interested in Bohr's ideas about how to improve his electron theory. Besides, he was then working on his "positive-ray apparatus," which would soon lead to the discovery of isotopes and the development of mass spectrometers (Holton, 1986).

Bohr therefore decided to leave Cambridge, and went to Manchester where Ernest Rutherford (1871–1937) was the Langworthy Professor of Physics. Perhaps as a test of Thomson's model of the atom, Rutherford had fired a beam of alpha particles at a piece of gold foil and had observed to his great surprise that the beam did not simply pass through the foil as expected. Instead, a significant number of alpha particles were scattered and some even seemed to bounce straight back as though they had struck a heavy object. He said later "It was quite the most incredible event that has ever happened to me in my life. It was almost as incredible as if you fired a 15-inch shell at a piece of tissue paper and it came back and hit you." (Jastrow, 1967, p. 9). After thinking about it, Rutherford decided that the massive objects had to be very small because they were struck only rarely by alpha particles and concluded that most

60

of the mass of the atom was concentrated in this small body, which he named the "nucleus."

Rutherford's model of the atom was inherently unstable because in classical physics an electron orbiting a positively charged nucleus loses energy by emitting an electromagnetic wave. As a result, the electron must speed up and the radius of its orbit must decrease until the electron crashes into the nucleus. Nevertheless, Bohr liked Rutherford's model because he had a hunch that he could improve it by extending the concept of quantization of electromagnetic energy to the mechanical energy of an electron orbiting the nucleus.

We all know that Bohr succeeded brilliantly, received the Nobel Prize in physics in 1922, and became one of the leaders in science in the 20th century. However, he had a hard time convincing even Rutherford that his model of the atom was right. When Bohr, who had briefly returned to Denmark, sent a manuscript copy of his paper, Rutherford agreed to submit it for publication but added, "I suppose you have no objection to my using my judgment to cut out any matter I may consider unnecessary in your paper?" (Holton, 1986, p. 240). Bohr did object and came back to Manchester from Copenhagen to discuss his paper with Rutherford point by point. In the end, Rutherford agreed to submit the entire manuscript, but grumbled that Bohr had been so obstinate.

When this paper and a second one were published (Bohr, 1913a, b), they were not well received. The prominent German physicist Otto Stern may have spoken for many of his colleagues when he reportedly said, "If this nonsense is correct, I will give up being a physicist." (Holton, 1986, p. 240). It was the end of an era. Physics was about to be shaken to its roots by quantum mechanics and the uncertainty principle.

5.2   Bohr's Theory of the Hydrogen Atom

The instability of an atom composed of one electron orbiting the nucleus can be demonstrated by means of Newtonian physics. We assume that the electron orbits the proton at a distance r and a velocity v. The condition for stability of such an atom is that the force of electrostatic attraction must be equal to the centrifugal force:

$$\frac{e^2}{r^2} = \frac{mv^2}{r} \qquad (5.1)$$

The energy of such an atom is the sum of the kinetic energy and the potential energy:

$$E = \tfrac{1}{2}mv^2 - \frac{e^2}{r} \qquad (5.2)$$

The potential energy term has a negative sign because the electron is attracted toward the proton. From equation 5.1 we obtain:

$$m = \frac{e^2}{rv^2} \qquad (5.3)$$

Substituting into equation 5.2 yields:

$$E = \frac{1}{2}\left(\frac{e^2}{r}\right) - \frac{e^2}{r} = -\frac{1}{2}\left(\frac{e^2}{r}\right) \qquad (5.4)$$

We see from equation 5.4 that the energy of the atom is negative and that it is inversely related to the radius of the orbit. In the limit, the energy approaches minus infinity as the radius goes to zero.

The instability arises because the atom is an "energy well." Energy is liberated as the radius of the electron orbit decreases. Since there are no restrictions on the magnitude of the electronic radius in this model, atoms will tend to achieve a condition of minimum energy and maximum stability by reducing the radius to zero.

Bohr avoided this problem by specifying that the energy of the atom is not continuously variable but is, in fact, quantized so that the electron must be restricted to orbits having specific radii. In order to accomplish this, he postulated that the angular momentum of the electron is restricted to multiples of $h/2\pi$, where h is Planck's constant, which is equal to 6.62517×10^{-27} erg/sec (see also the inside back cover). Therefore, according to Bohr:

$$mvr = \frac{nh}{2\pi} \qquad (5.5)$$

where $n = 1, 2, 3, \ldots$, and is known as the first quantum number. However, the atom must still satisfy the condition expressed by equation 5.1. Accordingly, the angular momentum of the electron must also be:

$$mvr = \frac{e^2}{v} \qquad (5.6)$$

and a stable atom must satisfy both equations 5.5 and 5.6; thus:

$$\frac{e^2}{v} = \frac{nh}{2\pi} \qquad (5.7)$$

The velocity of the electron is derivable from equation 5.5:

$$v = \frac{nh}{2\pi mr} \qquad (5.8)$$

Substituting into equation (5.7) yields:

$$\frac{2\pi mre^2}{nh} = \frac{nh}{2\pi} \qquad (5.9)$$

which leads to:

$$r = \frac{n^2 h^2}{4\pi^2 me^2} \qquad (5.10)$$

Equation 5.10 indicates that the radius of the electron now depends on the square of the first quantum number multiplied by a constant a whose magnitude is:

$$a = \frac{h^2}{4\pi^2 me^2} \qquad (5.11)$$

The smallest radius permitted by Bohr's theory is obtained from equation 5.10 by setting $n = 1$, in which case:

$$r = a_0 = 0.529 \times 10^{-8} \text{ cm} \qquad (5.12)$$

where a_0 is the so-called *Bohr radius*.

The energy of the atom can now be calculated by substituting equation 5.10 for the radius into equation 5.4 for the energy of the atom:

$$E = -\tfrac{1}{2}e^2\left(\frac{4\pi^2 me^2}{n^2 h^2}\right) = -\frac{2\pi^2 me^4}{n^2 h^2} \qquad (5.13)$$

We see that in Bohr's model the energy of the atom no longer depends on the radius of the electronic orbit but varies as the reciprocal of the square of the first quantum number. Evidently, the quantum number plays a decisive role because it controls both the radius of the electronic orbit and the energy of the atom.

In Bohr's model of the atom the electron can occupy only certain orbits whose radii depend on the first quantum number. The energy of the atom varies in increments depending on which orbit the electron occupies. When the electron jumps from a higher orbit to a lower one, the energy of the atom decreases. The energy liberated in this way is radiated away as an electromagnetic wave. Bohr used an equation derived by Albert Einstein to represent this energy change.

If E_1 is the energy of the atom when the electron is in a particular orbit and E_2 is the energy when the electron is in a *smaller* orbit, then the change in the energy of the atom as the electron moves from orbit 1 to orbit 2 is:

$$E_1 - E_2 = \Delta E = h\nu \qquad (5.14)$$

where ν is the frequency of the electromagnetic radiation and h is Planck's constant. The velocity of electromagnetic radiation c is related to its wavelength λ and the frequency ν by:

$$c = \lambda\nu \qquad (5.15)$$

Therefore, the energy of the radiation emitted by an atom as a result of an electronic "transition" is:

$$\Delta E = \frac{hc}{\lambda} \qquad (5.16)$$

In other words, the energy of the radiation is inversely proportional to its wavelength.

When Bohr developed this model of the atom between 1912 and 1913, it was already known that hydrogen atoms absorb and emit light at discrete wavelengths. These wavelengths had been expressed by the equation:

$$\frac{1}{\lambda} = R\left(\frac{1}{a_1^2} - \frac{1}{a_2^2}\right) \qquad (5.17)$$

where a_1 and a_2 are whole numbers, $a_1 > a_2$, and R is the Rydberg constant, whose value was known to be 109,678.18 cm^{-1}. (Note that we use a in equation 5.17 instead of n in order to avoid confusion with the first quantum number.) By applying quantum mechanics to Rutherford's model of the hydrogen atom Bohr was able to calculate the wavelengths of light that hydrogen atoms can emit as a result of transitions among the orbits the electron can occupy. By combining equations 5.13, 5.14, and 5.16 we have:

$$\frac{2\pi^2 me^4}{h^2}\left(\frac{1}{n_2^2} - \frac{1}{n_1^2}\right) = \frac{hc}{\lambda} \qquad (5.18)$$

where n_1 and n_2 are the quantum numbers, and $n_2 > n_1$ as required by the Rydberg formula. It follows that:

$$\frac{1}{\lambda} = \frac{2\pi^2 me^4}{h^3 c}\left(\frac{1}{n_2^2} - \frac{1}{n_1^2}\right) \qquad (5.19)$$

By comparing equations 5.19 (Bohr's model) and 5.17 (Rydberg formula) we see that:

$$R = \frac{2\pi^2 me^4}{h^3 c} \qquad (5.20)$$

Bohr substituted appropriate values into equation 5.20 and calculated a value of 109,677.76 cm^{-1} for the Rydberg constant. The result agreed very well with the observed value and thereby confirmed Bohr's model for the hydrogen atom. Niels Bohr was justifiably pleased with this result and subsequently devoted himself wholeheartedly to the study of quantum mechanics.

The Royal Danish Academy of Science, which receives a large portion of its funding from the Carlsberg Brewery, gave Bohr a grant of money to build an Institute for Theoretical Physics in Copenhagen. Bohr's institute became the center for research in quantum mechanics and nuclear physics. It was famous not only for its intellectual excitement but also for its relaxed atmosphere created by the "Professor" (Gamow, 1966).

5.3 Emission of X-rays

Bohr's model of the atom provided a direct explanation for the emission of x-rays, which had been discovered in 1895 by Wilhelm Konrad Röntgen (1845–1923). Although Bohr's model has been replaced by the wave mechanics of W. K. Heisenberg (1901–1976) and E. Schrödinger

(1887–1961), some aspects of it are still in use in the physics of x-rays.

In a commercial x-ray tube a stream of electrons is accelerated by a voltage difference in a vacuum before striking a target composed of a metal such as chromium, copper, molybdenum, or tungsten. The energetic electrons interact with the electrons of the target atoms and may knock them out of their orbits. X-rays are then generated when the resulting vacancies are filled by other electrons.

The wavelengths of these x-rays depend on the difference in the energies of the orbit in which a vacancy has been created and the orbit from which the replacing electron originated, as required by equation 5.16. These x-rays therefore have discrete wavelengths and give rise to a wavelength spectrum that is characteristic of the target element. The characteristic x-ray spectra were discovered in 1911 by Charles G. Barkla (1877–1944), who received the Nobel Prize for physics in 1917. The relationship between the characteristic x-ray spectrum and the atomic number of the target element was discovered in 1913 by H. G. J. Moseley (1877–1915).

In addition to the characteristic wavelength spectrum, a continuous x-ray spectrum is produced by electrons that pass through the target atoms without actually colliding with any of their electrons. Nevertheless, some energy is transferred in this type of interaction, and this energy is emitted as x-rays having continuously varying wavelengths.

The characteristic x-ray spectrum of the target atoms is divided into several series of wavelengths referred to by the letters K, L, M, etc. The "K x-rays" are emitted when a vacancy in the lowest energy level (K; $n = 1$) is filled. The electron filling the vacancy in the K-level may originate from the next higher level (L) or from others associated with electronic orbits having still larger radii. The L-series of characteristic x-rays is generated by transitions to the L-level ($n = 2$), and so on. The general scheme of identifying the characteristic x-rays is more complicated than that shown in Figure 5.1 because electrons in each of the K-, L-, and M-energy levels have slightly

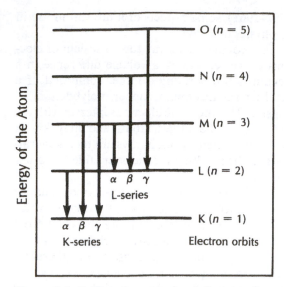

Figure 5.1 Schematic energy level diagram of atoms based on Bohr's model. X-rays of the K-series are generated by electronic transitions to the K-energy level ($n = 1$), whereas the L-series involves transitions to the L-level ($n = 2$). Greek letters are used to indicate the orbit from which the electrons originate. In reality the characteristic wavelength spectrum is more complicated than shown here because of small differences in the energies of the electrons that populate the M-, N-, and O-energy levels.

different energies. These differences are explainable in terms of wave mechanics and are not predicted in Bohr's model of the atom.

We see in Figure 5.1 that the K-series of characteristic x-rays is subdivided according to a simple scheme.

Electronic transition	Characteristic x-ray
L to K	K-alpha
M to K	K-beta
N to K	K-gamma
O to K	K-delta

The L-series is subdivided similarly, but the details in both series get more complicated

because of the "splitting" of the wavelengths caused by small energy differences of the electrons. In general, K-alpha x-rays are produced in greatest abundance because vacancies in the K-energy level are most often filled by electrons from the overlying L-level. Therefore, K-alpha x-rays are usually selected for analytical purposes both in x-ray-diffraction and in x-ray-fluorescence studies. The K-beta and K-gamma x-rays are more energetic, but their intensities are less than those of the K-alpha x-rays because the respective transitions occur less frequently.

X-ray diffraction and x-ray fluorescence have become indispensable tools in geochemical research. The former is based on the diffraction of x-rays by crystals (Klug and Alexander, 1954), whereas the latter arises from the emission of characteristic x-ray by atoms. X-ray fluorescence is now used to determine the chemical compositions of rocks including not only the major elements but also many trace elements (Norrish and Hutton, 1969).

5.4 Schrödinger's Model of the Atom

Bohr's model worked for the hydrogen atom and for ions having only one electron such as He^+, Li^{2+}, and Be^{3+}. However, it did not work well for atoms having two or more electrons. Arnold Sommerfeld (1868–1951) made some improvements by introducing elliptical orbits like those of the planets in the solar system. In addition, he allowed the planes of the orbits to include all the free space around the nucleus, thereby making the atom look more like a sphere than a flat disk. These refinements of Bohr's model required the introduction of a second quantum number but still did not make it applicable to multielectron atoms.

The problem with Bohr's model is that is treats electrons as particles whose positions in space can be determined. The validity of this idea was questioned by Louis-Victor 7th Duke de Broglie

(1892–1987) in 1923 who demonstrated that particles such as electrons also have the properties of waves. The wavelength λ of particles of mass m and velocity v is:

$$\lambda = \frac{h}{mv} \qquad (5.21)$$

where h is Planck's constant. For example, the wavelength of an electron having a kinetic energy of 10 electron volts is 0.12×10^{-8} cm, corresponding to fairly energetic x-rays (Moore, 1955, p. 271).

If electrons have wave properties, then it makes no sense to think of one as being in a particular place at a particular time. Werner Heisenberg expressed this condition in the form of the *uncertainty principle,* which implies that the motion of an electron around the nucleus of an atom cannot be described in terms of specific orbits regardless of whether they are circular or elliptical.

These strange ideas were incorporated into a new model of the atom proposed in 1926 by Werner Heisenberg and Erwin Schrödinger. Heisenberg's method was abstract and relied on matrix algebra, whereas Schrödinger (1926) used de Broglie waves to represent the distribution of electrons in three-dimensional space. Schrödinger's equation is generally stated in the form:

$$\frac{\partial^2 \psi}{\partial x^2} + \frac{\partial^2 \psi}{\partial y^2} + \frac{\partial^2 \psi}{\partial z^2} + \frac{8\pi^2 m}{h^2}(E - V)\psi = 0 \quad (5.22)$$

where ψ is a "wave function," m is the mass of the electron, E is the total energy of the atom, and V is the potential energy (Fyfe, 1964).

The wave equation can be understood by comparing it to a violin string vibrating between two fixed points. Such a string can develop nodes whose number may be 0, 1, 2, etc. The number of nodes in the one-dimensional case is determined by the quantum number n. If $n = 1$, the number of nodes is zero; when $n = 2$ there is one node, and so on. In the three-dimensional electron wave, nodes may form along the three principal directions; therefore at least three quantum numbers are needed to describe the motions of an electron in the space around the nucleus of an atom. These quantum numbers are:

$$n = 1, 2, 3, \ldots, \infty$$
$$\text{(principal quantum numbers)}$$
$$l = n - 1, n - 2, n - 3, \ldots, 0$$
$$\text{(azimuthal quantum number)}$$
$$m = 0, \pm 1, \pm 2, \ldots, \pm (l - 1), \pm l$$
$$\text{(magnetic quantum number)}$$
$$s = \pm \tfrac{1}{2} \quad \text{(spin quantum number)}$$

The fourth quantum number, s, indicates the direction of spin of the electron, which controls the polarity of the magnetic moment exerted by the electron. Note that the electron is again regarded as a charged particle in space in this context.

Solutions to the wave equation are obtained for wave functions having certain types of symmetry in space. For example, the wave function can have spherical symmetry around the nucleus, or it may be symmetrical with respect to the x, y, z directions or the xy, xz, yz planes. In general, four different types of wave functions are possible, which are identified by the letters s, p, d, and f. These wave functions describe the motions of the electron to which they apply and are therefore called "orbitals" in order to distinguish them from the "orbits" of the Bohr–Sommerfeld model. The s-orbital has spherical symmetry about the nucleus. The p-type orbital is symmetrical about the x-, y-, and z-axes, and the d-orbitals are symmetrical about the xy, xz, and yz planes. These orbitals are pictured in Figure 5.2. The f-type orbitals cannot be pictured in three dimensions. In the case of the s-type wave function there is only one solution; but there are three p-orbitals corresponding to the principal axes, five d-orbitals, and seven f-orbitals.

The wave functions of the different types of orbitals convey the probability that the electron of the hydrogen atom will be found in a certain volume of space. Consequently, we can illustrate the distribution of the electron around the nucleus by plotting the function $4\pi r^2 \psi^2$ versus

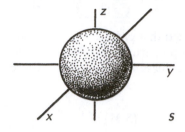

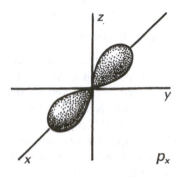

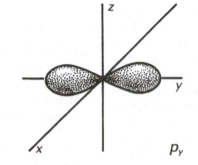

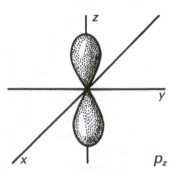

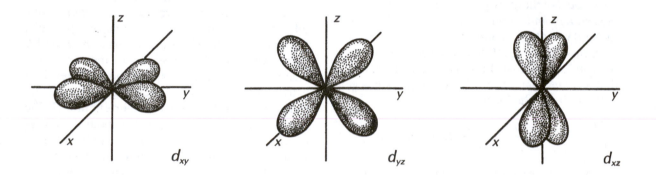

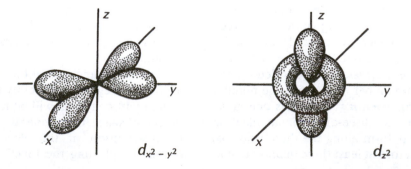

Figure 5.2 Electron clouds representing the *s*-, *p*-, and *d*-orbitals of the hydrogen atom according to the wave mechanical model of Schrödinger.

the distance r from the nucleus. Figure 5.3 is such a cross section of an s-orbital for the hydrogen atom. We see that the profile has a maximum at a distance $r = a_0$, the radius of the first electron orbit in Bohr's model. However, we also see that the electron can be found closer to or farther away from the nucleus some of the time. In other words, the electron actually occupies *all* of the space around the nucleus but can be found most often at a distance of a_0 from the nucleus. The same is true for all orbitals, and the orbital balloons shown in Figure 5.2 therefore have a "fuzzy" skin. The electrons occupying these orbitals do not spend all of their time on the skins of the balloons but move in and out in a random pattern predictable only in terms of probability. This aspect of quantum mechanics was very disturbing to Albert Einstein and even to Erwin Schrödinger. Einstein objected to the element of chance implied by the probabilistic description of the electronic structure of atoms by saying "Der liebe Gott würfelt nicht!" (the good Lord does not throw dice).

5.5 The Aufbau Principle

We are now ready to begin the construction of atoms of increasing atomic number by adding electrons to the possible orbitals of the hydrogen atom. This procedure is known as the *Aufbau principle* by means of which the orbitals of the hydrogen atoms are used to make multielectron atoms. In doing so we assume that the number of protons in the nucleus increases in step with the electrons.

The energy of an atom having a certain number of electrons depends on the orbitals the electrons occupy. Therefore, the electrons occupy only those orbitals that *minimize* the resulting energy of the atom because the *lowest* energy state of an atom is also its *most stable* or "stationary" state. This requirement controls the *sequence* in which orbitals are filled. The sequence can be derived by calculating the energy of an atom for different electron configurations. However, for multielectron atoms these calculations become very difficult and must be supplemented by spectroscopic studies.

The four quantum numbers play an important role in this process because, according to Pauli's exclusion principle, no two electrons in one atom can have the same set of quantum numbers. This principle therefore limits the number of electrons that can enter a particular orbital. The relationship between the quantum numbers and the different orbitals is indicated in Table 5.1.

We see there that the quantum numbers allow only the s-orbital for $n = 1$; s-, and p-orbitals for $n = 2$; s-, p-, d-orbitals for $n = 3$; and s-, p-, d-, and f-orbitals when $n = 4$. The table also indicates that there are three different p-orbitals, five d-orbitals (the seven f-orbitals are not listed). Moreover, each individual orbital can accommodate two electrons provided their spin numbers are different. As a result, only two electrons can be accommodated in orbitals having

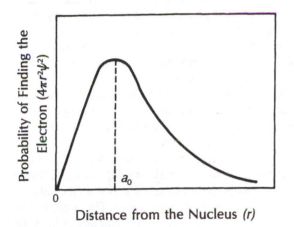

Figure 5.3 Variation of the radial distribution function $(4\pi r^2 \psi^2)$ for the s-orbital with increasing distance from the nucleus (r). The radial distribution function is a measure of the probability of finding the electron at a certain distance from the nucleus. In the s-orbital this probability has a maximum at a distance $r = a_0$, where a_0 is the radius of the smallest orbit $(n = 1)$ permitted by Bohr's model. We see that the electron in an s-orbital can also be found inside as well as outside that radius. The electron in effect occupies all of the space around the nucleus but can be found most often at the distance from the nucleus predicted by Bohr.

Table 5.1 Quantum Numbers and Possible Orbitals of the Hydrogen Atom

Quantum numbers				
n	l	m	s	Type of orbital
1	0	0	$+\frac{1}{2}, -\frac{1}{2}$	$1s$
2	0	0	$+\frac{1}{2}, -\frac{1}{2}$	$2s$
2	1	0	$+\frac{1}{2}, -\frac{1}{2}$	$2p$
2	1	1	$+\frac{1}{2}, -\frac{1}{2}$	$2p$
2	1	−1	$+\frac{1}{2}, -\frac{1}{2}$	$2p$
3	0	0	$+\frac{1}{2}, -\frac{1}{2}$	$3s$
3	1	0	$+\frac{1}{2}, -\frac{1}{2}$	$3p$
3	1	1	$+\frac{1}{2}, -\frac{1}{2}$	$3p$
3	1	−1	$+\frac{1}{2}, -\frac{1}{2}$	$3p$
3	2	0	$+\frac{1}{2}, -\frac{1}{2}$	$3d$
3	2	1	$+\frac{1}{2}, -\frac{1}{2}$	$3d$
3	2	−1	$+\frac{1}{2}, -\frac{1}{2}$	$3d$
3	2	2	$+\frac{1}{2}, -\frac{1}{2}$	$3d$
3	2	−2	$+\frac{1}{2}, -\frac{1}{2}$	$3d$
			etc.	

$n = 1$, eight electrons for $n = 2$, and 18 electrons for $n = 3$.

The sequence in which electronic orbitals are actually filled is:

$$1s, 2s, 2p, 3s, 3p, 4s, 3d, 4p, 5s,$$
$$4d, 5p, 6s, 4f, 5d, 6p, 7s, 6d$$

The Aufbau principle, by means of which we can predict the electronic configuration of the elements in the periodic table, is governed by a set of rules we state here in plain language without regard to their historical origins.

1. The first electron enters the orbital having the lowest energy. Additional electrons enter orbitals that minimize the energy of the atom (see above).

2. Each orbital can accommodate two electrons provided their spins are opposite.

3. When electrons enter a particular set of orbitals, they distribute themselves such that each orbital at first aquires only one electron.

Subsequently, a second electron may enter an orbital that already contains an electron provided its spin is opposite (rule 2).

4. Atoms achieve a state of decreased energy and enhanced stability when the available orbitals are either completely filled, half-filled, or empty.

We are now ready to construct the atoms of the elements in order of increasing atomic number by specifying their electronic structures. We do this by stating the value of the first quantum number, the type of orbital, and the number of electrons in each of those orbitals. For example, $1s^2 2s^2 2p^3$ is the electronic structure of nitrogen (atomic number 7). It has two electrons in the $1s$ orbital, two in the $2s$ orbital, and three in the $2p$ orbitals. Evidently, each of the three p-orbitals contains one electron. We can anticipate from rule 4 that nitrogen atoms may attract three additional electrons to themselves in order to fill the p-orbitals. In addition, they may also empty the p-orbitals by giving up three electrons, or lose five electrons by emptying both the p- and s-orbitals having $n = 2$. Consequently, nitrogen is likely to have different *oxidation states* or *valences:* −3, 0, +3, and +5. Evidently, we are already able to make predictions about the chemical properties of nitrogen based entirely on its electronic structure and the principles of wave mechanics.

Table 5.2 lists the electronic "formulas" of the elements in order of increasing atomic number from 1 to 96. The electrons are inserted into the orbitals in the proper sequence. The highest positive oxidation state of each element can be anticipated by means of rule 4. We have no difficulty in assigning hydrogen an oxidation state or valence of +1, although −1 is also possible and occurs in hydrides such as LiH. In natural systems hydrogen has a valence of +1, or 0 when it occurs as the diatomic molecule H_2.

Helium ($1s^2$) has filled the only orbital available to it, and therefore it neither accepts nor gives up additional electrons. Its valence is 0.

Lithium ($1s^2 2s^1$) begins a series of elements that fill orbitals having $n = 2$. The lone electron in the 2s orbital of lithium exists outside the closed $1s^2$ orbital and is therefore readily lost. Hence lithium has a valence of +1.

Beryllium ($1s^2 2s^2$) has two electrons in the 2s orbital, but it gives them up easily, thereby acquiring a valence of +2.

Boron ($1s^2 2s^2 2p^1$) starts a sequence of six elements in which the 2p orbitals are filled. The positive valences of these elements follow logically: boron +3, carbon +4, nitrogen +5. The next element, oxygen ($1s^2 2s^2 2p^4$), should have a valence of +6; but its valence in nature actually is −2 because, after the p-orbitals are half-full (nitrogen), atoms of oxygen attract electrons to themselves in order to *fill* their orbitals. Even nitrogen can acquire three electrons and have a valence of −3 as, for example, in ammonia (NH_3). Fluorine ($1s^2 2s^2 2p^5$) has the strongest tendency of all the elements to attract electrons to itself and, consequently, has a valence of −1. Neon ($1s^2 2s^2 2p^6$) completes the sequence by filling the 2p orbital and thereby ends the series of elements that started with lithium. In Table 5.2 square brackets indicate that the noble gases have filled orbitals. Neon is an inert or "noble" chemical element that does not form bonds with other elements.

In the next sequence the s- and p-orbitals of the third shell are filled. The valences of these elements (sodium, magnesium, aluminum, silicon, phosphorus, sulfur, chlorine, and argon) follow predictably from their electronic structures. Sodium has a $3s^1$ electron outside of a neon core and therefore has a valence of +1, just like lithium and hydrogen. Magnesium follows with +2, like beryllium before it, and so on. The electronic configuration of sulfur ($[Ne]3s^2 3p^4$) resembles that of oxygen, which has a valence of −2 because it strongly attracts electrons to itself. Sulfur atoms likewise attract two electrons to form S^{2-}, but they can also release their p-electrons and then assume a valence of +6, as in the sulfate ion (SO_4^{2-}). Chlorine strongly prefers the valence of −1 but takes a valence of +7 in the perchlorate (ClO_4^-) ion. Argon ($[Ne]3s^2 3p^6$) closes this sequence with a valence of 0.

The third energy shell includes d-type orbitals; however, potassium ($[Ar]4s^1$) puts the next electron into the 4s orbital rather than one of the 3d orbitals. Calcium ($[Ar]4s^2$) does likewise, so these elements have valences of +1 and +2, respectively. The d-orbitals are filled next, as we progress from scandium to zinc. In this sequence, chromium and copper are starred in Table 5.2 to draw attention to a significant anomaly in their electronic structures. Chromium should be $[Ar]3d^4 4s^2$ but is actually $[Ar]3d^5 4s^1$, and copper is $[Ar]3d^{10} 4s^1$. Both exemplify the benefit of half-filling or filling a set of electronic orbitals. Irregularity also occurs in the cases of iron, cobalt, and nickel because their valences do not follow the expected pattern. All three elements have valences of +2 and +3, even though we might have predicted +4 for cobalt and +5 for nickel.

Next, we have another sequence of six elements, starting with gallium and ending with krypton, in which the 4p orbitals are filled. These elements are similar to the aluminum–argon series in which 3p orbitals are filled.

Table 5.2 continues with another 18 elements: rubidium and strontium (5s), yttrium to cadmium (4d), and indium to xenon (5p). After that come elements with electrons in 6s, 4f, 5d, and 6p orbitals ending with radon (atomic number 86). The sequence continues with 7s, 6d, and 5f through the transuranium series. In general, the chemical properties of elements having partially filled d- and f-orbitals do not vary as much as those in which s- and p-orbitals are being filled.

5.6 Summary

Our concept of the internal structure of the atom has undergone many changes since J. J. Thomson formulated the "plum-pudding model." Some of the important milestones in this evolution are the scattering of alpha particles, discovered by E. Rutherford; the quantization of the energy of the hydrogen atom by N. Bohr; and the demonstration by L. V. de Broglie that small particles have wave properties. Ultimately W. K. Heisenberg and E. Schrödinger described the energy states of

Table 5.2 Electronic Structure and Valences of the Elements

Z^a	Element	Electronic structure	Valences
1	Hydrogen	$1s^1$	+1, −1
2	Helium	$1s^2$	0
3	Lithium	$1s^2 2s^1$	+1
4	Beryllium	$1s^2 2s^2$	+2
5	Boron	$1s^2 2s^2 2p^1$	+3
6	Carbon	$1s^2 2s^2 2p^2$	+4, +2, −4
7	Nitrogen	$1s^2 2s^2 2p^3$	+5, +4, +3, +2, −3
8	Oxygen	$1s^2 2s^2 2p^4$	+6, −2
9	Fluorine	$1s^2 2s^2 2p^5$	+7, −1
10	Neon	$1s^2 2s^2 2p^6$	0
11	Sodium	$[Ne]3s^1$	+1
12	Magnesium	$[Ne]3s^2$	+2
13	Aluminum	$[Ne]3s^2 3p^1$	+3
14	Silicon	$[Ne]3s^2 3p^2$	+4, −1
15	Phosphorus	$[Ne]3s^2 3p^3$	+5, +4, +3, −3
16	Sulfur	$[Ne]3s^2 3p^4$	+6, +4, +2, −2
17	Chlorine	$[Ne]3s^2 3p^5$	+7, +5, +3, +1, −1
18	Argon	$[Ne]3s^2 3p^6$	0
19	Potassium	$[Ar]4s^1$	+1
20	Calcium	$[Ar]4s^2$	+2
21	Scandium	$[Ar]3d^1 4s^2$	+3
22	Titanium	$[Ar]3d^2 4s^2$	+4, +3
23	Vanadium	$[Ar]3d^3 4s^2$	+5, +4, +3, +2
24*	Chromium	$[Ar]3d^5 4s^1$	+6, +3, +2
25	Manganese	$[Ar]3d^5 4s^2$	+7, +6, +4, +3, +2
26	Iron	$[Ar]3d^6 4s^2$	+3, +2
27	Cobalt	$[Ar]3d^7 4s^2$	+3, +2
28	Nickel	$[Ar]3d^8 4s^2$	+3, +2
29*	Copper	$[Ar]3d^{10} 4s^1$	+2, +1
30	Zinc	$[Ar]3d^{10} 4s^2$	+2
31	Gallium	$[Ar]3d^{10} 4s^2 4p^1$	+3, +1
32	Germanium	$[Ar]3d^{10} 4s^2 4p^2$	+4, +2
33	Arsenic	$[Ar]3d^{10} 4s^2 4p^3$	+5, +3, −3
34	Selenium	$[Ar]3d^{10} 4s^2 4p^4$	+6, +4, −2
35	Bromine	$[Ar]3d^{10} 4s^2 4p^5$	+7, +5, +1, −1
36	Krypton	$[Ar]3d^{10} 4s^2 4p^6$	0
37	Rubidium	$[Kr]5s^1$	+1
38	Strontium	$[Kr]5s^2$	+2
39	Yttrium	$[Kr]4d^1 5s^2$	+3
40	Zirconium	$[Kr]4d^2 5s^2$	+4
41	Niobium	$[Kr]4d^3 5s^2$	+5, +3
42*	Molybdenum	$[Kr]4d^5 5s^1$	+6, +5, +4, +3, +2
43	Technetium	$[Kr]4d^5 5s^2$	+7
44	Ruthenium	$[Kr]4d^6 5s^2$	+8, +6, +4, +3, +2
45	Rhodium	$[Kr]4d^7 5s^2$	+4, +3, +2
46	Palladium	$[Kr]4d^8 5s^2$	+4, +2
47*	Silver	$[Kr]4d^{10} 5s^1$	+1
48	Cadmium	$[Kr]4d^{10} 5s^2$	+2
49	Indium	$[Kr]4d^{10} 5s^2 5p^1$	+3, +1

Table 5.2 (continued)

Z^a	Element	Electronic structure	Valences
50	Tin	$[Kr]4d^{10}5s^25p^2$	+4, +2
51	Antimony	$[Kr]4d^{10}5s^25p^3$	+5, +3, −3
52	Tellurium	$[Kr]4d^{10}5s^25p^4$	+6, +4, −2
53	Iodine	$[Kr]4d^{10}5s^25p^5$	+7, +5, +1, −1
54	Xenon	$[Kr]4d^{10}5s^25p^6$	0
55	Cesium	$[Xe]6s^1$	+1
56	Barium	$[Xe]6s^2$	+2
57	Lanthanum	$[Xe]5d^16s^2$	+3
58*	Cerium	$[Xe]4f^25d^06s^2$	+3, +4
59	Praseodymium	$[Xe]4f^35d^06s^2$	+3, +4
60	Neodymium	$[Xe]4f^45d^06s^2$	+3
61	Promethium	$[Xe]4f^55d^06s^2$	+3
62	Samarium	$[Xe]4f^65d^06s^2$	+3, +2
63	Europium	$[Xe]4f^75d^06s^2$	+3, +2
64*	Gadolinium	$[Xe]4f^75d^16s^2$	+3
65	Terbium	$[Xe]4f^95d^06s^2$	+3, +4
66	Dysprosium	$[Xe]4f^{10}5d^06s^2$	+3
67	Holmium	$[Xe]4f^{11}5d^06s^2$	+3
68	Erbium	$[Xe]4f^{12}5d^06s^2$	+3
69	Thullium	$[Xe]4f^{13}5d^06s^2$	+3, +2
70	Ytterbium	$[Xe]4f^{14}5d^06s^2$	+3, +2
71	Lutetium	$[Xe]4f^{14}5d^16s^2$	+3
72	Hafnium	$[Xe]4f^{14}5d^26s^2$	+4
73	Tantalum	$[Xe]4f^{14}5d^36s^2$	+5
74	Tungsten	$[Xe]4f^{14}5d^46s^2$	+6, +5, +4, +3, +2
75	Rhenium	$[Xe]4f^{14}5d^56s^2$	+7, +6, +4, +2, −1
76	Osmium	$[Xe]4f^{14}5d^66s^2$	+8, +6, +4, +3, +2
77	Iridium	$[Xe]4f^{14}5d^76s^2$	+6, +4, +3, +2
78*	Platinum	$[Xe]4f^{14}5d^96s^1$	+4, +2
79	Gold	$[Xe]4f^{14}5d^{10}6s^1$	+3, +1
80	Mercury	$[Xe]4f^{14}5d^{10}6s^2$	+2, +1
81	Thallium	$[Xe]4f^{14}5d^{10}6s^16p^1$	+3, +1
82	Lead	$[Xe]4f^{14}5d^{10}6s^26p^2$	+4, +2
83	Bismuth	$[Xe]4f^{14}5d^{10}6s^26p^3$	+5, +3
84	Polonium	$[Xe]4f^{14}5d^{10}6s^26p^4$	+4, +2
85	Astatine	$[Xe]4f^{14}5d^{10}6s^26p^5$	+7, +5, +3, +1, −1
86	Radon	$[Xe]4f^{14}5d^{10}6s^26p^6$	0
87	Francium	$[Rn]7s^1$	+1
88	Radium	$[Rn]7s^2$	+2
89	Actinium	$[Rn]6d^17s^2$	+3
90	Thorium	$[Rn]5f^06d^27s^2$	+4
91*	Protactinium	$[Rn]5f^26d^17s^2$	+5, +4
92	Uranium	$[Rn]5f^36d^17s^2$	+6, +5, +4, +3
93	Neptunium	$[Rn]5f^46d^17s^2$	+6, +5, +4, +3
94	Plutonium	$[Rn]5f^66d^07s^2$	+6, +5, +4, +3
95	Americium	$[Rn]5f^76d^07s^2$	+6, +5, +4, +3
96	Curium	$[Rn]5f^76d^17s^2$	+3

^{a}The electron configuration of starred elements is irregular.

atoms in terms of electron orbitals based on the wave-mechanical treatment of the motions of electrons in the space around the nucleus. As a result, we can only describe the motion of an electron in terms of the probability of finding it at a particular point in space. Bohr's original model of the atom has been replaced by the wave-mechanical treatment except in the explanation for the generation of the characteristic x-ray spectra of the elements.

The electronic structure of the elements can be deduced from the possible orbitals of the hydrogen atom by means of the Aufbau principle together with a set of rules that tell us, in effect, how electrons distribute themselves in the available orbitals. We find that atoms change the number of electrons in such a way that partially filled orbitals are emptied completely, filled completely, or are occupied by only one electron per orbital in a set. This property of atoms can be used to explain the valences of the elements and hence their chemical properties.

We conclude that the chemical and physical properties of the elements depend on the number of protons in the nuclei of their atoms (atomic number) and on the distribution of the electrons around the available orbitals. Knowledge of the electronic structure of the elements therefore enhances our understanding of their chemical properties and provides a rational explanation for the construction of the periodic table.

Problems

1. Familiarize yourself with the names, chemical symbols, atomic numbers, and electronic structures of the first 36 elements.

2. How many electrons can be accommodated in orbitals having the quantum number $n = 4$?
(Answer: 32)

3. Write a complete set of quantum numbers for $n = 4$.

4. Using the electronic formulas in Table 5.2, explain the valences of the following elements: Co, S, Br, V, and Cu.

5. What do the elements of each group have in common
- **(a)** B, Al, Ga, and In
- **(b)** Be, Mg, Ca, Sr, Ba, and Ra
- **(c)** N, P, As, Sb, and Bi
- **(d)** K^+, Ca^{2+}, Sc^{3+}, Ti^{4+}, V^{5+}, Cr^{6+}

6. Without consulting Table 5.2, write the electron formula for a neutral atom having 14 electrons and predict its highest positive valence.

7. Table 5.2 suggests that the valences of the elements vary systematically with increasing atomic number. Determine how many sequences there are that end with a valence of zero.

8. Elements having similar chemical properties tend to be associated in nature. For the following pairs of elements indicate whether or not you expect them to be associated with each other. Base your answer on the electronic structures of these elements.
- **(a)** K and Rb
- **(b)** Al and Ge
- **(c)** Sc and Cu
- **(d)** S and Se
- **(e)** Li and Mg

References

BOHR, N., 1913a. Binding of electrons by positive nuclei. *Phil. Mag. J. Sci., Ser. 6,* **26**(151):1–25.

BOHR, N., 1913b. Systems containing only a single nucleus. *Phil. Mag. J. Sci., Ser. 6,* **26**(153):476–507.

FYFE, W. S., 1964. *Geochemistry of Solids.* McGraw-Hill, New York, 199 pp.

GAMOW, G., 1966. *Thirty Years That Shook Physics.* Doubleday, Garden City, NY, 224 pp.

HOLTON, G., 1986. Niels Bohr and the integrity of science. *Amer. Scientist,* **74:**237–243.

JASTROW, R., 1967. *Red Giants and White Dwarfs.* Harper & Row, New York, 176 pp.

KLUG, H. P., and L. E. ALEXANDER, 1954. *X-ray Diffraction Procedures.* Wiley, New York, 716 pp.

MOORE, W. J., 1955. *Physical Chemistry,* 2nd ed. Prentice-Hall, Upper Saddle River, NJ, 633 pp.

NORRISH, K., and J. T. HUTTON, 1969. An accurate x-ray spectrographic method for the analysis of a wide range of geological samples. *Geochim. Cosmochim. Acta,* **33:**431–453.

PLANCK, M., 1900. Zur Theorie des Gesetzes der Energieverteilung im Normalspectrum. *Verh. Deutschen Phys. Gesellschaft,* **2**(170):237–245.

SCHRÖDINGER, E., 1926. Quantisierung als Eigenwertproblem, Part 1. *Ann. Phys.,* **79**(4):361–376; Part 2, **79**(6):489–527.

6

The Periodic Table and Atomic Weights

In the early part of the 19th century chemists were almost overwhelmed by the discovery of new elements. By 1830 the number of elements had increased to 55, and there was no way of knowing how many more remained to be discovered. This uncomfortable situation motivated efforts to organize the elements in some systematic ways that might bring order to this increasingly chaotic state of affairs.

6.1 Mendeleev's Periodic Table

Although many chemists contributed to the effort to organize the elements, Dmitri Ivanovich Mendeleev (1834–1907) deserves to be recognized as the inventor of the periodic table (Asimov, 1965; van Sprosen, 1969).

Mendeleev was a graduate student in Germany when Friedrich August Kekulé von Stradonitz (known as Kekulé) organized a meeting of the leading chemists of Europe in 1860 in Karlsruhe. The purpose of this "First International Chemical Congress" was to debate the merits of the structural formulas of organic compounds Kekulé had proposed in 1858. These formulas were based on the idea that carbon has a valence of 4 and bonds with hydrogen, and even with itself, to form increasingly complex molecular chains. During the Congress in Karlsruhe the Italian chemist Stanislao Cannizzaro urged the delegates to recognize the importance of the atomic weights of the elements and to distinguish them from the equivalent weights and the molecular weights

based on an hypothesis his colleague Amadeo Avogadro had advanced in 1811.

When Mendeleev returned to Russia he used both the valences and the atomic weights to organize the elements into periods. In 1869 he published a table in the *Journal of the Russian Chemical Society* in which he identified several periods within which the valences rose and fell in regular patterns. In the following year (1870), the German chemist Julius Lothar Meyer published a graph of atomic volumes plotted versus atomic weights that also revealed the periodicity of these properties. However, Mendeleev (1871) boldly reordered some elements in his table on the basis of their valences and left gaps where, he claimed, elements existed that had not yet been discovered. For example, he predicted the existence of an element similar to silicon and specified many of its physical and chemical properties by comparison with those of its neighbors in his table. His predictions were confirmed when C. A. Winkler discovered germanium in 1886 during the analysis of the rare mineral argyrodite (Ag_8GeS_6) (Weeks, 1956).

6.2 The Modern Periodic Table

The periodic table has assumed many forms since it was first presented by Mendeleev. Mazurs (1974) described over 700 different graphic representations. Today we use the so-called long form depicted in Figure 6.1.

The elements are arranged into horizontal rows called *periods* and vertical columns called

New notation

VIII CAS version

Periods	1 / I	2 / II	3 / III	4 / IV	5 / V	6 / VI	7 / VII	8	9 VIII	10	11 / I	12 / II	13 / III	14 / IV	15 / V	16 / VI	17 / VII	18 / VIII
1	1 H																	2 He
2	3 Li	4 Be											5 B	6 C	7 N	8 O	9 F	10 Ne
3	11 Na	12 Mg											13 Al	14 Si	15 P	16 S	17 Cl	18 Ar
4	19 K	20 Ca	21 Sc	22 Ti	23 V	24 Cr	25 Mn	26 Fe	27 Co	28 Ni	29 Cu	30 Zn	31 Ga	32 Ge	33 As	34 Se	35 Br	36 Kr
5	37 Rb	38 Sr	39 Y	40 Zr	41 Nb	42 Mo	43 Tc	44 Ru	45 Rh	46 Pd	47 Ag	48 Cd	49 In	50 Sn	51 Sb	52 Te	53 I	54 Xe
6	55 Cs	56 Ba	57 La	72 Hf	73 Ta	74 W	75 Re	76 Os	77 Ir	78 Pt	79 Au	80 Hg	81 Tl	82 Pb	83 Bi	84 Po	85 At	86 Rn
7	87 Fr	88 Ra	89 Ac															

A Groups — B Groups

s — d — p — f

Rare earth elements

58 Ce	59 Pr	60 Nd	61 Pm	62 Sm	63 Eu	64 Gd	65 Tb	66 Dy	67 Ho	68 Er	69 Tm	70 Yb	71 Lw

Actinide elements

90 Th	91 Pa	92 U	93 Np	94 Pu	95 Am	96 Cm	97 Bk	98 Cf	99 Es	100 Fm	101 Md	102 No	103 Lr

Figure 6.1 The periodic table of the elements. The numbering of the groups has been changed to run consecutively from 1 to 18 across the table. The old method is used here to emphasize the relationship between the electronic structures and highest positive valences of atoms.

groups. The periods are numbered in accordance with the first quantum number of the orbitals that are being filled with increasing atomic number. For example, the elements of the first period contain electrons in the 1s orbital, the elements of the second period have electrons in the 2s and 2p orbitals, those in third period are filing 3s and 3p orbitals, and in the fourth period electrons enter 4s and 4p orbitals. However, the fourth period contains an additional group of 10 elements in which the 3d orbitals are filling up. Similarly, the fifth and sixth periods each contain elements in which d-orbitals are being filled. These elements together form the so-called d-block in the periodic table. The sixth period contains 14 extra elements known as the *rare earths* with electrons in 4f orbitals. The rare earths are pulled out of the table and placed separately below it together with the *actinide elements* in which 5f orbitals are being filled and which include the *transuranium* or "super-heavy" elements. We see, therefore, that the number of the elements in each period is a direct consequence of the progressive filling of electronic orbitals. These relationships are evident in Table 6.1.

The numbering of the vertical columns or groups has been confusing to say the least. They used to be numbered from left to right in accordance with the highest positive valence of the ele-ments in that group. In this version of the periodic table the elements in the s- and p-blocks formed the "A groups," and the elements in the d-block composed the "B groups." However, the identification of A and B groups has not been consistent. The valences of the elements in the A groups, as used in Figure 6.1, can be deduced directly from the electronic configurations of the atoms. For example, group IA consists of H, Li, Na, K, Rb, Cs, and Fr, each of which has one electron in an s-orbital outside of a noble gas core and thus a valence of +1. Excluding hydrogen, the elements in group IA are known as the *alkali metals*. The elements in group IIA, called the *alkaline earths*, include Be, Mg, Ca, Sr, Ba, and Ra. These elements have two electrons in an s-orbital outside a noble gas core and therefore have valences of +2.

The numbering of the A groups as defined above continues with the elements of the p-block on the right side of the periodic table. Once again the numbering corresponds to the highest positive valence consistent with the electronic configuration. Thus the elements in group 7A can have valences of +7 because they all have five electrons in p-orbitals and two electrons in s-orbitals. The elements in this group are called the *halogens* and include F, Cl, Br, I, and At. In nature the atoms of these elements attract an electron and actually form ions with a charge of −1. Nevertheless, they have electronic structures that make a valence of +7 at least technically possible. The last group of elements are the *noble gases* He, Ne, Ar, Kr, Xe, and Rn, all of which have filled orbitals. Consequently, they neither gain nor lose electrons under natural conditions and therefore have a valence of 0. In spite of that fact, this group used to be given the number 8, although 0 is more appropriate.

Partly because of these kinds of inconsistencies, chemists have decided to number the groups consecutively from left to right. In this new scheme the alkali metals are still in group 1 but the noble gases are now in group 18 (Figure 6.1). Nevertheless, we will continue the old numbering system outlined above because it emphasizes the

Table 6.1 Relationship Between Periods and Electron Orbitals of Atoms of the Elements

Period	Orbitals being filled	Number of elements
1	1s	2
2	2s, 2p	8
3	3s, 3p	8
4	4s, 3d, 4p	18
5	5s, 4d, 5p	18
6	6s, 4f, 5d, 6p	32
7	7s, 5f, 6d, 7p	32[a]

[a]Only 17 are known.

close relationship between the valence and the electronic structure of the elements.

The elements with incompletely filled *d*-orbitals are referred to as the *transition metals*. The groups in this block of elements are numbered in accordance with the valences of the elements just like the A groups. Therefore, group IB consists of the precious metals Cu, Ag, and Au, which have electron configurations of $d^{10}s^1$ and therefore can have a valence of +1. However, copper can also be +2, and gold can be +3. The numbering of the B groups continues with IIB composed of Zn, Cd, and Hg with $d^{10}s^2$ electron configurations that form +2 ions. Next, we move to the left to IIIB (Sc, Y, and La, d^1s^2, valence +3) and progress from there to group VIIB (Mn, Tc, and Re, d^5s^2, valence +7). However, the next three groups are anomalous. The iron group (Fe, Ru, and Os) should have a valence of +8 in accordance with their electronic structure (d^6s^2), but iron actually has valences of +2 and +3. Cobalt and nickel also have valences of +2 and +3. Therefore, these three groups are lumped together as group VIIIB. This is not a very elegant solution, but it has tradition on its side. The *congeners* of Fe, Co, and Ni form the *platinum group elements* (PGE) consisting of Ru, Os, Rh, Ir, Pd, and Pt. These elements can occur in the native state and, in addition, have valences of +2, +3, +4, +6, and even +8 in the case of Ru and Os.

The periodic table is founded on the recurrence of similar physical and chemical properties of the elements with increasing atomic number. This statement is known as the *periodic law* of the elements. In addition, elements in *groups* have the same valences because they have similar electronic structures. As a result, their physical and chemical properties tend to be similar.

In geochemistry this means that the elements of a group tend to occur together and display *geochemical coherence*. For example, the alkali metals, the alkaline earths, the halogens, the noble gases, and the precious metals display well-developed geochemical coherence in their distribution in nature.

6.3 Basic Principles of Atomic Physics

We know that atoms have nuclei composed of protons and neutrons. The number of protons in the nucleus determines the positive charge and therefore the number of extranuclear electrons required to form a neutral atom. The number of electrons and their distribution in the available orbitals give atoms their characteristic chemical properties. Therefore, the atomic number (Z) indirectly determines the chemical properties of an atom. Since all atoms of a chemical element must have the same (or nearly the same) chemical properties, it follows that all atoms of a given element must have the same atomic number.

The number (N) of neutrons in the nuclei of atoms affects atomic masses but has no direct bearing on chemical properties. Therefore, atoms of the same chemical element may contain different numbers of neutrons in their nuclei. Atoms that have the same atomic number but different neutron numbers are called *isotopes*. The word means "same place" and indicates that the isotopes of an element all occupy the same place in the periodic table because they have the same (or very similar) chemical properties.

The internal composition of the nuclei of isotopes is indicated by the shorthand notation used to identify them. This notation consists of the chemical symbol of the element, preceded by the atomic number (Z) written as a subscript, and the mass number (A) written as a superscript, where:

$$A = Z + N \qquad (6.1)$$

Therefore, the notation $^{12}_{6}C$ indicates that the atom or *nuclide* is an isotope of carbon, that its nucleus contains 6 protons, and that the sum of protons and neutrons is 12. From equation 6.1 we find that the nucleus also contains 6 neutrons.

Most chemical elements have two or more stable isotopes. However, some elements, including Be, F, Na, Al, P, Mn, Co, and As have only one stable isotope. Several other elements, including K, Rb, Sm, Lu, Re, Th, and U, have long-lived radioactive isotopes that decay so slowly that they still occur in the solar system even though they formed more than 6×10^9 years ago by nucleosynthesis reactions in ancestral stars. In addition, more than 2000 short-lived radioactive isotopes are known that do not occur in nature because they have decayed since the time of nucleosynthesis.

A fairly large group of short-lived radioactive isotopes *does* occur in nature because these isotopes are continuously produced either by the decay of long-lived parents (U and Th) or by nuclear reactions involving cosmic rays. The unstable daughters of uranium and thorium are polonium, astatine, radon, francium, radium, actinium, and protactinium. Another large number of radioactive isotopes is produced by nuclear reactions in the atmosphere: tritium, $_1^3H$, $_4^{10}Be$, $_6^{14}C$, $_{17}^{36}Cl$, and many others.

The abundances of the naturally occurring isotopes (stable or long-lived radioactive) are determined with mass spectrometers and are expressed in percent by *number*. For example, boron has two stable isotopes whose abundances are $_5^{10}B = 19.8\%$ and $_5^{11}B = 80.2\%$. This means that out of 1000 boron atoms 198 are the isotope $_5^{10}B$ and 802 are $_5^{11}B$. Actually, it turns out that the isotopic compositions of boron and of other elements of low atomic number (H, C, N, O, and S) may be changed by *isotope fractionation* because mass differences among the isotopes of an element affect the kinetics of chemical reactions and the strengths of chemical bonds. These effects are part of the subject of isotope geoscience presented in detail in a textbook by Faure (1986) and will be discussed in Chapter 18 of this book.

6.4   Atomic Weights

The concept of the *atomic weight* is fundamental to all forms of chemistry but is not always clearly understood. One reason for the uncertainty is that the scale by means of which atomic weights are expressed has changed over the years. Originally, chemists used oxygen as the standard and gave it an atomic weight of 16.0000. However, after the discovery of isotopes, physicists adopted a different scale based on $_8^{16}O = 16.0000$. All of the resulting confusion has now been eliminated by the adoption of $_6^{12}C$ as the standard. Therefore, the masses of atoms are now expressed in the atomic mass unit (amu), defined as 1/12 of the mass of $_6^{12}C$, which is the most abundant stable isotope of carbon. The masses of the naturally occurring isotopes have also been determined by mass spectrometry, and the results are tabulated in standard reference books such as the *CRC Handbook of Chemistry and Physics* (Weast et al., 1986).

We are now ready to define the atomic weight of an element.

The atomic weight of an element is the sum of the masses of its naturally occurring isotopes weighted in accordance with their abundances.

An example will clarify what this means. Silicon has three naturally occurring stable isotopes whose abundances and masses are

Isotope	Abundance, %	Mass, amu
$_{14}^{28}Si$	92.23	27.976 927
$_{14}^{29}Si$	4.67	28.976 495
$_{14}^{30}Si$	3.10	29.973 770

The atomic weight of silicon is found by multiplying the masses of the isotopes by their abundances expressed as decimal fractions and adding the resulting products:

$$\begin{aligned}
_{14}^{28}Si: & \quad 0.9223 \times 27.976\,927 = 25.803\,12 \\
_{14}^{29}Si: & \quad 0.0467 \times 28.976\,495 = 1.353\,20 \\
_{14}^{30}Si: & \quad 0.0310 \times 29.973\,770 = \underline{0.929\,18} \\
& \qquad\qquad\qquad\qquad\quad \text{Sum} = 28.085\,50
\end{aligned}$$

Therefore, the atomic weight of silicon, rounded to five significant figures, is 28.086. Chemists regard the atomic weights of the elements as dimensionless numbers because they are expressed relative to the mass of $^{12}_6C$ (Greenwood and Earnshaw, 1984).

The importance of the atomic weights of the elements arises from their use in the definition of the gram-atomic weight.

The *gram-atomic weight* of an element is equal to the atomic weight in grams.

Similarly, the gram-molecular weight or the gram-formula weight of a compound is defined as

The *gram-molecular weight* (or the *gram-formula weight*) of a compound is the molecular weight (or formula weight) in grams.

Both the gram-atomic weight and gram-molecular weight are referred to as the *mole* (mol). This is the basic unit of mass of elements and compounds in chemistry.

The reason why the mole is so important in chemistry is that one mole of an element or a compound always contains a fixed number of atoms or molecules. That number is:

$$N_A = 6.022\ 045 \times 10^{23} \text{ atoms or molecules}$$
$$\text{per mole}$$

which is known as *Avogadro's number*. It is a consequence of the hypothesis proposed by Amadeo Avogadro in 1811 that equal volumes of gases at the same pressure and temperature contain equal numbers of atoms or molecules. Therefore, the concept of the mole together with Avogadro's number enables us to convert an amount of an element or compound from grams to the corresponding number of atoms or molecules. In addition, we can now appreciate the formal definition of the mole stated on the inside back cover of this book.

The *mole* is the amount of a system that contains as many elementary entities as there are atoms in 0.012 kg of carbon-12.

In other words, 6.022×10^{23} atoms, molecules, ions, electrons, or other specified particles make up one mole.

The relationship between moles and the corresponding number of atoms and molecules enables us to represent chemical reactions by means of equations. For example, the equation:

$$2\,H_2 + O_2 \rightarrow 2\,H_2O \qquad (6.2)$$

indicates that two molecules of hydrogen (H_2) must combine with one molecule of oxygen (O_2) in order to form two molecules of water (H_2O). If we mix 2 g of H_2 with 1 g of O_2 we would *not* achieve the desired proportions of molecules. However, when the amounts of H_2 and O_2 are expressed in terms of moles, then the H_2 and O_2 molecules will be present with the necessary 2:1 ratio required for the reaction.

The atomic weights of the elements listed in Table 6.2 enable us to calculate molecular weights of compounds based on their chemical formulas and to convert amounts of such compounds from grams to the corresponding number of moles. For example, the molecular weight of $BaSO_4$ (barite) is calculated from the atomic weights of the elements:

Ba:	137.33
S:	32.06
O:	15.9994

The molecular weight of barite is 233.366, and one mole of barite weighs 233.366 g. The solubility of barite in cold water is 2.22×10^{-4} g/100 mL (Weast et al., 1986). Therefore, one liter of a saturated solution contains 2.22×10^{-3} g of barite, which corresponds to:

$$\frac{2.22 \times 10^{-3}}{233.366} = 9.513 \times 10^{-6} \text{ mol}$$

When barite dissolves in water, it dissociates into ions:

$$BaSO_4 \rightarrow Ba^{2+} + SO_4^{2-} \qquad (6.3)$$

The equation indicates that each mole of barite that dissolves produces one mole of Ba^{2+} and one mole of SO_4^{2-}. Since 9.512×10^{-6} mol of barite dissolves in 1 L of a saturated solution, the concentration of

Table 6.2 Atomic Weights of the Elements Relative to $^{12}_6C$

Z	Element	Atomic Weight	Z	Element	Atomic Weight
1	H	1.00794	48	Cd	112.41
2	He	4.00260	49	In	114.82
3	Li	6.941	50	Sn	118.71
4	Be	9.01218	51	Sb	121.75
5	B	10.81	52	Te	127.60
6	C	12.011	53	I	126.905
7	N	14.0067	54	Xe	131.29
8	O	15.9994	55	Cs	132.905
9	F	18.9984	56	Ba	137.33
10	Ne	20.179	57	La	138.906
11	Na	22.9898	58	Ce	140.12
12	Mg	24.305	59	Pr	140.908
13	Al	26.9815	60	Nd	144.24
14	Si	28.0855	61	Pm	(145)[a]
15	P	30.9738	62	Sm	150.36
16	S	32.06	63	Eu	151.96
17	Cl	35.453	64	Gd	157.25
18	Ar	39.948	65	Tb	158.925
19	K	39.0983	66	Dy	162.50
20	Ca	40.08	67	Ho	164.930
21	Sc	44.9559	68	Er	167.26
22	Ti	47.88	69	Tm	168.934
23	V	50.9415	70	Yb	173.04
24	Cr	51.996	71	Lu	174.967
25	Mn	54.9380	72	Hf	178.49
26	Fe	55.847	73	Ta	180.948
27	Co	58.9332	74	W	183.85
28	Ni	58.69	75	Re	186.207
29	Cu	63.546	76	Os	190.2
30	Zn	65.39	77	Ir	192.22
31	Ga	69.72	78	Pt	195.08
32	Ge	72.59	79	Au	196.967
33	As	74.9216	80	Hg	200.59
34	Se	78.96	81	Tl	204.383
35	Br	79.904	82	Pb	207.2
36	Kr	83.80	83	Bi	208.980
37	Rb	85.4678	84	Po	(209)[a]
38	Sr	87.62	85	At	(210)[a]
39	Y	88.9059	86	Rn	(222)[a]
40	Zr	91.224	87	Fr	(223)[a]
41	Nb	92.9064	88	Ra	226.025
42	Mo	95.94	89	Ac	227.028
43	Tc	(98)[a]	90	Th	232.038
44	Ru	101.07	91	Pa	231.036
45	Rh	102.906	92	U	238.029
46	Pd	106.42	93	Np	(237)[a]
47	Ag	107.868	94	Pu	(239)[a]

[a]*Mass number* of longest-lived isotope in nature.

both Ba^{2+} and SO_4^{2-} is also 9.512×10^{-6} mol/L. The concentration of Ba^{2+} in 1 L of saturated barite solution can also be expressed as:

$$9.512 \times 10^{-6} \times 137.33 = 1.306 \times 10^{-3} \text{ g/L}$$

Alternatively, we calculate the number of Ba^{2+} ions in one liter of a saturated barite solution by means of Avogadro's number:

$$9.512 \times 10^{-6} \times 6.022\,045 \times 10^{23}$$
$$= 5.728 \times 10^{18} \text{ ions/L}$$

Such computations involving conversions of amounts of elements or compounds in grams to the corresponding number of moles by means of the atomic or molecular weights occur commonly in geochemistry.

Before we close this chapter we define the gram-equivalent weight.

> A *gram-equivalent weight of an ion* is the gram-atomic or gram-molecular weight divided by the valence. For acids and bases the gram-equivalent weight is the gram-molecular weight of the compound divided by the number of hydrogen ions or hydroxyl ions the acid or base *can* produce when dissolved in water.

Equivalent weights are used in geochemistry when electrical charges if ions are an important consideration. In addition, they are traditionally used to express the concentrations of acids or bases in a way that indicates the concentrations of hydrogen or hydroxyl ions that may be available for reaction.

In practice, the equivalent weight of an element is calculated by dividing its atomic weight by the charge the atoms of that element acquire when the element is dissolved in water. Therefore, the equivalent weight of a divalent element like calcium is equal to the atomic weight divided by two. Conversely, one mole of Ca^{2+} contains two equivalents of Ca^{2+}. In the case of molecules of an acid such as H_2SO_4, the equivalent weight is equal to one half of the molecular weight because one mole of H_2SO_4 contains two equivalents of hydrogen ions. Similarly, phosphoric acid (H_3PO_4) can yield three moles of hydrogen ions per mole of the compound, and therefore its equivalent weight is equal to its molecular weight divided by three. In this case, it makes no difference that phosphoric acid actually gives up its hydrogen ions reluctantly and releases all three of them only under special conditions.

6.5 Summary

The periodic table was originally invented by D. I. Mendeleev and other chemists in the 19th century based on the relationship between the atomic weights of the elements and their chemical and physical properties. We now know that these relationships are caused by the composition of the nuclei of the atoms and by the distribution of their electrons in the available orbitals.

The so-called long form of the periodic table is organized into seven horizontal rows called periods and eighteen vertical columns called groups. The physical and chemical properties of elements in the periods change systematically with increasing atomic number. The elements in the groups tend to have similar properties because they have similar electronic structures.

The concept of the atomic weight is now based on the masses and abundances of the naturally occurring isotopes of the elements. The atomic weights are used to define the gram-atomic weight of elements or the gram-molecular weight of compounds, also known collectively as the mole. The importance of the mole in chemistry arises from the fact that one mole of an element or compound always contains the same number of atoms or molecules. This number is known as Avogadro's number.

Chemical reactions are represented by equations in which the quantities of reactants and products are specified in terms of moles, which can be converted to the numbers of atoms or molecules by virtue of Avogadro's number.

The gram-equivalent weight takes into account the electrical charges of ions or the number of hydrogen or hydroxyl ions that acids or bases can release into an aqueous solution.

Problems

1. Familiarize yourself with the names, atomic numbers, and positions in the periodic table of the elements having atomic numbers from 1 to 36.

2. Calculate the atomic weight of magnesium given the abundances and masses of its naturally occurring isotopes.

	Abundance, %	*Mass, amu*
$^{24}_{12}Mg$	78.99	23.985 042
$^{25}_{12}Mg$	10.00	24.985 837
$^{26}_{12}Mg$	11.01	25.982 593

(Answer: 24.305)

3. Calculate the formula weight of orthoclase feldspar ($KAlSi_3O_8$) using the atomic weights in Table 6.2.
(Answer: 278.332)

4. How many moles of Na^+ are present in one liter of an aqueous solution of Na_2SO_4 containing 4.760 g of the compound? (Answer: 0.0670 mol)

5. What is the weight of one atom of $^{238}_{92}U$ in grams given that the mass of this isotope is 238.050 784 amu?
(Answer: 3.953×10^{-22} g)

6. How many atoms of iron are present in 5.00 g of hematite (Fe_2O_3)? (Answer: 3.77×10^{22} atoms)

7. What is the concentration of Cr in the mineral chromite ($FeCr_2O_4$)? (Answer: 46.5%)

8. If the concentration of SO_4^{2-} is 2.5×10^{-3} mol/L, how many grams of SO_4^{2-} are present in one liter?
(Answer: 0.24 g)

9. A solution of $Al_2(SO_4)_3$ contains 2×10^{-4} mol/L of Al^{3+}. How many grams of SO_4^{2-} does it contain per liter? (Answer: 0.029 g/L)

10. How many gram-equivalent weights of Al^{3+} and SO_4^{2-} are in 2 mol of $Al_2(SO_4)_3$?
(Answer: Al^{3+}: 12; SO_4^{2-}: 12)

References

ASIMOV, I., 1965. *A Short History of Chemistry.* Doubleday, Garden City, NY, 263 pp.

FAURE, G., 1986. *Principles of Isotope Geology.* 2nd ed. Wiley, New York, 589 pp.

GREENWOOD, N. N., and A. EARNSHAW, 1984. *Chemistry of the Elements.* Pergamon Press, Oxford, England, 1542 pp.

MAZURS, E. G., 1974. *Graphic Representation of the Periodic System During One Hundred Years.* University of Alabama Press, Tuscaloosa.

VAN SPROSEN, J. W., 1969. *The Periodic System of the Elements.* Elsevier, Amsterdam, 368 pp.

WEAST, R. C., M. J. ASTLE, and W. H. BEYER, 1986. *CRC Handbook of Chemistry and Physics.* CRC Press, Boca Raton, FL.

WEEKS, M. E., 1956. *Discovery of the Elements,* 6th ed. J. Chem. Ed. Pub., 910 pp.

7

Chemical Bonds, Ionic Radii, and Crystals

The physical and chemical properties of the elements vary with increasing atomic number in accordance with the *periodic law*. Some of these properties are neatly summarized in periodic tables available in university book stores. It will be helpful to use such a table to study the topics presented in this chapter.

7.1 Electron Donors Versus Acceptors

We can divide the elements in the periodic table into two groups based on their tendency to give up electrons or to attract them. In general, the elements on the left side of the periodic table are the *electron donors* because their electronic orbitals are largely empty. The elements on the right side of the periodic table have orbitals that are nearly full, which causes them to be *electron acceptors*. The elements that are located between the alkali metals on the left and the halogens on the right vary progressively in their tendencies to be electron donors or acceptors, depending on how close they are to filling their orbitals with electrons. The electron donors are the *metals*, whereas the electron acceptors are the *non-metals*. The metallic character of the elements generally decreases from left to right in the periods as the orbitals are filling up with electrons.

The properties of metals and nonmetals are complementary in the sense that the electrons given up by metallic elements are picked up by the nonmetals. If a complete transfer of electrons is achieved, the electron-donor atoms become positively charged ions called *cations*, whereas the electron-acceptor atoms acquire a negative charge and form *anions*. Oppositely charged ions are attracted to each other and may form an electrostatic or *ionic bond*. Most minerals are chemical compounds composed of cations (metals) and anions (nonmetals) held together by ionic bonds.

When two elements having similar metallic or nonmetallic character interact, electrons are not actually transferred, but are shared by the atoms. The sharing of electrons among the atoms of two or more different elements establishes a *covalent bond* between them. The elements do not necessarily share the electrons equally, which causes covalent bonds to have a certain amount of ionic character. Similarly, the transfer of electrons among atoms of elements on opposite sides of the periodic table is never complete. As a result, ionic bonds have covalent character whose magnitude depends on the difference in the tendency of the atoms to attract electrons to themselves. As atoms become more equal in their ability to attract electrons, the covalent character of the bond increases. The distinction between ionic and covalent bonds is therefore an over-simplification of a very complex interaction among atoms of different elements (Pauling, 1960). The only truly covalent bonds occur when two atoms of the same element combine to form diatomic molecules such as H_2, O_2, or N_2.

7.2 Measures of Metallic Character

Although we can appreciate the difference between electron donors and electron acceptors *qualitatively,* we need to measure these tendencies *quantitatively* in order to be able to predict what kind of a bond the atoms of two elements are likely to form. Several parameters have been defined that, in fact, permit us to do this. They are identified and defined as follows.

> *First ionization potential* is the energy required to remove one electron from a neutral atom in a vacuum and to place it at rest an infinite distance away.
>
> *Electronegativity* is a measure of the ability of an atom in a molecule to attract electrons to itself (Pauling, quoted in Cotton and Wilkinson, 1962, p. 89).
>
> *Standard electrode potential* is the voltage generated when one mole of electrons is removed from one mole of an element or ion "in the standard state."

We deal with the standard electrode potential in Section 14.3 and consider here how the first ionization potentials and the electronegativities can be used to compare the metallic character of different elements.

The first ionization potential indicates how strongly the nucleus of a neutral atom attracts an electron in a partially filled orbital. Figure 7.1A indicates that the first ionization potentials *increase* with increasing atomic number in each period. The increase coincides with the progressive filling of the electronic orbitals and reflects the reluctance of atoms to give up electrons from orbitals that are nearly full. For example, we see that in periods 2 and 3 the first ionization potentials increase sharply with increasing atomic number toward the noble gas elements that terminate each period. We can even recognize the effect on the first ionization potential of filling the *s*-orbitals or half-filling the *p*-orbitals. This explains why the first ionization potential of B is less than that of Be and why it takes less energy to remove an electron from O than from N. The

same effect can be seen in period 3 in the first ionization potentials of Mg and Al and of P and S.

The first ionization potentials of the transition metals of period 4, which have partially filled 3*d* orbitals, vary much *less* than those in the first three periods. Period 4 starts out well enough with K and Ca, but then the first ionization potentials change little as we complete the sequence ending with Zn. Gallium starts to fill the 4*p* orbitals, which causes it to have a lower ionization potential than Zn. However, the first ionization potentials then rise rapidly with increasing atomic number until period 4 ends with Kr. The next period, 5, starting with Rb and ending with Xe, likewise illustrates the lack of differentiation among the elements with electrons in the 4*d* orbitals. The first ionization potentials for these and all other elements are listed in Table 7.1.

Evidently, the systematic variation of the first ionization potentials confirms that the *metallic character* of the elements in each period *decreases* with *increasing* atomic number. The first ionization potentials even reflect the reduction in the energy of atoms that have completely filled or half-filled orbitals and, in general, permit us to rank the elements throughout the periodic table in terms of their metallic character, which arises from their tendency to be electron donors.

Ionization potentials required to remove additional electrons rise to high values because it takes extraordinary amounts of energy to remove electrons from atoms that have acquired a positive charge because they have already lost one electron or because the remaining electrons completely fill the orbital they occupy. Therefore, during chemical reactions in nature elements generally lose only their "valence electrons," that is, those that reside outside of filled orbitals.

The problem with ionization potentials as predictors of chemical properties is that the chemical reactions we want to study do not occur in a vacuum but take place in mixtures of gases, in aqueous solutions, in silicate melts, and even in the solid state. Therefore, we need a parameter that is appropriate for the environments in which chemical reactions actually take place.

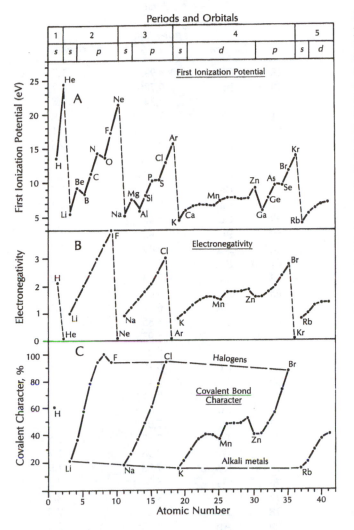

Figure 7.1 A: Variation of the first ionization potential with increasing atomic number in the first four periods. The first ionization potentials increase as the electronic orbitals are filling up in each period. Note that Be and N have exceptionally high ionization potentials consistent with filling or half-filling of electronic orbitals. The ionization potentials of elements in period 4 with partially filled d-orbitals vary little with atomic number. In general, the data show that the character of the elements in each period changes from metallic (low ionization potential, hence electron donor) to nonmetallic (high ionization potential, hence electron acceptor).

B: Variation of the electronegativity of the elements with increasing atomic number. The diagram shows that the electronegativity increases as the available orbitals become filled with electrons. This pattern, therefore, reflects the increasing tendency of elements in the periods to fill their orbitals by attracting electrons to themselves. Note that metallic elements have low electronegativities, whereas nonmetallic elements have high values. Therefore, the electronegativities also reflect the progressive decrease of the metallic character of elements with increasing atomic number in each period.

C: Variation of the covalent character of single bonds with *oxygen* based on electronegativity differences (Table 7.2). As the elements lose their metallic character and become increasingly reluctant electron donors, the bonds they form with oxygen are increasingly covalent in character. In general, the alkali metal–oxygen bonds are highly *ionic,* whereas the halogen–oxygen bonds are highly *covalent.* All other elements in the periodic table form bonds with oxygen whose character is intermediate between these extremes.

The electronegativities, proposed by Pauling (1960), are a set of dimensionless numbers calculated from the known strengths of bonds between atoms in molecules. They are a measure of the ionic character of covalent bonds and therefore indicate the extent to which two atoms in a molecule actually share their valence electrons equally. Elements having a low electronegativity act as electron donors, whereas elements with high electronegativity act as electron acceptors. Cesium has the lowest electronegativity (0.7) and therefore has the strongest tendency to be an electron donor, whereas fluorine has the highest elec-tronegativity (4.0) and is the strongest electron acceptor. The electronegativities of all naturally occurring elements are listed in Table 7.1 and are displayed in Figure 7.1B.

As expected, we find that the electronegativities rise with increasing atomic number in each period. The electronegativities of the noble gases are shown as "zero" because these elements do not attract electrons to themselves. The alkali metals and the halogens demonstrate that the electronegativities of group A elements *decrease* with increasing atomic number. The first ionization potentials of these elements also decrease with

Table 7.1 First Ionization Potentials and Electronegativities of the Elements

Z	Element	First ionization potential,[a] eV	Electronegativity [b]	Z	Element	First ionization potential,[a] eV	Electronegativity [b]
1	H	13.598	2.1	48	Cd	8.993	1.7
2	He	24.587		49	In	5.786	1.7
3	Li	5.392	1.0	50	Sn	7.344	1.8
4	Be	9.322	1.5	51	Sb	8.641	1.9
5	B	8.298	2.0	52	Te	9.009	2.1
6	C	11.260	2.5	53	I	10.451	2.5
7	N	14.534	3.0	54	Xe	12.130	
8	O	13.618	3.5	55	Cs	3.894	0.7
9	F	17.422	4.0	56	Ba	5.212	0.9
10	Ne	21.564		57	La	5.577	1.1
11	Na	5.139	0.9	58	Ce	5.47	1.1
12	Mg	7.646	1.2	59	Pr	5.42	1.1
13	Al	5.986	1.5	60	Nd	5.49	1.2
14	Si	8.151	1.8	61	Pm	5.55	
15	P	10.486	2.1	62	Sm	5.63	1.2
16	S	10.360	2.5	63	Eu	5.67	
17	Cl	12.967	3.0	64	Gd	6.14	1.1
18	Ar	15.759		65	Tb	5.85	1.2
19	K	4.341	0.8	66	Dy	5.93	
20	Ca	6.113	1.0	67	Ho	6.02	1.2
21	Sc	6.54	1.3	68	Er	6.10	1.2
22	Ti	6.82	1.5	69	Tm	6.18	1.2
23	V	6.74	1.6	70	Yb	6.254	1.1
24	Cr	6.766	1.6	71	Lu	5.426	1.2
25	Mn	7.435	1.5	72	Hf	7.0	1.3
26	Fe	7.870	1.8	73	Ta	7.89	1.5
27	Co	7.86	1.8	74	W	7.98	1.7
28	Ni	7.635	1.8	75	Re	7.88	1.9
29	Cu	7.726	1.9	76	Os	8.7	2.2
30	Zn	9.394	1.6	77	Ir	9.1	2.2
31	Ga	5.999	1.6	78	Pt	9.0	2.2
32	Ge	7.899	1.8	79	Au	9.225	2.4
33	As	9.81	2.0	80	Hg	10.437	1.9
34	Se	9.752	2.4	81	Tl	6.108	1.8
35	Br	11.814	2.8	82	Pb	7.416	1.8
36	Kr	13.999		83	Bi	7.289	1.9
37	Rb	4.177	0.8	84	Po	8.42	2.0
38	Sr	5.695	1.0	85	At		2.2
39	Y	6.38	1.3	86	Rn	10.748	
40	Zr	6.84	1.4	87	Fr		0.7
41	Nb	6.88	1.6	88	Ra	5.279	0.9
42	Mo	7.099	1.8	89	Ac	6.9	1.1
43	Tc	7.28	1.9	90	Th		1.3
44	Ru	7.37	2.2	91	Pa		1.5
45	Rh	7.46	2.2	92	U		1.7
46	Pd	8.34	2.2	93	Np		1.3
47	Ag	7.576	1.9	94	Pu	5.8	1.3

[a] Weast et al. (1986).
[b] Pauling (1960).

increasing atomic number in each group. This indicates that the *binding energy* of the first valence electron of elements in a given group decreases as the volume of the atoms in that group increases. In other words, the larger an atom, the more loosely it holds its valence electrons. Consequently, the metallic character of the group A elements *increases* with increasing atomic number.

In the B groups, where *d*- and *f*-orbitals are being filled, the electronegativities change much less dramatically with increasing atomic number than they do in the A groups. As a result, the chemical properties of these elements are more uniform than those of the elements in the A groups. Moreover, the electronegativities of elements in B groups do not decrease with increasing atomic number as they generally do in the A groups. For example, the electronegativities of the congeners of group VIIIB (Fe, Co, and Ni) actually rise with increasing atomic number.

The concept of electronegativity has merit because the *differences* in the electronegativities of elements are related to the ionic character of the bond formed by their atoms. Accordingly, two elements with similar electronegativities form covalent bonds with little ionic character, whereas elements having different electronegativities form bonds whose ionic character is proportional to the magnitude of the electronegativity difference. The relationship between the electronegativity difference and the ionic character of chemical bonds is expressed quantitatively by the data in Table 7.2. In Figure 7.1C we see how the bond character varies with increasing atomic number when elements form bonds with oxygen. Note that the graph shows the percent *covalent* character in order to emphasize the relationship between bond character and electronegativity (Figure 7.1B) or first ionization potential (Figure 7.1A). We see that the oxygen bonds of the elements become increasingly covalent in each period. The halogens form the most covalent oxygen bonds, whereas the alkali metals form oxygen bonds that are largely ionic. The transition metals in the fourth period show less variation in bond character than elements with valence electrons in *s*- and *p*-orbitals.

The data in Tables 7.1 and 7.2 permit us to predict the ionic character of the bond between any two elements in the periodic table. In general, we find that bonds between neighboring elements in the periodic table are highly covalent. For example, the nonmetals on the right side of the periodic table form many familiar compounds and anion complexes that are covalently bonded. The H—O bond in the H_2O, the C—O bond in CO_3^{2-}, the N—O bond in NO_3^-, the S—O bond in SO_4^{2-}, and the Si—O bond in SiO_4^{4-} are all strongly covalent.

In addition, we can demonstrate that bonds between elements from opposite ends of the periodic table are highly ionic. The best examples are the alkali halides (e.g., NaCl, KCl, and CsF). In fact, the Cs—F bond is the most ionic bond we know because Cs has the strongest tendency to be an electron donor (lowest electronegativity) and F is the strongest electron acceptor (highest electronegativity). Bonds with well-developed ionic character also occur between the cations of metals and complex anions exemplified by $CaCO_3$, $NaNO_3$, $MgSO_4$, and Fe_2SiO_4, all of which occur naturally as minerals.

Although the electronegativities of complex anions are not defined, we nevertheless know that the bonds between the cation of the metals and the complex anions are highly ionic because these compounds dissociate into ions when they dissolve in water.

7.3 Bonding in Molecules

We now come to the question: What difference does it make in geochemistry whether chemical bonds are covalent or ionic? The answer is that the physical and chemical properties of all compounds depend on the character of the bonds that hold them together.

An important property of most minerals is that they dissociate into ions when they dissolve in water. For example, when calcite ($CaCO_3$) dissolves in water, it forms Ca^{2+} ions and CO_3^{2-} ions:

$$CaCO_3 \rightarrow Ca^{2+} + CO_3^{2-} \qquad (7.1)$$

Table 7.2 Percent Ionic Character of a Single Chemical Bond with Oxygen

Difference in electronegativity	Ionic character, %	Difference in electronegativity	Ionic character, %
0.1	0.5	1.7	51
0.2	1	1.8	55
0.3	2	1.9	59
0.4	4	2.0	63
0.5	6	2.1	67
0.6	9	2.2	70
0.7	12	2.3	74
0.8	15	2.4	76
0.9	19	2.5	79
1.0	22	2.6	82
1.1	26	2.7	84
1.2	30	2.8	86
1.3	34	2.9	88
1.4	39	3.0	89
1.5	43	3.1	91
1.6	47	3.2	92

SOURCE: Sargent–Welch (1980).

The carbonate anion does *not* dissociate into carbon and oxygen ions because the C — O bonds are highly covalent, whereas the Ca—CO$_3$ bond is largely ionic. Why should that make a difference? The answer has to do with the interactions between the molecules of the solvent (water in this case) and the calcite "molecules."

The H — O bond in the water molecule is about 60% covalent as can be seen from Figure 7.1C. In addition, the electronegativity of oxygen (3.5) is larger than that of hydrogen (2.1). Consequently, we conclude that O attracts the electrons of the H atoms to itself. The unequal sharing of the electron between H and O therefore causes a net negative charge to develop on the O atoms of water molecules, whereas the H atoms have a net positive charge. In other words, the water molecule has electrical polarity and is therefore said to be "polar."

The manner in which the covalent H — O bonds form also determines the shape of the water molecule. A quick review of the electronic structure of O indicates that it has four electrons in the 2p orbitals, whereas H has one electron in the 1s orbital. The bond between them forms by sharing of the *unpaired* electrons. In accordance with rule 3 of the Aufbau principle (Chapter 5), one of the p-orbitals of oxygen contains two electrons with paired spins, whereas the other two p-orbitals of oxygen contain one electron each. We can represent the p-orbitals of oxygen as circles and the electrons they contain by arrows that point up or down depending on the spin.

The angle between these *p*-orbitals of oxygen is 90° because they are symmetrical about the *x*-, *y*-, and *z*-axes. Therefore, we expect that the angle between the H—O bonds in the water molecule should also be 90°. However, the observed angle is 104.45° (Fyfe, 1964).

The discrepancy arises because the *s*- and *p*-orbitals of O are merged in such a way that the angles between the "hybridized" orbitals are no longer 90°. Hybridization of orbitals is a common phenomenon in covalently bonded molecules and is the key to understanding the shapes or *stereochemistry* of covalently bonded molecules (Greenwood and Earnshaw, 1984; Pauling, 1960).

The polarity of the water molecule therefore arises not only because of the unequal sharing of the electrons between H and O but also because one of the hybridized *p*-orbitals of O containing the two paired electrons points away from the other two, which are involved in the bonding. In addition, the H_2O molecule is inherently polar because the O atom has more electrons than the H atoms.

Because of their electrical polarity, water molecules are attracted to electrical charges on the surface of an ionic crystal such as $CaCO_3$. The resulting electrostatic interactions weaken the ionic bonds between the Ca^{2+} and the CO_3^{2-} ions, but have no effect on the covalent C—O bonds of the carbonate ions.

The ions that form by dissociation attract water molecules to themselves and thereby become "hydrated." The number of water molecules bound to an ion depends primarily on its radius but also on its charge. For example, Be^{2+} (radius = 0.35 Å; Ahrens, 1952) holds four water molecules to form the hydrated ion $Be(OH_2)_4^{2+}$ whereas Al^{3+} (radius = 0.51 Å) holds six water molecules and forms $Al(OH_2)_6^{3+}$. Note that we represent the hydrated Be^{2+} ion as $Be(OH_2)_4^{2+}$ because the Be^{2+} is attracted to the negatively charged O atoms of the water molecule. Anions are also hydrated by attracting the positively charged hydrogen atoms of water. In some cases the water molecules are held so

strongly that they are incorporated into crystals as water of hydration, as, for example, in $CaSO_4\,2H_2O$ (gypsum).

Therefore, the answer to the original question is that water is able to dissolve ionic solids because it dissociates them into ions, which are then separated from each other by becoming hydrated. However, covalent bonds do not dissociate in water or in other polar solvents. Therefore, water is generally not an effective solvent for covalently bonded compounds.

7.4   Ionic Crystals

When ions come together to form a crystal they assemble in regular patterns dictated by the requirement of "closest packing." This requirement arises because the strength of the electrostatic bond (F) between cations and anions is inversely proportional to the square of the distance (r) between them:

$$F = \frac{e_1 e_2}{r^2} \qquad (7.2)$$

where e_1 and e_2 are the charges of the ions. Evidently, the magnitude of F increases as r decreases. Therefore, the ions in a crystal arrange themselves in such a way that the interionic distances are minimized, hence the tendency for closest packing. In this context, ions are assumed to be incompressible spheres like billiard balls.

A second requirement is that crystals must be electrically neutral. This requirement determines the ratio of cations to anions and is reflected in the chemical formulas of minerals. In addition, the requirement for electrical neutrality must be obeyed whenever substitutions occur in ionic crystals.

The arrangement of ions in a crystal can be understood in terms of the magnitudes of the radii of the cations and anions expressed by the *radius ratio,* which is defined as the ratio of the radius of the cation to that of the anion. In calculating radius ratios one commonly uses O^{2-} as the anion

Coordination

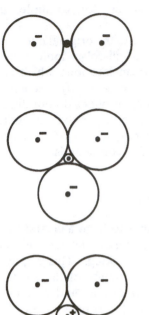

2

3

$$\frac{R_c}{R_a} = 0.155$$

4

$$\frac{R_c}{R_a} = 0.414$$

Figure 7.2 Examples of closest packing of anions around cations at increasing values of the radius ratio. When $R_c/R_a > 0.155$, three anions fit around the cation and when $R_c/R_a > 0.414$, fourfold coordination occurs. Coordination numbers 6, 8, and 12 are also important in mineral structures and are pictured in Figure 7.4.

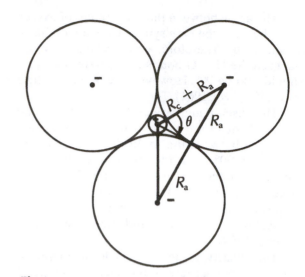

Figure 7.3 Derivation of the critical radius ratio for threefold coordination. R_c and R_a are the ionic radii of the cation and anion, respectively. The critical relationship is $\sin 60° = R_a/(R_a + R_c)$ from which it follows that $R_c/R_a = 0.155$.

because oxygen is the most abundant chemical element in the crust of the Earth.

The geometrical arrangements of ions in a crystal are described by the *coordination number,* which is the number of anions that surround a particular cation in an ionic crystal. It turns out that the coordination numbers of elements in ionic crystals can be predicted based on their radius ratios. We see from Figure 7.2 that only two large anions can fit around a very small cation. The smallest coordination number therefore is 2. If the cation is larger than in the case just considered and the radius ratio exceeds a critical value,

the coordination number rises to 3. At a still higher value of the radius ratio, four anions can be packed around the cation.

The critical value of the radius ratio at which threefold coordination is possible in a planar structure can be derived from geometrical considerations pictured in Figure 7.3. Let R_c and R_a be the radii of the cation and anion, respectively. By examining Figure 7.3 we find that:

$$\sin \theta = \frac{R_a}{R_c + R_a} \qquad (7.3)$$

since $\theta = 60°$ and $\sin 60° = 0.8660$:

$$\frac{R_a}{R_c + R_a} = 0.8660 \qquad (7.4)$$

By cross multiplying and collecting terms we obtain:

$$\frac{R_c}{R_a} = \frac{0.134}{0.866} = 0.155 \qquad (7.5)$$

Therefore, threefold coordination becomes possible in a planar structure when the radius ratio is equal to or greater than 0.155. A similar argument indicates that fourfold coordination occurs when the radius ratio is 0.414.

All of the packing arrangements shown in Figure 7.2 are two dimensional or planar. Since crystals are actually three-dimensional bodies, we must also consider three-dimensional patterns of fitting anions around cations. For example, four-fold coordination occurs when the anions occupy the corners of a tetrahedron with the cation in its center. Similarly, sixfold coordination is realized in an octahedron where six anions occupy the corners and the cation resides in the center. Higher coordination numbers of 8 and 12 are common and involve the cubic structures shown in Figure 7.4. Other coordination numbers, for example, 5, 7, 9, 10, and 11, occur only rarely and are not important in the crystal structures of minerals (Fyfe, 1964).

7.5   Ionic Radii

The radii of ions in crystals were first determined by V. M. Goldschmidt and Linus Pauling in the late 1920s by means of x-ray diffraction and are expressed in angstroms ($Å; 1Å = 10^{-10}$ m). Ionic radii are determined from the distances between ions in crystals and depend on the structure of the crystal in which they were measured. Most ionic radii were originally derived from crystals in which each cation is surrounded by six anions (Heydemann, 1969). A complete set of such radii was published by Ahrens (1952), but these have now been superseded by the work of Shannon and Prewitt (1969). Subsequently, Whittacker and Muntus (1970) published a new set of ionic radii (listed in Table 7.3) that takes into account the different coordination numbers that may occur in different kinds of crystals. Another set of radii was later published by Shannon (1976). The dependence of ionic radii on the coordination number occurs because the electron cloud of an ion can be deformed by the electrical charges that surround it in a crystal lattice. For example, the radius of Na^+ varies from 1.07 to 1.40 Å

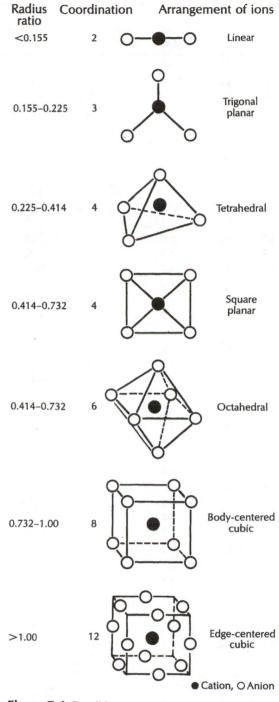

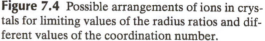

Figure 7.4 Possible arrangements of ions in crystals for limiting values of the radius ratios and different values of the coordination number.

Table 7.3 Radii of the Ions of the Elements in Angstrom Units (Å) for Different Charges and Coordination Numbers

Z	Symbol	Valence	II	III	IV[a]	V	VI[a,b]	VII	VIII	IX	X	XII
1	H	+1					extremely small					
3	Li	+1			0.68		0.82					
4	Be	+2		0.25	0.35							
5	B	+3		0.10	0.20							
6	C	−4					2.60 (P)					
		+4					0.16 (A)					
7	N	+3					0.16 (A)					
		+5					0.13 (A)					
8	O	−2	1.27	1.28	1.30		1.32		1.34			
9	F	−1	1.21	1.22	1.23		1.25					
11	Na	+1			1.07	1.08	1.10	1.21	1.24	1.40		
12	Mg	+2			0.66	0.75	0.80		0.97			
13	Al	+3			0.47	0.56	0.61					
14	Si	+4			0.34		0.48					
15	P	+3					0.44 (A)					
		+5			0.25		0.35 (A)					
16	S	−2			1.56		1.72		1.78			
		+6			0.20		0.30 (A)					
17	Cl	−1			1.67		1.72		1.65			
		+5		0.20			0.34 (A)					
		+7			0.28		0.27 (A)					
19	K	+1					1.46	1.54	1.59	1.63	1.67	1.68
20	Ca	+2					1.08	1.15	1.20	1.26	1.36	1.43
21	Sc	+3					0.83		0.95			
22	Ti	+2					0.94					
		+3					0.75					
		+4				0.61	0.69					
23	V	+2					0.87					
		+3					0.72					
		+4					0.67					
		+5			0.44	0.54	0.62					
24	Cr	+2					0.81 (L)					
							0.90 (H)					
		+3					0.70					
		+4			0.52		0.63					
		+5			0.43							
		+6			0.38		0.52 (A)					
25	Mn	+2					0.75 (L)			1.01		
							0.92 (H)					
		+3				0.66	0.66 (L)					
							0.73 (H)					
		+4					0.62					
		+6			0.35							
		+7			0.34		0.46 (A)					

Table 7.3 (continued)

							Coordination numbers					
Z	Symbol	Valence	II	III	IV[a]	V	VI[a,b]	VII	VIII	IX	X	XII
26	Fe	+2			0.71 (H)		0.69 (L)					
							0.86 (H)					
		+3			0.57 (H)		0.63 (L)					
							0.73 (H)					
27	Co	+2			0.65 (H)		0.73 (L)					
							0.83 (H)					
		+3					0.61 (L)					
							0.69 (H)					
28	Ni	+2					0.77					
		+3					0.64 (L)					
							0.68 (H)					
29	Cu	+1	0.54									
		+2			0.70	0.73	0.81					
30	Zn	+2			0.68	0.76	0.83		0.98			
31	Ga	+3			0.55	0.63	0.70					
32	Ge	+4			0.48		0.62					
33	As	+5			0.42		0.58					
34	Se	−2					1.88		1.90			
		+6			0.37		0.42 (A)					
35	Br	−1			1.88		1.95 (P)		1.84			
		+7			0.34		1.39 (A)					
37	Rb	+1					1.57	1.64	1.68		1.74	1.81
38	Sr	+2					1.21	1.29	1.33		1.40	1.48
39	Y	+3					0.98		1.10	1.18		
40	Zr	+4					0.80	0.86	0.92			
41	Nb	+2					0.79					
		+3					0.78					
		+4					0.77					
		+5			0.40		0.72	0.74				
42	Mo	+3					0.75					
		+4					0.73					
		+5					0.71					
		+6			0.50	0.58	0.68	0.79				
43	Tc	+4					0.72					
44	Ru	+3					0.76					
		+4					0.70					
45	Rh	+3					0.75					
		+4					0.71					
46	Pd	+1	0.67									
		+2			0.72		0.94					
		+3					0.84					
		+4					0.70					
47	Ag	+1	0.75		1.10	1.20	1.23	1.32	1.38			
		+3			0.73							
48	Cd	+2			0.88	0.95	1.03	1.08	1.15			1.39
49	In	+3					0.88		1.00			
50	Sn	+2					0.93 (A)		1.30			
		+4					0.77					

Table 7.3 (continued)

Z	Symbol	Valence	II	III	IV[a]	V	VI[a,b]	VII	VIII	IX	X	XII
							Coordination numbers					
51	Sb	+3			0.85	0.88	0.76 (A)					
		+5					0.69					
52	Te	+4		0.60			0.70 (A)					
		+6					0.56 (A)					
53	I	−1					2.13		1.97			
		+5					1.03					
55	Cs	+1					1.78		1.82	1.86	1.89	1.96
56	Ba	+2					1.44	1.47	1.50	1.55	1.60	1.68
57	La	+3					1.13	1.18	1.26	1.28	1.36	1.40
58	Ce	+3					1.09		1.22	1.23		1.37
		+4					0.88		1.05			
59	Pr	+3					1.08		1.22			
		+4					0.86		1.07			
60	Nd	+3					1.06		1.20	1.17		
61	Pm	+3					1.04					
62	Sm	+3					1.04		1.17			
63	Eu	+2					1.25		1.33			
		+3					1.03	1.11	1.15			
64	Gd	+3					1.02	1.12	1.14			
65	Tb	+3					1.00	1.10	1.12			
		+4					0.84		0.96			
66	Dy	+3					0.99		1.11			
67	Ho	+3					0.98		1.10			
68	Er	+3					0.97		1.08			
69	Tm	+3					0.96		1.07			
70	Yb	+3					0.95		1.06			
71	Lu	+3					0.94		1.05			
72	Hf	+4					0.79		0.91			
73	Ta	+3					0.75					
		+4					0.74					
		+5					0.72		0.77			
74	W	+4					0.73					
		+6			0.50		0.68					
75	Re	+4					0.71					
		+5					0.60					
		+6					0.60					
		+7			0.48		0.65					
76	Os	+4					0.71					
77	Ir	+3					0.81					
		+4					0.71					
78	Pt	+2			0.68		0.80 (A)					
		+4					0.71					
79	Au	+1					1.37 (A)					
		+3			0.78		0.85 (A)					
80	Hg	+1		1.05								
		+2	0.77		1.04		1.10		1.22			
81	Tl	+1					1.58		1.68			1.84
		+3					0.97		1.08			

Table 7.3 (continued)

Z	Symbol	Valence	II	III	IV[a]	V	VI[a,b]	VII	VIII	IX	X	XII
82	Pb	+2[c]			1.02		1.26		1.37	1.41		1.57
		+4					0.86		1.02			
83	Bi	+3				1.07	1.10		1.19			
84	Po	+4							1.16			
		+6					0.67 (A)					
85	At	+7					0.62 (A) ·					
86	Fr	+1					1.80 (A)					
87	Ra	+2					1.43 (A)		1.56			1.72
89	Ac	+3					1.18 (A)					
90	Th	+4					1.08		1.12	1.17		
91	Pa	+3					1.13 (A)					
		+4					0.98 (A)		1.09			
		+5					0.89 (A)		0.99	1.03		
92	U	+3					1.12					
		+4					0.97 (A)	1.06	1.08	1.13		
		+5					0.84	1.04				
		+6	0.53			0.56	0.81	0.96				
93	Np	+2					1.18					
		+3					1.10					
		+4					0.95 (A)		1.06			
94	Pu	+3					1.09					
		+4					0.88		1.04			

[a]Low spin and high spin are indicated by L and H, respectively.
[b]A = Ahrens (1952); P = Pauling (1927).
[c]Pb^{2+}, coordination number II, radius = 1.47 Å.
SOURCE: Whittacker and Muntus (1970).

depending on the coordination number on the lattice site. Evidently, ions are not actually hard spheres, although they are treated this way in some aspects of crystal chemistry.

The ionic radii listed in Table 7.3 vary in regular patterns depending on the atomic number of the element and the charge of the ion as shown in Figure 7.5, provided all radii are based on the same coordination number. We take note first of the existence of *isoelectronic series* consisting of sequences of ions in which the charge increases with increasing atomic number and which therefore contain the same number of electrons. Such isoelectronic series exist in all

but the first period. For example, in the second period we find:

$$Li^+, Be^{2+}, B^{3+}, C^{4+}, N^{5+}, O^{6+}, \text{ and } F^{7+}$$

Each of these ions has only two electrons. Similarly, the ions:

$$C^{4-}, N^{3-}, O^{2-}, \text{ and } F^-$$

form an isoelectronic series in which each ion contains 10 electrons. Other isoelectronic series in periods 3, 4, and 5 are indicated in Figure 7.5.

We will find it helpful later to have a mental image of the ionic radii and to recall some gener-

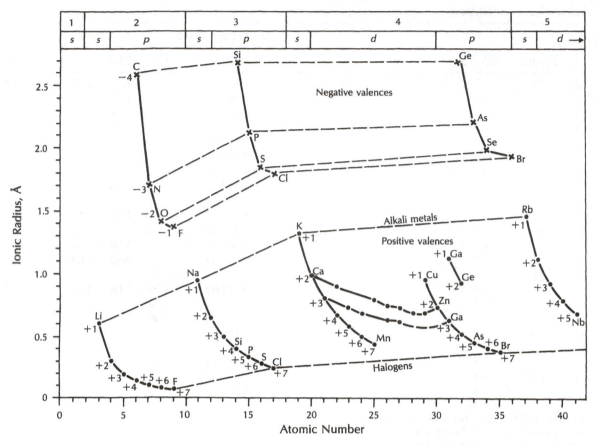

Figure 7.5 Variation of ionic radii with atomic number and valence. Note that ions having negative charges are larger than those having positive charges. Note also that the radii of ions forming isoelectronic series decrease with increasing atomic number (e.g., Li^+ to F^{7+}, Na^+ to Cl^{7+}, etc). The radii of ions in the same group that have the same charge increase with atomic number (e.g., Li^+ to Rb^+, F^- to Br^-, etc) [based on the ionic radii reported by Ahrens (1952) for sixfold coordination].

alizations about the way they vary within the periodic table. We therefore derive a number of conclusions about the ionic radii of the elements by examining Figure 7.5.

1. The ionic radii of isoelectronic series *decrease* with increasing atomic number for both positively and negatively charged ions.

2. The radii of ions with the same charge in a group *increase* with increasing atomic number, that is, downward in the periodic table.

3. The radii of ions of the same element *decrease* with increasing positive charge and *increase* with increasing negative charge.

4. The radii of ions with charges of +2 and +3 among the transition metals of the fourth period *decrease* with increasing atomic number, thereby implying a contraction of the electron cloud as the 3*d* orbitals are filled.

5. Ions of different elements may have similar ionic radii.

Examples to illustrate the last observation can be found in Figure 7.5 or by consulting Table 7.3.

$Na^+, Ca^{2+}, Cd^{2+}, Y^{3+},$	~ 1.00 Å
$Tl^{3+}, REE(+3), Th^{4+}$	(sixfold coordination)
Hf^{4+}, Zr^{4+}	~ 0.80 Å
	(sixfold coordination)
$Fe^{3+}, Co^{3+}, Ni^{3+}$	~ 0.65 Å
	(sixfold coordination)

The similarity of the ionic radii of different elements is an important phenomenon because it may permit the ion of one element to substitute for an ion of another element in ionic crystals (Goldschmidt et al., 1926).

The radii of ions in aqueous solution may be significantly changed by hydration. In this connection it is helpful to define the *ionic potential* as the ratio of the charge of an ion in electronic charges to its radius measured in angstrom units. The ionic potential is a measure of the density of the electrical charge on the surface of an ion and reflects the strength of the electrostatic bond the ion can form with another ion of the opposite charge or with a polar molecule like water. Ions with high ionic potential attract more water molecules and form hydrated ions that are larger than those of ions having lower ionic potentials. For example, the radii of the hydrated ions of the alkali metals *decrease* from Li^+ to Cs^+ even though their ionic radii in crystals increase with atomic number (Table 7.3). This can be explained by considering the ionic potentials of the alkali metals: $Li^+ = 1.2$. $Na^+ = 0.91$. $K^+ = 0.68$. $Rb^+ = 0.64$. and $Cs^+ = 0.56$. Therefore, even though Li^+ (radius = 0.82 Å) is much smaller than Cs^+ (radius = 1.78 Å), it forms a *larger* hydrated ion than Cs^+ because of its higher ionic potential.

7.6 Summary

The chemical elements can be divided into electron donors and electron acceptors, depending on whether their valence orbitals are nearly empty or nearly full. The tendencies of elements to be electron donors, and therefore to act like metals, is reflected by their first ionization potentials and electronegativities. Elements with low ionization potentials and low electronegativities have strongly developed metallic character. In general, the metallic character of the elements decreases from left to right across the periodic table and increases downward in those groups in which the elements have valence electrons in *s*- and *p*-orbitals.

Chemical bonds can be classified into ionic or covalent bonds, depending on whether the valence electrons are effectively transferred or shared, respectively. In reality, all ionic bonds have varying degrees of covalent character as indicated by the magnitude of their electronegativity *difference*. Polar solvents such as water are able to break ionic bonds because of electrostatic interactions between ions in the solid phase and charges on the water molecules. The hydration of cations and anions further enhances the solubility of ionic solids in water.

The radius ratio of a cation with respect to the surrounding anions determines its coordination number and hence affects the three-dimensional patterns the ions form in ionic crystals. The ionic radii of the elements in the periodic table vary predictably, depending on the atomic numbers of the elements, the electrical charges of the ions, and on the coordination number.

The metallic character of the elements, the kinds of bonds they form with each other, and the radii of ions are all related to their position in the periodic table and thus to their electronic configurations.

Problems

1. Determine the percent ionic character of the following bonds, based on the electronegativity differences: (a) NaCl; (b) $FeCl_2$; (c) CuCl; (d) $AlCl_3$; (e) CCl_4.

2. Look up the ionization potentials of silicon in a recent edition of the *CRC Handbook of Chemistry and Physics* (Weast et al., 1986) and plot them in coordinates of electron volts (eV) versus the number of electrons removed.

3. Explain why the third and fifth ionization potentials of silicon are significantly higher than those that immediately precede them.

4. Explain why the first ionization potential of sulfur ($Z = 16$) is less than that of phosphorus ($Z = 15$).

5. Write an equation to represent the dissociation of $(NH_4)_2SO_4$ into ions in an aqueous solution.

6. Calculate the volume of an ion of Na^+ in cubic meters (assume sixfold coordination).

7. Calculate the radius ratio for a planar structure with fourfold coordination.

8. Determine the ionic potentials of the ions in the isoelectronic series K^+ to Mn^{7+} and derive a conclusion about the size of the hydrated ions in the isoelectronic series (use sixfold and fourfold coordination depending on the availability of data in Table 7.3).

References

AHRENS, L. H., 1952. The use of ionization potentials Part 1. Ionic radii of the elements. *Geochim. Cosmochim. Acta,* **2:**155.

COTTON, F. A., and G. WILKINSON, 1962. *Advanced Inorganic Chemistry.* Wiley, New York, 959 pp.

FYFE, W. S., 1964. *Geochemistry of Solids.* McGraw-Hill, New York, 199 pp.

GOLDSCHMIDT, V. M., T. BARTH, G. LUNDE, and W. ZACHARIASEN, 1926. Geochemische Verteilungsgesetze der Elemente. VII. *Skrifter Norske Videnskaps.-Akad. Oslo, I. Mat.-Naturv. Klasse,* No. 2.

GREENWOOD, N. N., and A. EARNSHAW, 1984. *Chemistry of the Elements.* Pergamon Press, Oxford, England, 1542 pp.

HEYDEMANN, A., 1969. Tables. In K. H. Wedepohl (Ed.), *Handbook of Geochemistry,* vol. 1, 376–412. Springer-Verlag, Berlin, 442 pp.

PAULING, L., 1927. The sizes of ions and the structure of ionic crystals. *J. Amer. Chem. Soc.,* **49:**763

PAULING, L., 1960. *Nature of the Chemical Bond,* 3rd ed. Cornell University Press, Ithaca, NY.

SARGENT-WELCH SCIENTIFIC CO., 1980. *Periodic Table of the Elements.* 7300 Linder Ave., Skokie, IL 60076.

SHANNON, R. D., and C. T. PREWITT, 1969. Effective crystal radii in oxides and fluorides. *Acta Crystallogr., Sect. B,* **25:**925–946.

SHANNON, R. D., 1976. Revised effective ionic radii and systematic studies of interatomic distances in halides and chalcogenides. *Acta Crystallogr. Sect. A,* **32:**751–767.

WEAST, R. C., M. J. ASTLE, and W. H. BEYER, 1986. *CRC Handbook of Chemistry and Physics.* CRC Press, Boca Raton, FL.

WHITTACKER, E. J. W., and R. MUNTUS, 1970. Ionic radii for use in geochemistry. *Geochim. Cosmochim. Acta,* **34:**945–956.

8

Ionic Substitutions in Crystals

One of the principal objectives of geochemistry has been to explain the distribution of chemical elements in the Earth (Chapter 1). Before this objective could be achieved, it was first necessary to determine how the elements are actually distributed by analyzing large numbers of rocks and minerals from all over the Earth. F. W. Clarke took a giant step in that direction by his own efforts as an analytical chemist and by presenting the data of geochemistry in his book, first published in 1908. In subsequent years, geochemists continued to analyze geological materials in order to determine the chemical composition of the Earth and to understand the reasons for the observed variation in chemical composition of different kinds of rocks.

This was by no means an idle exercise because the information contributed directly to the understanding of the origin of metallic and nonmetallic ore deposits. The search for ore deposits of all kinds continues to be one of the principal tasks of geologists in society. In a very real sense, geologists are still expected to know where to find "the good rocks" that are the foundation of our modern industrial society.

8.1 Goldschmidt's Rules of Substitution

An important objective of Goldschmidt's work in geochemistry was to discover the laws of distribution of the chemical elements (Goldschmidt, 1937). He and Pauling (1927) saw perhaps more clearly than their contemporaries that the internal structure of crystals could be understood in terms of the size and charge of the ions and that an ion of one element can replace the ion of another if it is similar in size and charge. These insights are summarized by a set of generalizations known as *Goldschmidt's rules of substitution.*

1. The ions of one element can extensively replace those of another in ionic crystals if their radii differ by less than about 15%.

2. Ions whose charges differ by one unit substitute readily for one another provided electrical neutrality of the crystal is maintained. If the charges of the ions differ by more than one unit, substitution is generally slight.

3. When two different ions can occupy a particular position in a crystal lattice, the ion with the higher ionic potential forms a stronger bond with the anions surrounding the site.

4. Substitution may be limited, even when the size and charge criteria are satisfied, when the competing ions have different electronegatives and form bonds of different ionic character.

The fourth rule was actually formulated by Ringwood (1955) in order to explain discrepancies that arose from the three rules proposed by Goldschmidt. For example, Na^+ and Cu^+ have the same charge and their radii are virtually identical (Ahrens, 1952). Hence, according to Goldschmidt's rules, Cu^+ should replace Na^+ in sodium minerals such as albite ($NaAlSi_3O_8$) or halite (NaCl). We know, of course, that this substitution does not occur. The reason is that copper forms more covalent bonds than sodium, as indicated by their electronegatives.

The occurrence of elements in minerals and rocks either as major or minor constituents depends on their abundances and chemical properties. In general, the most abundant elements form the mineral compounds within which the minor elements may be accommodated by ionic substitution, in interstitial lattice positions, in fluid inclusions, or as exsolved mineral phases. Some minor elements also occur in accessory minerals such as zircon or apatite.

The tendencies of elements to form minerals are most clearly displayed during the crystallization of a cooling magma. Therefore, Goldschmidt's rules primarily apply during this process, which was thought to "sort" the ions according to their size and charge. Those ions that do not fit into the major rock-forming minerals are said to be *incompatible* and therefore accumulate in the residual magma. Consequently, the incompatible elements are concentrated in late-stage differentiates of magmas, including aplite dikes, pegmatites, and hydrothermal veins. The elements in this category include K^+, Rb^+, Cs^+, Sr^{2+}, Ba^{2+}, the rare earth elements (REE), Zr^{4+}, Hf^{4+}, Nb^{5+}, Ta^{5+}, Th^{4+}, and U^{4+}. All of these elements ultimately associate themselves with rocks composed of silicate minerals and, for that reason, are also referred to as the *large ion lithophile* (LIL) group.

Goldschmidt's rules were critically reviewed by many geochemists, including Shaw (1953) and Burns and Fyfe (1967). The rules are at best a first approximation to which there are many exceptions that arise because the replacement of one ion by another is generally site-specific, especially in complex compounds. The problem is magnified because the ionic radii, used as a criterion in Goldschmidt's rules, depend on the site where the replacement takes place. In spite of these limitations, Goldschmidt's rules still serve a useful purpose by identifying some of the major factors that must be considered in the possible replacement of ions in crystals forming in a cooling magma or from an aqueous solution.

8.2   Camouflage, Capture, and Admission

According to Goldschmidt's first rule, ions that have similar radii and equal charges replace each other extensively in ionic crystals. The extent of substitution that actually takes place depends on the concentration of the ions in the medium in which the crystals are forming, on the temperature, and on the compatibility of their bonds and coordination numbers. Crystals forming at high temperature are more tolerant of foreign ions than crystals forming at a low temperature. Therefore, the concentration of trace elements in crystals can be used to estimate the temperature of formation of certain minerals. For example, the concentration of Fe^{2+} in sphalerite (Fe, ZnS) increases with the temperature provided that enough iron was available to saturate the sphalerite (Kullerud, 1959).

The second rule applies to substitution of ions of similar size but having different charges. When the charge *difference* is greater than one, substitution is limited because of the difficulty in maintaining electrical neutrality. Charge deficiencies that result from substitution of ions of unequal charge must be compensated by a second substitution involving an ion having a different charge. This process of *coupled substitution* contributes to the diversification of chemical compositions of many minerals. An alternative to coupled substitution displayed by the clay minerals is *adsorption* of ions on the charged surfaces of small crystals.

The third rule describes the effect of the ionic potential on the relative strengths of ionic bonds of ions competing for the same site. Ions having a higher ionic potential (charge/radius ratio) form a stronger bond than their competitors and are therefore preferentially incorporated into the crystal. Similarly, competing ions with lower ionic potentials are discriminated against and are initially excluded from crystals forming from a magma. Consequently, ions with high ionic potentials are concentrated in early-formed crystals in a cooling magma, whereas those with low ionic potentials are concentrated in the residual magma and enter late-forming crystals.

According to the fourth rule, a minor ion can replace a major ion only if their electronegativities are similar. When two minor ions that are similar in size and charge compete for the same site, the one whose electronegativity is more similar to that of the major ion is preferred because the bonds are more compatible.

We see that Goldschmidt's rules of substitution provide criteria for predicting the extent to which minor elements can replace major elements in the minerals they form. As a result, we can predict how minor elements distribute themselves when minerals crystallize from a cooling magma or from a supersaturated aqueous solution. Some elements are selectively concentrated into the solid phase, whereas others remain in the liquid phase. The different ways in which the ions of minor elements are *partitioned* between the solid and liquid phases are described by the terms that form the heading of this section.

Camouflage occurs when the minor element has the same charge and a similar ionic radius as the major element it is replacing.

In this case, the minor element does not form its own minerals but is hidden in the crystals of another element. Camouflage is displayed by Zr^{4+} (0.80 Å) and Hf^{4+} (0.79 Å) because hafnium rarely forms its own minerals and is always present in the mineral zircon ($ZrSiO_4$). We could say that zircon crystals do not distinguish between Zr^{4+} and Hf^{4+} ions and accept both with equal ease.

Capture takes place when a minor element enters a crystal preferentially because it has a higher ionic potential than the ions of the major element.

Examples of capture occur in the formation of feldspar crystals, which may capture Ba^{2+} (1.44 Å) or Sr^{2+} (1.21 Å) in place of K^+ (1.46 Å). As a result, the concentrations of these elements in the residual magma *decrease* during the crystallization of K-feldspar. However, the replacement of the univalent K^+ ion by a divalent ion requires a coupled substitution of Al^{3+} for Si^{4+}

in order to preserve the electrical neutrality of the crystal lattice.

Admission involves the entry of a foreign ion that has a lower ionic potential than the major ion because it has either a lower charge or a larger radius, or both.

The occurrence of Rb^+ in K-feldspar and other potassium minerals is an example of admission because Rb^+ (1.57 Å) has a smaller ionic potential than K^+ (1.46 Å). Other examples are the replacement of Ca^{2+} (1.08 Å) by Sr^{2+} (1.21 Å) in calcite, and the substitution of Cl^- (1.72 Å) by Br^- (1.88 Å) in chlorides. The extent to which ions are admitted into a particular lattice site *decreases* as the difference in the radii of competing ions *increases*. For example, I^- (2.13 Å) replaces Cl^- (1.72 Å) much less than Br^- (1.88 Å) and Ba^{2+} (1.44 Å) is less abundant in calcite than Sr^{2+} (1.21 Å). Evidently, admission of foreign ions into the crystal of a major element is ultimately controlled by the size criterion expressed in Goldschmidt's first rule.

The third rule makes a statement about the relationship between the ionic potentials of cations and the strength of the bonds they form with anions. This relationship manifests itself in *some cases* in the melting temperatures of the compounds they form. For example, the Mg silicate forsterite (Mg_2SiO_4) has a higher melting temperature (1910 °C) than the Fe silicate fayalite (Fe_2SiO_4), which melts at 1503 °C. The two minerals form a solid solution known as the mineral olivine ($(Mg, Fe)_2 SiO_4$), which crystallizes from cooling magmas of basaltic composition. Early formed olivine is enriched in the forsterite (Mg) end member, whereas olivine forming at lower temperatures is enriched in the fayalite (Fe) end member. A possible explanation is that early formed olivine captures Mg^{2+} (0.67 Å) in favor of Fe^{2+} (0.74 Å) because it has a higher ionic potential (Ahrens, 1952). Similarly, the feldspar anorthite ($CaAl_2Si_2O_8$) has a higher melting temperature than albite ($NaAlSi_3O_8$) presumably because Ca^{2+} (1.08 Å) has a higher ionic potential than Na^+ (1.10 Å) (Ahrens, 1952). As a result, early-formed plagioclase is

Table 8.1 Classification of Silicate Minerals Based on Their Structures as Illustrated in Figure 8.1

Class	Shared corners	Silicate anion	Si/O	Examples
Neso	0	SiO_4^{4-}	1 : 4	fayalite Fe_2SiO_4
Soro	1	$Si_2O_7^{6-}$	2 : 7	akermanite $Ca_2MgSi_2O_7$
Cyclo	2	SiO_3^{2-}	1 : 3	benitoite $BaTiSi_3O_9$
Ino single	2	SiO_3^{2-}	1 : 3	enstatite $MgSiO_3$
Ino double	$2\frac{1}{2}$	$Si_4O_{11}^{6-}$	4 : 11	anthophyllite $Mg_7(Si_4O_{11})_2(OH)_2$
Phyllo	3	$Si_2O_5^{2-}$	2 : 5	kaolinite $Al_2Si_2O_5(OH)_4$
Tekto	4	SiO_2	1 : 2	quartz SiO_2

initially enriched in calcium relative to sodium. The ionic potential by itself, however, is *not* a reliable predictor of the melting temperatures of solid compounds and the examples presented above are more the exception than the rule (Burns and Fyfe, 1967).

8.3 Coupled Substitution: Key to the Feldspars

When the ion of a major element is replaced by a foreign ion having a different charge, the electrical neutrality of the crystal must be preserved by a complementary substitution elsewhere in the crystal lattice. This phenomenon helps to explain the chemical diversity of silicate minerals such as the feldspars, zeolites, and micas.

Silicate minerals contain a framework of *silica tetrahedra* linked together to form different kinds of structures. The silicate anion (SiO_4^{4-}) forms a tetrahedron in which the Si atom is located in the center and is bonded to the four O atoms that occupy the corners of the tetrahedron. The Si—O bonds are about 50% covalent based on the difference in their electronegativities. Each O atom at the corner of a tetrahedron has an unpaired electron in a hydridized *p*-orbital that can form ionic bonds with cations in orthosilicates, such as Mg_2SiO_4, Fe_2SiO_4, or Li_4SiO_4. The silica tetrahedra can also form bonds with each other by sharing O atoms at the corners to form

rings, chains, double chains, sheets, and three-dimensional networks listed in Table 8.1 and illustrated in Figure 8.1.

The feldspars are "tektosilicates" composed of a three-dimensional network of silica tetrahedra linked at their corners. In this arrangement, the two unpaired electrons of O atoms form bonds with different Si atoms and thereby link the tetrahedra together. When all of the O atoms are shared in this manner, each silicon atom "owns" only half of each of the four O atoms located at the corners of its tetrahedron. Therefore, in a three-dimensional network the ratio of Si to O is reduced from 1 : 4 to 1 : 2 and its formula is SiO_2.

This framework of linked silica tetrahedra can admit a limited number of Al^{3+} ions in place of Si^{4+}. The substitution is restricted because in fourfold coordination Al^{3+} (0.47 Å) is significantly *larger* than Si^{4+} (0.34 Å) and actually violates Goldschmidt's first rule. On the other hand, the radius ratio of Al^{3+} to O^{2-} is 0.36, which allows it to have a coordination number of 4 like Si^{4+}, whose radius ratio with O^{2-} is 0.26 (see Figure 7.4). As a result, one in four Si atoms is replaced to form the aluminosilicate anion $AlSi_3O_8^-$. This anion does not actually exist in discrete form because the lattice extends continuously in all directions. However, the formula of the aluminosilicate anion allows us to recognize that the replacement of one out of every four Si atoms by Al^{3+} causes a charge imbalance of -1

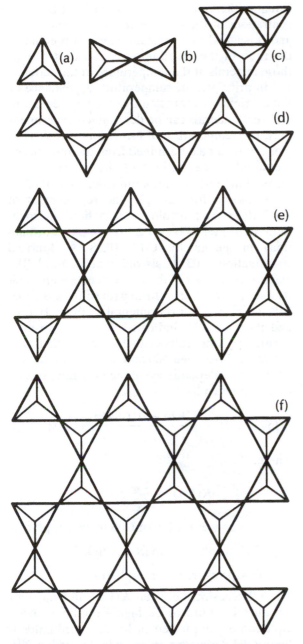

Figure 8.1 Two-dimensional networks of silica tetrahedra connected in different ways by sharing oxygen atoms at their corners. (a) Nesosilicates, (b) sorosilicates, (c) cyclosilicates, (d) inosilicates (single chain), (e) inosilicates (double chain), and (f) phyllosilicates. The tektosilicate structure, consisting of a three-dimensional network of tetrahedra in which all four oxygen atoms are shared, is not shown. More details are given in Table 8.1.

in the lattice. This excess negative charge is neutralized by the introduction of Na^+ or K^+. The resulting compounds are known to us as the alkali feldspars:

$$KAlSi_3O_8 \qquad NaAlSi_3O_8$$
orthoclase, microcline   albite

When two out of four Si atoms are replaced by Al^{3+}, the aluminosilicate anion has a charge of -2, which is neutralized by the addition of Ca^{2+} or Ba^{2+} to form:

$$CaAl_2Si_2O_8 \qquad BaAl_2Si_2O_8$$
anorthite   celsian

Albite and anorthite form a solid solution known as plagioclase, which is subdivided into six mineral species, depending on the molar concentrations of albite (Ab) or anorthite (An).

Mineral	Ab, %	An, %
Albite	100–90	0–10
Oligoclase	90–70	10–30
Andesine	70–50	30–50
Labradorite	50–30	50–70
Bytownite	30–10	70–90
Anorthite	10–0	90–100

The series is divided into 20% increments of Ab or An, except at the ends, where the increments are 10%. The names of the plagioclase series can be remembered by noting that the first letters spell the name of the fictitious geochemist A. O. Alba.

8.4   Distribution Coefficients and Geothermometers

Goldschmidt's rules provide a rational basis for predicting how minor elements may enter crystals forming from a cooling melt or from an aqueous solution. These effects are described qualitatively as "camouflage," "capture," and "admission." However, words alone are not sufficient to describe the partitioning of minor elements between crystals and the liquid from which they formed.

We therefore define the *distribution coefficient* (*D*):

$$D = \frac{C^x}{C^l} \qquad (8.1)$$

where C^x is the concentration of a minor element in the crystal (*x*) of a mineral and C^l is the concentration of that element in the liquid (l) from which the crystal formed under equilibrium conditions. The distribution coefficient of a particular element in a specific mineral may be > 1, < 1, or $= 1$. The magnitude of *D* is related to the verbal descriptors as follows:

$$D > 1 \quad \text{capture}$$
$$D < 1 \quad \text{admission}$$
$$D = 1 \quad \text{camouflage}$$

The numerical values of *D* must be determined experimentally in the laboratory or by analysis of natural systems in which both crystals and liquid can be sampled.

The observed distribution of trace elements in crystals can be used to estimate the temperature of formation of coexisting minerals. Recall that the extent of substitution is temperature dependent because crystals become more tolerant of foreign ions as the temperature increases. Therefore, distribution coefficients are, in general, temperature dependent and are also affected by the compositions of the liquid and the crystals, as well as by pressure. If two minerals A and B coprecipitate or crystallize from the same solution or magma, a minor element (*y*) can enter both minerals A and B at a particular temperature. The distribution coefficients of element *y* are:

$$D_A = \left(\frac{C_y^x}{C_y^l} \right)_A \quad \text{and} \quad D_B = \left(\frac{C_y^x}{C_y^l} \right)_B \qquad (8.2)$$

The ratio of the distribution coefficients for element *y* is:

$$\frac{D_A}{D_B} = \frac{(C_y^x)_A}{(C_y^x)_B} = K \text{ (constant)} \qquad (8.3)$$

Equation 8.3 indicates that the ratio of the concentrations of the trace element *y* in minerals A and B is equal to a constant *K*, which is equal to the ratio of the distribution coefficients of *y* in those minerals at the temperature of formation.

In principle, the temperature dependence of the distribution coefficients of a trace element in different minerals can be determined experimentally. The temperature of formation of rock samples can then be determined from the ratio of the measured concentrations of the selected trace element in these two minerals (McIntyre, 1963).

Consider, for example, the replacement of Zn^{2+} (0.68 Å, fourfold coordination) in sphalerite (ZnS) and Pb^{2+} (1.26 Å, sixfold coordination) in galena by Cd^{2+} (0.88 Å, fourfold coordination, 1.03 Å, sixfold coordination). The electronegativities of the three elements are similar and the bonds they form with sulfur are about 85% covalent. An experimental study by Bethke and Barton (1971) indicated that Cd is strongly partitioned into sphalerite (sp) coexisting with galena (gn) between 600 and 800 °C and that the temperature dependence of the distribution coefficients is given by:

$$\log K(Cd) = \frac{(2080 - 0.0264P)}{T} - 1.08 \qquad (8.4)$$

where:

$$K(Cd) = \frac{C_{sp}}{C_{gn}} \text{ for Cd}$$

C = concentration of CdS in weight percent

T = absolute temperature in kelvins

P = pressure in atmospheres

The relationship between $K(Cd)$ and the temperature is found to be highly nonlinear when equation 8.4 is plotted in linear coordinates in Figure 8.2. However, in coordinates of $\log K$ (*y*-coordinate) and $1/T$ (*x*-coordinate) equation 8.4 is a straight line in the slope-intercept form:

$$y = mx + b \qquad (8.5)$$

The slope $m = 2080 - 0.0264P$ and the intercept on the *y*-axis is $b = -1.08$. The pressure can be dis-

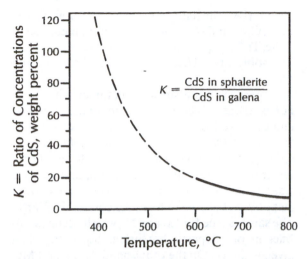

Figure 8.2 Geothermometer based on the distribution of CdS between sphalerite and galena (equation 8.4). The constant K is the ratio of the concentrations of CdS in sphalerite and galena in weight percent. The curve implies that the Cd content of sphalerite decreases relative to that of galena with increasing temperature of formation of these minerals. The solid part of the line is based on the experimental data of Bethke and Barton (1971). The dashed line is an extrapolation of those data.

regarded for low values of P. Other geothermometers studied by Bethke and Barton (1971) are based on the distributions of Mn (600–800°C) and Se (390–595°C) between sphalerite, galena, and chalcopyrite. The results indicate than Mn^{2+}, like Cd^{2+}, is preferentially incorporated into sphalerite rather than galena, whereas the order of preference for Se^{2-} is galena, chalcopyrite, and sphalerite, listed in order of decreasing preference.

The applicability of these trace element geothermometers is limited by the requirement that the minerals must have formed at the same temperature from the same solution under equilibrium conditions and that interfering effects caused by the presence of other ions or pressure variations were either negligible or cancel. These are serious limitations because minerals generally precipitate sequentially rather than simultaneously, which means that coexisting minerals do not necessarily form at the same temperature or from the same solution.

8.5   Geochemical Classification of the Elements

The geochemical classification of the elements is based on the way in which the elements actually distribute themselves between different kinds of liquids and a gas phase. During the smelting of oxide and sulfide ores three different liquids are encountered that are immiscible in each other and that segregate into layers depending on their density. These liquids are composed of molten Fe, molten sulfides (matte), and molten silicates (slag). Goldschmidt and his contemporaries considered it likely that the Earth was initially completely molten and that these liquids separated from each other under the influence of gravity to form an Fe core, a sulfide layer, and a silicate layer in the interior of the Earth. The gases formed the atmosphere, which subsequently produced the hydrosphere by condensation of water vapor.

The heat required for melting was provided by the impacts of the "planetesimals," by compression caused by the gravitational contraction of the Earth, by the migration of dense phases toward the center of the Earth, and by the decay of the radioactive isotopes of U, Th, and K, which were more abundant at the time of formation of the Earth 4.5×10^9 years ago than they are today and therefore generated more heat than they do at present. Although the Earth does not actually contain a sulfide shell, Goldschmidt's geochemical classification still conveys useful information about the tendencies of the elements to enter liquids of different composition or to be concentrated in the gas phase.

The information on which this classification is based came from the study of meteorites and from the smelting of the Kupferschiefer ore (copper slate) at Mansfeld in Germany. Meteorites are fragments of larger parent bodies that formed between the orbits of Mars and Jupiter (Mason, 1962) (see Table 3.2). Some of the parent bodies were large enough to retain sufficient heat for melting and differentiated into metallic Fe, silicate rocks, and sulfide minerals. The parent bodies were subsequently broken up by tides caused by the gravitational fields of Jupiter and Mars and by collisions among themselves. The remnants of the parent bodies still occupy the space between Mars and Jupiter. Fragments resulting from collisions among them continue to be deflected into Earth-crossing orbits and impact on the Earth as meteorites. Goldschmidt, as well as Ida and Walter Noddack in Berlin, analyzed the metallic, sulfide, and silicate phases of many meteorites and determined from the results how the elements had been partitioned into the three immiscible liquids during the geochemical differentiation of the parent bodies of meteorites (Noddack and Noddack, 1930). The information derived from the study of meteorites was consistent with the chemical compositions of the silicate slag, Fe–Cu sulfide matte, and metallic Fe, all of which form during the smelting of ore.

The resulting classification of the elements in Table 8.2 contains four groups that Goldschmidt named *siderophile* (iron liquid), *chalcophile* (sulfide liquid), *lithophile* (silicate liquid), and *atmophile* (gas phase). Note that several elements occur in more than one group and that such secondary affinity is indicated by parentheses. In general, we see from Figure 8.3 that the elements in group VIIIB are *siderophile* together with C, Mo, Au, Ge, Sn, and P. The congeners of groups IB and IIB are joined by the elements Ga, In, Tl, Pb, As, Sb, and Bi, as well as by S, Se, and Te, to form the *chalcophile* group. The *lithophile* elements include the alkali metals (group IA), alkaline earths (group IIA), the halogens (group VIIA), as well as B, Al, O, Si, and some of the transition metals such as Sc, Ti, V, Cr, Mn, and some of their congeners. The noble gases, H, and N make up the *atmophile* group.

There are a few surprises in this classification. For example, O is lithophile rather than atmophile, and C, as well as P, dissolves in metallic Fe in the absence of O. Thallium is a chalcophile element although it commonly substitutes for K^+ in silicate minerals. We note also that Au, Sn, and Mo are siderophile, and presumably have been carried away by metallic Fe to form the core of the Earth. The same is true for Co and Ni, which occur as sulfides in ore deposits but prefer liquid Fe when given a choice. On the other hand, Ti, Cr, and Mn, which are commonly associated with Fe in igneous rocks, are not siderophile but lithophile elements.

8.6   Summary

The ions of different elements can substitute for each other provided their radii, charges, and electronegativities are similar. In addition, they must be compatible with the lattice site in terms of their radius ratios and coordination numbers. Goldschmidt's rules provide a rational basis for understanding the distribution of trace elements but do *not* give enough emphasis to the character of the site of replacement in the crystal lattice.

Table 8.2   Geochemical Classification of the Elements[a]

Siderophile	*Chalcophile*	*Lithophile*	*Atmophile*
Fe, Co, Ni	Cu, Zn, Ga	Li, Be, B, (C), —, O, F	H, He
Ru, Rh, Pd	Ag, Cd, In	Na, Mg, Al, Si, P, —, Cl	(C), N, (O), Ne
Os, Ir, Pt	—, Hg, Tl	K, Ca, (Ga), (Ge), —, —, Br	Ar, Kr, Rn
plus	plus	Rb, Sr, —, I	
Mo, Re, Au	(Ge), As, S	Cs, Ba, (Tl), plus	
(W), and	(Sn), Sb, Se	Sc, Ti, V, Cr, Mn, (Fe)	
C, P, Ge	Pb, Bi, Te	Y, Zr, Nb	
Sn, (As), (Pb)	and (Cr), (Fe)	La, Hf, Ta, W	
	(Mo)	REE, Th, —, U, and	
		(H)	

[a]The elements in parentheses occur primarily in another class.

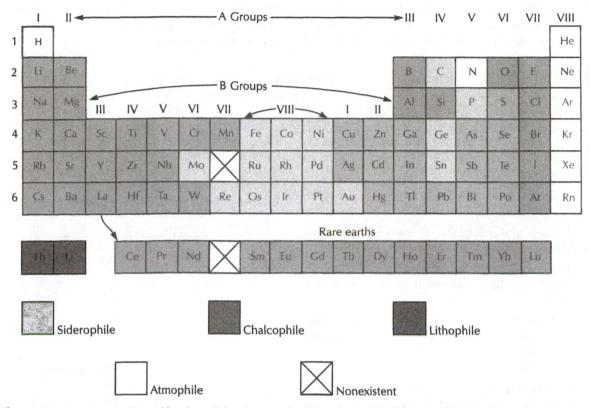

Figure 8.3 Geochemical classification of the elements in the periodic table. The classification is based on the way the elements distribute themselves between an iron liquid (siderophile), a sulfide liquid (chalcophile), a silicate liquid (lithophile), and a gas phase (atmophile).

The extent to which replacement of ions takes place is described by the terms *camouflage, capture,* and *admission.* Trace elements whose ions closely resemble those of major elements in terms of size, charge, and electronegativity are *camouflaged* in the crystals formed by the major elements. Trace elements that are *captured* during the crystallization of a magma are concentrated into early-formed crystals and are depleted in the residual liquid. Ions that are *admitted* into the crystals of a major element are initially enriched in the residual magma and subsequently enter late-forming crystals.

Substitution of Al^{3+} for Si^{4+} in silicate structures causes a charge imbalance in the lattice that is neutralized by the complementary introduction of certain cations that are compatible with the particular lattice sites. This phenomenon of *coupled substitution* accounts for the chemical composi-

tion of the feldspars and of other aluminosilicate mineral groups.

The *distribution coefficient* is a quantitative measure of ionic substitution. The numerical values of distribution coefficients must be determined experimentally as a function of temperature, chemical compositions of liquids and crystals, and of pressure. The distribution of trace elements between two coexisting minerals that formed under equilibrium conditions from the same liquid can be used to determine the temperature of formation. Such geothermometers are of great interest in geochemistry but their application is restricted by the conditions that must be imposed.

The geochemical properties of the elements are reflected by their distribution between a gas phase and among natural liquids composed of metallic Fe, transition metal sulfides, and silicates. The geochemical classification of the elements, which is based on this phenomenon, was derived from the chemical compositions of meteorites and from metallurgical studies in the smelting of sulfide and oxide ores.

Problems

1. Determine the coordination numbers of Ca^{2+} and Sr^{2+} relative to O^{2-}, and use the result to predict the substitution of Ca^{2+} by Sr^{2+} in *calcite* (coordination : 6) and aragonite (coordinate : 8).

2. Predict whether Hg^{2+} can replace Sr^{2+} in the mineral *strontianite* ($SrCO_3$).

3. Explain the formula of *leucite* (feldspathoid, $KAlSi_2O_6$) in terms of coupled substitution.

4. Examine the formula of the feldspathoid *sodalite* ($Na_8Al_6Si_6O_{24}Cl_2$) and explain it on the basis of coupled substitution.

5. Silver (Ag^+) is a common trace element in galena where it replaces Pb^{2+}. Identify the ion of a chalcophile element that is best suited to enter galena with Ag^+ in a coupled substitution.

6. Lithium (Li^+) and Mg^{2+} have similar radii and electronegativities, yet Li^+ does not replace Mg^{2+} in olivine. Explain the reason for this occurrence and suggest another host mineral for Li^+ in which it does replace Mg^{2+}.

7. Zircon crystals ($ZrSiO_4$) commonly admit uranium (U^{4+}) but strongly exclude Pb. Deduce the valence of Pb in magma from this observation.

8. The melting temperatures of the fluorides of the alkali metals and alkaline earths are listed below. Consider the observed variation in terms of both the ionic character of the bonds and the ionic potentials. How should the melting temperatures vary according to Goldschmidt's rules? Which compounds conform to this prediction and which do not?

Compound	Melting temperature °C
LiF	845
NaF	993
KF	858
RbF	795
CsF	682
BeF_2	800 (sublimes)
MgF_2	1261
CaF_2	1423
SrF_2	1573
BaF_2	1355

References

AHRENS, L. H., 1952. The use of ionization potentials Part 1. Ionic radii of the elements. *Geochim. Cosmochim. Acta,* **2**:155.

BETHKE, P. M., and P. B. BARTON, JR., 1971. Distribution of some minor elements between coexisting sulfide minerals. *Econ. Geol.,* **66**:140–163.

BURNS, R. G., and W. S. FYFE, 1967. Trace element distribution rules and their significance. *Chem. Geol.,* **2**:89–104.

GOLDSCHMIDT, V. M., 1937. The principles of distribution of chemical elements in minerals and rocks. *J. Chem. Soc. London,* **1937**:655–673.

KULLERUD, G., 1959. Sulfide systems as geological thermometers. In P. H. ABELSON (Ed.), *Researches in Geochemistry,* vol. 1, 301–335. Wiley, New York, 511 pp.

MASON, B., 1962. *Meteorites.* Wiley, New York, 274 pp.

MCINTYRE, W. L., 1963. Trace element partition coefficients—a review of theory and applications to geology. *Geochim. Cosmochim. Acta,* **27**:1209–1264.

NODDACK, I., and W. NODDACK, 1930. Die Häufigkeit der chemischen Elemente. *Naturwissenschaften,* **18**:757–764.

PAULING, L., 1927. The sizes of ions and the structure of ionic crystals, *J. Amer. Chem. Soc.,* **49**:765–790.

RINGWOOD, A. E., 1955. The principles governing trace element distribution during magmatic crystallization. *Geochim. Cosmochim. Acta,* **7**:189–202.

SHAW, D. M., 1953. The camouflage principle and trace element distribution in magmatic minerals. *J. Geol.,* **61**:142–151.

III

AQUEOUS GEOCHEMISTRY AND THE STABILITY OF MINERALS

Important chemical reactions occur at the surface of the Earth when minerals are exposed to water, carbon dioxide, and oxygen. The principles that govern such reactions can be used to explain the interactions of ions in aqueous solutions and to constrain the stabilities of solid compounds with which the ions are in equilibrium. The information derivable from such studies is conveniently presented in graphical form because Earth Scientists are accustomed to using maps and diagrams in their work.

9

Acids and Bases

Chemical reactions in nature commonly take place in the presence of water, which acts as the medium within which ions and molecules can interact. When two ions or molecules approach each other closely, they have an opportunity to establish a bond between them. When they do so, we say that a *chemical reaction* has taken place and a product has formed. Such chemical reactions are represented by means of algebraic equations that express the balance of masses and charges of the reactants and products. We use such equations to understand chemical reactions in nature, and therefore need to develop the skill of writing and interpreting them. For this reason, we begin the third part of this book with a very simple exercise.

9.1 Chemical Reactions and Equilibria

In order to understand how chemical reactions work we first consider what happens when sodium chloride is dissolved in pure water at room temperature and atmospheric pressure. The apparatus consists of a beaker containing continuously stirred water, into which is placed an ion-sensitive electrode capable of measuring the concentration of Na^+ ions. The electrode is connected to a strip-chart recorder operating at a known chart speed. We ignore all practical limitations and complications and concentrate our attention entirely on the events that occur when we place a quantity of solid NaCl into the water in the beaker. The results of this experiment are entirely predictable based on past experience: the salt will dissolve in the water.

However, we deliberately added a *larger* quantity of salt to the beaker than will dissolve in the water it contains.

As soon as the solid NaCl comes in contact with the water in the beaker, it begins to dissolve and the concentration of Na^+ in the water increases with time. After a few minutes the Na^+ concentration becomes constant regardless of how long we continue the experiment. At this time the beaker still contains some solid NaCl that did not dissolve in the water. In order to interpret the results of this experiment we turn to Figure 9.1, which is the recorder trace of our experiment.

We see that the Na^+ concentration increased rapidly at first but eventually leveled out and became constant. When no more NaCl would dissolve in the water at the temperature and pressure of the experiment, the solution became *saturated* with respect to NaCl. When we reached that condition, the reaction achieved a state of *equilibrium* in which the rate of solution of NaCl is equal to the rate at which it precipitates. Therefore, at equilibrium the amount of excess solid NaCl and the concentrations of Na^+ and Cl^- ions in the saturated solution are *invariant* with respect to time. Once the reaction reaches equilibrium, it stops being productive because the forward and backward reactions are occurring simultaneously and at the same rates. We can indicate this condition by means of equations. At *equilibrium* the forward reaction:

$$NaCl(s) \rightarrow Na^+ + Cl^- \qquad (9.1)$$

and the backward reaction:

$$Na^+ + Cl^- \rightarrow NaCl(s) \qquad (9.2)$$

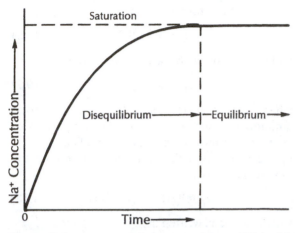

Figure 9.1 Increase of the concentration of Na⁺ with time after the addition of solid NaCl to pure water at 25°C. This experiment illustrates the point that chemical reactions progress toward a state of equilibrium in which the concentrations of reactants and products are invariant with time. In this illustration the reaction reaches equilibrium when the solution becomes saturated with respect to NaCl. While reactions are at disequilibrium they effectively convert reactants into products. When reactions reach equilibrium, the rate at which products are converted back to reactants equals the forward reaction rate and, as a result, the net change in the concentrations of reactants and products is zero.

occur at the same rates. We represent this state by means of double arrows:

$$NaCl(s) \rightleftharpoons Na^+ + Cl^- \qquad (9.3)$$

That is, double arrows mean that the reaction expressed by an equation is in a state of equilibrium.

Our experiment illustrates the important point that chemical reactions have a natural tendency to progress toward equilibrium. In order to reach this condition the reactants are consumed and products are formed such that the amounts of *both* change with time. We can think of this *disequilibrium condition* as the *active mode* of a reaction. After a reaction has reached equilibrium it becomes *inactive* in the sense that the amounts of reactants and products per unit weight or unit volume of water become constant. In fact, when a chemical reaction at equilibrium is disturbed by

changes in the physical and chemical conditions, it responds by reestablishing a new state of equilibrium. This property of chemical reactions was expressed by *Henry Le Châtelier* in 1888 in these words: Any change in one of the variables that determine the state of a system in equilibrium causes a shift in the position of equilibrium in a direction that tends to counteract the change in the variable under consideration (Moore, 1955, pp. 79–80). In other words, when a reaction at equilibrium is disturbed, it reestablishes equilibrium by counteracting the disturbance. This is the essence of *Le Châtelier's principle*. This principle will be very useful in the interpretation of equations representing chemical reactions.

Our experiment leads us also to understand the meaning of the concept of *solubility*, which is defined as the amount of a compound that dissolves to form a *saturated* solution. Therefore, in order to measure or calculate the solubility of a compound the reaction by which it dissolves must be at equilibrium. This is an important point to which we will return when we discuss solubility calculations.

The reaction we used in our experiment involves the dissociation of a solid into ions. Therefore, in this reaction no chemical bonds are actually formed by close encounters between ions or molecules. We chose that reaction mainly because it is familiar to all of us. Another reaction that is familiar to geologists is the decomposition of calcium carbonate by hydrochloric acid. This reaction can be represented by the following equation:

$$CaCO_3 + 2\,HCl \rightarrow Ca^{2+} + 2\,Cl^- + H_2O + CO_2 \qquad (9.4)$$

The carbon dioxide that is produced escapes in the form of bubbles when the reaction is carried out in contact with the atmosphere. Under these conditions the reaction cannot reach equilibrium because the CO_2 does not accumulate. Therefore, the reaction will continue in the direction indicated by the arrow in equation 9.4 until either the HCl or the $CaCO_3$ is used up. In this case, the reaction is prevented from reaching equilibrium and instead runs *to completion*.

Here we encounter the first of many differences between geochemistry and chemistry. In the chemistry laboratory the dissolution of $CaCO_3$ by HCl can be carried out in such a way that the CO_2 cannot escape. In this case, the reaction will achieve equilibrium and *excess* $CaCO_3$ and HCl will then coexist without appearing to react with each other. However, when this reaction occurs under natural conditions, it does not achieve equilibrium because the CO_2 escapes into the atmosphere. Similarly, chemical reactions on the surface of the Earth may not achieve equilibrium because some of the products continually escape either because they are gases or because ions and molecules are carried away by the movement of groundwater.

Another reason why some reactions in nature fail to achieve equilibrium is that their reaction rates are *slow*. This is especially true of reactions involving the transformation of a solid compound into another solid. Such reactions require a complex chain of events and their rates are determined by the slowest or the *rate-determining* step in the process. However, we will see that the failure of some reactions in nature to achieve equilibrium does not diminish the importance of this condition as an aid to understanding the geochemistry of the surface of the Earth.

9.2   The Law of Mass Action

The alchemists originally believed that chemical reactions occur becuse certain elements and compounds *love* each other. Robert Boyle expressed reservations about this idea in 1661 and Claude Louis de Berthollet pointed out about 150 years later that the direction of chemical reactions can be reversed by adding excess amounts of one of the products (Moore, 1955). The apparent reversibility of chemical reactions ultimately led to the conclusion that reactions achieve a state of equilibrium when the rate of forward reaction is equal to the rate of the backward reaction. For a simple reaction such as:

$$A + B \rightarrow C + D \tag{9.5}$$

The rate of the forward reaction is:

$$v_f = k_f(A)(B) \tag{9.6}$$

and the rate of the backward reaction is:

$$v_b = k_b(C)(D) \tag{9.7}$$

where (A), (B), (C), and (D) are the molar amounts or concentrations of the compounds or elements symbolized by A, B, C, and D and k_f and k_b are proportionality constants. At equilibrium $v_f = v_b$. Therefore:

$$k_f(A)(B) = k_b(C)(D) \tag{9.8}$$

which can be rewritten as:

$$\frac{k_f}{k_b} = \frac{(C)(D)}{(A)(B)} = K \tag{9.9}$$

where K is the *equilibrium constant*. In the general case of a reaction at equilibrium represented by:

$$a\,A + b\,B \rightleftharpoons c\,C + d\,D \tag{9.10}$$

the *Law of Mass Action*, first formulated by Guldberg and Waage in 1863, takes the form:

$$\frac{(C)^c(D)^d}{(A)^a(B)^b} = K \tag{9.11}$$

where *a, b, c,* and *d* are the molar coefficients taken from a balanced equation representing the reaction, and (A), (B), (C), and (D) are the concentrations of reactants and products of the reaction *at equilibrium*. The left side of equation 9.11 is the *reaction quotient,* which changes continuously as a reaction proceeds toward equilibrium. When a reaction has *achieved equilibrium,* the concentrations of reactants and products do not change and the reaction quotient becomes a *constant* known as the equilibrium constant K (Moore, 1955).

Although it is true that at equilibrium the rate of the forward reaction is equal to that of the backward reaction, the Law of Mass Action expressed in equation 9.11 cannot be derived from kinetics. The reason is that chemical reactions consist of sequences of discrete steps occurring at their own rates. Therefore, the overall rate of a reaction is not necessarily a linear function of the molar concentrations of reactants and products, but may vary as the square, the cube, or even the square root of the concentrations. Therefore, the Law of Mass Action was initially stated only as a generalization of experimental results and did not have the force of a scientific law. It turns out, however, that the Law of Mass Action can be derived from the first and second laws of thermodynamics for reactions among ideal gases at equilibrium. In order to apply it to reactions among ions and molecules in aqueous solutions we must replace their molar concentrations by their *activities*. The activity *a* of an ion in solution is related to its molar concentration *c* by the *activity coefficient* γ such that:

$$a = \gamma c \qquad (9.12)$$

The activity coefficients correct the molar concentration of ions for the interference by other ions in *real* solutions. In most cases, the values of activity coefficients are *less than one,* which indicates that the *effective concentration* (or activity) of ions is *less* than their actual concentration. Numerical values of activity coefficients can be calculated from the Debye–Hückel theory and its extensions, which we will discuss later.

The Law of Mass Action stated in equation 9.11 must now be restated in terms of activities: If a reaction represented by equation 9.10 is at equilibrium, then:

$$\frac{[C]^c[D]^d}{[A]^a[B]^b} = K \qquad (9.13)$$

where the brackets [ ] symbolize *activities,* wheras parentheses ( ) symbolize concentrations of ions and molecules. Unfortunately, there is no uniform convention to represent activities. We use brackets for activities and parentheses for concentrations following the example of Garrels and Christ (1965) and Krauskopf (1979), whereas Stumm and Morgan (1970) used the opposite convention and Drever (1988) symbolized activities by a lowercase *a* with appropriate subscripts. The difference between activity and concentration is unimportant in very *dilute* solutions and vanishes at infinite dilution when $\gamma = 1.0$. We will treat the activity coefficient as a dimensionless number, which enables us to express activities in the same units as concentration.

The units of concentration of ions and molecules in aqueous solutions are based on amounts expressed in moles but differ in terms of weight or volume of solvent or solution.

Molality (*m*) is the number of moles of solute per kg of water.
Formality (F) is the number of moles of solute per kg of solution.
Molarity (M) is the number of moles of solute per liter of solution.
Normality (N) is the number of equivalent weights of solute per liter of solution.

The molarity and normality are expressed in terms of volumes of solution, whereas molality and formality are based on weight of the solvent or solution, respectively. The latter are to be preferred because they are independent of the temperature, whereas molarity and normality vary with temperature because of the expansion of water with increasing temperature. The data in Table 9.1 indicate that the volume of 1 g of water increases from 1.000 13 mL at 0°C to 1.043 42 mL at 100°C or about 4.3%. Moreover, the specific volume of water has a minimum value of 1.000 00 mL at 3.98°C. Therefore, concentrations based on volumes must specify the temperature at which the volume was measured. On the other hand, it is more convenient to measure volumes of liquid than it is to weigh them. Therefore, both kinds of concentration units are in use.

When a chemical reaction between a solid and its ions and molecules in solution has achieved equilibrium, the Law of Mass Action applies with the following conventions:

1. The activities of ions and molecules must be expressed in terms of moles but may be referred to a unit weight of solvent (molality), a unit of weight of solution (formality), or a unit of volume of solution (molarity).

Table 9.1 Density and Specific Volume of Pure Water Free of Air

T, °C	Density, g/mL	Volume, mL/g
0	0.99987	1.00013
3.98	1.00000	1.00000
5	0.99999	1.00001
10	0.99973	1.00027
15	0.99913	1.00087
20	0.99823	1.00177
25	0.99707	1.00293
30	0.99567	1.00434
35	0.99406	1.00597
40	0.99224	1.00782
45	0.99025	1.00984
50	0.98807	1.01207
55	0.98573	1.01447
60	0.98324	1.01704
65	0.98059	1.01979
70	0.97781	1.02269
75	0.97489	1.02575
80	0.97183	1.02898
85	0.96865	1.03236
90	0.96534	1.03590
95	0.96192	1.03958
100	0.95838	1.04342

SOURCE: Weast et al. (1986), p. F-10.

2. The activities of pure solids and of water are equal to one.

3. The concentration of gases are expressed as partial pressures in atmospheres.

4. Reactions are assumed to take place at the standard temperature (25 °C) and pressure (1 atm) unless otherwise indicated.

The use of *moles* as the unit of mass is required because chemical reactions result from the interactions of individual ions or molecules whose number per mole is specified by Avogadro's number. The convention that solids and water have unit activities eliminates them from the equation of the Law of Mass Action. The activity of water may, however, be less than unity in very concentrated solutions in which the ratio of water to solute is significantly reduced. The convention to express the partial pressures of gases in units of atmospheres is arbitrary but affects the numerical values of equilibrium constants. The temparature and pressure at equilibrium must be specified because equilibrium constants vary with temperature. The effect of pressure on equilibrium constants is negligible under Earth-surface conditions.

9.3 Dissociation of Weak Acids and Bases

Following the tradition of the Swedish chemist Svante Arrhenius, we define an acid as a compound that releases hydrogen ions when it is dissolved in water. Similarly, a base is a compound that releases hydroxyl ions in aqueous solution. These definitions are no longer accepted by chemists who now prefer Brønsted's definition that an acid is any substance that can donate a proton to another substance and a base is a substance that can accept a proton. Strictly speaking, protons or hydrogen ions (H^+) cannot exist unattached in water and interact with water molecules to form hydronium ions (H_3O^+). We adopt the Arrhenius concept of acids and bases because in geochemistry we deal only with aqueous solutions of eletrolytes and we will represent protons as H^+ regardless of whether they actually occur in this form.

Some acids are said to be *strong* because they release all or most of the available hydrogen ions when they are dissolved in water, whereas *weak* acids release only a small fraction of their hydrogen ions. Bases are classified similarly by the extent to which they release hydroxyl ions (OH^-) in aqueous solutions. However, the behavior of bases is more complicated because some bases do not readily dissolve in water. For example, $Mg(OH)_2$ is so insoluble in water that it occurs naturally as the mineral *brucite*. Therefore, a saturated solution of brucite has a low concentration of OH^- even though $Mg(OH)_2$ is a strong base and dissociates extensively into Mg^{2+} and OH^-. Some common strong acids and bases are:

Strong acids	Strong bases
Hydrochloric acid (HCl)	hydroxides of alkali metals
Nitric acid (HNO_3)	hydroxides of alkaline earths (except Be)
Sulfuric acid (H_2SO_4)	lanthanum hydroxide ($La(OH)_3$)

The common weak acids include acetic acid (CH_3COOH), carbonic acid (H_2CO_3), phosphoric acid (H_3PO_4), and silicic acid (H_4SiO_4). Among the weak bases we mention ammonium hydroxide (NH_4OH), nickel hydroxide ($Ni(OH)_2$), copper hydroxide ($Cu(OH)_2$) and the hydroxides of the REEs except La. A clear distinction must be made between the *strength* of an acid or base and its *concentration*. For example, hydrochloric acid remains a strong acid even in dilute solution because it is completely dissociated. Similarly, acetic acid remains a weak acid even in a concentrated solution because it is only partially dissociated.

Many elements form hydroxides that can release either H^+ or OH^- depending on the concentration of H^+ in the water in which they dissolve. These hydroxides are said to be *amphoteric*. Among the more that 20 elements whose hydroxides are amphoteric we find Be, Al, Si, Ti, V, Fe, Co, Zn, Ag, Au, Sn, Pb, and U.

When a weak acid such as acetic acid is dissolved in water, it dissociates into ions:

$$CH_3COOH \rightarrow CH_3COO^- + H^+ \quad (9.14)$$

Dissociation reactions like this one reach equilibrium very quickly and are then subject to the Law of Mass Action:

$$\frac{[H^+][CH_3COO^-]}{[CH_3COOH]} = K \quad (9.15)$$

where $K = 1.76 \times 10^{-5}$ at 25 °C and 1 atm. The small value of K indicates that at equilibrium most of the acetic acid will remain in the undissociated molecular form. In order to solve equation 9.15 we now specify that 0.1 mol of the compound was dissolved in 1 L of pure water

(i.e., $\gamma = 1.0$) and that an unknown fraction x dissociated into ions. Therefore, at equilibrium:

$$[CH_3COOH] = 0.1 - x$$

$$[CH_3COO^-] = [H^+] = x$$

Substituting into equation 9.15 we obtain a quadratic equation:

$$\frac{x^2}{0.1 - x} = 1.76 \times 10^{-5} \quad (9.16)$$

This equation can be solved by putting it into the standard form:

$$ax^2 + bx + c = 0 \quad (9.17)$$

which then yields the quadratic formula:

$$x = \frac{-b \pm (b^2 - 4ac)^{1/2}}{2a} \quad (9.18)$$

In our case, we find that $a = 1$, $b = 1.76 \times 10^{-5}$, and $c = -1.76 \times 10^{-6}$ and that equation 9.18 yields $x_1 = +1.32 \times 10^{-3}$ and $x_2 = -1.33 \times 10^{-3}$. We discard the negative root of the quadratic equation and conclude that at equilibrium a 0.1 molar solution of acetic acid contains:

$$[H^+] = 1.32 \times 10^{-3} \text{ mol/L}$$

$$[CH_3COO^-] = 1.32 \times 10^{-3} \text{ mol/L}$$

$$[CH_3COOH] = 0.1 - 1.32 \times 10^{-3} \approx 0.1 \text{ mol/L}$$

The degree of dissociation (D) of a 0.1 molar solution of acetic acid is:

$$D = \frac{1.32 \times 10^{-3} \times 100}{0.1} = 1.32\% \quad (9.19)$$

Evidently, acetic acid is a weak acid because only 1.32% of the molecules in solution are dissociated into ions.

The first application of the Law of Mass Action has already taught us that even simple problems can result in complicated algebraic equations that are tedious to solve. We therefore resolve to make *approximations* whenever possible in order to reduce the effort required to work through to the answer. For example, going back to equation 9.16, we note that K is a very small number, which implies that only a *small fraction* of the acetic acid molecules will dissociate. Therefore, we assume that $x \ll 0.1$ and that $0.1 - x \approx 0.1$. Equation 9.16 now reduces to:

$$\frac{x^2}{0.1} = 1.76 \times 10^{-5} \quad (9.20)$$

from which it follows immediately that:

$$x = (1.76 \times 10^{-6})^{1/2} = 1.32 \times 10^{-3} \quad (9.21)$$

Therefore, x really is much less than 0.1 and our assumption was justified. The final result is indistinguishable from the value we obtained by solving a quadratic equation.

Another lesson we can learn here is that most numerical problems in geochemistry are based on conceptual models whose validity as representations of reality is limited. For this reason, it is inappropriate to extend numerical solutions of problems to more than two or three significant figures. Similarly, a difference of 10% between approximate and exact solutions is acceptable in many cases. In fact, approximate solutions have a special virtue because they force the investigator to develop some insight into a problem before attempting to solve it.

We have seen that acetic acid releases hydrogen ions when it is dissolved in water. However, water is itself dissociated into ions:

$$H_2O \rightleftharpoons H^+ + OH^- \quad (9.22)$$

and the dissociation constant at $25\,^{\circ}C$ is, to a very good approximation, $K_w = 1 \times 10^{-14}$. According to equation 9.22, pure water contains equal concentrations of hydrogen and hydroxyl ions whose activities are derivable from the Law of Mass Action:

$$\frac{[H^+][OH^-]}{[H_2O]} = 1 \times 10^{-14} \quad (9.23)$$

Since $[H_2O] = 1.0$ by convention and since $[H^+] = [OH^-]$, we find that in pure water:

$$[H^+] = 1 \times 10^{-7}\,mol/L \quad (9.24)$$

The activity of hydrogen ions is a useful parameter because it reflects directly or indirectly the progress of chemical reactions that may be occurring in an aqueous solution and because it can be measured by eletrical methods (Stumm and Morgan, 1970). Therefore, the activity of hydrogen ions is commonly expressed as the pH value, defined by:

$$pH = -\log_{10}[H^+] \quad (9.25)$$

It follows from equation 9.24 that the pH of pure water at $25\,^{\circ}C$ is equal to 7.0. When the hydrogen ion activity of a solution is greater than $10^{-7}\,mol/L$, its pH is less than 7, and when its activity is less than $10^{-7}\,mol/L$, the pH is greater that 7. These statements therefore define the *pH scale* according to which pH < 7.0 at $25\,^{\circ}C$ indicates acidic conditions and pH > 7.0 indicates basic conditions.

In conclusion, we can now calculate the pH of a 0.1 molar solution of acetic acid. According to equation 9.21, $[H^+] = 1.32 \times 10^{-3}\,mol/L$ from which we calculate the pH by applying equation 9.25:

$$pH = -\log(1.32 \times 10^{-3}) = 2.88 \quad (9.26)$$

The hydrogen ions released by the acetic acid affect the activity of hydroxyl ions that must remain in equilibrium with water. Therefore, from equation 9.23:

$$[OH^-] = \frac{1 \times 10^{-14}}{1.32 \times 10^{-3}}$$

$$= 0.757 \times 10^{-11}\,mol/L \quad (9.27)$$

The hydroxyl ion activity can be expressed as pOH defined by:

$$pOH = -\log_{10}[OH^-] \quad (9.28)$$

In the solution of acetic acid pOH $= -\log(0.757 \times 10^{-11}) = 11.12$. Evidently, it follows from equation 9.23 that at $25\,^{\circ}C$:

$$pH + pOH = 14.0 \quad (9.29)$$

The dissociation constant of water, like all other equilibrium constants, varies with temperature, as shown in Table 9.2. Therefore, the "neutral

Table 9.2 Dissociation Constant (K_w) of Water at Different Temperatures

T, °C	$-log\ K_w$	T, °C	$-log\ K_w$
0	14.9435	30	13.8330
5	14.7338	35	13.6801
10	14.5346	40	13.5348
15	14.3463	45	13.3960
20	14.1669	50	13.2617
24	14.0000	55	13.1369
25	13.9965	60	13.0171

SOURCE: Weast et al. (1986), p. D-164.

point" of the pH scale is at pH = 7.0 only at 25 °C. Actually, the data in Table 9.2 indicate that $K_w = 10^{-14.0}$ at 24 °C and at 25 °C it is slightly *larger* than this value.

Several weak acids are capable of yielding two or more hydrogen ions or protons per molecule of acid. When such acids dissolve in water they dissociate stepwise such that each step has a different degree of dissociation. Many bases behave exactly the same way when they dissociate in water to form OH^-. In order to illustrate the application of the Law of Mass Action to weak diprotic acids we choose hydrosulfuric acid (H_2S). This compound is a gas that dissolves in water to form H_2S molecules in aqueous solution, represented by $H_2S(aq)$, which then dissociate in two steps:

$$H_2S(aq) \rightleftharpoons H^+ + HS^- \quad K_1 \quad (9.30)$$

$$HS^- \rightleftharpoons H^+ + S^{2-} \quad K_2 \quad (9.31)$$

where $K_1 = 10^{-7.0}$ and $K_2 = 10^{-12.9}$ (Krauskopf, 1979). The reactions rapidly achieve equilibrium and then satisfy the Law of Mass Action:

$$\frac{[H^+][HS^-]}{[H_2S]} = 10^{-7.0} \quad (9.32)$$

$$\frac{[H^+][S^{2-}]}{[HS^-]} = 10^{-12.9} \quad (9.33)$$

Here we encounter for the first time two simultaneous equilibria that have two ions in common (H^+ and HS^-). Obviously, these ions must each

have the same concentration in both of the equilibria in which they occur. However, the amount of H^+ produced by the first equilibrium (9.30) is much greater than that contributed by the second equilibrium (9.31) because K_1 is about 1 million times larger than K_2.

The problem is to calculate the activities of all ionic and molecular species in a solution containing 0.1 mol of H_2S dissolved in 1 L of pure water at 25 °C. We will demonstrate *two* strategies for solving problems involving two or more simultaneous equilibria:

1. Stepwise solution.
2. Simultaneous equations.

The stepwise method applies to the dissociation of weak acids and bases, where the dissociation constants of successive steps differ by orders of magnitude. However, the use of simultaneous equations is more exact and applies to all situations in which several reactions are involved.

The stepwise method is based on simplifying assumptions that the reactions reach equilibrium in sequence. Therefore, we turn first to reaction 9.30 and let x be the number of moles of $H_2S(aq)$ that dissociate. At equilibrium we then have:

$$[H_2S(aq)] = 0.1 - x$$

$$[H^+] = x$$

$$[HS^-] = x$$

From the Law of Mass Action we obtain:

$$\frac{x^2}{0.1 - x} = 10^{-7.0} \quad (9.34)$$

Since K_1 is a very small number, we predict that only a small fraction of $H_2S(aq)$ will be dissociated at equilibrium, which justifies the assumption that $x \ll 0.1$ and that $0.1 - x \approx 0.1$. Therefore, equation 9.34 becomes:

$$\frac{x^2}{0.1} = 10^{-7.0} \quad (9.35)$$

from which it follows that $x = 10^{-4.0}$, $[H^+] = [HS^-] = 10^{-4.0}$ mol/L, and pH = 4.0. The degree of dissociation of $H_2S(aq)$ is:

$$D_1 = \frac{10^{-4.0} \times 100}{0.1} = 0.1\% \qquad (9.36)$$

Therefore, the assumption that $x \ll 0.1$ was justified in this case.

A certain number of HS^- molecules dissociate further to form S^{2-} and additional H^+. We let y be the number of moles of HS^- that dissociate and set:

$$[HS^-] = 10^{-4} - y$$

$$[H^+] = 10^{-4} + y$$

$$[S^{2-}] = y$$

Then, from the equation 9.33:

$$\frac{(10^{-4} + y)(y)}{(10^{-4} - y)} = 10^{-12.9} \qquad (9.37)$$

We again simplify the problem with the assumption that $y \ll 10^{-4.0}$ and that $10^{-4.0} + y \approx 10^{-4.0}$ and $10^{-4.0} - y \approx 10^{-4.0}$. It follows that equation 9.37 reduces to $y = 10^{-12.9}$, which confirms the validity of the assumption that $y \ll 10^{-4.0}$ and indicates that the degree of dissociation is:

$$D_2 = \frac{10^{-12.9} \times 100}{10^{-4.0}} = 10^{-6.9}\% \qquad (9.38)$$

These calculations therefore indicate that a 0.1 molar solution of $H_2S(aq)$ contains the following ions and molecules:

$$[H_2S] = 0.1 - 10^{-4.0} = 0.1 \text{ mol/L}$$

$$[HS^-] = 10^{-4.0} - 10^{-12.9} = 10^{-4.0} \text{ mol/L}$$

$$[H^+] = 10^{-4.0} + 10^{-12.9} = 10^{-4.0} \text{ mol/L}$$

$$[S^{2-}] = 10^{-12.9} \text{ mol/L}$$

The second dissociation of H_2S contributes only a negligible amount of H^+, but it provides *all* of the S^{2-} ions whose activity is equal to the value of the second dissociation constant.

The stepwise method presented above is inexact because it treats the two simultaneous equilibria as though they were independent of each other. If we look again at equations 9.30 and 9.31, we can see that the dissociation of HS^- (equation 9.31) causes more H_2S to dissociate

(equation 9.30) in accordance with Le Châtelier's principle. However, the stepwise solution disregards this interdependence of simultaneous equilibria having common ions.

The dissociation of $H_2S(aq)$ can be represented exactly by means of *simultaneous equations* derived from applications of the Law of Mass Action and from the requirement of electrical neutrality and mass balance. The number of equations required to solve the problem must be equal to the number of unknown parameters. In this case we have five unknowns: $H_2S(aq)$, HS^-, S^{2-}, H^+, and OH^-, and therefore we require five equations. The first two arise from the application of the Law of Mass Action to the two dissociation equilibria and were stated previously as equations 9.30 and 9.31. To these we add an equation based on the dissociation of water:

$$[H^+][OH^-] = 10^{-14.0} \qquad (9.39)$$

Next, we recall that all sulfur species in the solution arise by dissociation of $H_2S(aq)$ whose initial concentration was 0.1 mol/L. Therefore:

$$(H_2S) + (HS^-) + (S^{2-}) = 0.1 \qquad (9.40)$$

Note that the mass balance equation must be stated in terms of concentrations rather than activities. However, in dilute solutions the activity coefficients of all ions are close to unity and the difference between activities and concentration therefore vanishes. The fifth equation is derived from the requirement of electrical neutrality:

$$(H^+) = 2(S^{2-}) + (HS^-) + (OH^-) \qquad (9.41)$$

which states that the sum of the positive charges in the solution must be equal to the sum of the negative charges. The molar concentration of S^{2-} is multiplied by 2 because each S^{2-} ion contributes two charges. Once again we see the importance of Avogadro's law because it allows us to relate the quantity of an ion or molecule, expressed in moles, to the corresponding number of ions or molecules.

We now have five independent equations that express relationships among the five unknowns and therefore we can solve the problem. The procedure is to express the activities of the ions in terms of one of the ions, and then to substitute these relationships (after converting them to concentrations) into the equation for electrical neutrality because that equation contains all of the ions in the solution. We then solve this equation, by trial and error if necessary, and use the result to calculate the activities of all of the other ions and molecules in the solution.

Once again, the sensible approach is to make simplifying assumptions that reduce the computational labor without compromising the validity of the results. In the case at hand, we assume that $[OH^-] \ll [H^+]$ because we are discussing the dissociation of an *acid*. Therefore, we may omit the hydroxyl ion in summations such as equation 9.41 but *not* from multiplications such as equation 9.39. In addition, we know from past experience that $[S^{2-}] \ll [HS^-]$ because $K_2 \ll K_1$. This assumption simplifies both equations 9.40 and 9.41. From equation 9.40 we obtain:

$$(H_2S) + (HS^-) = 0.1 \qquad (9.42)$$

and from 9.41:

$$(H^+) = (HS^-) \qquad (9.43)$$

The latter can be used to revise equations 9.32 and 9.33. From equation 9.32 we obtain:

$$\frac{[HS^-]^2}{[H_2S]} = 10^{-7.0} \qquad (9.44)$$

and from equation 9.33:

$$[S^{2-}] = 10^{-12.9} \text{ mol/L} \qquad (9.45)$$

Next, we use equation 9.44 to express $[H_2S]$ as $[HS^-]^2/10^{-7.0}$ and substitute this value into equation 9.42:

$$\frac{[HS^-]^2}{10^{-7.0}} + HS^- = 0.1 \qquad (9.46)$$

Multiplying by $10^{-7.0}$ and rearranging terms yields:

$$[HS^-]^2 + 10^{-7.0}[HS^-] - 10^{-8.0} = 0 \qquad (9.47)$$

We solve this equation by means of the quadratic formula (equation 9.18) and obtain:

$$[HS^-] = \frac{-10^{-7.0} \pm [10^{-14.0} - 4(-10^{-8.0})]^{1/2}}{2}$$
$$(9.48)$$

which reduces to $[HS^-] = 1 \times 10^{-4.0}$ mol/L.

Evidently, in this case, the stepwise approach yields the same results as the solution of simultaneous equations because the dissociation constants are not only small numbers but also differ from each other by a factor of nearly 1 million. These circumstances do not apply to all of the acids and bases whose dissociation constants are listed in Table 9.3. Moreover, we will encounter sets of simultaneous equilibria in which the sequential treatment is inappropriate and which can be solved only by means of a set of independent equations.

The dissociation constants listed in Table 9.3 apply *only* to specific equations. In the case of a polyprotic acid such as arsenic acid the appropriate equations are:

$$H_3AsO_4(aq) \rightleftharpoons H_2AsO_4^- + H^+ \quad K_1 \quad (9.49)$$

$$H_2AsO_4^- \rightleftharpoons HAsO_4^{2-} + H^+ \quad K_2 \quad (9.50)$$

$$HAsO_4^{2-} \rightleftharpoons AsO_4^{3-} + H^+ \quad K_3 \quad (9.51)$$

Note that the acid molecule H_3AsO_4 is already dissolved in water in the first dissociation step. The situation is more complicated for bases because some of them are insoluble and because they may also be amphoteric (A). We choose $Cd(OH)_2$ to define the appropriate equations:

$$Cd(OH)_2(s) \rightleftharpoons Cd(OH)^+(aq) + OH^- \quad K_1$$
$$(9.52)$$

$$Cd(OH)^+(aq) \rightleftharpoons Cd^{2+} + OH^- \quad K_2 \quad (9.53)$$

$$Cd(OH)_2(s) \rightleftharpoons Cd(OH)_2(aq) \quad K_{aq} \quad (9.54)$$

$$Cd(OH)_2(s) + OH^- \rightleftharpoons Cd(OH)_3^-(aq)$$
$$K_A \text{ (amphoteric)} \qquad (9.55)$$

The two dissociation steps may be combined by adding equations 9.52 and 9.53. In this operation $Cd(OH)^+$ cancels, and we obtain:

$$Cd(OH)_2(s) \rightleftharpoons Cd^{2+}(aq) + 2\,OH^- \quad K_T \quad (9.56)$$

where $K_T = K_1 \times K_2$. Equation 9.54 represents the solution of solid $Cd(OH)_2$ to form molecular $Cd(OH)_2$ in aqueous solution. We can incorporate that step into the dissociation equilibria by

Table 9.3 Dissociation Constants of Acids and Bases at 25 °C[a]

Name	Formula	pK_1	pK_2	pK_3	pK_A	pK_{aq}	Ref.[b]
Acids							
Acetic	CH_3COOH	4.75	—	—	—	—	1
Arsenious	$H_3AsO_3(aq)$	9.2	—	—	—	—	2
Arsenic	$H_3AsO_4(aq)$	2.2	7.0	11.5	—	—	2
Boric	$H_3BO_3(aq)$	9.2	—	—	—	—	2
Carbonic	$H_2CO_3(aq)$	6.35	10.3	—	—	—	2
Hydrofluoric	$HF(aq)$	3.2	—	—	—	—	2
Phosphoric	H_3PO_4	2.1	7.2	12.4	—	—	2
Hydrosulfuric	$H_2S(aq)$	7.0	12.9?	—	—	—	2
Sulfuric	$H_2SO_4(aq)$	—	2.0	—	—	—	2
Hydroselenic	$H_2Se(aq)$	3.9	15.0	—	—	—	2
Selenic	$H_2SeO_4(aq)$	—	1.9	—	—	—	2
Silicic	H_4SiO_4	9.71	13.28	9.86	13.1 (pK_4)	—	3
Bases (Hydroxides)							
Ammonium	$NH_4OH(aq)$	4.7	—	—	—	—	2
Aluminum, amorph.	$Al(OH)_3$	12.3	10.3	9.0	−1.1	—	2
Aluminum, gibbsite	$Al(OH)_3$	14.8	10.3	9.0	1.4	—	2
Beryllium, amorph.	$Be(OH)_2$	—	—	—	2.2	—	2
Cadmium	$Cd(OH)_2$	10.5	3.9	—	4.1?	6.7	2
Cobalt	$Co(OH)_2$	10.6	4.3	—	5.2	6.5	2
Copper	$Cu(OH)_2$	13.0	6.3	—	2.9	—	2
Ferrous	$Fe(OH)_2$	10.6	4.5	—	5.1	8.4?	2
Ferric, amorph.	$Fe(OH)_3$	16.5	10.5	11.8	4.4	—	2
Lead, red	$PbO + H_2O$	9.0	6.3	—	1.4	4.4	2
Magnesium	$Mg(OH)_2$	8.6	2.6	—	—	—	2
Manganese	$Mn(OH)_2$	9.4	3.4	—	5.1	—	2
Mercurous, red	$HgO + H_2O$	14.8	10.6	—	4.5	3.6	2
Nickel	$Ni(OH)_2$	11.1	4.1	—	4	7	2
Silver	$\frac{1}{2}Ag_2O + \frac{1}{2}H_2O$	—	—	—	3.4	5.7	2
Thorium, amorph.	$Th(OH)_4$	—	—	10.3	5.8	10.8 (pK_4)	2
UO_2	$UO_2(OH)_2$	14.2	8.2	—	3.6	—	2
Vanadium	$V(OH)_3$	—	—	11.7	—	—	2
VO	$VO(OH)_2$	15.2	8.3	—	—	—	2
Zinc, amorph.	$Zn(OH)_2$	10.5	5.0	—	1.9	4.4?	2

[a]Explanations: $pK = -\log K$

$$Zn(OH)_2(s) \rightleftharpoons Zn(OH)^+ + OH^- \quad K_1$$
$$Zn(OH)^+ \rightleftharpoons Zn^{2+} + OH^- \quad K_2$$
$$Zn(OH)_2(s) \rightleftharpoons Zn(OH)_2(aq) \quad K_{aq}$$
$$Zn(OH)_2(s) + OH^- \rightleftharpoons Zn(OH)_3^- \quad K_A \text{ (amphoteric)}$$

"—" means "does not apply" or "unknown".

[b](1)Weast et al. (1986); (2) Krauskopf (1979); (3) calculated from thermodynamic data of Lindsay (1979).

combining equation 9.54 and 9.52. In order to accomplish this, we first invert equation 9.54:

$$Cd(OH)_2(aq) \rightleftharpoons Cd(OH)_2(s) \quad 1/K_{aq} \quad (9.57)$$

and we then add it to equation 9.52, thereby eliminating $Cd(OH)_2(s)$:

$$Cd(OH)_2(aq) \rightleftharpoons Cd(OH)^+ + OH^- \quad K_1/K_{aq} \quad (9.58)$$

Evidently, these equations can be manipulated algebraically with the understanding that:

1. When equations are added, their equilibrium constants are multiplied.
2. When equations are subtracted, their equilibrium constants are divided.
3. When an equation is reversed, its equilibrium constant is inverted.

Finally, we note that equation 9.55 reflects the amphoteric nature of $Cd(OH)_2$ because, when it takes on an additional hydroxyl ion to form $Cd(OH)_3^-$, a hydrogen ion is released by the dissociation of water. Therefore, $Cd(OH)_2$ in this case acts as an acid because it releases H^+ into solution. We predict, in accordance with Le Châtelier's principle, that the amphoteric character of $Cd(OH)_2$ is *enhanced* in basic solutions because a high activity of OH^- causes the equilibrium to make more $Cd(OH)_3^-$ in order to reduce the activity of OH^-. Conversely, in acidic solutions, which have a low activity of OH^-, the reaction shifts to the left by converting $Cd(OH)_3^-$ to OH^- and $Cd(OH)_2$. Therefore, amphoteric hydroxides act like acids in basic environments and like bases in acidic environments.

The behavior of amphoteric hydroxides teaches us two important lessons:

1. Weak acids and bases do *not* control the pH of natural environments but instead respond to it.
2. The solubility of sparingly soluble bases is pH dependent.

In geochemistry the pH is an *environmental parameter* that is determined by all of the simultaneous equilibria existing in a given environment. Therefore, the degree of dissociation of weak acids and bases and the resulting solubility of otherwise insoluble weak bases are controlled by the activity of H^+ ions in the environment.

9.4   Solubility of Sparingly Soluble Bases

We have defined the solubility of a compound as the amount of the compound that dissolves to form a saturated solution. Since many bases are only sparingly soluble, we may use the Law of Mass Action to calculate their solubilities. We choose $Mg(OH)_2$ for this demonstration because it has a low solubility and therefore occurs as the mineral *brucite*.

The first and most important step toward the solution of a problem involving the Law of Mass Action is to write the equations that represent the relevant equilibria. The dissociation of $Mg(OH)_2$ proceeds in two steps:

$$Mg(OH)_2(s) \rightleftharpoons Mg(OH)^+(aq) + OH^- \quad K_1 \quad (9.59)$$

$$Mg(OH)^+(aq) \rightleftharpoons Mg^{2+}(aq) + OH^- \quad K_2 \quad (9.60)$$

According to Table 9.3, $K_1 = 10^{-8.6}$ and $K_2 = 10^{-2.6}$. Evidently, in this case, K_2 is 1 million times *larger* than K_1, which may cause a problem if we attempt a stepwise solution. Therefore, we decide to treat the problem by means of a set of simultaneous equations. We have four unknowns [$Mg(OH)^+$, Mg^{2+}, OH^-, and H^+] and therefore need four independent equations to determine the activities of all ions in a saturated solution of brucite in pure water. We elect not to specify the pH of the solution at this time in order to demonstrate the pH dependence of the solubility of $Mg(OH)_2$, but we do set all *activity coefficients* equal to *one* and thereby equate activities and concentrations.

The necessary equations are derived from the dissociation equilibria of $Mg(OH)_2$ and H_2O and from the requirement of electrical neutrality:

$$[Mg(OH)^+][OH^-] = 10^{-8.6} \qquad (9.61)$$

$$\frac{[Mg^{2+}][OH^-]}{[Mg(OH^+)]} = 10^{-2.6} \qquad (9.62)$$

$$[H^+][OH^-] = 10^{-14.0} \qquad (9.63)$$

$$(OH^-) = 2(Mg^{2+}) + (MgOH^+) + (H^+) \qquad (9.64)$$

In order to simplify the problem, we assume that $(H^+) \ll (OH^-)$ and therefore drop (H^+) from equation 9.64. Equations 9.61 and 9.62 allow us to relate the Mg-bearing ions to $[OH^-]$:

$$[Mg(OH)^+] = \frac{10^{-8.6}}{[OH^-]} \qquad (9.65)$$

$$[Mg^{2+}] = \frac{10^{-2.6}[Mg(OH^+)]}{[OH^-]}$$

$$= \frac{10^{-11.2}}{[OH^-]^2} \qquad (9.66)$$

Substituting into equation 9.64 yields:

$$[OH^-] = \frac{2 \times 10^{-11.2}}{[OH^-]^2} + \frac{10^{-8.6}}{[OH^-]} \qquad (9.67)$$

After multiplying by $[OH^-]^2$ and rearranging terms, we obtain one equation in one unknown:

$$[OH^-]^3 - 10^{-8.6}[OH^-] = 2 \times 10^{-11.2}$$

$$= 1.26 \times 10^{-11} \qquad (9.68)$$

The equation can be solved by trial and error. We assume that $[OH^-] > 10^{-7}$ because $Mg(OH)_2$ is a base. When we try $[OH^-] = 10^{-4.0}$ in equation 9.68, we obtain:

$$10^{-12.0} - 10^{-12.6} = 7.5 \times 10^{-13} \qquad (9.69)$$

which is less than 1.26×10^{-11}. Further trials indicate that equation 9.68 is solved satisfactorily when $[OH^-] = 10^{-3.63}$ mol/L. Therefore, the activities of the Mg-bearing ions from the equations 9.65 and 9.66 are:

$$[Mg(OH)^+] = \frac{10^{-8.6}}{10^{-3.63}} = 10^{-4.97}$$

$$= 1.07 \times 10^{-5} \text{ mol/L} \qquad (9.70)$$

$$[Mg^{2+}] = \frac{10^{-11.2}}{10^{-7.26}} = 10^{-3.94}$$

$$= 11.48 \times 10^{-5} \text{ mol/L} \qquad (9.71)$$

The solubility of $Mg(OH)_2$ in pure water at 25°C is the *sum* of the concentrations of the two Mg-bearing ions and therefore equals 12.55×10^{-5} mol/L. Since the molecular weight of $Mg(OH)_2$ is 58.3267, we can express its solubility as $58.3267 \times 12.55 \times 10^{-5} = 7.32 \times 10^{-3}$ g/L, which is not a large amount. The activity of H^+ in a saturated solution of $Mg(OH)_2$ in pure water is $10^{-14.0}/10^{-3.63} = 10^{-10.37}$ mol/L and the pH = 10.37. Evidently a saturated solution of $Mg(OH)_2$ is strongly basic.

The relationship between the activities of the Mg ions in a saturated solution of $Mg(OH)_2$ and the pH is indicated by equations 9.65 and 9.66. By taking logarithms and converting pOH to pH by means of equation 9.69, we obtain from equation 9.65:

$$\log[Mg(OH)^+] = -8.6 + pOH \qquad (9.72)$$

or

$$\log[Mg(OH)^+] = 5.4 - pH \qquad (9.73)$$

Similarly, from equation 9.66:

$$\log[Mg^{2+}] = 16.8 - 2pH \qquad (9.74)$$

Each of these equations has been plotted in Figure 9.2 in coordinates of the log of the activity and pH. The total solubility of $Mg(OH)_2$ is the sum of the concentrations of $Mg(OH)^+$ and Mg^{2+} at different values of the pH.

Equations 9.73 and 9.74 are straight lines that intersect at pH = 11.4. We see by inspection of Figure 9.2 that Mg^{2+} is the dominant ion at pH values <11.4, whereas $Mg(OH)^+$ becomes dominant in more basic solutions. Such extremely basic environments occur only in highly saline lakes such as Lake Natron in East Africa. Therefore, Mg^{2+} is the dominant ion in most geological environments. Brucite is very soluble in acidic solutions and can occur in contact with water only when the activity of Mg^{2+} is large

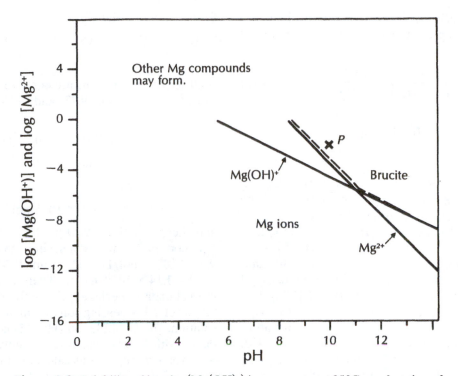

Figure 9.2 Solubility of brucite $(Mg(OH)_2)$ in pure water at 25 °C as a function of pH. $Mg(OH)_2$ dissociates to form $Mg(OH)^+$ and Mg^{2+} ions, whose activities in a saturated solution are strongly pH dependent, as given by equations 9.73 and 9.74. The resulting solubility of brucite is the sum of the concentrations of the Mg-bearing ions in the solution and is indicated by the dashed line. The coordinates of point P on the diagram represent conditions that will cause brucite to precipitate. The straight lines representing activities of Mg^{2+} and $Mg(OH)^+$ have been arbitrarily terminated at unit activities because in nature other Mg compounds will precipitate at elevated concentrations of Mg ions.

enough to maintain equilibrium. Therefore, the solubility of brucite in water can be used to define its stability as a solid phase in a geological environment.

Consider point P in Figure 9.2 whose coordinates are $[Mg^{2+}] = 10^{-2}$ mol/L and pH = 10. The coordinates of P do not satisfy equation 9.74, which implies that brucite cannot be in equilibrium with a solution of that composition because, at this value of $[Mg^{2+}]$, the activity of OH^- is too high. The application of Le Châtelier's principle to equations 9.59 and 9.60 indicates that the presence of excess OH^- will result in the precipitation of solid $Mg(OH)_2$ until both $[Mg^{2+}]$ and $[OH^-]$ have been reduced sufficiently to place point P on

the equilibrium line. We conclude, therefore, that brucite will precipitate in all environments that lie to the right of the solubility limits in Figure 9.2. In other words, the activities of the ions formed by $Mg(OH)_2$ constrain its stability in contact with water containing Mg^{2+} and $Mg(OH)^+$.

9.5   pH Control of Dissociation Equilibria

The dissociation of weak acids and bases is controlled by the pH of the geochemical environment. In order to examine this concept we return to the dissociation of $H_2S(aq)$ represented by equations 9.30 and 9.31. However, here we rewrite the mass-action equations (9.32 and 9.33) as follows:

$$\frac{[HS^-]}{[H_2S]} = \frac{10^{-7.0}}{[H^+]} \quad (9.75)$$

and:

$$\frac{[S^{2-}]}{[HS^-]} = \frac{10^{-12.9}}{[H^+]} \quad (9.76)$$

We see that the activity ratios of $[HS^-]/[H_2S]$ and $[S^2]/[HS^-]$ depend on the activity of the hydrogen ion. According to equation 9.75, $[HS^-]/[H_2S] = 1.0$ when $[H^+] = 10^{-7.0}$ mol/L. If $[H+] > 10^{-7.0}$, say $10^{-6.0}$, then $[HS^-]/[H_2S] = 0.1$, which means that the activity of H_2S is 10 times greater than the activity of $[HS^-]$. Similarly, if $[H^+] < 10^{-7.0}$, say $10^{-8.0}$, then $[HS^-]/[H_2S] = 10$ and $[HS^-]$ is 10 times greater than that of H_2S. Evidently, $[H^+] = 10^{-7.0}$ (pH = 7.0) is an important boundary at which the abundances of H_2S and HS^- are equal. H_2S dominates at pH < 7.0, whereas HS^- dominates at pH > 7.0

Similar arguments, applied to equation 9.76, indicate that pH = 12.9 is another boundary in the system. At pH < 12.9 we find that HS^- is more abundant than S^{2-}, whereas at pH > 12.9 the sulfide ion (S^{2-}) dominates. Such relationships occur in all weak acids and bases because their dissociation into ions is controlled by the pH of the environment.

The control of dissociation equilibria by the pH has an important consequence because it means that we can identify the dominant ions at different values of the pH. In order to demonstrate how this works we fix the total amount of sulfur-bearing species in the solution at 10^{-2} mol/L and calculate the activities of H_2S, HS^-, and S^{2-} as a function of pH. At pH = 6.0 we obtain from equation 9.75:

$$[H_2S] = \frac{10^{-6.0}[HS^-]}{10^{-7.0}} = 10[HS^-] \quad (9.77)$$

and from equation 9.76:

$$[S^{2-}] = \frac{10^{-12.9}[HS^-]}{10^{-6.0}} = 10^{-6.9}[HS^-] \quad (9.78)$$

From the mass balance:

$$(H_2S) + (HS^-) + (S^{2-}) = 1 \times 10^{-2} \text{ mol/L} \quad (9.79)$$

If all activity coefficients are equal to one, we can substitute equations 9.77 and 9.78 into 9.79, obtaining:

$$10[HS^-] + [HS^-] + 10^{-6.9}[HS^-] = 1 \times 10^{-2} \text{ mol/L} \quad (9.80)$$

which reduces to:

$$11[HS^-] = 1 \times 10^{-2} \text{ mol/L} \quad (9.81)$$

and hence $[HS^-] = 9.09 \times 10^{-4}$ mol/L. From equation 9.77 we determine that $[H_2S] = 9.09 \times 10^{-3}$ mol/L and from 9.78 we have $[S^{2-}] = 1.14 \times 10^{-10}$ mol/L. Therefore, H_2S is the most abundant species at pH = 6.0 and constitutes 90.9% of all S-bearing ions or molecules in the solution. The bisulfide ion (HS^-) is next in abundance with 9.09%, which leaves only 0.01% for S^{2-}. When we repeat this calculation for different values of the pH, we generate three curves that represent the changing abundances of H_2S, HS^-, and S^{2-} in the solution. As expected, we see from Figure 9.3 that H_2S dominates at pH < 7.0, that HS^- is the dominant ion for pH values from 7.0 to 12.9, and that S^{2-} is dominant at pH > 12.9. Although all of the ions or molecules are present throughout the range of pH values, most of the S in the solution is associated with a particular species except at the boundaries at which the abundances of two species are equal.

9.6 Solubility of Amorphous Silica

The oxides of many metals react with water to form bases and the oxides of nonmetals react with water to form acids. For example, CO_2 reacts with water to form carbonic acid:

$$CO_2 + H_2O \rightarrow H_2CO_3 \quad (9.82)$$

and SO_2 forms hydrosulfurous acid:

$$SO_2 + H_2O \rightarrow H_2SO_3 \quad (9.83)$$

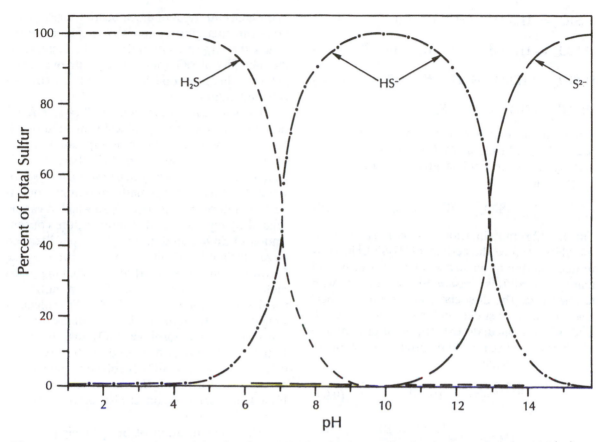

Figure 9.3 The pH dependence of the abundances of S-bearing ions and molecules in a solution of $H_2S(aq)$ containing a fixed amount of S in solution (equation 9.79). Note that H_2S is dominant at pH < 7.0, HS^- dominates between pH = 7.0 and 12.9, and that S^{2-} is dominant only at pH > 12.9. However, all of the ions are present at all pH values even though their abundances may be small compared to that of the dominant ion.

In a very similar manner, silicon dioxide reacts with water to form silicic acid, such that at equilibrium:

$$SiO_2(\text{amorph.}) + 2\,H_2O \rightleftharpoons H_4SiO_4$$

$$K = 10^{-2.74} \quad (9.84)$$

SiO_2 is a solid compound that can either be amorphous or assume several different polymorphic crystalline forms, including quartz, cristobalite, and tridymite. Silicic acid actually forms as a result of chemical weathering of the common rock-forming aluminosilicate minerals such as the feldspars and micas rather than by solution of crystalline or amorphous SiO_2. However, when its solubility is exceeded, it forms amorphous silica, which

settles out slowly as a gelatinous precipitate (Iler. 1979). Given sufficient time, the precipitate expels water and begins to crystallize, forming *opal A* and *opal CT* as intermediate phases (Kastner et al., 1977). The process ends with the crystallization of a cryptocrystalline variety of quartz called *chalcedony,* which forms *chert* or *flint* and occurs in *geodes* as *agate.* Quartz is highly insoluble, and dissolves in water only sparingly, even during long periods of geologic time. Amorphous silica, however, is much more reactive and generally maintains equilibrium with silicic acid.

We are now in a position to discuss the solubility of amorphous silica based on equation 9.84 and on the dissociation of the silicic acid:

$$H_4SiO_4 \rightleftharpoons H_3SiO_4^- + H^+ \quad K_1 = 10^{-9.71} \qquad (9.85)$$

$$H_3SiO_4^- \rightleftharpoons H_2SiO_4^{2-} + H^+ \quad K_2 = 10^{-13.28} \qquad (9.86)$$

$$H_2SiO_4^{2-} \rightleftharpoons HSiO_4^{3-} + H^+ \quad K_3 = 10^{-9.86} \qquad (9.87)$$

$$HSiO_4^{3-} \rightleftharpoons SiO_4^{4-} + H^+ \quad K_4 = 10^{-13.10} \qquad (9.88)$$

When amorphous silica and pure water have reached equilibrium at 25°C (equation 9.84), the activity of silicic acid is given by the Law of Mass Action:

$$[H_4SiO_4] = 10^{-2.74} \text{ mol/L} \qquad (9.89)$$

The equilibrium (equation 9.84) is *independent* of the pH; therefore, the activity of $[H_4SiO_4]$ in a saturated solution of silicic acid in contact with solid amorphous SiO_2 is *constant* and varies only with temperature. The dissociation constants of silicic acid are all very small, making this a very weak acid. We can calculate the degree of dissociation of silicic acid at pH = 7.0 by proceeding stepwise from equation 9.85:

$$\frac{[H^+][H_3SiO_4^-]}{[H_4SiO_4]} = 10^{-9.71} \qquad (9.90)$$

$$[H_3SiO_4^-] = \frac{10^{-9.71} \times 10^{-2.74}}{10^{-7.0}}$$

$$= 10^{-5.45} \text{ mol/L} \qquad (9.91)$$

$$D_1 = \frac{10^{-5.45} \times 10^2}{10^{-2.74}}$$

$$= 0.19\% \qquad (9.92)$$

Therefore, the activity of $H_3SiO_4^-$ at pH = 7.0 is about 500 times less than that of H_4SiO_4 and does not contribute appreciably to the solubility of amorphous SiO_2.

At pH = 8.0 the activity of $H_3SiO_4^-$ increases to $10^{-4.45}$ mol/L (equation 9.91) and D_1 rises to 1.95%. At pH = 9.0 $[H_3SiO_4^-]$ increases to $10^{-3.45}$ mol/L and D_1 reaches 19.5%. We see that the degree of dissociation of silicic acid is very low at pH < 8.0 and rises rapidly as the pH increases above that value. As a result, the solubility of amorphous silica also rises because it is based on the *sum*

of the concentrations of the Si-bearing ions in solution. The first dissociation step (equation 9.85) makes the largest contribution to the increase in the solubility of SiO_2 (amorphous) with increasing pH. The other ions come into play only in extremely basic solution.

The solubility of amorphous silica as a function of pH is shown graphically in Figure 9.4. Note that the concentration is expressed in milligrams per liter (mg/L) of SiO_2 because it is commonly stated this way in chemical analyses of water. The SiO_2 is not actually present in molecular form but represents H_4SiO_4 to which it can be related by equation 9.84. Therefore, the concentration of SiO_2 stated in milligrams per liter can be recalculated as H_4SiO_4 in moles per liter by the following consideration. If SiO_2 = 25.0 mg/L, its molar concentration is 25.0/60.08 mmol/L or 0.416×10^{-3} mol/L, where 60.08 is the molecular weight of SiO_2. Equation 9.84 indicates that for every mole of amorphous SiO_2 that dissolves 1 mol of H_4SiO_4 is produced. Therefore, since 0.416×10^{-3} mol of SiO_2 dissolved to give a concentration of 25.0 mg/L, the concentration of H_4SiO_4 in that solution is also 0.416×10^{-3} or 4.16×10^{-4} mol/L.

The concentration of SiO_2 in Figure 9.4 is plotted on a linear rather than a logarithmic scale in order to emphasize that the solubility of amorphous SiO_2 at 25°C increases steeply at pH > 8 because of the dissociation of H_4SiO_4 to $H_3SiO_4^-$, which introduces additional Si-bearing ions into the water. Therefore, even a small *decrease* in the pH of *basic* solutions containing H_4SiO_4 may *reduce* the solubility of SiO_2 enough to saturate the solution and to cause amorphous SiO_2 to be deposited. For example, a decrease of 0.1 pH units from 8.5 to 8.4 of a *saturated* solution can result in the deposition of 1.37 mg of amorphous SiO_2 per liter of solution. Such seemingly small amounts quickly increase when we apply them to the large volume of groundwater that may flow through an aquifer in the course of geologic time. Subtle changes in the geochemical environment can be magnified by time into large-scale transformations.

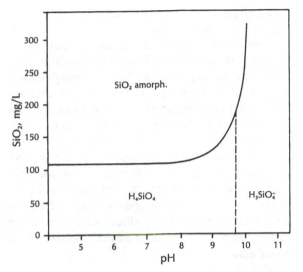

Figure 9.4 Solubility of amorphous SiO_2 expressed in units of mg of SiO_2 per liter of pure water at 25°C based on equations 9.84 and 9.85. Note that H_4SiO_4 begins to dissociate appreciably at about pH = 8.0 and that $H_3SiO_4^-$ becomes dominant at pH = 9.71. The second dissociation (equation 9.85) is negligible in the pH range shown here and, at pH = 10.0, it increases the solubility of amorphous SiO_2 by only 0.03%. Silicic acid is derived primarily by chemical weathering of the common rock-forming Al-silicate and silicate minerals, but may be precipitated from saturated solutions as gelatinous amorphous silica when the pH decreases. As amorphous silica ages, it recrystallizes to form opal A and opal CT until it ultimately forms chalcedony, which is a variety of quartz.

Surprisingly, quartz is not an important source of silicic acid in natural solution because it is quite insoluble. The equilibrium constant of the reaction:

$$SiO_2 + 2\,H_2O \rightleftharpoons H_4SiO_4 \qquad (9.93)$$
$$\text{Quartz}$$

is $K = 10^{-4.01}$. As a result, the activity of H_4SiO_4 in a saturated solution in contact with quartz at 25°C is 9.76×10^{-5} mol/L, which is equivalent to 5.9 mg/L of SiO_2 in solution. The concentrations of Si in average river water and seawater (Table 4.7) are 6.5 and 2.8 μg/g, respectively. These concentrations can be transformed into SiO_2 in milligrams

per liter by converting first to micromoles of Si and then to micrograms of SiO_2. For average river water:

$$SiO_2 = \frac{6.5 \times 60.08 \times 10^3}{28.086 \times 10^3} = 13.9 \text{ mg/L} \qquad (9.94)$$

Therefore, average river water is *supersaturated* with respect to quartz but is undersaturated with respect to amorphous silica. However, quartz does not precipitate from aqueous solutions directly but forms only by recrystallization of amorphous SiO_2 as discussed previously. The concentration of SiO_2 in seawater is 5.8 mg/L (assuming a density of 1.025 g/cm^3), which is close to the solubility of quartz at 25°C. Seawater has a lower concentration of SiO_2 than river water because certain organisms (sponges, diatoms, and radiolarians) form skeletons composed of opal A. When these organisms die, their siliceous skeletons accumulate on the bottom of the ocean and may ultimately form either chert, composed of cryptocrystalline quartz, or deposits of diatomite (Cressman, 1962; Wedepohl, 1972).

The geochemistry of silica illustrates the point that some reactions in nature are essentially *irreversible*. When NaCl dissolves in water the reaction is *reversible* because the compound does precipitate from supersaturated solutions. When amorphous silica dissolves in water, it can be made to precipitate again, although with some difficulty. However, when quartz dissolves in water at 25°C to form silicic acid, the reaction cannot be reversed at the same temperature. We will encounter this phenomenon again in Chapter 10 when we discuss the solubility of minerals, some of which form only by crystallization of magma at elevated temperatures.

9.7   Summary

Chemical reactions have a natural tendency to achieve a state of equilibrium in which the rates of the forward and backward reactions are equal. As a result, the amounts of reactants and products that coexist at equilibrium are invariant with time. This observational evidence was originally used to formulate the Law of Mass Action, which was later derived from the principles of thermodynamics.

Chemical reactions in the natural environment may fail to reach equilibrium because the products escape from the site of the reaction or because reactions involving the transformation of one solid into another are very slow. In addition, some natural reactions are irreversible.

Acids and bases are defined as compounds that release H^+ or OH^- ions in aqueous solutions, respectively. The bases of some elements are amphoteric and act like bases in acidic environments and like acids in basic environments. Weak acids and bases dissociate only partially depending on the pH of the environment. In general, the pH of natural environments results from the com-

bined effect of all chemical reactions that are taking place simultaneously.

Some bases are insoluble and actually occur as minerals in nature. For example, $Mg(OH)_2$ forms the mineral brucite. The solubility of such insoluble bases can be calculated by solving a set of simultaneous equations derived from the Law of Mass Action and from the requirement of conservation of mass and electrical charge.

Weak acids play an important role in geochemistry because the abundances of the ions they form by dissociation are controlled by the pH of the environment. This phenomenon affects the solubility of amorphous silica, which increases rapidly at pH > 8 because of the dissociation of silicic acid.

Problems

1. If the concentration of an ion in a solution is 5.0×10^{-2} mol/L at 25 °C, what is its concentration in the same solution at 45 °C? Use the data in Table 9.1. (Answer: 4.96×10^{-2} mol/L)

2. Derive a relationship between the molarity (M), the formality (F), and the density (d) of a solution.

3. Calculate the normality of hydrochloric acid (HCl) from the following information provided by the label on the bottle. Assay: 37%; net weight: 2.72 kg; molecular weight: 36.46; specific gravity: 1.18. (Answer: 12.0)

4. Calculate the normality of sulfuric acid (H_2SO_4) based on the information provided by the label on the

bottle. Assay: 98.0%; net weight: 4.08 kg; molecular weight: 98.08; specific gravity: 1.84.   (Answer 36.8)

5. What volume of 35 N (normal) H_2SO_4 is required to make 250 mL of 1.5 N H_2SO_4?

6. Calculate the pH of hydrofluoric acid containing 0.1 mol of HF per liter of solution. Find the dissociation constant in Table 9.3.

7. Calculate the activities of all ions and the pH of a solution containing 0.1 mol of phosphoric acid per liter of solution. Find the equilibrium constants in Table 9.3.

8. Calculate the solubility of gibbsite ($Al(OH)_3$) at pH = 5.0. Find the dissociation constants in Table 9.3.

References

CRESSMAN, E. R., 1962. Nondetrital siliceous sediments. Chapter T, 1–23. In M. Fleischer (Ed.), *Data of Geochemistry,* U.S. Geol. Surv. Paper 440-T.

DREVER, J. I., 1988. *The Geochemistry of Natural Waters,* 2nd ed. Prentice-Hall, Upper Saddle River, NJ, 435 pp.

GARRELS, R. M., and C. L. CHRIST, 1965. *Minerals, Solutions and Equilibria.* Harper & Row, New York, 450 pp.

ILER, R. K., 1979. *The Chemistry of Silica.* Wiley, New York, 866 pp.

KASTNER, M., J. B. KEENE, and J. M. GIESKES, 1977. Diagenesis of siliceous oozes—I. Chemical controls on the rate of opal A and opal CT transformation: an experimental study. *Geochim. Cosmochim. Acta,* **41**:1041–1060.

KRAUSKOPF, K. B., 1979. *Introduction to Geochemistry,* 2nd ed. McGraw-Hill, New York, 617 pp.

LINDSAY, W. L., 1979. *Chemical Equilibria in Soils.* Wiley, New York, 449 pp.

MOORE, W. J., 1955. *Physical Chemistry,* 2nd ed. Prentice-Hall, Upper Saddle River, NJ, 633 pp.

STUMM, W., and J. J. MORGAN, 1970. *Aquatic Chemistry. An Introduction Emphasizing Chemical Equilibria in Natural Waters.* Wiley, New York, 583 pp.

WEAST, R. C., M. J. ASTLE, and W. H. BEYER (Eds.), 1986. *CRC Handbook of Chemistry and Physics,* 66th ed. CRC Press, Boca Raton, FL.

WEDEPOHL, K. H. (Ed.), 1972. *Silicon. Handbook of Geochemistry,* vol. II-2. Springer-Verlag, Berlin.

10

Salts and Their Ions

When an acid and a base are mixed, the hydrogen ion of the acid combines with the hydroxyl ion of the base to form water. The remaining anion of the acid and the cation of the base form a salt, which may precipitate or remain in solution depending on its solubility. Salts are named after the acids that provided the anions. For example, hydrochloric acid forms *chlorides,* sulfuric acid forms *sulfates,* carbonic acid forms *carbonates,* and silicic acid forms *silicates.* The naturally occurring salts of these and other acids are *minerals* that form the rocks of the crust of the Earth. The only minerals that are not salts are the *oxides, hydroxides,* and *native elements.* Evidently, the study of salts and their properties is an important subject in geochemistry. In this chapter we will study the solubility of salts in water and the interactions of the resulting ions with molecules of water.

10.1   Solubility of Salts

When salts dissolve in water, they dissociate into the anions and cations of the acid and the base from which they were derived. Neutral molecules of salts may also form, but their abundance is low in most cases. When a salt has formed a *saturated* solution, a state of equilibrium exists between the ions in the solution and any excess salt remaining in the solid state. Therefore, the Law of Mass Action applies and can be used to calculate the activities of the ions in a saturated solution and hence the solubility of the salt.

We choose aluminum sulfate $[Al_2(SO_4)_3]$ to illustrate how the Law of Mass Action can be used for this purpose. Aluminum sulfate, a salt of sulfuric acid and aluminum hydroxide, is quite soluble in water and dissociates readily into ions:

$$Al_2(SO_4)_3(s) \rightleftharpoons 2\,Al^{3+} + 3\,SO_4^{2-} \quad (10.1)$$

We assume for the time being that these ions do not interact with water, that this reaction is the only source of Al^{3+} and SO_4^{2-} ions, that no other ions are present, and that the water temperature is 25 °C. We see from equation 10.1 that each mole of aluminum sulfate that dissolves produces two moles of Al^{3+} and three moles of SO_4^{2-} in the solution. Therefore, if x is the number of moles of salt that dissolve in a saturated solution, the concentrations of the resulting ions will be $(Al^{3+}) = 2x$ and $(SO_4^{2-}) = 3x$. According to the Law of Mass Action (equation 9.13):

$$[Al^{3+}]^2[SO_4^{2-}]^3 = K_{sp} \quad (10.2)$$

where K_{sp} is the *solubility product constant,* and the activity of excess solid aluminum sulfate is equal to one. If the concentrations of the ions are equal to the activities in the solution, $[Al^{3+}] = 2x$ and $[SO_4^{2-}] = 3x$. Substituting into the equation 10.2 we obtain:

$$(2x)^2(3x)^3 = K_{sp} \quad (10.3)$$

where $K_{sp} = 69.19$ (calculated from its solubility given by Weast et al., 1986). The solubility product constant is a large number, indicating that aluminum sulfate is, in fact, a very soluble salt. For this reason, it does not occur naturally as a mineral. Equation 10.3 reduces to $108x^5 = 69.19$, which yields $x = 0.9147$. Therefore, a saturated solution of aluminum sulfate contains $2x = 1.829$ mol/L of Al^{3+} and $3x = 2.744$ mol/L of SO_4^{2-}. The Al^{3+} reacts with water to form insoluble Al hydroxide.

The amount of aluminum sulfate that dissolves in a saturated solution is equal to $x = 0.9147$ mol/L. Since the molecular weight of aluminum sulfate is 342.1478, its solubility is $0.9147 \times 342.1478 = 3.13 \times 10^2$ g/L of water.

Most minerals are much less soluble in water than aluminum sulfate, and their solubility product constants have correspondingly smaller values. Table 10.1 contains the solubility product constants of some common minerals taken primarily from a compilation by Krauskopf (1979) but augmented with data from Lindsay (1979). These constants can be used to calculate the activities of the ions in saturated solutions of minerals by means of the Law of Mass Action.

We are also interested in finding out whether a solution is actually saturated with respect to a specific mineral. For example, if a sample of water contains 5.00×10^{-2} mol/L of Ca^{2+} and 7.00×10^{-3} mol/L of SO_4^{2-}, we may want to know whether this solution is saturated with respect to calcium sulfate (anhydrite). In order to find out, we first write the equation to represent the dissociation of $CaSO_4$ into its ions:

$$CaSO_4 \rightleftharpoons Ca^{2+} + SO_4^{2-} \qquad (10.4)$$

At equilibrium the Law of Mass Action applies; thus:

$$[Ca^{2+}][SO_4^{2-}] = 10^{-4.5} \qquad (10.5)$$

If the solution is saturated with respect to anhydrite, the product of activities of the ions in the solution must equal $10^{-4.5}$. We assume for the sake of argument that the measured concentrations are equal to the activities of the ions and therefore calculate the *ion activity product* (IAP):

$$(5.00 \times 10^{-2})(7.00 \times 10^{-3}) = 35.00 \times 10^{-5}$$

$$= 10^{-3.45} \qquad (10.6)$$

Since the IAP ($10^{-3.45}$) in this case is *larger* than the solubility product constant, the solution is *supersaturated* with respect to calcium sulfate. Therefore, anhydrite should *precipitate*. If the IAP is found to be *less* than the solubility product

constant, the solution is *undersaturated* and therefore can *dissolve* the mineral. If IAP $= K_{sp}$, the solution is *saturated* and a state of equilibrium exists between the solid and its ions.

In the case we are considering, it turns out that $CaSO_4$ does *not* actually precipitate from a supersaturated solution of its ions. Instead, *gypsum* ($CaSO_4 \cdot 2H_2O$) forms and subsequently crystallizes to anhydrite. This is another example of the *irreversibility* of some geochemical reactions we first mentioned in Section 9.6. In this case, $CaSO_4$ dissolves in water but does *not* precipitate at 25°C. Similarly, most silicate minerals dissolve in water but do not precipitate from aqueous solutions of their ions at low temperature. Some silicate minerals precipitate from aqueous solutions only at elevated temperatures or require a specific starting material like volcanic glass, and the silicate minerals of igneous rocks crystallize only from silicate melts at temperatures close to 1000°C.

In the case of the calcium sulfate solution we are considering, *gypsum* precipitates until the activities of the ions are reduced sufficiently so that the IAP is equal to K_{sp} and solid gypsum is in equilibrium with its ions:

$$CaSO_4 \cdot 2H_2O(s) \rightleftharpoons Ca^{2+} + SO_4^{2-} + 2H_2O$$
$$(10.7)$$

Therefore, at equilibrium between gypsum and its ions:

$$[Ca^{2+}][SO_4^{2-}] = 10^{-4.6} \qquad (10.8)$$

If x moles of gypsum precipitate, the activities of the ions at equilibrium will be:

$$[Ca^{2+}] = 5.00 \times 10^{-2} - x \qquad (10.9)$$

$$[SO_4^{2-}] = 7.00 \times 10^{-3} - x \qquad (10.10)$$

Substituting into equation 10.8 yields:

$$(5.00 \times 10^{-2} - x)(7.00 \times 10^{-3} - x) = 10^{-4.6}$$
$$(10.11)$$

which reduces to a quadratic equation:

$$x^2 - (5.7 \times 10^{-2})x + 3.25 \times 10^{-4} = 0 \qquad (10.12)$$

Table 10.1 Solubility Product Constants for Different Salts at 25°C

		$pK^{a,b}$
Carbonates		
Ag_2CO_3		11.09 (L)
$BaCO_3$	witherite	8.3 (K)[c]
$CaCO_3$	calcite	8.35 (K)
$CaCO_3$	aragonite	8.22 (K)
$CaCO_3 \cdot 6H_2O$	ikaite	6.38 (L)
$CdCO_3$	otavite	13.7 (K), 12.0 (L)
$CoCO_3$	sphaerocobaltite	10.0 (K)
$CuCO_3$		9.63 (L)
$Cu_2(OH)_2CO_3$	malachite	33.8 (K), 33.2 (L)
$Cu_3(OH)_2(CO_3)_2$	azurite	66.5 (L)
$FeCO_3$	siderite	10.7 (K), 10.2 (L)
$HgCO_3$		22.6 (L)
Hg_2CO_3	$[Hg_2^{2+}][CO_3^{2-}]$	14.0 (L)
$MgCO_3$	magnesite	7.5 (K), 7.46 (L)
$MgCO_3 \cdot 3H_2O$	nesquehonite	5.6 (K), 4.67 (L)
$MgCO_3 \cdot 5H_2O$	lansfordite	4.54 (L)
$MgCa(CO_3)_2$	dolomite	17.9 (L)
$MnCO_3$	rhodochrosite	9.3 (K), 10.1 (L)
$NiCO_3$		6.9 (K)
$PbCO_3$	cerussite	13.1 (K), 13.5 (L)
$Pb_2CO_3Cl_2$	phosgenite	20.0 (L)
$Pb_3(CO_3)_2(OH)_2$		46.8 (L)
$SrCO_3$	strontianite	9.0 (K)[c]
UO_2CO_3	$[UO_2^{2+}][CO_3^{2-}]$	10.6 (K)
$ZnCO_3$	smithsonite	10.0 (20°C, K), 10.2 (L)
Sulfates		
Ag_2SO_4		4.81 (K, L)
$KAl_3(SO_4)_2(OH)_6$	alunite	80.95 (L)
$BaSO_4$	barite	10.0 (K)
$CaSO_4$	anhydrite	4.5 (K), 4.41 (L)
$CaSO_4 \cdot 2H_2O$	gypsum	4.6 (K), 4.63 (L)
$CdSO_4$		0.044 (L)
$CdSO_4 \cdot 2H_2O$		1.59 (L)
$CuSO_4$		−3.72 (L)
$CuSO_4 \cdot 5H_2O$		2.61 (L)
$Cu_4(OH)_6SO_4$	brochantite	68.6 (L)
$FeSO_4$		−2.65 (L)
$FeSO_4 \cdot 7H_2O$		2.46 (L)
$Fe_2(SO_4)_3$		−2.89 (L)
$KFe_3(SO_4)_2(OH)_6$	jarosite	96.5 (L)
$HgSO_4$		3.34 (L)
Hg_2SO_4		6.20 (L)
$MgSO_4$		−8.18 (L)
$MnSO_4$		−3.43 (L)
$MnSO_4 \cdot H_2O$		−0.47 (L)

Table 10.1 (continued)

		$pK^{a,b}$
$Mn_2(SO_4)_3$		11.6 (L)
$PbSO_4$	anglesite	7.8 (K), 7.79 (L)
$SrSO_4$	celestite	6.5 (K)
$ZnSO_4$	zinkosite	-3.41 (L)
Sulfides		
α-Ag_2S		49.01 (L)
β-Ag_2S		48.97 (L)
Ag_2S		50.1 (K)
Bi_2S_3	bismuthinite	100.0 (K)
CaS	oldhamite	0.78 (L)
CdS	greenockite	27.0 (K, L)
$CoS\,(\alpha)$		21.3 (K)
$CoS(\beta)$		25.6 (K)
Cu_2S	chalcocite	48.5 (K, L)
CuS	covellite	36.1 (K)
α-$Fe_{0.95}S$	pyrrhotite	17.4 (L)
α-FeS	troilite	16.2 (L)
FeS	troilite	18.1 (K)
FeS_2	pyrite	42.5 (L)
FeS_2	marcasite	41.8 (L)
Fe_2S_3		88.0 (L)
α-HgS	cinnabar (red)	53.0 (K), 52.0 (L)
β-HgS	metacinnabar (black)	52.7 (K), 51.7 (L)
Hg_2S	$[Hg_2^{2+}][S^{2-}]$	54.8 (L)
MgS		-5.59 (L)
MnS	(green)	13.5 (K), 11.7 (L)
MnS	(pink) alabandite	10.5 (K), 12.8 (L)
MnS_2	haurite	81.4 (L)
MoS_2	molybdenite	?
α-NiS		19.4 (K)
γ-NiS		26.6 (K)
PbS	galena	17.5 (K, L)
Sb_2S_3	stibnite	90.8 (K)
SnS		25.9 (K)[c]
α-ZnS	sphalerite	24.7 (K, L)
β-ZnS	wurtzite	22.5 (K, L)
Phosphates		
Ag_3PO_4		16.0 (L)
$AlPO_4$	berlinite	19.0 (L)
$AlPO_4 \cdot 2H_2O$	variscite	22.1 (K, L)
$H_6K_3Al_5(PO_4)_8 \cdot 18H_2O$	K-taranakite	178.7 (L)
$CaHPO_4 \cdot 2H_2O$	brushite	6.57 (K, L)
$CaHPO_4$	monetite	6.89 (L)
α-$Ca_3(PO_4)_2$		25.5 (L)
β-$Ca_3(PO_4)_2$	whitlockite	28.7 (K), 28.9 (L)

Table 10.1 (continued)

		$pK^{a,b}$
$Ca_8H_2(PO_4)_6 \cdot 5H_2O$	octacalcium phosphate	80.6 (L)
$Ca_{10}F_2(PO_4)_6$	fluorapatite	119.2 (L)
$Ca_{10}(OH)_2(PO_4)_6$	hydroxyapatite	116.4 (L)
$Cd_3(PO_4)_2$		38.1 (L)
$Cu_3(PO_4)_2$		36.9 (K, L)
$Cu(PO_4)_2 \cdot 2H_2O$		38.8 (L)
$FePO_4$ (amorph.)		21.6 (K)
$FePO_4 \cdot 2H_2O$	strengite	26.4 (K, L)
$Fe_3(PO_4)_2 \cdot 8H_2O$	vivianite	36.0 (L)
Hg_2HPO_4		12.4 (L)
KH_2PO_4		0.21 (L)
K_2HPO_4		−3.60 (L)
$MgHPO_4 \cdot 3H_2O$	newberyite	5.82 (L)
$MgKPO_4 \cdot 6H_2O$		10.6 (L)
$MgNH_4PO_4 \cdot 6H_2O$	struvite	13.2 (L)
$Mg_3(PO_4)_2$		25.2 (K), 14.6 (L)
$Mg_3(PO_4)_2 \cdot 8H_2O$	bobierrite	25.0 (L)
$Mg_3(PO_4)_2 \cdot 22H_2O$		23.1 (L)
$MnHPO_4$		12.9 (L)
$Mn_3(PO_4)_2$		27.4 (L)
$PbHPO_4$		11.4 (K), 11.5 (L)
$Pb(H_2PO_4)_2$		114.7 (L)
$Pb_3(PO_4)_2$		43.5 at 38°C (K), 44.4 (L)
$Pb_5(PO_4)_3Br$	bromopyromorphite	78.2 (L)
$Pb_5(PO_4)_3Cl$	chloropyromorphite	83.7 (L)
$Pb_5(PO_4)_3F$	fluoropyromorphite	71.6 (L)
$Pb_5(PO_4)_3OH$	hydroxypyromorphite	76.8 (L)
$(UO_2)_3(PO_4)_2$	$[UO_2^{2+}]^3[PO_4^{3-}]^2$	49.7 (K)
UO_2HPO_4	$[UO_2^{2+}][HPO_4^{2-}]$	12.2 (K)
$Zn_3(PO_4)_2 \cdot 4H_2O$	hopeite	35.3 (L)
$Zn_3(PO_4)_2$		35.3 (K)

Molybdates

Ag_2MoO_4		11.55 (L)
$CaMoO_4$	powellite	7.94 (L)
$CuMoO_4$		6.48 (L)
$FeMoO_4$		7.70 (L)
$MgMoO_4$		0.62 (L)
$MnMoO_4$		4.13 (L)
$PbMoO_4$	wulfenite	16.0 (L)
$ZnMoO_4$		4.49 (L)

Chlorides

$AgCl$	cerargyrite	9.75 (K, L)
$CuCl$		6.7 (K)
$PbCl_2$	cotunnite	4.8 (K)
Hg_2Cl_2	$[Hg_2^{2+}][Cl^-]^2$	17.9 (K)

Table 10.1   (continued)

		p$K^{a,b}$
Fluorides		
AgF		100.0 (L)
BaF$_2$		5.8 (K)
CaF$_2$	fluorite	10.4 (K)
MgF$_2$	sellaite	8.2 (K)^c
PbF$_2$		7.5 (K)
SrF$_2$		8.5 (K)

^{a}Expressed as $-\log K_{sp}$.
^{b}K = Krauskopf (1979), pp. 552–553; L = Lindsay (1979), calculated from standard free energies.
^{c}Uncertain by more than 0.8 (Krauskopf, 1979).

We solve this equation by means of the quadratic formula (equation 9.18) and obtain two roots: $x_1 = 5.05 \times 10^{-2}$ and $x_2 = 6.45 \times 10^{-3}$. We choose x_2 because x_1 reduces $[SO_4^{2-}]$ in equation 10.10 to a negative value. Therefore, the amount of gypsum that will precipitate *per liter* of solution is $6.45 \times 10^{-3} \times 172.17 = 1.11$ g. When equilibrium has been established after gypsum has precipitated, the IAP $(5.00 - 0.645) \times 10^{-2} \times (7.00 - 6.45) \times 10^{-3} = 10^{-4.6}$, as required for equilibrium. Note that the $[Ca^{2+}]/[SO_4^{2-}]$ ratio has increased by more than a factor of 10 from its initial value of $(5.00 \times 10^{-2})/(7.00 \times 10^{-3}) = 7.1$ to $(4.35 \times 10^{-2})/(0.055 \times 10^{-2}) = 79.2$. Evidently, the precipitation of a salt from a supersaturated solution not only reduces the concentrations of the ions but actually *changes* the *chemical composition* of the remaining solution.

If gypsum continues to precipitate from this solution, the activity of SO_4^{2-} approaches zero and the $[Ca^{2+}]/[SO_4^{2-}]$ ratio rises toward infinity. This phenomenon, which is illustrated in Figure 10.1, occurs during the evolution of brines by evaporative concentration and constitutes a *geochemical divide*. The solution we have been discussing is depleted in sulfate and is enriched in Ca^{2+} by the progressive precipitation of gypsum because its initial molar Ca^{2+}/SO_4^{2-} ratio was greater than one. If $Ca^{2+}/SO_4^{2-} < 1.0$, precipitation of gypsum causes depletion in Ca^{2+} and enrichment in sulfate. Therefore, the value of the molar Ca^{2+}/SO_4^{2-} ratio of a solution from which gypsum precipitates deter-

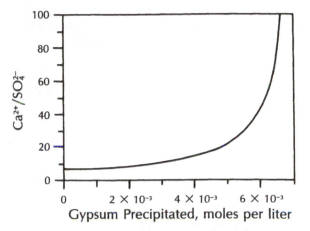

Figure 10.1   Changes in the molar Ca^{2+}/SO_4^{2-} ratio of a solution from which gypsum is precipitating based on equations 10.9 and 10.10. In this example, the solution contains more Ca^{2+} than SO_4^{2-}, and its Ca^{2+}/SO_4^{2-} ratio therefore *increases* as equal amounts of Ca^{2+} and SO_4^{2-} are removed from it. As a result, the solution becomes enriched in Ca^{2+} and depleted in SO_4^{2-}. If the solution initially contains more SO_4^{2-} than Ca^{2+}, it becomes enriched in SO_4^{2-} and depleted in Ca^{2+} as a result of progressive gypsum precipitation. This phenomenon constitutes a *geochemical divide* that affects the chemical evolution of brines by progressive evaporative concentration.

mines whether the solution becomes enriched in Ca^{2+} or SO_4^{2-}. Similar geochemical divides operate when other salts precipitate from isolated volumes of water undergoing evaporative concentration in a desert lake (Hardie and Eugster, 1970).

Natural water samples are much more complex than the solution we have been discussing because they contain a mixture of cations and anions and therefore may precipitate several different salts. For example, a solution may be saturated with respect to both gypsum and barium sulfate (barite). In this case we have two simultaneous equilibria with a common ion:

$$CaSO_4 \cdot 2H_2O \rightleftharpoons Ca^{2+} + SO_4^{2-} + 2H_2O \quad (10.13)$$

$$BaSO_4 \rightleftharpoons Ba^{2+} + SO_4^{2-} \quad (10.14)$$

Both reactions contribute SO_4^{2-} ions to the solution, but the resulting activities must be the same for both equilibria. We can calculate the $[Ca^{2+}]/[Ba^{2+}]$ ratio of a solution that is saturated with respect to both gypsum and barite by applying the Law of Mass Action to equations 10.13 and 10.14:

$$[Ca^{2+}][SO_4^{2-}] = 10^{-4.6} \quad (10.15)$$

$$[Ba^{2+}][SO_4^{2-}] = 10^{-10.0} \quad (10.16)$$

Since $[SO_4^{2-}]$ has the *same* value in both equilibria, we eliminate it by substituting $[SO_4^{2-}] = 10^{-4.6}/[Ca^{2+}]$ into equation 10.16:

$$\frac{[Ba^{2+}] \times 10^{-4.6}}{[Ca^{2+}]} = 10^{-10.0} \quad (10.17)$$

from which it follows that $[Ba^{2+}]/[Ca^{2+}] = 10^{-5.4}$ or $[Ca^{2+}]/[Ba^{2+}] = 2.5 \times 10^5$. Evidently, the activity of Ca^{2+} is 250,000 times that of Ba^{2+} when the solution is saturated with respect to both gypsum and barite.

We can determine the $[SO_4^{2-}]$ concentration of such a solution from the requirement for electrical neutrality:

$$2(Ba^{2+}) + 2(Ca^{2+}) + (H^+) = 2(SO_4^{2-}) + (OH^-) \quad (10.18)$$

From equations 10.15 and 10.16 we have, in terms of concentrations:

$$(Ca^{2+}) = \frac{10^{-4.6}}{\gamma_{\pm 2}[SO_4^{2-}]} \quad (10.19)$$

and:

$$(Ba^{2+}) = \frac{10^{-10.0}}{\gamma_{\pm 2}[SO_4^{2-}]} \quad (10.20)$$

where $\gamma_{\pm 2}$ is the activity coefficient of Ca^{2+} and Ba^{2+} defined by equation 9.12. By substituting into equation 10.18, setting $\gamma_{\pm 2} = 1.0$, and dropping (H^+) and (OH^-), we obtain:

$$\frac{2 \times 10^{-4.6}}{[SO_4^{2-}]} + \frac{2 \times 10^{-10.0}}{[SO_4^{2-}]} = 2(SO_4^{2-}) \quad (10.21)$$

If we stipulate that $[SO_4^{2-}] = (SO_4^{2-})$, equation 10.21 reduces to:

$$10^{-4.6} + 10^{-10.0} = (SO_4^{2-})^2 \quad (10.22)$$

Evidently, the barite contributes a negligible amount of sulfate to the solution and $[SO_4^{2-}] = (SO_4^{2-}) = 10^{-2.3}$ mol/L.

The results of this calculation indicate that gypsum is able to force barite to precipitate when a saturated solution of barite comes in contact with gypsum. In other words, barite can replace gypsum or anhydrite because barite is less soluble than gypsum or anhydrite. The calcium sulfate can dissolve in a solution that is already saturated with respect to barite and thereby it can increase the sulfate concentration. As a result, the solution becomes *supersaturated* with respect to barite. Therefore, barite precipitates as gypsum or anhydrite dissolves until the $[Ca^{2+}]/[Ba^{2+}]$ ratio of the solution approaches 250,000 which is required for equilibrium. When that value is reached, the replacement of gypsum or anhydrite by barite stops.

Replacement of one mineral by another is a common phenomenon in geology and affects not only sulfates but also sulfide minerals, carbonates, and other types of compounds. In each case, the introduction of a common ion causes the solution to become supersaturated with respect to the *less soluble* compound. The effective ion may be the anion, as we have seen, or the cation itself. For example, if $BaCl_2$, which is quite soluble, is added to a saturated solution of $BaSO_4$, the sulfate precipitates until most of it is removed from the solution. In this case barite replaces $BaCl_2$. The process always involves the replacement of the *more* soluble compound by the *less* soluble one.

Solutions in nature may become supersaturated with respect to a specific compound in several different ways, including (1) the introduction of a common ion, as just described; (2) a change in the pH, exemplified in Chapter 9 by the precipitation of amorphous silica; (3) evaporative concentration of water in closed basins and in soil or sediment drying in the sun; and (4) temperature variations.

The solubility of compounds does depend on the temperature, but not always as we might expect. In general, solubilities *increase* with increasing temperature, but for some compounds the opposite is true. For example, the solubility of $CaCO_3$ and other carbonates *decreases* with increasing temperature because the solubility product constants of *calcite* and *aragonite* actually *decrease* with increasing temperature. We will take up the temperature dependence of the solubilities of minerals in Chapter 11 after we learn to calculate equilibrium constants from thermodynamic data.

10.2   Hydrolysis

Weak acids are only partly dissociated in water because the anion of the acid has a strong affinity for hydrogen ions. When a salt of a *weak acid* is dissolved in water, it releases anions into solution that have the same strong affinity for hydrogen ions. Therefore, these anions immediately bond to hydrogen ions and thereby form the parental acid. The hydrogen ions that bond with anions may already exist in the environment, or they may originate from the dissociation of water. In either case, when the salt of a weak acid is dissolved in water, the removal of H^+ causes the solution to become more basic.

The process we have just described is *hydrolysis*. It is defined as the interaction between water and one or both ions of a salt that results in the formation of the parental acid or base, or both. We can classify salts by the strength of the acid and base from which they form:

1. Strong acid + strong base.
2. Strong acid + weak base.
3. Weak acid + strong base.
4. Weak acid + weak base.

Salts derived from strong acids and strong bases (category 1), such as NaCl or $CaSO_4$, do not hydrolyze because neither the anion nor the cation has a particular affinity for H^+ or for OH^-. Salts of strong acids and weak bases (category 2), such as $FeCl_3$ or $CuSO_4$, release cations into solution that combine with OH^- to form the parental base. The removal of OH^- causes the solution to become more *acidic*. Salts of weak acids and strong bases (category 3) release anions that hydrolyze and make the solution more *basic*. Salts of weak acids and weak bases (category 4), like $CuCO_3$ or FeS_2, release anions and cations both of which hydrolyze. In this case, the effect on the pH of a neutral solution must be determined by calculation because it depends on a comparison of the strengths of the parental acid and base.

Hydrolysis is a very important process because most of the common rock-forming minerals of the crust of the Earth are salts of *weak acids* and *strong bases*. For example, the carbonates and silicates of the alkali metals and alkaline earths are salts of this type. Therefore, the anions of these minerals split water molecules in order to form the weak parental acid and thereby cause the hydroxyl ion concentration of the solution to increase. This is why groundwater in carbonate aquifers is commonly basic. Even silicate minerals, such as the feldspars, hydrolyze when they interact with water by releasing Na^+ or K^+ from the surfaces of mineral grains and replacing it with H^+.

The dissociation of salts into ions and subsequent hydrolysis of the ions takes place very rapidly. As a result, these kinds of reactions are usually at equilibrium. Therefore, if we dissolve a small quantity of potassium carbonate in pure water at $25°C$, it will dissolve completely and dissociate into ions:

$$K_2CO_3 \rightarrow 2K^+ + CO_3^{2-} \qquad (10.23)$$

The carbonate ions react with water to form bicarbonate:

$$CO_3^{2-} + H_2O \rightleftharpoons HCO_3^- + OH^- \qquad K_{H1} \quad (10.24)$$

The bicarbonate ion (HCO_3^-) is itself an acid because it can release H^+ into solution. However, it also hydrolyzes:

$$HCO_3^- + H_2O \rightleftharpoons H_2CO_3 + OH^- \quad K_{H2} \quad (10.25)$$

By applying the Law of Mass Action to reaction 10.24 we have:

$$\frac{[HCO_3^-][OH^-]}{[CO_3^{2-}]} = K_{H1} \quad (10.26)$$

and from equation 10.25:

$$\frac{[H_2CO_3][OH^-]}{[HCO_3^-]} = K_{H2} \quad (10.27)$$

To find the value of the first hydrolysis constant (K_{H1}) we replace $[OH^-]$ in equation 10.26 by:

$$[OH^-] = \frac{K_w}{[H^+]} \quad (10.28)$$

where K_w is the dissociation constant of water. Therefore, from equation 10.26:

$$\frac{[HCO_3^-]K_w}{[CO_3^{2-}][H^+]} = K_{H1} \quad (10.29)$$

Since the dissociation of carbonic acid proceeds stepwise:

$$H_2CO_3 \rightleftharpoons HCO_3^- + H^+ \quad K_{A1} \quad (10.30)$$

$$HCO_3^- \rightleftharpoons CO_3^{2-} + H^+ \quad K_{A2} \quad (10.31)$$

we see that in equation 10.29:

$$\frac{[HCO_3^-]}{[CO_3^{2-}][H^+]} = \frac{1}{K_{A2}} \quad (10.32)$$

By substituting equation 10.32 into equation 10.29 we obtain:

$$\frac{K_w}{K_{A2}} = K_{H1} \quad (10.33)$$

The same procedure yields:

$$\frac{K_w}{K_{A1}} = K_{H2} \quad (10.34)$$

from equation 10.27.

We are now able to calculate the pH of a solution that was prepared by dissolving 0.1 mol of

K_2CO_3 in one liter of pure water at 25 °C. In addition, we specify that CO_2 gas does not escape from the solution. Since $K_{A1} = 10^{-6.35}$ and $K_{A2} = 10^{-10.3}$ (Table 9.3) and $K_w = 10^{-14}$, we find that $K_{H1} = 10^{-3.7}$ and $K_{H2} = 10^{-7.65}$. Since K_{H1} is nearly 9000 times larger than K_{H2}, the first reaction is much more productive than the second. We therefore attempt a stepwise solution. If x moles of the carbonate ion hydrolyze in the first step, we have at equilibrium:

$$[CO_3^{2-}] = 0.1 - x$$

$$[HCO_3^-] = [OH^-] = x$$

Substituting into equation 10.26 yields:

$$\frac{x^2}{0.1 - x} = 10^{-3.7} \quad (10.35)$$

If $x \ll 0.1$, $0.1 - x \approx 0.1$ and, from equation 10.35, $x = 10^{-2.35}$, which is only 4.5% of 0.1. Therefore, the approximation that $x \ll 0.1$ is justified. The second hydrolysis step produces a negligible amount of additional hydroxyl ion equal to $[OH^-] = 10^{-7.65}$ mol/L. The pH of a solution obtained by dissolving 0.1 mole of K_2CO_3 in one liter of pure water at $T = 25$ °C is:

$$pH = 14 - 2.35 = 11.65 \quad (10.36)$$

Clearly, the solution is quite basic, as predicted.

The problem can also be solved by means of a set of independent equations that arise from the simultaneous equilibria in the solution, from the requirement of electrical neutrality, and from the stipulation that the total amount of molecular and ionic carbonate species in the solution is 0.1 mol. We could have set ourselves a more general problem by allowing the solution to be open to the atmosphere, permitting the carbonate species in the solution to equilibrate with the CO_2 of the atmosphere. We choose not to do so here because we will treat carbonate equilibria in Section 10.4.

Salts derived from strong acids and *weak bases* also hydrolyze, and the necessary hydrolysis constants can be derived as demonstrated above. For example, $Cu(OH)_2$ is a weak base that dissociates in two steps:

$$Cu(OH)_2(s) \rightleftharpoons Cu(OH)^+ + OH^- \quad K_{B1} \quad (10.37)$$

$$Cu(OH)^+(aq) \rightleftharpoons Cu^{2+} + OH^- \quad K_{B2} \quad (10.38)$$

$Cu(OH)_2$ forms the salt $CuSO_4$ by reaction with H_2SO_4. When $CuSO_4$ is dissolved in water, Cu^{2+} is released and hydrolyzes, as shown:

$$Cu^{2+} + H_2O \rightleftharpoons Cu(OH)^+ + H^+ \quad K_{H1} \quad (10.39)$$

$$Cu(OH)^+ + H_2O \rightleftharpoons Cu(OH)_2(s) + H^+ \quad K_{H2}$$
$$(10.40)$$

where $K_{H1} = K_w/K_{B2}$ and $K_{H2} = K_w/K_{B1}$. Therefore, we predict that a solution of $CuSO_4$ in pure water becomes acidic because of the hydrolysis of the cupric ion (Cu^{2+}).

Hydrolysis is an important phenomenon because it enhances the solubility of salts formed from weak acids and bases and because it tends to stabilize the pH of salt solutions. Hydrolysis affects the solubility of salts by producing additional ionic and molecular species that permit more of the salt to dissolve. In fact, the solubility of salts derived from weak acids is greatly enhanced when they are dissolved in strong acids that provide an ample supply of hydrogen ions. Therefore, all carbonates, silicates, phosphates, and sulfides are more soluble in acids than in pure water because of the affinity of their anions for hydrogen ions. The same is true in principle of salts derived from weak bases, except that the bases they form may themselves be quite insoluble. For example, if we place ferric chloride into a basic solution, the Fe^{3+} ion reacts with OH^- to form insoluble $Fe(OH)_3$:

$$FeCl_3 \rightarrow Fe^{3+} + 3 Cl^- \quad (10.41)$$

$$Fe^{3+} + 3 OH^- \rightarrow Fe(OH)_3(s) \quad (10.42)$$

Solutions that contain a weak acid and a salt of that acid are *buffered* because they resist changes in their pH. Consider, for example, a solution containing carbonic acid and K_2CO_3. The carbonic acid is only partially dissociated, as indicated by equations 10.30 and 10.31. By adding K_2CO_3, which releases additional CO_3^{2-}, the dissociation of the acid is even more inhibited and the concentration of molecular H_2CO_3

increases in accordance with Le Châtelier's principle. Therefore, the solution has effectively *stored* hydrogen ions, and they are released when a reaction occurs in this solution that consumes H^+. Similarly, if additional hydrogen ions were to be released into the solution, they would be consumed by reacting with the carbonate and bicarbonate ions to form more carbonic acid.

10.3 Activities and Concentrations

We have assumed so far that the ions in electrolyte solutions do not interfere with each other. This can only be true in very dilute solutions because, in general, oppositely charged ions in a solution attract each other. Therefore, cations in an electrolyte solution are surrounded by anions and vice versa. The result is that the ions of a particular compound are less able to interact with each other than expected from their concentrations. We express this condition by saying that the *activity* of the ions in electrolyte solutions is *less* than their *concentration*. It is plausible to expect that the interference by the other ions increases with their concentrations and charges. Similarly, the effect of a given suite of ions on a particular ion depends on the charge of that ion and its radius, which together determine the charge density on the surface of the ion. In addition, the molecules of the solvent play a role because, if they have electrical polarity, they also interact with both anions and cations. All of these factors are included in the *Debye–Hückel theory,* on the basis on which the relationship between activity and concentration of ions in a solution of electrolytes can be estimated.

The concentrations and charges of ions in a solution are expressed by means of its *ionic strength,* defined as:

$$I = \tfrac{1}{2}\Sigma\, m_i z_i^2 \quad (10.43)$$

where m_i are the concentrations expressed in moles and z_i are the charges of the ions. The ionic strength of aqueous solutions on the surface of the Earth generally ranges from about 1×10^{-3} in rivers and lakes to 1×10^{-1} in old groundwater. Seawater has an ionic strength of 7×10^{-1} and oilfield brines and brine lakes may have ionic strengths of 5 or more. In order to calculate the ionic strength of a water sample, we must have a complete chemical analysis, such as the analysis of water from the Mississippi River at New Orleans in Table 10.2. The concentrations of the ions are converted from parts per million (ppm) to moles per kilogram before they are entered into equation 10.43. The result is $I = 4.4 \times 10^{-3}$. Note that SiO_2 does not contribute to the ionic strength of this water because its charge is zero and that H^+ and OH^- are omitted because their concentrations are very low (about 10^{-7} mol/kg) unless the water is highly acidic or basic. Note also that the concentrations of H_2CO_3 and CO_3^{2-} were not reported in the analysis in Table 10.2 because HCO_3^{2-} is the dominant ion when the pH is between 6.35 and 10.3.

The ionic strength is used in the Debye–Hückel theory to estimate values of the activity coefficient γ defined by the equation:

$$[\text{activity}] = \gamma(\text{concentration}) \quad (10.44)$$

For dilute solutions having ionic strengths $< 5 \times 10^{-3}$ the relationship is, to good approximation:

$$-\log \gamma = Az^2 I^{1/2} \quad (10.45)$$

where $A = 0.5085$ for water at $25°C$, z is the charge of the ion whose activity coefficient is being calculated, and I is the ionic strength of the solution as defined by equation 10.43. Note that the activity coefficient of molecules is equal to one because their charge is equal to zero.

A more complete statement, valid for $I < 0.1$, is given by:

$$-\log \gamma = \frac{Az^2 I^{1/2}}{1 + aBI^{1/2}} \quad (10.46)$$

where A and B are constants whose values depend on the dielectric constant of the solvent and the temperature, and a is the effective diameter of the ion in the solution in Ångstrom units. Values of A and B for water are listed in Table 10.3 as a function of temperature, and values of a of several ions in aqueous solution are compiled in Table 10.4. Equation 10.46 was used to calculate the activity

Table 10.2 Chemical Composition of Water of the Mississippi River at New Orleans

Ion	Concentration	
	ppm	*mol/kg*
HCO_3^-	116	1.90×10^{-3}
SO_4^{2-}	25.5	2.65×10^{-4}
Cl^-	10.3	2.90×10^{-4}
NO_3^-	2.7	4.3×10^{-5}
Ca^{2+}	34	8.5×10^{-4}
Mg^{2+}	8.9	3.7×10^{-4}
Na^+	11.9	5.17×10^{-4}
K^+	1.9[a]	4.9×10^{-5}
Fe^{2+}	0.14	2.5×10^{-6}
SiO_2	11.7	1.95×10^{-4}

[a]Based on $Na^+/K^+ = 6.4$ (concentration ratio) in average North American river water (Livingstone, 1963, Table 81, p. G41).
SOURCE: Livingstone (1963), Table 15, entry H, p. G15.

Table 10.3 List of Values for Constants Used in the Debye–Hückel Theory Expressed in Equation 10.46

Temperature, °C	A	B
0	0.4883	0.3241
5	0.4921	0.3249
10	0.4960	0.3258
15	0.5000	0.3262
20	0.5042	0.3273
25	0.5085	0.3281
30	0.5130	0.3290
35	0.5175	0.3297
40	0.5221	0.3305
45	0.5271	0.3314
50	0.5319	0.3321
55	0.5371	0.3329
60	0.5425	0.3338

SOURCE: Garrels and Christ (1965).

Table 10.4 Values of Parameter a of Ions in Aqueous Solution

Ions	$a, \text{Å}$
$Rb^+, Cs^+, NH_4^+, Tl^+, Ag^+$	2.5
$K^+, Cl^-, Br^-, I^-, NO_3^-$	3
$OH^-, F^-, HS^-, BrO_3^-, IO_4^-, MnO_4^-$	3.5
$Na^+, HCO_3^-, H_2PO_4^-, HSO_3^-, Hg_2^{2+}, SO_4^{2-}, SeO_4^{2-}, CrO_4^{2-}, HPO_4^{2-}, PO_4^{3-}$	4.0–4.5
$Sr^{2+}, Ba^{2+}, Ra^{2+}, Cd^{2+}, Hg^{2+}, S^{2-}, WO_4^{2-}$	5.0
$Li^+, Ca^{2+}, Cu^{2+}, Zn^{2+}, Sn^{2+}, Mn^{2+}, Fe^{2+}, Ni^{2+}, Co^{2+}$	6
Mg^{2+}, Be^{2+}	8
$H^+, Al^{3+}, Cr^{3+}, REE^{3+}$	9
$Th^{4+}, Zr^{4+}, Ce^{4+}, Sn^{4+}$	11

SOURCE: Garrels and Christ (1965).

coefficients for ions of different charges and hydrated radii shown in Figure 10.2. A third equation developed by Davies yields reliable results up to $I = 0.5$:

$$-\log \gamma = Az^2 \left[\frac{I^{1/2}}{1 + I^{1/2}} - 0.2I \right] \quad (10.47)$$

Further extensions of these methods by Pitzer (1973) enabled Harvie and Weare (1980) to calculate the solubilities of minerals in brines having ionic strengths greater than 20.

Activity coefficients for ions of different charges in solutions of varying ionic strengths, as determined from the Davies equation (10.47), are listed in Table 10.5. Note that the values of the activity coefficients in Table 10.5 *decrease* with increasing ionic strength of the solution and with increasing charge of the ion. For example, we find by interpolating in Table 10.5 that the activity of Na^+ in water of the Mississippi River ($I = 4.4 \times 10^{-3}$) is about 0.936, whereas that of Ca^{2+} is 0.759, and that of La^{3+} is only 0.543. In other words, the presence of other ions reduces the effective concentration of Na^+ in the Mississippi River by 6.4%, that of Ca^{2+} by 24.1%, and that of La^{3+} by 45.7%. In general, activity coefficients *decrease* with increasing ionic strength, but *increase* again at high values of I and may become larger than one (Garrels and Christ, 1965). The increase of the activity coefficient in highly saline

electrolyte solutions can be attributed to "crowding" of the ions, which counteracts the "interference" that occurs at lower ionic strengths.

The interference of other ions in the solution with the ions of a salt causes the solubility of the salt to increase. For example, the solubility of $CaSO_4$ in pure water at 25°C ($\gamma = 1.0$, $K_{sp} = 10^{-4.5}$) is $10^{-2.25}$ or 5.6×10^{-3} mol/L. However, in water of the Mississippi River ($I = 4.4 \times 10^{-3}$, $\gamma_{\pm 2} = 0.759$) the *concentration* of Ca^{2+} in equilibrium with anhydrite is:

$$(Ca^{2+}) = \frac{[Ca^{2+}]}{\gamma_{\pm 2}} = \frac{10^{-2.25}}{0.759} = 7.4 \times 10^{-3} \text{ mol/L} \quad (10.48)$$

which implies an *increase* of the solubility of anhydrite of 32.1%.

This result illustrates the point that the Law of Mass Action yields inaccurate results *unless* the concentrations of reactants and products of a reaction at equilibrium are converted to activities by means of activity coefficients in order to correct them for the nonideality of electrolyte solutions. Conversely, the results of calculations based on the Law of Mass Action are expressed in terms of *activities* and must be converted to concentrations before they are interpreted. The solubility calculation of anhydrite in Mississippi River water indicates that the presence of other ions in a solution must not be ignored.

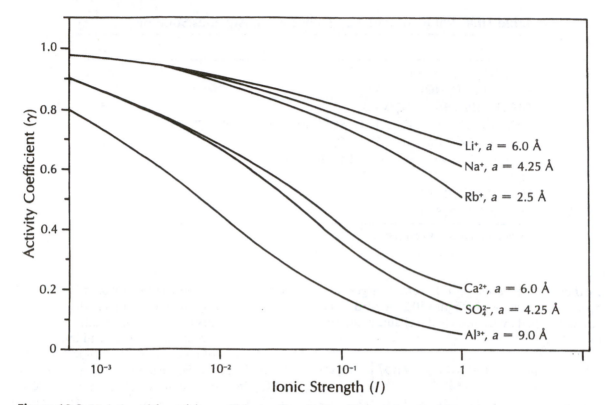

Figure 10.2 Variation of the activity coefficients of ions having different charges and hydrated radii with the ionic strength of aqueous solutions at 25°C. The curves were calculated from the Debye–Hückel theory expressed by equation 10.46 with data from Tables 10.3 and 10.4. Note that the magnitude of hydrated radii of ions of equal charge becomes important at $I > 10^{-2}$.

Table 10.5 Activity Coefficients for Ions of Different Charges in Electrolyte Solutions of Varying Ionic Strengths Based on Equation 10.47 by Davies

I	z		
	± 1	± 2	± 3
1×10^{-3}	0.97	0.87	0.73
5×10^{-3}	0.93	0.74	0.51
1×10^{-2}	0.90	0.66	0.40
5×10^{-2}	0.82	0.45	0.16
1×10^{-1}	0.78	0.36	0.10
2×10^{-1}	0.73	0.28	0.06
5×10^{-1}	0.69	0.23	0.04

SOURCE: Krauskopf (1979).

10.4   Solubility of Calcium Carbonate

We are now ready to calculate the solubility of calcite, which is one of the most common minerals on the face of the Earth. The calculation requires all of the skills we have developed in applying the Law of Mass Action and is by no means trivial (Butler, 1982; Garrels and Christ, 1965). We start with the dissociation of calcite into ions at 25°C and the subsequent hydrolysis of the carbonate and bicarbonate ions to form carbonic acid:

$$CaCO_3 \rightleftharpoons Ca^{2+} + CO_3^{2-} \qquad K_{sp} = 10^{-8.35} \quad (10.49)$$

$$CO_3^{2-} + H_2O \rightleftharpoons HCO_3^- + OH^- \qquad K_{H1} \quad (10.50)$$

$$HCO_3^- + H_2O \rightleftharpoons H_2CO_3 + OH- \qquad K_{H2} \quad (10.51)$$

If the solution is open to the atmosphere, carbonic acid is in equilibrium with carbon dioxide, which has a partial pressure of 3×10^{-4} atmosphere:

$$H_2CO_3 \rightleftharpoons CO_2(g) + H_2O \quad K = 32.2 \quad (10.52)$$

When calcite crystals are in equilibrium with their ions in pure water open to the atmosphere at 25°C, seven ions and molecules are present:

$$Ca^{2+}, CO_3^{2-}, HCO_3^-, H_2CO_3, CO_2, H^+, OH^-$$

However, the partial pressure of CO_2 is fixed by the composition of the atmosphere, although soil gases may contain up to ten times as much CO_2 as the open atmosphere. Therefore, we need only six independent equations in order to specify the concentrations of all ions and molecules in a saturated solution of calcite. We obtain five equations by applying the Law of Mass Action to equations 10.49 through 10.52 and to the dissociation equilibrium of water. From equation 10.49:

$$[Ca^{2+}][CO_3^{2-}] = 10^{-8.35} \quad (10.53)$$

From equations 10.50 and 10.51:

$$\frac{[HCO_3^-][OH^-]}{[CO_3^{2-}]} = K_{H1} = \frac{10^{-14.0}}{10^{-10.3}} = 10^{-3.7} \quad (10.54)$$

$$\frac{[H_2CO_3][OH^-]}{[HCO_3^-]} = K_{H2} = \frac{10^{-14.0}}{10^{-6.35}} = 10^{-7.65} \quad (10.55)$$

Finally, from equation 10.52:

$$\frac{[CO_2]}{[H_2CO_3]} = 32.2 \quad (10.56)$$

and for water:

$$[H^+][OH^-] = 10^{-14.0} \quad (10.57)$$

The sixth equation arises from the requirement for electrical neutrality:

$$2(Ca^{2+}) + (H^+) = 2(CO_3^{2-}) + (HCO_3^-) + OH^-) \quad (10.58)$$

We notice that according to equation 10.56:

$$[H_2CO_3] = \frac{[CO_2]}{32.2} = \frac{3 \times 10^{-4}}{32.2} \quad (10.59)$$
$$= 9.3 \times 10^{-6} \, mol/L$$

Evidently, the activity of carbonic acid is fixed by the partial pressure of CO_2. We can use this result in equation 10.55 to express $[HCO_3^-]$ as a function of $[OH^-]$:

$$[HCO_3^-] = \frac{[H_2CO_3][OH^-]}{10^{-7.65}}$$
$$= \frac{(9.3 \times 10^{-6})[OH^-]}{10^{-7.65}} \quad (10.60)$$
$$= 10^{2.618}[OH^-]$$

From equation 10.54 we derive a relationship between $[CO_3^{2-}]$ and $[OH^-]$:

$$[CO_3^{2-}] = \frac{[HCO_3^-][OH^-]}{10^{-3.7}} = \frac{10^{2.618}[OH^-]^2}{10^{-3.7}} \quad (10.61)$$
$$= 10^{6.318}[OH^-]^2$$

From equations 10.61 and 10.53 we obtain:

$$[Ca^{2+}] = \frac{10^{-8.35}}{[CO_3^{2-}]} = \frac{10^{-8.35}}{10^{6.318}[OH^-]^2}$$
$$= \frac{10^{-14.668}}{[OH^-]^2} \quad (10.62)$$

Finally, from equation 10.57:

$$[H^+] = \frac{10^{-14.0}}{[OH^-]} \quad (10.63)$$

We now convert the activities of the ions to their corresponding concentrations for substitution into equation 10.58. In order to do so we must know the ionic strength of the solution, which depends on the concentrations of the ions it contains. Strictly speaking, we cannot determine the ionic strength without first completing the calculation. However, we cannot complete the calculation without knowing the ionic strength. We deal with this Catch-22 situation by *iteration*. That is, we assume that all activity coefficients are equal to one, make the calculation, and then use the results to determine the ionic strength and hence the activity coefficients from the Davies equation (10.47) or from Table 10.5. We then return to equation 10.58, convert the activities of all ions to the appropriate concentrations, and solve the problem a second time. If necessary, the process can be repeated several times to achieve a desired level of accuracy.

Therefore, we now proceed with the solution of the problem by substituting equations 10.60, 10.61, 10.62, and 10.63 into equation 10.58, assuming that all activity coefficients are equal to one:

$$\frac{2 \times 10^{-14.668}}{[OH^-]^2} + \frac{10^{-14.0}}{[OH^-]} = (2 \times 10^{6.318})[OH^-]^2$$
$$+ 10^{2.618}[OH^-] + [OH^-] \tag{10.64}$$

We simplify equation 10.64 by eliminating the term for (H^+) because the solution will be basic and by combining $10^{2.618}[OH^-] + [OH^-]$ as $10^{2.619}[OH^-]$. Next, we clear the fractions by multiplying all remaining terms by $[OH^-]^2$ and collect terms:

$$10^{6.619}[OH^-]^4 + 10^{2.619}[OH^-]^3 = 10^{-14.366} \tag{10.65}$$

Dividing each term by $10^{6.619}$ yields:

$$[OH^-]^4 + 10^{-4.000}[OH^-]^3 = 10^{-20.985} \tag{10.66}$$

We solve this equation by trial and error starting with $[OH^-] = 10^{-6.0}$ because we predict that the solution will be basic. Substituting into equation 10.66 yields $10^{-24} + 10^{-22}$, which is about 10 times smaller than $10^{-20.985}$. Therefore, $[OH^-]$ must be larger than $10^{-6.0}$. We note that $10^{-24} \ll 10^{-22}$ and therefore eliminate the first term from equation 10.66 to get:

$$10^{-4.000}[OH^-]^3 = 10^{-20.985} \tag{10.67}$$

which yields $[OH^-] = 10^{-5.66}$ mol/L. By substituting this value into equation 10.66 we find that the term we omitted is only 2.2% of $10^{-4.000}[OH-]^3$, which is acceptable.

We can now calculate the activities of the other ions in the solution by substituting $[OH^-] = 10^{-5.66}$ into equations 10.60–10.63. The results are:

$$[HCO_3^-] = 10^{2.618}[OH^-] = 10^{2.618} \times 10^{-5.66}$$
$$= 10^{-3.042} = 9.08 \times 10^{-4} \text{ mol/L}$$

$$[CO_3^{2-}] = 10^{6.318}[OH^-]^2 = 10^{6.318} \times 10^{-11.32}$$
$$= 10^{-5.002} = 9.95 \times 10^{-6} \text{ mol/L}$$

$$[Ca^{2+}] = \frac{10^{-14.668}}{[OH^-]^2} = \frac{10^{-14.668}}{10^{-11.32}} = 10^{-3.348}$$
$$= 4.48 \times 10^{-4} \text{ mol/L}$$

$$[H^+] = \frac{10^{-14.0}}{[OH^-]} = \frac{10^{-14.0}}{10^{-5.66}} = 10^{-8.34} \text{ mol/L}$$

We check the correctness of these results by calculating the IAP of calcite in this solution:

$$IAP = (9.95 \times 10^{-6}) \times (4.48 \times 10^{-4}) = 10^{-8.35} \tag{10.68}$$

which is identical to K_{sp} for calcite and therefore confirms the correctness of the calculation. The ionic strength of the solution is $I = 1.4 \times 10^{-3}$, and the activity coefficients of the ions from Table 10.5 are $\gamma_{\pm 1} = 0.966$ and $\gamma_{\pm 2} = 0.857$.

Returning to equations 10.60–10.63, we now convert the activities to concentrations:

$$(HCO_3^-) = \frac{10^{2.618}[OH^-]}{\gamma_{\pm 1}} = \frac{10^{2.618}[OH^-]}{0.966}$$
$$= 10^{2.633}[OH^-]$$

$$(CO_3^{2-}) = \frac{10^{6.318}[OH^-]^2}{\gamma_{\pm 2}} = \frac{10^{6.318}[OH^-]^2}{0.857}$$
$$= 10^{6.385}[OH^-]^2$$

$$(Ca^{2+}) = \frac{10^{-14.668}}{\gamma_{\pm 2}[OH^-]^2} = \frac{10^{-14.668}}{0.857[OH^-]^2}$$
$$= \frac{10^{-14.600}}{[OH^-]^2}$$

Substituting into equation 10.58, dropping (H^+), and combining the $[OH^-]$ terms as before, we obtain:

$$\frac{2 \times 10^{-14.600}}{[OH^-]^2} = (2 \times 10^{6.385})[OH^-]^2$$
$$+ 10^{2.634}[OH^-] \qquad (10.69)$$

Note that this equation is stated in terms of the *activities* of OH^-, which express the *concentrations* of the other ions. By solving equation 10.69 as before, we obtain $[OH^-] = 10^{-5.644}$ mol/L. This value yields for the other ions:

$$[HCO_3^-] = 9.42 \times 10^{-4} \text{ mol/L}$$

$$[CO_3^{2-}] = 1.07 \times 10^{-5} \text{ mol/L}$$

$$[Ca^{2+}] = 4.17 \times 10^{-4} \text{ mol/L}$$

$$[H^+] = 10^{-8.356} \quad \text{or} \quad pH = 8.36$$

The ionic strength, based on these results, is $I = 1.3 \times 10^{-3}$, which does not differ enough from the previous estimate to justify a repetition of the calculation. Therefore, the concentration of Ca^{2+} in the solution is:

$$(Ca^{2+}) = \frac{4.17 \times 10^{-4}}{0.857} = 4.86 \times 10^{-4} \text{ mol/L}$$

or 19.5 mg/L. The solubility (S) of calcite in water in equilibrium with CO_2 of the atmosphere at $25\,^\circ$C is:

$$S = 4.86 \times 10^{-4} \times 100.0787 = 4.86 \times 10^{-2} \text{ g/L}$$

In addition, note that the pH of this solution is 8.36, thus confirming our prediction based on the effect of hydrolysis of the carbonate ion of calcite.

The chemical analysis in Table 10.2 indicates that the water in the Mississippi River has a Ca concentration of 34 ppm, which is nearly 75% *larger* than the value we obtained above. The difference is partly due to the higher ionic strength of water in the Mississippi River $(I = 4.4 \times 10^{-3})$ and to the correspondingly lower activity coefficients $(\gamma_{\pm 1} = 0.936, \quad \gamma_{\pm 2} = 0.759)$. However, allowance for the higher ionic strength increases

the predicted Ca concentration in the Mississippi River only to 22.0 mg/L and therefore does not resolve the discrepancy.

The water in the Mississippi River, like all natural solutions, contains a mixture of ions and molecules derived from different sources. Its Ca^{2+} content originates from the solution of calcium carbonate (calcite and aragonite), calcium sulfate (anhydrite and gypsum), and calcium phosphate (apatite) and from the transformation of plagioclase to kaolinite or other clay minerals. In addition, Ca^{2+} may be adsorbed on clay minerals and on the surfaces of other small mineral particles in suspension in the water. Finally, calcium occurs in many complex ions and molecules whose presence permits the calcium concentration of natural waters to exceed the concentration of Ca^{2+} based on the solubility of its principal minerals. A partial list of the complex ions of Ca includes $CaHCO_3^+$, $CaCO_3^0$, $CaCl^+$, $CaCl_2^0$, $CaNO_3^+$, $Ca(NO_3)_2^0$, $CaOH^+$, $Ca(OH)_2^0$, $CaPO_4^-$, $CaHPO_4^0$, $CaH_2PO_4^+$, $CaP_2O_7^{2-}$, $CaHP_2O_7^-$, $CaOHP_2O_7^{3-}$, $CaSO_4^0$ (Lindsay, 1979). The point is that natural waters are very complex and their compositions are not explainable by the solubility of a few common minerals. We therefore abandon the effort to explain the concentration of calcium or any other element in the Mississippi River. Instead, we use the discrepancy between the Ca^{2+} concentration of a saturated calcite solution and the calcium concentration in the water of the Mississippi River to consider how the solubility of calcite is affected by variations in the partial pressure CO_2, the pH, and the temperature.

The reaction of calcite with carbonic acid between pH values of 6.35 and 10.3 can be represented by the equation:

$$CaCO_3 + H_2CO_3 \rightleftharpoons Ca^{2+} + 2HCO_3^- \quad (10.70)$$

where carbonic acid is also in equilibrium with aqueous CO_2, which in turn is in equilibrium with CO_2 gas:

$$H_2CO_3 \rightleftharpoons CO_2(aq) + H_2O \quad (10.71)$$
$$CO_2(aq) \rightleftharpoons CO_2(g) \quad (10.72)$$

Note that equation 10.52 is the sum of 10.71 and 10.72 and that molecular CO_2 dissolved in water is thereby eliminated from consideration. We can use these equations to predict *qualitatively* how the solubility of calcite is affected by changes in the partial pressure of CO_2. An *increase* in the partial pressure of CO_2 at constant temperature increases the concentration of carbonic acid in the solution. Consequently, more calcite *dissolves* by equation 10.70 in accordance with Le Châtelier's principle. Similarly, a *decrease* in the partial pressure of CO_2 causes a saturated solution of calcite to become supersaturated and results in the *precipitation* of calcite until equilibrium is restored.

The partial pressure of CO_2 at a site on the surface of the Earth may *decrease* as a result of photosynthesis of aquatic plants, which has the effect of combining CO_2 and H_2O to produce glucose ($C_6H_{12}O_6$) and O_2 with the help of ultraviolet radiation. Consequently, communities of aquatic plants can lower the concentration of $CO_2(aq)$ in the water during daylight hours. This may cause calcite to precipitate if the water becomes supersaturated with respect to calcite. Certain colonial algae are capable of precipitating calcite by this mechanism, thereby forming thinly laminated calcite mounds called *stromatolites* (Walter, 1977).

A similar process in limestone caverns results in the deposition of calcite *speleothems* in the form of *stalactites* and *stalagmites*. Limestone caverns form by the solution of calcite by water containing carbonic acid in accordance with equation 10.70. This process occurs primarily below the water table, and the solution cavities are initially filled with water. However, if the groundwater table is lowered because of uplift of the crust or a decrease in meteoric precipitation, the solution cavities above the water table become filled with air. Meteoric water percolating through the soil may equilibrate with CO_2 at a *higher* partial pressure than exists in the atmosphere and thus becomes a more effective solvent of calcite than it would have been had it equilibrated with CO_2 of the atmosphere. When this water forms a drop on the ceiling of a limestone cavern, the excess CO_2 escapes, the solution therefore may become supersaturated, and calcite may then precipitate

to form a *stalactite* hanging from the roof of the cavern. When the rate of flow of groundwater is so rapid that the solution does not reach equilibrium on the roof of the cavern, additional calcite may precipitate after the water has dripped onto the floor of the cavern, thereby forming a *stalagmite*. Ultimately, stalactites and stalagmites may join to become pillars or curtains of calcite.

The effect of the partial pressure of CO_2 on the solubility of calcite is shown quantitatively in Figure 10.3. Here we see that the concentration of Ca^{2+} in a saturated solution of calcite at 25°C increases from about 20 mg/L to about 45 mg/L when the partial pressure of CO_2 increases from 3×10^{-4} atm in the air to $\times 10^{-3}$ atm in soil gases. Consequently, each liter of water appearing on the roof of a cavern could, under ideal conditions, deposit 25 mg of calcite. The amount of calcite that is actually deposited depends on the partial pressure of CO_2 in the soil above a given cavern, on the temperature, and on the establishment of equilibrium between calcite and the groundwater before the water reaches the cavern. Nevertheless, stalactites and stalagmites can grow

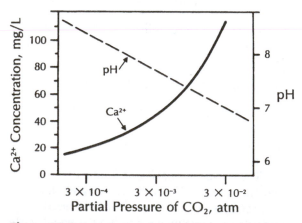

Figure 10.3 Increase of the solubility of calcite in water at 25°C with increasing partial pressure of CO_2 from 3×10^{-4} to 3×10^{-2} atm. The solubility of calcite is expressed as the concentration of Ca^{2+} in a saturated solution based on ionic strengths, which rise from 1.3×10^{-3} to 6.3×10^{-3}, whereas the pH decreases, as more CO_2 dissolves.

to large size over thousands or tens of thousands of years.

The solubility of calcite increases very significantly with increasing acidity of the solution, if the pH is controlled independent of the carbonate equilibria. Equation 10.62 provides a relationship between the activity of Ca^{2+} in a saturated solution of calcite in equilibrium with CO_2 of the atmosphere of 25°C and the activity of OH^-. Replacing $[OH^-]$ by $K_w/[H^+]$ gives:

$$[Ca^{2+}] = \frac{10^{-14.668}}{[OH^-]^2} = \frac{10^{-14.668}[H^+]^2}{(K_w)^2} \quad (10.73)$$
$$= 10^{13.332}[H^+]^2$$

Taking logarithms of both sides and converting $[H^+]$ to pH, we obtain:

$$\log[Ca^{2+}] = 13.332 - 2pH \quad (10.74)$$

This equation indicates that the solubility of calcite changes by a factor of 100 when the pH is varied by one unit. For example, at pH = 7, $[Ca^{2+}] = 2.15 \times 10^{-1}$ mol/L, whereas at pH = 6, $[Ca^{2+}] = 2.15 \times 10^{+1}$ mol/L. Evidently, when calcite reacts with an excess amount of a strong acid, its solubility is very high. Similarly, calcite cannot persist as a stable phase in natural environments that are even mildly acidic. For example, calcite in till of Wisconsin age in the midcontinent area of North America has been removed by leaching to depths of about one meter or more in only 15,000 years.

The solubility of calcite is also affected by the temperature because of changes in the numerical values of all of the equilibrium constants. Garrels and Christ (1965) compiled the set of values of the relevant equilibrium constants between 0 and 50°C that is listed in Table 10.6. We see by inspection that the dissociation constants of carbonic acid *increase* with increasing temperature, whereas the solubility product constants of calcite and CO_2 *decrease*. As a result, the solubility of calcite in pure water in equilibrium with CO_2 of the atmosphere actually *decreases* with increasing temperature, as shown in Figure 10.4. For example, a saturated solution of calcite in equilibrium with CO_2 at 3×10^{-3} atm contains about 75 mg/L

of Ca^{2+} at 5°C but only 40 mg/L at 30°C. Such changes in the environmental conditions may occur when cold water from the bottom of the oceans rises to the surface and is warmed in a shallow lagoon. As a result, seawater may become supersaturated with respect to calcite (or aragonite) because of the *increase* in temperature as well as the *decrease* in the partial pressure of CO_2. However, carbonate sediment in the oceans is primarily of biogenic origin and results from the secretion of carbonate skeletons by plants and animals living in the water.

The opposite environmental changes are experienced by skeletal carbonate grains sinking through the ocean toward the bottom. Oceanic bottom water is colder than surface water and contains more dissolved CO_2 produced by decay of organic material. Consequently, the solubility of calcite (or aragonite) rises with increasing depth, and skeletal carbonate particles dissolve as they sink in the oceans. Whether such particles survive the trip through the water column and reach the bottom of the oceans depends on the

Table 10.6 Equilibrium Constants of Carbonate Equilibria as a Function of Temperature[a]

Temperature, °C	pK_{A1}	pK_{A2}	pK_{sp}	pK_{CO_2}
0	6.58	10.62	8.02	1.12
5	6.52	10.56	8.09	(1.21)[b]
10	6.47	10.49	8.15	(1.28)
15	6.42	10.43	8.22	(1.35)
20	6.38	10.38	8.28	(1.41)
25	6.35	10.33	8.34	1.47
30	6.33	10.29	8.40	(1.52)
40	6.30	10.22	8.52	1.64
50	6.29	10.17	8.63	(1.74)

[a]$pK_{A1} = -\log K_{A1}$; $pK_{A2} = -\log K_{A2}$; $pK_{sp} = -\log K_{sp}$; $pK_{CO_2} = -\log K_{CO_2}$ for $CO_2 + H_2O \rightleftharpoons H_2CO_3$
[b]Values in parentheses were obtained by graphical interpolation.
SOURCE: Garrels and Christ (1965), Table 3.2.

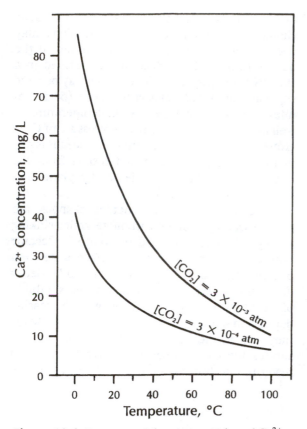

Figure 10.4 Decrease of the concentration of Ca^{2+} in a saturated solution of calcite in pure water with *increasing* temperature at two partial pressures of CO_2. Activity coefficients were taken from Table 10.5 for the ionic strengths of the solutions, which range from about 5×10^{-3} ($P_{CO_2} = 3 \times 10^{-3}$ atm, 5°C) to 0.5×10^{-3} ($P_{CO_2} = 3 \times 10^{-4}$ atm, 80°C).

are the dissociation of the compound into ions as it dissolves in the water, accompanied by hydrolysis of the carbonate ion. The point here is that calcite and many other naturally occurring chemical compounds dissolve *congruently,* that is, without forming another compound. However, the Al silicate minerals generally dissolve *incongruently;* that is, they react with water to form another solid compound as well as ions and molecules.

The process of chemical weathering not only includes the congruent and incongruent solution of minerals but also encompasses all kinds of interactions of water and atmospheric gases with minerals at or near the surface of the Earth. Chemical weathering also includes oxidation–reduction reactions, which we will take up in Chapter 14. The reactions that constitute chemical weathering are not restricted to the surface of the Earth where minerals are actually exposed to the weather, but also operate effectively in soils and below the groundwater table where minerals are continuously in contact with water. The products of chemical weathering consist of (1) new minerals such as clay minerals, oxides, and hydroxides; (2) ions and molecules that dissolve in the water; and (3) grains of minerals that are unreactive such as quartz, garnet, zircon, muscovite, and native gold.

The solid weathering products, including new compounds and resistant minerals, are a major source of "sediment" that may accumulate in residual deposits at the site of weathering or is transported to the oceans where it is deposited and ultimately forms clastic sedimentary rocks. The ions and molecules in solution determine the chemical composition or "quality" of water and make up the solution load that is transported to the oceans by streams. Ultimately, as noted in Chapter 4, these ions and molecules are removed from the oceans by precipitation of compounds or by adsorption on solid particles. Evidently, chemical weathering is part of the *geochemical machine* that operates on the surface of the Earth. Chemical and mechanical forms of weathering cause the destruction of rocks, thereby releasing the elements for a trip to the ocean. We recognize, of course, that the trip to the ocean may be delayed by deposition of sediment or of dissolved constituents in basins on the continents and

rates at which they sink and dissolve. If the water is deep enough, carbonate particles dissolve before they reach the bottom. Therefore, no carbonate sediment can accumulate in the oceans below this critical depth, which is known as the *carbonate compensation depth* (CCD).

10.5 Chemical Weathering

The solution of calcite as limestone or as mineral grains in soil is a form of chemical weathering. The chemical reactions that take place in this case that chemical weathering occurs also in the oceans themselves. Some of these complexities will become the subject of more detailed consideration in Part V of this book. For the time being, we attempt to apply the Law of Mass Action to the chemical weathering of a common Al silicate such as K-feldspar.

10.6 Transformation of Potassium Feldspar to Kaolinite

When K-feldspar (orthoclase or microcline) in igneous or high-grade metamorphic rocks is exposed to chemical weathering, it develops a "chalky" appearance caused by the formation of *kaolinite*. This mineral belongs to the large group of clay minerals (Chapter 13) that can form by chemical weathering of Al-silicate minerals of igneous and metamorphic rocks, by reactions involving volcanic glass, by hydrothermal processes, and by transformations from other clay minerals. In order to apply the Law of Mass Action to the reaction that causes K-feldspar to form kaolinite, we must first write the equation that represents this reaction. The construction of such equations requires an understanding of the chemical properties of the elements that may participate in the reaction and of the environment in which the reaction takes place. Therefore, we set up a list of guidelines that will govern the way in which we construct equations for chemical reactions in nature.

We start by placing the reactant on the left and the known product on the right side of the equation and then proceed to balance the equation based on the following conventions and requirements, provided that the valence numbers of the elements do *not* change during the reaction.

1. In reactions involving Al-silicates the Al is conserved; that is, it does not dissolve appreciably but is transferred into the solid product of the reaction.

2. Excess Si forms silicic acid, which does not dissociate unless the environment is very basic (pH > 9.0).

3. Alkali metals and alkaline earth elements form ions in solution.

4. Any H^+ ions consumed by the reaction originate from the environment.

5. After the equation has first been balanced with respect to Al, Si, and soluble cations, we balance oxygen by adding H_2O as necessary.

6. We then balance hydrogen by adding H^+ as needed.

7. The equation must be balanced both in terms of matter and in terms of electrical charges.

Note that oxygen cannot be balanced by O_2 unless the reaction involves oxidation–reduction. In balancing oxidation–reduction reactions the electrons that are exchanged must be balanced first before a balance of matter and electrical charges is accomplished.

The rules listed above now permit us to construct an equation for the conversion of K-feldspar to kaolinite. We start by writing down the known reactant and its product:

$$KAlSi_3O_8 \rightarrow Al_2Si_2O_5(OH)_4 \quad (10.75)$$
$$\text{K–feldspar} \qquad \text{kaolinite}$$

Next, we balance Al and then take care of K and Si:

$$2\,KAlSi_3O_8 \rightarrow Al_2Si_2O_5(OH)_4$$
$$+\ 2\,K^+ + 4\,H_4SiO_4 \qquad (10.76)$$

We now have 16 oxygens on the left and $9 + 16 = 25$ oxygens on the right. Therefore, we add 9 H_2O to the left side of the equation to balance oxygen:

$$2\,KAlSi_3O_8 + 9\,H_2O \rightarrow Al_2Si_2O_5(OH)_4$$
$$+\ 2\,K^+ + 4\,H_4SiO_4 \qquad (10.77)$$

Finally, we have 18 hydrogens on the left and $4 + 16 = 20$ hydrogens on the right. Therefore, equation 10.77 is balanced in all respects by adding two H^+ to the left side:

$$2\,KAlSi_3O_8 + 9\,H_2O + 2\,H^+ \rightarrow$$
$$Al_2Si_2O_5(OH)_4 + 2\,K^+ + 4\,H_4SiO_4 \qquad (10.78)$$

This equation also satisfies the requirement for electrical neutrality because we have two positive charges on the left and two on the right side.

We now examine the equation we have constructed to learn by the application of Le Châtelier's principle how the reaction at equilibrium responds to changes in certain environmental parameters. We recall that reactions in nature may not achieve equilibrium but run to completion if one of the products escapes from the system. In the case we are considering, this could occur if K^+ and H_4SiO_4 are removed by the movement of the water in which the reaction is taking place. Therefore, the conversion of K-feldspar to kaolinite is favored by the movement of groundwater or surface water in response to a hydraulic or topographic gradient. We also note that nine moles of water are consumed for each mole of kaolinite that is produced. Therefore, the formation of kaolinite is favored by an abundance of water, which, in addition, must be acidic to provide the H^+ needed by the reaction. We conclude, therefore, that K-feldspar is converted into kaolinite in places where an abundance of acidified water is available and where the water flows at a sufficient rate to remove the soluble products of the reaction. Also, since reaction rates generally double for every 10°C increase in temperature, the formation of kaolinite is likely to be favored by tropical climatic conditions near the equator of the Earth.

The reaction between K-feldspar and water involves the hydrolysis of a salt derived from a weak acid and strong base. We predict that such a reaction should remove H^+ from the solution and increase its pH. This is indeed the case here because H^+ is consumed by the reaction. However, the predicted increase of the pH can only be demonstrated in the laboratory because natural waters are buffered (Wollast, 1967). Therefore, when this reaction occurs in nature, the H^+ originates from the environment by the dissociation of carbonic or hydrosulfuric acid, or some other reaction. If the supply of H^+ is large and if the soluble products are removed, the reaction can run to completion. If, however, the water is basic and the soluble products remain at the site, the reaction can achieve equilibrium. In that case, the conversion of K-feldspar to kaolinite stops and the reaction becomes unproductive. Therefore, the establishment of chemical equilibrium may *prevent* the formation of a potentially valuable mineral deposit and, in that sense, is not desirable from our point of view.

The probability that a chemical reaction in nature, such as equation 10.78, can achieve equilibrium depends, among other things, on the water/rock ratio. If there is an abundance of water compared to the volume of rock, reactions are likely to run to *completion* because the environment controls the reaction. If, on the other hand, the water/rock ratio is small, the reactions control the environment and *equilibrium* is possible, provided reaction rates are adequate. Therefore, reactions taking place in large bodies of water and in aquifers with high percolation rates having large water/rock ratios are more likely to run to completion than to achieve equilibrium, whereas reactions in small volumes of pore water (small water/rock ratios) may achieve a localized state of equilibrium between ions and their solids. Such restricted environments occur in the pore spaces of rocks in which mineral cements may be deposited by reactions that maintain a state of equilibrium.

The Law of Mass Action enables us to determine the activities of ions and molecules required for equilibrium in a chemical reaction represented by a balanced equation. Therefore, we return to equation 10.78 and place that reaction at equilibrium. By applying the Law of Mass Action we obtain:

$$\frac{[H_4SiO_4]^4[K^+]^2}{[H^+]^2} = K \qquad (10.79)$$

We "linearize" this equation by taking logarithms:

$$4\log[H_4SiO_4] + 2\log[K^+] - 2\log[H^+] = \log K \qquad (10.80)$$

In order to reduce the number of variables to two, we combine K^+ and H^+ as the $[K^+]/[H^+]$ ratio:

$$4\log[H_4SiO_4] + 2\log\frac{[K^+]}{[H^+]} = \log K \qquad (10.81)$$

We rewrite this equation as:

$$\log\frac{[K^+]}{[H^+]} = \tfrac{1}{2}\log K - 2\log[H_4SiO_4] \qquad (10.82)$$

which is the equation of a straight line in coordinates of $\log[K^+]/[H^+]$ and $\log[H_4SiO_4]$ having a slope $m = -2$ and an intercept of $b = \frac{1}{2}\log K$. It is the locus of all points whose coordinates are the logarithms of the $[K^+]/[H^+]$ ratio and of $[H_4SiO_4]$ required for equilibrium of reaction 10.78. Thus, any point P on the line in Figure 10.5 (equation 10.82) represents an environment in which K-feldspar and kaolinite coexist in equilibrium with the ions and molecules in equation 10.78. Further, any point that is *not* on the line represents an environment in which K-feldspar and kaolinite are *not* at equilibrium. In that case, the reaction

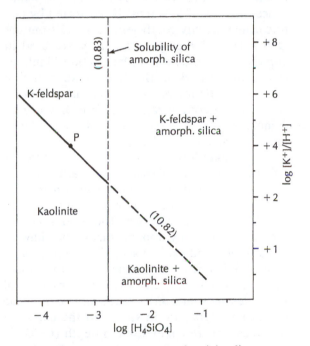

Figure 10.5 Partially completed activity diagram depicting the environmental conditions required to convert K-feldspar into kaolinite, and vice versa, in the presence of water at 25 °C. The line labeled 10.82 represents conditions at which K-feldspar and kaolinite coexist in equilibrium. Point P on the line was chosen arbitrarily to help determine the direction of reaction 10.78 when $[H_4SiO_4]$ is increased or decreased. Line 10.83 represents the solubility of amorphous silica, which limits the activity of $[H_4SiO_4]$ to values less than $10^{-2.74}$ mol/L, as shown in equation 9.89. (Line 10.82 was plotted from equation 10.16 of Garrels and Christ, 1965.)

will proceed in the direction that moves it closer to a state of equilibrium. If the water/rock ratio is small, the environmental parameters can change until they satisfy equation 10.82 and equilibrium is established. If the water/rock ratio is large, the environmental parameters may not be affected appreciably by the reaction we are now considering. The reaction therefore runs to completion and does *not* achieve equilibrium.

The straight line we have drawn in Figure 10.5 (equation 10.82) is a boundary that separates the environments in which K-feldspar is produced by reaction 10.78 from those in which kaolinite is produced. We assume here that the reaction represented by equation 10.78 is, in fact, reversible. Although this may not be true in all cases, a K-feldspar known as *adularia* does form from kaolinite at low temperature (Mensing and Faure, 1983; Kastner and Siever, 1979). In any case, if we move from point P on the equilibrium line in Figure 10.5 toward higher activities of H_4SiO_4, the reaction will respond by combining kaolinite with H_4SiO_4 to produce K-feldspar. Therefore, K-feldspar is the stable phase in all environments represented in Figure 10.5 that have an excess of H_4SiO_4 above that required for equilibrium. Similarly, kaolinite is stable in all environments containing less H_4SiO_4 than is required for equilibrium of reaction 10.78.

The activity of silicic acid is limited by its solubility, as shown in Figure 9.4. Therefore, amorphous silica precipitates when its solubility limit is reached and the activity of $[H_4SiO_4]$ in Figure 10.5 cannot rise above the value required for equilibrium in the reaction:

$$SiO_2(\text{amorph.}) + 2\ H_2O \rightleftharpoons H_4SiO_4 \qquad (10.83)$$

At equilibrium, $[H_4SiO_4] = 10^{-2.74}$ mol/L (equation 9.89). Therefore, a line has been drawn in Figure 10.5 at $\log [H_4SiO_4] = -2.74$ which is an impenetrable boundary because $[H_4SiO_4]$ cannot rise to higher values. The two lines in Figure 10.5, therefore, divide the compositional plane into three fields that represent environmental conditions needed to stabilize K-feldspar, kaolinite, and amorphous silica.

The procedure we have used to determine the environmental conditions necessary for the conversion of K-feldspar to kaolinite can serve to define the stabilities of all minerals that may be exposed to water and atmospheric gases at or near the surface of the Earth. Unfortunately, we cannot proceed in this direction because the equilibrium constants of the reactions we want to study have not been measured owing to experimental difficulties. For this reason, we next undertake a brief excursion into thermodynamics because it is possible to calculate equilibrium constants for chemical reactions from the change in the standard free energy of reactants and products. Thermodynamics will enable us to evaluate most chemical reactions we may care to investigate and will also provide a proof for the validity of the Law of Mass Action.

10.7 Summary

When salts dissolve in water, they dissociate into cations and anions. The solubility of a salt can be determined by calculating the activities of its ions in a saturated solution based on the Law of Mass Action. A salt may precipitate from a supersaturated solution in which the ion activity product (IAP) of its ions exceeds its solubility product constant. When a salt does precipitate, the concentration of its ions remaining in the solution change and their ratio tends toward infinity or zero. The resulting change in the chemical composition of the water contributes to the evolution of brines formed by progressive evaporation of water in desert lakes or in sediment drying in the sun.

The precipitation of salt A may also be caused by the addition of another, more soluble, salt B that shares a common ion with salt A. As a result, salt A precipitates as salt B dissolves until equilibrium between the ions and both salts is established. This process can account for the replacement of a mineral (salt B) by another (salt A) that is less soluble.

Salts can be classified on the basis of the strength or weakness of the acids and bases from which they originated. When the ions of a weak acid or weak base are released into solution by the dissociation of a salt, they react with water to form the parental weak acid or weak base. This phenomenon is called hydrolysis and causes the pH of a salt solution to change. Hydrolysis of carbonates, silicates, and phosphates of the alkali metals and alkaline earths tends to make groundwater basic, thereby enhancing their solubility. In addition, salt solutions that also contain the parental acid (or base) of the salt resist change in their pH and are therefore said to be buffered.

The ability of the ions of a salt in solution to interact with each other is diminished by the presence of other ions and by interactions with the molecules of the solvent. Therefore, the molar concentrations of ions must be corrected by factors called activity coefficients derived from the Debye–Hückel theory before they are used in applications of the Law of Mass Action. Similarly, calculations based on the Law of Mass Action yield the activities of ions, which must then be converted to concentrations before they can be compared with observed concentrations determined by chemical analyses.

The solubility of calcium carbonate is enhanced by the hydrolysis of the carbonate ion and also rises with increasing partial pressure of CO_2, with decreasing pH of the environment, and decreasing temperature. These relationships are all derivable by applications of the Law of Mass Action and explain such diverse phenomena as the deposition of stromatolites, the formation of limestone caverns, the growth of stalactites and stalagmites, leaching or deposition of calcium carbonate in soils, and the existence of the carbonate compensation depth (CCD) in the oceans.

The Law of Mass Action can also be used to study chemical weathering of Al silicates such as K-feldspar, which is converted by incongruent solution into kaolinite. These kinds of reactions can be represented by equations that are constructed on the basis of certain assumptions and conventions. The equations indicate the necessary conditions for the complete conversion of one mineral into another, although the decomposition of Al silicates is not reversible at surface temperatures in many cases. The equations also yield the activities of ions and molecules required for equilibrium and thereby define the stability limits of the solids. By plotting the relevant equations derived from the Law of Mass Action in log–log coordinates, useful diagrams can be constructed that depict the environmental conditions at which the solids are stable. The numerical values of the equilibrium constants required for this purpose can be calculated from thermodynamics, which is the topic of the next chapter.

Problems

1. Calculate activities of Ca^{2+} and F^- and the solubility of fluorite (CaF_2) in pure water at 25°C without regard to possible hydrolysis. Assume that all activity coefficients are equal to one. Express the solubility of fluorite in terms of mol/L and g/100 mL.

2. A quantity of 0.50 g of crystalline NaCl is added to one liter of a solution containing $10^{-3.0}$ mol of dissolved Ag^+. Demonstrate that AgCl precipitates from the solution and calculate the weight of AgCl in grams that forms per liter. Assume that all activity coefficients are equal to one.

3. What is the ratio of the activities of Zn^{2+} and Pb^{2+} in a solution that is in equilibrium with respect to sphalerite (ZnS) and galena (PbS)? (Disregard hydrolysis effects.)

4. Predict the outcome of reactions that occur when a solution having $[Cu^{2+}]/[Fe^{2+}] = 10$ comes in contact with hydrotroilite (FeS). What will be the ratio of $[Cu^{2+}]/[Fe^{2+}]$ in the solution after equilibrium is established? (Disregard hydrolysis effects.)

5. Calculate the ionic strengths of water in Lake Superior (A) and Lake Erie (B) and explain the difference (Livingstone,1963, Table 7). The concentrations are in mg/L.

	A	*B*
HCO_3^-	50.0	121
SO_4^{2-}	4.8	28
Cl^-	1.5	17

NO_3^-	0.52	1.2
Ca^{2+}	14.1	39
Mg^{2+}	3.7	8.7
Na^+	2.9	8.2
K^-	0.5	1.4
Fe^{2+}	0.36	0.03
SiO_2	4.1	2.1

6. Calculate the activity coefficient of Mg^{2+} in an aqueous solution having $I = 5 \times 10^{-2}$ at 15°C based on the Debye–Hückel theory represented by equation 10.46.

7. Calculate the solubility of strontianite $(SrCO_3)$ in pure water at 25°C in equilibrium with CO_2 having a partial pressure of 3×10^{-4} atm. Specify the *concentrations* of all ions in a saturated solution in units of mol/L.

8. Calculate the solubility of calcite at 15°C in pure water in equilibrium with CO_2 at 3×10^{-3} atm pressure. Set up the equations based on the dissociation of carbonic acid and solve for the activity of H^+. Specify the concentrations of all ions and express (Ca^{2+}) in mg/L.

9. Construct an equation to represent the incongruent solution of K-feldspar $(KAlSi_3O_8)$ to "illite" $(KAl_3Si_3O_{10}(OH)_2)$.

References

BUTLER, J. N., 1982. *Carbon Dioxide Equilibria and Their Applications.* Addison-Wesley, Reading, MA, 259 pp.

GARRELS, R. M., and C. L. CHRIST, 1965. *Solutions, Minerals and Equilibria.* Harper & Row, New York, 450 pp.

HARDIE, L. A., and H. P. EUGSTER, 1970. The evolution of closed-basin brines. *Mineral. Soc. Amer. Spec. Pub.* 3, 273–290.

HARVIE, C. E., and J. H. WEARE, 1980. The prediction of mineral solubilities in natural waters: the Na-K-Mg-Ca-Cl-SO$_4$-

H$_2$O system from zero to high concentration at 25 °C. *Geochim. Cosmochim. Acta,* **44**:981–997.

KASTNER, M., and R. SIEVER, 1979. Low temperature feldspars in sedimentary rocks. *Amer. J. Sci.,* **279**:435–479.

KRAUSKOPF, K. B., 1979. *Introduction to Geochemistry,* 2nd ed. McGraw-Hill, New York, 617 pp.

LINDSAY, W. L., 1979. *Chemical Equilibria in Soils.* Wiley, New York, 449 pp.

LIVINSTONE, D. A., 1963. Chemical composition of rivers and lakes. In M. Fleischer (Ed.), Data of Geochemistry, 6th ed. Ch. G., *U.S. Geol. Surv. Prof. Paper* 440–G, 64 pp.

MENSING, T. M., and G. FAURE, 1983. Identification and age of neoformed Paleozoic feldspar (adularia) in a Precambrian basement core from Scioto County, Ohio, USA. *Contrib. Mineral. Petrol.,* **82**:327–333.

PITZER, K. S., 1973. Thermodynamics of electrolytes. I. Theoretical basis and general equations. *J. Phys. Chem.,* **77**:268–277.

WALTER, M. R., 1977. *Stromatolites.* Elsevier, Amsterdam, 790 pp.

WEAST, R. C., M. J. ASTLE, and W. H. BEYER (Eds.), 1986. *CRC Handbook of Chemistry and Physics,* 66th ed. CRC Press, Boca Raton, FL.

WOLLAST, R., 1967. Kinetics of the alteration of K-feldspar in buffered solutions at low temperature. *Geochim. Cosmochim. Acta,* **31**:635–648.

11

Thermodynamics

The science of thermodynamics originally evolved from a variety of observational evidence indicating that mechanical energy is transformable into heat. For example, Benjamin Thompson (1753–1814) of Woburn, Massachusetts, better known as Count Rumford of the Holy Roman Empire, was impressed by the amount of heat generated during the boring of cannon at the arsenal in Munich. In 1798 he proposed that the heat was formed by conversion of the mechanical energy expended by the boring and actually calculated the amount of heat generated by one horse turning the borer for one hour. Count Rumford's ideas about the equivalence of work and heat were supported by Sir Humphry Davy (1778–1829), who demonstrated in 1799 that two pieces of ice could be made to melt by rubbing them together by a clockwork mechanism in a vacuum. Nevertheless, the scientific community did not accept this idea for nearly 50 years until James P. Joule (1818–1889), working in his father's brewery in Manchester, England, carefully measured the rise in temperature caused by stirring water with a mechanical paddle wheel. His research between 1840 and 1849 clearly demonstrated that the expenditure of a certain amount of work always produced the same amount of heat (Moore, 1955).

The observation that work is transformable into heat is an extension of the principle of conservation of energy. It can be restated by saying that whenever work is done, heat energy is consumed, implying that one must always pay a price by the expenditure of energy if work is to be accomplished.

Thermodynamics is founded on three basic principles, which arise from common experience but cannot be proved or disproved. They must be accepted as articles of faith. The science of thermodynamics has evolved from these so-called laws by careful deductive reasoning aided by mathematics.

11.1 Definitions

Thermodynamics describes the world the way it appears to an observer in terms of certain measurable properties of matter, such as volume, pressure, temperature, and chemical composition. Matter is regarded as continuous and indivisible. In this regard thermodynamics is the opposite of *statistical mechanics* in which matter is treated as a collection of particles in motion. In order to enter the world of thermodynamics we must first become familiar with certain definitions and conventions that may differ from common usage.

The basic entity for consideration in thermodynamics is the *system,* which is that part of the universe whose properties are under consideration. The system is separated from the rest of the universe, known as the *surroundings,* by a *boundary* whose properties can be defined. A system is said to be *open* when matter can pass across the boundary and *closed* when matter is prevented from entering or leaving the system. The boundary may prevent not only matter but also heat and any other form of energy from passing through it. Systems that have such impervious boundaries are said to be *isolated.* The investigator must identify the system to be studied and specify the properties of its boundary.

The properties of a system are classified as being either *intensive* or *extensive.* Intensive properties, such as pressure, temperature, and density, are independent of the amount of matter that is present, whereas the extensive properties, such as volume and mass, depend on the mass and are therefore additive. When the properties of a system are changed, it experiences a change in its *state*. For example, when a quantity of gas in a cylinder is compressed by moving a frictionless piston, the system undergoes a *change in state*.

The chemical composition of thermodynamic systems is expressed in terms of *components* and *phases*. Nordstrom and Munoz (1986, p. 67) defined a *phase* as "a uniform, homogeneous, physically distinct, and mechanically separable portion of a system." When the system under consideration is a rock, the minerals of which the rock is composed are the phases. Phases need not be solids, but can also be liquids or gases. For example, a system containing ice, water, and water vapor has three phases (solid, liquid, and gas). The aqueous phase in such a system may also contain a number of ionic or molecular *species* in solution.

The *components* of a system are the chemical constituents by means of which the chemical compositions of the phases of a system can be completely described. The choice of components is arbitrary. For example, the chemical composition of a rock specimen can be expressed in terms of oxides or in terms of chemical elements. However, the *number* of components required to describe the compositions of all of the phases present in a system at equilibrium is constrained by the *Gibbs phase rule*.

The number of variables that must be specified in order to define the state of a system is equal to the number of components plus the number of intensive properties, such as pressure and temperature. The state of the system is completely defined when the number of variables is equal to the number of independent relations between them. If the number of independent equations is *less* than the number of variables, the system has *variance* or *degrees of freedom*. In this case, the state of the system cannot be completely constrained. When the number of variables *exceeds* the number of independent equations by one, the system is *univariant* and has one degree of freedom. Univariant systems can be con-

strained by specifying the value of *one* of the variables. When the number of variables exceeds the number of equations by *two*, the system is *divariant*, and so on.

When a chemical system is at equilibrium, the number of phases (p), the number of components (c), and the number of degrees of freedom (f) must satisfy the Gibbs phase rule (Nordstrom and Munoz, 1986):

$$p + f = c + 2 \qquad (11.1)$$

Therefore, systems in which the number of phases is *equal* to the number of components have two degrees of freedom. In other words, pressure and temperature must be specified in this case in order to completely define the state of the system.

All systems contain a certain amount of *heat*, which is defined as that which flows across the boundary of a closed system during a change in its state by virtue of a difference in temperature between the system and its surroundings. Moreover, heat always flows from a point of higher to a point of lower temperature. Thermodynamics defines heat as though it were an invisible fluid whose *direction* of flow is indicated by its algebraic sign. Heat is *positive* when it flows from the surroundings to the system and *negative* when it flows from the system to the surroundings. We can remember this important convention by recognizing that systems are "selfish." When a system gains heat, the added heat has a positive sign, and when it loses heat the sign is negative.

Work is defined in mechanics as the product of force times distance, whereas in thermodynamics work is defined as a quantity that flows across the boundary of a system during a change in its state and is completely convertible into the lifting of a weight in the surroundings. Work, like heat, is an algebraic quantity, but the sign convention of work is opposite to that of heat. When work flows to the surroundings it has a positive sign, and when it flows to the system it has a negative sign. Another way to visualize the sign convention for work is to say that when the system does work, the work is positive, but when the surroundings do it, the work is negative.

With these definitions and conventions we have now set the stage to introduce the first law of thermodynamics. The presentation will be selective and based on plausibility rather than mathematical rigor. More complete and rigorous presentations can be found in college level textbooks on physical chemistry. Thermodynamics for use by geologists is presented in books by Kern and Weisbrod (1967), Wood and Fraser (1976), Fraser (1977), Nordstrom and Munoz (1986), and Anderson (1996).

11.2   The First Law

Every system whose state has been defined by specifying the numerical values of its properties contains a certain amount of energy. This energy arises from the heat content of the system as well as from other internal sources such as gravitational, electrical, magnetic, molecular, atomic, and even nuclear phenomena. The amount of internal energy (E) of a particular system is difficult to determine. However, its magnitude depends on the state of the system. When a system undergoes a change in its state in a reversible manner, its internal energy also changes. The magnitude of the change in the internal energy ΔE of a system that undergoes a reversible change in state from state 1 to state 2 is given by:

$$\Delta E = E_2 - E_1 \qquad (11.2)$$

where E_2 and E_1 are the internal energies of the system in states 2 and 1, respectively. It makes no difference *how* the change in state is carried out, provided that it is *reversible,* which requires that the system is never far from being in equilibrium. If the system is returned to its original state, the net change in its internal energy is zero. We conclude therefore that the change in the internal energy of a system undergoing a change in state along a circular path is zero:

$$\Delta E = (E_2 - E_1) + (E_1 - E_2) = 0 \qquad (11.3)$$

The *first law of thermodynamics* states that the increase in the internal energy of a system during a reversible change in its state is equal to the heat that flows across its boundary from the surroundings (q) minus the work done by the system (w).

Therefore, the net change in the internal energy of the system is:

$$\Delta E = q - w \qquad (11.4)$$

The first law of thermodynamics implies that work and heat are related and that some of the heat that flows into the system during a change in state is converted into work and therefore does not augment the internal energy.

We can imagine that a reversible change in state can be carried out in small incremental steps such that the internal energy changes by small amounts designed by dE. Now we can restate the first law in differential form:

$$dE = dq - dw \qquad (11.5)$$

where dq and dw are very small increments of heat and work, respectively. The most common form of work a system can do during a change in its state is to expand against the constant pressure of the surroundings. Thus we can specify that:

$$dw = PdV \qquad (11.6)$$

where P is the external pressure and dV is an incremental change in the volume of the system. A simple demonstration indicates that PdV does have the dimensions of work (force × distance):

$$PdV = \frac{\text{force} \times (\text{distance})^3}{(\text{distance})^2} = \text{force} \times \text{distance} \qquad (11.7)$$

Therefore, we can express the first law by the equation:

$$dE = dq - PdV \qquad (11.8)$$

11.3   Enthalpy

The statement of the first law represented by equation 11.8 enables us to predict the change in the internal energy of a system during a change in its state. The equation indicates that the increase in the internal energy of the system is equal to the heat absorbed from the surroundings minus the work done by the expansion of the system against the pressure exerted by the surroundings. We can carry out such a change in state by integrating equation 11.8 from state 1 to state 2:

$$\int_1^2 dE = \int_1^2 dq - P \int_1^2 dV \qquad (11.9)$$

The integration yields:

$$E_2 - E_1 = (q_2 - q_1) - P(V_2 - V_1) \qquad (11.10)$$

We simplify this result by replacing $q_2 - q_1$ by q_P, which is an *amount* of heat added at constant pressure. In addition, we multiply the PV terms:

$$E_2 - E_1 = q_P - PV_2 + PV_1 \qquad (11.11)$$

By rearranging the terms of this equation we obtain:

$$(E_2 + PV_2) - (E_1 + PV_1) = q_P \qquad (11.12)$$

It is advantageous at this point to define a new kind of energy called the *enthalpy* (H):

$$H = E + PV \qquad (11.13)$$

The enthalpy is a function of the state of the system, as is the internal energy. Therefore, the enthalpy of the system in states 1 and 2 is:

$$H_1 = E_1 + PV_1 \qquad (11.14)$$

$$H_2 = E_2 + PV_2 \qquad (11.15)$$

Substituting equations 11.14 and 11.15 into equation 11.12 yields:

$$H_2 - H_1 = q_P = \Delta H \qquad (11.16)$$

In other words, the change in the enthalpy of a system (ΔH) during a reversible change in its state at constant pressure is equal to the heat absorbed by the system during that change in state.

This is a very important and concrete result because q_P is a *measurable* quantity. For example, when a chemical reaction is carried out in a well-insulated vessel called a calorimeter, the heat evolved or consumed is indicated by the change in temperature inside the calorimeter. Therefore, equation 11.16 enables us to measure the *change* in the enthalpy of a system as a result of a chemical reaction, but we still do not know the *amount* of enthalpy the system contains.

In order to establish a *scale* for enthalpy we now define a *reference state* for chemical elements and their compounds. The reference state is defined in terms of certain *standard properties* that define the so-called *standard state*. The temperature of the standard state is equal to 25 °C (298.15 K), the pressure is 1 atm (recently redefined as 1 bar, which is equal to 10^6 dyne/cm^2, or 0.987 atm, 29.53 in. Hg), and the activity of ions and molecules in aqueous solution is equal to one. In addition, we specify that the enthalpies of the pure elements in their stable state of aggregation in the standard state are equal to zero. Thus we have the definition:

The *enthalpy of formation* of compounds and their ions and molecules in aqueous solution is the heat absorbed or given off by chemical reactions in which the compounds, ions, and molecules form from the elements in the standard state.

We represent this important thermodynamic parameter by the symbol H_f°, where the superscript $^\circ$ identifies the standard state and the subscript f symbolizes "formation."

The standard enthalpies of formation and other thermodynamic constants of many compounds, ions, and molecules have been measured experimentally and are tabulated in the *CRC Handbook of Chemistry and Physics* (Weast et al., 1986), in Technical Notes of the National Bureau of Standards, and in compilations by Kelley (1962), Robie et al. (1978), Barin et al. (1989), and Woods and Garrels (1987). However, it is important to remember that standard enthalpies of formation and other thermodynamic constants are *measurements* and are subject to experimental errors. Consequently, the values reported by different investigators for a given compound commonly differ from each other by amounts that may or may not be within the error of the measurements. Such discrepancies in thermodynamic constants affect the results of calculations in which they are used and cause internal inconsistencies. Some of the standard enthalpies of formation listed in Appendix B are average values of two or more determinations that appear to be compatible within the errors of measurement. However, no assurance can be given that these values are accurate to within the stated number of significant figures or that the listed values are internally consistent.

11.4   Heats of Reaction

When two compounds or elements, A and B, react to form a product A_2B_3 by the reaction:

$$2\,A + 3\,B \to A_2B_3 \qquad (11.17)$$

a certain amount of heat is either used up or given off. We can now calculate the amount of this heat (q_P) from the enthalpy change of the system when the reaction takes place in the standard state. In other words, we treat the reaction as a change in state and calculate the difference in the enthalpy (ΔH_R°) between the final state (A_2B_3) and the initial state ($2\,A + 3\,B$). Therefore, the change in enthalpy of the reaction in the standard state is:

$$\Delta H_R^\circ = H_f^\circ(A_2B_3) - [2H_f^\circ(A) + 3H_f^\circ(B)] \qquad (11.18)$$

where ΔH_R° is the heat of the reaction in the standard state and H_f° is the standard enthalpy of formation of A_2B_3, A, and B, respectively. Note that the heat is evolved or absorbed as a result of, or during, the reaction and *not* when the reaction is at equilibrium. After a reaction has reached equilibrium, the heat of the reaction (ΔH_R) is zero because the rates of the forward and backward reactions are exactly equal. Thus, we can state that at equilibrium $\Delta H_R = 0$. However, the reference state is a defined condition in which equilibrium does not occur. Therefore ΔH_R° is never zero. In general, ΔH_R° is calculated by summing the standard enthalpies of the products and by subtracting the enthalpies of the reactants:

$$\Delta H_R^\circ = \sum_i n_i H_{f_i}^\circ(\text{products}) - \sum_i n_i H_{f_i}^\circ(\text{reactants}) \qquad (11.19)$$

where n is the molar coefficient of each reactant and product taken from a balanced equation representing the reaction and i identifies the compounds or ions that participate in the reaction.

Standard enthalpies are expressed in units of kilocalorie per mole (kcal/mol), where the normal *calorie* is defined as the amount of heat required to raise the temperature of 1 g of water from 14.5 to 15.5 °C. The calorie is equivalent to

4.1840 joules (J) so that 1 kcal/mol is equal to 4.1840 kJ/mol. At the present time both units are in use.

When ΔH_R° of a reaction is positive, the reaction is *endothermic*, which means that heat flows from the surroundings to the system. Conversely, when ΔH_R° is negative, the reaction is *exothermic* and heat flows from the system to the surroundings. Endothermic reactions consume heat, whereas exothermic reactions produce heat.

In order to illustrate how heats of reaction are calculated, we choose the reaction between hydrogen and oxygen to form water in the gaseous state, symbolized by (g):

$$2\,H_2(g) + O_2(g) \to 2\,H_2O(g) \qquad (11.20)$$

The amount of heat absorbed or given off when this reaction occurs at 1 atm and 25 °C is given by:

$$\Delta H_R^\circ = 2H_f^\circ(H_2O(g)) \\ - [2H_f^\circ(H_2(g)) + H_f^\circ(O_2(g))] \qquad (11.21)$$

The standard enthalpies of H_2 and O_2 are both equal to zero by definition, so that in this case:

$$\Delta H_R^\circ = 2H_f^\circ(H_2O(g)) \qquad (11.22)$$

The standard enthalpy of formation of water vapor is -57.80 kcal/mol (Appendix B). Therefore:

$$\Delta H_R^\circ = 2(-57.80) = -115.6 \text{ kcal} \qquad (11.23)$$

We conclude that this reaction is strongly *exothermic*. In fact, this reaction is commonly used to demonstrate the explosive nature of some chemical reactions. Note that the reaction is *endothermic* when equation 11.20 is *reversed*.

Heat effects are also associated with the dissociation of salts into ions, with phase transformations, and with mixing or dilution of solutions. For example, heat is absorbed from the surroundings when water evaporates to form water vapor:

$$H_2O(l) \to H_2O(g) \qquad (11.24)$$

When the evaporation occurs in the standard state:

$$\Delta H_R^\circ = H_f^\circ(H_2O(g)) - H_f^\circ(H_2O(l)) \qquad (11.25)$$

From Appendix B:

$$\Delta H_R^\circ = (-57.80) - (-68.32) = +10.52 \text{ kcal} \qquad (11.26)$$

Evidently, the evaporation of water at 25 °C and 1 atm pressure is an *endothermic* process that removes heat from the surroundings and thereby causes the temperature to decrease. The heat required to evaporate one mole of a liquid is called the *latent heat of evaporation*. When the vapor condenses to form the liquid, the heat given off is called the *latent heat of condensation*. In other words, evaporation is endothermic, whereas condensation of water vapor is exothermic, and the latent heat of evaporation is equal in magnitude to the heat of condensation. Similarly, fusion or melting of solids is endothermic, whereas crystallization of liquids is exothermic, and the *latent heat of fusion* is equal to the *latent heat of crystallization*.

Note that we are unable to calculate the latent heat of fusion or crystallization of water because it takes place at 0 °C rather than at 25 °C. Therefore, in order to calculate the latent heat of fusion of water, we must first extrapolate the enthalpies of formation from 25 to 0 °C. In general, chemical reactions on the surface of the Earth occur at temperatures between 0 and 100 °C, and their heats of reaction therefore commonly differ from those calculated at 25 °C.

The response of a reaction *at equilibrium* to a change in temperature can be predicted on the basis of Le Châtelier's principle. If the forward reaction is exothermic, an *increase* in the temperature favors the backward reaction because it consumes heat and therefore counteracts the increase in the temperature. Consequently, the equilibrium will shift to the left making the equilibrium constant smaller. Similarly, a *decrease* in temperature applied to a reaction at equilibrium in which the forward reaction is exothermic causes the equilibrium to shift to the right and increases the value of the equilibrium constant. Evidently, the enthalpy change of a chemical reaction is related to the way in which the numerical value of the equilibrium constant varies with temperature. This relationship is formally expressed by the van't Hoff equation, which will be derived in Section 11.10.

We may also expect that exothermic reactions take place spontaneously because they do not require heat from the surroundings. However, this generalization is not correct. It turns out that not all of the heat that enters a system during a reversible change in its state is available to do work. Similarly, the direction in which a reaction proceeds spontaneously is not determined solely by the enthalpy change because systems have an additional property called *entropy*. We return to this phenomenon in Section 11.6.

11.5   Heat Capacity

When heat is added to a solid, a liquid, or a gas, the temperature of the substance increases. This familiar observation can be stated as an equation:

$$dq = C\,dT \qquad (11.27)$$

where dq is an increment of heat added, C is a constant known as the *heat capacity,* and dT is the corresponding incremental increase in the temperature expressed in *kelvins* (K). In thermodynamics, temperatures are always expressed on the "absolute" or Kelvin scale, which has its zero point at -273.15 °C, when all molecular motion stops and the heat content of pure crystalline materials becomes equal to zero.

The heat capacity is a characteristic property of chemical elements and their compounds that must be measured experimentally. It turns out that heat capacities actually vary with temperature and that it makes a difference whether the heat is added at constant pressure or at constant volume. On the surface of the Earth pressure is constant, and we therefore work with the "heat capacity at constant pressure" symbolized by C_P. Consequently, we restate equation 11.27 as:

$$dq_P = C_P\,dT \qquad (11.28)$$

We now recall that the heat added to a closed system during a reversible change in its state at constant pressure is equal to the increase in the enthalpy of the system. This conclusion was expressed by equation 11.16, which we now restate in terms of differentials:

$$dq_P = dH \qquad (11.29)$$

Substituting equation 11.29 into 11.28 yields:

$$dH = C_P\,dT \qquad (11.30)$$

This equation can be integrated between appropriate temperature limits and therefore enables us to extrapolate enthalpies of formation of compounds from the standard temperature ($T°$) to some other temperature (T) above or below it:

$$\int_{T°}^{T} dH = C_P \int_{T°}^{T} dT \qquad (11.31)$$

By carrying out the integration we obtain:

$$H_T - H_{T°} = C_P(T - T°) \qquad (11.32)$$

Equation 11.32 enables us to calculate the enthalpy of formation of a compound at temperature T if we know its standard enthalpy and its heat capacity, *provided* the temperature dependence of C_P can be neglected.

In reality, use of equation 11.32 to extrapolate enthalpies is limited because the heat capacity is in fact a function of the temperature and therefore cannot be treated as a constant in the integration. Moreover, the range of temperatures over which the integration is carried out must not include phase transformations, such as boiling or freezing, which cause a discontinuity in the heat capacity and involve their own enthalpy changes. These restrictions generally are not severe provided the environmental conditions remain between 0 and 100 °C at 1 atm pressure. Therefore, the most straightforward procedure is to recalculate the enthalpies of formation of reactants and products to the desired temperature (between 0 and 100 °C) and then to calculate the enthalpy change of the reaction at that temperature.

A more accurate and elegant procedure is based on equations that express the variation of C_P as a function of temperature in the form:

$$C_P = a + (b \times 10^{-3})T + (c \times 10^{-6})T^2 \qquad (11.33)$$

where $a, b,$ and c are constants derived by fitting an algebraic equation to experimentally determined data in coordinates of C_P and T. Substituting equation (11.33) into equation 11.31 yields:

$$\int_{T°}^{T} dH = a \int_{T°}^{T} dT + b \times 10^{-3} \int_{T°}^{T} T dT$$

$$+ c \times 10^{-6} \int_{T°}^{T} T^2 dT \qquad (11.34)$$

By integrating this equation we obtain:

$$H_T - H_{T°} = a(T - T°) + \frac{b \times 10^{-3}}{2}[T^2 - (T°)^2]$$

$$+ \frac{c \times 10^{-6}}{3}[T^3 - (T°)^3] \qquad (11.35)$$

Equation 11.35 can be applied to the reactants and products of a chemical reaction as specified by equation 11.19. As a result, we calculate Δa, Δb, and Δc for the reaction and then evaluate the enthalpy change of the reaction at temperature T ($\Delta H_{R,T}$) from the equation:

$$\Delta H_{R,T} - \Delta H_R° = \Delta a(T - T°) \qquad (11.36)$$

$$+ \frac{\Delta b \times 10^{-3}}{2}[T^2 - (T°)^2]$$

$$+ \frac{\Delta c \times 10^{-6}}{3}[T^3 - (T°)^3]$$

The relationship between C_P and T is sometimes expressed in ways that differ from equation 11.33, but the application of these equations is limited by the availability of experimental data.

11.6   The Second Law

In the 19th century much work was done by means of steam engines in which hot steam expands in a cylinder and pushes a piston. The steam cools as it expands and is returned to the reservoir at a lower temperature. An important scientific problem of that time was to understand the factors that limit the *efficiency* of steam engines. This problem was solved in 1824 by the French engineer Sadi Carnot (1796–1832). Without going into the details, we can say that the efficiency of steam engines can never be 100% because some of the heat added to the system must be discharged when the expanded steam returns to the reservoir.

The work of Carnot was a very important contribution to thermodynamics and from it arose a new principle known as the *second law of thermodynamics*. The essence of the second law is that all systems possess a property called *entropy*, named by the German physicist Rudolf Clausius (1822–1888). One way to explain entropy is to say that a certain fraction of the enthalpy of a system is not convertible into work because it is consumed by an increase in the entropy. Another way to explain the meaning of entropy is to compare it to the property of "randomness" recognized in *statistical mechanics*. When one tries to predict the behavior of a collection of particles, one comes to the conclusion that they will most probably assume a condition of maximum randomness. This conclusion leads to the generalization that every system that is left to itself will, on the average, change toward a condition of maximum randomness. The implication of this statement is that the entropy of a system increases spontaneously and that energy must be spent to reverse this tendency.

When entropy is thought of as randomness, it can be recognized in many natural phenomena. We have all learned from the common experiences of life that work must be done to counteract the natural tendency toward disorder. Many popular science books have been written about entropy and its manifestations in everyday life including one by Rifkin (1980) who claimed that "entropy is the supreme law of nature and governs everything we do."

The second law introduces entropy by the statement:

In any *reversible* process the change in the entropy of the system (dS) is equal to the heat received by the system (dq) divided by the absolute temperature (T):

$$dS = \frac{dq}{T} \quad \text{(reversible process)} \quad (11.37)$$

In any spontaneous *irreversible* process the change in entropy is greater than this amount:

$$dS > \frac{dq}{T} \quad \text{(irreversible process)} \quad (11.38)$$

Entropy, like internal energy and enthalpy, is a property whose magnitude depends on the state of the system. However, unlike internal energy and enthalpy, the absolute value of the entropy of a system can be determined by virtue of the third law of thermodynamics.

Measurements indicate that the heat capacities of pure crystalline materials *decrease* with decreasing temperature and approach zero at absolute zero. This evidence has been elevated into the *third law of thermodynamics,* which states:

The heat capacities of pure crystalline substances become zero at absolute zero.

Since $dq = CdT$ (equation 11.28) and $dS = dq/T$ (equation 11.37), it follows that:

$$dS = C\left(\frac{dT}{T}\right) \quad (11.39)$$

If the heat capacity is equal to zero at absolute zero, then $dS = 0$. Therefore, when a system composed of pure crystalline substances undergoes a reversible change in state at absolute zero, the entropy change ΔS is equal to zero. The German physicist Walther Hermann Nernst (1864–1941) expressed these insights by stating that at absolute zero the molar entropies of pure crystalline solids are equal to zero. Therefore, the third law enables us to calculate entropies by integrating equation 11.39 from absolute zero to some higher temperature (T):

$$\int_0^T dS = \int_0^T C\left(\frac{dT}{T}\right) \quad (11.40)$$

In order to calculate the entropy of an element or compound in the standard state from equation 11.40, the temperature dependence of its heat capacity must be determined experimentally. The standard molar entropies of many elements and compounds have been determined and are listed in reference books including the *CRC Handbook of Chemistry and Physics* (Weast et al., 1986). Molar entropies are expressed in "entropy units" or calories per degree (cal/deg). When the standard molar entropies of the reactants and products of a chemical reaction are known, the change in the entropy (ΔS_R°) of a chemical reaction or phase transformation can be calculated from the statement:

$$\Delta S^\circ_R = \sum n_i S^\circ_i(\text{products}) - \sum n_i S^\circ_i(\text{reactants})$$
(11.41)

When ΔS°_R is positive, the entropy of the system increases as a result of the change in state, whereas when ΔS°_R is negative, it decreases. For example, when liquid water evaporates in the standard state to form water vapor:

$$H_2O(l) \rightarrow H_2O(g)$$
(11.42)

the change in entropy is (Krauskopf, 1979):

$$\Delta S^\circ_R = 45.10 - 16.71 = +28.39 \text{ cal/deg}$$
(11.43)

The increase in the entropy is consistent with our expectation that water molecules in the gas phase are more randomly distributed than molecules of liquid water. Similarly, when solid NaCl dissociates to form ions:

$$NaCl(s) \rightarrow Na^+ + Cl^-$$
(11.44)

the change in entropy is (Krauskopf, 1979):

$$\Delta S^\circ_R = 14.0 + 13.5 - 17.2 = +10.3 \text{ cal/deg}$$
(11.45)

Evidently, the entropy increases when NaCl dissolves to form ions in aqueous solution because the ions occupy fixed lattice positions in the crystal whereas the ions in the solution are mobile. However, for other reactions ΔS°_R may be negative, which implies a *reduction* in the entropy of the system.

11.7   Gibbs Free Energy

The second law of thermodynamics implies that the *increase* in the *enthalpy* of a system during a reversible change in its state at constant temperature is diminished because a certain amount of the enthalpy is consumed by an increase in the *entropy* of the system. This insight was used by J. Willard Gibbs (1839–1903) to define a new form of energy, now known as the *Gibbs free energy* (G):

$$G = H - TS$$
(11.46)

where H is the enthalpy and S is the entropy. The change in the Gibbs free energy of a chemical reaction in the standard state is:

$$\Delta G^\circ_R = \Delta H^\circ_R - T\Delta S^\circ_R$$
(11.47)

The Gibbs free energy is a function of the state of the system like the internal energy and the enthalpy. It is therefore subject to the same conventions regarding the standard state as enthalpy and is measured in kilocalories per mole:

The *standard Gibbs free energy of formation* of a compound (G°_f) is the change in the free energy of the reaction by which it forms from the elements in the standard state.

Numerical values of G°_f for compounds and ions in aqueous solution are listed in Appendix B together with values of H°_f. The value of ΔG°_R for a reaction is calculated by summing the values of G°_f of the products, each multiplied by its molar coefficient, and by subtracting from it the sum of the G°_f values of the reactants:

$$\Delta G^\circ_R = \sum n_i G^\circ_{fi}(\text{products}) - \sum n_i G^\circ_{fi}(\text{reactants})$$
(11.48)

The algebraic sign and magnitude of ΔG°_R depend on the sign and magnitude of ΔH°_R and ΔS°_R in equation 11.47. If the forward reaction is exothermic (ΔH°_R is negative) and the entropy increases (ΔS°_R is positive), then the two terms of equation 11.47 combine to make ΔG°_R negative. If, however, the entropy decreases (ΔS°_R is negative), then $T\Delta S^\circ_R$ in equation 11.47 is positive. If ΔH°_R is negative, ΔG°_R will be less negative than ΔH°_R and may even become positive if $T\Delta S > \Delta H$. Note that $T\Delta S$ is in fact a form of enthalpy because $\Delta S = \Delta H/T$ at constant pressure and temperature (equations 11.16 and 11.37). Therefore, $T\Delta S$ has the dimensions of ΔH and the two terms can be added or subtracted from each other.

When ΔG°_R of a reaction is *negative*, the forward reaction has excess energy when it occurs in the standard state. When ΔG°_R is *positive*, there is a deficiency of energy for the forward reaction in the standard state and the reaction will therefore run in the backward direction. In general, ΔG_R° is a measure of the *driving force* of a chemical reaction such as equation 11.17:

$$2\,A + 3\,B \rightarrow A_2B_3$$

If ΔG_R° of this reaction is negative, the reaction proceeds as written from left to right in the standard state. When the reaction ultimately reaches equilibrium, the product is more abundant than the reactants so that the equilibrium constant K is large:

$$\frac{[A_2\,B_3]}{[A]^2\,[B]^3} = K \tag{11.49}$$

If, however, ΔG_R° for reaction 11.17 in the standard state is positive, the reaction runs the other way:

$$2\,A + 3\,B \leftarrow A_2\,B_3 \tag{11.50}$$

Therefore, the equilibrium constant is small because the reactants are more abundant than the product. Evidently, the sign and magnitude of ΔG_R° of a reaction can be used to predict both the *direction* in which a reaction runs in the standard state and the magnitude of the equilibrium constant.

In conclusion, we calculate ΔG_R° for the familiar reaction:

$$\underset{\text{calcite}}{CaCO_3} \rightarrow Ca^{2+} + CO_3^{2-} \tag{11.51}$$

using the data in Appendix B:

$$\Delta G_R^\circ = [(-132.3) + (-126.17)] - [(-269.9)]$$

$$= +11.43 \text{ kcal} \tag{11.52}$$

Since ΔG_R° is positive, the reaction runs from right to left in the standard state and calcite precipitates. The equilibrium constant for reaction 11.51 should therefore be a small number, as we know it to be. It is not surprising that the reaction (equation 11.51) runs to the left in the standard state because the activities of the ions in the standard state are each equal to one and the ion activity product is therefore much larger than the equilibrium constant.

We can also calculate ΔH_R° and ΔS_R° for reaction 11.51 using values of H_f° from Appendix B and of S_f° from Krauskopf (1979):

$$\Delta H_R^\circ = [(-129.7) + (-161.84)] - [(-288.6)]$$

$$= -2.94 \text{ kcal} \tag{11.53}$$

$$\Delta S_R^\circ = [(-12.7) + (-13.6)] - [(+22.2)]$$

$$= -48.5 \text{ cal/deg} \tag{11.54}$$

Note that ΔS_R° for the dissociation of calcite into ions is negative, implying a *decrease* in the entropy. Note also that the entropy change is expressed in calories per degree (cal/deg) rather than in kilocalories (kcal). We can now combine ΔH_R° and ΔS_R° for reaction 11.51 and calculate ΔG_R°, thereby testing the internal consistency of the thermodynamic data:

$$\Delta G_R^\circ = -2.94 - \frac{298.15(-48.5)}{1000} \tag{11.55}$$

where 298.15 is the temperature of the standard state and we divide the $T\Delta S$ term by 1000 in order to convert calories to kilocalories.

The result is:

$$\Delta G_R^\circ = +11.5 \text{ kcal} \tag{11.56}$$

which is in satisfactory agreement with $\Delta G_R^\circ = +11.43$ kcal calculated from equation 11.52. Note that ΔG_R° for the dissociation of calcite into ions is positive even though the reaction is exothermic, that is, ΔH_R° is negative. This occurs because more heat energy is consumed to decrease the entropy than is produced by the dissociation. Therefore, in this case ΔH_R° is *not* a reliable predictor of the direction in which the reaction runs in the standard state.

The sign of ΔH_R° does, however, enable us to use Le Châtelier's principle to predict how the equilibrium constant of reaction 11.51 varies with temperature. Since the reaction is exothermic as written, it produces heat which increases the temperature. Therefore, if reaction 11.51 is at equilibrium and the temperature is *decreased,* the reaction moves to the right because that tends to *increase* the temperature. When equilibrium is reestablished at a lower temperature, the equilibrium constant is *larger* than it was before. Therefore, the solubility-product constant of calcite *increases* with *decreasing* temperature and calcite becomes *more* soluble. As we discussed in Section 10.4, thermodynamics provides a rational explanation for the temperature dependence of the solubility of calcite and allows us to predict the course of other chemical reactions provided the necessary data are available.

11.8   Derivation of the Law of Mass Action

Because the Gibbs free energy is defined in terms of both the first and second laws of thermodynamics, we can express it in terms of directly measurable properties. We can replace H in equation 11.46 by equation 11.13 and obtain:

$$G = E + PV - TS \qquad (11.57)$$

For an infinitesimal change in state at constant temperature $dT = 0$ and therefore:

$$dG = dE + PdV + VdP - TdS \qquad (11.58)$$

Substituting equation 11.8 for dE yields:

$$dG = dq - PdV + PdV + VdP - TdS \qquad (11.59)$$

Since $dS = dq/T$, $dq = TdS$ and equation 11.59 therefore reduces to:

$$dG = VdP \qquad (11.60)$$

We can use equation 11.60 to calculate the change in the Gibbs free energy of one mole of an ideal gas as a function of pressure at constant temperature. Since the volume of gases varies with pressure, we must express the volume in equation 11.60 by an equation that relates volume to pressure. According to the ideal gas law, for one mole of gas:

$$V = \frac{RT}{P} \qquad (11.61)$$

where R is the gas constant.

Substituting into equation 11.60 yields:

$$dG = RT\left(\frac{dP}{P}\right) \qquad (11.62)$$

Integrating from the standard pressure $(P°)$ to some other pressure (P):

$$\int_{P°}^{P} dG = RT \int_{P°}^{P} \frac{dP}{P} \qquad (11.63)$$

yields:

$$G_P - G° = RT(\ln P - \ln P°) \qquad (11.64)$$

Since $P° = 1$ atm, $\ln P° = 0$. Therefore, equation 11.64 reduces to:

$$G_P - G° = RT \ln P \qquad (11.65)$$

where G_P is the molar free energy of the ideal gas at pressure P, $G°$ is the molar free energy in the standard state, and T is the temperature of the standard state in kelvins. The molar free energy is also called the chemical potential (μ), and equation 11.65 can be written as:

$$\mu = \mu° + RT \ln P \qquad (11.66)$$

The free energy of n moles of an ideal gas can be calculated by multiplying equation 11.65 by n:

$$nG_P - nG° = nRT \ln P \qquad (11.67)$$

According to Dalton's law, the total pressure of a mixture of ideal gases is the sum of the partial pressures exerted by each gas in the mixture. Therefore, for a mixture of i ideal gases:

$$\sum n_i G_i - \sum n_i G_i° = RT \sum n_i \ln P_i \qquad (11.68)$$

Next, we assume that the mixture of ideal gases is the result of a chemical reaction and specify that the molar coefficients (n) of the *products* are *positive* and those of the *reactants* are *negative*. The provision is not new or arbitrary but goes back to the convention regarding the algebraic sign of heat. If we choose to represent the chemical reaction by the equation:

$$a\,A + b\,B \rightarrow c\,C + d\,D \qquad (11.69)$$

then we can evaluate what equation 11.68 tells us to do. For example, $\sum n_i G_i$ means:

$$\sum n_i G_i(\text{products}) - \sum n_i G_i(\text{reactants}) = \Delta G_R$$
$$(11.70)$$

Similarly, $\sum n_i G_i^\circ$ means:

$$\sum n_i G_i^\circ(\text{products}) - \sum n_i G_i^\circ(\text{reactants}) = \Delta G_R^\circ$$
$$(11.71)$$

and $RT \sum n_i \ln P_i$ means:

$$RT(c \ln P_C + d \ln P_D - a \ln P_A - b \ln P_B)$$
$$= RT(\ln P_C^c P_D^d - \ln P_A^a P_B^b)$$
$$= RT \ln \left(\frac{P_C^c P_D^d}{P_A^a P_B^b} \right) = RT \ln Q$$
$$(11.72)$$

where Q is the *reaction quotient*. Therefore, we now restate equation 11.68 as:

$$\Delta G_R - \Delta G_R^\circ = RT \ln Q \qquad (11.73)$$

which is valid for a reacting mixture of ideal gases.

The reaction quotient (Q) varies continuously as the reaction proceeds until the reaction reaches a state of equilibrium. After equilibrium has been established, the partial pressures of the reactant and product gases become constant because the forward and backward reactions are occurring at the same rate. Therefore, at equilibrium the reaction quotient becomes constant and is known to us as the *equilibrium constant.*

Next, we consider a very *important point:* the free energy change of a chemical reaction *at equilibrium* is equal to zero. This statement becomes obvious if we recall that ΔG_R is regarded as the driving force of a chemical reaction. When a reaction is in equilibrium, there is no driving force pushing it one way or the other. However, note that ΔG_R° is *not* zero because it is the change in the Gibbs free energy when the reaction takes place in the standard state. Therefore, ΔG_R° is a *constant,* whereas ΔG_R varies as the reaction proceeds from the standard state toward equilibrium.

When the reaction reaches equilibrium, $\Delta G_R \rightarrow 0$, but ΔG_R° remains the same.

Returning to equation 11.73 we can now state that for a mixture of ideal gases at equilibrium $\Delta G_R = 0$ and therefore:

$$\Delta G_R^\circ = -RT \ln K \qquad (11.74)$$

where K is the equilibrium constant at the standard temperature. This remarkable equation establishes the functional relationship that exists between ΔG_R° and K, both of which are constants that apply to chemical reactions in the standard state and at equilibrium, respectively. In the process of deriving this relationship we have also proved the validity of the Law of Mass Action for reactions among ideal gases. These two results are among the most important contributions thermodynamics has made to science. Equation 11.74 allows us to calculate the numerical values of equilibrium constants at 25°C for all chemical reactions for which we can calculate ΔG_R° from existing measurements of the standard free energies of formation.

We reduce equation 11.74 to a formula by converting the equation to logarithms to the base 10 and by substituting $R = 1.987$ cal/deg · mol and $T = 298.15$ K:

$$\Delta G_R^\circ = -\frac{2.3025 \times 1.987 \times 298.15}{1000} \log K \quad (11.75)$$

$$= -1.364 \log K \qquad (11.76)$$

The right side of equation 11.75 must be divided by 1000 because R is expressed in calories, whereas ΔG_R° is in kilocalories. Therefore, the equilibrium constant K is related to ΔG_R° by the formula:

$$K = 10^{-\Delta G_R^\circ / 1.364} \qquad (11.77)$$

This is a very useful result, but it applies only to reactions among ideal gases unless we take steps to make it applicable to real gases and to ions and molecules in aqueous solutions.

11.9 Fugacity and Activity

In order to make equations 11.74 and 11.77 applicable to reactions among *real gases* we define the *fugacity* as the partial pressure a real gas would have if it were ideal. Therefore, fugacities satisfy equation 11.74. In order to apply equation 11.74 to *real solutions* of any kind we define the *activity* (*a*) of a solute as:

$$a = \frac{f}{f^\circ} \qquad (11.78)$$

where *f* is the fugacity of the solute in the solution and f° is its fugacity when the vapor is in equilibrium with the pure substance in the standard state. Since fugacities are measured in atmospheres or other units of pressure, the activity as defined by equation 11.78 is a dimensionless number. It also follows from equation 11.78 that for a pure substance:

$$a = 1.0 \qquad (11.79)$$

because the fugacity *f* of a pure substance is equal to f°.

Binary mixtures of two nonelectrolytes A and B obey Raoult's law, which states that the fugacity of component A is given by:

$$f_A = f_A^\circ N_A \qquad (11.80)$$

where N_A is the mole fraction of A defined as:

$$N_A = \frac{n(A)}{n(A) + n(B)} \qquad (11.81)$$

where $n(A)$ and $n(B)$ are the numbers of moles of A and B, respectively. When the solution is dilute, that is, $n(A) \ll n(B)$, the vapor obeys Henry's law:

$$f_A = kN_A \qquad (11.82)$$

By dividing both sides of equation 11.82 by f_A°, we obtain:

$$\frac{f_A}{f_A^\circ} = \frac{k}{f_A^\circ} N_A = a_A \qquad (11.83)$$

Next, we note that for a dilute solution of solute A in solvent B the number of moles of A is much smaller than the number of moles of B. Therefore, if $n(A) \ll n(B)$, equation 11.81 can be restated to a very good approximation as:

$$N_A = \frac{n(A)}{n(B)} \qquad (11.84)$$

The mole fraction of A in a dilute solution can be transformed into the molal concentration of A (m_A) by converting $n(B)$ to kilograms of B:

$$m_A = \frac{n(A) \times 1000}{n(B) \times MW(B)} \qquad (11.85)$$

where $MW(B)$ is the gram-molecular weight of B. Since $n(A)/n(B) = N_A$ in dilute solutions of A in B:

$$m_A = \frac{N_A \times 1000}{MW(B)} = N_A k' \qquad (11.86)$$

and k' is a constant equal to $1000/MW(B)$. When we substitute equation 11.86 into equation 11.83, we obtain:

$$a_A = \frac{k}{f_A^\circ} N_A = \frac{km_A}{f_A^\circ k'} \qquad (11.87)$$

By combining the three constants in equation 11.87 we obtain:

$$a_A = \gamma_A m_A \qquad (11.88)$$

where γ_A is the *activity coefficient.*

This relationship applies to all solutions regardless of the vapor pressures of the solutes or of the solvent. If the solution is a mixture of volatile nonelectrolytes such as acetone dissolved in water, we have no difficulty detecting the presence of acetone vapor and the definition of the activity of acetone in the solution given by equation 11.78 makes sense. Ions produced by the dissociation of salts, acids, and bases in water also produce vapors, but their partial pressures or fugacities are so low that they may be undetectable. However, this experimental difficulty does not alter the validity of relationships derived by consideration of vapors in equilibrium with solutes and solvents regardless of whether they are electrolytes or nonelectrolytes.

The concept of *activity* presented above is quite different from that discussed in Sections 9.2 and 11.3 We can reconcile the two representations by saying that activities are introduced in thermodynamics in order to extend the Law of Mass Action from reacting mixtures of ideal gases to chemical reactions among molecules and ions in nonideal aqueous solutions. This is accomplished by means of activity coefficients that convert molal concentrations of ions and molecules into activities. When activity coefficients are regarded as dimensionless conversion factors, then activities have the same units as concentrations.

11.10 The van't Hoff Equation

Thermodynamics enables us to calculate the equilibrium constants of chemical reactions at 25°C. In order to extrapolate the equilibrium constant from the standard temperature to other temperatures between 0 and 100°C, we return briefly to equation 11.74:

$$\Delta G_R^\circ = -RT \ln K$$

By solving this equation for $\ln K$ we obtain:

$$\ln K = -\frac{\Delta G_R^\circ}{RT} \qquad (11.89)$$

Next, we differentiate equation 11.89 with respect to T:

$$\frac{d \ln K}{dT} = -\left(\frac{1}{R}\right) d\left(\frac{\Delta G_R^\circ / T}{dT}\right) \qquad (11.90)$$

Since $\Delta G_R^\circ = \Delta H_R^\circ - T\Delta S_R^\circ$, it follows that:

$$\frac{\Delta G_R^\circ}{T} = \frac{\Delta H_R^\circ}{T} - \Delta S_R^\circ \qquad (11.91)$$

and:

$$d\frac{\Delta G_R^\circ / T}{dT} = -\frac{\Delta H_R^\circ}{T^2} - 0 \qquad (11.92)$$

Substituting into equation 11.90 yields:

$$\frac{d \ln K}{dT} = \left(-\frac{1}{R}\right)\left(-\frac{\Delta H_R^\circ}{T^2}\right) = \frac{\Delta H_R^\circ}{RT^2} \qquad (11.93)$$

Multiplying by dT gives:

$$d \ln K = \left(\frac{\Delta H_R^\circ}{R}\right)\left(\frac{dT}{T^2}\right) \qquad (11.94)$$

We integrate equation 11.94 from the standard temperature (T°) to some other temperature (T), which may be higher or lower than T°:

$$\int_{T^\circ}^{T} d \ln K = \frac{\Delta H_R^\circ}{R} \int_{T^\circ}^{T}\left(\frac{dT}{T^2}\right) \qquad (11.95)$$

The result of the integration is:

$$\ln K_T - \ln K_{T^\circ} = \left(-\frac{\Delta H_R^\circ}{R}\right)\left(\frac{1}{T} - \frac{1}{T^\circ}\right) \qquad (11.96)$$

Equation 11.96 is the important van't Hoff equation, which we now restate in terms of logarithms to the base 10:

$$\log K_T = \log K_{T^\circ} - \frac{\Delta H_R^\circ}{2.3025R}\left(\frac{1}{T} - \frac{1}{298.15}\right) \qquad (11.97)$$

Note that equation 11.97 can be used to calculate K at any temperature between 0 and 100°C if we know the value of K at the standard temperature. However, if the reference temperature *differs* from 25°C, then ΔH_R must be evaluated at that temperature by means of heat capacities, as shown in equation 11.32 or 11.35.

We have now completed this selective presentation of thermodynamics, which has enabled us to prove the validity of the Law of Mass Action and to calculate equilibrium constants of chemical reactions not only at the standard temperature but also at other temperatures between 0 and 100°C.

11.11 Solubility of Amorphous Silica between 0 and 100°C

At the end of Chapter 9 we calculated the solubility of amorphous silica in water as a function of pH at 25°C. We are now in a position to investigate how the solubility of amorphous silica varies as a function of temperature at different values of the pH. The thermodynamic constants required for this calculation are listed in Appendix B.

Amorphous silica reacts with water to form silicic acid:

$$SiO_2(\text{amorph.}) + 2\,H_2O(l) \rightarrow H_4SiO_4^\circ \qquad (11.98)$$

The change in Gibbs free energy of this reaction in the standard state is:

$$\Delta G_R^\circ = [(-312.66)] - [(-203.33) + 2(-56.687)]$$
$$= +4.044 \text{ kcal} \qquad (11.99)$$

and the equilibrium constant K_{T° is:

$$K_{T^\circ} = 10^{-(4.044/1.364)} = 10^{-2.96} \quad (11.100)$$

The change in enthalpy in the standard state is:

$$\Delta H_R^\circ = -349.1 - [-215.94 + 2(-68.315)]$$

$$= +3.47 \text{ kcal} \quad (11.101)$$

The reaction is *endothermic,* suggesting that its equilibrium constant *increases* with temperature. From the van't Hoff equation (11.97):

$$\log K_T =$$

$$-2.96 - \frac{3.47 \times 10^3}{2.3025 \times 1.987}\left(\frac{1}{T} - \frac{1}{298.15}\right) \quad (11.102)$$

where $R = 1.987$ cal/deg·mol. Note that ΔH_R° must be converted to calories in order to be compatible with R. Equation 11.102 reduces to:

$$\log K_T = -2.96 - 758.4\left(\frac{1}{T} - 0.00335\right)$$

$$(11.103)$$

When $T > 298.15$ K, $1/T - 0.003\,35$ is negative and the algebraic sign of the second term in equation 11.103 changes to plus. Therefore, at $T = 273.15$ K (0°C) equation 11.103 yields $K_T = 10^{-3.20}$, whereas at 373.15 K (100°C) $K_T = 10^{-2.45}$. Evidently, the equilibrium constant of reaction 11.98 *increases* with *increasing* temperature as predicted.

The first dissociation of silicic acid forms $H_3SiO_4^-$ and H^+:

$$H_4SiO_4 \rightarrow H_3SiO_4^- + H^+ \quad (11.104)$$

The standard free energy change is $\Delta G_R^\circ = +13.24$ kcal, and the standard enthalpy change is $\Delta H_R^\circ = +7.0$ kcal. Note that the G_f° and H_f° of H^+ are both equal to zero. The equilibrium constant of reaction 11.104 at 25°C is $K = 10^{-9.71}$ and varies with temperature in accordance with the van't Hoff equation:

$$\log K_T = -9.71 - 1530.03\left(\frac{1}{T} - 0.00335\right)$$

$$(11.105)$$

Consequently at 0°C, $K_T = 10^{-10.19}$, whereas at 100°C, $K_T = 10^{-8.68}$. Once again, the equilibrium constant *increases* with rising temperature because reaction 11.104 is endothermic.

These calculations illustrate the point that thermodynamics not only confirms our qualitative predictions based on Le Châtelier's principle but actually permits us to determine the numerical values of the equilibrium constants of all kinds of chemical reactions between temperatures of 0 and 100°C or other appropriate temperature limits.

Although this is a very important advance in geochemistry, the calculations can become quite tedious. Therefore, it is helpful to summarize the results of thermodynamic calculations by means of suitable diagrams. In the case at hand, we want to depict the concentration of SiO_2 in saturated solutions of amorphous silica at different values of the pH and temperature. We therefore carry out the calculations first made in Section 9.6 at a series of temperatures for selected pH values and plot the results in Figure 11.1 as a set of curves in coordinates of the concentration of SiO_2 and temperature. The diagram depicts the increase in the solubility of amorphous SiO_2 with increasing temperature at specified values of the pH. Such diagrammatic representations of chemical reactions are widely used in geochemistry and therefore are the subject of Chapter 12.

11.12 Summary

Thermodynamics is based on three statements regarding the relationship between heat and other forms of energy that do not require proof because they are self-evident. Starting from these basic premises, thermodynamics has grown into a large body of knowledge by rigorous deductive reasoning aided by mathematics. The products of this development include the concepts of enthalpy and Gibbs free energy and a proof of the Law of Mass Action for reacting mixtures of ideal gases.

These results of thermodynamics have great practical value in geochemistry because they permit us to calculate the equilibrium constants of chemical reactions at any desired temperature, limited only by the availability of the necessary thermodynamic parameters and constrained by the stabilities of the reactants and products. Consequently, we can now explore the behavior of chemical reactions on the surface of the Earth and use the results to specify the stability limits of the common rock-forming minerals. Many of the reactions that take place in the natural environment can be studied in the laboratory only with great difficulty because these reactions are so slow. Thermodynamics enables us to make predictions about the outcome of these reactions that can be tested by evidence in the field.

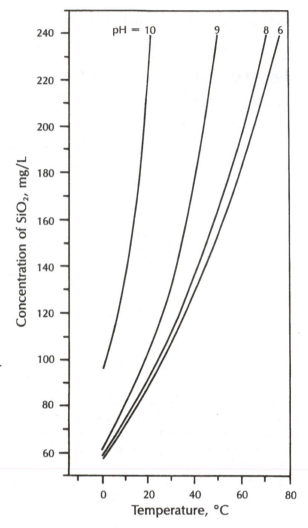

Figure 11.1 Solubility of amorphous silica as a function of temperature at different pH values. The equilibrium constants at 25°C were calculated from equation 11.77 using the standard free energies of formation from Appendix B. Equilibrium constants at other temperatures were derived from the van't Hoff equation (11.97) based on K and ΔH_R°. The activity coefficients of all ions and molecules were assumed to be equal to one. The resulting curves indicate that the concentration of SiO_2 in milligrams per liter in saturated solutions of amorphous silica in water increases both with increasing temperature and pH.

Problems

1. Calculate the enthalpy change when fluorite (CaF_2) dissolves in water in the standard state.

2. Based on the result in Problem 1, predict how the solubility of fluorite varies with temperature.

3. Calculate the standard free energy change of the reaction by which fluorite dissociates into ions.

4. Calculate the solubility of fluorite in water at 10, 20, and 30°C and express each in terms of the concentration of Ca^{2+} in mg/L. (Assume $\gamma = 1$ for all species.)

5. Calculate K_{T° for the reaction between albite $(NaAlSi_3O_8)$, water, and H^+ to form kaolinite $(Al_2Si_2O_5(OH)_4)$ and silicic acid.

6. Apply the Law of Mass Action to the reaction in Problem 5, take logarithms to the base 10, and express the resulting equation in terms of $\log [Na^+]/[H^+]$ and $\log [H_4SiO_4]$.

7. The water of the Mississippi River at Baton Rouge, Louisiana, has the following chemical composition at pH = 7.2.

	ppm		ppm
HCO_3^-	101	Mg^{2+}	7.6
SO_4^{2-}	41	Na^+	11
Cl^-	15	K^+	3.1
NO_3^-	1.9	SiO_2	5.9
Ca^{2+}	34		

Calculate $\log [Na^+]/[H^+]$ and $\log [H_4SiO_4]$ for this water using the necessary activity coefficients.

8. Use the results of Problems 6 and 7 to determine the direction of the reaction in Problem 5 in the Mississippi River at Baton Rouge; that is, determine whether albite or kaolinite is produced.

9. Make the necessary calculation (assuming $T = 25\,°C$) to determine whether the Mississippi River at Baton Rouge is saturated or undersaturated with respect to (a) calcite, (b) gypsum, (c) amorphous silica, (d) quartz, and (e) dolomite.

10. Calculate the value of $G_f^°$ for BaF_2 given that the concentration of Ba^{2+} in a saturated solution at $25\,°C$, having an ionic strength of 0.01, is 1.12×10^{-2} m/L. Assume that Ba^{2+} and F^- are derived entirely from BaF_2 and that hydrolysis is negligible.

References

ANDERSON, G. M., 1996. *Thermodynamics of Natural Systems.* Wiley, New York, 382 pp.

BARIN, I., F. SAUERT, E. SCHULTZE-RHONHOF, and W. S. SHENG, 1989. *Thermochemical Data of Pure Substances,* VCH Publishers, New York, 1739 pp.

FRASER, D. G. (Ed.), 1977. *Thermodynamics in Geology.* Reidel, Amsterdam, 410 pp.

KELLEY, K. K., 1962. Heats of free energies of formation of anhydrous silicates. *U.S. Bureau of Mines Rept. Invest.* 5901.

KERN, R., and A. WEISBROD, 1967. *Thermodynamics for Geologists.* W. H. Freeman, New York, 304 pp.

KRAUSKOPF, K. B., 1979. *Introduction to Geochemistry,* 2nd ed. McGraw-Hill, New York, 617 pp.

MOORE, W. J., 1955. *Physical Chemistry,* 2nd ed. Prentice-Hall, Upper Saddle River, NJ, 633 pp.

NORDSTROM, D. K., and J. L. MUNOZ, 1986. *Geochemical Thermodynamics.* Blackwell Scientific, Palo Alto, 477 pp.

RIFKIN, J., 1980. *Entropy.* Bantam, New York, 302 pp.

ROBIE, R. A., B. S. HEMINGWAY, and J. R. FISHER, 1978. Thermodynamic properties of minerals and related substances at 298.15 K and 1 bar pressure and at higher temperatures. *U.S. Geol. Survey Bull.* 1452.

WEAST, R. C., M. J. ASTLE, and W. H. BEYER (Eds.), 1986. *CRC Handbook of Chemistry and Physics.* CRC Press, Boca Raton, FL.

WOOD, B. J., and D. G. FRASER, 1976. *Elementary Thermodynamics for Geologists.* Oxford University Press, Oxford, England, 303 pp.

WOODS, T. L., and R. S. GARRELS, 1987. *Thermodynamic Values at Low Temperature for Natural Inorganic Materials.* Oxford University Press, New York, 242 pp.

12

Mineral Stability Diagrams

When minerals participate in chemical reactions as reactants or products, they are either consumed or produced depending on the direction in which the reaction is moving. We can therefore say that *reactant* minerals are *unstable,* whereas *product* minerals are *stable* while the reaction is in progress. If the reaction ultimately reaches equilibrium, *both* reactant and product minerals become stable. Thus, we can investigate the stability of minerals under two kinds of conditions:

1. We can determine the environmental conditions necessary for minerals to coexist at equilibrium.
2. We can determine which minerals are stable and which are unstable in a particular geochemical environment.

The theoretical basis for these kinds of studies is provided by the Law of Mass Action and by the relationship between the standard free energy change of a chemical reaction and its equilibrium constant at 25 °C.

The activity diagrams we want to construct are based on the *incongruent* solution of certain aluminosilicate minerals in igneous and high-grade metamorphic rocks. The reactions on which the diagrams are based contribute to chemical weathering and result in the formation of oxides, hydroxides, clay minerals, or zeolites, depending on the geochemical environment. Transformations among minerals that form only at elevated temperature and pressure are inappropriate in activity diagrams drawn at the temperature and pressure of the Earth's surface. However, the stability of minerals can also be defined by equations arising from their *congruent* solution in dilute aqueous solutions. Both kinds of activity diagrams present complex chemical interactions in a form that is especially appropriate for geologists accustomed to using phase diagrams in igneous petrology.

12.1 Chemical Weathering of Feldspars

The feldspars in igneous rocks crystallize from silicate melts at high temperature, but can also form on the surface of the Earth by reactions of aqueous solutions with other aluminosilicates such as the clay minerals. In Section 10.6 we discussed how to construct the equations that represent the reactions among aluminosilicates, but we were unable to proceed because we did not know the equilibrium constants of the reactions we wanted to investigate.

Now we can restate the equation for the conversion of microcline to kaolinite:

$$2\,KAlSi_3O_8 + 9\,H_2O + 2\,H^+ \rightleftharpoons$$
$$Al_2Si_2O_5(OH)_4 + 2\,K^+ + 4\,H_4SiO_4 \quad (12.1)$$

Using the standard free energies of formation from Appendix B we obtain:

$$\Delta G_R^\circ = [-(906.84) + 2(-67.70) + 4(-312.66)]$$
$$-[2(-894.9) + 9(-56.687)]$$
$$= +7.103 \text{ kcal} \quad (12.2)$$

Therefore, the equilibrium constant of the reaction at 25 °C is:

$$K = 10^{-(7.103/1.364)} = 10^{-5.21} \quad (12.3)$$

Applying the Law of Mass Action when the reaction is at equilibrium gives:

$$\frac{[H_4SiO_4]^4[K^+]^2}{[H^+]^2} = 10^{-5.21} \quad (12.4)$$

where the square brackets signify activities or fugacities of reactants and products. Next, we take logarithms of equation 12.4:

$$4 \log [H_4SiO_4] + 2 \log [K^+] - 2 \log [H^+]$$
$$= -5.21 \quad (12.5)$$

Collecting terms:

$$4 \log [H_4SiO_4] + 2(\log [K^+] - \log [H^+])$$
$$= -5.21 \quad (12.6)$$

from which it follows that:

$$4 \log [H_4SiO_4] + 2 \log \frac{[K^+]}{[H^+]} = -5.21 \quad (12.7)$$

and:

$$\log \frac{[K^+]}{[H^+]} = -2 \log [H_4SiO_4] - 2.60 \quad (12.8)$$

Equation 12.8 is a straight line, shown in Figure 12.1, along which $\log [K^+]/[H^+]$ and $\log [H_4SiO_4]$ have the values required for equilibrium between microcline and kaolinite. If the two parameters have values that place the environment *off* the line, the reaction is *not* at equilibrium. In that case, we can determine the direction in which the reaction proceeds by applying Le Châtelier's principle to equation 12.1. If $\log [H_4SiO_4]$ of the environment is *greater* than required for equilibrium at a particular value of $\log [K^+]/[H^+]$, reaction 12.1 moves to the left, thereby converting kaolinite into microcline. Therefore, microcline is *stable* under these conditions, whereas kaolinite is *unstable*. The same reasoning indicates that if $\log [H_4SiO_4]$ is *less* than required for equilibrium, kaolinite is stable and microcline is unstable at 25 °C.

We might have chosen to write an equation for the conversion of microcline to gibbsite instead of kaolinite. The resulting line plots to the left of the microcline–kaolinite equilibrium in Figure 12.1. This means that microcline is reacting to form gibbsite under conditions in which kaolinite, not microcline, is stable. Consequently, the transformation of microcline to gibbsite is *metastable* and must be excluded from the diagram.

When reaction 12.1 occurs in a closed environment having a small water/rock ratio, $\log [H_4SiO_4]$ and $\log [K^+]/[H^+]$ may change as the reaction progresses until equilibrium is established. In that case, kaolinite and microcline coexist in equilibrium with the aqueous solution, provided the temperature remains constant and the activities of the ions and molecules remain unchanged. On the other hand, if the environment is open or the water/rock ratio is large, the reaction may not achieve equilibrium because it is unable to change the activity of H_4SiO_4 and the $[K^+]/[H^+]$ ratio of the environment enough to make them compatible with equation 12.8. In this case, the reaction continues until either kaolinite or microcline is used up, depending on the direction in which it is moving; that is, the reaction goes to *completion*.

We can investigate the effect of a temperature change on reaction 12.1 by calculating ΔH_R°:

$$\Delta H_R^\circ = [(-983.5) + 2(-60.32) + 4(-348.3)]$$
$$- [2(-948.7) + 9(68.315)]$$
$$= +14.895 \text{ kcal} \quad (12.9)$$

We see that the reaction is *endothermic* when it occurs in the standard state, which means that it consumes heat in the forward direction. Therefore, an *increase* in temperature favors the conversion of microcline to kaolinite and causes the equilibrium constant to *increase*. The relationship between K_{T° and K_T is given by the van't Hoff equation:

$$\log K_T = \log K_{T^\circ} - \frac{\Delta H_R^\circ \times 10^3}{2.3025R} \left(\frac{1}{T} - 0.00335\right)$$
$$(12.10)$$

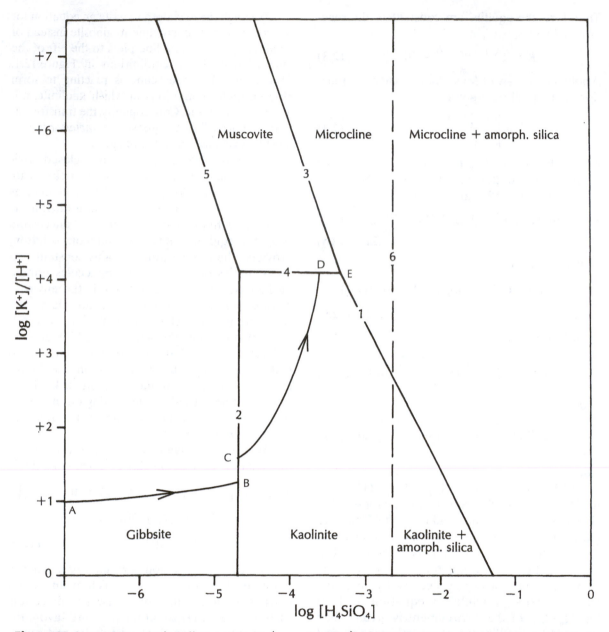

Figure 12.1 Stabilities of microcline, muscovite (proxy for illite), kaolinite, gibbsite, and amorphous silica in equilibrium with K^+, H^+, and H_4SiO_4 in aqueous solution at 25 °C and 1 atm pressure. The equations for the stability boundaries are based on the standard free energy values from Appendix B and are listed in Table 12.1. Note that K-smectite, pyrophyllite, and quartz are omitted from the diagram in order to keep it simple. Note also that the diagram lies in the upper left quadrant of the coordinate system. The reaction path was adapted from Steinmann et al. (1994).

Since $K_{T^\circ} = 10^{-5.21}$, $\Delta H_R^\circ = +14.895$ kcal, and if $T = 308.15$ K (35 °C):

$$\log K_T = -5.21 - \frac{14{,}895}{4.575}(0.003\,24 - 0.003\,35)$$

$$K(35\,°C) = 10^{-4.85} \tag{12.11}$$

Since $10^{-4.85} > 10^{-5.21}$, the equilibrium *does* shift to the right in favor of kaolinite with increasing temperature, as predicted.

When reaction 12.1 proceeds from left to right as written, it produces K^+ ions and H_4SiO_4 molecules while consuming H^+ and H_2O. Therefore, if the system is closed, the chemical composition of the water changes in such a way that $\log[H_4SiO_4]$ and $\log[K^+]/[H^+]$ become more compatible with the values required for equilibrium. Therefore, chemical weathering alters the chemical composition of water and hence its *quality* as a natural resource. Moreover, reactions such as 12.1 take place not only on the surface of the Earth but also *below* the surface where groundwater is in contact with minerals that are unstable in the presence of water. Chemical "weathering" is *not* restricted to places that are exposed to the weather, but occurs at varying depth below the surface depending on the location of the water table and the porosity and permeability of the rocks.

We can now complete the construction of the stability diagram for microcline (Figure 12.1) by adding other crystalline or amorphous compounds that can occur in the system composed of SiO_2–Al_2O_3–K_2O–HCl–H_2O. These include gibbsite, muscovite (proxy for illite), K-smectite, pyrophyllite, and amorphous silica. Note that "HCl" serves as a source of hydrogen ions and is not actually present in natural environments. Note also that the stability fields of K-smectite and pyrophyllite are not shown in Figure 12.1, but they do occur between kaolinite and microcline. The equations required to define the stability limits of the solid phases in this system are derivable by the same procedure we demonstrated above for reaction 12.1 and are therefore listed in Table 12.1.

The boundaries in Figure 12.1 are the loci of points whose coordinates have the values required

for equilibrium between the solids in adjoining fields. Each field is labeled to indicate which solid is stable within its boundaries. The solids are stable only within their stability fields. When they are exposed to an environment that lies outside their own stability field, they are converted into the solid that is stable in that environment. For example, if a grain of microcline is placed into an environment within the stability field of kaolinite, reaction 12.1 will run from left to right and convert the microcline into kaolinite, provided enough time is available to allow the reaction to run to completion. Similarly, if kaolinite is placed into an environment within the stability field of microcline, it forms secondary K-feldspar known as adularia (Kastner, 1971; Kastner and Siever, 1979; Mensing and Faure, 1983).

Most natural waters plot within the stability field of kaolinite in Figure 12.1, whereas seawater straddles the boundary between muscovite (illite) and microcline. Therefore, microcline and other feldspars in igneous and metamorphic rocks on the continents weather to form clay minerals (kaolinite and smectite). When these clay minerals are deposited in the oceans, they tend to take up K^+ and may be converted into illite (represented by muscovite in Figure 12.1) or low-temperature feldspar, given sufficient time.

The evolution of the chemical composition of water as a result of chemical weathering of minerals such as microcline can be predicted by numerical modeling (Helgeson et al., 1969). The progress of such reactions produces curved reaction paths on activity diagrams. An example of such a path is shown in Figure 12.1 based on modeling by Steinmann et al. (1994). Their calculations were based on the assumptions that:

1. The system is closed and the water/rock ratio is small.

2. The secondary minerals are in equilibrium with ions in the solution even though the primary minerals are not in equilibrium.

3. Temperature and pressure remain constant.

4. Aluminum is conserved in the reactions.

5. The solutions remain dilute such that all activity coefficients are close to unity.

Table 12.1 Stability Relations Among Microcline, Kaolinite, Gibbsite, and Amorphous Silica in the Presence of Water at 25 °C and 1 atm pressure[a]

Microcline–kaolinite

$$2\,KAlSi_3O_8 + 9\,H_2O + 2\,H^+ \rightleftharpoons Al_2Si_2O_5(OH)_4 + 2\,K^+ + 4\,H_4SiO_4$$
$$\Delta G_R^\circ = +7.103\ kcal \quad K = 10^{-5.21}$$
$$\log \frac{[K^+]}{[H^+]} = -2\log[H_4SiO_4] - 2.60 \tag{1}$$

Kaolinite–gibbsite

$$Al_2Si_2O_5(OH)_4 + 5\,H_2O \rightleftharpoons 2\,Al(OH)_3 + 2\,H_4SiO_4$$
$$\Delta G_R^\circ = +12.755\ kcal \quad K = 10^{-9.35}$$
$$\log[H_4SiO_4] = -4.68 \tag{2}$$

Microcline–muscovite

$$3\,KAlSi_3O_8 + 12\,H_2O + 2\,H^+ \rightleftharpoons KAl_3Si_3O_{10}(OH)_2 + 2\,K^+ + 6\,H_4SiO_4$$
$$\Delta G_R^\circ = +16.184\ kcal \quad K = 10^{-11.865}$$
$$\log \frac{[K^+]}{[H^+]} = -3\log[H_4SiO_4] - 5.93 \tag{3}$$

Muscovite–kaolinite

$$2\,KAl_3Si_3O_{10}(OH)_2 + 3\,H_2O + 2\,H^+ \rightleftharpoons 3\,Al_2Si_2O_5(OH)_4 + 2\,K^+$$
$$\Delta G_R^\circ = -11.059\ kcal \quad K = 10^{+8.11}$$
$$\log \frac{[K^+]}{[H^+]} = +4.05 \tag{4}$$

Muscovite–gibbsite

$$KAl_3Si_3O_{10}(OH)_2 + 9\,H_2O + H^+ \rightleftharpoons 3\,Al(OH)_3 + K^+ + 3\,H_4SiO_4$$
$$\Delta G_R^\circ = +13.603\ kcal \quad K = 10^{-9.97}$$
$$\log \frac{[K^+]}{[H^+]} = -3\log[H_4SiO_4] - 9.97 \tag{5}$$

Solubility limit of amorphous silica

$$SiO_2(amorph.) + 2\,H_2O \rightleftharpoons H_4SiO_4$$
$$\Delta G_R^\circ = +3.604\ kcal \quad K = 10^{-2.64}$$
$$\log[H_4SiO_4] = -2.64 \tag{6}$$

---

[a]Based on the thermodynamic data in Appendix B.

When microcline is exposed to an environment represented by the coordinates of point A in Figure 12.1 (i.e., log $[H_4SiO_4] = -6.0$, log $[K^+]/[H^+] = -2.0$, pH = 4), it will react to form gibbsite:

$$KAlSi_3O_8 + 7 H_2O + H^+ \rightarrow Al(OH)_3$$
$$+ 3 H_4SiO_4 + K^+ \qquad (12.12)$$

As a result, the $[K^+]/[H^+]$ ratio and $[H_4SiO_4]$ of the water both increase because the system is assumed to be closed and the water/rock ratio is small. Therefore, the chemical composition of the aqueous phase changes along the reaction path from A to B.

When the chemical composition of the aqueous phase reaches point B, kaolinite becomes the stable phase. Therefore, the previously formed gibbsite reacts with silicic acid in the water to form kaolinite:

$$2 Al(OH)_3 + 2 H_4SiO_4 \rightarrow \qquad (12.13)$$
$$Al_2Si_2O_5(OH)_4 + 5 H_2O$$

This reaction consumes silicic acid, but does not affect the $[K^+]/[H^+]$ ratio of the water. At the same time, microcline continues to dissolve incongruently to form kaolinite (equation 1, Table 12.1):

$$2 KAlSi_3O_8 + 9 H_2O + 2 H^+ \rightarrow \qquad (12.14)$$
$$Al_2Si_2O_5(OH)_4 + 2 K^+ + 4 H_4SiO_4$$

This reaction releases silicic acid and also causes the $[K^+]/[H^+]$ ratio of the water to increase. If equilibrium between gibbsite and kaolinite is maintained, the concentration of silicic acid in the water must remain constant until all of the previously formed gibbsite has been converted to kaolinite. Under these conditions, the reactions at point B in Figure 12.1 can be represented by one equation written in such a way that Si is conserved (Steinmann et al., 1994):

$$KAlSi_3O_8 + 2 Al(OH)_3 + H^+ \rightarrow \qquad (12.15)$$
$$1.5 Al_2Si_2O_5(OH)_4 + K^+ + 0.5 H_2O$$

This equation can be constructed by multiplying equation 12.14 by 1/2 and adding it to 12.13. The reactions represented by equation 12.15 cause the $[K^+]/[H^+]$ ratio of the water to rise along the gibbsite–kaolinite boundary from point B to point C, which is reached when all of the gibbsite has been converted to kaolinite.

Subsequently, microcline reacts with ions in the water to form kaolinite in accordance with equation 12.14. As a result, the $[K^+]/[H^+]$ ratio as well as $[H_4SiO_4]$ both increase along the reaction path between points C and D, provided that enough microcline is available to allow reaction 12.14 to continue.

If the chemical composition of the water reaches the stability boundary between kaolinite and muscovite (illite), kaolinite becomes unstable and reacts with the ions in the water to form muscovite (equation 4, Table 12.1):

$$3 Al_2Si_2O_5(OH)_4 + 2 K^+ \rightarrow \qquad (12.16)$$
$$2 KAl_3Si_3O_{10}(OH)_2 + 3 H_2O + 2 H^+$$

This reaction causes a decrease of the $[K^+]/[H^+]$ ratio of the water without affecting the activity of $[H_4SiO_4]$. However, if microcline is still available, it will now form muscovite (equation 3, Table 12.1):

$$3 KAlSi_3O_8 + 12 H_2O + 2 H^+ \rightarrow \qquad (12.17)$$
$$KAl_3Si_3O_{10}(OH)_2 + 2 K^+ + 6 H_4SiO_4$$

This reaction increases both the $[K^+]/[H^+]$ ratio and $[H_4SiO_4]$. Since the composition of the water must remain on the stability boundary between kaolinite and muscovite until all of the previously formed kaolinite has been converted to muscovite, the $[K^+]/[H^+]$ ratio of the solution must remain constant. Therefore, equations 12.16 and 12.17 can be combined to yield:

$$KAlSi_3O_8 + Al_2Si_2O_5(OH)_4 + 3 H_2O \rightarrow$$
$$KAl_3Si_3O_{10}(OH)_2 + 2 H_4SiO_4 \qquad (12.18)$$

This reaction increases $[H_4SiO_4]$ without affecting the $[K^+]/[H^+]$ ratio of the water. Therefore, the chemical composition of the water moves along the kaolinite–muscovite boundary in Figure 12.1 from point D toward point E.

If sufficient microcline is available to drive reaction 12.18 until the chemical composition of the water reaches point *E*, all three minerals (microcline, kaolinite, and muscovite) coexist in equilibrium with the ions in the solution. Consequently, the chemical weathering reactions stop at point *E* provided that the temperature remains constant and the system remains closed.

The reaction path shown in Figure 12.1 is one of an infinite number of possible paths whose course depends on the initial chemical composition and pH of the water. The work of Steinmann et al. (1994) indicated that the shape of reaction paths in Figure 12.1 is especially sensitive to the initial pH of the water. If pH > 5, the reaction paths proceed from point *A* to the gibbsite–muscovite boundary without entering the kaolinite field.

The purpose of this discourse is to emphasize that the reactions between microcline (or another mineral) and acidified water not only cause transformations among the solid phases but also affect the chemical composition of the water. These reactions therefore contribute significantly to the chemical evolution of water both on the surface and below the surface of the Earth. Therefore, activity diagrams like Figure 12.1 provide useful information about several geochemical questions:

1. What environmental conditions are required to allow a particular mineral to form?
2. What minerals are stable in a given geochemical environment?
3. What ions or molecules are consumed or produced when an unstable mineral reacts in a given geochemical environment?
4. How does the water evolve chemically when reactions occur in a closed system with a small water/rock ratio?

The answers to these kinds of questions provide information that is useful in hydrogeology, sedimentary petrology, soil science, geochemical prospecting, economic geology, clay mineralogy, and other important applications of inorganic aqueous geochemistry. A large number of activity diagrams for many different systems at different temperatures and pressures was published by Bowers et al. (1984).

The stability limits of albite ($NaAlSi_3O_8$) and anorthite ($CaAl_2Si_2O_8$) are shown in Figure 12.2, and the relevant equations are listed in Table 12.2. These diagrams can be interpreted in the same way as the microcline diagram. They are, in fact, complementary to Figure 12.1 because they not only introduce Na^+ and Ca^{2+}, respectively, but also involve H^+ and H_4SiO_4. In most natural environments all three minerals are likely to be present. Therefore, the activities of H^+ and H_4SiO_4 of natural solutions reflect the progress of all three chemical reactions occurring simultaneously. In addition, other reactions may be taking place involving other silicate minerals, as well as carbonates, phosphates, sulfides, or sulfates. As a result, predictions about the progress of one particular reaction occurring in such complex natural environments may be affected by the other reactions that are occurring simultaneously.

The stability diagram of albite in Figure 12.2A includes the Na-mica *paragonite* and therefore is similar to that of microcline. The solubility limit of magadiite (Surdam and Eugster, 1976) is listed in Table 12.2 but is not shown in Figure 12.2A. The stability of analcime is discussed in Section 12.2 with other zeolite minerals, whereas reactions among the clay minerals are discussed in Chapter 13.

The stability field of the Ca-mica *margarite* is not shown in Figure 12.2B because it is a rare mineral and its standard free energy of formation is not well known. In addition, the activity of Ca^{2+} in aqueous solutions reacting with anorthite is limited by the solubility of calcite, which depends on the fugacity of CO_2 of the environment and on the H^+ activity. The equation that governs the solubility limit of calcite can be constructed as follows. We know that calcite dissolves to form Ca^{2+} and HCO_3^- at pH values between 6.4 and 10.3 (Section 9.5). However, the bicarbonate ion is also in equilibrium with molecular $H_2CO_3^0$, which equilibrates with $CO_2(g) + H_2O$. Therefore, we express the solubility of calcite in terms of Ca^{2+} and $CO_2(g)$ because we

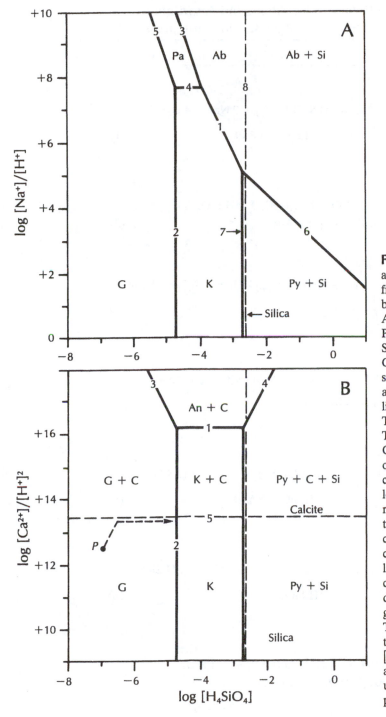

Figure 12.2 Stability of albite (**A**) and anorthite (**B**) in the presence of acidified water at 25°C and 1 atm pressure based on equations listed in Table 12.2. Ab = albite, K = kaolinite, G = gibbsite, Pa = paragonite, Py = pyrophyllite, Si = amorphous silica, An = anorthite, C = calcite. Na- and Ca-smectite and the solubility limits of quartz and magadiite are not shown, but the magadiite solubility limit is given by equation 9 in Table 12.2A. The solubility limit of calcite (equation 7, Table 12.2B) in **B** is based on a fugacity of $CO_2 = 3 \times 10^{-4}$ atm. The solubility limits of amorphous silica and calcite restrict the compositions of natural solutions to the lower left corner of **B**. Point P in that area represents an environment in which anorthite alters to gibbsite, a reaction that can cause the composition of the water to change. When the reaction path reaches line 5 (Table 12.2B), calcite begins to precipitate and the path turns to the right but does not advance beyond the gibbsite–kaolinite border (equation 2, Table 12.2B). Because of the way in which the solubility limit of calcite restricts the $[Ca^{2+}]/[H^+]^2$ ratio of water, authigenic anorthite cannot form by reactions of natural solutions with gibbsite, kaolinite, or pyrophyllite at 25°C.

Table 12.2 Stability Relations for Albite and Anorthite in Contact with Aqueous Solutions at 25°C and 1 atm pressure[a]

A. Albite

Albite–kaolinite

$$2\,NaAlSi_3O_8 + 9\,H_2O + 2\,H^+ \rightleftharpoons Al_2Si_2O_5(OH)_4 + 2\,Na^+ + 4\,H_4SiO_4$$

$$\log \frac{[Na^+]}{[H^+]} = -2\log [H_4SiO_4] - 0.19 \tag{1}$$

Kaolinite–gibbsite

$$Al_2Si_2O_5(OH)_4 + 5\,H_2O \rightleftharpoons 2\,Al(OH)_3 + 2\,H_4SiO_4$$

$$\log [H_4SiO_4] = -4.68 \tag{2}$$

Albite–paragonite

$$3\,NaAlSi_3O_8 + 12\,H_2O + 2\,H^+ \rightleftharpoons NaAl_3Si_3O_{10}(OH)_2 + 2\,Na^+ + 6\,H_4SiO_4$$

$$\log \frac{[Na^+]}{[H^+]} = -3\log [H_4SiO_4] - 4.10 \tag{3}$$

Paragonite–kaolinite

$$2\,NaAl_3Si_3O_{10}(OH)_2 + 3\,H_2O + 2\,H^+ \rightleftharpoons 3\,Al_2Si_2O_5(OH)_4 + 2\,Na^+$$

$$\log \frac{[Na^+]}{[H^+]} = +7.63 \tag{4}$$

Paragonite–gibbsite

$$NaAl_3Si_3O_{10}(OH)_2 + 9\,H_2O + H^+ \rightleftharpoons 3\,Al(OH)_3 + Na^+ + 3\,H_4SiO_4$$

$$\log \frac{[Na^+]}{[H^+]} = -3\log [H_4SiO_4] - 6.40 \tag{5}$$

Albite–pyrophyllite

$$2\,NaAlSi_3O_8 + 4\,H_2O + 2\,H^+ \rightleftharpoons Al_2Si_4O_{10}(OH)_2 + 2\,Na^+ + 2\,H_4SiO_4$$

$$\log \frac{[Na^+]}{[H^+]} = -\log [H_4SiO_4] + 2.48 \tag{6}$$

Pyrophyllite–Kaolinite

$$Al_2Si_4O_{10}(OH)_2 + 5\,H_2O \rightleftharpoons Al_2Si_2O_5(OH)_4 + 2\,H_4SiO_4$$

$$\log [H_4SiO_4] = -2.67 \tag{7}$$

Table 12.2 (continued)

Solubility limit of amorphous silica

$$SiO_2(amorph.) + 2\,H_2O \rightleftharpoons H_4SiO_4$$
$$\log[H_4SiO_4] = -2.64 \tag{8}$$

Solubility limit of magadiite

$$NaSi_7O_{13}(OH)_3 + 12\,H_2O + H^+ \rightleftharpoons Na^+ + 7\,H_4SiO_4$$
$$\Delta G_R^\circ = +18.88\ \text{kcal} \quad K = 10^{-13.84}$$
$$\log\frac{[Na^+]}{[H^+]} = -7\log[H_4SiO_4] - 13.84 \tag{9}$$

B. Anorthite

Anorthite–kaolinite

$$CaAl_2Si_2O_8 + H_2O + 2\,H^+ \rightleftharpoons Al_2Si_2O_5(OH)_4 + Ca^{2+}$$
$$\log\frac{[Ca^{2+}]}{[H^+]^2} = +16.17 \tag{1}$$

Kaolinite–gibbsite

$$Al_2Si_2O_5(OH)_4 + 5\,H_2O \rightleftharpoons 2\,Al(OH)_3 + 2\,H_4SiO_4$$
$$\log[H_4SiO_4] = -4.68 \tag{2}$$

Anorthite–gibbsite

$$CaAl_2Si_2O_8 + 6\,H_2O + 2\,H^+ \rightleftharpoons 2\,Al(OH)_3 + Ca^{2+} + 2\,H_4SiO_4$$
$$\log\frac{[Ca^{2+}]}{[H^+]^2} = -2\log[H_4SiO_4] + 6.82 \tag{3}$$

Anorthite–pyrophyllite

$$CaAl_2Si_2O_8 + 2\,H_4SiO_4 + 2\,H^+ \rightleftharpoons Al_2Si_4O_{10}(OH)_2 + Ca^{2+} + 4\,H_2O$$
$$\log\frac{[Ca^{2+}]}{[H^+]^2} = 2\log[H_4SiO_4] + 21.50 \tag{4}$$

Solubility limit of calcite

$$CaCO_3 + 2\,H^+ \rightleftharpoons Ca^{2+} + CO_2 + H_2O$$
$$\log\frac{[Ca^{2+}]}{[H^+]^2} = 13.38 \quad \text{if} \quad [CO_2] = 3 \times 10^{-4}\ \text{atm} \tag{5}$$

[a]Based on the thermodynamic data in Appendix B.

thereby avoid having to introduce the bicarbonate ion into the equations:.

$$CaCO_3(s) + 2 H^+ \rightleftharpoons Ca^{2+} + CO_2(g) + H_2O(l) \tag{12.19}$$

In the standard state, $\Delta G_R^\circ = -13.441$ kcal and hence $K = 10^{9.85}$. If $[CO_2] = 3 \times 10^{-4}$ atm:

$$\frac{[Ca^{2+}]}{[H^+]^2} = \frac{10^{9.85}}{3 \times 10^{-4}} = 10^{13.38} \tag{12.20}$$

and $\log [Ca^{2+}]/[H^+]^2 = 13.38$. Therefore, calcite precipitates when $\log [Ca^{2+}]/[H^+]^2 > 13.88$. Consequently, the decomposition of anorthite can result in the formation of gibbsite + calcite, kaolinite + calcite, or kaolinite + calcite + amorphous silica, depending on the environment to which anorthite is exposed. For example, if anorthite is placed into an environment represented by point P in the stability field of gibbsite, anorthite is converted to gibbsite in accordance with reaction 3 in Table 12.2B. If the system is closed and the water/rock ratio is small, the composition of the water changes because Ca^{2+} and H_4SiO_4 are released. If the fugacity of CO_2 is 3×10^{-4} atm, calcite begins to precipitate when the reaction path reaches the solubility boundary of calcite. As a result, the reaction path moves along the solubility boundary of calcite (line 5 in Figure 12.2B) because the activity of H_4SiO_4 increases as anorthite continues to be converted to gibbsite.

When the reaction path reaches the stability boundary between gibbsite and kaolinite (line 2 in Figure 12.2B), the remaining anorthite reacts to form kaolinite by reaction 1 in Table 12.2B. This reaction tends to increase the $[Ca^{2+}]/[H^+]^2$ ratio of the solution and therefore causes more calcite to precipitate. The reaction path does not advance beyond the gibbsite–kaolinite boundary because the conversion of anorthite to kaolinite does not increase the activity of H_4SiO_4 of the solution and because the $[Ca^{2+}]/[H^+]^2$ activity ratio cannot exceed the solubility limit of calcite. Therefore, the solubility limits of calcite and amorphous silica in Figure 12.2B *restrict* the compositions of natural waters to the lower left corner of the diagram.

Another consequence of the calcite solubility limit is that neither gibbsite nor kaolinite can be converted into anorthite because $\log [Ca^{2+}]/[H^+]^2$ of the solution cannot rise above 13.38. Therefore, the reactions between acidified water and anorthite that produce either kaolinite or gibbsite are not reversible at 25 °C. If the fugacity of CO_2 is increased by a factor of 10 to 3×10^{-3} atm, $\log [Ca^{2+}]/[H^+]^2 = 12.38$ (equation 12.20), which causes calcite to precipitate at even lower values of the $[Ca^{2+}]/[H^+]^2$ ratio. Therefore, anorthite is *not* formed from clay minerals or aluminum hydroxide at Earth-surface temperatures, whereas albite and K-feldspar *do* form because their carbonates, oxides, and hydroxides are much more soluble than those of Ca or Mg and therefore do not impose limits on the activities of their ions in aqueous solutions.

12.2   Formation of Zeolites

The zeolites are a large and highly diversified group of aluminosilicates that occur in certain igneous, metamorphic, and sedimentary rocks (Gottardi and Galli, 1985). They can form by direct precipitation from aqueous solution in vesicles and fractures of lava flow. However, zeolites also form in sedimentary rocks by reactions between volcanic glass, feldspars, feldspathoids, and other silicate minerals with saline solutions of high pH (Lisitzina and Butuzova, 1982; Petzing and Chester, 1979). The zeolites contain water of hydration, which is gradually forced out by increases in pressure and temperature. Hence, zeolites have been used to determine the depth of burial of lava flows and of volcaniclastic sedimentary rocks (Coombs et al., 1959; Walker, 1960; Sutherland, 1977). The crystal structures of zeolites contain channels and cavities that make these minerals useful as molecular sieves. Therefore, zeolites are used in the purification of water and other liquids and gases, for chemical separations, and for decontamination of radioactive waste (Barrer, 1978). Because of their many industrial uses, zeolites are synthesized for specific applications, but natural materials recovered from large sedimentary deposits are also used (Hay, 1966).

Zeolites are tektosilicates (Section 8.3) composed of silica tetrahedra linked at the corners by sharing oxygen atoms so that the ratio of Si to O is $1:2$. The replacement of some Si atoms by Al atoms creates a *deficiency* of *positive charge* in the lattice because Al is trivalent whereas Si is tetravalent. The resulting *excess negative charge* is neutralized by incorporation of Na^+ or Ca^{2+} and, more rarely, of K^+, Ba^{2+}, and Sr^{2+} ions. The zeolites have a more open lattice than other tektosilicates, such as the feldspars and feldspathoids, which allows them to contain water and makes them useful as molecular sieves. Some zeolites (chabazite) can be dehydrated by heating without permanently damaging the crystal lattice. Dehydrated chabazite can reabsorb the same amount of water it contained originally. Other zeolites (phillipsite and gismondine) experience major structural changes during dehydration below 200°C.

The chemical compositions of zeolites range widely because of variations in the number of Si atoms that are replaced by Al atoms, because the resulting charge deficiency is neutralized by varying proportions of Na^+, K^+, Ca^{2+}, Ba^{2+}, and Sr^{2+}, and because the amount of water may vary depending on pressure and temperature. Nevertheless, the chemical formulas of zeolites can be rationalized by normalizing the Al to two atoms, regardless of the numbers of SiO_2 "units" or water molecules. For example, the formula of stilbite ($CaAl_2Si_6O_{16} \cdot 6H_2O$) can be understood in terms of coupled substitution:

$$8(SiO_2) + 2\,Al^{3+} \rightarrow Al_2Si_6O_{16}^{2-} + 2\,Si^{4+} \quad (12.21)$$

$$Al_2Si_6O_{16}^{2-} + Ca^{2+} + 6\,H_2O \rightarrow$$

$$CaAl_2Si_6O_{16} \cdot 6\,H_2O \quad (12.22)$$

The composition of zeolites in sedimentary rocks may change continuously in response to changes in the physical and chemical state of their environment. The complexity of the chemical compositions of zeolites and their sensitivity to environmental change make it difficult to determine their standard free energies of formation (Johnson et al., 1982; Hemingway and Robie,

1984). Hence, thermodynamic constants of only a few zeolites (analcime, wairakite, laumontite, and leonhardite) are listed in Appendix B.

We begin by modifying the stability diagram of albite (Figure 12.2A) by including the zeolite analcime ($NaAlSi_2O_6 \cdot H_2O$). The reaction by which albite is transformed into analcime is easily constructed:

$$NaAlSi_3O_8 + 3\,H_2O \rightleftharpoons$$

$$NaAlSi_2O_6 \cdot H_2O + H_4SiO_4 \quad (12.23)$$

$$\Delta G_R^\circ = +5.741 \text{ kcal} \quad K = 10^{-4.21}$$

$$\log[H_4SiO_4] = -4.21 \quad (12.24)$$

When equation 12.24 is plotted as line 1 in Figure 12.3A, it becomes apparent that analcime must also react to form paragonite. Therefore, we construct the equation to represent this reaction:

$$3\,NaAlSi_2O_6 \cdot H_2O + 3\,H_2O + 2\,H^+ \rightleftharpoons$$

$$NaAl_3Si_3O_{10}(OH)_2 + 2\,Na^+ + 3\,H_4SiO_4 \quad (12.25)$$

$$\Delta G_R^\circ = -6.045 \text{ kcal} \quad K = 10^{+4.43}$$

$$\log\frac{[Na^+]}{[H^+]} = -1.5\log[H_4SiO_4] + 2.22 \quad (12.26)$$

The remainder of Figure 12.3A is identical to Figure 12.2A. We see that analcime is stable in environments having high activities of Na^+ but low activities of H^+, which is typical of saline brines formed by evaporative concentration of surface water. Therefore, analcime commonly occurs in playa-lake deposits and in soils and volcaniclastic sediment in arid regions of the world. It occurs in very small crystals ranging from 5 to 100 μm but can be detected by x-ray diffraction (Hay, 1966).

Next, we consider the Ca-zeolites:

$$CaAl_2Si_4O_{12} \cdot 2H_2O$$

wairakite

$$CaAl_2Si_4O_{12} \cdot 3.5H_2O$$

leonhardite

$$CaAl_2Si_4O_{12} \cdot 4H_2O$$

laumontite

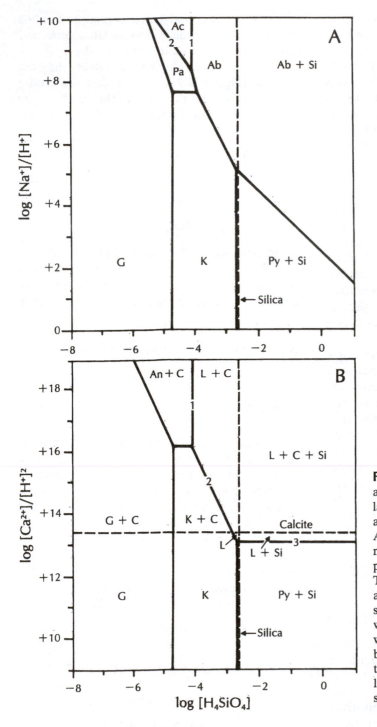

Figure 12.3 Stability of albite (**A**) and anorthite (**B**) with respect to analcime and laumontite, respectively, in the presence of acidified water at 25°C at 1 atm pressure. Ab = albite, Ac = analcime, Pa = paragonite, K = kaolinite, G = gibbsite, Si = amorphous silica, An = anorthite, L = laumontite. The equations for the numbered boundaries are given in the text and the others are the same as in Figure 12.2. Analcime can form when any of the solids in Figure 12.3A reacts with Na-rich brines of high pH. However, the brine compositions required to form laumontite at 25°C are restricted by the solubility limits of calcite and amorphous silica to the small area in Figure 12.3B labeled L.

all of which have identical molar Si/Al ratios of 2 : 1 but differ in the number of water molecules they contain. Therefore, we choose laumontite to represent these minerals on the stability diagram in Figure 12.3B because it forms in near-surface environments and may subsequently convert to leonhardite or wairakite. Leonhardite forms by dehydration of laumontite upon exposure to the atmosphere, whereas wairakite becomes stable between 150 and 300°C at 1 atm pressure (Bowers et al., 1984). The transformation of anorthite to laumontite proceeds as follows (line 1, Figure 12.3B):

$$CaAl_2Si_2O_8 + 2\,H_4SiO_4 \rightleftharpoons CaAl_2Si_4O_{12} \cdot 4H_2O \tag{12.27}$$

$$\Delta G_R^\circ = -11.32 \text{ kcal} \quad K = 10^{8.30}$$
$$\log [H_4SiO_4] = -4.15 \tag{12.28}$$

The decomposition of laumontite to kaolinite is represented by the equation:

$$CaAl_2Si_4O_{12} \cdot 4H_2O + H_2O + 2\,H^+ \rightleftharpoons$$
$$Al_2Si_2O_5(OH)_4 + Ca^{2+} + 2\,H_4SiO_4 \tag{12.29}$$

$$\Delta G_R^\circ = -10.73 \text{ kcal} \quad K = 10^{7.87}$$

$$\log \frac{[Ca^{2+}]}{[H^+]^2} = -2 \log [H_4SiO_4] + 7.87 \tag{12.30}$$

Equation 12.30 has been plotted as line 2 in Figure 12.3B. Finally, the laumontite–pyrophyllite equilibrium is represented by:

$$CaAl_2Si_4O_{12} \cdot 4H_2O + 2\,H^+ \rightleftharpoons$$
$$Al_2Si_4O_{10}(OH)_2 + Ca^{2+} + 4\,H_2O \tag{12.31}$$

$$\Delta G_R^\circ = -18.00 \text{ kcal} \quad K = 10^{+13.20}$$

$$\log \frac{[Ca^{2+}]}{[H^+]^2} = +13.20 \tag{12.32}$$

Equation 12.32 forms line 3 in Figure 12.3B. All other boundaries are the same as in Figure 12.2B.

We see that laumontite has a large stability field in Figure 12.3B, but only a very small part of it lies within the compositional range of natural solutions. Therefore, anorthite, or any other Al-bearing compound, can alter to laumontite at

25°C only in those environments that lie within the small triangular field labeled "L." In addition, the conversion of anorthite to laumontite is irreversible because the calcite solubility boundary at atmospheric values of the fugacity of CO_2 prevents the $[Ca^{2+}]/[H^+]^2$ ratio of natural solutions from increasing to values within the anorthite field.

We can combine the stability diagrams of analcime and laumontite (Figure 12.3) by specifying that the solutions are saturated with respect to amorphous silica. This presents no problem for laumontite, which can be in equilibrium with such a solution. However, analcime, paragonite, and gibbsite are not stable in contact with silica-saturated solutions. Therefore, we anticipate that *kaolinite* is the stable phase in silica-saturated solutions having *low* activities of Na^+ and Ca^{2+} and that kaolinite can coexist in equilibrium with albite and laumontite as the activities of Na^+ and Ca^{2+} in the solution increase. Actually, pyrophyllite forms from kaolinite just below the silica-saturation limit (Figure 12.2). However, we prefer kaolinite because it is far more abundant as a weathering product of aluminosilicate minerals than pyrophyllite. Accordingly, we begin the construction of this diagram with the conversion of kaolinite to albite in a silica-saturated solution:

$$Al_2Si_2O_5(OH)_4 + 2\,Na^+ + 4\,SiO_2(\text{amorph.}) \rightleftharpoons$$
$$2\,NaAlSi_3O_8 + H_2O + 2\,H^+ \tag{12.33}$$

$$\Delta G_R^\circ = +13.899 \text{ kcal} \quad K = 10^{-10.19}$$

$$\log \frac{[Na^+]}{[H^+]} = +5.09 \quad \text{(line 1, Figure 12.4)} \tag{12.34}$$

The reaction of kaolinite with Ca^{2+} to form laumontite is represented by:

$$Al_2Si_2O_5(OH)_4 + Ca^{2+} + 2\,SiO_2(\text{amorph.})$$
$$+ 3\,H_2O \rightleftharpoons CaAl_2Si_4O_{12} \cdot 4H_2O + 2\,H^+ \tag{12.35}$$

$$\Delta G_R^\circ = +17.941 \text{ kcal} \quad K = 10^{-13.15}$$

$$\log \frac{[Ca^{2+}]}{[H^+]^2} = +13.15 \quad \text{(line 2, Figure 12.4)} \tag{12.36}$$

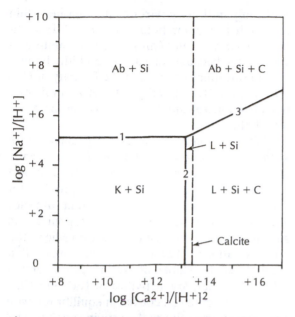

Figure 12.4 Stability diagram for albite (Ab), kaolinite (K), laumontite (L), and calcite (C) in contact with solutions saturated with respect to amorphous SiO_2 (Si) at 25°C. Analcime, paragonite, anorthite, and gibbsite were excluded from this diagram because they are unstable in silica-saturated solutions as shown in Figure 12.3. However, pyrophyllite is probably stable, but was excluded because kaolinite is more abundant as a weathering product of Al-silicates than pyrophyllite. The calcite-saturation limit is based on $[CO_2] = 3 \times 10^{-4}$ atm (Table 12.2B, equation 5). The numbered lines (1, 2, and 3) were derived from equations 12.33, 12.35, and 12.37, respectively.

Equilibrium between albite and laumontite in silica-saturated solutions is based on:

$$2\,NaAlSi_3O_8 + Ca^{2+} + 4\,H_2O \rightleftharpoons$$
$$CaAl_2Si_4O_{12} \cdot 4H_2O + 2\,Na^+$$
$$+ 2\,SiO_2(amorph.) \quad (12.37)$$

$$\Delta G_R^\circ = +4.042 \text{ kcal} \quad K = 10^{-2.96}$$

We note that reaction 12.37 is independent of the pH and therefore must be modified before it can be plotted in the coordinates of Figure 12.4.

Applying the Law of Mass Action to equation 12.37 yields:

$$\frac{[Na^+]^2}{[Ca^{2+}]} = 10^{-2.96} \quad (12.38)$$

We now divide the $[Na^+]^2$ and $[Ca^{2+}]$ in equation 12.38 by $[H^+]^2$:

$$\frac{[Na^+]^2/[H^+]^2}{[Ca^{2+}]/[H^+]^2} = 10^{-2.96} \quad (12.39)$$

Taking logarithms of equation 12.39 yields the desired equation:

$$2\log\frac{[Na^+]}{[H^+]} - \log\frac{[Ca^{2+}]}{[H^+]^2} = -2.96 \;(12.40)$$

and hence:

$$\log\frac{[Na^+]}{[H^+]} = 0.5\log\frac{[Ca^{2+}]}{[H^+]^2} - 1.48$$
$$\text{(line 3, Figure 12.4)} \quad (12.41)$$

The three lines derived from reactions 12.33, 12.35, and 12.37 have been plotted in Figure 12.4 together with the solubility limit of calcite (Table 12.2B, equation 5) for $[CO_2] = 3 \times 10^{-4}$ atm. The resulting diagram delineates the conditions under which albite decomposes to kaolinite + silica or laumontite + silica. Its virtue lies in the fact that the environmental conditions are described in terms of both Na^+ and Ca^{2+}, but it is limited by the instability of certain solids in the presence of silica-saturated solutions.

12.3   Magnesium Silicates

Magnesium forms a variety of silicate minerals, some of which include Al (phlogopite, cordierite, vermiculite, chlorite, and Mg-smectite), whereas others lack Al (forsterite, enstatite, serpentine, talc, and sepiolite). Some Mg silicates contain other cations, such as K^+ (phlogopite), Fe^{2+}, Ca^{2+}, and K^+ (vermiculite), and Fe^{2+} (Mg-montmorillonite). Therefore, these minerals can be shown only on diagrams that assume a specified activity of silicic acid or require other assumptions. For this reason, Figure 12.5A includes the stability fields only of chlorite (Ct) together with pyrophyllite (Py), kaolinite (K), gibbsite (G), amorphous silica (Si), and magnesite (Ms). Brucite and periclase have solubility limits of $\log [Mg^{2+}]/[H^+]^2 = 16.71$ and $\log [Mg^{2+}]/[H^+]^2 = 21.59$, respectively, and are unstable in Figure 12.5A. Therefore, magnesite effectively limits the $[Mg^{2+}]/[H^+]^2$ ratio of natural solutions to values less than $10^{13.52}$, assuming that $[CO_2] = 3 \times 10^{-4}$ atm. The equations for these reactions are listed in Table 12.3A.

The Al-bearing silicates of Mg dissolve *incongruently* because gibbsite, kaolinite, and pyrophyllite are highly insoluble. This situation does not occur when the Mg silicates that lack Al dissolve in acidified water. These minerals therefore dissolve *congruently* and form Mg^{2+} and H_4SiO_4. Consequently, the stabilities of the Mg-silicates are limited by the activities of Mg^{2+}, H^+, and H_4SiO_4 of the solution with which they come in contact. The equations defining the solubilities of the Mg silicates in coordinates of $\log [Mg^{2+}]/[H^+]^2$ and $\log [H_4SiO_4]$ are listed in Table 12.3B and have been plotted as dashed lines in Figure 12.5B.

The solubilities of magnesite, serpentine, sepiolite, and amorphous silica outline a region in Figure 12.5B in which all of the minerals are *soluble* and are therefore *unstable*. When the activity ratio $[Mg^{2+}]/[H^+]^2$ and/or $[H_4SiO_4]$ of solutions in that region increase, either magnesite, serpentine, sepiolite, or amorphous silica precipitates, depending on which solubility line is contacted. The composition of a solution may change either because a soluble Mg silicate such as forsterite, talc, enstatite, or anthophyllite is dissolving in the water in a closed system with a small water/rock ratio, or because of evaporative concentration of surface water, or both. For example, if enstatite dissolves in acidified water represented by the point P in Figure 12.5B, the reaction path intersects the serpentine solubility line. Therefore, serpentine precipitates from the solution as enstatite dissolves. In effect, serpentine *replaces* enstatite by the reactions:

$$MgSiO_3 + H_2O + 2 H^+ \rightleftharpoons Mg^2 + H_4SiO_4 \quad (12.42)$$

$$3 Mg^{2+} + 2 H_4SiO_4 + H_2O \rightleftharpoons \\ Mg_3Si_2O_5(OH)_4 + 6 H^+ \quad (12.43)$$

If Mg^{2+} is conserved by multiplying equation 12.42 by 3 before adding it to equation 12.43, the net result is:

$$3 MgSiO_3 + 4 H_2O \rightleftharpoons \\ Mg_3Si_2O_5(OH)_4 + H_4SiO_4 \quad (12.44)$$

Therefore, the conversion of enstatite to serpentine increases the activity of silicic acid. If enstatite continues to dissolve, the activity of silicic acid may increase until the solubility limit of sepiolite is reached. Continued solution of enstatite causes sepiolite to precipitate, and so on, until the solution becomes saturated with amorphous silica. However, the suggested sequence of events depends on assumptions about the reactions (i.e., Mg is conserved, enstatite continues to dissolve, no other reactions are taking place) and about the environment (closed, with small water/rock ratio, stable with respect to erosion, constant temperature) that are rarely satisfied in nature.

The replacement of enstatite by serpentine in reaction 12.44 can, in principle, achieve equilibrium at a specific value of $\log [H_4SiO_4]$. Since $\Delta G_R^\circ = -2.63$ kcal and $K = 10^{1.93}$, equilibrium between enstatite and serpentine occurs at $\log [H_4SiO_4] = 1.93$. This value *exceeds* the solubility limit of amorphous silica, which means that enstatite and serpentine cannot coexist in equilibrium in contact with natural solutions at 25°C. This is true also of enstatite–talc, serpentine–talc, forsterite–talc, serpentine–anthophyllite, and anthophyllite–talc because, in each case, the activity of H_4SiO_4 required for equilibrium is *greater* than that of a saturated solution of amorphous silica. Similarly, forsterite cannot coexist in equilibrium with enstatite, sepiolite, or anthophyllite because the $[Mg^{2+}]/[H^+]^2$ ratios required for equilibrium in each case exceed the solubility limit of magnesite.

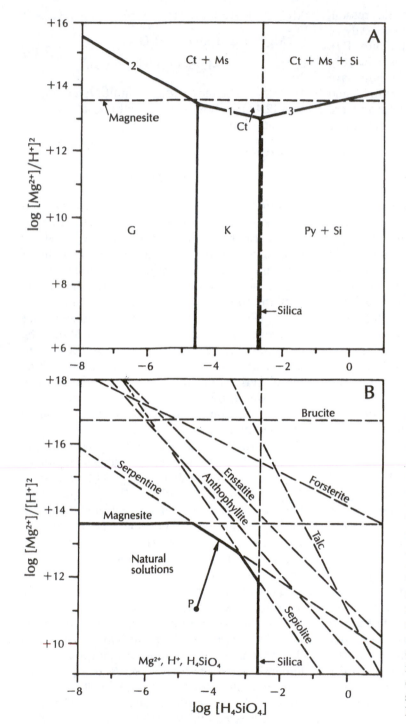

Figure 12.5 A: Stability diagram of Mg chlorite (Ct) (aluminosilicate) with respect to gibbsite (G), kaolinite (K), pyrophyllite (Py), amorphous silica (Si), and magnesite (Ms) at $[CO_2] = 3 \times 10^{-4}$ atm in the presence of acidified water at 25°C.

B: Solubility limits of pure Mg-silicates in acidified water at 25°C, based on equations in Table 12.3. Each of the minerals shown here dissolves in solutions whose $[Mg^{2+}]/[H^+]^2$ ratios and H_4SiO_4 activities are *less* than those of their respective solubility lines. Conversely, when these parameters increase in a solution in the lower left corner of the diagram, magnesite, serpentine, sepiolite, or amorphous silica precipitates, depending on which solubility line is reached by the solution. Consequently, these are the only stable minerals in contact with natural solutions on the surface of the Earth. The combination of solution and reprecipitation causes the replacement of forsterite, enstatite, talc, and anthophyllite by magnesite, serpentine, sepiolite, or amorphous silica. If amorphous silica precipitates first, and if one of the more soluble Mg silicates continues to dissolve, the $[Mg^{2+}]/[H^+]^2$ of the solution rises until sepiolite coprecipitates with silica. Similarly, the initial precipitation of magnesite may eventually result in the formation of serpentine, provided the activity of H_4SiO_4 increases sufficiently.

Table 12.3 Stability Relations among Mg Silicates in the Presence of Acidified Water at 25 °C and 1 atm pressure[a]

A. Mg–Al Silicates

Kaolinite–chlorite

$$Al_2Si_2O_5(OH)_4 + 5\,Mg^{2+} + H_4SiO_4 + 5\,H_2O \rightleftharpoons Mg_5Al_2Si_3O_{10}(OH)_8 + 10\,H^+$$
$$\Delta G_R^\circ = +85.235 \text{ kcal} \quad K = 10^{-62.49}$$
$$\log\frac{[Mg^{2+}]}{[H^+]^2} = -0.2\log[H_4SiO_4] + 12.50 \tag{1}$$

Gibbsite–chlorite

$$2\,Al(OH)_3 + 5\,Mg^{2+} + 3\,H_4SiO_4 \rightleftharpoons Mg_5Al_2Si_3O_{10}(OH)_8 + 10\,H^+$$
$$\Delta G_R^\circ = +72.48 \text{ kcal} \quad K = 10^{-53.14}$$
$$\log\frac{[Mg^{2+}]}{[H^+]^2} = -0.6\log[H_4SiO_4] + 10.62 \tag{2}$$

Pyrophyllite–chlorite

$$Al_2Si_4O_{10}(OH)_2 + 5\,Mg^{2+} + 10\,H_2O \rightleftharpoons Mg_5Al_2Si_3O_{10}(OH)_8 + H_4SiO_4 + 10\,H^+$$
$$\Delta G_R^\circ = +92.51 \text{ kcal} \quad K = 10^{-67.82}$$
$$\log\frac{[Mg^{2+}]}{[H^+]^2} = +0.2\log[H_4SiO_4] + 13.56 \tag{3}$$

Solubility limit of brucite

$$Mg(OH)_2 \rightleftharpoons Mg^{2+} + 2\,OH^-$$
$$\Delta G_R^\circ = +15.40 \text{ kcal} \quad K = 10^{-11.29}$$
$$\log\frac{[Mg^{2+}]}{[H^+]^2} = +16.71 \tag{4}$$

Solubility limit of periclase

$$MgO + 2\,H^+ \rightleftharpoons Mg^{2+} + H_2O$$
$$\Delta G_R^\circ = -29.45 \text{ kcal} \quad K = 10^{21.59}$$
$$\log\frac{[Mg^{2+}]}{[H^+]^2} = +21.59 \tag{5}$$

We conclude that the Mg-silicates whose solutibilities are shown as *dashed* lines in Figure 12.5B are *unstable* in the presence of acidified water and therefore dissolve in it. As a result, the $[Mg^{2+}]/[H^+]^2$ and $[H_4SiO_4]$ values of the solutions may rise until they reach the solubility limits of magnesite, serpentine, sepiolite, or amorphous silica, which are the only stable phases that can coexist with natural solutions at 25 °C.

Table 12.3 (continued)

B. Solubility of Mg Silicates

Forsterite

$$Mg_2SiO_4 + 4 H^+ \rightleftharpoons 2 Mg^{2+} + H_4SiO_4$$
$$\Delta G_R^\circ = -38.76 \text{ kcal} \quad K = 10^{+28.42}$$
$$\log \frac{[Mg^{2+}]}{[H^+]^2} = -0.5 \log [H_4SiO_4] + 14.2 \tag{1}$$

Enstatite

$$MgSiO_3 + H_2O + 2 H^+ \rightleftharpoons Mg^{2+} + H_4SiO_4$$
$$\Delta G_R^\circ = -15.31 \text{ kcal} \quad K = 10^{11.23}$$
$$\log \frac{[Mg^{2+}]}{[H^+]^2} = -\log [H_4SiO_4] + 11.23 \tag{2}$$

Serpentine

$$Mg_3Si_2O_5(OH)_4 + 6 H^+ \rightleftharpoons 3 Mg^{2+} + 2 H_4SiO_4 + H_2O$$
$$\Delta G_R^\circ = -43.31 \text{ kcal} \quad K = 10^{31.75}$$
$$\log \frac{[Mg^{2+}]}{[H^+]^2} = -0.67 \log [H_4SiO_4] + 10.58 \tag{3}$$

Anthophyllite

$$Mg_7Si_8O_{22}(OH)_2 + 8 H_2O + 14 H^+ \rightleftharpoons 7 Mg^{2+} + 8 H_4SiO_4$$
$$\Delta G_R^\circ = -93.95 \text{ kcal} \quad K = 10^{68.88}$$
$$\log \frac{[Mg^{2+}]}{[H^+]^2} = -1.14 \log [H_4SiO_4] + 9.84 \tag{4}$$

Talc

$$Mg_3Si_4O_{10}(OH)_2 + 4 H_2O + 6 H^+ \rightleftharpoons 3 Mg^{2+} + 4 H_4SiO_4$$
$$\Delta G_R^\circ = -29.91 \text{ kcal} \quad K = 10^{21.93}$$
$$\log \frac{[Mg^{2+}]}{[H^+]^2} = -2 \log [H_4SiO_4] + 10.96 \tag{5}$$

12.4   Solubility Diagrams

All minerals dissolve *congruently* when placed in *pure* water. For example, microcline dissolves congruently in pure water until the activities of

Al^{3+} ions in the solution increase sufficiently to stabilize gibbsite. Subsequently, microcline dissolves *incongruently* to form gibbsite plus K^+ and H_4SiO_4. We therefore need to know how gibbsite dissolves and under what conditions it is stable

Table 12.3 (continued)

Sepiolite (see Appendix B)

(a) $Mg_4Si_6O_{15}(OH)_2 \cdot 6H_2O + H_2O + 8\,H^+ \rightleftharpoons 4\,Mg^{2+} + 6\,H_4SiO_4$

$$\Delta G_R^\circ = -43.28 \text{ kcal} \quad K = 10^{31.73}$$

$$\log \frac{[Mg^{2+}]}{[H^+]^2} = -1.5 \log [H_4SiO_4] + 7.93 \tag{6}$$

(b) $Mg_2Si_3O_6(OH)_4 + 2\,H_2O + 4\,H^+ \rightleftharpoons 2\,Mg^{2+} + 3\,H_4SiO_4$

$$\Delta G_R^\circ = -21.26 \text{ kcal} \quad K = 10^{15.58}$$

$$\log \frac{[Mg^{2+}]}{[H^+]^2} = -1.5 \log [H_4SiO_4] + 7.79 \tag{7}$$

[a]Based on the thermodynamic data in Appendix B.

with respect to the different Al ions that can form in an aqueous solution.

a. Gibbsite

Since gibbsite is the insoluble hydroxide of Al, we recall the discussion of the dissociation of bases and their amphoteric character from Section 9.4. At that stage in our discussion we were restricted to known dissociation constants, whereas we can now calculate them from values of ΔG_R° for any reaction we wish to study. The Appendix B contains G_f° values for Al^{3+}, $Al(OH)^{2+}$, $Al(OH)_2^+$, $Al(OH)_3^0$, $Al(OH)_4^-$, and $Al(OH)_5^{2-}$, all of which can form when an Al-bearing mineral dissolves congruently at different pH values. The reactions by which these ions form from gibbsite and their equilibrium constants at 25°C are listed in Table 12.4.

The equations derived in this way outline the stability field of gibbsite in Figure 12.6. Outside this area, gibbsite dissolves congruently to form ions whose relative abundances become equal at certain pH values corresponding to the points of intersection of the solubility lines. For example, Al^{3+} is the dominant ion at pH < 4.65, and $Al(OH^+)_2$ dominates at pH values from 4.65 to 5.7, as shown in Figure 12.6. The *solubility* of gibbsite at a specified pH is the sum of the concentrations of all Al-bearing species that exist in a saturated solution in equilibrium with gibb-

site. The solubility of this mineral has a minimum at pH from 6 to 8 and rises at pH >8 and <6 in accordance with the amphoteric character of Al.

When microcline dissolves congruently in very dilute solutions having a pH <4.65, it releases K^+, Al^{3+}, and H_4SiO_4 according to the reaction:

$$KAlSi_3O_8 + 4\,H_2O + 4\,H^+ \rightleftharpoons$$
$$K^+ + Al^{3+} + 3\,H_4SiO_4 \tag{12.45}$$

When microcline dissolves congruently in water of composition P in Figure 12.6, both the activity of Al^{3+} and the pH rise along the reaction path. If the system is closed and its water/rock ratio is small, the solution may become saturated with respect to gibbsite, which then precipitates:

$$Al^{3+} + 3\,H_2O \rightleftharpoons Al(OH)_3 + 3\,H^+ \tag{12.46}$$

The net result, obtained by adding equations 12.45 and 12.46, is that microcline then dissolves *incongruently* to form gibbsite plus K^+ and H_4SiO_4:

$$KAlSi_3O_8 + 7\,H_2O + H^+ \rightleftharpoons$$
$$Al(OH)_3 + K^+ + 3\,H_4SiO_4 \tag{12.47}$$

If microcline continues to dissolve, the composition of the solution moves across the gibbsite stability field in Figure 12.1 until it reaches the gibbsite–kaolinite boundary (line 2). At that

Table 12.4 Congruent Solution of Gibbsite to Form Different Ionic and Molecular Species at 25 °C[a]

$Al(OH)_3(s) + 3 H^+ \rightleftharpoons Al^{3+} + 3 H_2O$
$\Delta G^\circ_R = -11.29 \text{ kcal} \quad K = 10^{8.28}$
$\log [Al^{3+}] = -3pH + 8.28 \qquad (1)$

$Al(OH)_3(s) + 2 H^+ \rightleftharpoons Al(OH)^{2+} + 2 H_2O$
$\Delta G^\circ_R = -4.44 \text{ kcal} \quad K = 10^{3.26}$
$\log [Al(OH)^{2+}] = -2pH + 3.26 \qquad (2)$

$Al(OH)_3(s) + H^+ \rightleftharpoons Al(OH)_2^+ + H_2O$
$\Delta G^\circ_R = +1.39 \text{ kcal} \quad K = 10^{-1.02}$
$\log [Al(OH)_2^+] = -pH - 1.02 \qquad (3)$

$Al(OH)_3(s) \rightleftharpoons Al(OH)_3^0$
$\Delta G^\circ_R = +9.16 \text{ kcal} \quad K = 10^{-6.72}$
$\log [Al(OH)_3^0] = -6.72 \qquad (4)$

$Al(OH)_3(s) + H_2O \rightleftharpoons Al(OH)_4^- + H^+$
$\Delta G^\circ_R = +20.5 \text{ kcal} \quad K = 10^{-15.1}$
$\log [Al(OH)_4^-] = pH - 15.1 \qquad (5)$

$Al(OH)_3(s) + 2 H_2O \rightleftharpoons Al(OH)_5^{2-} + 2 H^+$
$\Delta G^\circ_R = +35.41 \text{ kcal} \quad K = 10^{-26.0}$
$\log [Al(OH)_5^{2-}] = 2pH - 26.0 \qquad (6)$

[a]Based on the thermodynamic data in Appendix B.

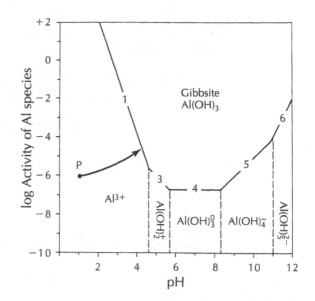

Figure 12.6 Congruent solubility of gibbsite as a function of the environmental pH at 25 °C, based on the equations in Table 12.4. The lines define the stability field of gibbsite where it precipitates from saturated solutions. The environments in which gibbsite is soluble have been subdivided by identifying the dominant Al species depending on the pH. Note that $Al(OH)^{2+}$ is not dominant at any pH according to the standard free energy values used to construct this diagram. The amphoteric character of Al causes the solubility of a gibbsite to increase with both decreasing and increasing pH. Its solubility is at a minimum at pH values between 6 and 8, which justifies the assumption that Al is conserved during the incongruent solution of aluminosilicates.

point, microcline forms kaolinite and the previously formed gibbsite is converted to kaolinite:

$2 KAlSi_3O_8 + 9 H_2O + 2 H^+ \rightleftharpoons$

$\qquad Al_2Si_2O_5(OH)_4 + 2 K^+ + 4 H_4SiO_4 \quad (12.48)$

$4 Al(OH)_3 + 4 H_4SiO_4 \rightleftharpoons$

$\qquad 2 Al_2Si_2O_5(OH)_4 + 10 H_2O \quad (12.49)$

If the conversion of gibbsite to kaolinite keeps pace with the incongruent solution of microcline to kaolinite, then $[H_4SiO_4]$ remains constant and only $[K^+]/[H^+]$ increases. Consequently, the reaction path moves "up" along the gibbsite-kaolinite boundary until all of the gibbsite is converted to kaolinite. If more microcline dissolves, the path

then traverses the kaolinite field as discussed before. Therefore the solubility diagram of gibbsite in Figure 12.6 complements the stability diagram of microcline in Figure 12.1.

The pH dependence of the solubility of gibbsite and of other Al-bearing minerals causes the abundance of Al species in aqueous solution to rise at both high and low environmental pH values. High concentrations of dissolved Al in soil tend to inhibit the growth of rooted plants for reasons that are not yet fully understood. The Al toxicity associated with strongly acidic or basic environments contributes to the "infertility" of such environments. In addition, we note that Al is mobile in strongly acidic and basic environments because it can be transported by the movement of water. Moreover, gibbsite can precipitate when strongly acidic or basic solutions are neutralized.

When other Al-bearing minerals such as kaolinite, pyrophyllite, feldspars, feldspathoids, zeolites, or clay minerals dissolve congruently in aqueous solutions, the abundances of the aqueous Al species depend on the environmental pH as shown in Figure 12.6. However, these minerals also produce H_4SiO_4 and the equilibria therefore involve three variables: the Al species, pH, and H_4SiO_4. The stabilities of these minerals in equilibrium with their ions can be represented either in three-dimensional diagrams or in coordinates of log [Al species] and log [H_4SiO_4] at a series of pH values.

b. Hematite

Iron, like Al, is amphoteric and forms hydroxyl complexes whose standard free energies are listed in Appendix B. Equations for the congruent solution of hematite as a function of pH are easily derived and have been plotted in Figure 12.7. The results define the stability field of hematite where it precipitates from aqueous solutions. The environments in which hematite dissolves have been subdivided according to the dominant Fe species in solution at different pH ranges. The solubility of hematite has a minimum between pH values of about 4.3–8.7 when $Fe(OH)_3^0$ is the dominant ion. The Fe^{3+} ion is dominant only at pH <2.2.

The solubilities of all oxides, hydroxides, carbonates, and sulfides of iron are strongly dependent on the pH of the environment. At pH <4.3 Fe becomes increasingly mobile in Earth-surface environments and is leached from soils, regolith, and rocks exposed to such conditions. When such acidic waters are neutralized, they may become saturated with respect to ferric hydroxide [$Fe(OH)_3$], which may then precipitate. The ferric hydroxide recrystallizes in time to form more stable crystalline phases such as lepidocrocite, goe-

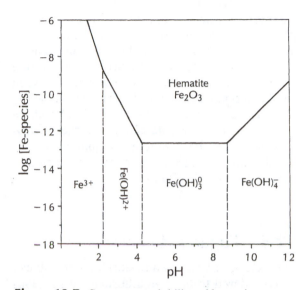

Figure 12.7   Congruent solubility of hematite as a function of pH based on equations like those for gibbsite in Table 12.4. The solubility of hematite, or other compounds of Fe, has a minimum at pH values from 4.3 to 8.7 and increases both in more acid and in more basic solutions because Fe, like Al, is amphoteric. As a result, Fe is mobile in acidic and basic environments and precipitates as amorphous ferric hydroxide when Fe-bearing solutions are neutralized. Ferric hydroxide recrystallizes through lepidocrocite and goethite to hematite.

thite, or hematite, which is the most stable form of Fe oxide on the surface of the Earth.

The geochemistry of Fe differs from that of Si and Al because it has two oxidation states (+2 and +3) both of which occur at Earth-surface conditions. The transformation from Fe^{2+} to Fe^{3+} occurs by the loss of an electron from Fe^{2+} in the presence of an electron acceptor. Loss of electrons is known to us as "oxidation," whereas a gain of electrons is "reduction." Oxidation and reduction must always occur together by a transfer of electrons from an electron donor to an electron acceptor. Oxidation–reduction reactions must be balanced *first* in terms of the electrons that are transferred before a mass balance is achieved. These matters are the subject of

Chapter 14, where we review some basic principles of electrochemistry and apply them to the construction of the so-called Eh–pH diagrams.

In Section 14.5 we take up the construction of fugacity diagrams based on equilibria between solid compounds and certain gases, including O_2, CO_2, and S_2. The solid–gas equilibria of both Fe and Cu include possible *changes* in their *oxidation states,* but the reactions are quite uncomplicated. Thus we can present fugacity diagrams here rather than in Chapter 14.

12.5   Fugacity Diagrams

The oxides, carbonates, and sulfides of metals can be considered to be in equilibrium with gaseous O_2, CO_2, and S_2, respectively. For example, the fugacity of O_2 ranges from about 2×10^{-1} atm in air at the surface of the Earth to much lower values in certain anoxic environments such as peat bogs, stagnant basins, and deep-sea trenches. Therefore, the first question we need to consider is the range of fugacities of O_2 that can occur on the surface of the Earth.

a. Stability Limits of Water
We know that liquid water dissociates into H^+ and OH^- ions and that the ion activity product of water at $25\,°C$ is about 1×10^{-14}. However, water must also maintain an equilibrium with the gases O_2 and H_2:

$$2\,H_2O(l) \rightleftharpoons O_2(g) + 2\,H_2(g) \quad (12.50)$$

The standard free energy change of this reaction is $\Delta G_R^\circ = +113.374$ kcal and its equilibrium constant at $25\,°C$ is $K = 10^{-83.1}$. Therefore, at equilibrium:

$$[O_2][H_2]^2 = 10^{-83.1} \quad (12.51)$$

The total pressure of all gases occurring naturally on the surface of the Earth must be 1 atm or less. If the pressure of gases rises to higher values, they expand against the atmosphere, which means that bubbles form in water exposed to the atmosphere

and the gases escape. Therefore, the fugacities of O_2 and H_2 must each remain <1.0 atm if liquid water is to be stable at sea level on the surface of the Earth. If the fugacity of H_2 alone is 1.0 atm, then the fugacity of O_2 must be as low as it can get in the presence of liquid water. Setting $[H_2] = 1.0$ in equation 12.51 yields:

$$[O_2] = 10^{-83.1} \text{ atm} \quad (12.52)$$

Therefore, the fugacity of O_2 in the presence of liquid water may vary from 1.0 to $10^{-83.1}$ atm. Similarly, the fugacity of H_2 can vary from 1.0 to $10^{-41.6}$ atm, based on equation 12.51.

b. Oxides and Carbonates of Iron
Metallic Fe exposed to the air is known to form the oxides wüstite (FeO), magnetite (Fe_3O_4), and hematite (Fe_2O_3). Therefore, we can represent the oxidation of metallic Fe by a series of equations starting with the formation of wüstite:

$$\text{Fe (metallic)} + \tfrac{1}{2}O_2 \rightleftharpoons FeO \quad (12.53)$$

For this reaction, $\Delta G_R^\circ = -58.7$ kcal and $K = 10^{+43.0}$. Therefore, at equilibrium:

$$\frac{1}{[O_2]^{1/2}} = 10^{43.0} \quad (12.54)$$

and $[O_2] = 10^{-86.1}$. Evidently, wüstite can coexist in equilibrium with metallic Fe only when the fugacity of O_2 is less than is permitted in normal terrestrial environments.

We can continue either by considering the transformation of wüstite to magnetite, which is a mixed oxide ($FeO \cdot Fe_2O_3$), or by relating magnetite directly to metallic Fe. The latter is preferable because wüstite does not occur in terrestrial rocks. Therefore, we consider the reaction:

$$3\,\text{Fe (metallic)} + 2\,O_2 \rightleftharpoons FeO \cdot Fe_2O_3$$
$$\Delta G_R^\circ = -242.6 \text{ kcal} \quad K = 10^{177.86} \quad (12.55)$$

It follows that at equilibrium the fugacity of O_2 must be $10^{-88.92}$ atm, which is also less than the lowest fugacity of O_2 in equilibrium with liquid water. Therefore, metallic Fe cannot be stable on the surface of the Earth, where the fugacity of O_2 must be $>10^{-83.1}$ atm. However, magnetite *does* occur in surface environments and, therefore, is stable at O_2 fugacities that are compatible with liquid water.

We consider next the reaction between magnetite and hematite in the hope that this equilibrium will provide a stability limit for magnetite that is within the range of permissible O_2 fugacities:

$$2\,FeO \cdot Fe_2O_3 + \tfrac{1}{2}O_2 \rightleftharpoons 3\,Fe_2O_3$$
$$\Delta G_R^\circ = -47.6 \text{ kcal} \quad K = 10^{34.89} \tag{12.56}$$

Thus, magnetite and hematite are in equilibrium at $[O_2] = 10^{-69.8}$ atm, which is within the permissible range of O_2 fugacities. Hematite is the most O-rich oxide of Fe and can coexist with water at O_2 fugacities from $10^{-69.8}$ to 1.0 atm.

Next, we add siderite ($FeCO_3$) to the system by reacting magnetite and hematite with CO_2 gas. Since Fe in siderite is divalent, we start with magnetite because it also contains divalent Fe. We construct the desired equation with magnetite as the reactant and siderite as the product. Then we add CO_2 to the reactant side to balance the C in siderite. Finally, we complete the equation by adding O_2 as needed for balance:

$$FeO \cdot Fe_2O_3 + 3\,CO_2 \rightleftharpoons FeCO_3 + \tfrac{1}{2}O_2 \tag{12.57}$$

Note that this is an oxidation–reduction reaction and that the equations *cannot* be balanced in terms of mass only, as we just did. In this case, two of the three divalent Fe ions in siderite give up one electron each for a total of two. These electrons are accepted by one atom of O_2 gas, which becomes one of the O^{2-} ions in the magnetite. Equation 12.57 is reversed because we have treated magnetite as the reactant rather than as the product. The standard free energy change of reaction 12.57 is $\Delta G_R^\circ = +44.48$ kcal and $k = 10^{-32.61}$. Therefore, at equilibrium:

$$\frac{[O_2]^{1/2}}{[CO_2]^3} = 10^{-32.61} \tag{12.58}$$

Taking logarithms to the base 10:

$$\tfrac{1}{2}\log[O_2] - 3\log[CO_2] = -32.61 \tag{12.59}$$
$$\log[O_2] = 6\log[CO_2] - 65.22 \tag{12.60}$$

Similarly, hematite reacts with CO_2 to form siderite according to:

$$Fe_2O_3 + 2\,CO_2 \rightleftharpoons 2\,FeCO_3 + \tfrac{1}{2}O_2$$
$$\Delta G_R^\circ = +45.52 \text{ kcal} \quad K = 10^{-33.37} \tag{12.61}$$

and:

$$\log[O_2] = 4\log[CO_2] - 66.74 \tag{12.62}$$

The equations representing solid–gas equilibria derived above have been plotted in Figure 12.8 in coordinates of $\log[O_2]$ and $\log[CO_2]$. The result is a *fugacity diagram* for the oxides and carbonates of Fe that indicates the fugacities of the two gases that are required to stabilize each of the minerals we have considered. The diagram also predicts the direction of solid–gas reactions that occur when a particular mineral is placed in an environment in which another mineral is stable.

For example, so-called *ironstone concretions* composed of siderite may be exposed to the atmosphere as a result of erosion of the enclosing sedimentary rocks. Figure 12.8 indicates that hematite is stable in contact with the atmosphere. Therefore, siderite is transformed into hematite in accordance with equation 12.61. At equilibrium between siderite and hematite:

$$\frac{[O_2]^{1/2}}{[CO_2]^2} = 10^{-33.37} \tag{12.63}$$

However, in the atmosphere $[O_2] = 2 \times 10^{-1}$ atm and $[CO_2] = 3 \times 10^{-4}$ atm. Therefore, siderite can coexist in equilibrium with hematite in contact with the atmosphere only when:

$$[CO_2] = \left[\frac{[O_2]^{1/2}}{K}\right]^{1/2} = \left[\frac{(O_2)^{1/2}}{10^{-33.37}}\right]^{1/2}$$
$$= 10^{16.51} \text{ atm} \tag{12.64}$$

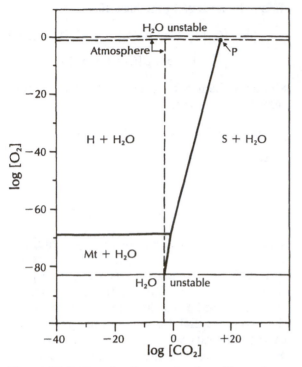

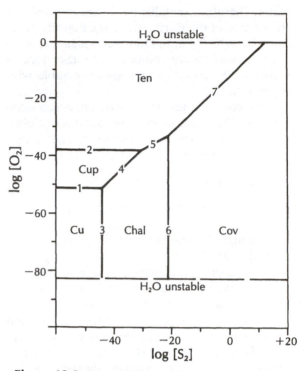

Figure 12.8 Fugacity diagram for the oxides and carbonate of Fe in terms of the fugacities of O_2 and CO_2, where Mt = magnetite, H = hematite, and S = siderite. Note that siderite and magnetite are not stable in contact with the atmosphere, where $[O_2] = 2 \times 10^{-1}$ atm and $[CO_2] = 3 \times 10^{-4}$ atm. Therefore, when siderite or magnetite is exposed to the atmosphere, it converts to hematite. The diagram is based on equations 12.52 (stability limit of water) and 12.56, 12.58, and 12.61. Point P represents conditions when siderite is stable in contact with O_2 of the atmosphere.

Figure 12.9 Fugacity diagram for the oxides and sulfides of Cu. Cu = metallic Cu, Cup = cuprite (Cu_2O), Ten = tenorite (CuO), Chal = chalcocite (Cu_2S), and Cov = covellite (CuS), based on equations in Table 12.5. Metallic Cu is stable in the presence of water, but when it is exposed to the atmosphere, it reacts to form tenorite. Note that G_f° of $S_2(g)$ is not zero because the stable form of S in the standard state is the rhombic solid.

which is identified by point P in the diagram. Since the fugacity of CO_2 in the atmosphere is only 3×10^{-4} atm, reaction 12.62 is not in equilibrium in contact with the atmosphere but runs to the left, thereby converting siderite to hematite. Therefore, ironstone concretions should weather to hematite upon exposure to the air. This prediction is confirmed by the presence of a hematite rind around ironstone concretions where the thickness of the rind depends on the rate of the reaction and the duration of the exposure.

c. Oxides and Sulfides of Copper

Copper forms the oxides cuprite (Cu_2O) and tenorite (CuO) and the sulfides chalcocite (Cu_2S) and covellite (CuS). Note that Cu is univalent in cuprite and chalcocite and divalent in tenorite and covellite. Therefore, when metallic Cu forms Cu^+ and when Cu^+ forms Cu^{2+}, electrons are transferred either to O_2 or to S_2, depending on the compound that is produced. The equations required to construct the fugacity diagram for Cu in Figure 12.9 are listed in Table 12.5.

We see that native Cu is stable in the presence of water, whereas native Fe is not. However,

Table 12.5 Fugacity Diagram of Copper Oxides and Sulfides at 25 °C[a]

$2\,Cu\,(metal) + \frac{1}{2}O_2 \rightleftharpoons Cu_2O$

$\Delta G_R^\circ = -35.1\ kcal \quad K = 10^{25.73}$

$$\log[O_2] = -51.5 \tag{1}$$

$Cu_2O + \frac{1}{2}O_2 \rightleftharpoons 2\,CuO$

$\Delta G_R^\circ = -26.3\ kcal \quad K = 10^{19.28}$

$$\log[O_2] = -38.6 \tag{2}$$

$2\,Cu\,(metal) + \frac{1}{2}S_2 \rightleftharpoons Cu_2S$

$\Delta G_R^\circ = -30.08\ kcal \quad K = 10^{22.05}$

$$\log[S_2] = -44.1 \tag{3}$$

$Cu_2O + \frac{1}{2}S_2 \rightleftharpoons Cu_2S + \frac{1}{2}O_2$

$\Delta G_R^\circ = +5.02\ kcal \quad K = 10^{-3.60}$

$$\log[O_2] = \log[S_2] - 7.4 \tag{4}$$

$2\,CuO + \frac{1}{2}S_2 \rightleftharpoons Cu_2S + O_2$

$\Delta G_R^\circ = +31.32\ kcal \quad K = 10^{-22.96}$

$$\log[O_2] = 0.5\log[S_2] - 23.0 \tag{5}$$

$Cu_2S + \frac{1}{2}S_2 \rightleftharpoons 2\,CuS$

$\Delta G_R^\circ = -14.48\ kcal \quad K = 10^{10.61}$

$$\log[S_2] = -21.2 \tag{6}$$

$CuO + \frac{1}{2}S_2 \rightleftharpoons CuS + \frac{1}{2}O_2$

$\Delta G_R^\circ = +8.42\ kcal \quad K = 10^{-6.17}$

$$\log[O_2] = \log[S_2] - 12.34 \tag{7}$$

[a]Based on the thermodynamic data in Appendix B.

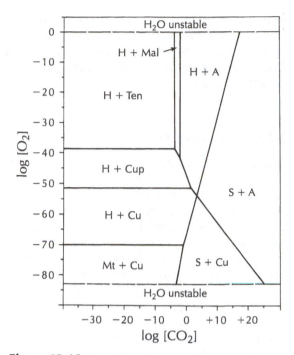

Figure 12.10 Combined fugacity diagram for Fe and Cu and their oxides and carbonates. H = hematite (Fe_2O_3); Mt = magnetite (Fe_3O_4); S = siderite ($FeCO_3$); Cu = native Cu; Cup = cuprite (Cu_2O); Ten = tenorite (CuO); Mal = malachite [$CuCO_3 \cdot Cu(OH)_2$]; A = azurite [$2CuCO_3 \cdot Cu(OH)_2$]. The boundaries between tenorite and malachite and between malachite and azurite were drawn after Garrels and Christ (1965). This diagram indicates the environmental conditions under which certain assemblages of Fe and Cu minerals can occur together.

when metallic Cu is exposed to O_2 at fugacities $>10^{-51.5}$ atm, it forms the oxide cuprite (Cu_2O) or tenorite (CuO), depending on the O_2 fugacity. Similarly, Cu reacts with S_2 to form the sulfides chalcocite (Cu_2S) and covellite (CuS). Copper also forms mixed sulfides with Fe including chalcopyrite ($CuFeS_2$), cubanite ($CuFe_2S_3$), and bornite (Cu_5FeS_4).

Fugacity diagrams of different metals can be superimposed on each other to form composite diagrams that indicate the stability limits of the compounds of both metals. Such a diagram has been constructed in Figure 12.10 for the oxides and carbonates of Fe and Cu. The diagram defines the fugacities of O_2 and CO_2 that permit certain assemblages of these minerals to coexist in equilibrium in the presence of water.

For example, the diagram indicates that the carbonates of Cu (malachite and azurite) form at lower fugacities of CO_2 than siderite and therefore can coexist in equilibrium with hematite and tenorite. Consequently, malachite, azurite, tenorite, and hematite are common weathering products of Fe–Cu sulfide minerals and occur in the oxidized part of sulfide ore bodies (gossans), whereas siderite is unstable at the surface of the Earth (see also Section 21.2). Figure 12.10 also points out that native Cu has a large stability field in the absence of S and can coexist with magnetite, hematite, and siderite, provided the fugacity of O_2 is low enough. When native Cu is exposed to the atmosphere, it *tarnishes* by forming a layer of tenorite or malachite or both, depending on the fugacity of CO_2.

12.6   Summary

When the silicate minerals of igneous and metamorphic rocks are exposed to acidified water, they may dissolve *incongruently* to produce clay minerals and zeolites as well as insoluble oxides and hydroxides, depending on the environmental conditions. The chemical reactions that take place during these transformations can be used to define the conditions required for equilibrium when both reactants and products are stable. Therefore, equations can be derived from the Law of Mass Action, which specify the conditions required for equilibrium between the solids.

The resulting mineral stability diagrams define the environmental conditions in which each of the solids in a given compositional system is stable. In addition, the diagrams predict what happens to a mineral when it is exposed to conditions outside of its own stability field. The reactions that occur in such cases not only cause transformations among the solids but also affect the chemical composition, or *quality,* of water both above and below the surface of the Earth. When the reactions occur in closed systems with small water/rock ratios, the chemical composition of the water may evolve along curved reaction paths. Such conditions exist in the pore spaces or fractures of rocks in subsurface where reactions between minerals and water not only contribute to the chemical evolution of the water but also cause the deposition of mineral cements.

The solubility limits of certain compounds, including calcite, magnesite, amorphous silica, and others, which can precipitate from natural solutions, significantly restrict the chemical compositions of aqueous solutions in nature.

Therefore, certain minerals cannot be in equilibrium in the presence of water and also cannot form by transformation from other minerals. This phenomenon is exemplified by anorthite, which does not form as an authigenic mineral at Earth-surface temperatures, whereas albite and K-feldspar do occur in places where alkali-rich brines can react with volcanic glass, clay minerals, or other kinds of solids.

The conditions required for stability of minerals can also be expressed by their congruent solution or by their interaction with gases such as O_2, CO_2, and S_2. For example, the non-aluminous Mg silicates dissolve congruently to form Mg^{2+} and H_4SiO_4, which may precipitate as a different Mg silicate whose solubility is exceeded. The combination of congruent solution of a "primary" Mg silicate and the precipitation of a "secondary" mineral results in the replacement of one mineral by another. The apparent conversion of enstatite to serpentine is an example of such paired reactions.

The equilibrium between water and its constituent gases, O_2 and H_2, defines the stability range of water within which all reactions on the surface of the Earth take place. Solid oxides, carbonates, and sulfides must similarly remain in equilibrium with their constituent gases O_2, CO_2, and S_2, respectively. These solid–gas reactions can be used to construct fugacity diagrams that contain the stability fields of the solids appropriate to a particular system. The fugacity diagrams of different metals can be superimposed on each other to form combination diagrams. These diagrams effectively subdivide the stability field of water into environments in which suites of minerals of different elements can coexist in equilibrium.

The virtue of activity and fugacity diagrams is that they depict the relationships between minerals and aqueous solutions or gases in a readily usable form. As a result, idle speculation is replaced by understanding and progress is made in understanding the chemical reactions that contribute to geological processes.

Problems

1. Construct the equations necessary to introduce pyrophyllite into the activity diagram for microcline in Figure 12.1.

2. Construct the equations necessary to introduce Na-smectite $(Na_{0.33}Al_{2.33}Si_{3.67}O_{10}(OH)_2$, $G_f^o = -1277.76$ kcal/mol) into the activity diagram for albite in Figure 12.2.

3. Construct a solubility diagram for goethite (FeOOH) as a function of pH.

4. What are the environmental conditions in terms of $[Mg^{2+}]/[H^+]^2$ and $[H_4SiO_4]$ at which enstatite and forsterite coexist in equilibrium?

5. Is equilibrium possible between enstatite and forsterite in the presence of aqueous solutions in nature?

6. What mineral precipitates first when forsterite dissolves congruently in water having the composition of point P in Figure 12.5?

7. Construct a fugacity diagram for the sulfides and carbonates of Pb.

8. Construct a fugacity diagram for the sulfides and carbonates of Zn.

9. Combine the two diagrams for Zn and Pb.

10. What are the fugacities of O_2 and CO_2 at which native Cu, cuprite, and azurite coexist in equilibrium?

11. Can hematite and siderite both join that assemblage in equilibrium with each other?

References

BARRER, R. M., 1978. *Zeolites and Clay Minerals as Sorbents and Molecular Sieves.* Academic Press, London, 497 pp.

BOWERS, T. S., K. J. JACKSON, and H. C. HELGESON, 1984. *Equilibrium Activity Diagrams.* Springer-Verlag, New York, 397 pp.

COOMBS, D. S., A. J. ELLIS, W. S. FYFE, and A. M. TAYLOR, 1959. The zeolite facies, with comments on the interpretation of hydrothermal syntheses. *Geochim. Cosmochim. Acta,* **17**: 52–107.

GARRELS, R. M., and C. L. CHRIST, 1965. *Solutions, Minerals and Equilibria.* Harper & Row, New York, 450 pp.

GOTTARDI, G., and E. GALLI, 1985. *Natural Zeolites,* Springer-Verlag, New York, 390 pp.

HAY, R. L., 1966. Zeolites and zeolitic reactions in sedimentary rocks. *Geol. Soc. Amer., Spec. Paper* No. 85, 130 pp.

HELGESON, H. C., R. M. GARRELS, and F. T. MACKENZIE, 1969. Evaluation of irreversible reactions in geochemical processes involving minerals and aqueous solutions. II. Application. *Geochim. Cosmochim. Acta,* **33**: 455–481.

HEMINGWAY, B. S., and R. A. ROBIE, 1984. Thermodynamic properties of zeolites: low-temperature heat capacities and thermodynamic functions for phillipsite and clinoptilolite. Estimates of the thermochemical properties of zeolitic water at low temperature. *Amer. Mineral.,* **69**: 692–700.

JOHNSON, G. K., H. E. FLOTOW, P. A. G. O'HARE, and W. S. WISE, 1982. Thermodynamic studies of zeolites: analcime and dehydrated analcime. *Amer. Mineral.,* **67**: 736–748.

KASTNER, M., 1971. Authigenic feldspars in carbonate rocks. *Amer. Mineral.,* **56**: 1413–1442.

KASTNER, M., and R. SIEVER, 1979. Low temperature feldspars in sedimentary rocks. *Amer. J. Sci.,* **279**: 435–479.

LISITZINA, N. A., and G. YU. BUTUZOVA, 1982. Authigenic zeolites in the sedimentary mantle of the world ocean. *Sed. Geol.,* **31**: 33–42.

MENSING, T. M., and G. FAURE, 1983. Identification and age of neoformed Paleozoic feldspar (adularia) in a Precambrian basement core from Scioto County, Ohio, U.S.A. *Contrib. Mineral. Petrol.,* **82**: 327–333.

PETZING, J., and R. CHESTER, 1979. Authigenic marine zeolites and their relationship to global volcanism. *Marine Geol.,* **29**: 253–272.

STEINMANN, P., P. C. LICHTNER, and W. SHOTYK, 1994. Reaction path approach to mineral weathering reactions. *Clays and Clay Minerals,* **42**: 197–206.

SURDAM, R. C., and H. P. EUGSTER, 1976. Mineral reactions in the sedimentary deposits of the Lake Magadi region, Kenya. *Geol. Soc. Amer. Bull.,* **87**: 1739–1752.

SUTHERLAND, F. L., 1977. Zeolite minerals in the Jurassic dolerites of Tasmania: Their use as possible indicators of burial depth. *Geol. Soc. Aust. J.,* **24**: 171–178.

WALKER, G. P. L., 1960. Zeolite zones and dike distribution in relation to the structure of the basalts of eastern Iceland. *J. Geol.,* **68**: 515–528.

13

Clay Minerals

Clay minerals form one of the largest and most highly diversified groups of minerals known. The majority of clay minerals are aluminosilicates and they crystallize as phyllosilicates. However, some clay minerals are Mg-silicates or Fe-silicates, others are *not* crystalline, or they are not phyllosilicates. The property common to all clay minerals is the small size of their crystals or grains with diameters less than 2 micrometers (μm). As a result, sediment in that size range is sometimes referred to as the *clay-size fraction* or even as *clay*. This usage is incorrect in geochemistry because clay minerals are defined on the basis of their crystal structure and chemical composition and not by the size of their crystals or grains. Moreover, the clay-size fraction of sediment frequently includes small grains of other minerals (quartz, feldspar, calcite, etc.), as well as grains of insoluble oxides or hydroxides of Fe, Al, Mn, and so on, which may be partly or completely amorphous. Nevertheless, clay minerals are typically fine grained and therefore cannot be identified solely by their physical or optical properties. Instead, the classification of clay minerals relies heavily on their crystallographic and certain physical properties that are revealed by x-ray diffraction.

Clay mineralogy has evolved into an important field with applications in geology, agronomy, ceramic engineering, soil mechanics, and geochemistry. We will be concerned primarily with the classification of clay minerals, with their stability in different geochemical environments, with ion exchange phenomena, and with the use of clay minerals for dating sedimentary rocks. The presence of clay minerals in natural environments frequently makes a significant difference to the geochemical processes that may be occurring. Since we encountered the clay minerals kaolinite, pyrophyllite, and smectite in Chapter 12, it is now necessary for us to examine these minerals in more detail.

13.1 Crystal Structure

Clay minerals are divided into three groups based on their crystallographic properties: (1) platy clay minerals (phyllosilicates), (2) fibrous clay minerals, and (3) amorphous clay. The platy clay minerals are much more abundant than the other two groups and include most of the minerals that traditionally make up the clay minerals. The fibrous clays *sepiolite* and *palygorskite* (also known as attapulgite), though not as common as the platy clays, do occur in many different kinds of depositional environments, including marine and lacustrine sediment, hydrothermal deposits, and soils of arid regions (Singer and Galan, 1984). So-called amorphous clay, referred to as allophane, is an aluminosilicate of variable chemical composition that is amorphous to x-rays in most cases but does diffract electrons. It may be an intermediate product in the formation of phyllosilicate clays from various kinds of materials, but its abundance is not well documented because it is not directly detectable by x-ray diffraction (Carroll, 1970).

Clay minerals are described in many textbooks on mineralogy and in numerous specialized books, including those by Grim (1968), Millot (1970), Weaver and Pollard (1973), Velde (1977), Sudo and Shimoda (1978), Sudo et al. (1981), and Cairns-Smith and Hartman (1987). In addition, several scientific journals are devoted to clay mineralogy, including *Clays and Clay Minerals, Clay Minerals Bulletin, Clay Science,* and *Clays and Clay Technology.* Articles on clay mineralogy are also published in the *American Mineralogist,* the *Journal of the Soil Science Society of America,* and the *Journal of Sedimentary Petrology.*

The phyllosilicate clay minerals consist of two kinds of layers of different chemical composition and coordination. One of these is the *tetrahedral layer* (T), which is composed of silica tetrahedra. The second layer consists of Al hydroxide with Al^{3+} ions located in the centers of octahedra and is therefore known as the *octahedral layer* (O). In some clay minerals, such as serpentine and talc, the octahedral layer is composed of Mg hydroxide, which forms the mineral *brucite.* Therefore, the octahedral layer is also referred to as the *brucite layer* or the *gibbsite layer,* depending on its composition.

When Al^{3+} occurs in the octahedral layer, only two of every three available sites are filled. Therefore, such clays are said to be dioctahedral. When Al^{3+} is replaced by Fe^{3+} or by other trivalent ions, the minerals remain dioctahedral. However, when the two Al^{3+} ions are replaced by three divalent ions of Fe^{2+}, Mg^{2+}, Zn^{2+}, etc., then all positions in the octahedral layer are filled and the clay is trioctahedral. Based on these distinctions we conclude that kaolinite is dioctahedral, whereas serpentine is trioctahedral.

The tetrahedral layer consists of silica tetrahedra that are linked together by sharing O atoms at three of the four corners. The linkages occur only at the bases of the tetrahedra, thus forming a sheet with the unshared O atoms all pointing in the same direction. If *all* four O atoms of silica tetrahedra are shared (as in feldspar), then each Si atom controls only one half of its O atoms. Therefore, the Si/O ratio in feldspar and other tektosilicates is 1 : 2 (Section 8.3). Since in phyllosilicates only the three *basal* O atoms are shared, the Si/O ratio is 1 : 1.5. When the *unshared* O atom is counted with the basal ones, the Si/O ratio becomes 1 : 2.5.

The octahedral layer has six hydroxyl ions located at the corners of an octahedron. Therefore, the ratio of Al to hydroxyl ions is basically 1 : 6. However, the octahedra that make up the gibbsite sheet are linked at the corners by

sharing hydroxyl ions and have an Al/OH ratio of 1 : 3 instead of 1 : 6.

The tetrahedral and octahedral layers are joined to form two-layer clays (T–O), three-layer clays (T–O–T), or mixed-layer clays that are mixtures of different two-layer and three-layer clays. The linking of the octahedral and tetrahedral sheets in T–O clay minerals takes place by sharing of O atoms and hydroxyl radicals, which reduces the number of hydroxyls by *one each time such a link is established.* Therefore, two-layer clays, such as kaolinite, are composed of Si, Al, O, and OH in the proportions $SiO_{2.5}Al(OH)_2$. We multiply the formula by two in order to clear the decimal fraction and obtain the formula of kaolinite and its polymorphs:

$$Al_2Si_2O_5(OH)_4 \qquad (13.1)$$

Similarly, the formula of three-layer clays (T–O–T), such as pyrophyllite, is $2(SiO_{2.5})Al(OH)$ or $Si_2O_5Al(OH)$. Again, we multiply by two and obtain:

$$Al_2Si_4O_{10}(OH)_2 \qquad (13.2)$$

The structures of two- and three-layer clays are illustrated in Figure 13.1.

The phyllosilicate clay minerals have widely varying chemical compositions because the Al^{3+} ion in the octahedral layers can be wholly or partly replaced by Fe^{3+}, Cr^{3+}, Fe^{2+}, Mg^{2+}, Zn^{2+}, Li^+, and many other ions. In addition, some of the Si^{4+} ions in the tetrahedral layer are replaced by Al^{3+}. This results in an excess of negative charge, which is neutralized by the adsorption of cations to the outer surfaces of the tetrahedral layers of adjoining clay units. The substitutions within the layers are constrained by the requirement for electrical neutrality, which is met when the cations in the *octahedral layer* have a combined charge of +6 and the *tetrahedral layers* contain four moles of Si + Al per formula weight. The two-layer clays permit very little substitution of either Al or Si, whereas the three-layer clays, with the exception of pyrophyllite, are characterized by extensive substitutions in either the octahedral layer or the tetrahedral layer, or in both.

(a) Two-layer clay: $AlSiO_{2.5}(OH)_2$

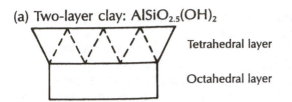

Tetrahedral layer

Octahedral layer

(b) Three-layer clay: $AlSi_2O_5(OH)$

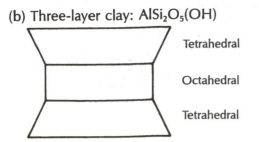

Tetrahedral

Octahedral

Tetrahedral

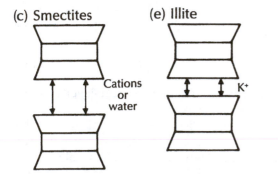

(c) Smectites

Cations or water

(e) Illite

K^+

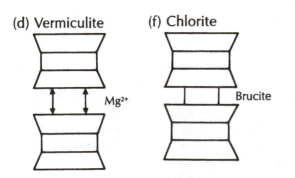

(d) Vermiculite

Mg^{2+}

(f) Chlorite

Brucite

Figure 13.1 Schematic diagrams of the structures of two-layer and three-layer clay minerals.

13.2 Classification and Chemical Composition

The classification of clay minerals is based on their crystal structure and chemical composition. In practice, clay minerals are identified by x-ray diffraction combined with treatment procedures that together serve to identify certain mineral species. In other words, some clay minerals are defined by the procedure that is used to identify them.

a. Two-Layer Clays

The phyllosilicate clays are subdivided into the two-layer, three-layer, and mixed-layer clays. The two-layer clays consist of two subgroups, each of which contains a number of mineral species, as listed in Table 13.1.

1. KAOLINITE The minerals in the *kaolinite subgroup* all have the familiar formula $Al_2Si_2O_5(OH)_4$ and include *kaolinite, dickite, nacrite,* and *halloysite.* These minerals differ from kaolinite only by having different crystal structures. For example, halloysite forms tubular crystals, whereas kaolinite crystals are platy, but all members of the kaolinite subgroup have identical chemical compositions. Therefore, halloysite and the other mineral species in the kaolinite subgroup are *polymorphs* of kaolinite but occur in certain geochemical environments in preference to kaolinite.

2. SERPENTINE AND GREENALITE *Serpentine* is a two-layer phyllosilicate like kaolinite in which the gibbsite layer is replaced by brucite. Consequently, the formula of serpentine $[Mg_3Si_2O_5(OH)_4]$ contains Si and O in the characteristic proportion of the tetrahedral layer, and the octahedral layer contains a positive charge of +6 as a result of the replacement of two Al^{3+} by three Mg^{2+}. The mineral species in that group are *chrysotile, lizardite,* and *antigorite,* all of which are polymorphs of serpentine. *Greenalite* $[Fe_3^{2+}Si_2O_5(OH)_4]$ is a two-layer clay in which three Fe^{2+} ions take the place of two Al^{3+} ions.

Table 13.1 Classification of Clay Minerals

Group	Subgroup	Species
A. Two-Layer Clays		
Kaolinite	kaolinite (dioctahedral)	kaolinite
		dickite
		nacrite
		halloysite
Kaolinite	serpentine (trioctahedral)	chrysotile
		lizardite
		antigorite
		greenalite
B. Three-Layer Clays		
Pyrophyllite	dioctahedral	pyrophyllite
	trioctahedral	talc
	dioctahedral	minnesotaite
Smectite	dioctahedral	montmorillonite
		beidellite
		nontronite
Smectite	trioctahedral	saponite
		hectorite
		sauconite
Vermiculite	dioctahedral	vermiculite
	trioctahedral	vermiculite
Mica	dioctahedral	illite
		muscovite
		paragonite
		glauconite
Mica	trioctahedral	biotite
		phlogopite
		lepidomelane
Brittle mica	dioctahedral	margarite
	trioctahedral	seybertite
		xanthophyllite
		brandisite
Chlorite	dioctahedral	cookeite (?)
	trioctahedral	Fe-rich varieties
		thuringite
		chamosite
		Mg-rich varieties
		clinochlore
		penninite

Table 13.1 (continued)

Group	Subgroup	Species
C. Mixed-Layer Clays		
Complex interstratified two-layer and three-layer clay minerals following either a regular or a random pattern.		
D. Fibrous Clay Minerals		
		palygorskite (attapulgite)
		sepiolite

Adapted from Grim (1968) and Carroll (1970).

b. Three-Layer Clays

The three-layer clays can be divided into the *pyrophyllite, smectite, vermiculite, mica, brittle mica,* and *chlorite groups.* All of the minerals included here have the basic three-layer structure of pyrophyllite but differ from it in chemical composition and physical properties.

1. PYROPHYLLITE AND TALC *Pyrophyllite* itself $(Al_2Si_4O_{10}(OH)_2)$ is the prototype of all three-layer clays and permits no substitutions in the octahedral or tetrahedral layers.

Talc is related to pyrophyllite in the same way that serpentine is related to kaolinite because in talc the octahedral layer is composed of brucite instead of gibbsite. Consequently, talc has the formula $Mg_3Si_4O_{10}(OH)_2$, where three Mg^{2+} ions replace two Al^{3+} and the Si/O ratio is derived from the presence of two tetrahedral layers. *Minnesotaite* is the Fe member in this group in which the octahedral layer contains Fe^{2+} instead of Al^{3+}. Consequently, its formula is $Fe_3^{2+}Si_4O_{10}(OH)_2$.

2. SMECTITES The *smectite group* contains a large number of mineral species that differ in chemical composition because of the complete or partial replacement of Al^{3+} in the octahedral layer and the partial replacement of Si^{4+} in the tetrahedral layer.

The *smectite group* is divided into dioctahedral and trioctahedral subgroups each of which contains mineral species that differ from each other in chemical composition. For example, *montmorillonite* is a prominent member of the *dioctahedral subgroup* of smectites in which Mg^{2+} occurs in the octahedral layer, but the replacement of Si^{4+} by Al^{3+} in the tetrahedral layer is quite limited. For example, a sample of montmorillonite from Germany has the formula (Carroll, 1970, Table 6):

$$[(Al_{1.77}Fe^{3+}_{0.03}Mg_{0.20})(Si_{3.74}Al_{0.26})O_{10}(OH)_2]$$

(octahedral)   (tetrahedral)

$$Ca_{0.16} \cdot Na_{0.07} \cdot Mg_{0.04} \qquad (13.3)$$

(exchangeable)

This formula can be readily interpreted in terms of the three-layer model if we determine the electrical charges in each of the layers. The octahedral layer contains $1.77\ Al^{3+} + 0.03\ Fe^{3+} + 0.20\ Mg^{2+}$, which adds up to $+5.8$ electronic charges. Since the octahedral layer is supposed to contain six positive charges, there is a deficiency of 0.2 positive charge, which is equivalent to an excess of 0.2 negative charge. In the tetrahedral layer replacement of some Si^{4+} by Al^{3+} causes a deficiency of positive charge equivalent to a charge of -0.26. Evidently, the substitutions in the two kinds of layers have caused a total charge of $-0.2 + (-0.26) = -0.46$. In the example before us, the negative charges are neutralized by adsorption of $0.16\ Ca^{2+} + 0.07\ Na^+ + 0.04\ Mg^{2+}$, which together provide 0.47 positive charge. The agreement between the negative charges in the layers (-0.46) and the positive charges of the additional ions ($+0.47$) is well within the uncertainty of the chemical analysis of this mineral. The overall charge distribution for this montmorillonite is given by adding up the ionic charges of all elements that make up this mineral:

$$1.77(Al^{3+}) + 0.03(Fe^{3+}) + 0.20(Mg^{2+})$$
$$+ 3.74(Si^{4+}) + 0.26(Al^{3+})$$
$$+ 0.16(Ca^{2+}) + 0.07(Na^+)$$
$$+ 0.04(Mg^{2+}) - 10(O^{2-}) - 2(OH^-)$$
$$= +0.01 \qquad (13.4)$$

The additional cations, which neutralize the negative charges caused by imperfect substitutions within the layers, are *adsorbed* on the surfaces of the mineral grains and on the tetrahedral layers of the clay units. Therefore, they are partly or completely *exchangeable* and can be replaced by other cations or by the polar molecules of water in the environment. The adsorbed ions located on the outer surfaces of the tetrahedral layers bond the structural units together. When the bonding involves the cations of metals, the interlayer distances between adjacent structural units are short. However, when certain clay minerals are immersed in water, the interlayer cations are replaced by molecules of water, which have much smaller net positive charges than metal ions. Consequently, a larger number of water molecules is required to neutralize excess negative charges generated within the layers. As a result, the interlayer distances *increase* as water molecules crowd in and the clay *expands*.

The swelling of clay minerals in the smectite group is a characteristic property that is used to identify them. For example, when natural montmorillonite is heated at 300°C for an hour or more, the interlayer distance *decreases* from about 15 angstroms (Å) to about 9 Å because adsorbed water molecules are expelled. When the clay is subsequently exposed to the vapor of ethylene glycol ($HOCH_2CH_2OH$) for several hours, the interlayer distance *expands* to 17 or 18 Å.

Montmorillonite commonly forms by alteration of volcanic ash and therefore is the principal mineral of *bentonite* beds, which form by alteration of volcanic ash deposits. The well-known swelling of bentonite, when it is placed in water, is actually caused by the expansion of the interlayer distances of montmorillonite. The mineral also occurs in soils of arid regions where leaching is limited by lack of water and the pH is above 7.

Beidellite is a dioctahedral smectite containing little if any Fe^{2+} or Mg^{2+} and in which negative charges originate primarily by replacement of Si^{4+} by Al^{3+} in the tetrahedral layer. According to Grim (1968), its formula can be written as:

$$[(Al_{2.17})(Si_{3.17}Al_{0.83})O_{10}(OH)_2] \cdot Na_{0.33} \quad (13.5)$$

Note that the replacement of Si^{4+} by Al^{3+} in the tetrahedral layer has caused a net charge of -0.83, which is partly compensated by the positive charge ($+0.51$) generated by an excess of 0.17 Al^{3+} in the octahedral layer. The remaining excess negative charge ($-0.83 + 0.51 = -0.32$) is neutralized by adsorption of 0.33 moles of Na^+ per formula weight of beidellite.

Nontronite, the third dioctahedral smectite in Table 13.1, contains Fe^{3+} in the octahedral layer in place of Al^{3+} with minor amount of Fe^{2+} and Mg^{2+}. The net charge imbalance arises from substitutions of Si^{4+} by Al^{3+} in the tetrahedral layer. Therefore, the formula of nontronite is:

$$[(Fe_2^{3+})(Si_{3.67}Al_{0.33})O_{10}(OH)_2] \cdot Na_{0.33} \quad (13.6)$$

As in beidellite, the negative charge generated in the tetrahedral layer may be partly compensated by the presence of additional Fe^{3+}, Fe^{2+}, or Mg^{2+} in the octahedral layer. When Al^{3+} in the octahedral layer is replaced by Cr^{3+}, the resulting dioctahedral smectite is known as *volkhonskoite.*

The *trioctahedral smectites* include the mineral species *saponite, hectorite,* and *sauconite.* In saponite, the octahedral layer is composed primarily of Mg^{2+} in place of Al^{3+} with minor amounts of Fe^{3+} and Fe^{2+}. Negative charges arise in the tetrahedral layer by replacement of Si^{4+} by Al^{3+} and may be partly compensated by additional cations in the octahedral layer. The formula of saponite according to Grim (1968) is:

$$[(Mg_3(Si_{3.67}Al_{0.33})O_{10}(OH)_2] \cdot Na_{0.33} \quad (13.7)$$

Hectorite is a Li-bearing smectite in which Mg^{2+} and Li^+ occur together in the octahedral layer, but no substitutions occur in the tetrahedral layer. Thus the formula for hectorite is:

$$[(Mg_{2.67}Li_{0.33})(Si_4)O_{10}(OH)_2] \quad (13.8)$$

The smectite mineral *stevensite* is similar to hectorite because it is Mg-rich and lacks Al in the tetrahedral layer. However, stevensite also lacks Li in the octahedral layer.

The smectite mineral *sauconite* is characterized by the presence of Zn^{2+} in the octahedral layer and by limited replacement of Si^{4+} by Al^{3+} in the tetrahedral layer. Carroll (1970) recalculated a chemical analysis of sauconite to obtain the following formula:

$$[(Zn_{2.40}Mg_{0.18}Al_{0.22}Fe_{0.17}^{3+})$$
$$(Si_{3.47}Al_{0.53})O_{10}(OH)_2]^{-0.20} \quad (13.9)$$

The charge imbalance in the octahedral layer of this clay mineral is $+0.33$, that of the tetrahedral layer is -0.53 for a net charge of -0.20, which is neutralized by the adsorption of exchangeable cations in interlayer sites.

The word *vermiculite* is derived from the Latin verb *vermiculari,* which means to *breed worms.* The clay mineral vermiculite occurs both in dioctahedral and trioctahedral varieties with other clay minerals and as an alteration product of micas. According to Gruner (1934), vermiculite from Bare Hills, Maryland, has the formula:

$$[(Fe_{0.24}^{3+}Mg_{2.70}Fe_{0.04}^{2+}Ni_{0.01})$$
$$(Si_{2.73}Al_{1.26})O_{10}(OH)_2]Mg_{0.55}Ca_{0.01} \quad (13.10)$$

Evidently, Mg^{2+} dominates in the octahedral layer, and the Si^{4+} in the tetrahedral layer is extensively replaced by Al^{3+}. The octahedral layer has an excess positive charge of $+0.22$, which partly compensates for the strong negative charge of -1.30 in the tetrahedral layer. The remaining negative charge of -1.08 is neutralized by adsorbed Mg^{2+} and Ca^{2+} in the interlayer position. Vermiculite characteristically contains both Mg^{2+} and water molecules in the interlayer sites. The water can be expelled by heating at 500°C, which causes the interlayer distance to decrease from about 14 Å to about 9 Å. After heating at 500°C or less, vermiculite rehydrates spontaneously; however, heating at 700°C causes irreversible dehydration. In addition, when vermiculite is treated with a solution of KCl, the interlayer Mg^{2+} ions are replaced by K^+, and it becomes nonexpandable. Vermiculite differs from smectites by having a higher charge in the tetrahedral layer, by commonly containing

Mg^{2+} in the exchangeable interlayer sites, and by being less expandable.

3. ILLITE The power of K^+ ions to stabilize clay minerals is well illustrated by *illite,* which is listed in Table 13.1 with the *mica* minerals. Actually, illite could also be regarded as a special member of the smectite group. However, it differs from typical smectites by having a *nonexpandable lattice* because of the presence of K^+ ions in interlayer sites. Illite was named by Grim et al. (1937) after the state of Illinois in the United States. They intended it to represent mica-like clay minerals as a group, rather than a specific clay mineral, and expressed the formula of illite as:

$$[(Al_2, Fe_2, Mg_2, Mg_3)(Si_{4-x}Al_x)O_{10}(OH)_2] \cdot K_x \tag{13.11}$$

The formula indicates that the octahedral layer may contain Al, Fe, or Mg and that illites may be dioctahedral (2 Al^{3+} or 2 Fe^{3+}) or trioctahedral (3 Mg^{2+}). The value of x is less than one and usually varies between 0.5 and 0.75. The negative charge generated in the tetrahedral layer is compensated by K^+ ions, which fit into hexagonal holes formed by the silica tetrahedra of the tetrahedral layers. As a result, the K^+ ions are not exchangeable, the spacing of the interlayers is fixed, and the mineral is not expandable. However, illites may be interlayered with montmorillonite or other clay minerals that are expandable.

Illite is a common weathering product of *muscovite* and differs from it primarily by the extent of substitution of Si^{4+} by Al^{3+} in the tetrahedral layer. Muscovite and its Na-analog *paragonite* are dioctahedral and have the formula:

$$[(Al_2)(Si_3Al)O_{10}(OH)_2]K, Na \tag{13.12}$$

The charge deficiency in the tetrahedral layer of -1 is neutralized by K^+ in muscovite and by Na^+ in paragonite. The stacking of adjacent mica units in muscovite can occur in different ways, giving rise to polymorphs designated as $1M$, $1Md$, $2M_1$, $2M_2$, and $3T$ (Smith and Yoder, 1956). Illite in the smallest grain-size fraction ($<0.2 \mu m$) of sediment or soil is of the $1Md$ or $1M$ type. Larger grains consist of the

2M polymorphs and are derived by mechanical weathering of igneous and metamorphic rocks. Therefore, the 2M illite in sedimentary rocks is *detrital* in origin, whereas the $1Md$ or $1M$ polymorphs may consist of *authigenic* illite. The x-ray diffraction spectra of the different muscovite polymorphs in the clay fraction of sedimentary materials were worked out by Yoder and Eugster (1955).

4. GLAUCONITE OR GLAUCONY *Glauconite* is a dioctahedral clay mineral in which Fe^{3+}, Fe^{2+}, and Mg^{2+} replace Al^{3+} in the octahedral layer. Negative charges in the octahedral and tetrahedral layers are neutralized by interlayer K^+, Ca^{2+}, and Na^+. The concentration of K^+ in glauconite is variable and increases with time depending on the availability of K^+ in the environment. Johnston and Cardile (1987) reported that glauconite from Point Jackson, Francosia, Wisconsin, has the formula:

$$[(Fe^{2+}_{1.097}Al_{0.849}Mg_{0.442}Ti_{0.003}Mn_{0.001})$$
$$(Si_{3.611}Al_{0.389})O_{10}(OH)_2]K_{0.725}Ca_{0.096} \tag{13.13}$$

In this sample of glauconite, Fe^{2+} dominates in the octahedral layer, and the deficiency of positive charges in both the octahedral and tetrahedral layers is largely compensated by interlayer K^+ and Ca^{2+} ions.

Glauconite is an *authigenic* clay mineral that forms in the marine environment by transformation from other kinds of material (McRae, 1972). It commonly occurs as greenish pellets about 1–2 mm in diameter comprised of aggregates of platy glauconite crystallites about $10 \mu m$ in diameter. Glauconite has been used for dating unmetamorphosed sedimentary rocks by the K–Ar and Rb–Sr methods (Chapter 16) because it contains both K and Rb and because it is authigenic in origin.

The status of glauconite as a discrete mineral species has become controversial, partly because of its importance for dating sedimentary rocks and the hoped-for refinement of the geological time scale. Burst (1958) and Hower (1961) regarded glauconite as an Fe-rich illite with variable K content that occurs in pure form as the $1Md$ or $1M$ polymorph. However, this definition included poorly crystallized materials with low K concentrations that are unsuitable for dating by isotopic methods. Therefore, Odin and Matter (1981) proposed the word "glaucony" for a spectrum of Fe-rich clay minerals with varying K concentrations that form the characteristic green pellets in sedimentary rocks of marine origin. Glaucony therefore represents clay minerals that are intermediate between Fe-rich and K-poor smectite and Fe-rich K-mica. As the K concentration of glaucony increases, the material becomes less expandable and more retentive with respect to the radiogenic daughters of K and Rb.

5. MICAS   The trioctahedral micas in igneous and metamorphic rocks include *biotite* and *phlogopite*. These differ from muscovite by the chemical composition of the octahedral layer, but both replace one Si^{4+} by one Al^{3+} in the tetrahedral layer. Actually, biotite is intermediate between phlogopite:

$$[(Mg_3)(Si_3Al)O_{10}(OH)_2] \cdot K$$

and annite or lepidomelane:

$$[(Fe_3^{2+})(Si_3Al)O_{10}(OH)_2] \cdot K$$

Therefore, the general formula for biotite can be written as:

$$[(Mg, Fe^{2+})_3(Si_3Al)O_{10}(OH)_2] \cdot K \quad (13.14)$$

Biotite is considerably more susceptible to chemical weathering than muscovite because the Fe^{2+} can oxidize to Fe^{3+} in the presence of an electron acceptor. As a result, the charge balance of the lattice is disturbed and the mineral alters to other aluminosilicates and ferric oxide, depending on environmental conditions.

The so-called *brittle micas* in Table 13.1 include the dioctahedral species *margarite* and other rare trioctahedral phyllosilicates listed by Grim (1968).

6. CHLORITE   The *chlorite* group contains primarily trioctahedral species composed of three-layer clay units bonded to brucite sheets. The chemical compositions of chlorite range from Mg-rich to Fe-rich varieties, but all contain varying amounts of Al, which replaces Si in the tetrahedral layers.

The only dioctahedral chlorite is an Al-rich variety known as *cookeite* described by Brown

and Bailey (1962) and by Eggleston and Bailey (1967). Dioctahedral chlorite has been identified in soil samples with pH <5.5 and may also crystallize metastably in sedimentary environments as an authigenic mineral (Carroll, 1970).

The trioctahedral chlorites have been classified on the basis of their chemical compositions by Hey (1954) and by Foster (1962). The latter specified compositional limits for eight varieties of chlorite including the Fe-rich species *thuringite* and *chamosite* and the Mg-rich species *clinochlore* and *penninite*.

The structural formula of trioctahedral chlorites includes the mica unit $(Mg, Fe^{2+})_3(Si, Al)_4O_{10}(OH)_2$, which carries a negative charge because of the replacement of up to 1.6 Si^{4+} by Al^{3+} in the tetrahedral layer. The negative charge is neutralized by a brucite sheet that carries a positive charge caused by replacement of Mg^{2+} by Al^{3+}: $(Mg, Al)_3(OH)_6$. The brucite layer occupies an interlayer position and bonds two mica units together as K^+ ions do in illite (Figure 13.1). The complete structural formula of chlorite is obtained by combining the components of the mica and brucite units:

$$[(Mg, Fe^{2+})_3(Si, Al)_4O_{10}(OH)_2](Mg, Al)_3(OH)_6 \quad (13.15)$$

In addition, chlorites may contain Fe^{3+} ions, which replace Al^{3+} in the brucite layer.

Chamosite, one of the Fe-rich chlorites, is an important constituent of Fe-rich sedimentary rocks of marine origin described by James (1966). Another Fe-rich phyllosilicate called *berthierine* has a 1 : 1 kaolinite-type lattice, but has also been called "chamosite" (Brindley, 1961, Brindley et al., 1968). This mineral has the formula:

$$(Fe^{2+}, Mg)_{2.2}(Fe^{3+}, Al)_{0.7}Si_{1.4}Al_{0.6}O_5(OH)_4 \quad (13.16)$$

and is structurally related to kaolinite but compositionally similar to serpentine. However, it differs from serpentine by containing Fe in the octahedral layer and by limited substitution of Si^{4+} by Al^{3+} in the tetrahedral layer.

Both chamosite and berthierine are authigenic minerals that form in the ocean and may recrystallize during low-grade regional metamorphism to the Fe-chlorite *thuringite*.

c. Mixed-Layer Clays

The clay minerals we have described occur as interstratified layers in the so-called mixed-layer clays, which are very common in soils and in about 70% of clay-rich sedimentary rocks (Weaver, 1956). To some extent the principal types of mixed-layer clays depend on climatic factors that affect the environment within which chemical weathering occurs. Under moist and tropical conditions the stratification in mixed-layer clays is montmorillonite–halloysite–kaolinite, whereas in humid temperature regions the mixed-layer clays consist of montmorillonite–chlorite–mica or mica–intermediate products–illite (Carroll, 1970). The crystallographic properties of mixed-layers clays have been discussed by Zen (1967), Reynolds (1967, 1980), and Corbató and Tettenhorst (1987).

d. Fibrous Clay Minerals

The fibrous clay minerals palygorskite and sepiolite are composed of double chains of silica tetrahedra with an Si/O ratio of 4 : 11. The channel between the double chains is occupied by water molecules that are lost stepwise during heating up to 850°C, at which temperature the structure is destroyed. Neither palygorskite nor sepiolite expand when treated with organic liquids.

Palygorskite is a Mg-silicate with some replacement of Mg^{2+} and Si^{4+} by Al^{3+}. Its structure was worked out by Bradley (1940), who reported the formula:

$$Mg_5Si_8O_{20}(OH)_2(OH_2)_4 \cdot 4H_2O \quad (13.17)$$

where OH_2 is bound water and H_2O is water in channels like that of zeolites. The structure presented above is electrically neutral but may contain exchangeable Ca^{2+} to compensate charge imbalances caused by limited substitutions in this structure. Palygorskite is called *attapulgite* in the United States after an occurrence of this clay in Attapulgus, Georgia.

Palygorskite is an authigenic mineral that occurs in alkaline, unleached soils and as a product of hydrothermal alteration of pyroxene and amphibole. One of the largest accumulations is in the Hawthorn Formation of Miocene age in Georgia and Florida, which was deposited in fresh to brackish water lagoons.

Sepiolite is also composed of chains of silica tetrahedra but it differs from palygorskite in specific details. Grim (1968) reported two different structural formulas for this mineral based on a review of the literature available at the time. The formulas currently in use (see Appendix B) are $Mg_2Si_3O_6(OH)_4$ and $Mg_4Si_6O_{15}(OH)_2 \cdot 6H_2O$. Carroll (1970) reported that a chemical analysis of sepiolite from Little Cottonwood, Utah, yields the formula:

$$[(Mg_{7.42}Mn^{2+}_{0.53}Fe^{3+}_{0.02}) (Si_{11.67}Al_{0.24}Fe^{3+}_{0.09})O_{32}]Cu^{2+}_{0.15} \quad (13.18)$$

Evidently, Mg^{2+} is partly replaced by Mn^{2+} and Fe^{3+}, whereas Si^{4+} is replaced by Al^{3+} and Fe^{3+}. The structure has a net charge of -0.37, which is largely neutralized by exchangeable Cu^{2+}.

13.3 Gibbs Free Energies of Formation

The thermodynamic properties of the three-layer and mixed-layer clay minerals are difficult to establish because of their highly variable chemical compositions. As a result, standard free energies or enthalpies of formation are available only in certain specific cases or apply only to idealized compositions that do not reflect the true chemical diversity of real clay minerals.

In order to overcome this problem, Tardy and Garrels (1974), Chen (1975), and Nriagu (1975) proposed empirical methods for estimating standard free energies of formation of clay minerals. Tardy and Garrels (1976, 1977), Tardy and Gartner (1977), Tardy and Vieillard (1977), and Tardy (1979) subsequently generalized this approach to estimate the standard free energies of formation of other kinds of compounds and their ions in aqueous solution.

The method of Tardy and Garrels (1974) is based on the premise that the standard free energy of a phyllosilicate can be treated as the sum of the free energies of its oxide and hydroxide components. The numerical values of the free energies of the components must reflect the fact that they exist in a silicate lattice and not in free form. Therefore, they must first be determined from phyllosilicate minerals whose standard free energies of formation are known from experimental determinations. The first step toward that end is to restate the structural formulas of phyllosilicates in terms of oxide components, except that Mg is converted to the hydroxide until all of the hydroxyl ion in the formula has been assigned to $Mg(OH)_2$. For example, serpentine is rewritten:

$$Mg_3Si_2O_5(OH)_4 = 2Mg(OH)_2 \cdot MgO \cdot 2SiO_2 \tag{13.19}$$

Similarly, talc becomes:

$$Mg_3Si_4O_{10}(OH)_2 = Mg(OH)_2 \cdot 2MgO \cdot 4SiO_2 \tag{13.20}$$

and sepiolite, which is not a phyllosilicate, nevertheless is recast into the form:

$$Mg_2Si_3O_6(OH)_4 = 2Mg(OH)_2 \cdot 3SiO_2 \tag{13.21}$$

The three Mg-silicates are composed of $Mg(OH)_2$, MgO, and SiO_2, whose standard free energies in the silicate lattice are to be determined. This is done by assuming that the known standard free energies of the three minerals can be expressed as the sums of the unknown standard free energies of the components in the silicate lattice. Therefore, we can write appropriate equations for each of the three minerals. For serpentine:

$$2G_f^\circ(Mg(OH)_2^s) + G_f^\circ(MgO^s) + 2G_f^\circ(SiO_2^s)$$
$$= -965.1 \text{ kcal/mol} \tag{13.22}$$

For talc:

$$G_f^\circ(Mg(OH)_2^s) + 2G_f^\circ(MgO^s) + 4G_f^\circ(SiO_2^s)$$
$$= -1320.38 \text{ kcal/mol} \tag{13.23}$$

And for sepiolite:

$$2G_f^\circ(Mg(OH)_2^s) + 3G_f^\circ(SiO_2^s)$$
$$= -1020.95 \text{ kcal/mol} \tag{13.24}$$

Equations 13.22–13.24 can be solved by substitution and yield the standard free energies of formation in the silicate lattice of the three components listed in Table 13.2. Other components can be added by considering the Al silicates kaolinite, pyrophyllite, and muscovite, which yield values for $G_f^\circ(Al_2O_3^s)$, $G_f^\circ(H_2O^s)$, and $G_f^\circ(K_2O^s)$. Tardy and Garrels (1974) also determined the standard free energies of formation in the silicate lattice of exchangeable cations and water. These values are listed in Table 13.2.

In order to illustrate their method of estimating standard free energies of formation of a real

Table 13.2 Standard Free Energies of Formation of Oxide and Hydroxide Components in the Lattice of Phyllosilicate Minerals at 25 °C

Component	G_f° (silicated), kcal/mol
K_2O	−188.0
Na_2O	−162.8
MgO	−149.2
FeO	−64.1
Fe_2O_3	−177.7
Al_2O_3	−382.4
SiO_2	−204.6
H_2O	−59.2
$Mg(OH)_2$	−203.3
$Al(OH)_3$	−280.0
FeOOH	−117.0
KOH	−123.6
NaOH	−111.0
Exchangeable Ions Expressed as Oxides	
K_2O	−188.0
CaO	−182.8
Na_2O	−175.4
MgO	−159.5
H_2O	−58.6
Li_2	−190.6

SOURCE: Tardy and Garrels (1974).

clay mineral, Tardy and Garrels (1974) chose a typical montmorillonite:

$$[(Al_{1.5}Fe^{3+}_{0.2}Mg_{0.3})(Si_{3.5}Al_{0.5})O_{10}(OH)_2] \cdot Mg_{0.4} \quad (13.25)$$

The formula is electrically neutral, which is important because the exchangeable cations do contribute to the free energy of the mineral and therefore cannot be omitted. In Table 13.3 the formula for this montmorillonite has been recast in terms of molar amounts of the oxide components. These are then multiplied by their respective G°_f (silicated) values from Table 13.2 and the products are summed. The result is -1282.5 kcal/mol, which is a reasonable value for the standard free energy of formation of montmorillonite. For example, Appendix B contains a value of -1258.84 kcal/mol listed by Lindsay (1979) for a Mg-montmorillonite.

More specific comparisons between free energies of formation, estimated by the method of Tardy and Garrels (1974), and experimentally determined values are presented in Table 13.4. The experimental determinations were made by

Table 13.3 Estimate of the Standard Free Energy of Formation of a Typical Montmorillonite with the Formula Stated in Equation 13.25

1 Component, moles	2 G°_f (silicated), kcal/mol	3 1 × 2, kcal
1.0 Al_2O_3	-382.4	-382.4
0.1 Fe_2O_3	-177.7	-17.77
0.3 $Mg(OH)_2$	-203.3	-60.99
3.5 SiO_2	-204.6	-716.1
0.7 H_2O[a]	-59.2	-41.44
0.4 MgO(exch.)	-159.5	-63.8
	Sum	-1282.5 kcal

[a] 0.3 $Mg(OH)_2$ uses up 0.6 hydrogen atom, leaving 1.4 hydrogen atoms to make 0.7 mole of H_2O.
SOURCE: Tardy and Garrels (1974).

Table 13.4 Standard Free Energies of Formation of Clay Minerals Based on Experimental Determinations and Estimates by the Method of Tardy and Garrels (1974)

Mineral or locality	G°_f (measured), kcal/mol	G°_f (estimated), kcal/mol
Illite		
Fithian	-1277.2[a]	-1278.8
Beavers Bend	-1274.7[a]	-1276.2
Goose Lake	-1272.1[a]	-1273.4
Montmorillonite		
Aberdeen	-1230.6[c]	-1228.3
Belle Fourche	-1240.6[b]	-1236.4
Colony, Wyoming	-1241.5[d]	-1240.6
Other Phyllosilicates		
Annite (1)	-1149.3	-1150.2
Annite (2)	-1151.7	-1149.1
Clinochlore	-1961.8	-1958.7
Phlogopite	-1410.9	-1400.7
Greenalite	—	-720.0
Minnesotaite	—	-1055.0

Illite formulas

Fithian,
$$[(Al_{1.54}Fe^{3+}_{0.29}Mg_{0.19})(Si_{3.51}Al_{0.49})O_{10}(OH)_2]K_{0.64}$$
Beavers Bend,
$$[(Al_{1.66}Fe^{3+}_{0.20}Mg_{0.13})(Si_{3.62}Al_{0.38})O_{10}(OH)_2]K_{0.53}$$
Goose Lake,
$$[(Al_{1.58}Fe^{3+}_{0.24}Mg_{0.15})(Si_{3.65}Al_{0.35})O_{10}(OH)_2]K_{0.59}$$

Montmorillonite formulas

Aberdeen,
$$[(Al_{1.29}Fe^{3+}_{0.335}Mg_{0.445})(Si_{3.82}Al_{0.19})O_{10}(OH)_2]K_{0.415}$$
Belle Fourche,
$$[(Al_{1.515}Fe^{3+}_{0.225}Mg_{0.29})(Si_{3.935}Al_{0.065})O_{10}(OH)_2]H_{0.29}$$
Colony,
$$[(Al_{1.52}Fe^{3+}_{0.22}Mg_{0.29})(Si_{3.81}Al_{0.19})O_{10}(OH)_2]K_{0.40}$$

[a] Routson and Kittrick (1971).
[b] Kittrick (1971a).
[c] Kittrick (1971b).
[d] Weaver et al. (1971).

dissolving the minerals and by analyzing the resulting saturated solutions. For example, Weaver et al. (1971) dissolved a sample of montmorillonite from Colony, Wyoming:

$$[(Al_{1.52}Fe^{3+}_{0.22}Mg_{0.29})(Si_{3.81}Al_{0.19})O_{10}(OH)_2]Mg_{0.20}$$
$$+ 6.10 H^+ + 3.57 H_2O \rightleftharpoons$$
$$1.71 Al^{3+} + 3.81 H_4SiO_4 \qquad (13.26)$$
$$+ 0.11 Fe_2O_3(s) + 0.49 Mg^{2+}$$

At equilibrium in a saturated solution the Law of Mass Action applies:

$$\frac{[Mg^{2+}]^{0.49}[H_4SiO_4]^{3.81}[Al^{3+}]^{1.71}}{[H^+]^{6.10}} = K$$
$$(13.27)$$

where $Fe_2O_3(s)$ and $H_2O(l)$ were given activities of one. The value of the equilibrium constant, calculated from equation 13.27 using the measured activities of the ions and molecules, was found to be $K = 10^{0.991}$. Since $\log K = -\Delta G^\circ_R/1.364$, it follows that $\Delta G^\circ_R = -1.352$ kcal. The standard free energy change that takes place during the dissociation of the montmorillonite from Colony, Wyoming, in accordance with equation 13.26 can be expressed as:

$$\Delta G^\circ_R = 1.71\, G^\circ_f(Al^{3+}) + 3.81\, G^\circ_f(H_4SiO_4)$$
$$+ 0.11\, G^\circ_f(Fe_2O_3) + 0.49\, G^\circ_f(Mg^{2+})$$
$$- 3.57\, G^\circ_f(H_2O) - 6.10\, G^\circ_f(H^+) - G^\circ_f(mont.)$$
$$(13.28)$$

Since ΔG°_R for this reaction is known from the measured equilibrium constant, G°_f of the montmorillonite can be calculated from the G°_f values of the other constituents. Tardy and Garrels (1974) obtained a value of $G^\circ_f = -1241.5$ kcal per mole for the Colony montmorillonite from the solubility data, whereas their own method yields $G^\circ_f = -1240.6$ kcal per mole. The agreement is excellent. However, Weaver et al. (1971) obtained $G^\circ_f(mont.) = -1255.8 \pm 0.6$ kcal/mol for the Mg-substituted montmorillonite. The

other direct comparisons in Table 13.4 are also impressive.

13.4 Stability Diagrams

The three-layer clays are more sensitive to the geochemical environment in which they exist than most other minerals because of the presence of exchangeable ions and molecules. Their approach to equilibrium takes place by exchange of the adsorbed ions with ions or molecules from the environment, by transformations involving the chemical composition of the layers and their stacking, and perhaps even by the formation of mixed-layer clays. Although ion exchange is rapid, reactions that require adjustments in the chemical compositions of the octahedral and tetrahedral layers, and in the bonding between individual clay units, are slow. Consequently, natural materials commonly contain mixtures of several clay minerals that cannot be in equilibrium with each other under one set of environmental conditions. For this reason, we now turn to the construction and use of activity diagrams for clay minerals in order to identify the stable phases and to indicate the direction of reactions among the clay minerals in a given geochemical environment.

a. Aluminosilicate Clay Minerals

In order to simplify the construction of stability diagrams, we restrict ourselves to a suite of montmorillonites in which negative charges in the tetrahedral layers are compensated by different adsorbed cations, whereas the octahedral layers consist of unaltered gibbsite sheets. These idealized montmorillonites and their standard free energies of formation were proposed by Helgeson (1969) and Helgeson et al. (1978), and are listed in Table 13.5 together with a standardized illite. These data permit us to explore the stabilities of montmorillonites in the same coordinates we used in Chapter 12 to depict the stability fields of other aluminosilicate minerals. Real montmorillonites and other smectites should react similarly after

Table 13.5 Formulas and Standard Free Energies of Formation of Standardized Montmorillonites and Illite

Formula	G_f°, kcal/mol
Montmorillonite	
$[(Al_{2.00})(Si_{3.67}Al_{0.33})O_{10}(OH)_2]Ca_{0.167}$	-1279.24
$[(Al_{2.00})(Si_{3.67}Al_{0.33})O_{10}(OH)_2]Na_{0.33}$	-1277.76
$[(Al_{2.00})(Si_{3.67}Al_{0.33})O_{10}(OH)_2]K_{0.33}$	-1279.60
$[(Al_{2.00})(Si_{3.67}Al_{0.33})O_{10}(OH)_2]Mg_{0.167}$	-1275.34
Illite	
$[(Al_{1.80}Mg_{0.25})(Si_{3.50}Al_{0.50})O_{10}(OH)_2]K_{0.33}$	-1300.98
	(-1307.485)[a]

[a]Estimated by the method of Tardy and Garrels (1974).
SOURCE: Helgeson (1969).

allowance is made for their more complex chemical compositions.

Based on the stability diagrams we derived in Chapter 12, we can say that kaolinite is the stable aluminosilicate clay mineral in environments where product ions are removed by leaching. If the activity of H_4SiO_4 is less than $10^{-4.7}$ mol/L, even kaolinite breaks down to form gibbsite. Therefore, we can reverse the decomposition of aluminosilicate minerals and inquire how montmorillonite and illite might form from kaolinite when the activities of their ions increase in solutions that are saturated with respect to amorphous silica. Figure 13.2 is an example of such a stability diagram in coordinates of $\log [K^+]/[H^+]$ and $\log [Na^+]/[H^+]$, assuming saturation with respect to amorphous silica. The equations for the boundaries are based on the thermodynamic data in Table 13.5 and in Appendix B. The reaction to form K-montmorillonite from kaolinite is easily constructed based on the conservation of Al:

$$7\,Al_2Si_2O_5(OH)_4 + 8\,SiO_2 + 2\,K^+ \rightleftharpoons$$

$$6[(Al_{2.00})(Si_{3.67}Al_{0.33})O_{10}(OH)_2]K_{0.33}$$

$$+ 7\,H_2O + 2\,H^+ \qquad (13.29)$$

If we multiply kaolinite by 7 and K-montmorillonite by 6, we can balance Al to a good approximation. As a result, we must add 8 moles of SiO_2 (amorphous)

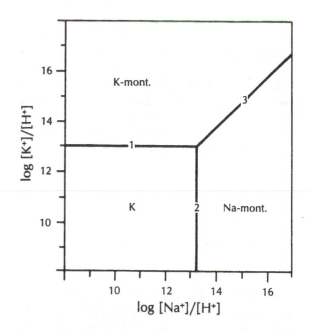

Figure 13.2 Stability of kaolinite, K-montmorillonite, and Na-montmorillonite in the presence of amorphous silica based on data in Table 13.5 and Appendix B.

to the left side of the equation to balance Si and 2 moles of K^+ to balance K. Finally, we need 7 moles of H_2O on the right side to balance O and 2 moles of H^+ to balance H. Equation 13.29 yields

$\Delta G_R^\circ = +35.511$ kcal and $K = 10^{-26.03}$. From the Law of Mass Action:

$$\frac{[H^+]^2}{[K^+]^2} = 10^{-26.03} \qquad (13.30)$$

and therefore (line 1, Figure 13.2):

$$\log \frac{[K^+]}{[H^+]} = 13.02 \qquad (13.31)$$

Equation 13.29 indicates that K-montmorillonite is favored over kaolinite in environments where $[H^+]$ is low and $[K^+]$ is high. Such conditions occur in poorly drained soils under arid climatic conditions or in the oceans where seawater provides a source of K^+ at a pH of about 8.2. The formation of Na-montmorillonite from kaolinite takes place by a reaction that is analogous to equation 13.29 (line 2, Figure 13.2, $\log [Na^+]/[H^+] = 13.32$).

The direct conversion of K-montmorillonite to Na-montmorillonite involves only an exchange of the two elements:

$$3[(Al_{2.00})(Si_{3.67}Al_{0.33})O_{10}(OH)_2]K_{0.33} + Na^+ \rightleftharpoons$$

$$3[(Al_{2.00})(Si_{3.67}Al_{0.33})O_{10}(OH)_2]Na_{0.33} + K^+ \qquad (13.32)$$

$$\Delta G_R^\circ = +0.413 \text{ kcal} \quad K = 10^{-0.302}$$

Therefore, at equilibrium:

$$\frac{[K^+]}{[Na^+]} = 10^{-0.302} \qquad (13.33)$$

and hence (line 3, Figure 13.2):

$$\log \frac{[K^+]}{[H^+]} = \log \frac{[Na^+]}{[H^+]} - 0.30 \qquad (13.34)$$

We can also construct a stability diagram in coordinates of $\log [K^+]/[H^+]$ and $\log [Mg^{2+}]/[H^+]^2$, which includes kaolinite, illite, muscovite, microcline, phlogopite, and chlorite in contact with amorphous silica. The equations for such a diagram are listed in Table 13.6 and have been plotted in Figure 13.3.

During the construction of Figure 13.3 we encountered difficulties caused by the lack of internal consistency of some of the thermodynamic data. A difference of 1.364 kcal in the calculated

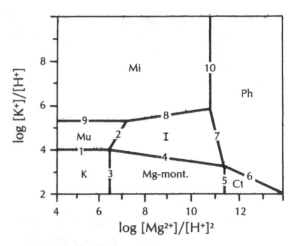

Figure 13.3 Stability of selected minerals in the system K_2O–MgO–Al_2O_3–SiO_2–H_2O–HCl at 25 °C at 1 atm pressure in the presence of amorphous silica. The positions of the lines were adjusted as indicated in the footnotes to Table 13.6 to overcome internal inconsistencies in the thermodynamic data. Nevertheless, the diagram shows that the clay minerals kaolinite (K), illite (I), montmorillonite (Mg-mont.), and chlorite (Ct) can form as weathering products of microcline (Mi), muscovite (Mu), and phlogopite (Ph), which crystallize from cooling magma or grow by metamorphic recrystallization of clay minerals at elevated temperatures and pressures.

value of ΔG_R° of a reaction changes its equilibrium constant by a factor of 10 and displaces the corresponding stability boundary in the activity diagram. In order to avoid some of the resulting inconsistencies, the standard free energy of formation of illite of Helgeson (1969) in Table 13.5 was recalculated by the method of Tardy and Garrels (1974), which increased it from -1300.98 to -1307.49 kcal/mol. In addition, the value of G_f° of Mg-montmorillonite was changed from -1275.34 to -1279.61 kcal/mol. These and other adjustments are listed at the bottom of Table 13.6.

The resulting stability diagram in Figure 13.3 indicates that in acidic low-K environments phlogopite (or biotite) alters to chlorite, Mg-montmorillonite, or kaolinite, depending on the activity of Mg^{2+}. Although phlogopite may be transformed directly into each of the alteration products, sys-

Table 13.6 Equations for Mineral Stability Diagram (Figure 13.3) in the System K_2O–MgO–Al_2O_3–SiO_2–HCl–H_2O at 25 °C, 1 atm Pressure, and in Equilibrium with Amorphous Silica

Kaolinite–muscovite

$$3\,Al_2Si_2O_5(OH)_4 + 2\,K^+ \rightleftharpoons 2\,KAl_2Si_3AlO_{10}(OH)_2 + 3\,H_2O + 2\,H^+$$

$$\Delta G_R^\circ = +10.959 \text{ kcal} \quad K = 10^{-8.03}$$

$$\log \frac{[K^+]}{[H^+]} = 4.02 \tag{1}$$

Illite–muscovite

$$1.3\,[(Al_{1.80}Mg_{0.25})(Si_{3.50}Al_{0.50})O_{10}(OH)_2]K_{0.60} + 0.22\,K^+ + 0.43\,H^+ \rightleftharpoons$$

$$KAl_2Si_3AlO_{10}(OH)_2 + 1.55\,SiO_2(\text{amorph.}) + 0.325\,Mg^{2+} + 0.5\,H_2O$$

$$\Delta G_R^\circ = -1.658^a \text{ kcal} \quad K = 10^{1.215}$$

$$\log \frac{[K^+]}{[H^+]} = 1.48 \log \frac{[Mg^{2+}]}{[H^+]^2} - 5.52 \tag{2}$$

Kaolinite–Mg-montmorillonite

$$7\,Al_2Si_2O_5(OH)_4 + 8\,SiO_2\,(\text{amorph.}) + Mg^{2+} \rightleftharpoons$$

$$6\,[(Al_{2.00})(Si_{3.67}Al_{0.33})O_{10}(OH)_2]Mg_{0.167} + 7\,H_2O + 2\,H^+$$

$$\Delta G_R^\circ = +8.75^b \text{ kcal} \quad K = 10^{-6.42}$$

$$\log \frac{[Mg^{2+}]}{[H^+]^2} = 6.42 \tag{3}$$

Illite–Mg-montmorillonite

$$1.013[(Al_{1.80}Mg_{0.25})(Si_{3.50}Al_{0.50})O_{10}(OH)_2K_{0.60} + 0.1245\,SiO_2(\text{amorph.}) + 0.781\,H^+ \rightleftharpoons$$

$$[(Al_{2.00})(Si_{3.67}Al_{0.33})O_{10}(OH)_2]Mg_{0.167} + 0.608\,K^+ + 0.086\,Mg^{2+} + 0.405\,H_2O$$

$$\Delta G_R^\circ = -3.28^{a,\,b} \text{ kcal} \quad K = 10^{2.41}$$

$$\log \frac{[K^+]}{[H^+]} = -0.14 \log \frac{[Mg^{2+}]}{[H^+]^2} + 3.96^c \tag{4}$$

Mg-montmorillonite–chlorite

$$[(Al_{2.00})(Si_{3.67}Al_{0.33})O_{10}(OH)_2]Mg_{0.167} + 5.658\,Mg^{2+} + 9.32\,H_2O \rightleftharpoons$$

$$1.165\,Mg_5Al_2Si_3O_{10}(OH)_8 + 0.175\,SiO_2(\text{amorph.}) + 11.316\,H^+$$

$$\Delta G_R^\circ = +87.67^d \text{ kcal} \quad K = 10^{-64.27}$$

$$\log \frac{[Mg^{2+}]}{[H^+]^2} = 11.36 \tag{5}$$

Table 13.6 (continued)

Chlorite–phlogopite

$$Mg_5Al_2Si_3O_{10}(OH)_8 + Mg^{2+} + 3\ SiO_2(amorph.) + 2\ K^+ \rightleftharpoons 2\ KMg_3Si_3AlO_{10}(OH)_2 + 4\ H^+$$

$$\Delta G_R^\circ = +26.60^e\ kcal \quad K = 10^{-19.50}$$

$$\log \frac{[K^+]}{[H^+]} = -0.5 \log \frac{[Mg^{2+}]}{[H^+]^2} + 9.75^f \tag{6}$$

Phlogopite–illite

$$2.30\ KMg_3Si_3AlO_{10}(OH)_2 + 15.0\ H^+ \rightleftharpoons [(Al_{1.80}Mg_{0.25})(Si_{3.50}Al_{0.50})O_{10}(OH)_2]K_{0.60}$$

$$+ 1.70\ K^+ + 6.65\ Mg^{2+} + 3.40\ SiO_2(amorph.) + 8.8\ H_2O$$

$$\Delta G_R^\circ = -113.98\ kcal \quad K = 10^{+83.57}$$

$$\log \frac{[K^+]}{[H^+]} = -3.91 \log \frac{[Mg^{2+}]}{[H^+]^2} + 49.15 \tag{7}$$

Microline–illite

$$2.3\ KAlSi_3O_8 + 0.25\ Mg^{2+} + 0.4\ H_2O + 1.2\ H^+ \rightleftharpoons$$

$$[(Al_{1.80}Mg_{0.25})(Si_{3.50}Al_{0.50})O_{10}(OH)_2]K_{0.60} + 1.7\ K^+ + 3.4\ SiO_2(amorph.)$$

$$\Delta G_R^\circ = -14.67\ kcal \quad K = 10^{+10.76}$$

$$\log \frac{[K^+]}{[H^+]} = 0.147 \log \frac{[Mg^{2+}]}{[H^+]^2} + 6.33 \tag{8}$$

Muscovite–microcline

$$KAl_2Si_3AlO_{10}(OH)_2 + 6\ SiO_2(amorph.) + 2\ K^+ \rightleftharpoons 3\ KAlSi_3O_8 + 2\ H^+$$

$$\Delta G_R^\circ = +14.43\ kcal \quad K = 10^{-10.58}$$

$$\log \frac{[K^+]}{[H^+]} = 5.29 \tag{9}$$

Microcline–phlogopite

$$KAlSi_3O_8 + 3\ Mg^{2+} + 4\ H_2O \rightleftharpoons KMg_3Si_3AlO_{10}(OH)_2 + 6\ H^+$$

$$\Delta G_R^\circ = +44.948\ kcal \quad K = 10^{-32.95}$$

$$\log \frac{[Mg^{2+}]}{[H^+]^2} = 10.98 \tag{10}$$

[a]Based on G_f° (illite) $= -1307.485$ kcal/mol, according to Tardy and Garrels (1974).

[b]G_f° (Mg-montmorillonite) $= -1279.61$ kcal/mol adjusted so that the kaolinite–Mg-montmorillonite line goes through log $[K^+]/[H^+] = 4.02$ and log $[Mg^{2+}]/[H^+]^2 = 6.42$.

[c]Increased intercept to 4.86 to force line through log $[K^+]/[H^+] = 4.02$ and log $[Mg^{2+}]/[H^+]^2 = 6.42$.

[d]G_f°(chlorite) $= -1974.0$ kcal/mol (Zen, 1972).

[e]G_f°(phlogopite) $= -1400.7$ kcal/mol (Tardy and Garrels, 1974).

[f]Reduced to 9.0 to improve the fit.

tematic variations in the chemical composition of the fluid in the immediate vicinity of a mineral grain may cause a sequence of reactions to occur such that phlogopite first alters to chlorite, which changes to Mg-montmorillonite, which ultimately decomposes to kaolinite or even gibbsite. Similarly, illite may form either microcline or muscovite by a reduction of $[K^+]$ or decrease in pH. Depending on the existence of the appropriate chemical gradients, muscovite flakes may alter to kaolinite or illite, which may transform into illite–montmorillonite or illite–kaolinite mixed-layer clays. However, predictions of reaction products based on thermodynamics alone are not always reliable because unfavorable kinetics may prevent the reactions from reaching equilibrium or from going to completion.

b. Solubility of Ferrous Silicates

The ferrous silicates include fayalite (Fe_2SiO_4), ferrosilite ($FeSiO_3$), and the two-layer phyllosilicate *greenalite*. In addition, the carbonate mineral siderite ($FeCO_3$) and the hydroxide [$Fe(OH)2$] may form. We assume that Fe is predominantly in the divalent state and that minerals containing Fe^{3+} do not occur in the system.

The ferrous silicates dissolve congruently much like the Mg-silicates we discussed in Chapter 12 (Figure 12.5). The solution of the ferrous silicates is expressed by the equation in Table 13.7, which have been plotted in Figure 13.4 in coordinates of log $[Fe^{2+}]/[H^+]^2$ and log $[H_4SiO_4]$. The lines are the boundaries between solutions that are undersaturated or supersaturated with respect to each mineral. For example, fayalite dissolves in solutions represented by points lying to the left of the equilibrium line (equation 1, Table 13.7). Evidently, greenalite (equation 3, Table 13.7) is less soluble, and therefore more stable, than either fayalite or ferrosilite in contact with water at 25°C. However, the composition of natural solutions in contact with the atmosphere is restricted by the solubility boundaries of siderite (equation 4, Table 13.7) and amorphous silica (equation 6, Table 13.7). Therefore, even greenalite dissolves in natural solutions and precipitates only from solutions that have low bicarbonate ion activities.

Table 13.7 Solubility Equations of Pure Fe-Silicate Minerals in Pure Water at 25°C[a]

Fayalite

$$Fe_2SiO_4 + 4\,H^+ \rightleftharpoons 2\,Fe^{2+} + H_4SiO_4$$
$$\Delta G^\circ_R = -20.76 \text{ kcal} \qquad K = 10^{15.22}$$

$$\log \frac{[Fe^{2+}]}{[H^+]^2} = -0.5 \log [H_4SiO_4] + 7.61 \qquad (1)$$

Ferrosilite

$$FeSiO_3 + H_2O + 2\,H^+ \rightleftharpoons Fe^{2+} + H_4SiO_4$$
$$\Delta G^\circ_R = -7.663 \text{ kcal} \qquad K = 10^{5.62}$$

$$\log \frac{[Fe^{2+}]}{[H^+]^2} = -\log [H_4SiO_4] + 5.62 \qquad (2)$$

Greenalite

$$Fe_3Si_2O_5(OH)_4 + 6\,H^+ \rightleftharpoons$$
$$3\,Fe^{2+} + 2\,H_4SiO_4 + H_2O$$
$$\Delta G^\circ_R = -18.577 \text{ kcal} \qquad K = 10^{13.60}$$

$$\log \frac{[Fe^{2+}]}{[H^+]^2} = -0.67 \log [H_4SiO_4] + 0.53 \qquad (3)$$

Siderite

$$FeCO_3 + 2\,H^+ \rightleftharpoons Fe^{2+} + CO_2 + H_2O$$
$$\Delta G^\circ_R = +0.509 \text{ kcal} \qquad K = 10^{-0.37}$$

$$\log \frac{[Fe^{2+}]}{[H^+]^2} = 3.15 \text{ at } [CO_2] = 3 \times 10^{-4} \text{ atm} \qquad (4)$$

Ferrous hydroxide

$$Fe(OH)_2(s) + 2\,H^+ \rightleftharpoons Fe^{2+} + 2\,H_2O$$
$$\Delta G^\circ_R = -15.924 \text{ kcal} \qquad K = 10^{11.67}$$

$$\log \frac{[Fe^{2+}]}{[H^+]^2} = 11.67 \qquad (5)$$

Amorphous silica

$$SiO_2(\text{amorph.}) + 2\,H_2O \rightleftharpoons H_4SiO_4$$
$$\Delta G^\circ_R = +4.044 \text{ kcal} \qquad K = 10^{-2.96}$$
$$\log [H_4SiO_4] = -2.96 \qquad (6)$$

[a]Based on thermodynamic constants in Appendix B.

The solubility data in Figure 13.4 indicate that greenalite could precipitate in place of siderite if log $[Fe^{2+}]/[H^+]^2$ is greater than about 7. The fugacity of CO_2 that would permit this can be calculated from the solubility equation for siderite in Table 13.7:

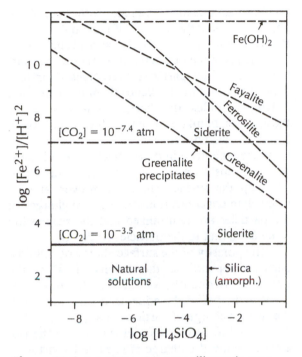

Figure 13.4 Solubility of ferrous silicates in pure water at 25 °C. Greenalite, which is a two-layer clay analogous to kaolinite and serpentine, is soluble in natural solutions open to the atmosphere. It can precipitate at 25 °C, only when $[CO_2] = 10^{-7.4}$ atm or less, from solutions that are nearly saturated with respect to amorphous silica. Greenalite is the most abundant primary Fe-silicate in the Gunflint Iron Formation of Proterozoic age in western Ontario, Canada, and its absence from the siderite facies implies that the siderite facies was deposited where the fugacity of CO_2 was greater than $10^{-7.4}$ atm.

$$\frac{[Fe^{2+}]}{[H^+]^2} = \frac{10^{-0.373}}{[CO_2]} = 10^{7.0} \qquad (13.35)$$

which yields $[CO_2] = 10^{-7.4}$ atm. In other words, greenalite can precipitate only from solutions that have very low carbonate ion concentrations and are close to being saturated with respect to amorphous silica. Such solutions may be discharged by volcanic hotsprings in shallow marine basins where cherty iron formations were deposited in Precambrian time. For example, Goodwin (1956) reported that greenalite is the

most common iron silicate mineral in unaltered parts of the Gunflint Formation in western Ontario, Canada. It occurs as light green to olive green granules up to 3 mm in diameter, except in the banded chert–siderite and tuffaceous shale facies. The absence of greenalite from siderite-bearing rocks is explained by the difference in their solubilities when the solutions are in contact with CO_2 of the atmosphere. Hydrothermal solutions discharged into the Gunflint basin could precipitate greenalite and amorphous silica if they had sufficiently low carbonate ion concentrations to prevent siderite from forming.

Minnesotaite is common in the Fe-silicate facies of the Biwabik Iron Formation of northern Minnesota, but is absent from the Gunflint Formation. James (1966) pointed out that this difference between the two iron formations may be related to the fact that the Biwabik Formation was intruded by the Embarras Granite over much of its length, whereas the Gunflint Formation was not intruded by granite. Consequently, he agreed with Tyler (1949) that minnesotaite is a product of metamorphism, which made the Biwabik Formation more susceptible to chemical weathering than the Gunflint Formation and resulted in the formation of hematite ore bodies in Minnesota.

13.5 Colloidal Suspensions and Ion Exchange

When clay minerals and other small particles with diameters of <1 μm are dispersed in pure water, they form a colloidal suspension called a *sol.* Suspensions of small droplets of a liquid that is immiscible in another liquid are known as *emulsions,* and solid particles or liquid droplets dispersed in a gaseous medium form *aerosols.* Such colloidal suspensions may appear to be homogeneous, and the particles do not settle out under the influence of gravity. Some kinds of particles in sols attract water molecules to themselves and are therefore said to be *hydrophilic,* whereas others lack this tendency and are classified as *hydrophobic.* A sol composed of hydrophilic particles may *set* to form a translucent or transparent *gel.* Some gels can be dispersed again into a sol, whereas in others the process of gel formation is irreversible. For example, concentrated solutions of silic acid can polymerize to form amorphous silica particles, and the resulting colloidal suspension can set to form a gel. However, the silica gel does not redisperse spontaneously, but slowly loses water as it crystallizes to form quartz (variety chalcedony) with opal as an intermediate product (see Section 9.6).

The properties of colloidal suspensions arise from electrical charges on the surface of the particles. The presence of electrical charges on colloidal particles is related to their small size and resultant large surface area. The atoms on the surfaces of solid particles of all kinds of materials are incompletely bonded because they are not surrounded by ions of the opposite charge. Therefore, such particles attract ions from the solution in which they are suspended and thereby acquire a characteristic charge. When *clay minerals* are suspended in water, some of the adsorbed interlayer ions may be lost, leaving the individual particles with a negatively charged surface. In addition, clay particles also have surface ions with unsatisfied charges and hence become charged not only because of internal crystallographic reasons but also because of the surface effects that occur in all small particles.

The surfaces of silicate, oxide, or hydroxide particles expose atoms of Si, Al, Fe, Mn, etc., which are bonded to O^{2-} or OH^- radicals within the solid. However, OH^- radicals on the surface exert an attractive force on positively charged ions in the solution, including H^+. Therefore, such particles acquire a *positive* charge in strongly acidic solutions by virtue of the reaction:

$$Al-OH^{(-)} + H^+ \rightarrow Al-OH_2^{(+)} \quad (13.36)$$

where the parentheses indicate "net" positive or negative charges that are not necessarily equal to one electronic unit. In strongly basic solutions the OH^- radicals may release H^+, thereby causing a net *negative* charge at the surface:

$$Al-OH^{(-)} \rightarrow Al-O^{(-)} + H^+ \quad (13.37)$$

Similar reactions may take place with atoms of Si, Fe, Mn, etc., on the surface of colloidal particles that contribute to their total electrical charges.

The presence of electrical charges on the surface of colloidal particles overcomes their tendency to settle out and keeps them in suspension indefinitely unless the charges are neutralized by the addition of an electrolyte or the passage of an electrical current. In general, colloidal suspensions are stabilized in dilute electrolyte solutions because ions from the solution are needed to build up the surface charge. However, at high electrolyte concentrations the surface charges of the particles are neutralized and the suspension *coagulates,* that is, the particles settle out.

The polarity of the surface charge of colloidal particles depends on the activity of H^+ in the water. Consequently, we define the *isoelectric point* as the pH of the aqueous medium at which the surface charge of particles of a specific material is equal to zero. If the pH is less than the isoelectric point, the charge is *positive* by virtue of equation 13.36; whereas if the pH is greater than the isoelectric points the charge is *negative* according to equation 13.37. The isoelectric points of different kinds of materials in Table 13.8 range from pH = 1.0 (amorphous silica) to pH = 9.1 (Al_2O_3, corundum). The isoelectric points can be used to deduce the polarity of the surface charge of colloidal particles of each material in acid and basic media. The data in Table 13.8 indicate that colloidal particles composed of amorphous silica, quartz, Mn-dioxide, montmorillonite, and kaolinite have a *negative* surface charge, except in highly acidic environments caused, for example, by local discharge of acid mine waters. The oxides of Fe form particles whose surface charges may change from positive to negative between pH values 5 and 7. However, Fe-oxide or hydroxide and TiO_2 particles in the oceans (pH = 8.2) have *negative* charges. The oxides and hydroxides of Al stand out from the others in Table 13.8 because they form *positively* charged particles in all but highly basic environments.

It turns out that the presence of electrolytes in a colloidal suspension affects the development

Table 13.8 Isoelectric Points of Various Natural Materials When Suspended in Water of Varying Acidity but Lacking Dissolved Electrolytes

Material	pH	Polarity in acid medium	Polarity in basic medium
SiO_2(amorph.)	1.0–2.5	$-$ [a]	$-$
Quartz	2.0	$-$	$-$
MnO_2 (different forms)	2.0–4.5	$-$	$-$
Albite	2.0	$-$	$-$
Montmorillonite	2.5	$-$	$-$
Kaolinite	4.6	$-$	$-$
Hematite	5–9	$\pm$	$\pm$
	commonly 6–7		
Magnetite	6.5	$+$	$-$
Goethite	6–7	$\pm$	$\pm$
Limonite ($Fe_2O_3 \cdot nH_2O$)	6–9	$\pm$	$\pm$
Anatase (TiO_2)	7.2	$+$	$-$
Gibbsite	~9	$+$	$+$
Corundum	9.1	$+$	$+$
Periclase (MgO)	12.4	$+$	$+$

[a]The polarity changes from negative to positive at pH values less than the isoelectric point. Thus, the charge is negative if the pH is greater than the isoelectric point. *Acid* and *basic* are defined relative to pH = 7.0 at 25 °C.

After Drever (1982), Table 4.2; Stumm and Morgan (1970).

of surface charges of the particles. Therefore, the pH at which surface charges of colloidal particles become equal to zero in the presence of electrolytes differs from the isoelectric point in ways that depend on the composition of the particles and the charges of the ions of the electrolyte. The pH at which the surface charge of colloidal particles in the presence of electrolytes is equal to zero is known as the *zero point of charge* (*ZPC*). The numerical values of the ZPC depend not only on the composition of the colloidal particles but also on the composition of the electrolytes in the suspension.

The charges of colloidal particles of different compositions affect geochemical processes in the oceans, in soils, and in the deposition of sediment because negatively charged particles, of colloidal size or larger, adsorb cations (metals), whereas positively charged particles adsorb anions (non-metals). Particles of all kinds that sink in the oceans scavenge metal ions from seawater and transfer them to the sediment accumulating at the bottom of the oceans. Consequently, deep-sea sediment is enriched in many metals compared to normal marine sediment, which is dominated by detrital sediment derived from nearby continental areas (Section 4.5).

Colloidal particles in soils act as temporary repositories of ions released by chemical weathering or by the addition of fertilizers. The resulting capacity of soils to store plant nutrients enhances their fertility. In addition, the fixation of ions in soils and sediment enhances the solubility of minerals and results in the purification of groundwater.

When clay minerals or other particles of colloidal or larger size are dispersed in water, their surfaces become charged and attract ions of the opposite charge, which are adsorbed by means of electrostatic bonds. With a few exceptions, mineral surfaces are not particularly selective and attract ions of different elements without much discrimination. This means that the adsorbed ions are *exchangeable* for others in the solution depending primarily on their activity. The resulting phenomenon of *ion exchange* can be represented by a simple exchange reaction:

$$AX + B^+ \rightleftharpoons BX + A^+ \qquad (13.38)$$

which indicated that ion A^+ on the adsorber is exchanged for ion B^+ in the solution. Such exchange reactions reach equilibrium very rapidly and are then subject to the Law of Mass Action; that is:

$$\frac{[BX][A^+]}{[AX][B^+]} = K_{ex} \qquad (13.39)$$

where K_{ex} is the exchange constant. If the exchanger accepts A^+ and B^+ with equal ease, then $K_{ex} = 1.0$. However, in general this is not the case because A^+ and B^+ do not have the same charge/radius ratio and their hydrated ions have different radii. Moreover, divalent or trivalent ions form stronger electrostatic bonds than univalent ions and may therefore be adsorbed preferentially.

Equation 13.39 can be restated in the form:

$$\frac{[A^+]}{[B^+]} = K_{ex}\left(\frac{(AX)}{(BX)}\right)^n \qquad (13.40)$$

where (AX) and (BX) are the molar concentrations of A and B on the exchanger and n is an exponent with values between 0.8 and 2.0 (Berner, 1971; Garrels and Christ, 1965) . These equations explain why the activities of ions in the aqueous phase of colloidal suspension are controlled by ion exchange. For example, if A^+ is added to a colloidal solution containing A^+ and B^+ in ion exchange equilibrium with clay minerals in suspension, A^+ will displace B^+ from exchange sites and thereby reestablish equilibrium in accordance with Le Châtelier's principle.

The capacity of clay minerals to act as ion exchangers varies widely, depending at least in part on the charge imbalances that originate within the crystal lattice. For example, kaolinite commonly has a low *cation exchange capacity* (*CEC*), whereas montmorillonite and vermiculite have large CECs. The CEC of a sample of clay is determined by means of a standard procedure: 100 g of dry clay is repeatedly placed into NaCl solutions at pH = 7.0 until all exchangeable cations are replaced by Na^+. The Na-saturated clay is then added to concentrated KCl solutions and the amount of Na^+ released by the clay at pH = 7.0 is determined by analysis and is expressed in units of milliequivalents per 100 g of dry clay. The CEC values of clay minerals are related to their chemical compositions and therefore vary widely. However, the values in Table 13.9 are fairly typical of the respective clay minerals.

The CEC of a clay mineral can be estimated from the net negative charge of its structural formula. For example, the montmorillonite represented by formula 13.3 has a charge of -0.46 electronic units per formula weight which is neutralized by exchangeable Ca^{2+}, Mg^{2+}, and Na^+. The CEC of this mineral caused only by the charge in the layers is:

Table 13.9 Cation Exchange Capacity (CEC) of Various Clay Minerals

Mineral	CEC, meq/100 g at pH = 7.0
Kaolinite	3–15
Chlorite	10–40?
Illite	10–40
Glauconite	11–20+
Palygorskite	20–30
Allophane	~70
Smectite (montmorillonite)	70–100
Vermiculite	100–150

SOURCE: Garrels and Christ (1965).

$$CEC = \frac{0.46 \times 100 \times 1000}{360.25} = 127.7 \text{ meq/100 g}$$

$$(13.41)$$

This value is probably an underestimate because it does not include charges caused by broken bonds. Nevertheless, the result indicates that 100 g of this clay can adsorb 64.6 mmol of Ca^{2+}, which amounts to $40.08 \times 64.6/1000 = 2.59$ g of Ca or 2.97 g of Na. These amounts are not only substantial, but the presence of Na^+, Ca^{2+}, or other cations has a significant effect on the mechanical properties of the clay.

The presence of Na^+ in exchangeable positions increases the plasticity of clay and can make soil containing such clay unsuitable for agriculture. The best soils contain clay minerals having a mixture of Ca^{2+}, H^+, Mg^{2+}, and K^+ as exchangeable cations. However, in the ceramic industry, where plasticity is a desirable property, clay may be converted to the Na-form by suitable treatment prior to its use.

Smectite clays expand when water molecules replace exchangeable ions, and slurries of smectite may set to form a solid that can revert to a liquid when it is disturbed. This property, called *thixotropy*, characterizes bentonite, which is composed largely of smectite clay minerals, and makes it useful as a sealant of building foundations and as a drilling mud in the petroleum industry.

13.6   Dating of Clay Minerals

One of the primary tasks in the Earth Sciences is to establish an accurate chronology of the history of the Earth, which is recorded in the sedimentary rocks that have survived to the Present. This has turned out to be a very difficult task because sedimentary rocks are generally not datable by methods based on radioactivity. One of several reasons for this difficulty is that sedimentary rocks are composed primarily of detrital particles of minerals that crystallized to form igneous or metamorphic rocks long before they were incorporated into clastic sedimentary rocks, such as sandstones or shales. One can make a virtue of this circumstance by dating detrital minerals such as muscovite, feldspar, or zircon to determine their *provenance* by identifying their source by its age (Faure and Taylor, 1981). However, the age of sedimentary rocks can be determined only by dating minerals that formed at the time of deposition (Clauer and Chaudhuri, 1995).

Unfortunately, most of the *authigenic* minerals in sedimentary rocks are not suitable for dating by isotopic methods because they do not contain radioactive parent elements in sufficient concentration or because they do not retain the radiogenic daughter isotope quantitatively. These shortcomings effectively eliminate several authingenic minerals that commonly occur in sedimentary rocks, including calcite, dolomite, quartz, gypsum, anhydrite, hematite, magnetite, pyrite, pyrolusite, halite, sylvite, and other evaporite minerals. However, certain clay minerals are authigenic and do contain K and Rb, which make them potentially datable by the K–Ar and Rb–Sr methods described in Chapter 16.

Glauconite and illite are best suited for dating of sedimentary rocks, but only under narrowly defined circumstances. For example, glaucony in the form of greenish pellets and encrustations in marine sedimentary rocks contains varying amounts of K, which stabilizes the lattice by strengthening the bonds between clay units. As a result, the retentivity of glaucony with respect to radiogenic ^{40}Ar improves with increasing K content and the mineral becomes closed to ^{40}Ar when $K_2O > 6.0\%$ (Odin and Matter, 1981). Therefore, only glauconitic micas with the required K concentration can yield reliable dates by the K–Ar method. The selection criteria for dating glaucony by the Rb–Sr method are not as well defined, partly because radiogenic ^{87}Sr is less mobile than ^{40}Ar. However, both methods are also sensitive to gain or loss of the radioactive parent element so that well-crystallized glauconitic mica is also preferable for dating by the Rb–Sr method. However, even well-crystallized glauconies may be appreciably altered by thermal metamorphism, tectonic deformation, and ion exchange with brines in the subsurface. Therefore, glauconies that remember their age are valuable exceptions rather than the rule (Odin, 1982).

The application of the Rb–Sr method to dating *clay-rich* sedimentary rocks has been summarized by Clauer (1982). The whole-rock isochron method sometimes yields useful results (Cordani et al., 1985), especially in unfossiliferous shales of Precambrian age containing large amounts of accumulated radiogenic ^{87}Sr. In general, however, it is better to use methods that rely on the properties of a specific authigenic mineral, such as illite. Morton (1985) demonstrated that coarse-grained illite (1–2 μm) in Upper Devonian black shales of Texas is composed of the 2M polymorph of detrital origin and yielded a "provenance date" by the Rb–Sr method that predates the time of deposition. However, 1Md illite in the <0.2-μm fraction in this case dated the time of diagenetic alteration of the illite by migrating brines. The fit of data points to a Rb–Sr isochron was dramatically improved by treating fine-grained 1Md illites with ammonium acetate, which removed Rb^+ and Sr^{2+} from exchangeable sites.

13.7   Summary

Clay minerals are classified according to their crystal structure into platy clays, fibrous clays, and x-ray amorphous clays. The platy clays are composed of sheets of silica tetrahedra joined at their bases (tetrahedral layer, T) and Al hydroxide octahedra joined by sharing hydroxyl ions at the corners (octahedral layer, O). The phyllosilicate clays are made up of two-layer (T–O) and three-layer (T–O–T) packages. The chemical diversity of clay minerals results from complete or partial replacement of Al^{3+} in the octahedral layer and of Si^{4+} in the tetrahedral layer.

In the octahedral layer, Al^{3+} can be replaced by Fe^{3+}, Fe^{2+}, Mg^{2+}, Zn^{2+}, Li^+, and Cr^{3+}, whereas in the tetrahedral layer Si^{4+} is partially replaced by Al^{3+}. The charge imbalance that results from this substitution in the tetrahedral layer may be partly compensated by additional cations in the octahedral layer. However, the net negative charge arising from either the tetrahedral or the octahedral layer, or both layers, is neutralized by cations that are adsorbed to the outer surfaces of the terahedral layer and thereby bond individual clay units to each other. Potassium ions are especially effective in this regard because they fit into the hexagonal rings of silica tetrahedra that make up the tetrahedral layer. Consequently, K-bearing clays like illite have short interlayer spacings and do not expand in water. Clays that have adsorbed Na^+, Ca^{2+}, or Mg^{2+} ions have wider interlayer spacings than K-clays and expand in water because the interlayer cations are displaced by water molecules.

The chemical diversity of clay minerals makes it difficult to ascertain their thermodynamic properties. Several empirical schemes have been proposed to overcome this difficulty, including one by Tardy and Garrels (1974). Their method for estimating standard free energies is based on the oxides and hydroxides that make up the layers of two- or three-layer clays and on the adsorbed ions, including H^+. However, in order to construct mineral stability diagrams of general validity, standardized smectite-type clay minerals are used. The results indicate that primary mica minerals (muscovite, biotite, and phlogopite) alter to a variety of clay minerals depending on environmental conditions. Mineral stability diagrams can be used to understand natural geochemical processes such as the formation of greenalite in cherty iron formations and the incompatibility of greenalite and siderite.

Clay minerals and other small particles can form colloidal suspensions in dilute electrolyte solutions because of electrical charges on their surfaces. Such suspensions may be stable for long periods of time but are coagulated by an increase in the concentration of electrolytes. The polarity of the electrical charges of colloidal particles of different compositions depends on the pH of the solution. Clay minerals, quartz, feldspar, and Mn oxides are negatively charged except in strongly acidic environments with pH < 2.0. The various oxides and hydroxides of Fe can have positive or negative surface charges in the pH range from 5 to 9, whereas Al oxides and hydroxide particles have positive charges except at pH > 9.

Cations that are adsorbed on the charged surfaces of small particles in contact with dilute electrolyte solutions are exchangeable for ions in the solution. These cation exchange reactions can achieve a state of equilibrium that causes water–particle mixtures to respond to changes in water compositions in accordance with Le Châtelier's principle. The ability of small particles of different minerals to adsorb exchangeable ions is expressed by its cation exchange capacity (CEC), which can be measured by means of a standardized procedure. The physical properties of clays are strongly affected by the kinds of adsorbed cations.

Certain authigenic K-bearing clay minerals are dateable by the K–Ar and Rb–Sr methods and therefore provide rare opportunities to determine the ages of sedimentary rocks. However, only clay samples whose chemical composition, crystallinity, or grain size has been evaluated should be used, and the resulting dates may be affected by thermal metamorphism, structural deformation, or ion exchange with circulating brines. Dating of sedimentary rocks by means of clay minerals requires a thorough understanding of clay mineralogy as well as of geochemistry.

Problems

1. Determine the electrical charges of the octahedral and tetrahedral layers and check whether the exchangeable ions balance the charge in the following clay mineral.

$$[(Fe^{3+}_{1.345}Mg_{0.595}Mn_{0.004})$$

$$(Si_{3.836}Al_{0.112}Fe^{3+}_{0.051})O_{10}(OH)_2]\,K_{0.779}Ca_{0.076}$$

2. A glauconite from Point Jackson, Francosia, Wisconsin, has the following chemical composition (Johnston and Cardile, 1987).

Element	Moles	Element	Moles
Si	3.611	Mn	0.001
Al	1.238	Ca	0.096
Fe^{2+}	1.097	K	0.725
Mg	0.442	H	2.000
Ti	0.003	O	12.000

Rewrite this analysis in the form of a structural formula for a three-layer clay and test it for electrical neutrality.

3. Recalculate the chemical analysis of an impure clay mineral from Macon, Georgia, into a structural formula for a two-layer clay (Grim, 1968).

Component	%	Component	%
SiO_2	45.20	K_2O	0.49
Al_2O_3	37.02	Na_2O	0.36
Fe_2O_3	0.27	TiO_2	1.26
FeO	0.06	H_2O-	1.55[a]
MgO	0.47	H_2O+	13.27
CaO	0.52	Sum	100.47

[a]H_2O- is water removed at 105 °C, whereas H_2O+ is lost only at high temperature.

4. Recalculate the chemical analysis of a clay mineral from Montmorillon, France, into a structural formula for a three-layer clay (Grim, 1968).

Component	%	Component	%
SiO_2	51.14	K_2O	0.11
Al_2O_3	19.76	Na_2O	0.04
Fe_2O_3	0.83	TiO_2	—
FeO	—	H_2O-	14.81
MgO	3.22	H_2O+	7.99
CaO	1.62	Sum	99.52

5. Recalculate the chemical analysis of a chlorite mineral from Schmiedefeld, Thuringia, into a structural formula (Grim, 1968).

Component	%	Component	%
SiO_2	20.82	MgO	4.15
Al_2O_3	17.64	H_2O+	10.30
Fe_2O_3	8.70	Sum	99.57
FeO	37.96		

6. Estimate the standard free energy of formation of "average" montmorillonite by the method of Tardy and Garrels (1974) based on the formula

$$[(Al_{1.50}Mg_{0.34}Fe^{3+}_{0.17}Fe^{2+}_{0.04})$$
$$(Si_{3.83}Al_{0.17})O_{10}(OH)_2]K_{0.40}$$

7. Estimate the standard free energy of formation of "pure" illite by the method of Tardy and Garrels (1974) based on the formula

$$[(Al_{1.50}Mg_{0.34}Fe^{3+}_{0.17}Fe^{2+}_{0.04})$$
$$(Si_{3.43}Al_{0.57})O_{10}(OH)_2]K_{0.80}$$

8. Calculate the solubility product constant for hectorite at 25 °C, assuming congruent solution based on the formula

$$[(Mg_{2.67}Li_{0.33})(Si_4)O_{10}(OH)_2]Na_{0.33}$$

References

BERNER, R. A., 1971. *Principles of Chemical Sedimentology.* McGraw-Hill, New York, 240 pp.

BRADLEY, W. F., 1940. The structural scheme of attapulgite. *Amer. Mineral.,* **25**:405–410.

BRINDLEY, G. S., 1961. Kaolin, serpentine, and kindred minerals. In G. Brown (Ed.), *The X-ray Identification and Crystal Structures of Clay Minerals,* 2nd ed., 51–131. Mineralogy Society London, Monograph.

BRINDLEY, G. S., S. W. BAILEY, G. T. FAUST, S. A. FORMAN, and C. I. RICH, 1968. Report of the Nomenclature Committee (1966–67) of the Clay Minerals Society. *Clays and Clay Minerals,* **16**:322–324.

BROWN, B. E., and S. W. BAILEY, 1962. Chlorite polytypisms, part 1, regular and semi-random one-layer structures. *Amer. Mineral.,* **47**:819–850.

BURST, J. F., 1958. Glauconite pellets, their mineral nature and application to stratigraphic interpretations. *Amer. Assoc. Petrol. Geologists Bull.,* **42**:310–327.

CAIRNS-SMITH, A. G., and H. HARTMAN (Eds.), 1987. *Clay Minerals and the Origin of Life.* Cambridge University Press, Cambridge, England, 202 pp.

CARROLL, D., 1970. Clay minerals: a guide to their x-ray identification. *Geol. Soc. Amer., Special Paper* 126, 80 pp.

CHEN, C.-H., 1975. A method of estimating standard free energies of formation of silicate minerals at 298.15 K. *Amer. J. Sci.,* **275**:801–817.

CLAUER, N., 1982. The rubidium–strontium method applied to sediments: certitudes and uncertainties. In G. S. Odin (Ed.), *Numerical Dating in Statigraphy,* vol 1, 245–276. Wiley, Chichester, England, 630 pp.

CLAUER, N., and S. CHAUDHURI, 1995. *Clays in Crustal Environments; Isotopic Dating and Tracing.* Springer-Verlag, Heidelberg, 359 pp.

CORBATÓ, C. E., and R. T. TETTENHORST, 1987. Analysis of illite-smectite interstratification. *Clay Miner.,* **22**:269–285.

CORDANI, U. G., A. THOMAZ-FILHO, B. B. BRITO-NEVES, and K. KAWASHITA, 1985. On the applicability of the Rb–Sr method to argillaceous sedimentary rocks: some examples from Precambrian sequences of Brazil. *Giornale di Geologia, ser. 3a.,* **47**:253–280.

DREVER, J. I., 1982. *The Geochemistry of Natural Waters.* Prentice-Hall, Upper Saddle River, NJ, 388 pp.

EGGLESTON, R. A., and S. W. BAILEY, 1967. Structural aspects of dioctahedral chlorite. *Amer. Mineral.,* **52**:673–689.

FAURE, G., and K. S. TAYLOR, 1981. Provenance of some glacial deposits in the Transantarctic Mountains based on Rb–Sr dating of feldspars. *Chem. Geol.,* **32**:271–290.

FOSTER, M. D., 1962. Interpretation of the composition of trioctahedral micas. *U. S. Geol. Surv. Prof. Paper* **354-B**:11–49.

GARRELS, R. M., and C. L. CHRIST, 1965. *Solutions, Minerals, and Equilibria,* Harper & Row, New York, 450 pp.

GOODWIN, A. M., 1956. Facies relations in the Gunflint Iron Formation. *Econ. Geol.,* **51**:588–595.

GRIM, R. E., 1968. *Clay Minerology,* 2nd ed. McGraw-Hill, New York, 596 pp.

GRIM, R. E., R. H. BRAY, and W. F. BRADLEY, 1937. The mica in argillaceous sediments. *Amer. Mineral., 22*:813–827.

GRUNER, J. W., 1934. The structure of vermiculites and their collapse by dehydration. *Amer. Mineral., 19*:557–575.

HELGESON, H.C., 1969. Thermodynamics of hydrothermal systems at elevated temperatures and pressures. *Amer. J. Sci., 267*:729–804.

HELGESON, H. C., J. M. DELANY, H. W., NESBITT, and D. K. BIRD, 1978. Summary and critique of the thermodynamic properties of rock-forming minerals. *Amer. J. Sci., 278A*:1–229.

HEY, M. H., 1954. New review of the chlorites. *Mineral. Mag., 30*:277–292.

HOWER, J., 1961. Some factors concerning the nature and origin of glauconite. *Amer. Mineral., 46*:313–334.

JAMES, H. L., 1966. Chemistry of the iron-rich sedimentary rocks. In M. L. Fleischer (Ed.), Data of geochemistry. *U. S. Geol. Surv. Prof. Paper*, 440-W; 60 pp.

JOHNSTON, J. H., and C. M. CARDILE, 1987. Iron substitution in montmorillonite, illite, and glauconite by ^{57}Fe Mössbauer spectroscopy. *Clays and Clay Minerals, 35*:170–176.

KITTRICK, J. A., 1971a. Stability of montmorillonites: I. Belle Fourche and Clay Spur montmorillonites. *Soil Sci. Soc. Amer. Proc., 35*:140–145.

KITTRICK, J. A., 1971b. Stability of montmorillonites: II. Aberdeen montmorillonite. *Soil Sci. Soc. Amer. Proc., 35*:820–823.

LINDSAY, W. L., 1979. *Chemical Equilibira in Soils.* Wiley, New York, 449 pp.

McRAE, S. G., 1972. Glauconite. *Earth Sci. Rev., 8(4)*:397–440.

MILLOT, G., 1970. *Geology of Clays.* Springer-Verlag, Berlin, 429 pp.

MORTON, J. P., 1985. Rb–Sr dating of diagenesis and source age of clays in Upper Devonian black shales of Texas. *Geol. Soc. Amer. Bull., 96*:1043–1049.

NRIAGU, J. O., 1975. Thermochemical approximations for clay minerals. *Amer. Mineral., 60*:834–39.

ODIN, G. S. (Ed.), 1982. *Numerical Dating in Stratigraphy*, vols. 1 and 2. Wiley, Chichester, England, 1040 pp.

ODIN, G. S., and A. MATTER, 1981. De glauconiarum origine. *Sedimentology, 28*:611–641.

REYNOLDS, R. C., 1967. Interstratified clay systems—calculation of the total one-dimensional diffraction function. *Amer. Mineral., 52*:661–672.

REYNOLDS, R. C., 1980. Interstratified clay minerals. In G. W. Brindley and G. Brown (Eds.), *Crystal Structures of Clay Minerals and Their X-ray Identification.* Mineral. Soc. London, Monograph 5, 495 pp.

ROUTSON, R. C., and J. A. KITTRICK, 1971. Illite solubility. *Soil Sci. Soc. Amer. Proc., 35*:714–718.

SINGER, A., and E. GALAN (Eds.), 1984. *Palygorskite-Sepiolite.*

Occurrences, Genesis and Uses. Elsevier, Amsterdam, 352 pp.

SMITH, J. V., and H. S. YODER, 1956. Studies of mica polymorphs. *Mineral. Mag., 31*:209–235.

STUMM, W., and J. J. MORGAN, 1970. *Aquatic Chemistry.* Wiley, New York, 583 pp.

SUDO, T., and S. SHIMODA (Eds.), 1978. *Clays and Clay Minerals in Japan.* Elsevier, Amsterdam.

SUDO, T., S. SHIMODA, H. YOTSUMOTO, and S. AITA, 1981. *Electron Micrographs of Clay Minerals.* Elsevier, Amsterdam, 204 pp.

TARDY, Y., and R. M. GARRELS, 1974. A method of estimating the Gibbs energies of formation of layer silicates. *Geochim. Cosmochim. Acta, 38*:1101–1116.

TARDY, Y., and R. M. GARRELS, 1976. Prediction of Gibbs energies of formation—I. Relationships among Gibbs energies of formation of hydroxides, oxides and aqueous ions. *Geochim. Cosmochim. Acta, 40*:1051–1056.

TARDY, Y., and R. M. GARRELS, 1977. Prediction of Gibbs energies of formation of compounds from the elements—II. Monovalent and divalent metal silicates. *Geochim. Cosmochim. Acta, 41*:87–92.

TARDY, Y., and L. GARTNER, 1977. Relationships among Gibbs energies of formation of sulfates, nitrates, carbonates, oxide, and aqueous ions. *Contrib. Mineral. Petrol., 63*:89–102.

TARDY, Y., and P. VIEILLARD, 1977. Relationships among Gibbs free energies and enthalpies of formation of phosphates, oxides and aqueous ions. *Contrib. Mineral. Petrol., 63*:75–88.

TARDY, Y., 1979. Relationships among Gibbs energies of formation of compounds. *Amer. J. Sci., 279*:217–224.

TYLER, S. A., 1949. Development of Lake Superior soft iron ores from metamorphosed iron formation. *Geol. Soc. Amer. Bull., 60*:1101–1124.

VELDE, B., 1977. *Clays and Clay Minerals in Natural and Synthetic Systems.* Elsevier, Amsterdam, 218 pp.

WEAVER, C. E., 1956. The distribution and identification of mixed-layer clays in sedimentary rocks. *Amer. Mineral., 41*:202–221.

WEAVER, R. M., M. L. JACKSON, and J. K. SYERS, 1971. Magnesium and silicon activities in matrix solutions of montmorillonite-containing soils in relation to clay mineral stability. *Soil Sci. Soc. Amer. Proc., 35*:823–830.

WEAVER, C. E., and L. D. POLLARD, 1973. *The Chemistry of Clay Minerals.* Elsevier, Amsterdam.

YODER, H. S., and H. P. EUGSTER, 1955. Synthetic and natural muscovites. *Geochim. Cosmochim. Acta, 8*:225–280.

ZEN, E-AN, 1967. Mixed-layer minerals as one-dimensional crystals. *Amer. Mineral., 52*:635–660.

ZEN, E-AN, 1972. Gibbs free energy, enthalpy and entropy of ten rock-forming minerals: calculations, discrepancies, implications. *Amer. Mineral., 57*:524–553.

14

Oxidation–Reduction Reactions

We now return to a subject we first considered in Chapter 7 and mentioned again in connection with fugacity diagrams in Section 12.5. In Chapter 7 we divided the elements of the periodic table into the *electron donors* (metals) and *electron acceptors* (nonmetals) based on their electron configurations. Metals are electron donors because their available orbitals are largely empty, whereas those of nonmetals are almost filled to capacity. As a result, the atoms of metals give up electrons from partially filled orbitals to form *cations,* whereas atoms of nonmetals form *anions* by attracting electrons to themselves until their available orbitals are filled. These tendencies of the elements are expressed numerically by means of the *electronegativity,* which is defined as the ability of an atom in a molecule to attract electrons to itself.

Some elements always give up or accept the same number of electrons and therefore have fixed valences. For example, the alkali metals and alkaline earths give up one or two electrons, respectively, whereas the halogens accept one electron under natural conditions. Other elements, such as Fe, Cu, and U can lose varying numbers of electrons depending on the strength of the electron acceptor they are interacting with. Similarly, nonmetals such as S, N, and C can take up varying numbers of electrons depending on their availability in the environment.

These concepts come into play in chemical reactions in which the valences of some of the participating elements change. The loss of electrons by the electron donor is called *oxidation* and the gain of electrons by the electron acceptor is called *reduction.* Oxidation must be accompanied by

reduction because the number of electrons released by the donor must equal the number of electrons gained by the acceptor. As a result of the transfer of electrons in an oxidation–reduction reaction the electron donor is *oxidized* and the electron acceptor is *reduced.* Evidently, the electron donor is the *reducing agent* and the electron acceptor is the *oxidizing agent.* Since the atoms of metals are electron donors, we can say that metals are reducing agents and that nonmetals are oxidizing agents.

14.1 Balancing Equations of Oxidation–Reduction Reactions

In order to balance oxidation–reduction reactions correctly, it is helpful to make use of the *valence number* (also known as the oxidation number), which is defined as the electrical charge an atom would acquire if it formed ions in aqueous solution (Hogness and Johnson, 1954, p. 90). This definition presents no difficulty in assigning valence numbers to atoms in ionically bonded compounds. However, molecules with predominantly covalent bonds do not dissociate in water and the valence numbers assigned to elements in covalently bonded compounds are therefore somewhat fictitious. We avoid these difficulties by adopting additional conventions regarding the assignment of valence numbers.

1. The valence number of all elements in pure form is equal to zero.

2. The valence number of H is +1, except in metal hydrides (e.g., LiH) in which it is −1.

3. The valence number of O is −2, except in peroxides (e.g., H_2O_2) in which it is −1.

4. Valence numbers are assigned to elements in molecules or complex ions in such a way that the algebraic sum of the valence numbers of the atoms of a neutral molecule is zero, or equals the charge of the complex ions.

The application of these rules is illustrated in Table 14.1. The examples listed there show that S can have valence numbers of +6, +4, 0, and −2. The valence numbers of C can be +4, 0, −2, and −4, those of Mn can be +7, +4, +3 and +2, and N can have valence numbers of +5 or −3. These values are derived from the chosen compounds or ions by the rules listed above, but do not necessarily represent all of the valence numbers these elements can assume.

The first step in balancing an equation representing an oxidation–reduction reaction is to assign valence numbers to all of the elements that are involved. For example, Cl_2 gas reacts with metallic Fe in the presence of water to form Cl^- and Fe^{3+} ions:

$$Fe + Cl_2 \rightarrow Fe^{3+} + Cl^- \qquad (14.1)$$

The valence numbers of metallic Fe and Cl_2 are zero in accordance with the convention, whereas those of the ions are equal to their charges. Evidently, Fe gives up three electrons per atom to form Fe^{3+}, whereas Cl_2 accepts two electrons per molecule to form $2\,Cl^-$. The loss and gain of electrons is conveniently indicated by arrows:

$$Fe + Cl_2 \rightarrow Fe^{3+} + 2Cl^- \qquad (14.2)$$
$$\downarrow \qquad \uparrow$$
$$3\,e^- \quad 2\,e^-$$

where an arrow pointing down means *loss* and an arrow pointing up means *gain*. In order to balance the loss and gain of electrons, we cross multiply:

$$2\,Fe + 3\,Cl_2 \rightarrow Fe^{3+} + Cl^- \qquad (14.3)$$

The Fe now gives up $2 \times 3 = 6$ electrons and the Cl_2 gains $3 \times 2 = 6$ electrons. Therefore, the elec-

Table 14.1 Assignment of Valence Numbers of Atoms in Chemical Compounds

Formula	Compound	Valence numbers
H_2SO_4	sulfuric acid	H = +1
		O = −2
		S = +6
SO_2	sulfur dioxide	O = −2
		S = +4
H_2S	hydrogen sulfide	H = +1
		S = −2
S_2	sulfur gas	S = 0
$CaCO_3$	calcium carbonate	O = −2
		Ca = +2
		C = +4
CO_2	carbon dioxide	O = −2
		C = +4
CO	carbon monoxide	O = −2
		C = +2
CH_4	methane	H = +1
		C = −4
C	graphite	C = 0
$FeCr_2O_4$	chromite	O = −2
		Fe = +2
		Cr = +3
MnO_4^-	permanganate ion	O = −2
		Mn = +7
MnO_2	pyrolusite	O = −2
		Mn = +4
Mn_3O_4	hausmannite	O = −2
		Mn = +3
		Mn = +2
HNO_3	nitric acid	O = −2
		H = +1
		N = +5
NH_3	ammonia	H = +1
		N = −3
NH_4^+	ammonium	H = +1
		N = −3

trons are in balance, and we can proceed with the mass balance of this equation:

$$2\,Fe + 3\,Cl_2 \rightarrow 2\,Fe^{3+} + 6\,Cl^- \qquad (14.4)$$

Note that the equation is electrically balanced as it must be. Equation 14.4 represents an oxidation–reduction reaction because both oxidation and reduction take place. The metallic Fe loses three electrons and is thereby *oxidized,* whereas the $Cl_2(g)$ gains two electrons and is *reduced.* The electrons picked up by Cl_2 originate from the Fe, which acts as the *reducing agent.* The $Cl_2(g)$ picks up the electrons given up by Fe and thereby acts as the *oxidizing agent.*

In general, metals are reducing agents (electron donors) and nonmetals are oxidizing agents (electron acceptors). However, the anions of nonmetals can also act as electron donors (reducing agents) because they may give up the electrons they acquired during their formation. Similarly, the cations of metals can be electron acceptors (oxidizing agents) to make up for the loss of electrons that occurred during their formation. These relationships are summarized in Table 14.2.

The oxidation–reduction reaction represented by equation 14.4 can be divided into two complementary half-reactions:

$$2\,Fe \rightarrow 2\,Fe^{3+} + 6\,e^- \qquad (14.5)$$

and

$$3\,Cl_2 + 6\,e^- \rightarrow 6\,Cl^- \qquad (14.6)$$

Equation 14.5 is the electron-donor (oxidation) half-reaction and equation 14.6 is the electron-acceptor (reduction) half-reaction. When the two half-reactions are added, the electrons cancel and we obtain equation 14.4 for the complete oxidation–reduction reaction.

Table 14.2 Schematic Representation of the Properties of Elements in Oxidation–Reduction Reactions

Elements	Electron donors, reducing agents	Electron acceptors, oxidizing agents
Metals	×	
Cations of metals		×
Nonmetals		×
Anions of nonmetals	×	
Nonmetals in complex anions $(NO_3^-, SO_4^{2-}, PO_4^{3-})$		×
Nonmetals in complex cations (NH_4^+)	×	

An important oxidation–reduction reaction in nature is the conversion of pyrite (FeS_2) into a precipitate of $Fe(OH)_3$ and SO_4^{2-} ions in solution. In order to construct the equation for this reaction we treat FeS_2 as a reactant and $Fe(OH)_3$ and SO_4^{2-} as products:

$$FeS_2 \rightarrow Fe(OH)_3 + SO_4^{2-} \qquad (14.7)$$

Equation 14.7 indicates that Fe experiences a change in its valence number from +2 in pyrite to +3 in ferric hydroxide. Therefore, the reaction liberates electrons that *must* be taken up by an electron acceptor (oxidizing agent). Natural environments may contain several different electron acceptors, which may be nonmetals like $O_2(g)$ of the atmosphere, or cations such as Fe^{3+}, or complex ions containing oxidized forms of nonmetals like NO_3^-, SO_4^{2-}, or PO_4^{3-}. Some of these potential electron acceptors may not be "strong" enough to take an electron away from Fe^{2+}, but we leave that question for later consideration.

Diatomic O_2 molecules are the most common oxidizing agents in all natural environments in contact with the atmosphere. Therefore, we add O_2 as a reactant to equation 14.7:

$$FeS_2 + O_2 \rightarrow Fe(OH)_3 + SO_4^{2-} \qquad (14.8)$$

Next, we assign valence numbers and determine the number of electrons released and taken up in the reaction.

Compound	Element	Valence number
FeS_2	Fe	+2
	S	−1
O_2	O	0
$Fe(OH)_3$	Fe	+3
	O	−2
	H	+1
SO_4^{2-}	S	+6
	O	−2

Note that the valence number of S in pyrite is −1, whereas in the SO_4^{2-} ion it is +6. Therefore, the S in the pyrite is an electron donor and provides seven electrons per atom. The transformation of Fe^{2+} to Fe^{3+} provides an additional electron per atom of Fe. Therefore, the conversion of one mole of pyrite into $Fe(OH)_3$ and SO_4^{2-} ions releases $(7 \times 2) + 1 = 15$ moles of electrons. The O_2 molecules accept four electrons per molecule because each O atom can take two electrons, which fill vacancies in its $2p$ orbital. The resulting electronic configuration of O^{2-} is $1s^2 2s^2 2p^6$, which is identical to that of Ne (Chapter 5). The electrons given off and accepted are indicated by means of arrows:

$$FeS_2 + O_2 \rightarrow Fe(OH)_3 + SO_4^{2-}$$
$$\begin{array}{cc} \downarrow & \uparrow \\ 15\,e^- & 4\,e^- \end{array} \qquad (14.9)$$

In order to balance the transfer of electrons, we cross multiply.

$$4\,FeS_2 + 15\,O_2 \rightarrow Fe(OH)_3 + SO_4^{2-} \qquad (14.10)$$

Next, we balance Fe and S:

$$4\,FeS_2 + 15\,O_2 \rightarrow 4\,Fe(OH)_3 + 8\,SO_4^{2-} \qquad (14.11)$$

In order to balance O, we add 14 H_2O to the left side:

$$4\,FeS_2 + 15\,O_2 + 14\,H_2O \rightarrow$$
$$4\,Fe(OH)_3 + 8\,SO_4^{2-} \qquad (14.12)$$

In order to balance H, we add 16 H^+ to the right side:

$$4\,FeS_2 + 15\,O_2 + 14\,H_2O \rightarrow$$
$$4\,Fe(OH)_3 + 8\,SO_4^{2-} + 16\,H^+ \qquad (14.13)$$

Equation 14.13 indicates that the conversion of one mole of pyrite into solid ferric hydroxide and sulfate ion releases four moles of H^+ into the environment. The reaction occurs naturally when pyrite weathers by exposure to oxygenated water at or near the surface of the Earth. This process is greatly accelerated when pyrite-bearing rocks are exposed to weathering as a result of mining or quarrying. Pyrite occurs not only in sulfide ores associated with igneous and metamorphic rocks but also in shales that may be interbedded with coal seams, especially those that formed in marine coastal swamps. Stripmining of such coal seams commonly causes acidification of local surface and groundwater because of the oxidation of pyrite (or its polymorph marcasite) by exposure to oxygenated water. The uncontrolled discharge of acid mine waters into streams or lakes is harmful to plant and animal life and may be accompanied by deposition of ferric hydroxide and even gypsum $(CaSO_4 \cdot 2H_2O)$ where sufficient Ca^{2+} is present.

Even a complicated oxidation–reduction reaction like 14.13 can be divided into two complementary half-reactions representing oxidation and reduction, respectively:

$$4\,FeS_2 + 44\,H_2O \rightarrow 4\,Fe(OH)_3 + 8\,SO_4^{2-}$$
$$+ 76\,H^+ + 60\,e^- \quad (14.14)$$
$$15\,O_2 + 60\,H^+ + 60\,e^- \rightarrow 30\,H_2O \qquad (14.15)$$

The Fe^{2+} and S^{1-} of pyrite are the electron donors (reducing agents) and are themselves oxidized to Fe^{3+} and S^{6+}. The O_2 is the electron acceptor (oxidizing agent) and is reduced to O^{2-}.

An oxidation–reduction reaction can, of course, achieve a state of chemical equilibrium, in which case it obeys the Law of Mass Action. The value of the equilibrium constant can be calculated from the standard free energy change of the reaction using data from Appendix B. However, in reality the oxidation of pyrite commonly does not reach equilibrium at or near the surface of the Earth because SO_4^{2-} and H^+ are lost from the system by the movement of water. Moreover, the amount of O_2 available to the reaction is commonly unlimited, allowing the reaction to continue until all of the available pyrite is consumed.

14.2   The Electromotive Series

In the preceding section we alluded to the fact that an oxidizing agent (electron acceptor) may not be strong enough to remove electrons from a particular reducing agent (electron donor). The ability of elements to act as electron donors or electron acceptors arises basically from the extent to which their orbitals are filled with electrons.

More specifically, this property depends on the decrease in the energy of the atoms that results from having only completely filled orbitals. Metallic elements empty partially filled orbitals, whereas nonmetals add electrons to orbitals that are nearly filled. For example, Na^+ and F^- both have the same electronic structure as Ne, but Na achieves this condition by losing an electron, whereas F must gain an electron.

Thermodynamics enables us to rank the elements in terms of their strengths as reducing agents or oxidizing agents. In order to maintain a consistent point of view in this presentation, we consider all elements to be *reducing agents* of varying strengths. Since metals have a natural tendency to be reducing agents, we consider how metals interact with each other. For example, suppose we place metallic Zn into a solution containing Fe^{2+}:

$$Zn + Fe^{2+} \rightarrow Zn^{2+} + Fe \quad (14.16)$$

If the reaction proceeds as written, Zn atoms are able to force Fe^{2+} to accept two electrons. Consequently Zn atoms dissolve as Zn^{2+} and the Fe^{2+} ions precipitate as metallic Fe. However, the reaction could also run in the opposite direction, in which case metallic Fe could displace Zn^{2+} from the solution. We resolve the question by calculating $\Delta G^\circ_R = -16.29$ kcal for reaction 14.16 as written. Since free energy is released by this reaction, we conclude that in the standard state *Zn is a stronger reducing agent than Fe* because it is able to force Fe^{2+} to accept its two electrons.

Next, we consider the reaction:

$$Fe + Cu^{2+} \rightarrow Fe^{2+} + Cu \quad (14.17)$$

and find that $\Delta G^\circ_R = -34.51$ kcal. Evidently, Fe displaces Cu^{2+} ions from solution in the standard state. Therefore Fe is a stronger reducing agent than Cu. Finally, we add metallic Cu to a solution containing Ag^+:

$$Cu + 2 Ag^+ \rightarrow Cu^{2+} + 2 Ag \quad (14.18)$$

In this case, two Ag^+ ions are needed to take the two electrons that Cu atoms can give up. For this reaction, $\Delta G^\circ_R = -21.21$ kcal, which means that Cu is a stronger reducing agent than Ag.

The displacement reactions we have considered (14.16–14.18) permit us to rank the metals in order of decreasing strength as reducing agents in the so-called *electromotive series*:

$$Zn \rightarrow Zn^{2+} + 2 e^- \quad \text{(strongest reducing agent)}$$

$$Fe \rightarrow Fe^{2+} + 2 e^-$$

$$Cu \rightarrow Cu^{2+} + 2 e^-$$

$$Ag \rightarrow Ag^+ + e^- \quad \text{(weakest reducing agent)}$$

Zinc is the strongest reducing agent in this group of metals and displaces the ions of all metals located below it in the electromotive series. Silver is the weakest reducing agent in this group of elements, but it does displace ions of elements below it in the electromotive series. Conversely, because Ag atoms are the weakest reducing agent in the series of metals listed, Ag^+ is the strongest oxidizing agent among the ions we are considering.

The displacement reactions 14.16–14.18 are oxidation–reduction reactions involving the transfer of electrons from an electron donor to an electron acceptor and can be expressed as two complementary half-reactions. For example, when metallic Zn displaces Cu^{2+} from solution:

$$Zn + Cu^{2+} \rightarrow Zn^{2+} + Cu \quad (14.19)$$

the reaction can be divided into two half-reactions:

$$Zn \rightarrow Zn^{2+} + 2 e^- \quad (14.20)$$

$$Cu^{2+} + 2 e^- \rightarrow Cu \quad (14.21)$$

These half-reactions can be separated from each other in an experimental setup called an *electrochemical cell*, which, in this case, consists of a Zn electrode and a Cu electrode dipping into an electrolyte solution containing Cu^{2+} ions. Figure 14.1 shows that the two electrodes are connected by a conducting wire through a voltmeter, which measures the electrical potential or voltage difference between the two electrodes. The voltage arises because electrons are produced at the Zn electrode according to equation 14.20, whereas electrons are consumed at the Cu electrode as shown by equation 14.21. The electrical potential generated by the half-reactions is called the *electromotive force*

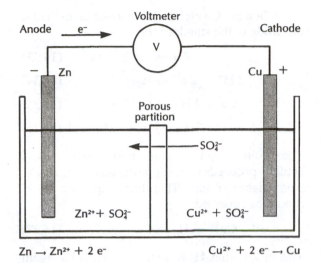

Voltmeter

Anode e^-

Cathode

V

$-$ | Zn

Cu | $+$

Porous partition

$\longleftarrow$ SO$_4^{2-}$

Zn^{2+} + SO$_4^{2-}$

Cu^{2+} + SO$_4^{2-}$

Zn $\rightarrow$ Zn^{2+} + 2 e$^-$

Cu^{2+} + 2 e$^-$ $\rightarrow$ Cu

Figure 14.1 Schematic diagram of a Zn–Cu electrochemical cell. The two metal electrodes dip into solutions of ZnSO$_4$ and CuSO$_4$, which are prevented from mixing by a porous partition. Metallic Zn gives up two electrons per atom and dissolves as Zn^{2+}. The electrons flow to the Cu electrode where Cu^{2+} ions pick up two electrons each and plate out as metallic Cu. Sulfate ions migrate through the porous partition to maintain the electrical neutrality of the solutions. As electrons flow from the Zn electrode to the Cu electrode, the activity of Zn^{2+} increases, whereas that of Cu^{2+} decreases until equilibrium is achieved at which time the flow of electrons stops.

(emf). The magnitude of the emf varies with time as the oxidation–reduction reaction composed of the two complementary electrode half-reactions approaches equilibrium. When the reaction is in the standard state ($[Zn^{2+}] = [Cu^{2+}] = 1.0$, $T = 25\,°C$ $P = 1$ atm) the value of the emf is fixed by the definition of the standard state. When the reaction proceeds from the standard state toward equilibrium, the emf decreases and becomes equal to zero when equilibrium is achieved.

At equilibrium the Zn–Cu electrochemical cell obeys the Law of Mass Action and:

$$\frac{[Zn^{2+}]}{[Cu^{2+}]} = K \qquad (14.22)$$

The magnitude of the equilibrium constant (K) can be calculated from the standard free energy change of the reaction. In this case, $\Delta G_R^{\circ} = -50.8$ kcal and $K = 10^{37.24}$. Evidently, when most of the Cu^{2+} ions have been driven out of the solution such that $[Zn^{2+}]/[Cu^{2+}] = 10^{37.24}$, the emf of the Zn–Cu cell becomes equal to zero. Note that at equilibrium the free energy change of the reaction is also equal to zero and that both ΔG_R and the emf change from fixed values in the standard state to zero at equilibrium.

The relationship between ΔG_R of an oxidation–reduction reaction in any state and the corresponding emf (E) is given by:

$$\Delta G_R = n\mathcal{F}E \qquad (14.23)$$

where $\mathcal{F}$ is the Faraday constant and n is the number of electrons transferred in the reaction. The value of $\mathcal{F}$ is 96,489 ± 20 coulombs per mole or 23.06 kcal per volt per gram equivalent. Since we can calculate the free energy change of an oxidation–reduction reaction in the standard state, we can use equation 14.23 to determine the corresponding *standard emf* ($E°$). For the Zn–Cu cell, $\Delta G_R^{\circ} = -50.8$ kcal and therefore:

$$E° = \frac{\Delta G_R^{\circ}}{n\mathcal{F}} = \frac{-50.8}{2 \times 23.06} = -1.10 \text{ volts (V)}$$
$$(14.24)$$

The minus sign originates from the sign of ΔG_R° and contains information about the electrical polarity of the electrodes.

The relationship between ΔG_R° and $E°$ can be used to rank the elements and their ions in terms of their strengths as reducing agents. In order to establish such a ranking, the emf generated by the dissociation of H$_2$ gas into H$^+$ in the standard state is arbitrarily set equal to zero:

$$H_2(g) \rightarrow 2\,H^+ + 2\,e^- \qquad E° = 0.00 \text{ V} \quad (14.25)$$

If $E° = 0.00$ V, then ΔG_R° for this half-reaction must also be zero, which requires that we adopt the convention that the *standard* free energies of formation H$^+$ and e$^-$ are equal to zero:

$$G_f^{\circ}(H^+) = G_f^{\circ}(e^-) = 0.00 \qquad (14.26)$$

The standard hydrogen electrode can be connected to any metal electrode in the standard state to form an electrochemical cell represented by an oxidation–reduction reaction that is obtained by adding the half-reactions. For example, if Zn is the metal electrode:

$$Zn \rightarrow Zn^{2+} + 2\,e^- \quad (14.27)$$

$$2\,H^+ + 2\,e^- \rightarrow H_2 \quad (14.28)$$

$$Zn + 2\,H^+ \rightarrow Zn^{2+} + H_2 \quad (14.29)$$

Equation 14.29 represents a displacement reaction and implies that Zn atoms can displace H^+ from solution and force it to form bubbles of H_2 gas. The value of ΔG_R° of equation 14.29 is calculated in the conventional manner:

$$\Delta G_R^{\circ} = G_f^{\circ}(Zn^{2+}) + G_f^{\circ}(H_2)$$
$$- G_f^{\circ}(Zn) - 2G_f^{\circ}(H^+) \quad (14.30)$$

However, $G_f^{\circ}(H_2)$ and $G_f^{\circ}(H^+)$ are both equal to zero by convention, so that ΔG_R° of reaction 14.29 arises *entirely* from the standard free energy change of the Zn electrode:

$$\Delta G_R^{\circ} = G_f^{\circ}(Zn^{2+}) - G_f^{\circ}(Zn)$$
$$= -35.14 - 0 = -35.14\,kcal \quad (14.31)$$

Hence, the emf generated by the Zn–H_2 cell in the standard state is based *only* on the magnitude of the standard free energy change of the Zn electrode:

$$E^{\circ} = \frac{\Delta G_R^{\circ}}{n\mathscr{F}} = -\frac{35.14}{2 \times 23.06} = -0.76\,V \quad (14.32)$$

The emf generated by an electrode in the standard state relative to the H_2 electrode in the standard state is called the *standard electrode potential*. The standard electrode potential of the Zn electrode is $-0.76\,V$, where the minus sign originates from the sign of ΔG_R° of the half-reaction. It indicates that when the Zn electrode is connected to the H_2 electrode in the standard state, the reaction proceeds in the direction indicated by the arrow in equation 14.29. In other words, Zn atoms force H^+ to accept its electrons; thus Zn is the *stronger reducing agent* and lies *above* H_2 in the electromotive series.

When the Cu electrode is connected to the H_2 electrode in the standard state:

$$Cu \rightarrow Cu^{2+} + 2\,e^- \quad (14.33)$$

$$2\,H^+ + 2\,e^- \rightarrow H_2 \quad (14.34)$$

$$Cu + 2\,H^+ \rightarrow Cu^{2+} + H_2 \quad (14.35)$$

$$\Delta G_R^{\circ} = G_f^{\circ}(Cu^{2+}) = +15.66\,kcal \quad (14.36)$$

The positive sign indicates that reaction 14.35 actually proceeds in the opposite direction, that is, from right to left. Therefore, equation 14.35 should be reversed:

$$Cu^{2+} + H_2 \rightarrow Cu + 2\,H^+ \quad (14.37)$$

This means that H_2 is a stronger reducing agent than Cu because it forces Cu^{2+} from solution in the standard state. Therefore, the Cu electrode lies *below* the hydrogen electrode in the electromotive series. The standard electrode potential of the Cu electrode is calculated from the value of ΔG_R° of equation 14.35 by means of equation 14.23:

$$E^{\circ} = \frac{+15.66}{2 \times 23.06} = +0.34\,V \quad (14.38)$$

Note again that the H_2 electrode is "inert" and does not enter into the calculation. We can now recombine the Zn and Cu electrodes knowing that Zn is a stronger reducing agent than Cu:

$$Zn \rightarrow Zn^{2+} + 2\,e^- \quad E_{Zn}^{\circ} = -0.76\,V \quad (14.39)$$

$$Cu^{2+} + 2\,e^- \rightarrow Cu \quad E_{Cu}^{\circ} = -0.34\,V \quad (14.40)$$

$$Zn + Cu^{2+} \rightarrow Zn^{2+} + Cu \quad (14.41)$$

$$E_{Zn+Cu}^{\circ} = E_{Zn}^{\circ} + E_{Cu}^{\circ} = -0.76 + (-0.34)$$
$$= -1.10\,V \quad (14.42)$$

Note that we change the sign of E_{Cu}° when we inverted the Cu electrode half-reaction because it acts as the electron acceptor in this case.

The manipulation of electrode half-reactions, illustrated above, indicates that the electromotive series can be derived from standard free energy changes of the half-reactions, provided these are written with the *electrons on the right side*. When this convention is obeyed, the algebraic sign of

ΔG_R° indicates the position of the electrode above or below the H_2 electrode in the electromotive series. A minus sign means that the electrode is a *stronger* reducing agent than H_2, whereas a positive sign indicates that it is *weaker* than the hydrogen electrode as a reducing agent. The numerical values of E° of the electrode half-reactions then permit us to rank elements such that the strongest reducing agent is at the top and the weakest reducing agent is at the bottom of the electromotive series.

The electromotive series in Table 14.3 is a listing of electrode half-reactions in order of decreasing strength of the reducing agents in the standard state. All of the reducing agents *above* the H_2 electrode are *stronger* than H_2, and their half-reactions have negative standard electrode potentials. Those *below* the H_2 electrode are *weaker* reducing agents than H_2. The algebraic sign and magnitude of the standard electrode potentials of the electrodes are derived from the sign and magnitude of ΔG_R°. Note that the alkali metals and alkaline earths are all strong reducing agents because these metals are good electron donors. Some members of the transition groups are moderate to weak reducing agents compared to H_2 (Zn, Cr, Fe, Ni, and Pb), whereas others are actually weaker than H_2 (Bi, Cu, Hg, Ag, and Pt).

The electrodes with positive standard electrode potentials act as *electron acceptors* in the standard state and therefore are *oxidizing agents*. For example, the standard electrode potential of the fluorine electrode in Table 14.3 is +2.88 V. The positive sign and large value indicate that F^- has only a weak tendency to give up electrons to form F_2 gas, whereas F_2 has a strong tendency to act as an electron acceptor or oxidizing agent:

$$F_2 + 2\,e^- \rightarrow 2\,F^- \qquad (14.43)$$

Evidently, the *standard electrode potentials* reflect the same property of the elements as *electronegativities* (Section 7.2), and both arise from the tendencies of atoms to have completely filled orbitals by either giving up or accepting electrons.

Table 14.3 Electromotive Series Presented as Electrode Half-reactions in Order of Decreasing Strengths as Reducing Agents[a]

Reducing agent	Oxidizing agent		Standard electrode potential, V
Cs	$\rightarrow Cs^+$	$+\,e^-$	−3.03
Li	$\rightarrow Li^+$	$+\,e^-$	−3.04
K	$\rightarrow K^+$	$+\,e^-$	−2.94
Ba	$\rightarrow Ba^{2+}$	$+\,2\,e^-$	−2.91
Sr	$\rightarrow Sr^{2+}$	$+\,2\,e^-$	−2.90
Ca	$\rightarrow Ca^{2+}$	$+\,2\,e^-$	−2.87
Na	$\rightarrow Na^+$	$+\,e^-$	−2.71
Rb	$\rightarrow Rb^+$	$+\,e^-$	−2.60
Y	$\rightarrow Y^{3+}$	$+\,3\,e^-$	−2.40
Mg	$\rightarrow Mg^{2+}$	$+\,2\,e^-$	−2.36
La	$\rightarrow La^{3+}$	$+\,3\,e^-$	−2.36
Ce	$\rightarrow Ce^{3+}$	$+\,3\,e^-$	−2.32
Sc	$\rightarrow Sc^{3+}$	$+\,3\,e^-$	−2.03
Be	$\rightarrow Be^{2+}$	$+\,2\,e^-$	−1.97
Th	$\rightarrow Th^{4+}$	$+\,4\,e^-$	−1.83
Al	$\rightarrow Al^{3+}$	$+\,3\,e^-$	−1.70
U	$\rightarrow U^{4+}$	$+\,4\,e^-$	−1.38
Mn	$\rightarrow Mn^{2+}$	$+\,2\,e^-$	−1.18
Nb	$\rightarrow Nb^{3+}$	$+\,3\,e^-$	−1.10
V	$\rightarrow V^{3+}$	$+\,3\,e^-$	−0.87
Zn	$\rightarrow Zn^{2+}$	$+\,2\,e^-$	−0.76
Cr	$\rightarrow Cr^{3+}$	$+\,3\,e^-$	−0.74
S^{2-}	$\rightarrow S$	$+\,2\,e^-$	−0.44
Fe	$\rightarrow Fe^{2+}$	$+\,2\,e^-$	−0.41
Cd	$\rightarrow Cd^{2+}$	$+\,2\,e^-$	−0.40
Co	$\rightarrow Co^{2+}$	$+\,2\,e^-$	−0.28
Ni	$\rightarrow Ni^{2+}$	$+\,2\,e^-$	−0.24
Mo	$\rightarrow Mo^{3+}$	$+\,3\,e^-$	−0.20
Sn	$\rightarrow Sn^{2+}$	$+\,2\,e^-$	−0.14
Pb	$\rightarrow Pb^{2+}$	$+\,2\,e^-$	−0.13
H_2	$\rightarrow 2\,H^+$	$+\,2\,e^-$	0.00
Bi	$\rightarrow Bi^{3+}$	$+\,3\,e^-$	+0.29
Cu	$\rightarrow Cu^{2+}$	$+\,2\,e^-$	+0.34
Cu	$\rightarrow Cu^+$	$+\,e^-$	+0.52
$2\,I^-$	$\rightarrow I_2$	$+\,2\,e^-$	+0.53
Se^{2-}	$\rightarrow Se$	$+\,2\,e^-$	+0.67
Ag	$\rightarrow Ag^+$	$+\,e^-$	+0.80
Hg	$\rightarrow Hg^{2+}$	$+\,e^-$	+0.85
Pd	$\rightarrow Pd^{2+}$	$+\,2\,e^-$	+0.92
$2\,Br^-$	$\rightarrow Br_2$	$+\,2\,e^-$	+1.08
Pt	$\rightarrow Pt^{2+}$	$+\,2\,e^-$	+1.19
$2\,Cl^-$	$\rightarrow Cl_2$	$+\,2\,e^-$	+1.36
Au	$\rightarrow Au^+$	$+\,e^-$	+1.69
Pt	$\rightarrow Pt^+$	$+\,e^-$	+2.64
$2\,F^-$	$\rightarrow F_2$	$+\,2\,e^-$	+2.88

[a]Based on G_f° values in Appendix B.

14.3   The Emf of Electrochemical Cells

The relationship between the free energy change of a chemical reaction and the activities of the reactants and products is given by an equation derived in Section 11.8 (equation 11.73):

$$\Delta G_R - \Delta G_R^\circ = RT \ln Q \qquad (14.44)$$

where ΔG_R is the free energy change of the reaction in any state, ΔG_R° is the free energy change in the standard state, R is the gas constant, T is the standard temperature in kelvins, and Q is the reaction quotient. We now convert ΔG values to emf by substituting equation 14.23 into equation 14.44. After rearranging terms and converting to logarithms to the base 10, we have:

$$E = E^\circ + \frac{2.303RT}{n\mathcal{F}} \log Q \qquad (14.45)$$

Since R $= 1.987 \times 10^{-3}$ kcal/deg·mol, $T = 298.15$ K, $\mathcal{F} = 23.06$ kcal/V · number of electrons transferred:

$$E = E^\circ + \frac{2.303 \times (1.987 \times 10^{-3}) \times 298.15}{n \times 23.06} \log Q \qquad (14.46)$$

$$E = E^\circ + \frac{0.05916}{n} \log Q \qquad (14.47)$$

Equation 14.47 provides a relationship between the activities of the reactants and products of an oxidation–reduction reaction, expressed by Q, and the emf that would be generated by the reaction if it were split into two half-reactions in an electrochemical cell. When the reaction achieves equilibrium, $\Delta G_R = 0$, $E = 0$, and $Q = K$ where K is the equilibrium constant. Therefore, at equilibrium:

$$E^\circ = -\frac{0.05916}{n} \log K \qquad (14.48)$$

Equation 14.48 relates the emf of an oxidation–reduction reaction in the standard state to the equilibrium constant of that reaction.

We can apply these relationships to the Zn–Cu electrochemical cell represented by equation 14.41. When this reaction is neither in the standard state nor at equilibrium, the emf it generates is given by:

$$E = E^\circ + \frac{0.05916}{2} \log \frac{[Zn^{2+}]}{[Cu^{2+}]} \qquad (14.49)$$

where E° is the emf generated by the Zn–Cu cell in the standard state, which has a value of -1.10 V as calculated in equation 14.24. Therefore, when the reaction is proceeding toward equilibrium, the emf is:

$$E = -1.10 + 0.02958 \log \frac{[Zn^{2+}]}{[Cu^{2+}]} \qquad (14.50)$$

In the standard state, $[Zn^{2+}] = [Cu^{2+}] = 1.0$ and therefore $\log [Zn^{2+}]/[Cu^{2+}] = 0$. Consequently, in the standard state the emf is:

$$E = E^\circ = -1.10 \text{ V} \qquad (14.51)$$

At equilibrium, $[Zn^{2+}]/[Cu^{2+}] = 10^{37.24}$ (equation 14.22). Therefore, at equilibrium:

$$E = -1.10 + 0.02958 \times 37.24 = 0.00 \text{ V} \quad (14.52)$$

The progress of the reaction in the Zn–Cu cell is illustrated by plotting equation 14.50 in Figure 14.2 in coordinates of E and log $[Zn^{2+}]/[Cu^{2+}]$. The straight line indicates how E changes from -1.10 V in the standard state (log $[Zn^{2+}]/[Cu^{2+}] = 0$) to 0.00 V at equilibrium (log $[Zn^{2+}]/[Cu^{2+}] = 37.24$). When an external voltage is applied to the Zn–Cu cell, the $[Zn^{2+}]/[Cu^{2+}]$ ratio changes until it is compatible with the applied voltage.

These concepts seem fairly removed from oxidation–reduction reactions in natural environments. However, they enable us to define a special kind of emf known as the Eh. The Eh is the emf generated between an electrode in any state and the H_2 electrode in the standard state. Since the H_2 electrode is in the standard state, it does not affect the magnitude of the Eh and acts only as a reference. We illustrate the usefulness of this concept by considering an electrode at which Fe^{2+} is oxidized to Fe^{3+}:

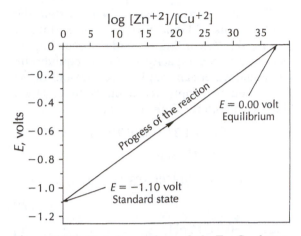

Figure 14.2 Variation of the emf of a Zn–Cu electro-chemical cell as the oxidation–reduction reaction on which it is based progresses from the standard state when $[Zn^{2+}] = [Cu^{2+}] = 1.0$ mol/L, and $E^\circ = -1.10$ V toward equilibrium when $[Zn^{2+}]/[Cu^{2+}] = 10^{37.5}$ and $E = 0.00$ V.

$$Fe^{2+} \rightarrow Fe^{3+} + e^- \qquad (14.53)$$

The complete oxidation–reduction reaction, when this electrode is combined with the H_2 electrode in the standard state, is obtained by adding the H_2 half-reaction:

$$H^+ + e^- \rightarrow \tfrac{1}{2} H_2 \qquad (14.54)$$

to equation 14.53, which yields:

$$Fe^{2+} + H^+ \rightarrow Fe^{3+} + \tfrac{1}{2} H_2 \qquad (14.55)$$

Since the H_2 electrode is in the standard state, $[H^+] = [H_2] = 1.0$. Therefore, the reaction quotient of reaction 14.55 is:

$$Q = \frac{[Fe^{3+}]}{[Fe^{2+}]} \qquad (14.56)$$

The *Eh* generated by the Fe^{2+} electrode is given by equation 14.47. When both electrodes are in the standard state, the change in free energy of equation 14.55 is

$$\Delta G_R^\circ = [G_f^\circ(Fe^{3+}) + \tfrac{1}{2}G_f^\circ(H_2)]$$
$$- [G_f^\circ(H^+) + G_f^\circ(Fe^{2+})] \qquad (14.57)$$

Since $G_f^\circ(H_2)$ and $G_f^\circ(H^+)$ are both equal to zero, ΔG_R° of equation 14.55 depends entirely on the Fe^{2+} electrode. Therefore, we can calculate it *directly* from the electrode half-reaction. In this case, $\Delta G_R^\circ = +17.75$ kcal and the standard electrode potential $E^\circ = +0.769$ V. The Eh generated by the Fe^{2+} electrode is related to the ratio of the activities of Fe^{3+} and Fe^{2+} by the equation:

$$Eh = 0.769 + \frac{0.05916}{1} \log \frac{[Fe^{3+}]}{[Fe^{2+}]} \qquad (14.58)$$

The equation establishes the relationship between the emf generated by the Fe^{2+} electrode connected to the H_2 electrode in the standard state for different values of the $[Fe^{3+}]/[Fe^{2+}]$ activity ratio.

In natural systems the Eh, like pH, is an *environmental parameter* whose value reflects the ability of the natural system to be an electron donor or acceptor relative to the standard H_2 electrode. Therefore, equation 14.58 leads to the important conclusion that the $[Fe^{3+}]/[Fe^{2+}]$ ratio of a natural geochemical system depends on the Eh of the environment. This example suggests that the concepts of electrochemistry presented in Section 14.2 are indeed applicable to natural systems.

14.4   Stability Limits of Water in Terms of Eh and pH

We showed in Section 12.5 that liquid water must maintain an equilibrium with O_2 and H_2 gas:

$$H_2O(l) \rightleftharpoons \tfrac{1}{2} O_2(g) + H_2(g) \qquad (14.59)$$

We used this equation to set limits on the fugacity of O_2 required for stability of liquid water on the surface of the Earth. Now we want to determine the stability limits of water in terms of Eh and pH. At first glance, this may appear to be a pretty hopeless proposition until we recall that we are dealing here with electrode half-reactions. Therefore, we invent the *water electrode*:

$$H_2O(l) \rightarrow 2 H^+ + \tfrac{1}{2} O_2 + 2 e^- \qquad (14.60)$$

In this electrode the O in the water molecule gives up two electrons and thereby changes its valence number from -2 to 0. Equation 14.60 therefore is an electron-donor half-reaction that can generate an Eh when it is connected to the hydrogen reference electrode, which is maintained in the standard state. It does not matter whether an H_2O–H_2 electrochemical cell can actually be constructed. Moreover, we do not need to complete the oxidation–reduction reaction by adding the standard H_2 electrode reaction to equation 14.60 because the Eh is generated entirely by the water electrode. Therefore, we can express the Eh of the water electrode by the equation:

$$Eh = E° + \frac{0.05916}{2} \log([O_2]^{1/2}[H^+]^2) \quad (14.61)$$

We determine $E°$ by calculating $\Delta G_R°$ for the water electrode as written in equation 14.60:

$$\Delta G_R° = [2G_f°(H^+) + \tfrac{1}{2}G_f°(O_2) + 2G_f°(e^-)]$$
$$- [G_f°(H_2O(l))] \quad (14.62)$$

Since

$$G_f°(H^+) = G_f°(O_2) = G_f°(e^-) = 0,$$
$$\Delta G_R° = -G_f°(H_2O(l)) = +56.687 \text{ kcal} \quad (14.63)$$

and hence:

$$E° = +\frac{56.687}{2 \times 23.06} = +1.23 \text{ V} \quad (14.64)$$

Therefore, the Eh generated by the water electrode is:

$$Eh = 1.23 + \frac{0.05916}{2} \log([O_2]^{1/2}[H^+]^2) \quad (14.65)$$

$$Eh = 1.23 + \frac{0.05916}{2}\left(\frac{1}{2}\right)\log[O_2]$$
$$+ \frac{0.05916 \times (2)}{2}\log[H^+] \quad (14.66)$$

$$Eh = 1.23 + 0.01479 \log[O_2] - 0.05916 \text{ pH}$$
$$(14.67)$$

Equation 14.67 is very important because it establishes the relationship between the Eh and pH of a geochemical environment and the corresponding fugacity of O_2.

We can also use equation 14.67 to state the stability limits of liquid water on the surface of the Earth. According to the discussion in Section 12.5, the fugacity of O_2 in equilibrium with liquid water can vary between 1.0 and $10^{-83.1}$ atm. When we apply these limits to equation 14.67, we obtain for $O_2 = 1.0$ atm:

$$Eh = 1.23 - 0.05916 \text{ pH} \quad (14.68)$$

and for $O_2 = 10^{-83.1}$ atm:

$$Eh = 1.23 + 0.01479(-83.1) - 0.05916 \text{ pH}$$
$$(14.69)$$

$$Eh = -0.05916 \text{ pH} \quad (14.70)$$

Equations 14.68 and 14.70 are straight lines in coordinates of Eh and pH, which define the stability field of liquid water in Figure 14.3. All natural chemical reactions that take place in the presence of water at 25°C and 1 atm pressure are restricted to that region of the diagram.

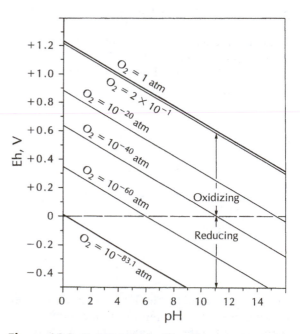

Figure 14.3 Stability field of liquid water at the surface of the Earth at 25°C and 1 atm pressure in coordinates of Eh and pH. The contours of O_2 fugacity are based on equation 14.67 and arise from the dissociation of liquid water into O_2, H^+, and electrons.

The stability field of liquid water in Figure 14.3 contains environments having positive as well as negative Eh values. A positive Eh implies that the environment is a *weaker* reducing agent than the standard H_2 electrode. Therefore, such environments actually function as *electron acceptors* (oxidizing agents) relative to the standard H_2 electrode. Hence, such environments are said to be *oxidizing environments,* meaning that they have the ability to *accept* electrons from electron donors.

Environments having negative Eh values are *stronger* reducing agents than the standard H_2 electrode. Therefore, such environments function as *electron donors* (reducing agents) and are said to be *reducing environments.*

An oxidizing environment contains elements or compounds that are capable of accepting electrons in case an electron donor becomes exposed to it. The most common oxidizing agent at or near the surface of the Earth is atmospheric molecular oxygen, which dissolves in water and may be transported by the movement of water to environments that are not directly exposed to the atmosphere. In addition, the equilibrium of water with O_2 and H_2 requires the presence of both gases wherever water is present on the Earth. Consequently, some O_2 is present even in environments described as *anoxic.* However, the fugacity of O_2 in anoxic environments may be much less than is required to sustain life, although certain kinds of bacteria thrive in such environments.

The relationship of Eh, pH, and O_2, expressed by equation 14.67, enables us to calculate the fugacity of O_2 from the Eh and pH of an environment. Equation 14.67 can also be used to derive a series of contours for selected values of the O_2 fugacity, as shown in Figure 14.3. These contours reflect the fact that the O_2 fugacity of natural environments can vary over an extremely wide range from 1 to $10^{-83.1}$ atm. Even environments classified as *oxidizing* in Figure 14.3 may contain only very small amounts of O_2.

Consider an environment with pH = 7.0 and Eh = +0.6 V. Equation 14.67 indicates that in this environment $[O_2] = 10^{-14.6}$ atm. The gaseous O_2 is in equilibrium with O_2 molecules dissolved in water:

$$O_2(g) \rightleftharpoons O_2(aq) \qquad (14.71)$$

At equilibrium 25 °C (Stumm and Morgan, 1996, p. 215):

$$\frac{[O_2(aq)]}{[O_2(g)]} = 10^{-2.9} \qquad (14.72)$$

Therefore, $[O_2(aq)] = 10^{-17.5}$ mol/L, which is equivalent to about $1 \times 10^{-13} \mu g/mL$ of dissolved O_2. Most fish require more than one million times more O_2. Nevertheless, the environment is classified as being *oxidizing.*

The Eh of an environment can be measured directly as the voltage between a shiny Pt electrode inserted into the solution to be tested and a reference electrode. The reference electrode used most frequently is the *calomel electrode* in which Hg reacts with Cl^- to form Hg_2Cl_2 plus electrons:

$$2\,Hg + 2\,Cl^- \rightarrow Hg_2Cl_2 + 2\,e^- \qquad (14.73)$$

The emf of this electrode is stabilized by maintaining the activity of Cl^- at a constant value by immersing it in a saturated solution of KCl. However, in practice, Eh measurements are difficult to make because the insertion of the electrode may disturb the environment, or because the evironment is inaccessible, and for other reasons discussed by Garrels and Christ (1965) and by Stumm and Morgan (1996). Nevertheless, sufficient Eh measurements exist to characterize common geological environments in terms of the range of Eh and pH values (Baas-Becking et al., 1960). A schematic representation of these data in Figure 14.4 indicates that environments in contact with the atmosphere (rainwater, streams, oceans, and mine waters) have positive Eh values, whereas environments that are isolated from the atmosphere (waterlogged soils, euxenic marine basins, and organic-rich brines) have negative Eh values. However, many environments are transitional between these extremes and have intermediate Eh values.

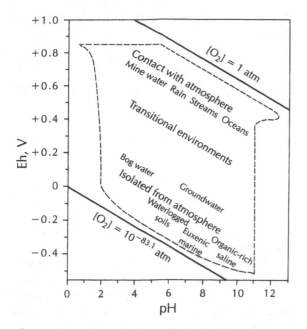

Figure 14.4 Range of Eh and pH conditions in natural environments based on data by Baas-Becking et al. (1960) and modified after Garrels and Christ (1965).

14.5 Stability of Iron Compounds

The stability of solid compounds of elements having different oxidation states can be represented in terms of Eh and pH when they participate in electrode half-reactions. We therefore demonstrate the construction of such *Eh–pH stability diagrams* by using the naturally occurring compounds of Fe as an example. Such diagrams were originally adapted by Garrels (1960) from electrochemical theory used in the study of the corrosion of metals (Pourbaix, 1949). Eh–pH diagrams are widely used in geochemistry to study the formation and weathering of ore deposits (Garrels and Christ, 1965), environmental geochemistry, and nuclear waste disposal (Brookins, 1978). A book devoted exclusively to this subject was published by Brookins (1988).

a. Oxides of Iron
The naturally occurring oxides of Fe include magnetite (Fe_3O_4) and hematite (α-Fe_2O_3). Other oxides such as maghemite (γ-Fe_2O_3) and oxyhy-

droxides goethite (α-FeOOH) and lepidocrocite (γ-FeOOH) occur metastably and, in time, recrystallize to hematite. Nevertheless, their stability in coordinates of Eh and pH can be delineated by the same methods that apply to hematite and magnetite.

We begin the presentation with the oxidation of metallic Fe to magnetite, which we express as an electrode half-reaction:

$$3\,Fe \rightarrow Fe_3O_4 + 8\,e^- \qquad (14.74)$$

Note that magnetite is a mixed oxide composed of FeO and Fe_2O_3. Therefore, one of the metallic Fe atoms loses two electrons to form FeO while the other two atoms lose three electrons each to form Fe_2O_3. Therefore, the three moles of Fe required to make one mole of magnetite give up a total of eight moles of electrons. In order to account for the O of magnetite we add four moles of H_2O to the left side of the equation, which requires the appearance of eight moles of H^+ on the right side:

$$3\,Fe + 4\,H_2O \rightarrow Fe_3O_4 + 8\,H^+ + 8\,e^- \qquad (14.75)$$

This electrode generates an emf when it is connected to the standard H_2 electrode:

$$Eh = E^\circ + \frac{0.05916}{8} \log[H^+]^8 \qquad (14.76)$$

Since ΔG_R° for this electrode is -15.85 kcal, $E^\circ = -0.086$ V. By substituting this value into equation 14.76 and after converting to pH, we obtain:

$$Eh = -0.086 - 0.0592\,pH \qquad (14.77)$$

The straight line represented by this equation relates the Eh of the Fe–Fe_3O_4 electrode to the activity of H^+. If the Eh is fixed by the environment, $[H^+]$ is uniquely determined by equation (14.77) in terms of the corresponding pH value. If the $[H^+]$ of the environment is greater than required by equation 14.77, the electrode half-reaction reduces magnetite to metallic Fe in accordance with equation 14.75. Therefore, metallic Fe is the stable phase under conditions represented by all points located left of the

Eh–pH line for the Fe–Fe_3O_4 electrode. If $[H^+]$ of the environment having a fixed Eh value is less than that required by the electrode (equation 14.75), metallic Fe is converted to Fe_3O_4. Therefore, magnetite is the stable phase in environments represented by points located to the right of the Eh–pH line for the Fe–Fe_3O_4 electrode. However, metallic Fe is not stable on the surface of the earth because its stability field lies *outside* the stability limits of water. We reached the same conclusion during the construction of the fugacity diagram for Fe oxides in Section 12.5b.

The conversion of magnetite to hematite can be expressed as an electrode half-reaction:

$$2\,Fe_3O_4 \rightarrow 3\,Fe_2O_3 + 2\,e^- \quad (14.78)$$

We require one mole of H_2O on the left side to balance O and then obtain $2\,H^+$ on the right:

$$2\,Fe_3O_4 + H_2O \rightarrow 3\,Fe_2O_3 + 2\,H^+ + 2\,e^-$$

$$\Delta G_R^\circ = +9.087\ kcal \qquad E^\circ = +0.20\ V \quad (14.79)$$

Therefore, the Eh generated by this electrode as a function of pH is:

$$Eh = 0.20 + \frac{0.05916}{2}\log\,[H^+]^2 \quad (14.80)$$

which reduces to:

$$Eh = 0.20 - 0.0592\ pH \quad (14.81)$$

Equations 14.77 and 14.81 define the stability limits of metallic Fe, magnetite, and hematite in Figure 14.5. Magnetite is stable and can precipitate from aqueous solution primarily in reducing environments having Eh < 0.0 V. Hematite is stable under oxidizing conditions, but its stability field does extend into reducing environments having pH > 4.6.

b. Solubility of Iron Oxides

Both magnetite and hematite are slightly soluble in water depending on the pH (Section 12.5b, Figure 12.7). Several different ionic and molecular species can form, but Fe^{2+} and Fe^{3+} dominate at low pH. The solubility of Fe-oxides limits their

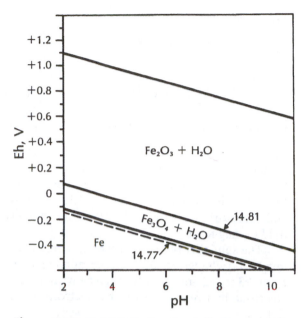

Figure 14.5 Stability limits of metallic Fe, magnetite, and hematite in the presence of water at 25 °C in coordinates of Eh and pH. Note that the pH scale starts at 2 rather than 0.

stability in natural environments because they may be removed by the movement of acidic solutions in the course of geologic time.

We consider first the solubility of magnetite and hematite with respect to Fe^{2+}. The solution of *magnetite* to form Fe^{2+} involves a change in the valence number of the trivalent Fe and therefore can be expressed as an electrode half-reaction:

$$Fe_3O_4 + 2\,e^- \rightarrow 3\,Fe^{2+} \quad (14.82)$$

We must reverse this equation in order to keep the electrons on the right side and, by adding H_2O and H^+ as required, we obtain:

$$3\,Fe^{2+} + 4\,H_2O \rightarrow Fe_3O_4 + 8\,H^+ + 2\,e^-$$

$$\Delta G_R^\circ = +40.698\ kcal \qquad E^\circ = +0.88\ V \quad (14.83)$$

Therefore,

$$Eh = 0.88 + \frac{0.05916}{2}\log\left(\frac{[H^+]^8}{[Fe^{2+}]^3}\right) \quad (14.84)$$

which reduces to:

$$Eh = 0.88 + \frac{0.05916 \times 8}{2} \log[H^+]$$

$$- \frac{0.05916 \times 3}{2} \log[Fe^{2+}] \qquad (14.85)$$

$$Eh = 0.88 - 0.237\ pH - 0.089 \log[Fe^{2+}] \quad (14.86)$$

Equation 14.86 represents a family of parallel lines in coordinates of Eh and pH corresponding to selected values of $[Fe^{2+}]$. These lines express the solubility of *magnetite* in the Eh–pH plane and therefore are valid only within the stability field of magnetite.

The solubility of *hematite* with respect to Fe^{2+} involves a change in the valence number of Fe from +3 to +2 and therefore can be written as an electrode half-reaction with the electrons on the right side:

$$2\ Fe^{2+} + 3\ H_2O \rightarrow Fe_2O_3 + 6\ H^+ + 2\ e^- \quad (14.87)$$

$$\Delta G^\circ_R = +30.161\ kcal \qquad E^\circ = +0.65\ V$$

$$Eh = 0.65 + \frac{0.05916}{2} \log\left(\frac{[H^+]^6}{[Fe^{2+}]^2}\right) \qquad (14.88)$$

$$Eh = 0.65 - 0.177\ pH - 0.0592 \log[Fe^{2+}] \qquad (14.89)$$

Equations 14.86 and 14.89 have been used to draw contours across the stability fields of magnetite and hematite in Figure 14.6 at activities of $Fe^{2+} = 10^{-4}, 10^{-6}$, and 10^{-8} mol/L. Evidently, the solubility of both oxides increases with decreasing environmental pH and Eh. Therefore, when hematite and magnetite are exposed to acidic environments with low Eh values, equivalent to low O_2 fugacities, they become *unstable* and go into solution.

Conversely, when a solution containing Fe^{2+} is neutralized (pH increases), or aerated (Eh increases), or both, hematite or magnetite may precipitate. For example, a solution containing $[Fe^{2+}] = 10^{-4}$ mol/L (5.6 μg/mL), whose Eh and pH are the coordinates of point P in Figure 14.6, is undersaturated with respect to hematite. If the environmental Eh and pH change along the

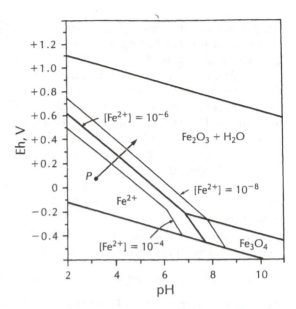

Figure 14.6 Solubility of hematite and magnetite with respect to Fe^{2+} represented by equations 14.89 and 14.86, respectively. An activity of 10^{-6} mol/L of ions is regarded as an effective limit on the stability of a solid phase because when $[Fe^{2+}] > 10^{-6}$ mol/L, the minerals can be removed from rocks by the movement of groundwater in the course of geologic time.

arrow in Figure 14.6, the solution becomes saturated with respect to hematite when the path reaches the 10^{-4} contour. If Eh and pH continue to change in the direction of the arrow, hematite precipitates continuously as the activity of Fe^{2+} decreases to 10^{-6} mol/L (0.056 μg/mL) and 10^{-8} mol/L (0.00056 μg/mL). The amount of hematite that precipitates during this process is 0.44 g/L as $[Fe^{2+}]$ in the solution decreases from 10^{-4} to 10^{-6} mol/L. If the environmental conditions continue to change in the direction of the arrow, the activity of Fe^{2+} in the solution decreases from 10^{-6} to 10^{-8} mol/L, but the amount of hematite that precipitates is only 0.0044 g/L. These figures justify the convention that an activity of 10^{-6} mol/L of ionic species is an effective limit on the stability of solids.

The solubility of hematite with respect to Fe^{3+} does not require a change in the valence number of Fe. Therefore, we can construct the desired equation purely on the basis of mass balance:

$$Fe_2O_3 + 6 H^+ \rightarrow 2 Fe^{3+} + 3 H_2O \quad (14.90)$$

$$\Delta G_R^\circ = +5.339 \text{ kcal} \qquad K = 10^{-3.91}$$

$$\frac{[Fe^{3+}]^2}{[H^+]^6} = 10^{-3.91} \quad (14.91)$$

$$2 \log [Fe^{3+}] - 6 \log [H^+] = -3.91 \quad (14.92)$$

$$\log [Fe^{3+}] + 3 \text{ pH} = -1.96 \quad (14.93)$$

If

$$[Fe^{3+}] = 10^{-6} \text{ mol/L},$$

$$\text{pH} = 1.35 \quad (14.94)$$

Evidently, the solubility of hematite with respect to Fe^{3+} is independent of Eh because it does not require an oxidation or reduction.

The solution of magnetite to form Fe^{3+} does require the oxidation of one divalent Fe^{2+} to the trivalent state and therefore must be expressed as an electrode half-reaction:

$$Fe_3O_4 + 8 H^+ \rightarrow 3 Fe^{3+} + 4 H_2O + e^- \quad (14.95)$$

$$\Delta G_R^\circ = +12.552 \text{ kcal} \qquad E^\circ = +0.54 \text{ V}$$

$$Eh = 0.54 + \frac{0.05916}{1} \log \left(\frac{[Fe^{3+}]^3}{[H^+]^8} \right) \quad (14.96)$$

$$Eh = 0.54 + \frac{0.05916 \times 3}{1} \log [Fe^{3+}]$$

$$- \frac{0.05916}{1} \times 8 \log [H^+] \quad (14.97)$$

$$Eh = 0.54 + 0.177 \log [Fe^{3+}] + 0.473 \text{ pH} \quad (14.98)$$

If $[Fe^{3+}] = 10^{-6}$ mol/L,

$$Eh = -0.52 + 0.473 \text{ pH} \quad (14.99)$$

Equations 14.94 and 14.99 have been plotted in Figure 14.7A by superimposing them on Figure 14.6. The results reveal a discrepancy.

Equation 14.94 ($\log [Fe^{3+}] = 10^{-6}$ mol/L) cuts off a corner of the hematite field in the upper left corner of the diagram and intersects the

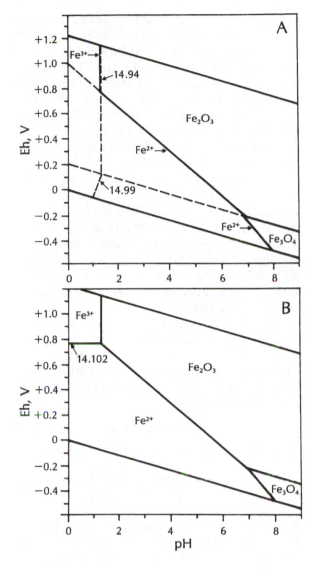

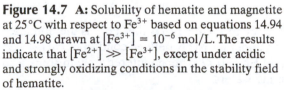

Figure 14.7 **A:** Solubility of hematite and magnetite at 25 °C with respect to Fe^{3+} based on equations 14.94 and 14.98 drawn at $[Fe^{3+}] = 10^{-6}$ mol/L. The results indicate that $[Fe^{2+}] \gg [Fe^{3+}]$, except under acidic and strongly oxidizing conditions in the stability field of hematite.

B: Completed Eh–pH diagram for the oxides of Fe and their dominant ions at an activity of 10^{-6} mol/L and 25 °C. Equation 14.102 fixes the boundary where $[Fe^{3+}] = [Fe^{2+}]$ in equilibrium with hematite. Note that the pH scale has been shifted compared to Figure 14.6.

$[Fe^{2+}] = 10^{-6}$ mol/L solubility contour at a point whose coordinates are pH = 1.35, Eh = +0.77 V. At the point of intersection, the activities of Fe^{3+} and Fe^{2+} are both equal to 10^{-6} mol/L. However, at lower Eh values the line (equation 14.94) enters the region of the diagram where $[Fe^{2+}] \gg 10^{-6}$ mol/L, whereas $[Fe^{3+}]$ remains at 10^{-6} mol/L. Therefore, the solubility of hematite with respect to Fe^{2+} is much greater than its solubility to form Fe^{3+} in all environments where Eh < +0.77 V.

The solubility of magnetite in its own stability field also favors Fe^{2+} over Fe^{3+}. Equation 14.99 (for log $[Fe^{3+}] = 10^{-6}$ mol/L) plots far to the left of the Fe^{2+} solubility boundary at 10^{-6} mol/L. Therefore, Fe^{3+} is much less abundant than Fe^{2+} in equilibrium with *hematite*, except if pH < 1.35 and Eh > 0.77 V, and it does *not* become dominant over Fe^{2+} in equilibrium with *magnetite* under any Eh–pH conditions.

The boundary along which $[Fe^{3+}]$ is equal to $[Fe^{2+}]$ in equilibrium with hematite can be drawn graphically by plotting contours at different activities and by connecting the points of intersection. A more elegant method is to write the electrode half-reaction (14.53):

$$Fe^{2+} \rightarrow Fe^{3+} + e^- \quad (14.100)$$

Since $\Delta G_R^\circ = +17.75$ kcal and $E^\circ = 0.77$ V,

$$Eh = 0.77 + \frac{0.05916}{1} \log\left(\frac{[Fe^{3+}]}{[Fe^{2+}]}\right) \quad (14.101)$$

We are interested in environments where $[Fe^{3+}] = [Fe^{2+}]$, and log($[Fe^{3+}]/[Fe^{2+}]$) = 0. Therefore, the boundary along which this condition occurs is given by:

$$Eh = 0.77 \text{ V} \quad (14.102)$$

We see in Figure 14.7B that Fe^{3+} is the dominant ion under narrowly defined conditions (pH < 1.35 and Eh > 0.77 V) encountered only in some acid mine-waters.

c. Stability of Ferrous Carbonate

The dominance of Fe^{2+} over Fe^{3+} in most environments on the surface of the Earth is reflected

in the fact that *siderite* ($FeCO_3$) rather than $Fe_2(CO_3)_3$ is the common carbonate of Fe. Siderite may precipitate from saturated solutions which are in equilibrium with CO_2 gas or contain a fixed amount of carbonate ions as a result of prior interactions with rocks or minerals.

When a solution is in equilibrium with CO_2 gas in the stability field of magnetite, siderite may form from magnetite by an electrode half-reaction:

$$Fe_3O_4 + 2 e^- \rightarrow 3 FeCO_3 \quad (14.103)$$

The two Fe^{3+} ions in magnetite must each accept one electron to convert them to the divalent state. This puts the electrons n equation 14.104 on the wrong side. Therefore, we reverse the equation and add CO_2, H_2O, ar 」 H^+ as required for balance:

$$3 FeCO_3 + H_2O \rightarrow Fe_3O_4 + 3 CO_2$$
$$+ 2 H^+ + 2 e^- \quad (14.104)$$

$$\Delta G_R^\circ = +12.225 \text{ kcal} \quad E^\circ = +0.265 \text{ V}$$

$$Eh = 0.265 + \frac{0.05916}{2} \log([H^+]^2[CO_2]^3)$$
$$(14.105)$$

$$Eh = 0.265 - 0.05916 \text{ pH} + 0.0887 \log[CO_2]$$
$$(14.106)$$

If $[CO_2] = 3 \times 10^{-4}$ atm (composition of the atmosphere):

$$Eh = -0.048 - 0.0592 \text{ pH} \quad (14.107)$$

Evidently, the siderite–magnetite boundary plots slightly *below* the stability limit of water when the solution is in contact with the atmosphere. Therefore, siderite is *not* expected to precipitate unless the fugacity of CO_2 is greater than 3×10^{-4} atm. If $[CO_2] = 10^{-2}$ atm, equation 14.106 yields:

$$Eh = +0.0876 - 0.0592 \text{ pH} \quad (14.108)$$

Under these conditions, siderite does have a small stability field and may precipitate under strongly reducing conditions from basic solutions as indicated in Figure 14.8

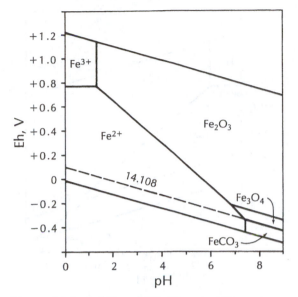

Figure 14.8 Stability of siderite, magnetite, and hematite in contact with water at 25 °C in coordinates of Eh and pH assuming a fugacity of $CO_2 = 10^{-2}$ atm and activities of 10^{-6} mol/L for Fe^{2+} and Fe^{3+}.

However, the stability of siderite is restricted by its solubility in water under reducing conditions:

$$FeCO_3 + 2\,H^+ \rightleftharpoons Fe^{2+} + CO_2 + H_2O \quad (14.109)$$

Since no oxidation or reduction takes place, equation 14.109 is based only on mass-balance considerations. At equilibrium:

$$\frac{[CO_2][Fe^{2+}]}{[H^+]^2} = K \quad (14.110)$$

Since $\Delta G_R^\circ = -9.491$ kcal and $K = 10^{6.958}$,

$$\log[CO_2] + \log[Fe^{2+}] - 2\log[H^+] = 6.958 \quad (14.111)$$

and

$$2\,pH = 6.958 - \log[CO_2] - \log[Fe^{2+}] \quad (14.112)$$

If $[CO_2] = 10^{-2}$ mol/L and $[Fe^{2+}] = 10^{-6}$ mol/L,

$$pH = 7.48 \quad (14.113)$$

Equation 14.113 limits the stability field of siderite in Figure 14.8, which also depicts the stability limits of hematite and magnetite in coordinates of Eh and pH at 25 °C and a fugacity of $CO_2 = 10^{-2}$ atm.

If a solution contains carbonate ions, but is not actually in contact with CO_2 gas, we must take into consideration the way in which the abundances of the carbonate species vary with the pH of the solution. Molecular carbonic acid dominates at pH < 6.35, bicarbonate dominates at pH = 6.35–10.3, and the carbonate ion dominates at pH > 10.3 (Section 9.5, Table 9.3). Therefore, if the total carbonate content of a solution is 10^{-2} mol/L, $[H_2CO_3] = 10^{-2}$ mol/L at pH < 6.35, $[HCO_3^-] = 10^{-2}$ at pH 6.35–10.3, and $[CO_3^{2-}] = 10^{-2}$ mol/L at pH > 10.3. The formation of siderite from magnetite in the different pH intervals can be expressed by electrode half-reactions modeled after equation 14.104. At pH < 6.35:

$$3\,FeCO_3 + 4\,H_2O \rightarrow Fe_3O_4 + 3\,H_2CO_3$$
$$+ 2\,H^+ + 2\,e^- \quad (14.114)$$

At pH 6.35–10.3:

$$3\,FeCO_3 + 4\,H_2O \rightarrow Fe_3O_4 + 3\,HCO_3^-$$
$$+ 5\,H^+ + 2\,e^- \quad (14.115)$$

At pH > 10.3:

$$3\,FeCO_3 + 4\,H_2O \rightarrow Fe_3O_4 + 3\,CO_3^{2-}$$
$$+ 8\,H^+ + 2\,e^- \quad (14.116)$$

Similarly, the electrode half-reactions involving the conversion of siderite to hematite in the specified pH intervals are:

(pH < 6.35): $2\,FeCO_3 + 3\,H_2O \rightarrow Fe_2O_3$
$$+ 2\,H_2CO_3 + 2\,H^+ + 2\,e^- \quad (14.117)$$

(pH 6.35–10.3): $2\,FeCO_3 + 3\,H_2O \rightarrow Fe_2O_3$
$$+ 2\,HCO_3^- + 4\,H^+ + 2\,e^- \quad (14.118)$$

(pH > 10.3): $2\,FeCO_3 + 3\,H_2O \rightarrow Fe_2O_3$
$$+ 2\,CO_3^{2-} + 6\,H^+ + 2\,e^- \quad (14.119)$$

Table 14.4 Eh–pH Equations for the Conversion of Siderite to Magnetite and Hematite at Specified pH Intervals in Solutions Having a Total Carbonate Content of 10^{-2} mol/L Without Being in Contact with CO_2 Gas

A. Siderite to Magnetite

pH < 6.35 (equation 14.114)

$$Eh = 0.217 - 0.0592\ pH \text{ (metastable) (1)}$$

pH 6.35–10.3 (equation 14.115)

$$Eh = 0.782 - 0.148\ pH \tag{2}$$

pH > 10.3 (equation 14.116)

$$Eh = 1.70 - 0.236\ pH \text{ (metastable) (3)}$$

B. Siderite to Hematite

pH < 6.35 (equation 14.117)

$$Eh = 0.211 - 0.0592\ pH \tag{4}$$

pH 6.35–10.3 (equation 14.118)

$$Eh = 0.587 - 0.118\ pH \tag{5}$$

pH > 10.3 (equation 14.119)

$$Eh = 1.20 - 0.177\ pH \text{ (metastable) (6)}$$

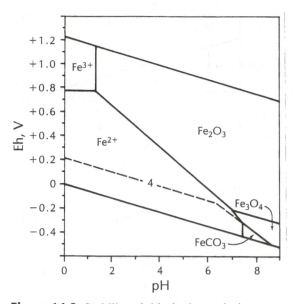

Figure 14.9 Stability of siderite in a solution containing 10^{-2} mol/L of dissolved carbonate ion at 25 °C without being in contact with CO_2 gas. The dashed lines are extensions of equations 2, 4 and 5 in Table 14.4, which represent metastable transformations because at pH < 7.48 the solubility of siderite is greater than 10^{-6} mol/L.

The corresponding Eh–pH equations are listed in Table 14.4.

A comparison of equations 1 and 4 in Table 14.4 indicates that they have identical slopes but different intercepts, namely, 0.217 (siderite → magnetite) and 0.211 (siderite → hematite). This means that the conversion of siderite to magnetite (equation 1) is metastable because it occurs within the stability field of hematite where magnetite is not stable. For this reason, equation 4 in Table 14.4 forms the boundary between siderite and hematite at pH < 6.35 in Figure 14.9.

Above pH = 6.35, the transformation of siderite to hematite is represented by equation 5 in Table 14.4 until the boundary enters the magnetite field where equation 2 is valid. The siderite field delineated in this way actually pinches out at pH < 10.3 when equation 2 intersects the lower stability limit of water. Therefore, the reactions represented by equations 3 and 6 (valid at pH > 10.3) are also metastable and are not shown in Figure 14.9.

The stability field of siderite delineated by these equations is restricted by the solubility of siderite with respect to Fe^{2+} expressed by equation 14.109. When the limit is taken to be $[Fe^{2+}] = 10^{-6}$ mol/L, the boundary between siderite and Fe^{2+} lies at pH = 7.48 (equation 14.113). The stability field of siderite *expands* as the total carbonate content of the system increases, but it *decreases* with increasing pH because the activity of CO_3^{2-} rises up to a pH value of about 10.5 and remains constant at higher pH values. When the system is in contact with CO_2, the stability field of siderite increases with increasing fugacity of CO_2 and does not shrink as the pH rises (Figure 14.8) because, in this case, the activity of CO_3^{2-} increases with increasing pH. (See Problem 3 at the end of this chapter.) However, regardless of which case we consider (fixed carbonate content or CO_2 gas), siderite precipitates only when the carbonate content or the fugacity of CO_2 are greater than is consistent with the present atmosphere of the Earth.

d. Stability of Sulfur Species

The dominant S-bearing ions and molecules in natural solutions are SO_4^{2-}, HSO_4^-, S^{2-}, HS^-, and H_2S^0. The abundances of these ions in solutions containing a fixed amount of dissolved S depend both on pH and Eh because the transformation of sulfide to sulfate species requires a change in valence number from -2 to $+6$. In addition, native S (valence number $= 0$) may precipitate when the amount of S in the system exceeds its solubility.

The boundary between SO_4^{2-} and HSO_4^- can be calculated from the equilibrium:

$$SO_4^{2-} + H^+ \rightleftharpoons HSO_4^- \qquad (14.120)$$

$$\Delta G_R^\circ = -2.94 \text{ kcal} \qquad K = 10^{2.155}$$

Therefore, when $[HSO_4^-] = [SO_4^{2-}]$, $[H^+] = 10^{-2.155}$ mol/L and pH = 2.16. The boundaries between H_2S^0, HS^-, and S^{2-} are also independent of Eh and occur at pH = 7.0 ($[H_2S] = [HS^-]$) and pH = 12.9 ($[HS^-] = [S^{2-}]$) as shown in Figure 9.3 (Section 9.5). Therefore, H_2S can be oxidized to HSO_4^- at pH < 2.16 and to SO_4^{2-} at pH > 2.16 but < 7.0. In basic environments (pH > 7.0 but < 12.9) HS^- is oxidized to SO_4^{2-}, but S^{2-} becomes dominant only in extremely basic environments, which are rare in nature.

The electrode half-reactions that establish the boundaries between H_2S–HSO_4^-, H_2S–SO_4^{2-}, and HS^-–SO_4^{2-} are listed in Table 14.5A and have been plotted in Figure 14.10. The boundaries are drawn where the activities of reacting S species are equal to each other and are therefore independent of the total S content of the system.

Native S can precipitate from aqueous solution when the total S content of the system exceeds its solubility. This process is Eh-dependent because the valence numbers of S change from -2 to 0 and from $+6$ to 0 during the transformation of sulfide and sulfate to native S, respectively. Therefore, we establish the stability field of native S by a series of electrode half-reactions written for the Eh–pH conditions in which H_2S^0, HS^-, HSO_4^-, and SO_4^{2-}

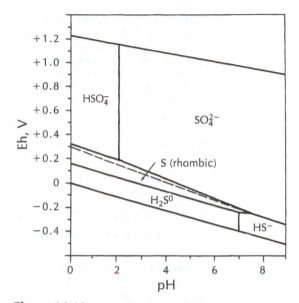

Figure 14.10 Abundance of aqueous S-species and stability field of native S (rhombic) in a system containing 10^{-1} mol/L of total S. The stability field of native S is based on the assumption that the aqueous species have activities of 10^{-1} mol/L in the Eh–pH fields in which they are the dominant ion or molecule.

dominate. Moreover, we specify that the total S content of the system is 10^{-1} mol/L in order to assure that the stability field of native S occupies a significant area in the diagram. The relevant equations are listed in Table 14.5B and have been used to draw the stability field of native S in Figure 14.10.

The diagram indicates that SO_4^{2-} and HSO_4^- are the dominant stable S species in well-oxygenated surface water, whereas H_2S^0 and HS^- predominate under reducing conditions corresponding to low O_2 fugacities. Native S has a wedge-shaped stability field that expands with the increasing total-S content of the system but disappears when $\Sigma S < 10^{-4.8}$ mol/L.

e. Stability of Ferrous Sulfides

A comparison of Figures 14.7B and 14.10 indicates that the Eh–pH area in which Fe^{2+} is the dominant Fe ion in equilibrium with hematite and magnetite overlaps the fields of H_2S^0,

Table 14.5 Electrode Half-reactions Among Dissolved Molecular and Ionic S Species in Aqueous Solution at 25 °C

A. Ionic and Molecular Species in Solution

$H_2S^0 + 4 H_2O \rightarrow HSO_4^- + 9 H^+ + 8 e^-$
$\Delta G_R^\circ = +52.718$ kcal $E^\circ = +0.285$ V

$$Eh = 0.285 - 0.0665 \text{ pH} \quad \text{when } [H_2S^0] = [HSO_4^-] \tag{1}$$

$H_2S^0 + 4 H_2O \rightarrow SO_4^{2-} + 10 H^+ + 8 e^-$
$\Delta G_R^\circ = +55.658$ kcal $E^\circ = +0.302$ V

$$Eh = 0.302 - 0.0739 \text{ pH} \quad \text{when } [H_2S^0] = [SO_4^{2-}] \tag{2}$$

$HS^- + 4 H_2O \rightarrow SO_4^{2-} + 9 H^+ + 8 e^-$
$\Delta G_R^\circ = +46.068$ kcal $E^\circ = +0.249$ V

$$Eh = 0.249 - 0.0665 \text{ pH} \quad \text{when } [HS^-] = [SO_4^{2-}] \tag{3}$$

B. Solubility of Native Sulfur ($\Sigma S = 10^{-1}$ mol/L)

$H_2S^0 \rightarrow S \text{ (rhombic)} + 2 H^+ + 2 e^-$
$\Delta G_R^\circ + 6.66$ kcal $E^\circ = +0.144$ V

$$Eh = 0.174 - 0.0592 \text{ pH} \quad \text{when } [H_2S^0] = 10^{-1} \text{ mol/L} \tag{4}$$

$HS^- \rightarrow S \text{ (rhombic)} + H^+ + 2 e^-$
$\Delta G_R^\circ = -2.93$ kcal $E^\circ = -0.0635$ V

$$Eh = 0.034 - 0.0296 \text{ pH} \quad \text{when } [HS^-] = 10^{-1} \text{ mol/L} \tag{5}$$

$S \text{ (rhombic)} + 4 H_2O \rightarrow HSO_4^- + 7 H^+ + 6 e^-$
$\Delta G_R^\circ = +46.058$ kcal $E^\circ = +0.333$ V

$$Eh = 0.323 - 0.0690 \text{ pH} \quad \text{when } [HSO_4^-] = 10^{-1} \text{ mol/L} \tag{6}$$

$S \text{ (rhombic)} + 4 H_2O \rightarrow SO_4^{2-} + 8 H^+ + 6 e^-$
$\Delta G_R^\circ = +48.998$ kcal $E^\circ = +0.0354$ V

$$Eh = 0.344 - 0.0789 \text{ pH} \quad \text{when } [SO_4^{2-}] = 10^{-1} \text{ mol/L} \tag{7}$$

HSO_4^-, and SO_4^{2-}. Therefore, ferrous sulfides may be stable under a wide range of Eh and pH conditions, whereas ferric sulfides cannot form because Fe^{3+} is dominant *only* under highly oxidizing and acidic conditions in the area of HSO_4^- dominance.

The ferrous sulfides that may form include FeS (amorphous), pyrrhotite, mackinawite, troilite, pyrite, and marcasite. However, only pyrrhotite ($Fe_{0.95}S$) and pyrite (FeS_2) are stable and need to be considered here. The others may form as intermediate metastable products that recrystallize to form either pyrrhotite or pyrite in the course of geologic time.

In order to introduce ferrous sulfides into the Eh–pH diagram for the oxides of Fe, we guess that pyrite can occur with hematite in the SO_4^{2-} field. The transformation of pyrite into hematite plus SO_4^{2-} requires changes in the valence numbers of both Fe and S and therefore can be expressed as an electrode half-reaction:

$$2 FeS_2 + 19 H_2O \rightarrow Fe_2O_3 + 4 SO_4^{2-}$$
$$+ 38 H^+ + 30 e^- \tag{14.121}$$

In this electrode, the valence numbers of S change from -1 in pyrite to $+6$ in sulfate, which causes the release of 28 electrons. In addition, each of the Fe^{2+} ions in pyrite gives up one electron for a total of 30 electrons. For this reaction $\Delta G^{\circ}_R = +266.453$ kcal and $E^{\circ} = +0.385$ V. If we set $[SO_4^{2-}] = 10^{-1}$ mol/L, the Eh–pH boundary equation is:

$$Eh = +0.377 - 0.0749 \text{ pH} \quad (14.122)$$

This line establishes the boundary between pyrite and hematite in that part of the stability field of hematite where SO_4^{2-} is dominant.

The further development of this Eh–pH diagram is best done by plotting each new line on Figure 14.11, which already contains the stability fields of the Fe-oxides as well as the distribution of S species. The pyrite–hematite boundary intersects the magnetite–hematite boundary, which implies that pyrite and magnetite can also coexist. Therefore, we construct a pyrite–magnetite electrode reaction in which the S appears as SO_4^{2-}:

$$3 \text{ FeS}_2 + 28 \text{ H}_2\text{O} \rightarrow \text{Fe}_3\text{O}_4 + 6 \text{ SO}_4^{2-}$$
$$+ 56 \text{ H}^+ + 44 \text{ e}^- \quad (14.123)$$

$$\Delta G^{\circ}_R = +395.136 \text{ kcal} \quad E^{\circ} = +0.389 \text{ V}$$

$$Eh = 0.381 - 0.0753 \text{ pH}$$
$$([SO_4^{2-}] = 10^{-1.0} \text{ mol/L}) \quad (14.124)$$

By plotting equation 14.124 in Figure 14.11 we find that magnetite retains a small stability field in strongly reducing (Eh < −0.30 V) and basic (pH > 11.5) environments where it can precipitate from sulfate-dominated solutions.

The pyrite–hematite boundary continues into the HSO_4^- field by virtue of the electrode half-reaction and the corresponding Eh–pH equation:

$$2 \text{ FeS}_2 + 19 \text{ H}_2\text{O} \rightarrow \text{Fe}_2\text{O}_3 + 4 \text{ HSO}_4^-$$
$$+ 34 \text{ H}^+ + 30 \text{ e}^- \quad (14.125)$$

$$\Delta G^{\circ}_R + 254.693 \text{ kcal} \quad E^{\circ} = +0.368 \text{ V}$$

$$Eh = +0.360 - 0.0670 \text{ pH}$$
$$([HSO_4^-] = 10^{-1} \text{ mol/L}) \quad (14.126)$$

Equation 14.126 extends the pyrite–hematite boundary from SO_4^{2-}-dominated environments into HSO_4^--dominated environments with only a slight change in slope. However, we do not

plot equation 14.126 because we have entered the field where hematite is soluble with respect to Fe^{2+}.

Therefore, we must now consider the solubility of pyrite with respect to Fe^{2+} and the various S species. In the SO_4^{2-}-dominated Eh–pH region, the electrode-half reaction is:

$$\text{FeS}_2 + 8 \text{ H}_2\text{O} \rightarrow \text{Fe}^{2+} + 2 \text{ SO}_4^{2-}$$
$$+ 16 \text{ H}^+ + 14 \text{ e}^- \quad (14.127)$$

$$\Delta G^{\circ}_R = +118.146 \text{ kcal} \quad E^{\circ} = +0.366 \text{ V}$$

$$Eh = 0.366 + \frac{0.05916}{14}$$
$$\times \log \left([H^+]^{16}[SO_4^{2-}]^2[Fe^{2+}] \right) \quad (14.128)$$

By converting $[H^+]$ to pH, setting $[SO_4^{2-}] = 10^{-1}$ mol/L and $[Fe^{2+}] = 10^{-6}$ mol/L, equation 14.128 reduces to:

$$Eh = 0.322 - 0.067 \text{ pH} \quad (14.129)$$

In the HSO_4^--dominated field the electrode half-reaction is:

$$\text{FeS}_2 + 8 \text{ H}_2\text{O} \rightarrow \text{Fe}^{2+} + 2 \text{ HSO}_4^-$$
$$+ 14 \text{ H}^+ + 14 \text{ e}^- \quad (14.130)$$

$$\Delta G^{\circ}_R = 112.266 \text{ kcal} \quad E^{\circ} = +0.348 \text{ V}$$

$$Eh = 0.316 - 0.0592 \text{ pH} \quad (14.131)$$

where $[HSO_4^-] = 10^{-1}$ mol/L and $[Fe^{2+}] = 10^{-6}$ mol/L.

Because this line (equation 14.131) enters the stability field of native S at a very low angle we must consider the solubility of pyrite with respect to Fe^{2+} and native S.

$$\text{FeS}_2 \rightarrow \text{Fe}^{2+} + 2 \text{ S} + 2 \text{ e}^- \quad (14.132)$$

$$\Delta G^{\circ}_R = +20.15 \text{ kcal} \quad E^{\circ} = +0.437 \text{ V}$$

$$Eh = +0.437 + \frac{0.05916}{2} \log [Fe^{2+}] \quad (14.133)$$

If $[Fe^{2+}] = 10^{-6}$ mol/L, Eh = 0.259 V.

The construction of Figure 14.11 indicates that pyrite is stable in the Eh–pH region lying below the boundaries based on equations 14.122, 14.124, 14.129, 14.131, and 14.133. Therefore,

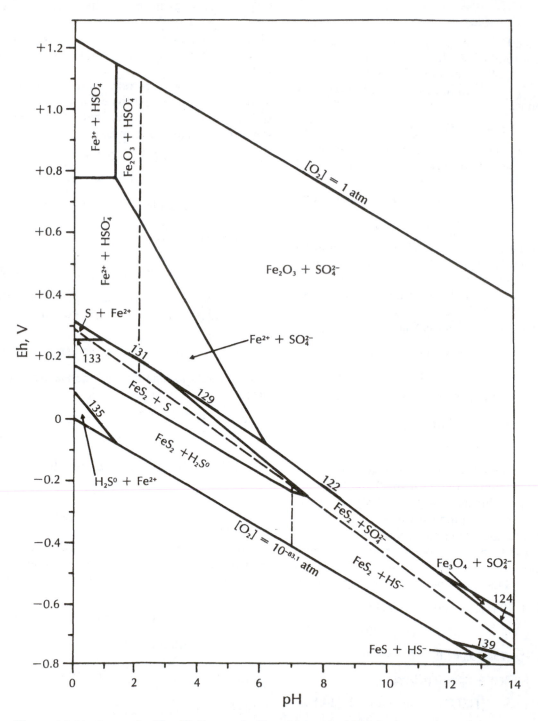

Figure 14.11 Composite Eh–pH diagram for Fe oxides and sulfides in the presence of water at 25 °C containing $\Sigma S = 10^{-1}$ mol/L. In addition, the dashed lines indicate the conditions under which the different S species become dominant. The numbers adjacent to some of the boundaries identify equations in the text. An even more comprehensive diagram including oxides, sulfides, and the carbonate of Fe was published by Garrels and Christ (1965).

pyrite occurs in the area where H_2S^0 is dominant and could be soluble with respect to Fe^{2+} and H_2S^0. The relevant electrode half-reaction is:

$$2 H_2S^0 + Fe^{2+} \rightarrow FeS_2 + 4 H^+ + 2 e^-$$
$$(14.134)$$

$$\Delta G_R^\circ = -6.83 \text{ kcal} \qquad E^\circ = -0.148 \text{ V}$$

$$Eh = +0.088 - 0.118 \text{ pH} \quad (14.135)$$

where $[H_2S^0] = 10^{-1}$ mol/L and $[Fe^{2+}] = 10^{-6}$ mol/L.

Since pyrite also occurs where HS^- is the dominant S species, we write the electrode half-reaction:

$$2 HS^- + Fe^{2+} \rightarrow FeS_2 + 2 H^+ + 2 e^-$$
$$(14.136)$$

$$\Delta G_R^\circ = 26.01 \text{ kcal} \qquad E^\circ = -0.564 \text{ V}$$

$$Eh = -0.327 - 0.0591 \text{ pH} \quad (14.137)$$

where $[HS^-] = 10^{-1}$ mol/L and $[Fe^{2+}] = 10^{-6}$ mol/L. Evidently this line lies outside the stability field of water, which implies that pyrite is not appreciably soluble with respect to Fe^{2+} under conditions where HS^- is the dominant S species.

Finally, we consider the relationship between pyrite and pyrrhotite, which does require a change in the valence number of S from -1 in pyrite to -2 in pyrrhotite. The relevant electrode reaction in the HS^--dominated Eh–pH region is:

$$FeS + HS^- \rightarrow FeS_2 + H^+ + 2 e^- \quad (14.138)$$

$$\Delta G_R^\circ = -17.93 \text{ kcal} \qquad E^\circ = -0.389 \text{ V}$$

$$Eh = -0.359 - 0.0296 \text{ pH}$$

$$([HS^-] = 10^{-1} \text{ mol/L}) \quad (14.139)$$

Note that the formula for pyrrhotite in equation 14.138 is given as FeS for the sake of simplicity even though its standard free energy value in Appendix B is for $Fe_{0.95}S$. Equation 14.139 delineates a small stability field for pyrrhotite under extremely basic (pH > 12) and reducing (Eh < -0.52 V) conditions, which explains why this mineral is rarely, if ever, found as an authigenic mineral in sedimentary rocks.

Figure 14.11 contains a great deal of information about the geochemistry of Fe on the surface of the Earth. It identifies the environments in which hematite, magnetite, pyrite, pyrrhotite, and native S are stable in the presence of water containing 10^{-1} mol/L of S species. In addition, the diagram delineates environmental conditions that cause the Fe-minerals to dissolve significantly to form Fe^{2+} or Fe^{3+} and which therefore favor the transport of Fe in aqueous solutions at 25 °C. In general, we see that Fe is mobile primarily under acidic and oxidizing conditions and that it becomes immobile when Fe-bearing solutions are neutralized or enter reducing environments containing dissolved S. The minerals that form by precipitation from aqueous solutions depend on the Eh and pH conditions as shown in Figure 14.11. However, intermediate compounds may form initially, which may subsequently recrystallize to form the stable phases having the lowest standard free energies of formation.

All natural solutions contain not only Fe and S but also carbonate ions. Therefore, natural systems may precipitate siderite, which is not shown in Figure 14.11 in order to avoid overcrowding. An Eh–pH diagram including the oxides, sulfides, and the carbonate of Fe was worked out by Garrels and Christ (1965, Fig. 7.21) for a system at 25 °C containing $\Sigma S = 10^{-6}$ mol/L and Σ carbonates $= 1$ mol/L.

The methods illustrated here can be used to develop Eh–pH diagrams for all elements in the periodic table with the exception of the noble gases, which do not change their oxidation state. Such diagrams are useful in the study of metallic mineral deposits, in geochemical prospecting, and in environmental geochemistry. The most comprehensive treatment of this subject is by Brookins (1988).

14.6 Summary

The elements and their ions can be ranked by their tendency to give up electrons as expressed in displacement reactions that involve the transfer of electrons from the electron donor (reducing agent) to the electron acceptor (oxidizing agent). The resulting oxidation–reduction reactions must be balanced by use of valence numbers to assure that the number of electrons given up by the reducing agent is equal to the number of electrons gained by the oxidizing agent.

Oxidation–reduction reactions can be divided into two complementary half-reactions that may be separated physically to form an electrochemical cell. The electromotive force (emf) exerted by such a cell is measurable as a voltage between the electrodes dipping into an electrolyte solution. The emf of an electrochemical cell in the standard state is related to the standard free energy change of the reaction by Faraday's constant. When the cell is *not* in the standard state, the emf is related to the reaction quotient and decreases to zero as the oxidation–reduction reaction approaches equilibrium.

A systematic scale of emf values is created by arbitrarily assigning a value of zero to the emf of the H_2 electrode in the standard state. This is accomplished by setting the standard free energies of formation of H^+ and e^- equal to zero. The standard electrode potentials of all other electrodes can be calculated from their standard free energy change alone when they are connected to the standard H_2 electrode. When the electrode half-reactions are written with the electrons on the right side of the equation, the algebraic sign and magnitude of their standard electrode potentials become the basis for ranking them relative to the standard H_2 electrode. Such a ranked list of electrode half-reactions is called the electromotive series. A negative sign of the standard electrode potential indicates that the half-reaction is a stronger electron donor (reducing agent) than the standard H_2 electrode, whereas a positive sign indicates that the electrode is a weaker reducing agent and is therefore a stronger oxidizing agent than the standard H_2 electrode.

As expected from their electronic configurations, the alkali metals and alkaline earths are strong reducing agents, whereas the halogens and other nonmetallic elements are strong oxidizing agents. However, the cations of a metal can act as electron acceptors (oxidizing agents) and the anions of a nonmetal can function as electron donors (reducing agents).

In order to facilitate the application of electrochemistry to natural environments a special emf is defined as the voltage generated by any electrode in any state relative to the H_2 electrode in the standard state (Eh). Natural environments on the surface of the Earth are restricted to a range of Eh and pH conditions determined by the dissociation of liquid water into H^+ and O_2, which releases electrons and therefore constitutes an electrode half-reaction. The limits on the stability of water result from the requirement that the fugacity of O_2 in an environment open to the atmosphere must be less than 1.0 but greater than $10^{-83.1}$ atm. The Eh and pH values of natural environments depend on their chemical compositions and on the reactions that are taking place within them. Therefore, the Eh and the pH are both *environmental parameters* that are imposed on any specific reaction occurring in the environment.

The algebraic sign of measured environmental Eh values gives meaning to the widely used concept of *oxidizing* and *reducing* conditions. Oxidizing environments have positive Eh values, so they are weaker reducing agents than the standard H_2 electrode. Reducing environments have negative Eh values and are therefore stronger reducing agents than the standard H_2 electrode. The capacity of an environment to be a reducing agent results from the presence of potential electron donors that can force the standard H_2 electrode to accept electrons. Organic compounds of biological origin containing reduced forms of C, such as CH_4, where the C has a valence number of -4, are a common source of electrons in natural environments. Similarly, the oxidizing character of an environment results from the presence of potential electron acceptors or oxidizing agents. A common oxidizing agent in natural environments is molecular O_2, which can accept four electrons per molecule. However, many other compounds and ions have an effect on the Eh of a particular environment.

The Eh and pH of natural environments can be used to define relationships between solids, between solids and ions, and among ions in aqueous solution of elements that can have different valence numbers. The Eh–pH stability diagrams of Fe compounds, including oxides, carbonates, and sulfides, are constructed from appropriate electrode half-reactions that define compatible relationships between Eh and pH for these electrodes. The resulting diagrams can be used to understand the geochemistry of Fe and S and to predict the effect of environmental changes on the compounds and ions of these elements. Such diagrams have been constructed for most of the elements. The Eh–pH diagrams are useful in studying the origin of metallic mineral deposits and the subsequent dispersal of metals during chemical weathering. For this reason they are relevant also in hydrogeology, environmental geochemistry, and soil chemistry.

Problems

1. Write electronic formulas for $S^{2-}, Ti^{4+}, P^{3-}, Zr^{4+}$, and Cl^{7+}. Note the tendency of these elements to achieve the electronic configuration of noble gases.

2. Balance the following oxidation–reduction reactions.

 (a) $MnO_4^- + Cl^- \rightarrow Mn^{2+} + Cl_2$
 (b) $As_2S_3 + NO_3^- \rightarrow HAsO_3 + S + NO$
 (c) $Cr_2O_7^{2-} + I^- \rightarrow Cr^{3+} + I_2$
 (d) $CrO_2^- + ClO^- \rightarrow CrO_4^{2-} + Cl^-$
 (e) $CH_4^0 + SO_4^{2-} \rightarrow HCO_3^- + HS^-$

3. Calculate the abundance of CO_3^{2-} in moles per liter at pH = 9, 10, 11 in two different systems.

 (a) Having a fixed total carbonate content of 10^{-2} mol/L.
 (b) In equilibrium with CO_2 gas at a fugacity of 10^{-2} atm.

4. Complete the following electrode half-reactions in the conventional form and calculate their standard electrode potentials.

 (a) PbO (red)$\rightarrow PbO_2$
 (b) $Cu_2O \rightarrow Cu$
 (c) $UO_2^{2+} \rightarrow U^{4+}$
 (d) $Cr^{3+} \rightarrow Cr_2O_7^{2-}$
 (e) $Si \rightarrow SiF_6^{2-}$

5. Combine the Ag and Cu electrode half-reactions and calculate the emf when $[Cu^{2+}]/[Ag^+]^2 = 10^{-4}$.

6. Combine the Al and Ni electrodes and calculate the emf when $[Ni^{2+}]^3/[Al^{3+}]^2 = 10^{-2}$.

7. Manganese forms the following oxide minerals: manganite ($MnOOH$), manganosite (MnO), pyrolusite (MnO_2), and hausmannite (Mn_3O_4). Determine the valence numbers of Mn in each compound and order them in terms of increasing valence number of Mn.

8. (a) Derive Eh–pH equations based on the progressive oxidation of Mn to form oxides starting with metallic Mn.

 (b) Plot these equations in coordinates of Eh and pH using a full sheet of graph paper, add the stability limits of water, and label all fields.

 (c) Is metallic Mn stable at the surface of the Earth?

9. (a) Derive equations in terms of Eh and pH for the solution of the oxide minerals of Mn to form Mn^{2+}.

 (b) Plot the equations for $[Mn^{2+}] = 10^{-6}$ mol/L on the diagram constructed in Problem 8b.

10. What is the fugacity of O_2 of an environment having pH = 7.0 and Eh = +0.2 V?

11. A solution containing 10^{-6} mol/L of Fe^{2+} and Mn^{2+} at Eh = +0.2 V is neutralized such that its pH changes from 4 to 6. What effect will this have on the concentrations of Fe^{2+} and Mn^{2+} in the solution? Refer to the appropriate Eh–pH diagrams.

12. Based on the insight gained above, explain why Fe-oxide deposits in near shore marine environments have low Mn concentrations and speculate about the fate of Mn in the oceans.

References

BAAS-BECKING, L. G. M., I. R. KAPLAN, and O. MOORE, 1960. Limits of the natural environment in terms of pH and oxidation–reduction potentials. *J. Geol.,* **68**:243–284.

BROOKINS, D. G., 1978. Eh–pH diagrams for elements from $Z = 40$ to $Z = 52$: Application to the Oklo natural reactor. *Chem. Geol.,* **23**:324–342.

BROOKINS, D. G., 1988. *Eh–pH Diagrams in Geochemistry.* Springer-Verlag, New York, 200 pp.

GARRELS, R. M., 1960. *Mineral Equilibria.* Harper & Row, New York, 244 pp.

GARRELS, R. M., and C. L. CHRIST, 1965. *Solutions, Minerals and Equilibria.* Harper & Row, New York, 450 pp.

HOGNESS, T. R., and W. C. JOHNSON, 1954. *Qualitative Analysis and Chemical Equilibrium,* 4th ed. Henry Holt, New York, 621 pp.

POURBAIX, M. J. N., 1949. *Thermodynamics of Dilute Aqueous Solutions.* Arnold, London.

STUMM, W., and J. J. MORGAN, 1996. *Aquatic Chemistry,* 3rd ed. Wiley, New York, 1022 pp.

15

Rates of Geochemical Processes

The progress of chemical reactions depends on their *rates,* which are controlled by a variety of factors, including the temperature, the concentrations of reactants and products, and the rate of mass transfer into and out of a fixed reaction site. In some cases, the transformation that we observe is the end product of several intermediate reactions each of which proceeds at its own rate. As a result, the rate of the complete reaction may be controlled by the slowest of the intermediate reactions, which becomes the rate-determining step. Therefore, the study of the kinetics of chemical reactions leads to insights about what *really* happens during such familiar processes as the formation of a crystalline precipitate in aqueous solution, the incongruent solution of minerals, or the formation of concretions during diagenesis of sediment. The explanations of *reaction mechanisms* and the mathematical descriptions of the rates of chemical transformations are often complex and are, in many cases, not as well understood as the thermodynamic treatment of chemical reactions. In general, thermodynamics allows us to predict what *should* happen, whereas kinetics determines whether it actually does occur in a reasonable interval of time.

15.1 Rates of Chemical Reactions

The first quantitative study of chemical kinetics was undertaken by L. Wilhelmy in 1850. It was followed in 1863 by the important work of Guldberg and Waage, who derived the Law of Mass Action from the concept that at equilibrium the rate of the forward reaction is equal to the rate of the backward reaction. The principles of chemical kinetics and other rate-dependent processes such as diffusion and advection have been combined with thermodynamics only recently in comprehensive mathematical descriptions of the rates of geochemical processes (Berner, 1971, 1980; Hofmann et al., 1974; Lerman, 1979; Lasaga and Kirkpatrick, 1981).

The rate of a chemical reaction can be expressed in terms of the consumption of one of the reactants or in terms of the formation of one of the products. Whatever method is used, the rates of chemical reactions *decline* with time from some initial value until they become virtually equal to zero as the reaction is either reaching *equilibrium* or is going to *completion.* We first encountered this idea in Section 9.1, where we described the dissolution of NaCl(s) in water as a function of time (Figure 9.1).

a. Order of the Reaction

The rate of change of the concentration of a reactant or product of a chemical reaction generally varies as some power of its concentration. For example, in the reaction:

$$X \rightarrow Y + Z \qquad (15.1)$$

the rate of consumption of reactant X is given by an equation of the form:

$$-\frac{d(X)}{dt} = k(X)^n \qquad (15.2)$$

where (X) is the concentration of reactant X at any time t, n is a number whose value indicates the *order* of the reaction, k is the specific rate constant, and the minus sign is needed to indicate that the rate *decreases* with time. The numerical values of n and k for a specific reaction must be determined experimentally. If $n = 1.0$, the reaction is said to be a *first-order* reaction, and so on. The value of n may also be a fraction like $\frac{1}{2}$ in which case the order is *one half*.

The *order* of the reaction represented by equation 15.1 can be determined by measuring how the *initial* rate of the reaction (R_0^X) varies with the initial concentration of reactant X represented by $(X)_0$. For this purpose we rewrite equation 15.2 as:

$$R_0^X = \text{constant} \times ((X)_0)^n \qquad (15.3)$$

which we linearize in the form:

$$\log R_0^X = \log \text{constant} + n \log (X)_0 \quad (15.4)$$

Therefore, a plot of $\log R_0^X$ versus $\log (X)_0$ yields a straight line whose slope is equal to n.

The rate equation of reactions such as:

$$2\,X \rightarrow Y + Z \qquad (15.5)$$

may take the form:

$$-\frac{d(X)}{dt} = k(X)^2 \qquad (15.6)$$

which makes it a second-order reaction. If the reaction is:

$$X + Y \rightarrow Z \qquad (15.7)$$

the rate equation may take the form:

$$-\frac{d(X)}{dt} = k(X)^1(Y)^1 \qquad (15.8)$$

This is also a second-order reaction, but it is said to be first order with respect to X, first order with respect to Y, and second order overall. In general, the rate equation of a chemical reaction:

$$a\,X + b\,Y \rightarrow c\,Z \qquad (15.9)$$

may be expressed as:

$$-\frac{d(X)}{dt} = k(X)^{n_1}(Y)^{n_2} \qquad (15.10)$$

and the order of such a reaction is $n_1 + n_2$. Note that the exponents of the rate equation *do not* necessarily agree with the molar coefficients of the balanced equation of the reaction.

b. Rate Constant

The value of the specific rate constant (k) must also be determined experimentally by measuring the concentration of reactant X at known intervals of time as the reaction runs its course. For this purpose we rearrange equation 15.2 for a first-order reaction:

$$-\frac{d(X)}{(X)} = k\,dt \qquad (15.11)$$

which we can integrate:

$$\int \frac{d(X)}{(X)} = -k \int dt \qquad (15.12)$$

$$\ln (X) = -kt + \text{constant of integration} \qquad (15.13)$$

If $t = 0$, the integration constant is $\ln (X)_0$ where $(X)_0$ is the initial concentration of X. Therefore,

$$\ln (X) = \ln (X)_0 - kt \qquad (15.14)$$

$$\ln \frac{(X)}{(X)_0} = -kt \qquad (15.15)$$

$$(X) = (X)_0 e^{-kt} \qquad (15.16)$$

Equation 15.14 is a straight line in coordinates of $\ln (X)$ and t whose slope is negative because the concentration of reactant X *decreases* with time. The absolute value of the slope is the specific rate constant (k) of the reaction. Equation 15.16 has the same form as equation 16.9 in Chapter 16, which expresses the *law of radioactive decay*.

The rate of decay of a *radioactive* nuclide can be stated in terms of the *half-life*, which is the time required for one half of an initial number of radioactive atoms to decay. We therefore define the *reaction half-life* $(t_{1/2})$ in a similar manner and can show (see Chapter 16) that for a first-order reaction:

$$t_{1/2} = \frac{\ln 2}{k} \qquad (15.17)$$

When time is measured in seconds and concentrations are expressed in moles/liter, the units of the rate constant (k) of a first-order reaction are:

$$k = \frac{d(X)/dt}{(X)} = \frac{\text{mol/L}}{\text{mol/L} \times \text{sec}} = \text{sec}^{-1} \quad (15.18)$$

For second-order reactions the units of k are:

$$k = \frac{d(X)dt}{(X)^2} = \frac{\text{mol/L}}{(\text{mol/L})^2 \times \text{sec}}$$

$$= \frac{\text{L}}{\text{mol} \times \text{sec}} = \text{L mol}^{-1} \text{sec}^{-1} \quad (15.19)$$

The rate of a first-order reaction may also be expressed in terms of the *increase* of the concentration of one of the *products*. For example, for the reaction represented by equation 15.1 the rate of formation of product Y can be expressed as follows. If x moles of reactant X have reacted at time t, the concentration of products Y and Z is equal to x and the concentration of reactant X is $(X)_0 - x$. Therefore, the rate of formation of Y is:

$$\frac{dx}{dt} = k[(X)_0 - x] \qquad (15.20)$$

By rearranging this equation and by integration we obtain:

$$-\ln [(X)_0 - x] = kt + \text{constant} \quad (15.21)$$

If $t = 0$, then $x = 0$, and the constant of integration is $-\ln (X)_0$. Therefore:

$$\ln \left(\frac{(X)_0}{(X)_0 - x} \right) = kt \qquad (15.22)$$

$$\frac{(X)_0}{(X)_0 - x} = e^{kt} \qquad (15.23)$$

$$\frac{(X)_0 - x}{(X)_0} = e^{-kt} \qquad (15.24)$$

and thus

$$x = (X)_0(1 - e^{-kt}) \qquad (15.25)$$

Equation 15.22 is a straight line in coordinates of $\ln (X)_0/[(X)_0 - x]$ and t whose slope is equal to k.

Equation 15.25 has the same form as equation 16.19 for the growth of a stable radiogenic daughter. Therefore, the consumption of a reactant and the formation of a product in a *first-order* chemical reaction are governed by the same equations as the decay of a radioactive parent nuclide and the resulting growth of a radiogenic daughter. Both are illustrated in Figure 15.1 for a hypothetical reaction whose rate constant $k = 0.0693$ sec^{-1}, which corresponds to a reaction half-life of 10 sec.

c. Temperature Dependence of Rate Constants

Before a reaction of the form:

$$X + Y \rightarrow Z \qquad (15.26)$$

can take place, the atoms or molecules of the reactants X and Y must collide with *sufficient* energy to allow them to form a bond. The minimum energy required for two atoms or molecules to react is called the *activation energy*. When X and Y are molecules, the chemical bonds they contain may have to break before the atoms can recombine to form an intermediate *activated complex,* which subsequently decomposes to from the product. The activation energy (E_a) acts as a barrier, illustrated in Figure 15.2, between the reactants and products that must be overcome before the reactants can form the desired product. The rates of chemical reaction ultimately depend on the *frequency* with which atoms or molecules of the reactants acquire the necessary activation energy that enables them to form bonds. Therefore, the rates of chemical reactions *increase* with increasing temperature because the fraction of atomic or

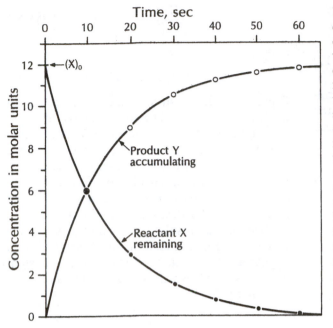

Figure 15.1 Kinetics of a hypothetical first-order chemical reaction represented by X→Y + Z. The consumption of reactant X follows equation 15.16 where the specific rate constant $k = 0.0693$ sec^{-1}, corresponding to a reaction half-life of 10 sec. The formation of products Y and Z is given by equation 15.25. The graph shows that the rate of consumption of reactant X and the rate of formation of product Y both *decrease* with time from their initial values at $t = 0$. In this example the concentration of reactant X apporaches zero, indicating that the reaction is going to *completion*. If a finite amount of X remains when the reaction rates approach zero, then the reaction is in *equilibrium*. Note that both curves represent the forward reaction. The consumption of a reactant and the formation of a product in a first-order chemical reaction are governed by the same equations as the decay of a radioactive parent nuclide to a stable radiogenic daughter discussed in Chapter 16.

molecular collisions that are sufficiently energetic to provide the activation energy increases with temperature.

The relationship between the specific rate constant (k) of a chemical reaction, the temperature (T) in kelvins, and the activation energy (E_a) in kilocalories is given by an equation derived by Svante Arrhenius in 1889 (Moore, 1955):

$$k = Ae^{-E_a/RT} \qquad (15.27)$$

where A is a constant that has a characteristic value for each reaction, and R is the gas constant. The constant A is called the *frequency factor* and $e^{-E_a/RT}$ is known as the *Boltzmann factor*, which is the fraction of atoms or molecules that manages to acquire the necessary activation energy from the energy liberated by collisions. By taking base 10 logarithms of equation 15.27 we can transform it into a straight line in coordinates of log k and $1/T$:

$$\log k = \log A - \frac{E_a}{2.303R}\left(\frac{1}{T}\right) \qquad (15.28)$$

A plot of experimentally determined values of the rate constant of a chemical reaction at differ-

ent temperatures should therefore yield a straight line whose slope is equal to:

$$m = -\frac{E_a}{2.303R} \qquad (15.29)$$

from which the activation energy of the reaction can be calculated. The intercept on the log k axis is equal to the logarithm of the frequency factor A. The activation energy is itself slightly dependent on temperature, but the slope of equation 15.28 is not perceptibly changed by this effect.

The formation of the activated complex and its deactivation to form the product of a chemical reaction each have their own rate constants. As a result, even a straightforward first-order reaction involves internal mechanisms that are difficult to express mathematically. The problem can be solved only for the condition of *steady state* when the rate of formation of the activated complex is equal to the rate of its deactivation and the number of activated complexes remains constant. Although the concept of the activated complex was initially developed for reactions in the gas phase, it also applies to reactions among ions in aqueous solution.

$$F = dv \qquad (15.30)$$

When the density is expressed in grams per cubic centimeter and the velocity in centimeters per second, the units of the flux are:

$$\text{units of flux } (F) = \frac{g \times cm}{cm^3 \times sec}$$

$$= g\ cm^{-2}\ sec^{-1} \quad (15.31)$$

When the material in motion contains another substance X in solution at a concentration (X) expressed in grams per cubic centimeter, then the flux of X is:

$$F_X = (X)v \quad g\ cm^{-2}\ sec^{-1} \qquad (15.32)$$

In general, typical velocities of advective transport processes are (Lerman, 1979):

	Meters per year
Particles settling in water	3×10^2 to 3×10^4
Groundwater flow	1–100
Upwelling in the oceans	1–2
Sedimentation in oceans and lakes	3×10^{-6} to 1×10^{-6}

The flow of water through the pore spaces of sediment or rocks in the subsurface (i.e., an aquifer) is an important advective transport mechanism that affects chemical reactions occurring within the aquifer because the water may supply some of the reactants or remove some of the products. The movement of water, or of another liquid such as petroleum, through a porous medium is governed by a law that was stated by Henry Darcy in 1856. *Darcy's law* expresses the flux of water (F) flowing through unit area (a) of an aquifer in unit time as a function of the hydrostatic pressure difference, which is equivalent to the hydraulic gradient (S):

$$F = KS \qquad (15.33)$$

where K is the *coefficient of hydraulic conductivity* whose dimensions are centimeters per second. The flux F has the dimensions of volume/unit area/unit time (cm^3 cm^{-2} sec^{-1}), which reduces to

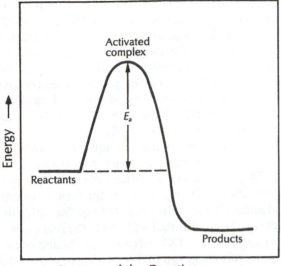

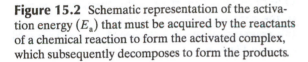

Figure 15.2 Schematic representation of the activation energy (E_a) that must be acquired by the reactants of a chemical reaction to form the activated complex, which subsequently decomposes to form the products.

15.2   Transport of Matter: Advection

The equations describing the rates of chemical reactions discussed above apply to stationary systems such as a cylinder of gas or an aqueous electrolyte solution in a beaker. However, chemical reactions in nature commonly require the movement of reactants or products and therefore depend on the applicable transport mechanism.

Transport of a solute in a solution can occur primarily by advection or by diffusion. *Advection* involves the displacement of matter in response to the action of a force. Examples of advection include the flow of water in the channel of a stream or in subsurface in response to a hydraulic gradient. In addition, rain falling from a cloud, sediment settling out of water, and the movement of lithospheric plates on the surface of the Earth are all examples of advection.

The displacement of matter produces a *flux*(F), which is the product of the density of the material (d) and its velocity (v):

centimeter per second and is therefore equivalent to the velocity of the water. The hydraulic gradient S is given by:

$$S = \frac{\Delta h}{L} \qquad (15.34)$$

where Δh is the difference in elevation between two points within the aquifer and L is the distance between these points measured along the flow path. Equation 15.33 can be modified to yield the *discharge* (D) in units of cubic centimeters per second:

$$D = KSa = Fa \qquad (15.35)$$

where a is the cross-sectional area of the aquifer shown schematically in Figure 15.3. When the dip of the aquifer is zero $(\Delta h = 0)$, flow occurs only when a pressure difference is imposed on the aquifer.

The hydraulic conductivity (K) is related to the *coefficient of hydraulic permeability* $(\mathbf{k})$ by the equation:

$$K = \frac{\mathbf{k} d g}{\eta} \qquad (15.36)$$

where g is the acceleration due to gravity and η is the viscosity of the fluid. For water at 20 °C, $\eta = 0.01$ g cm^{-1} sec^{-1}, $d = 1.0$ g/cm^3, and since $g = 980$ cm/sec^2:

$$K \approx 10^5 \mathbf{k} \qquad (15.37)$$

The coefficient of hydraulic permeability has the dimension of length squared and is related to the *porosity* of the aquifer and to the radius of the particles of which it is composed (Lerman, 1979). Values of $\mathbf{k}$ range over 12 orders of magnitude from 10^{-3} cm^2 for clean gravel to 10^{-15} cm^2 for granite and unweathered clay. The hydraulic conductivity (K) varies similarly. Good aquifers composed of clean sand or sand-gravel mixtures characteristically have $K \approx 10^{-1}$ cm/sec and $\mathbf{k} \approx 10^{-6}$ cm^2 (Lerman, 1979).

We can now consider an aquifer within which a chemical reaction is occurring (Moore, 1955, p. 544). A solution containing reactant X at concentration (X) mol/L flows through a volume element dV (the reactor) of the aquifer with a discharge rate D (L/sec) and leaves the reactor with a concentration $(X) - d(X)$. If no mixing occurs, then the net change in the number of moles of X with time in the reactor is the sum of two terms. One of these is the effect of the chemical reaction in the reactor, and the other is the difference in the concentration of X between the inflow and outflow. Hence:

$$\frac{dn_x}{dt} = R\,dV - D\,d(X) \qquad (15.38)$$

where R is the reaction rate per unit volume. For a first-order reaction with respect to reactant X:

$$-R = k(X) \qquad (15.39)$$

After the system has operated for some time, it achieves a *steady state* such that:

$$\frac{dn_x}{dt} = 0 \qquad (15.40)$$

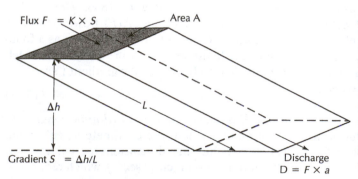

Figure 15.3 Schematic representation of a volume increment of a porous bed (aquifer) in subsurface. The movement of water through the aquifer is governed by Darcy's law expressed by equation 15.33 in terms of the flux F and the gradient S. The flux has units of mass (or volume) per unit area per unit time, which reduce to cm/sec. The flux F times the area a yields the discharge D, which is the volume of water leaving the system in unit time.

Under conditions of steady state and after substituting the rate equation for R, equation 15.38 becomes:

$$-k(X)dV = Dd(X) \qquad (15.41)$$

which can be integrated between the inlet and outlet of the reactor (dV):

$$-\frac{k}{D}\int_0^V dV = \int_{(X)_2}^{(X)_1}\frac{d(X)}{(X)} \qquad (15.42)$$

where $(X)_1$ and $(X)_2$ are the molar concentrations of X at the entrance and exit of volume dV, respectively. The result is:

$$-\frac{kV}{D} = \ln\frac{(X)_2}{(X)_1} \qquad (15.43)$$

Note that V/D has the dimensions of time called the *contact time:*

$$\frac{V}{D} = \frac{\text{volume} \times \text{time}}{\text{volume}} = \text{time} \qquad (15.44)$$

and therefore represents the average time molecule of X spends in the reactor. If we let $V/D = t$, equation 15.43 becomes:

$$\ln\frac{(X)_2}{(X)_1} = -kt \qquad (15.45)$$

$$(X)_2 = (X)_1 e^{-kt} \qquad (15.46)$$

Thus, equation 15.43 reduces to a first-order rate equation when it is applied to a static system.

15.3   Transport of Matter: Diffusion

Molecules or ions of a solute in a solution tend to disperse throughout a stationary solvent by a process known as *diffusion*. If the medium is itself in turbulent motion, then the dispersion of the solute is called *turbulent* or *eddy diffusion*. The flux (F) of ions or molecules diffusing through a stationary medium is proportional to the concentration gradient (dc/dx):

$$F = -D\left(\frac{dc}{dx}\right) \qquad (15.47)$$

where D here represents the *diffusion coefficient* and the minus sign indicates that the concentration gradient has a negative slope; that is, the ions and molecules move from points of high concentration toward points of low concentration within the medium. Equation 15.47 was formulated by the German chemist Adolf Fick in the 1850s and is known as *Fick's first law*. The diffusion coefficient for ions and molecules in water are of the order of 10^{-5} cm^2/sec, whereas for diffusion of ions in solids at temperatures around 1000°C, D has values around 10^{-10} cm^2/sec. Diffusion coefficients for selected ions in dilute aqueous solutions are listed in Table 15.1.

In case a solute is transported both by advection and by diffusion, the total flux (F) of an ion or molecule X is the sum of the advective flux (F_a) and the diffusive flux (F_d) (Lerman, 1979, p. 58):

$$F = F_a + F_d \qquad (15.48)$$

If the solvent moves with velocity v and the concentration of component X is (X), the advective flux according to equation 15.32 is:

$$F_a = (X)v \qquad (15.49)$$

Table 15. 1 Diffusion Coefficients for Ions in Dilute Aqueous Solutions at 25 °C

Cation	D, cm²/sec	Anion	D, cm²/sec
Li$^+$	10.3×10^{-6}	F$^-$	14.6×10^{-6}
Na$^+$	13.3×10^{-6}	Cl$^-$	20.3×10^{-6}
K$^+$	19.6×10^{-6}	Br$^-$	20.1×10^{-6}
Rb$^+$	20.6×10^{-6}	I$^-$	20.0×10^{-6}
Mg^{2+}	7.05×10^{-6}	HS$^-$	17.3×10^{-6}
Ca^{2+}	7.93×10^{-6}	HSO$_4^-$	13.3×10^{-6}
Sr^{2+}	7.94×10^{-6}	SO$_4^{2-}$	10.7×10^{-6}
Ba^{2+}	8.48×10^{-6}	NO$_3^-$	19.0×10^{-6}
Mn^{2+}	6.88×10^{-6}	HCO$_3^-$	11.8×10^{-6}
Fe^{2+}	7.19×10^{-6}	CO$_3^{2-}$	9.55×10^{-6}
Co^{2+}	6.99×10^{-6}	H$_2$PO$_4^-$	8.46×10^{-6}
Ni^{2+}	6.79×10^{-6}	HPO$_4^{2-}$	7.34×10^{-6}
Cu^{2+}	7.33×10^{-6}	PO$_4^{3-}$	6.12×10^{-6}
Zn^{2+}	7.15×10^{-6}	MoO$_4^{2-}$	9.91×10^{-6}

SOURCE: Lerman (1979), Table 3.1.

The diffusive flux is given by equation 15.47. Therefore, the total flux of component X is:

$$F_X = (X)v - D\frac{d(X)}{dx}$$

We may want to compare the fluxes caused by advection and by diffusion. For that purpose we define the advective transport distance (L_a) by the equation:

$$L_a = vt \qquad (15.50)$$

Since the diffusion coefficient (D) has the dimensions of square centimeters per second, Dt has the dimensions of square centimeters and $(Dt)^{1/2}$ has dimensions of length only. Therefore, the diffusional transport distance (L_d) is:

$$L_d = (Dt)^{1/2} \qquad (15.51)$$

and the ratio of transport distances is:

$$\frac{L_a}{L_d} = \frac{vt}{(Dt)^{1/2}} = v\left(\frac{t}{D}\right)^{1/2} \qquad (15.52)$$

Evidently, for fixed values of v and D the ratio L_a/L_d increases with the square root of time. The time required to make the respective transport distances equal to each other is obtained by solving equation 15.52 for t in the case that $L_a/L_d = 1.00$:

$$v\left(\frac{t}{D}\right)^{1/2} = 1.00 \qquad (15.53)$$

$$t = \frac{D}{v^2} \qquad (15.54)$$

For example, let $D = 1 \times 10^{-5}$ cm^2/sec and $v = 1.0$ m/yr, which is fairly typical for the movement of groundwater through an aquifer. Converting the velocity to centimeters per second:

$$v = \frac{1(100)}{1(365.25)(24)(60)(60)} = 3.2 \times 10^{-6} \text{ cm/sec}$$

$$(15.55)$$

and from equation 15.54 for the case that $L_a = L_d$:

$$t = \frac{1 \times 10^{-5}}{(3.2 \times 10^{-6})^2} = 9.8 \times 10^5 \text{ sec}$$

$$= 11.3 \text{ days} \qquad (15.56)$$

Evidently, after 11.3 days advection (based on a velocity of 1.0 m/yr) can transport the solute as far as diffusion could alone. Therefore, transport of solute in this case is primarily by advection rather than by diffusion. However, in a case where the velocity of the water is only $v = 1$ cm/1000 yr, which is representative of the rate of sediment deposition in the oceans, then $v = 1/1000(365.25)(24)(60)(60) = 3.16 \times 10^{-11}$ cm/sec and:

$$t = \frac{1 \times 10^{-5}}{(3.16 \times 10^{-11})^2} = 1.00 \times 10^{16} \text{ sec}$$

$$= 320 \times 10^6 \text{ yr} \qquad (15.57)$$

In this case, it takes a long time before transport by advection becomes equivalent to diffusion. This means that diffusion is the dominant transport mechanism in deep-sea sediment.

Fick's first law relates the flux of component X to the concentration gradient of X but does not indicate how the concentration of X varies with time. We therefore consider a volume increment having a unit cross section and a length dx extending from x to $x + dx$. The change in the concentration of X with time in this volume increment is the difference between the flux entering at x and leaving at $x + dx$, divided by dx. This is dimensionally correct because:

$$\frac{\text{flux}}{\text{length}} = \frac{g}{(\text{cm}^2)(\text{cm})(\text{sec})} = \frac{g}{(\text{cm}^3)(\text{sec})}$$

$$= \frac{\text{concentration}}{\text{time}} \qquad (15.58)$$

Therefore, based on Fick's first law (equation 15.47):

$$\frac{\partial c}{\partial t} = \frac{1}{dx}\left[-D\left(\frac{\partial c}{\partial x}\right)_x + D\left(\frac{\partial c}{\partial x}\right)_{x+dx}\right] \qquad (15.59)$$

where the equation is written in terms of partial derivatives because we are differentiating only with respect to the x-direction while holding the other two directions, y and z, constant. The outgoing flux at $x + dx$ has a positive sign because it is being subtracted from the flux entering at x. Since:

$$\left(\frac{\partial c}{\partial x}\right)_{x+dx} = \left(\frac{\partial c}{\partial x}\right)_x + \frac{\partial}{\partial x}\left(\frac{\partial c}{\partial x}\right)dx \quad (15.60)$$

we substitute into equation 15.59 (Moore, 1955, p. 448) and obtain:

$$\frac{\partial c}{\partial t} = -\frac{D}{dx}\left(\frac{\partial c}{\partial x}\right)_x + \frac{D}{dx}$$

$$\times \left[\left(\frac{\partial c}{\partial x}\right)_x + \frac{\partial}{\partial x}\left(\frac{\partial c}{\partial x}\right)dx\right] \quad (15.61)$$

which reduces to:

$$\frac{\partial c}{\partial t} = D\left(\frac{\partial^2 c}{\partial x^2}\right) \quad (15.62)$$

Equation 15.62 is *Fick's second law*, which states that the rate of change of the concentration of a component X is equal to the second derivative of the linear concentration gradient multiplied by the diffusion coefficient. Equation 15.62 has the same form as the equation that governs heat conduction and has been solved for many different boundary conditions (Crank, 1975).

The application of Fick's second law is illustrated by the case where a thin layer of the diffusing substance is placed between two layers of a stationary medium. According to Crank (1975), the solution of equation 15.62 is:

$$\frac{c}{c_0} = (\pi Dt)^{-1/2}e^{-x^2/4Dt} \quad (15.63)$$

When the concentration c is plotted as a function of distance x for different values of t using an appropriate value of the diffusion coefficient (D), a series of Gaussian error curves is obtained that show how the diffusing substance spreads out as time increases (Moore, 1955, p. 449). Three such curves are shown in Figure 15.4, where the diffusion takes place both in the $+x$ and in the $-x$ direction and the concentration must therefore be divided by 2 because diffusion occurs in two directions.

Equation 15.63 can be used to evaluate diffusion data. If we take natural logarithms of equation 15.63, we obtain:

$$\ln\left(\frac{c}{c_0}\right) = -\tfrac{1}{2}\ln(\pi Dt) - \frac{x^2}{4Dt} \quad (15.64)$$

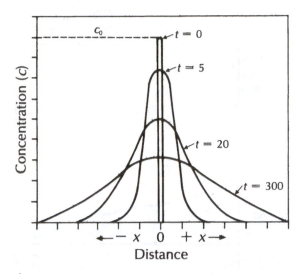

Figure 15.4 Schematic representation of the distribution of a diffusing substance that starts as a thin film in a stationary medium. The distribution of the diffusing substance at different times (t) is governed by Fick's second law expressed by equation 15.63.

which is a straight line in coordinates of $\ln(c/c_0)$ and x^2. Therefore, measurements of the concentration of an ion in a body of water through which the ion is diffusing define a straight line in this coordinate system. The slope of this line is:

$$m = -\frac{1}{4Dt} \quad (15.65)$$

and its intercept on the y-axis is:

$$b = -\tfrac{1}{2}\ln(\pi Dt) \quad (15.66)$$

An example of diffusion of ions occurs in Lake Vanda in Wright Valley, southern Victoria Land, Antarctica. This lake contains a layer of dense brine from which ions have diffused into an overlying layer of less saline water. Green and Canfield (1984) explained the variation in the chemical composition of water taken at different depths in Lake Vanda in terms of diffusion of ions. The variation of the concentration of Ca^{2+} ions with depth is illustrated in Figure 15.5, which also contains a plot of $\ln Ca^{2+}/Ca_0^{2+}$ versus x^2,

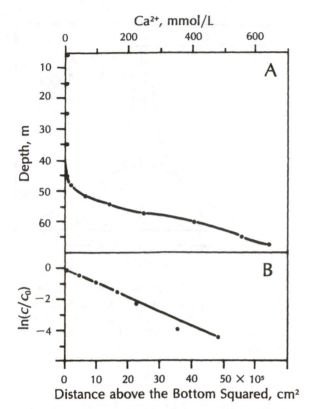

Figure 15.5 A: Variation of the concentration of Ca^{2+} as a function of depth in Lake Vanda, Wright Valley, southern Victoria Land, Antarctica. The increase of Ca^{2+} and other ions is caused by the presence of a layer of dense brine whose temperature is about 25 °C, although Lake Vanda has a permanent ice cover that is 4 m thick (Green and Canfield, 1984).

B: Plot of Ca^{2+} concentrations in Lake Vanda in accordance with equation 15.64, which is based on Fick's second law of diffusion. The data points define a straight line whose slope was used in equation 15.68 to calculate an age of 1290 years for the dilute layer above the brine.

where x is the distance in centimeters from the bottom of the lake to each analyzed sample. The data points below a depth of 45 m in the lake fit a straight line whose slope is $m = -0.090\,96 \times 10^{-5}$. Since the water temperature *increases* with depth from ~10 °C at 45 m to ~25 °C at the bottom, the diffusion coefficient of Ca^{2+} was taken to be $D = 6.73 \times 10^{-6}$ cm^2/sec at an average tempera-

ture of 18 °C (Li and Gregory, 1974). We can now solve equation 15.65 for time in years. Since:

$$-\frac{1}{4Dt} = -0.090\,96 \times 10^{-5} \qquad (15.67)$$

$$t = \frac{10.99 \times 10^5}{4(6.73 \times 10^{-6})(3.155 \times 10^7)} = 1290 \text{ yr} \qquad (15.68)$$

This result indicates that the observed distribution of Ca^{2+} concentrations in Lake Vanda below a depth of ~45 m is attributable to ionic diffusion from the dense brine layer in the past 1290 years.

When diffusion involves the movement of charged particles, electrical neutrality throughout the system must be maintained either by the movement of positively and negatively charged species in the same direction or by the movement of charged particles in the opposite direction. Moreover, the magnitude of the diffusion coefficient of an ionic species is affected by electrostatic interactions with other ions depending on its ionic potential, that is, the charge-to-radius ratio (Lasaga, 1979). Therefore, the magnitudes of diffusion coefficients depend to some extent on the composition of the environment in which diffusion is taking place. In general, several different kinds of diffusion coefficients have been defined that apply to specific types of diffusion (Crank, 1975).

The magnitudes of diffusion coefficients are also strongly dependent on the temperature. The temperature dependence of D is expressed by the Arrhenius equation:

$$D = D_0 e^{-E_a/RT} \qquad (15.69)$$

where D_0 is the *frequency factor*, E_a is the activation energy, R is the gas constant, and T is the temperature in kelvins. In this context, E_a is a measure of the energy required to start the movement of the diffusing substance or ion. We can linearize equation 15.69 by taking natural logarithms:

$$\ln D = \ln D_0 - \frac{E_a}{RT} \qquad (15.70)$$

This is the equation of a straight line in coordinates of ln D and the reciprocal of the temperature whose slope is:

$$m = -\frac{E_a}{R} \tag{15.71}$$

Since the slope is negative, ln D *increases* with *decreasing* values of $1/T$, that is, with *increasing* temperature. The temperature dependence of diffusion coefficients is especially important in igneous petrology because it affects the distribution of cations in minerals that form at elevated temperatures by crystallization of silicate melts. Values of relevant diffusion coefficients of ions in melts and in minerals at 1000°C were compiled by Henderson (1982, Tables 8.1 and 8.2). Diffusion coefficients of ions and molecules through solids at 25°C are listed in Table 15.2.

15.4   Growth of Concretions During Diagenesis

Diagenesis includes all of the chemical, physical, and biological processes that take place in sediment after it was deposited. The *chemical* processes may include dissolution of mineral particles in the pore water, precipitation of insoluble compounds from the pore water, and ion exchange reactions between aqueous species and the surfaces of solids. *Biological* activity may stir up the sediment (bioturbation), and bacteria can bring about chemical changes, such as the reduction of sulfate to bisulfide ions. In addition, the presence of certain kinds of biogenic compounds may inhibit dissolution of mineral grains by coating their surfaces or enhance it by adsorbing some of the ions that are released. The chemical reactions and biological activity take place in an environment controlled by *physical* processes, including the deposition of sediment and its compaction as well as the flow of pore water in response to pressure gradients.

The description of these processes can be approached in three different ways all of which are useful in the study of diagenesis (Berner, 1980):

1. Observations regarding changes in the chemical composition of the pore water and the sediment with depth and hence with time.

2. Experimentation in the laboratory or thermodynamic calculations both of which may permit qualitative predictions about changes that should occur with depth as a result of diagenesis.

3. A combination of transport mechanisms (advection and diffusion) and measured rates of chemical reactions from which a transport-dependent reaction model of diagenesis can be developed.

Such a model was presented by Berner (1980) to describe the growth of a concretion during diagenesis.

Minerals that precipitate from the pore water of sediment may either form *within* the pore spaces as a cement or replace preexisting minerals that dissolve. Both processes may cause the formation of "localized mineral segregations" called *concretions* (Berner, 1980, p. 108). The reactants required to form a mineral cement may originate by dissolution of other minerals within the sediment by a process called *diagenetic redistribution,* or they may be transported to the site of deposition from distant sources. Similarly, the ions released by dissolution of the preexisting minerals may participate in the formation of the authigenic minerals at the same site (replacement), or they may be transported away.

After nucleation of an authigenic mineral has occurred, the continued growth of crystals depends on (1) the transport of ions to the surface of the crystal; (2) surface reactions such as adsorption, dehydration, and ion exchange; and (3) removal of the products of chemical reactions. The rate of growth of the crystals may be controlled by each of the three processes or by combinations of them. For the sake of simplicity, we consider only transport and surface chemistry as the rate-controlling processes of crystal growth.

In case crystal growth is controlled primarily by *transport mechanisms,* the concentrations of

Table 15.2 Diffusion Coefficients and Activation Energies of Ions in Solids at 25 °C

Matrix	Species	D, (25°C) cm²/sec	E_a kcal/mol	Direction or composition
Quartz	Na	1.1×10^{-15}	20.2	parallel to c-axis
Quartz	Na	1.8×10^{-28}	41	90° to c-axis
Quartz	Ca	1.4×10^{-45}	68	
Na_2SiO_3 glass	Na	8.2×10^{-17}	18.5	
K_2SiO_3 glass	K	5.9×10^{-17}	17.5	
SiO_2 glass	H_2O	3.4×10^{-12}	11.3	
Obsidian	H_2O	1.0×10^{-16}	13	H_2O = 0 wt.%
Obsidian	H_2O	8.0×10^{-16}	13	H_2O = 0.5 wt.%
Obsidian	H_2O	1.0×10^{-14}	13	H_2O = 6.0 wt.%
Feldspar	K	1.6×10^{-49}	68	orthoclase, Or_{94}
Feldspar	Na	2.1×10^{-38}	53	orthoclase, Or_{94}
Feldspar	^{40}Ar	3.6×10^{-34}	43.1	orthoclase, Or_{94}
Feldspar	O_2	5.6×10^{-28}	29.6	microcline, Or_{100}
Feldspar	O_2	1.1×10^{-28}	29.6	adularia, Or_{100}
Feldspar	O_2	7.7×10^{-27}	25.6	adularia, Or_{98}
Feldspar	O_2	5.6×10^{-25}	21.3	albite, Ab_{97-99}
Feldspar	O_2	8.7×10^{-27}	26.2	anorthite, An_{96}
Mica	Ar	1.7×10^{-43}	57.9	phlogopite, 4% annite, P = 2 kbar
Mica	Ar	1.4×10^{-22}	23.3	phlogopite in vacuum
Zeolite	H_2O	10^{-7} to 10^{-8}		heulandite, parallel to structure
Zeolite	H_2O	10^{-11}		heulandite, perpendicular to structure
Zeolite	H_2O	2.1×10^{-8}		Ca-heulandite
Zeolite	H_2O	1.3×10^{-7}		Ca-chabazite
Zeolite	H_2O	2.0×10^{-13}		Na-analcime
Zeolite	H_2O	1.5×10^{-12}		Na-analcime
Zeolite	Na	10^{-12} to 10^{-13}		analcime, chabazite
Zeolite	Cs	$10^{-24}, 10^{-13}$		analcime, chabazite
Zeolite	Ca	4×10^{-16}		chabazite
Calcite	CO_2	1.4×10^{-46}	58.0	$CaCo_3$ = 99.5%
				$MgCO_3$ = 0.5%
				$FeCO_3$ = trace

SOURCE: Lerman (1979), Table 3.9.

the reactants in the solution adjacent to the crystal are maintained at levels required for equilibrium and the crystals grow as rapidly as the reactants arrive at the surface of the crystal by advection or diffusion. If the rate of crystal growth is dominated by *surface reactions,* then the concentrations of the reactants in the solution adjacent to the crystal may be in excess of equilibrium values and may remain uniform throughout the aqueous phase. In reality, *both* processes may affect the growth rate of crystals in sediment during diagenesis, but rapid growth of authigenic crystals is generally associated with transport-controlled reactions.

The rate of dissolution of minerals is similarly controlled by the transport mechanism, by surface chemistry, or by a combination of the two. If dissolution of preexisting minerals and simultaneous precipitation of authigenic minerals are both transport controlled, then a zone of depletion develops around each growing crystal. In this process of diagenetic redistribution the thickness of the depleted zone around each crystal depends on the abundance of the dissolving mineral in the sediment and on the radius of the crystal as required by mass balance. If, on the other hand, the rate of dissolution and crystal growth are both entirely controlled by surface processes, then dissolution of preexisting minerals is selective—that is, smaller crystals dissolve faster than larger ones because they have more surface area—and growing crystals are surrounded only by a poorly defined zone of depletion whose thickness cannot be predicted by a mass-balance calculation.

Therefore, the presence of a well-defined zone of depletion in the sediment around a concretion indicates that it formed as a result of local diagenetic redistribution by a transport-dominated process. In this case, it is possible to determine how much time was required for the concretion to form.

We assume that a concretion is forming, as shown in Figure 15.6, by dissolution of minerals in the sediment at the outer edge of a spherical zone of depletion whose width is L and that the transport of reactants to the surface of the concretion is entirely by diffusion. Under these conditions Fick's first law (equation 15.47) applies, and the flux (F) of reactants arriving on the surface of the spherical concretion is (Berner, 1980; Nielsen, 1961; Frank 1950):

$$F_c = -\frac{\phi D}{L}(c_d - c_p) \qquad (15.72)$$

where c is the molar concentration of the material in the pore water in equilibrium with the solid that is dissolving (d) or precipitating (p) in the zone of depletion and on the surface of the concretion, respectively. Therefore, $(c_d - c_p)/L$ is the concentration gradient across the zone of deple-

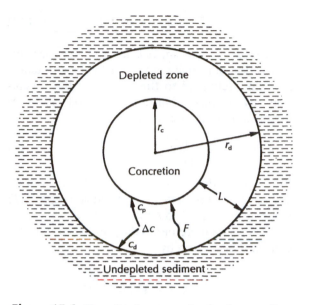

Figure 15.6 Growth of a concretion having a radius of r_c by diagenetic redistribution caused by dissolution of a mineral in the sediment, diffusion of the resulting ions to the surface of the concretion, and precipitation of a new mineral in the pore spaces of the sediment. The process is assumed to be transport dominated and therefore causes the formation of a depleted zone having a radius of r_d and a width L. In addition, V is the molar volume of the dissolving or precipitating minerals, ϕ is the porosity of the sediment, D is the diffusion coefficient, f is the volume fraction of the dissolving or precipitating mineral, and F is the flux of ions to the surface of the concretion and leaving the outer surface of the depleted zone. The model discussed here was presented by Berner (1980).

tion. We can simplify equation 15.72 by replacing $c_d - c_p$ by Δc:

$$F_c = -\frac{\phi D \Delta c}{L} \qquad (15.73)$$

The flux due to diffusion is diminished by a factor ϕ, which is the *porosity* of the sediment defined as the fraction of a unit volume of sediment that consists of water-filled pore spaces.

The concretion grows as a function of time by the addition of a certain amount of mass (dM_p) to its surface:

$$\frac{dM_p}{dt} = -4\pi(r_c)^2 F_c \qquad (15.74)$$

where $4\pi r_c^2$ is the surface area of the spherical concretion of radius r_c and F_c is the flux of material being delivered to the surface of the concretion by diffusion alone. Equation 15.74 is dimensionally correct because:

$$4\pi(r_c)^2 F_c = \frac{\text{length}^2 \times \text{mass}}{\text{length}^2 \times \text{time}} = \frac{\text{mass}}{\text{time}} \qquad (15.75)$$

In addition, the minus sign is required because dM/dt must remain positive when equation 15.73 is substituted for F_c in equation 15.74.

The mass increment (dM) added to the concretion is also given by:

$$dM_p = \frac{4\pi(r_c)^2}{V_p} f_p \, dr_c \qquad (15.76)$$

where f_p is the volume fraction of the precipitating material in the concretion and V_p is the molar volume of the precipitation material. Molar volumes of common minerals are listed in Table 15.3. The dimensions of equation 15.76 are in terms of *mass* because f_p is a dimensionless number and:

$$\frac{4\pi}{V_p}(r_c)^2 f_p \, dr_c = \frac{\text{length}^2 \times \text{length}}{\text{length}^3/\text{mass}} = \text{mass} \qquad (15.77)$$

We now substitute equation 15.73 for F_c and equation 15.76 for dM_p into equation 15.74. The result is:

$$\frac{4\pi(r_c)^2}{V_p} f_p\left(\frac{dr_c}{dt}\right) = 4\pi(r_c)^2 \phi D\left(\frac{\Delta c}{L}\right) \qquad (15.78)$$

which reduces to:

$$\frac{dr_c}{dt} = \frac{\phi V_p D}{f_p}\left(\frac{\Delta c}{L}\right) \qquad (15.79)$$

Unfortunately, we cannot integrate this equation because the width (L) of the depleted zone also increases with time as the concretion grows. Therefore, we must express L as a function of time before we can use equation 15.79 to derive a relationship between the radius of a concretion and the time required for it to grow to a given size.

The thickness (L) of the zone of depletion increases with time at a rate that depends on the rate of growth of the concretion:

Table 15.3 Molar Volumes of Common Minerals in Sedimentary Rocks

Mineral	Formula	Molar volume, cm^3
Troilite (hex.)	FeS	18.20
Pyrrhotite (hex.)	$Fe_{0.98}S$	18.11
Alabandite (cubic)	MnS	21.457
Pyrite (cubic)	FeS_2	23.940
Marcasite (orth.)	FeS_2	24.579
Boehmite (orth.)	AlO(OH)	19.535
Diaspore (orth.)	AlO(OH)	17.760
Gibbsite (mon.)	$Al(OH)_3$	31.956
Hematite (hex.)	Fe_2O_3	30.274
Magnetite (cubic)	Fe_3O_4	44.524
Goethite (orth.)	α-FeO(OH)	20.82
Manganosite (cubic)	MnO	13.221
Pyrolusite (tet.)	MnO_2	16.61
Bixbyite (cubic)	Mn_2O_3	31.37
Hausmannite (tet.)	Mn_3O_4	46.95
Quartz (α) (hex.)	SiO_2	22.688
Quartz (β) (hex.)	SiO_2	23.718
Cristobalite (α) (tet.)	SiO_2	25.739
Cristobalite (β) (cubic)	SiO_2	27.381
Tridymite (β) (hex.)	SiO_2	27.414
Uraninite (cubic)	UO_2	24.618
Calcite (hex.)	$CaCO_3$	36.934
Aragonite (orth.)	$CaCO_3$	34.15
Siderite (hex.)	$FeCO_3$	29.378
Magnesite (hex.)	$MgCO_3$	28.018
Rhodochrosite (hex.)	$MnCO_3$	31.073
Dolomite (hex.)	$MgCa(CO_3)_2$	64.341
Barite (orth.)	$BaSO_4$	52.10
Anhydrite (orth.)	$CaSO_4$	45.94
Gypsum (mon.)	$CaSO_4 \cdot 2H_2O$	74.69

SOURCE: Weast et al. (1986).

$$\frac{dL}{dt} = \frac{dr_d}{dt} - \frac{dr_c}{dt} \qquad (15.80)$$

where the subscripts identify the zone of depletion (d) and the concretion (c) as before and r_d is the radius of the depleted zone. The mass of the dissolving material is:

$$\frac{dM_d}{dt} = 4\pi (r_d)^2 F_d \qquad (15.81)$$

The rate of dissolution (dM_d/dt) in equation 15.81 is a negative quantity because the material is lost from the system. Therefore, the two minus signs associated with F and with the loss of material cancel each other.

The amount of material lost from the zone of depletion is also expressible by an equation that is similar to 15.76:

$$dM_d = -\frac{4\pi}{V_d}(r_d)^2 f_d\, dr_d \qquad (15.82)$$

where f_d is the volume fraction of the dissolving mineral in the sediment and V_d is the molar volume of the dissolving material. We can now substitute equation 15.82 for dM_d and equation 15.73 for F_d in equation 15.81. The result is:

$$-\frac{4\pi}{V_d}(r_d)^2 f_d\left(\frac{dr_d}{dt}\right) = -4\pi (r_d)^2 \phi D\left(\frac{\Delta c}{L}\right) \quad (15.83)$$

which reduces to:

$$\frac{dr_d}{dt} = \frac{\phi V_d D}{f_d}\left(\frac{\Delta c}{L}\right) \qquad (15.84)$$

Next, we substitute the rate of growth of the concretion (equation 15.79) and the rate of growth of the depleted zone (equation 15.84) into equation 15.80, which expresses the rate of change of the width (L) of the depleted zone:

$$\frac{dL}{dt} = \frac{\phi V_d D}{f_d}\left(\frac{\Delta c}{L}\right) - \frac{\phi V_p D}{f_p}\left(\frac{\Delta c}{L}\right) \quad (15.85)$$

which yields:

$$\frac{dL}{dt} = \phi D\left(\frac{\Delta c}{L}\right)\left(\frac{V_d}{f_d} - \frac{V_p}{f_p}\right) \qquad (15.86)$$

Equation 15.85 can be integrated in the form:

$$\int L\, dL = \text{constant} \int dt \qquad (15.87)$$

If $L = 0$ when $t = 0$, the result is $\frac{1}{2}L^2 = \text{con-stant} \times t$ from which it follows that:

$$L = \left[2\phi D\Delta c\left(\frac{V_d}{f_d} - \frac{V_p}{f_p}\right)t \right]^{1/2} \quad (15.88)$$

We are now able to integrate equation 15.79 and thereby obtain the desired relationship between the radius of the growing concretion (r_d) and time by substituting equation 15.88 into equation 15.79. In order to simplify the algebra we write equation 15.88 as:

$$L = G^{1/2}\, t^{1/2} \qquad (15.89)$$

and equation 15.79 as:

$$\frac{dr_c}{dt} = \frac{H}{L} \qquad (15.90)$$

where G and H are constants defined by equations 15.88 and 15.79, respectively. Therefore:

$$dr_c = \left(\frac{H}{G^{1/2}}\right)t^{-1/2}\, dt \qquad (15.91)$$

By integrating equation 15.91 for the condition that $r_c = 0$ when $t = 0$, we obtain:

$$r_c = \left(\frac{2H}{G^{1/2}}\right)t^{1/2} \qquad (15.92)$$

or

$$t = \left(\frac{r_c}{2}\right)^2 \frac{G}{H^2} \qquad (15.93)$$

The ratio of the constant G/H^2 can now be simplified:

$$\frac{G}{H^2} = \frac{2\phi D\Delta c(V_d/f_d - V_p/f_p)f_p^2}{\phi^2 V_p^2 D^2 \Delta c^2} \quad (15.94)$$

$$\frac{G}{H^2} = \frac{2(V_d/f_d - V_p/f_p)f_p^2}{\phi V_p^2 D\Delta c}$$

$$= \frac{2(V_d f_p - V_p f_d)f_p^2}{\phi V_p^2 D\Delta c f_d f_p} \qquad (15.95)$$

Next, we assume that the precipitating mineral fills the available pore spaces in the sediment ($f_p = \phi$) and that the molar volumes of the dissolving and precipitating minerals are approximately equal ($V_d \approx V_p$):

$$\frac{G}{H^2} = \frac{2(\phi - f_d)}{V_p F_d D\Delta c} \qquad (15.96)$$

Therefore, equation 15.93 becomes:

$$t = \frac{r_c^2(\phi - f_d)}{2V_p f_d D \Delta c} = \frac{r_c^2}{2V_p D \Delta c}\left(\frac{\phi}{f_d} - 1\right) \quad (15.97)$$

This equation, derived by Berner (1980), is applicable to the growth of calcite concretions by dissolution of aragonite crystals disseminated throughout a body of sediment. If transport of ions is by diffusion and if $D = 2 \times 10^{-6}$ cm^2/sec, $\phi = f_p = 0.60$, and V (calcite) $\approx V$ (aragonite) $= 35$ cm^3/mol, then equation 15.97 can be solved for t for selected values of the abundance of aragonite crystals in the sediment (f_d). The concentration gradient (Δc) is the difference in the equilibrium concentrations of Ca^{2+} or CO$_3^{2-}$ ions at the surfaces of aragonite and calcite crystals. At 25°C this difference is about 1.0×10^{-5} mol/kg or 1×10^{-8} mol/g, based on the solubility product constants in Table 10.1. Therefore, if the aragonite content of the sediment is 15% ($f_d = 0.15$), a calcite concretion having a radius of 5.0 cm will form in:

$$t = \frac{(5.0)^2(0.60/0.15 - 1)}{2 \times 35 \times 2 \times 10^{-6} \times 1 \times 10^{-8}}$$

$$= 0.5357 \times 10^{14} \text{ sec} \quad (15.98)$$

Since 1 yr $= 3.155 \times 10^7$ sec,

$$t = \frac{0.5357 \times 10^{14}}{3.155 \times 10^7} = 1.7 \times 10^6 \text{ yr} \quad (15.99)$$

We see from this result that a calcite concretion the size of a small cannonball can grow in ~1.7 million years, which is quite fast on a geological time scale. In general, the *rate of growth* of concretions increases with the abundance of the dissolving mineral in the sediment. The relationship between the radius of a concretion and time, expressed by equation 15.97, is shown graphically in Figure 15.7 for the case where the aragonite content of the sediment is 5, 10, and 20% by volume (i.e., $f_d = 0.05, 0.10,$ and 0.20). The resulting curves can be used to determine the time when a particular layer within a concretion was deposited. Therefore, we can regard carbonate concretions as *geochemical recorders* that have preserved a record of changes in the environment in which they formed. Concretions can reveal not only the

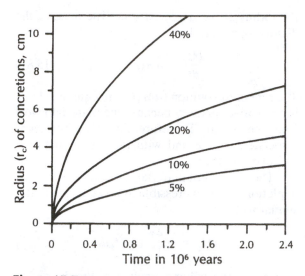

Figure 15.7 Growth of calcite concretions by dissolution of aragonite during diagenesis of sediment containing varying initial concentrations of aragonite (see equation 15.97 after Berner, 1980). Transport of ions is by diffusion, the concentration gradient across the zone of depletion is 1×10^{-5} mol/kg, the diffusion coefficient $D = 2 \times 10^{-6}$ cm^2/sec, the porosity of the sediment in the depleted zone is 60%, which is assumed to be equal to the abundance of calcite in the concretions, and the molar volumes of calcite and aragonite are both taken to be 35 cm^3/mol. The curves indicate that the size of calcite concretions that grow in unit time increases with increasing abundance of aragonite in the sediment.

sequence of geochemical changes but also the length of the time intervals between them based on equation 19.97. Since concretions form soon after deposition of the sediment, they can provide a time scale for geochemical events that may have affected their mineral compositions, their growth rate, trace element concentrations, and the isotope compositions of O, C, Sr, Pb, and Nd they contain.

15.5 Growth of Monomineralic Layers

The sediment that is accumulating in deep-sea basins and in some lakes becomes anoxic because O$_2$ is consumed by organisms and by the oxidation of organic matter. As a result, the solubility of the oxides in Fe and Mn increases as the environmental Eh decreases, as shown in Figures 14.6 and 14.7B. (The stability of Mn oxides in terms of Eh and pH is the subject of Problems 7–9 in Chapter 14.) Therefore, a concentration gradient of Fe^{2+} and Mn^{2+} ions may develop, which causes these ions to diffuse upward toward the sediment–water interface. When the ions return to the more oxygenated environment, they precipitate as oxyhydroxides of Fe^{3+} and Mn^{4+} because of the decrease in the solubilities of these compounds at the higher Eh (and pH) conditions in the water above the sediment. As a result, a layer of Fe and Mn oxides and hydroxides accumulates at the sediment–water interface, provided that the process is not terminated prematurely by burial caused by rapid sedimentation. The dissolution of scattered particles of Fe- and Mn-bearing minerals and the precipitation of authigenic oxides at the sediment–water interface is an example of diffusion-controlled diagenetic redistribution. The rate of growth of the monomineralic layer at the sediment–water interface can therefore be expressed by the equations derived in the preceding section for the growth of concretions.

The flux (F) at the sediment–water interface is given by Fick's first law (Berner, 1980):

$$F = -\phi D \frac{c_d - c_p}{L} \qquad (15.100)$$

where c_d and c_p are the concentrations of Fe^{2+} and Mn^{2+} in the water in equilibrium with the dissolving mineral grains (c_d) and in equilibrium with the authigenic minerals at the sediment–water interface (c_p). Since L is the thickness of the depleted zone shown in Figure 15.8 $(c_d - c_p)/L$ is the concentration gradient across the depleted zone. In addition, ϕ is the porosity and D is the diffusion coefficient defined as in Section 15.4.

In order to constrain the model to the simplest set of circumstances we specify that:

1. The authigenic minerals form within the oxygenated water on top of the sediment.
2. Transport of ions by advection is negligible.
3. The ions are not adsorbed on the surfaces of particles in the sediment.

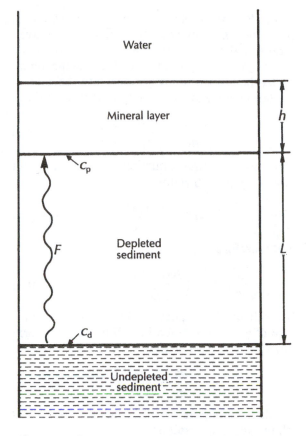

Figure 15.8 Formation of a layer of minerals at the sediment–water interface by diffusion of ions such as Fe^{2+} and Mn^{2+} from the sediment. The ions form by dissolution of particles in the sediment, which is anoxic compared to the water above; that is, it has a lower Eh. The ions precipitate when they reach the sediment–water interface to form a layer whose thickness (h) increases with time in accordance with equation 15.110. The thickness of the depleted layer (L) also increases with time (equation 15.104). However, calculations show that a layer of Fe or Mn oxyhydroxides can accumulate only when the rate of deposition of detrital sediment is virtually zero.

4. Sediment deposition from above is negligible.

By analogy with equation 15.79, the rate of growth of the monomineralic layer of thickness h is:

$$\frac{dh}{dt} = \frac{\phi V_p D}{f_p} \left(\frac{\Delta c}{L} \right) \qquad (15.101)$$

where V_p is the molar volume of the precipitating phase and f_p is its volume fraction in the layer. Similarly, the rate of growth of the zone of depletion is given by equation 15.84 because the monomineralic layer is not expanding into the zone of depletion but is growing away from it. Therefore:

$$\frac{dL}{dt} = \frac{\phi V_d D}{f_d}\left(\frac{\Delta c}{L}\right) \qquad (15.102)$$

Integration of this equation for the case that $L = 0$ when $t = 0$ yields:

$$\tfrac{1}{2}L^2 = \left(\frac{\phi V_d D \Delta c t}{f_d}\right) \qquad (15.103)$$

and therefore:

$$L = \left(\frac{2\phi V_d D \Delta c}{f_d}\right)^{1/2} t^{1/2} \qquad (15.104)$$

By substituting equation 15.104 in the form $L = G'^{1/2} t^{1/2}$ into equation 15.101 in the form $dh/dt = H'/L$, we obtain:

$$\frac{dh}{dt} = \frac{H't^{-1/2}}{G'^{1/2}} \qquad (15.105)$$

which we integrate for the condition that $h = 0$ when $t = 0$. The result is:

$$h = \frac{2H'}{G'^{1/2}} t^{1/2} \qquad (15.106)$$

or

$$t = \left(\frac{h}{2}\right)^2 \frac{G'}{H'^2} \qquad (15.107)$$

which is similar to equation 15.93. Next, we evaluate the ratio of the constant terms:

$$\frac{G'}{H'^2} = \frac{2\phi V_d D \Delta c f_p^2}{f_d \phi^2 V_p^2 D^2 (\Delta c)^2} = \frac{2V_d f_p^2}{f_d \phi V_p^2 D \Delta c} \qquad (15.108)$$

If $V_p \approx V_d$, we obtain:

$$\frac{G'}{H'^2} = \frac{2f_p^2}{f_d \phi V_p D \Delta c} \qquad (15.109)$$

and therefore equation 15.107 becomes:

$$t = \frac{h^2 f_p^2}{2f_d \phi V_p D \Delta c} \qquad (15.110)$$

Note that in this case the volume fraction of the precipitating phase (f_p) is not equal to the porosity (ϕ) of the underlying sediment because the layer is forming in the water above the sediment.

Berner (1980) evaluated equation 15.110 for the case that $h = 1$ cm, $f_p = 0.2$ (flocculant precipitate), $\phi = 0.8$, $V_p \approx V_d = 20$ cm^3/mol, $D = 3 \times 10^{-6}$ cm^2/sec or 100 cm^2/yr. According to equation 15.110, the time required to form a monomineralic layer of specified thickness increases with *decreasing* abundance (f_d) of the dissolving mineral in the sediment and with *decreasing* concentration gradient (Δc). If $f_d = 0.002$ (0.2% by volume) and $\Delta c = 10^{-4}$ mol/kg, equation 15.110 yields:

$$t = \frac{(1)^2(0.2)^2(10^3)}{2(0.002)(0.8)(20)(100)(10^{-4})}$$

$$= \frac{40 \times 10^4}{6.4} = 62{,}500 \text{ yr} \qquad (15.111)$$

The time required to form such a monomineralic layer is apparently quite *long* compared to normal sedimentation rates in the oceans or in lakes. Even on the abyssal plains of the oceans the sedimentation rate is of the order of 0.1 cm/1000 yr. Therefore, the layer of *detrital* sediment deposited in 62,500 yr is more than six times thicker than the 1-cm monomineralic layer; thus such a layer cannot form under normal circumstances in the oceans or in lakes. The existence of such monomineralic layers of Fe and Mn oxides in sediment therefore requires either that a period of sediment starvation occurred at the site, or that the upward diffusive flux of ions was augmented by an advective flux caused by the movement of water, or both.

The mathematical approach to the study of geochemical processes involving transport of reactants in different kinds of environments on the surface of the Earth is presented in the textbooks by Berner (1971, 1980) and by Lerman (1979). These books combine elegant mathematical descriptions of complex geological processes with analytical measurements and with understanding based on thermodynamics. In this way we can actually achieve the high level of *under-*

standing of geochemical processes, which is the principal objective of the science of geochemistry. The grandeur of the processes that convert sediment at the bottom of the oceans into sedimentary rocks was appreciated more than 100 years ago by the geologist K. von Gümbel (1888), who wrote (freely translated):

> There is no doubt that magnificent chemical processes are taking place on the bottom of the ocean. They explain the partial reconstitution (diagenesis) and solidification of sediment by the formation of cements as exemplified by the marine deposits of the geologic past. (From a quotation in Berner, 1980.)

15.6 Summary

Geochemical processes, such as weathering or diagenesis, consist of sequences of chemical reactions, which are commonly dependent on the transport of reactants and products. The chemical reactions themselves involve complex mechanisms, each of which takes place at a characteristic rate. As a result, the rates of geochemical processes depend both on the rates of the chemical reactions and on the rates and methods of transport that contribute reactants and remove products from the reaction site.

Geochemical processes can be studied by three methods. The first consists of analyzing appropriately chosen samples representing different stages or aspects of the process. Second, predictions about the outcome of geochemical processes can be made from thermodynamic considerations using various graphical representations or computations. In the third method, the entire process may be expressed in the form of equations that combine information about reaction mechanisms and their rates with appropriate transport models.

All three methods of study need to be employed because without analytical data and without a sense of direction provided by thermodynamics, mathematical modeling of geochemical processes cannot succeed. However, when done correctly, mathematical representations of geochemical processes achieve the highest level of understanding because they provide not only an explanation of what has already happened, but also permit predictions of what may happen in the future.

Geochemistry acquires a special importance among the sciences because it helps us to understand the world we live in and to make use of the resources it contains. The understanding we have of the environment we live in today permits us to anticipate how it may respond in the future to various anthropogenic perturbations.

Problems

1. Determine the *order* and write the rate equation of the reaction between ammonium ion and nitrite ion in aqueous solution at 30°C based on the following information about the initial rates of this reaction (Brown, 1968).

$$NH_4^+ + NO_2^- \rightarrow N_2 + 2 H_2O$$

The concentration of NH_4^+ is kept constant so that the initial rates of the reaction depend only on the concentration of the nitrite ion.

Concentration of NO_2^- mol/L	Initial rate, mol L^{-1} sec $^{-1}$
2.4×10^{-3}	18×10^{-8}
4.9×10^{-3}	33×10^{-8}
10.0×10^{-3}	65×10^{-8}
24.9×10^{-3}	156×10^{-8}
50.7×10^{-3}	338×10^{-8}
94.0×10^{-3}	643×10^{-8}

(Answer: $n \approx 1$)

2. The decomposition of gaseous nitrogen pentoxide (N_2O_5) is a first-order reaction where the concentration of N_2O_5 remaining is measurable by its pressure (Moore, 1955). Given the data below, determine the rate constant of this reaction.

Pressure of N_2O_5, mm	Time, sec
348.4	0
247	600
185	1200
140	1800
105	2400
78	3000
58	3600
44	4200
33	4800
24	5400
18	6000
10	7200
5	8400
3	9600

3. Calculate the reaction half-life of the decomposition of N_2O_5 based on the data in Problem 2.

4. The rate constants for the formation of hydrogen iodide ($H_2 + I_2 \rightarrow 2\ HI$) vary with temperature in the following manner (Moore, 1955).

k, sec^{-1}	T, °C
316.2	473
100	441
31.6	412
14.1	394
3.16	362
2.20	352
0.316	320
0.251	315
0.0354	282

Determine the activation energy (E_a) of this reaction.

5. A major artesian aquifer in Florida has a vertical drop of 10 m over a map distance of 100 km. The hydraulic conductivity of this aquifer is $K = 10^{-1}$ cm/sec (Lerman, 1979).

(a) Calculate the rate of flow of water in this aquifer in centimeters per year.

(b) Calculate the residence time of water in this aquifer, that is, the time required for a "particle" of water to traverse the entire aquifer.

6. The chloride concentrations of Lake Vanda, Wright Valley, southern Victoria Land, Antarctica, varies with depth as shown below (Green and Canfield, 1984).

Depth, m	Cl$^-$, mmol/L
5	8.26
15	10.04
25	13.91
35	14.39
45	29.34
48	43.72
51	231.88
54	496.5
57	846.26
60	1390.7
65	1900.2
67	2268.0

Interpret these data as in Figure 15.5 to determine the age of the dilute layer, assuming that $D = 17.1 \times 10^{-6}$ cm^2/sec for Cl$^-$ at 18°C.

7. Calculate the time required for a calcite concretion to grow by dissolution of disseminated aragonite and diffusion of the resulting Ca^{2+} ions at 10°C, based on the following information. The radius of the concretion is 8.0 cm, the diffusion coefficient of Ca^{2+} is 5.4×10^{-6} cm^2/sec, the porosity of the sediment is 0.75, the volume fraction of aragonite in the sediment is 0.15, the difference in equilibrium concentration of Ca^{2+} in the pore water on the surfaces of calcite and aragonite crystals is 2.0×10^{-5} mol/kg, the molar volume of calcite is 36.934 cm^3/mol, and the volume fraction of the authigenic calcite is equal to the porosity of the sediment.

(Answer: 1.0×10^6 yr)

8. What is the width of the depleted zone around the concretion described in Problem 7? Obtain the molar volume of aragonite from Table 15.3.

(Answer: 30 cm)

9. Calculate the time required for a siderite concretion to grow and the width of the depleted zone around it given the following information: $r_c = 50$ cm, $\Delta c = 2.4 \times 10^{-7}$ mol/g, $\phi = f_p = 0.85$, D (Fe^{2+}) = 7.2×10^{-6} cm^2/sec, $V_d \approx V_p = 30$ cm^3/mol, $f_d = 0.30$.

(Answer: $t = 2.5 \times 10^6$ yr, $L = 120$ cm)

References

BERNER, R. A., 1971. *Principles of Chemical Sedimentology.* McGraw-Hill, New York, 240 pp.

BERNER, R. A., 1980. *Early Diagenesis. A Theoretical Approach.* Princeton University Press, Princeton, NJ, 241 pp.

BROWN, T. L., 1968. *General Chemistry,* 2nd ed. Merrill, Columbus, OH, 668 pp.

CRANK, J., 1975. *The Mathematics of Diffusion,* 2nd ed. Oxford University Press, London, 414 pp.

FRANK, F. C., 1950. Radial symmetric growth controlled by diffusion. *Proc. Roy. Soc. London,* **A-201**:48–54.

GREEN, W. J., and D. E. CANFIELD, 1984. Geochemistry of the Onyx River (Wright Valley, Antarctica) and its role in the chemical evolution of Lake Vanda. *Geochim. Cosmochim. Acta,* **48**(12):2457—2468.

HENDERSON, P., 1982. *Inorganic Geochemistry.* Pergamon Press, Oxford, England, 353 pp.

HOFMANN, A. W., B. J. GILETTI, H. S. YODER, JR., and R. A. YUND, 1974. *Geochemical Transport and Kinetics.* Carnegie Institution of Washington, Pub. 634, 353 pp.

LASAGA, A. C., 1979. The treatment of multicomponent diffusion and ion pairs in diagenetic fluxes. *Amer. J. Sci.,* **279**:324–346.

LASAGA, A. C., and R. J. KIRKPATRICK (Eds.), 1981. Kinetics of geochemical processes. *Rev. Mineral.,* **8,** Mineral. Soc. Amer., Washington, DC, 398 pp.

LERMAN, A., 1979. *Geochemical Processes. Water and Sediment Environments.* Wiley, New York, 481 pp.

LI, Y.-H., and S. GREGORY, 1974. Diffusion of ions in sea water and in deep-sea sediments. *Geochim. Cosmochim. Acta,* **38**: 703–714.

MOORE, W. J., 1955. *Physical Chemistry,* 2nd ed. Prentice-Hall, Upper Saddle River, NJ, 633 pp.

NIELSEN, A. E., 1961. Diffusion-controlled growth of a moving sphere. The kinetics of crystal growth in potassium perchlorate precipitation. *J. Phys. Chem.,* **65**:46–49.

WEAST, R. C., M. J. ASTLE, and W. H. BEYER (Eds.), 1986. *CRC Handbook of Chemistry and Physics.* CRC Press, Boca Raton, FL.

IV

ISOTOPE GEOCHEMISTRY
AND MIXING

The decay of long-lived radioactive isotopes of certain elements generates heat in the crust and mantle of the Earth and enables us to measure the ages of rocks and minerals. In addition, the fractionation of the stable isotopes of certain elements of low atomic number provides useful information about processes occurring in the hydrosphere, biosphere, and lithosphere. The isotope compositions of certain elements and the chemical compositions of water and rocks are also being altered by mixing of two, three, or more components. The chapters in Part IV deal with all of these subjects because they are important aspects of geochemistry.

16

Isotopic
Geochronometers

The discovery of the radioactivity of uranium salts by Henry Becquerel in 1896 started a revolution in physics. Important advances soon resulted from the pioneering work of Marie Sklodowska-Curie and her husband Pierre in Paris and by the studies of Ernest Rutherford and Frederick Soddy in Montreal. By 1902 radioactivity was recognized as a process associated with several elements, some of which were newly discovered. It was also recognized that the process results in the emission of alpha and beta particles and of gamma radiation. The rate of emission is proportional to the number of radioactive atoms remaining and therefore decreases exponentially with time. These are the roots of atomic physics that, in the course of the 20th century, led to nuclear physics and to the inquiry into the most primitive forms of matter (Romer, 1971).

The importance of radioactivity to geology became apparent after Curie and Laborde (1903) reported that radioactivity is an exothermic process. This fact invalidated Lord Kelvin's calculations of the age of the Earth, which were based on the assumption that the Earth had cooled from its original molten state and does not have a source of heat of its own (Burchfield, 1975). By 1905 Rutherford was using the amount of He (alpha particles) in U ore to calculate geologic ages of the order of 500 million years, which were far in excess of Lord Kelvin's estimates of the age of the Earth (Rutherford, 1906; Thomson, 1899). The rest, as they say, is history. In 1909 John Joly summarized his measurements of the radioactivity of rocks in a small book in which he also calculated the resulting rate of heat production. In 1911 Arthur Holmes published the first of his many papers on the dating of rocks based on the decay of U to Pb. In 1913 he published a book entitled *The Age of the Earth,* which contained the first geologic time scale based on age determinations of rocks and minerals.

These are the beginnings of a new discipline in the Earth Sciences called *Isotope Geology,* which seeks to use natural variations in the abundances of the isotopes of certain elements to study geological processes. The growth of this discipline has been closely linked to the design and subsequent improvement of analytical equipment, such as mass spectrometers and sensitive detectors of ionizing radiation. The entire subject has been presented in a textbook by Faure (1986). In addition, many earth science journals now publish papers based on or containing information derived from isotopic data. For this reason, we will briefly review the principles of isotope geology so that we can refer to them later as we take up different topics in geochemistry.

16.1 Decay Modes

The nucleosynthesis process in the ancestral stars (Chapter 2) produced more than 2200 nuclides, most of which were unstable and decomposed spontaneously in a short time. Only those nuclides that are stable or decay slowly have survived to the present. In addition, certain short-lived radioactive nuclides (*radionuclides*) exist naturally because:

1. They are the short-lived daughters of U and Th, both of which consist of long-lived radioactive isotopes.
2. They are produced by nuclear reactions between cosmic rays and atoms in the atmosphere, as well as in rocks exposed at the surface.

A third group of radionuclides now exists on the Earth because of nuclear reactions caused by humans. The reactions consist primarily of induced fission of ^{235}U in nuclear reactors, but also include the detonation of fission and fusion devices in the atmosphere and in the subsurface. The anthropogenic radioactive contaminants produced by these activities include a very large number of radionuclides: $^{3}_{1}H$ (tritium), $^{14}_{6}C$, $^{90}_{38}Sr$, $^{129}_{53}I$, $^{137}_{55}Cs$, and many others. The containment of this radioactive waste has become a serious problem in environmental geochemistry to be discussed in Chapter 24.

The large number of nuclides, whether stable or unstable, can be accommodated on a diagram called the *chart of the nuclides* (Chapter 2). This diagram is a plot of the atomic number (Z) versus the neutron number (N) and consists of squares, each of which is identified by the values of its N and Z coordinates. Figure 16.1 is such a chart for elements of low atomic number starting with hydrogen. The stable isotopes of each element are *shaded* and form a band that extends diagonally across the diagram. The short-lived radioactive isotopes are *unshaded* and form horizontal rows on either side of the stable isotopes. The ratio of neutrons to protons of the stable isotopes is approximately equal to one, but rises to three with increasing atomic number. The unstable isotopes of each element that lie to the left of the stable isotopes are *deficient* in neutrons, whereas those on the right side of the band of stable isotopes have an *excess* of neutrons. The existence of an excess or deficiency of neutrons in the nuclei of unstable atoms is reflected in the way they decay.

The phenomenon known to us as *radioactivity* consists of spontaneous nuclear transformations that change the number of protons and neutrons in the nucleus. The process continues until a stable daughter is produced whose values of Z and N place it within the band of stable nuclides. The transformations are accompanied by the emission of alpha and beta particles and of gamma rays from the nucleus. Some isotopes of the light elements in Figure 16.1 actually disintegrate into alpha particles or emit protons or neutrons. Most radionuclides undergo *beta decay,* a much smaller number are subject to *alpha decay,* and a few of the heaviest isotopes decay by *spontaneous nuclear fission.*

a. Beta Decay

Beta decay can take place in three different ways, depending on whether the radionuclide has an excess or deficiency of neutrons. Radionuclides having an excess of neutrons emit a negatively charged beta particle from the nucleus. This has the effect of reducing the number of neutrons and increasing the number of protons:

$$\text{neutron} \rightarrow \text{proton} + \beta^- + \bar{\nu} \quad (16.1)$$

where β^- is a negatively charged beta particle, which is identical to an extranuclear electron, and $\bar{\nu}$ is an antineutrino. The decay energy takes the form of kinetic energies of the beta particle and the antineutrino. The kinetic energy of the beta particle is complementary to that of the associated antineutrino, such that their sum is equal to the total decay energy. The product nuclide is the isotope of a different element because its atomic number differs from that of the parent. The nucleus of the product nuclide may have excess energy, which is emitted in the form of a gamma ray having a discrete energy. In some cases, the product nuclide may have two or more different excited states, which enables it to emit several gamma rays. If the product nuclide is unstable, it decays to form a second daughter by emitting another β^- particle, and so on, until at last a stable nuclide is formed. Each of the radionuclides in such a decay series emits β^- particles and antineutrinos with complementary kinetic energies equal to the total decay energy. In addition, each transition from one successive daughter to the next may be

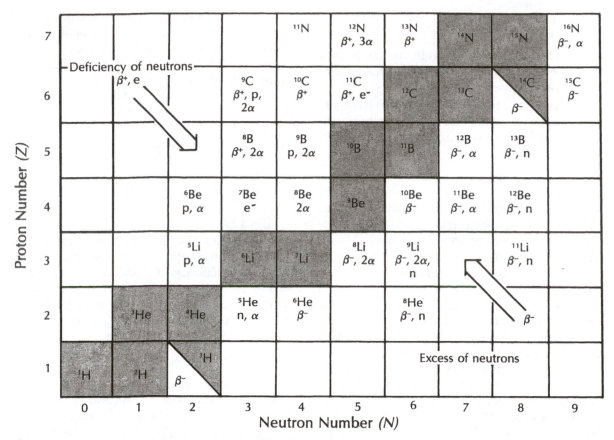

Figure 16.1 A segment of the chart of the nuclides. The stable isotopes of the elements are shaded and form a diagonal band across the diagram. The unstable isotopes are unshaded. Tritium (^{3}H) and ^{14}C are unstable but occur naturally because they are produced continuously in the atmosphere by cosmic rays. Nuclides having an excess of neutrons decay by β^- emission, whereas those with a deficiency of neutrons decay by β^+ emission or electron (e) capture. Some of the nuclides disintegrate into one or two alpha particles (α), neutrons (n), and protons (p).

accompanied by the emission of gamma rays. The energies of the β^- particles and the gamma rays are dissipated by collisions with neighboring atoms and electrons and are thereby converted into thermal energy. The antineutrinos interact *sparingly* with matter and carry their kinetic energies with them into the universe. Their ultimate fate is still unknown.

Radionuclides located on the left side of the band of stable nuclides in Figure 16.1 are deficient in neutrons and decay by position emission (β^+) or by electron capture. These forms of beta decay increase the number of neutrons and decrease the

number of protons. In positron decay a proton is converted into a neutron by the transformation:

$$\text{proton} \rightarrow \text{neutron} + \beta^+ + \nu \quad (16.2)$$

The same result is achieved when the nucleus captures an extranuclear electron:

$$\text{proton} + e^- \rightarrow \text{neutron} + \nu \quad (16.3)$$

The positrons emitted by the nucleus are slowed by collisions until they come to rest next to an electron. Both are then annihilated in the matter–antimatter reaction, which converts their rest masses into two gamma rays with a combined energy of 1.02 million electron volts (MeV). The neutrinos produced during positron decay have a range of kinetic energies complementary to those of the positrons, whereas those emitted during electron-capture decay are monoenergetic. Both escape into the universe and carry their energy with them.

A few radionuclides can decay by all three kinds of beta decay and, therefore, give rise to two different daughter products. For example, $^{40}_{19}K$ decays by positron decay and electron capture to stable $^{40}_{18}Ar$ and also decays by β^- (negatron) decay to stable $^{40}_{20}Ca$. Such *branched decay* also occurs among the unstable daughters of U and Th, some of which can decay both by β^- emission and by alpha decay.

b. Alpha Decay

Since the alpha particle is composed of two protons and two neutrons, it is identical to the nucleus of ^4_2He. The emission of an alpha particle lowers both the proton number and the neutron number of the product nucleus by two:

$$^A_Z P \rightarrow ^{A-4}_{Z-2}D + ^4_2He + Q \qquad (16.4)$$

where P is the parent nuclide, D is the daughter, and Q is the total decay energy. Alpha decay may be followed by emission of gamma rays, which reduces the product nucleus to the ground state. Some of the kinetic energy of the alpha particles is imparted to the product nucleus as recoil energy, which may displace the nucleus from its position in the crystal lattice. The alpha particles are slowed by collisions and their energies are thereby converted into heat. These collisions also cause radiation damage in crystals containing alpha emitters.

Alpha decay is available to a large group of radionuclides having atomic numbers >58 and to a few nuclides of low atomic number, including ^5_2He, ^5_3Li, ^6_4Be, and others shown in Figure 16.1. The naturally occurring radioactive isotopes of U ($^{235}_{92}U$ and $^{238}_{92}U$) and Th ($^{232}_{90}Th$) decay to stable isotopes of Pb through series of short-lived radioactive daughters that form chains linked by alpha and beta decays. The decay chain of $^{238}_{92}U$ is illus-

trated in Figure 16.2. It contains 18 intermediate nuclides, all of which ultimately form stable $^{206}_{82}Pb$, regardless of the sequence of decay events. The isotopes $^{235}_{92}U$ and $^{232}_{90}Th$ each give rise to decay chains ending with $^{207}_{82}Pb$ and $^{208}_{82}Pb$, respectively.

All of the intermediate daughters of U and Th are short-lived radioactive nuclides that occur in the solar system only because they are continuously produced by decay of their long-lived parents. Some chemical elements including Po, At, Rn, Ra, and Pa occur only because their isotopes are members of these decay chains. The occurrence of Rn can be a hazard when it is inhaled since the alpha particles it emits can damage lung tissue.

c. Nuclear Fission

Some of the isotopes of U and of the transuranium elements decay by spontaneous fission. In this process the nucleus breaks into two fragments of unequal weight, which form the nuclei of two fission-product nuclides. These nuclides have an excess of neutrons, and therefore decay by emitting β^- particles and gamma rays until a stable nuclide is produced. Spontaneous fission liberates large amounts of energy and results in the emission of neutrons, alpha particles, and other fragments of nuclear material. The large amount of energy liberated by the spontaneous fission of $^{238}_{92}U$ in U-bearing minerals causes damage in the form of tracks about 10 μm in length. The tracks can be made visible by etching, so that they can be counted with a microscope, and are used to determine the ages and thermal histories of U-bearing minerals. The resulting *fission-track method* of dating has been described by Faure (1986).

Nuclear fission can also be *induced* when the nuclei of certain isotopes of U and Pu (plutonium) absorb neutrons or other nuclear particles. Induced fission of $^{235}_{92}U$ results in the release of neutrons, which can cause additional nuclei of $^{235}_{92}U$ to fission in a self-sustaining *chain reaction* that can liberate very large amounts of energy. In nuclear fission *reactors* such a chain reaction is maintained under controlled conditions and the heat generated by the reaction is used to generate electricity by means of conventional generators.

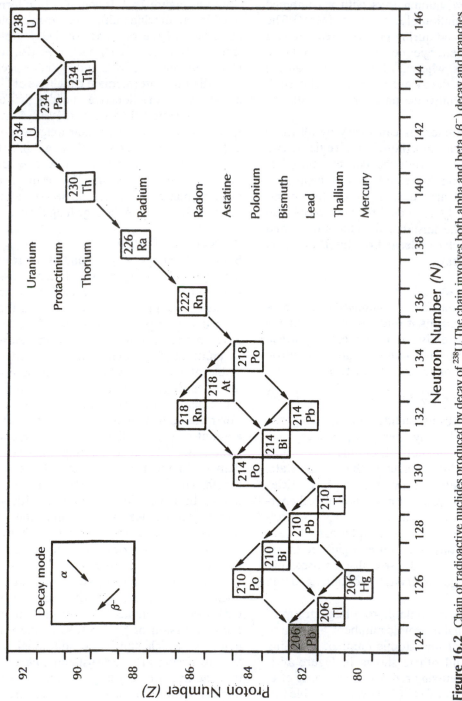

Figure 16.2 Chain of radioactive nuclides produced by decay of ^{238}U. The chain involves both alpha and beta (β^-) decay and branches repeatedly. However, all possible paths lead to stable ^{206}Pb. Note that this decay series includes 11 different chemical elements including Ra and Rn, which are important in environmental geochemistry.

The fission products accumulate in the *fuel rods,* which must be replaced when most of the $^{235}_{92}U$ has been consumed. The spent fuel rods are extremely radioactive because of the emission of β^- particles and gamma rays by the fission-product nuclides and their radioactive daughters. In addition, the spent fuel rods contain isotopes of Np (neptunium) and Pu (plutonium), that form from $^{238}_{92}U$ by neutron capture and subsequent beta decay leading to the formation of $^{239}_{94}Pu$ and $^{240}_{94}Pu$. These isotopes are *fissionable* and are used in the manufacture of nuclear weapons. Plutonium is also poisonous, and its dispersal creates an environmental health hazard (Chapter 24).

16.2   Law of Radioactivity

The *law of radioactivity* was discovered by Rutherford and Soddy (1902) and is expressed by the statement that the rate of decay of a radioactive nuclide is proportional to the number of atoms of that nuclide remaining at any time. If N is the *number* of atoms remaining, then:

$$-\frac{dN}{dt} = \lambda N \qquad (16.5)$$

where λ is a proportionality constant known as the *decay constant* and the minus sign indicates that the rate of decay decreases with time. By integrating, we obtain:

$$\ln N = -\lambda t + C \qquad (16.6)$$

where ln is the logarithm to the base e and C is the constant of integration. When $t = 0$, $C = \ln N_0$, where N_0 is the number of radioactive atoms at $t = 0$. Therefore, the integrated form of the equation is:

$$\ln N = -\lambda t + \ln N_0 \qquad (16.7)$$

$$\ln \left(\frac{N}{N_0}\right) = -\lambda t \qquad (16.8)$$

$$N = N_0 e^{-\lambda t} \qquad (16.9)$$

It is convenient to define the *half-life* ($T_{1/2}$) as the time required for one half of a given number of radioactive atoms to decay. Therefore, if $t = T_{1/2}$,

then $N = N_0/2$. Substituting into equation 16.9, we obtain:

$$\frac{N_0}{2} = N_0 e^{-\lambda T_{1/2}} \qquad (16.10)$$

$$\ln 1 - \ln 2 = -\lambda T_{1/2} \qquad (16.11)$$

$$\ln 2 = \lambda T_{1/2} \qquad (16.12)$$

$$T_{1/2} = \frac{\ln 2}{\lambda} \qquad (16.13)$$

Every radionuclide has a characteristic decay constant that must be determined experimentally. This can be done by measuring its rate of decay at intervals over a period of time and by plotting the resulting *decay curve.* According to equation 16.5, the rate of disintegration (A) is equal to λN. Therefore, we can restate the law of radioactivity in terms of *disintegration rates* by multiplying both sides of equation 16.9 by λ:

$$\lambda N = \lambda N_0 e^{-\lambda t} \qquad (16.14)$$

Since $A = \lambda N$, we obtain:

$$A = A_0 e^{-\lambda t} \qquad (16.15)$$

Taking logarithms to the base e gives:

$$\ln A = \ln A_0 - \lambda t \qquad (16.16)$$

Equation 16.16 is the equation of a straight line in the slope–intercept form where $\ln A_0$ is the intercept on the ordinate and $-\lambda$ is the slope. Evidently, the decay constant of a radionuclide can be determined from the slope of the straight line fitted to measurements of the disintegration rate at known intervals of time. The decay curve can also be plotted in terms of the number of radionuclides remaining (N) versus time measured in multiples of the half-life. Both forms of the decay curve are illustrated in Figure 16.3.

If the decay of a radionuclide gives rise to a stable daughter nuclide, we can say that the number of radiogenic daughter nuclides (D^*) that have accumulated at any time is equal to the number of parent nuclides that have decayed. Therefore,

$$D^* = N_0 - N \qquad (16.17)$$

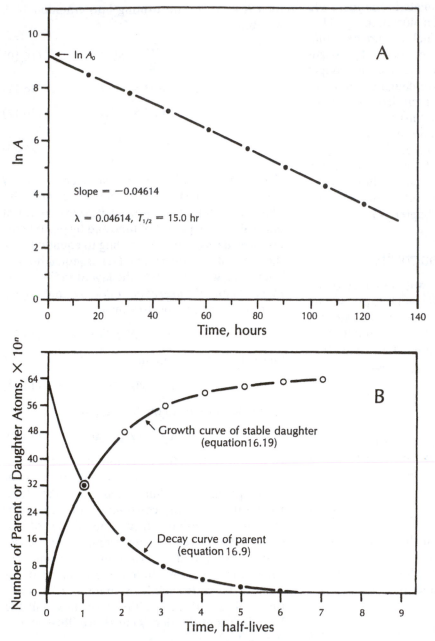

Figure 16.3 **A.** Decay of a radionuclide ($^{24}_{11}$Na) in ln-normal coordinates according to equation 16.16 where A is the disintegration rate and time t is measured in hours. The numerical value of the slope is equal to the decay constant λ from which the half-life $T_{1/2}$ can be calculated by means of equation 16.13.

B. Decay curve of a radionuclide and growth curve of its stable daughter in linear coordinates, where time is expressed in multiples of the half-life. Both curves can be constructed by using the definition of the half-life. Note that, in general, the number of parent atoms is reduced by a factor of 2^{-n} after n half-lives have elapsed.

where N_0 is the number of parent atoms at $t = 0$ and N is the number of parent atoms remaining at any time t. Substituting equation 16.9 yields:

$$D^* = N_0 - N_0 e^{-\lambda t} \qquad (16.18)$$

$$D^* = N_0(1 - e^{-\lambda t}) \qquad (16.19)$$

Equation 16.19 is the *growth curve* in Figure 16.3B, which relates the number of *radiogenic* daughters to the time that has elapsed and to the number of parent atoms that were present initially when t was equal to zero. Notice that the decay curve of the parent and the growth curve of the stable daughter are complementary. This condition arises from equation 16.17, which is based on the assumption that the numbers of parent and daughter atoms change *only* as a result of decay and that no daughter or parent atoms are gained or lost by any other process.

The relationship between the radiogenic daughter and remaining parent atoms can also be expressed by replacing N_0 in equation 16.17 by:

$$N_0 = Ne^{\lambda t} \qquad (16.20)$$

which we obtain from equation 16.9. Hence:

$$D^* = Ne^{\lambda t} - N \qquad (16.21)$$

$$D^* = N(e^{\lambda t} - 1) \qquad (16.22)$$

This is a more useful relationship than equation 16.19 because it relates the number of radiogenic daughter atoms (D^*) to the number of parent atoms (N) remaining, where both are *measurable* quantities. Therefore, equation 16.22 is the *geochronometry equation* used to determine the age of a mineral or rock that contains a long-lived naturally occurring radionuclide. The total number of radiogenic daughters (D) per unit weight of rock or mineral is:

$$D = D_0 + D^* \qquad (16.23)$$

where D_0 is the number of daughters that entered a unit weight of mineral or rock at the time of its formation and D^* is the number of radiogenic daughters that formed by decay of the parent in the same unit weight of mineral or rock. A complete description of the number of daughter atoms is obtained by combining equations 16.22 and 16.23:

$$D = D_0 + N(e^{\lambda t} - 1) \qquad (16.24)$$

Notice again that D and N are measurable quantities expressed in terms of numbers of atoms per unit weight of mineral or rock, D_0 is the number of daughters present *initially*, and t is the time elapsed since a sample of mineral or rock came

into existence. In other words, t is the age of the mineral or rock provided D and N changed only because of decay of the parent to the radiogenic daughter D^*.

In order to date a rock or mineral by means of equation 16.24, the measured concentrations of the parent and daughter elements must be recalculated into numbers of atoms of the appropriate isotopes by converting weights to numbers of moles and by using Avogadro's number and the abundances of the appropriate isotopes. These matters were discussed in Sections 6.3 and 6.4. However, in order to solve equation 16.24 for t we must also know the decay constant λ and the value of D_0. The decay constants of the radionuclides have been measured and accurate values are available for most of them. However, how do we determine the value of D_0?

The number of daughter atoms present initially is determined in one of two ways. The least desirable but simplest method is to *assume* a value whose magnitude depends on the particular decay scheme and on the circumstances. For example, in the K–Ar method of dating we may assume that $D_0 = 0$ because Ar is a noble gas that is not likely to enter K-bearing minerals crystallizing from a magma. In the Rb–Sr method we may select a value for D_0 based on the assumption that the Sr originated from the mantle of the Earth and had an isotopic composition similar to that of Sr in midocean ridge basalt (MORB). The error caused by this assumption depends on the magnitude of the term $N(e^{\lambda t} - 1)$ in equation 16.24. If that term is much larger than D_0, then the calculated date is not sensitive to the choice of D_0. Therefore, as a rule, *old Rb-rich* minerals such as muscovite or biotite can be reliably dated using an assumed value of D_0.

The preferred procedure is based on the analysis of several samples from the same body of rock such that all samples have the same age (t) and the same initial daughter content (D_0). For such a suite of samples D_0 and t are constants and equation 16.24 is a straight line in coordinates of D and N. This straight line is called an *isochron* beçause it is the locus of all points representing mineral or rock samples that have the same age (t) and the same initial abundance of the daughter nuclide (D_0). The slope and intercept of the isochron are determined by the least-squares method using the analytical data from at least three *cogenetic* samples. The slope (m) of the isochron is related to the common age of the suite of cogenetic samples by:

$$m = e^{\lambda t} - 1 \qquad (16.25)$$

and the age t is:

$$t = \frac{1}{\lambda} \ln(m + 1) \qquad (16.26)$$

The *date* so calculated is the *age* of the rocks, provided the assumptions implied by the calculation are actually satisfied:

1. The number of parent and daughter atoms per unit weight of mineral or rock changed only by decay of the parent to the daughter.

2. The isotopic composition of the parent element was not altered by fractionation of isotopes at the time of formation of the mineral.

3. The decay constant of the parent is known accurately.

4. The isochron is not a mixing line.

5. The analytical data are accurate.

The units of time used in geochronology are years (a), thousands of years (ka), millions of years (Ma), or billions of years (Ga). In the ideal case described by these assumptions, the isochron technique yields not only the age of the samples but also indicates the abundance of the daughter isotope at the time of their formation (D_0). This value is of great interest because it contains information about the previous history of the daughter element before it was incorporated into the rocks in which it now resides (Dickin, 1995).

16.3 Methods of Dating

The Earth contains several naturally occurring long-lived radioactive isotopes, which are being used for dating by the methods outlined above. The parents and their daughters are listed in Table 16.1,

Table 16.1 Physical Constants Required for Dating Based on Naturally Occurring Long-Lived Radioactive Nuclides

Parent	Daughter	$T_{1/2}$,[a] a	λ,[a] a^{-1}	Isotopic abundance of parent, %
$^{40}_{19}K$	$^{40}_{18}Ar$	1.19×10^{10}	0.581×10^{-10}	0.01167
$^{40}_{19}K$	$^{40}_{20}Ca$	1.40×10^{10}	4.962×10^{-10}	0.01167
$^{40}_{19}K$	$^{40}Ar + {}^{40}Ca$	1.250×10^{9}	5.543×10^{-10}	0.01167
$^{87}_{37}Rb$	$^{87}_{38}Sr$	48.8×10^{9}	1.42×10^{-11}	27.8346
$^{147}_{62}Sm$	$^{143}_{60}Nd$	1.06×10^{11}	0.654×10^{-12}	15.0
$^{138}_{57}La$	$^{138}_{58}Ce$	2.70×10^{11}	2.57×10^{-12}	0.09
$^{138}_{57}La$	$^{138}_{56}Ba$	1.51×10^{11}	4.59×10^{-12}	0.09
$^{138}_{57}La$	$^{138}_{58}Ce + {}^{138}_{56}Ba$	96.8×10^{9}	7.16×10^{-12}	0.09
$^{176}_{71}Lu$	$^{176}_{72}Hf$	35.7×10^{9}	1.94×10^{-11}	2.600
$^{187}_{75}Re$	$^{187}_{76}Os$	45.6×10^{9}	1.52×10^{-11}	62.602
$^{238}_{92}U$	$^{206}_{82}Pb$	4.468×10^{9}	1.55125×10^{-10}	99.2743
$^{235}_{92}U$	$^{207}_{82}Pb$	0.7038×10^{9}	9.8485×10^{-10}	0.7200
$^{232}_{90}Th$	$^{208}_{82}Pb$	14.010×10^{9}	4.9475×10^{-11}	100.00

[a] a = year, derived from Latin word *annum*.

which also contains information about their decay constants, half-lives, and relevant isotopic abundances (Steiger and Jäger, 1977). Each method is applicable to those minerals and rocks that contain the parent element in significant concentrations and that retain the parent and daughter isotopes quantitatively. The applicability of the different dating methods is therefore governed, at least in part, by geochemical considerations.

a. The K–Ar Method

The positron and electron capture decay of radioactive $^{40}_{19}K$ to stable $^{40}_{18}Ar$ in K-bearing minerals causes the abundance of the radiogenic daughter to increase as a function of time in accordance with equation 16.24:

$$^{40}Ar = {}^{40}Ar_0 + \left(\frac{\lambda e}{\lambda}\right) {}^{40}K(e^{\lambda t} - 1) \quad (16.27)$$

where ^{40}Ar and ^{40}K are the numbers of atoms of these isotopes per unit sample weight, λ_e is the decay constant for electron capture and positron decay ($\lambda_e = 0.581 \times 10^{-10}$ a^{-1}), and λ is the total decay constant of ^{40}K ($\lambda = 5.543 \times 10^{-10}$ a^{-1}). Note that $^{40}Ar_0 = 0$ and that the factor λ_e/λ is needed to identify the fraction of ^{40}K atoms that decay to ^{40}Ar.

The decay of $^{40}_{19}K$ to ^{40}Ca by β^- decay having a decay constant λ_β is expressed by a complementary equation:

$$^{40}Ca = {}^{40}Ca_0 + \left(\frac{\lambda_\beta}{\lambda}\right) {}^{40}K(e^{\lambda t} - 1) \quad (16.28)$$

This method of dating is rarely used because ^{40}Ca is the most abundant isotope of Ca (96.94%) and because its isotopic composition is difficult to measure by mass spectrometry.

The K–Ar method is widely used to date K-bearing minerals that retain ^{40}Ar quantitatively such as muscovite, biotite, and hornblende in igneous and metamorphic rocks. The dates calculated from equation 16.27 represent the time elapsed since the mineral being dated cooled through its *blocking temperature* below which ^{40}Ar is quantitatively retained. Therefore, K–Ar dates of slowly cooled minerals *underestimate* their ages. The K–Ar method has also been used

to date authigenic glauconite and illite in sedimentary rocks (Section 13.6) as well as whole-rock samples of glassy basalt. However, such applications require very careful selection of samples and are certainly not routine.

The $^{40}Ar/^{39}Ar$ method is a variation of the K–Ar method in which stable $^{39}_{19}K$ is partially converted to unstable but long-lived $^{39}_{40}Ar$ ($T_{1/2} = 269$ a) by irradiating the sample with neutrons in a nuclear reactor. The sample is then heated stepwise and the $^{40}Ar/^{39}Ar$ ratio of Ar, released at each step, is used to calculate a date. The resulting spectrum of dates provides information about the cooling age of the sample as well as about the distribution of K and Ar within the mineral grains (Dalrymple and Lanphere, 1969).

b. The Rb–Sr Method

The β^- decay of naturally occurring $^{87}_{37}Rb$ to stable $^{87}_{38}Sr$ is the basis for the Rb–Sr method of dating. The geochronometry equation is written in terms of the isotopic ratio $^{87}Sr/^{86}Sr$, which is conveniently measured by mass spectrometry:

$$\frac{^{87}Sr}{^{86}Sr} = \left(\frac{^{87}Sr}{^{86}Sr}\right)_0 + \frac{^{87}Rb}{^{86}Sr}(e^{\lambda t} - 1) \quad (16.29)$$

The Rb–Sr method is used to date Rb-rich minerals such as muscovite, biotite, and K-feldspar in igneous and metamorphic rocks based on assumed values of the initial $^{87}Sr/^{86}Sr$ ratio, that is, 0.704 ± 0.001, which is representative of Sr in the source regions of oceanic basalt. The resulting dates may indicate the time elapsed since the minerals cooled through their respective blocking temperatures for radiogenic ^{87}Sr, which lie *above* those for radiogenic ^{40}Ar. Therefore, Rb–Sr dates of slowly cooled minerals tend to be *older* than their K–Ar dates and are better estimates of their crystallization ages.

The Rb–Sr method is most often used to date suites of cogenetic igneous and high-grade metamorphic rocks by means of the *isochron* technique illustrated in Figure 16.4. Whole-rock samples of igneous and metamorphic rocks may retain Rb and Sr quantitatively even though radiogenic ^{87}Sr may have been redistributed among their constituent minerals by diffusion during

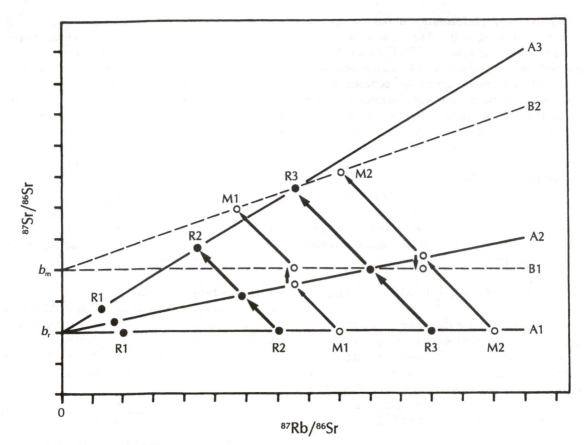

Figure 16.4 Rb–Sr isochron diagram for three whole-rock samples of igneous rocks (R1, R2, and R3) and two minerals (M1 and M2) of rock R3. The two axes in this diagram have equal scale to enhance the clarity of this illustration. Initially, all rock and mineral samples lie on isochron A1, which has a slope of zero and an intercept equal to b_r. As time passes, all rocks and minerals on A1 move along straight lines indicated by arrows and, at some time after crystallization, they form isochron A2. The slope of A2 is a measure of the amount of time that has passed since crystallization, but the intercept is still at b_r. The isotopic compositions of Sr in minerals M1 and M2 are changed at this time by diffusion of radiogenic ^{87}Sr caused by thermal metamorphism so that both minerals and R3 have the same ^{87}Sr/^{86}Sr ratio equal to b_m at the end of this episode. As a result, the rock R3 and its constituent minerals M1 and M2 now lie on isochron B1, but rock R3 also lies on isochron A2. After the passage of additional time, the whole-rock samples form isochron A3 whose slope is a function of the time elapsed since crystallization and whose intercept is still equal to b_r. However, minerals M1 and M2 now form isochron B2 whose slope is less than that of isochron A3 and measures the time elapsed since their isotopic reequilibration. The intercept of isochron B2 at b_m is greater than b_r because it is the ^{87}Sr/^{86}Sr ratio of R3 and its constituent minerals after isotopic reequilibration. This hypothetical example suggests that whole-rock Rb–Sr isochrons can yield the crystallization age of igneous and high-grade metamorphic rocks, whereas their constituent minerals form isochrons that indicate the time elapsed since the last episode of isotopic reequilibration.

thermal metamorphism. The date derived from the slope of a whole-rock isochron is therefore regarded as a reliable indicator of the crystallization age of igneous and metamorphic rocks. The intercept is the initial $^{87}Sr/^{86}Sr$ ratio, which reflects the geochemical history of the Sr before it was incorporated into the rocks that form the isochron. The initial $^{87}Sr/^{86}Sr$ is, therefore, an important parameter for the study of the petrogenesis of igneous rocks and of the geochemical differentiation of the continental crust from the underlying mantle (Faure and Powell, 1972).

c. The Sm–Nd, La–Ce, and Lu–Hf Methods

These rare earth geochronometers are useful for dating Ca-rich igneous and high-grade metamorphic rocks of Precambrian age. The relevant equations (equations 16.30–16.32), follow directly from equation 16.24 and are written in terms of isotopic ratios:

$$\frac{^{143}Nd}{^{144}Nd} = \left(\frac{^{143}Nd}{^{144}Nd}\right)_0 + \frac{^{147}Sm}{^{144}Nd}(e^{\lambda t} - 1) \qquad (16.30)$$

$$\frac{^{138}Ce}{^{140}Ce} = \left(\frac{^{138}Ce}{^{140}Ce}\right)_0 + \left(\frac{\lambda_\beta}{\lambda}\right)\frac{^{138}La}{^{140}Ce}(e^{\lambda t} - 1) \quad (16.31)$$

$$\frac{^{176}Hf}{^{177}Hf} = \left(\frac{^{176}Hf}{^{177}Hf}\right)_0 + \frac{^{176}Lu}{^{177}Hf}(e^{\lambda t} - 1) \qquad (16.32)$$

All three methods are used to construct isochrons by analysis of suites of cogenetic mafic or even ultramafic rocks. These geochronometers can be used to date rocks that are not datable by the K–Ar and Rb–Sr methods. In addition, the initial isotope ratios are important for studies of the petrogenesis of igneous rocks and of crustal evolution.

d. The Re–Os Method

The Re–Os method has become easier to use because of improvements in the isotopic analysis of Os (Creaser et al., 1991). The average concentrations of Re and Os in rocks are very low and amount to only fractions of parts per billion. Both elements are siderophile and are strongly concentrated in the Fe–Ni phases of meteorites. In addition, Re occurs in *molybdenite* (MoS_2), whereas

Os is concentrated with other Pt-group metals in the naturally occurring alloy *osmiridium*.

The geochronometry equation takes the form:

$$\frac{^{187}Os}{^{186}Os} = \left(\frac{^{187}Os}{^{186}Os}\right)_0 + \frac{^{187}Re}{^{186}Os}(e^{\lambda t} - 1) \,(16.33)$$

It is used to construct isochrons for dating of meteorites, ultramafic rocks, and molybdenite-bearing ore samples. An advantage of the Re–Os decay scheme is that crustal rocks have high Re/Os ratios. For example, the average Re/Os ratio of tholeiite basalts is 28 and that of granites is about 10. This means that crustal rocks of Precambrian age have high $^{187}Os/^{186}Os$ ratios, which is useful to igneous petrologists who study the contamination of basalt magma by assimilation of crustal rocks.

e. The U, Th–Pb Methods

The decay of ^{238}U, ^{235}U, and ^{232}Th runs through separate series of intermediate radioactive daughters to stable isotopes of Pb. These decay chains can achieve a condition of secular equilibrium when the rates of decay of the parents are imposed on all of their intermediate daughters. Under these conditions the geochronometry equations are:

$$\frac{^{206}Pb}{^{204}Pb} = \left(\frac{^{206}Pb}{^{204}Pb}\right)_0 + \frac{^{238}U}{^{204}Pb}(e^{\lambda_1 t} - 1) \qquad (16.34)$$

$$\frac{^{207}Pb}{^{204}Pb} = \left(\frac{^{207}Pb}{^{204}Pb}\right)_0 + \frac{^{235}U}{^{204}Pb}(e^{\lambda_2 t} - 1) \qquad (16.35)$$

$$\frac{^{208}Pb}{^{204}Pb} = \left(\frac{^{208}Pb}{^{204}Pb}\right)_0 + \frac{^{232}Th}{^{204}Pb}(e^{\lambda_3 t} - 1) \qquad (16.36)$$

The three decay schemes can yield three independent age determinations of minerals or rocks containing both U and Th, provided they did not gain or lose U, Th, or Pb after the date of their formation. In most cases, however, the dates indicated by these geochronometers do *not* agree with each other and are said to be *discordant*. Most commonly, the discordance results from loss of Pb or of intermediate daughters. The loss of daughter nuclides from U, Th-bearing minerals is caused in part by the radiation damage resulting from the energy released during alpha decay.

In such cases, a fourth date can be calculated by combining equations 16.34 and 16.35 for the U–Pb geochronometers:

$$\frac{\frac{^{207}Pb}{^{204}Pb} - \left(\frac{^{207}Pb}{^{204}Pb}\right)_0}{\frac{^{206}Pb}{^{204}Pb} - \left(\frac{^{206}Pb}{^{204}Pb}\right)_0} = \frac{^{235}U}{^{238}U}\left(\frac{e^{\lambda_2 t} - 1}{e^{\lambda_1 t} - 1}\right) \quad (16.37)$$

This equation represents a family of straight lines in coordinates of $^{207}Pb/^{204}Pb$ (y) and $^{206}Pb/^{204}Pb$ (x) that intersect in a point whose coordinates are the initial $^{207}Pb/^{204}Pb$ and $^{206}Pb/^{204}Pb$ ratios. Therefore, the left side of equation 16.37 is the slope of these straight lines, which are known as *Pb–Pb isochrons*:

$$\frac{\frac{^{207}Pb}{^{204}Pb} - \left(\frac{^{207}Pb}{^{204}Pb}\right)_0}{\frac{^{206}Pb}{^{204}Pb} - \left(\frac{^{206}Pb}{^{204}Pb}\right)_0} = \left(\frac{^{207}Pb}{^{206}Pb}\right)^* = \text{slope} \quad (16.38)$$

where the asterisk (*) indicates that the slope of the Pb–Pb isochrons is the ratio of the *radiogenic* Pb isotopes. Equation 16.37 relates the ratio of the radiogenic Pb isotopes to the age t of the rock or mineral based *only* on the isotopic composition of Pb. However, the equation is transcendental and cannot be solved for t by ordinary algebraic methods. Solutions can be obtained most conveniently by interpolating in a table of calculated $(^{207}Pb/^{206}Pb)^*$ ratios and t (Faure, 1986, Table 18.3, p. 290). In cases where equations 16.34–16.36 yield discordant dates, the *207/206 date* derived from equation 16.37 is commonly older than the U–Pb and Th–Pb dates because the Pb–Pb date is not affected by recent Pb loss that does not change the isotopic composition of the remaining Pb.

Although the number of known U- and Th-bearing minerals is very large, only a few are suitable for dating by the U, Th–Pb methods because they retain radiogenic Pb and because they are *common*. Zircon ($ZrSiO_4$) meets these criteria better than all other minerals, although monazite (rare earth phosphate) and apatite (Ca phosphate) are used occasionally. Zircon permits the replacement of Zr^{4+} by U^{4+} and Th^{4+} in accordance with Goldschmidt's first rule (Table 7.3 and

Section 8.1) because their radii in sixfold coordinated sites are similar ($Zr^{4+} = 0.80$ Å, $U^{4+} = 0.97$ Å, and $Th^{4+} = 1.08$ Å). However, zircon strongly *excludes* Pb, which can form both Pb^{2+} (1.26 Å) and Pb^{4+} (0.86 Å) ions. Since Pb is excluded from zircon, it must occur as Pb^{2+}, which cannot replace Zr^{4+} because it is too large and its charge differs too much from that of Zr^{4+}. Therefore, Goldschmidt's rules help to explain why zircon crystals exclude Pb but accept U and Th, thereby making zircon one of the most useful geological chronometers.

However, not even zircon is perfectly retentive and, in most cases, U, Th–Pb dates of zircons are *discordant*. This problem was solved brilliantly by Wetherill (1956) by means of the *concordia diagram*, which is constructed by calculating the $^{206}Pb^*/^{238}U$ and $^{207}Pb^*/^{235}U$ ratios of U–Pb systems that yield concordant dates. From equation 16.34 we have:

$$\frac{\frac{^{206}Pb}{^{204}Pb} - \left(\frac{^{206}Pb}{^{204}Pb}\right)_0}{\frac{^{238}U}{^{204}Pb}} = \frac{^{206}Pb^*}{^{238}U} = e^{\lambda_1 t} - 1 \quad (16.39)$$

Similarly,

$$\frac{^{207}Pb^*}{^{235}U} = e^{\lambda_2 t} - 1 \quad (16.40)$$

Compatible sets of $^{206}Pb^*/^{238}U$ and $^{207}Pb^*/^{235}U$ ratios, calculated from equations 16.39 and 16.40 for selected values of t, are the coordinates of points that form the *concordia curve*. In other words, the concordia curve is the locus of all points representing U–Pb systems that yield concordant U–Pb dates. Any U-bearing rock or mineral that does *not* plot on concordia yields *discordant* U–Pb dates.

Zircon crystals or other U–Pb systems that remain closed to U and Pb, and to the intermediate daughters, move along the concordia curve in Figure 16.5 as they age. When they lose Pb as a result of an episode of thermal metamorphism, they leave concordia and move along a straight line toward the origin. When all accumulated radiogenic Pb is lost, the $^{206}Pb/^{238}U$ and

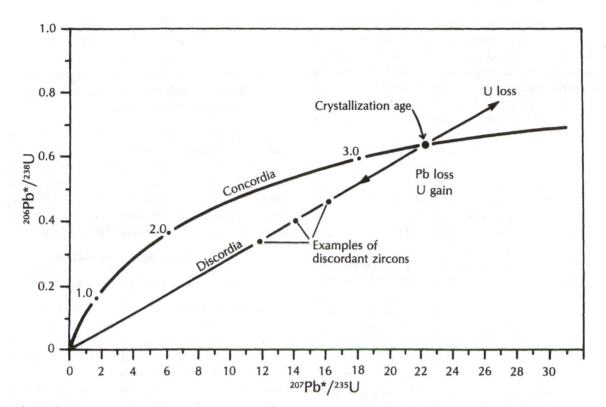

Figure 16.5 Concordia diagram used in dating minerals such as zircon, monazite, apatite, and others by the U–Pb method. The coordinates of points on concordia are generated by solving equations 16.38 and 16.39 for selected values of t. During episodic Pb loss or U gain, minerals are displaced from concordia and move along the discordia line toward the origin. Zircons, or other U-bearing minerals, which yield discordant U–Pb dates by solution of equations 16.33 and 16.34, are represented by points that define discordia. The upper point of intersection of discordia with concordia is the crystallization age of the zircons that define discordia. Loss of U causes U–Pb minerals to be represented by points on discordia *above* concordia. The effect of Pb gain is not predictable unless the isotopic composition of the added Pb is known. U–Th–Pb concordias are possible but have been rarely used because differences in the geochemical properties of U and Th may produce discordias that do not pass through the origin.

$^{207}Pb/^{235}U$ ratios approach zero along a chord directed toward the origin. When only a fraction of Pb is lost from zircons, they are represented by points on the straight line. Such U–Pb systems yield discordant dates, which is why the chord is called *discordia.*

When zircon crystals from a large sample of granitic rock are sorted according to their color, magnetic properties, or shape prior to analysis, they commonly form points that define a straight line on the concordia diagram. The upper point of intersection of this discordia line with concordia

yields the time elapsed since crystallization because, if the zircons had not lost Pb, they would have resided on concordia at that point. The lower point of intersection *may* be the time elapsed since the end of episodic Pb loss, because this is where zircon crystals would now be located on concordia if they had lost all of the accumulated Pb during the episode of thermal metamorphism (see Section 19.5).

The significance of the lower point of intersection of discordia with concordia is uncertain because Pb loss may be *continuous* rather than *episodic,* or it may be caused by the relaxation of lithostatic pressure as a result of erosion of the overlying rocks following crustal uplift. In addition, zircon crystals and other U-bearing minerals may suffer not only loss of Pb but may also lose or gain U. Gain of U has the same effect on U–Pb systems as loss of Pb, whereas loss of U causes them to move *away* from the origin and places them on discordias *above* concordia. This kind of behavior is sometimes seen in monazite and other U-bearing phosphate minerals. In such cases, the age of the minerals is indicated by the upper point of intersection as before, but the lower point of intersection does not have a unique interpretation.

Use of highly selected and purified samples consisting of only a small number of zircon crystals (or fragments of a single crystal) can reduce the statistical error of ages determined by this method to ±2 million years (Ma) for zircons whose age is 2700 Ma (Krogh, 1982a,b).

The decay of U and Th to stable isotopes of Pb permits still more elaborate interpretations that go beyond the limits of this book. We have already mentioned the fission-track method of dating, which arises from the spontaneous fission decay of $^{238}_{92}$U. In addition, the growth of radiogenic Pb in the Earth is the subject of *plumbology,* which seeks to interpret the isotopic composition of Pb in Pb-bearing minerals that do *not* contain U or Th. Therefore, plumbology is concerned with the study of Pb in ore deposits containing galena (PbS) and minerals of other metals (Stacey and Kramers, 1975). Before we leave this subject, however, we consider briefly the so-called U-series disequilibrium methods of dating.

f. The Ionium (^{230}Th) Methods

The daughters of ^{238}U, shown in Figure 16.2, form a decay chain that can be broken by geological processes because the daughters of U are isotopes of different elements and therefore have different geochemical properties. Two different situations may arise:

1. A member of the decay chain is separated from its parent and therefore "dies" at a rate determined by its half-life.

2. A member of the decay chain is separated from its daughter and subsequently regenerates its own daughter.

Both conditions occur between ^{234}U and its daughter ^{230}Th (ionium) in the decay series of ^{238}U.

Uranium can assume valences of +4 and +6, whereas Th has only one valence of +4. The reason for this difference is that U atoms ($Z = 92$) have two more electrons than Th atoms ($Z = 90$). Their electron configurations (Table 5.2) are:

$$U: \quad [Rn]5f^3 6d^1 7s \rightarrow +4, +6$$

$$Th: \quad [Rn]5f^0 6d^2 7s^2 \rightarrow +4$$

The tetravalent (+4) ions of U and Th form compounds that are insoluble in water. In addition, their high ionic potentials (charge-to-radius ratio) cause them to be strongly adsorbed on the charged surface of small particles. However, in the presence of molecular O at or near the surface of the Earth, U is oxidized to the hexavalent state (+6) and forms the *uranyl* ion (UO_2^{2+}), which has a smaller ionic potential than U^{4+} and therefore is not scavenged from aqueous solutions by small particles as is Th^{4+}. Therefore, when U-bearing minerals decompose as a result of chemical weathering, ^{234}U forms the uranyl ion and remains in solution, whereas ^{230}Th forms a tetravalent ion and is strongly adsorbed on sedimentary particles. As a result, ^{230}Th (ionium) is separated from its parent by being concentrated in sediment deposited in the oceans or in lakes.

After deposition of the sediment, the adsorbed "excess" ^{230}Th decays by emitting alpha particles with its characteristic half-life ($T_{1/2} = 7.52 \times 10^4$ a, $\lambda = 9.217 \times 10^{-6}$ a^{-1}) in accordance with the law of radioactivity. The number of ^{230}Th atoms remaining in a unit weight of dry sediment is measured by observing the rate of disintegration (A) using an alpha spectrometer of known efficiency. When samples are taken from a piston core, the ^{230}Th activity $(^{230}Th)_A$ of a particular sample is related to its age (t) by equation 16.15:

$$\left(^{230}\text{Th}\right)_A = \left(^{230}\text{Th}\right)_A^0 e^{-\lambda t} \qquad (16.41)$$

where $\left(^{230}\text{Th}\right)_A^0$ is the activity of ^{230}Th at the top of the core. For practical reasons, the activity of ^{230}Th is divided by the activity of adsorbed ^{232}Th, whose activity is virtually independent of time because of its long half-life (Table 16.1). When the rate of sedimentation s is defined as:

$$s = \frac{h}{t} \qquad (16.42)$$

where h is the depth below the top of the core, $t = h/s$, and the ratio of the activities of the Th isotopes is:

$$\left(\frac{^{230}\text{Th}}{^{232}\text{Th}}\right)_A = \left(\frac{^{230}\text{Th}}{^{232}\text{Th}}\right)_A^0 e^{-\lambda h/s} \qquad (16.43)$$

We linearize the equation by taking natural logarithms:

$$\ln\left(\frac{^{230}\text{Th}}{^{232}\text{Th}}\right)_A = \ln\left(\frac{^{230}\text{Th}}{^{232}\text{Th}}\right)_A^0 - \frac{\lambda h}{s} \qquad (16.44)$$

Figure 16.6 illustrates how sediment samples taken at different depths from a core generate an array of points in coordinates of $\ln\left(^{230}\text{Th}/^{232}\text{Th}\right)_A$ and h to which a straight line can be fitted. The intercept of that line with the $\ln\left(^{230}\text{Th}/^{232}\text{Th}\right)_A$ (y) axis is the $^{230}\text{Th}/^{232}\text{Th}$ activity ratio at the sediment–water interface and the slope of the line is:

$$m = -\frac{\lambda}{s} \qquad (16.45)$$

Since the slope also has a negative value, it follows that the sedimentation rate is:

$$s = \frac{\lambda}{m} \qquad (16.46)$$

This method of dating works best when it is applied to sediment samples taken from known depths below the sediment–water interface. It can yield not only the ages of individual samples based on the extrapolated value of $\left(^{230}\text{Th}/^{232}\text{Th}\right)_A^0$ but also indicates sedimentation rates. The method is limited by the eventual regrowth of ^{230}Th from small amounts of ^{234}U adsorbed on the sediment particles, which causes data points to

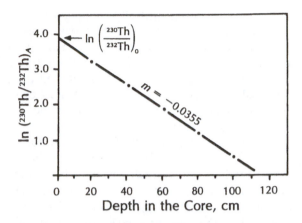

Figure 16.6 Decay of excess ^{230}Th in a core of detrital sediment at increasing depths below the water–sediment interface according to equation 16.44. The slope of the line fitted to the hypothetical data points is $m = -0.0355$. According to equation 16.46, the sedimentation rate is $s = \lambda/m = (9.217 \times 10^{-6})/0.0355 = 2.60 \times 10^{-4}$ cm/a or 2.60 mm/10^3 yr. This value is fairly typical of the rate of deposition of detrital sediment in the oceans.

deviate from the linear relationship required by equation 16.44.

The differences in geochemical properties of U and Th come into play in yet another way. The water-soluble uranyl ion (UO_2^{2+}) may enter calcite or aragonite by replacement of Ca^{2+}, whereas Th^{4+} is strongly excluded. Therefore, freshly deposited calcium carbonate contains ^{234}U that has been separated from its daughter ^{230}Th. The daughter is then regenerated in the calcite or aragonite at a rate determined by the difference between the rate of decay of its parent and its own decay rate. The differential equation governing the decay of a radionuclide to a radioactive daughter was solved by Bateman (1910). When the solution of that equation is adapted to the decay of ^{230}Th that is forming by decay of ^{234}U, we obtain (Faure, 1986, ch. 21, sec. 3):

$$\left(^{230}\text{Th}\right)_A = \left(^{238}\text{U}\right)_A\left(1 - e^{-\lambda_{230} t}\right) \qquad (16.47)$$

This equation is similar to equation 16.19, which relates the number of daughters to the number of parent atoms present initially. In equation 16.47 the parent is ^{238}U, which is assumed to be in secular equilibrium with its daughter ^{234}U. Since ^{238}U has a very long half-life ($T_{1/2} = 4.468 \times 10^9$ a, Table 16.2), its activity is virtually constant on a time scale of 1 million years. Therefore, the growth curve of the activity of ^{230}Th in calcium carbonate in Figure 16.7 has the same shape as the one in Figure 16.3B, which is a plot of equation 16.19.

The ^{234}U–^{230}Th method of dating is useful for dating calcium carbonate deposited in the oceans in the past 150,000 years or less. It has been used to date coral reefs related to sea-level fluctuations and the results have supported the *Milankovitch hypothesis* that the global climate is affected by periodic variations in the amount and timing of solar radiation received by the Earth

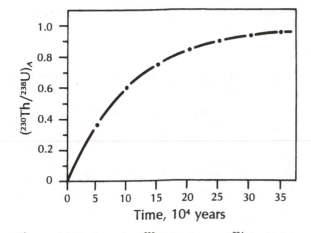

Figure 16.7 Growth of ^{230}Th by decay of ^{234}U, which is in secular radioactive equilibrium with ^{238}U, in accordance with equation 16.47. During the formation of calcium carbonate from aqueous solution U is admitted as the uranyl ion (UO_2^{2+}), whereas Th is excluded. Therefore, the decay chain of ^{238}U is thereby broken and ^{234}U in the carbonate regenerates the daughter ^{230}Th until their activity ratio reaches a value of one and secular equilibrium is reestablished. The diagram assumes that ^{234}U is in equilibrium with ^{238}U, but in reality the oceans and water on the continents contain excess ^{234}U because it is leached preferentially from U-bearing minerals. A derivation of the equation representing cases where $(^{234}U/^{238}U)_A > 1$ was given by Faure (1986, pp. 370–371).

because of the dynamics of its orbit around the Sun (Broecker et al., 1968).

Equation 16.47 is inexact because ^{234}U is actually *not* in secular equilibrium with ^{238}U in water on the surface of the Earth. The reason is that ^{234}U is leached preferentially from U-bearing minerals because it resides in lattice sites that were damaged by the decay of ^{238}U and of the two daughters that precede ^{234}U (Figure 16.2). As a result, the disintegration rate of ^{234}U in the oceans is 15% *greater* than that of ^{238}U. The discrepancy is even greater in groundwater, where the activity ratio ^{234}U/^{238}U may be as high as 10. A complete description of the growth of ^{230}Th in systems containing excess ^{234}U can be found in Faure (1986, p. 370) and in the references cited therein.

The decay of ^{238}U and ^{235}U through chains of unstable daughters has permitted the development of several other geochronometers. All of them are used to study the rates of sedimentation and other geological processes on the surface of the Earth. These applications of *radiochemistry* to geology have grown into a large subdivision of isotope geochemistry.

16.4 Cosmogenic Radionuclides

A large group of radionuclides is produced by nuclear reactions between cosmic rays and the stable atoms of the atmosphere and in rocks exposed at the surface of the Earth. Cosmic rays originate primarily from the Sun and consist predominantly of highly energetic protons. The products of these nuclear reactions may be removed from the atmosphere by meteoric precipitation or they may remain in the atmosphere for long periods of time, depending on their chemical properties. Many of the product nuclides have half-lives measured in hours or days and therefore are not useful for geologic research. However, a surprisingly large number of these nuclides decay sufficiently slowly to be of interest in geology.

The radionuclides listed in Table 16.2 are becoming increasingly useful because of dramatic improvements in the equipment used to detect them. The traditional measurement technique is low-level counting using heavily shielded radia-

Table 16.2 Physical Constants of Some Cosmogenic Radionuclides

Nuclide	$T_{1/2}$, a	λ, a^{-1}
$^3_1$H (tritium)	12.26	5.653×10^{-2}
$^{10}_4$Be	1.5×10^6	0.462×10^{-6}
$^{14}_6$C	5730	0.1209×10^{-3}
$^{26}_{13}$Al	0.716×10^6	0.968×10^{-6}
$^{32}_{14}$Si	276	0.251×10^{-2}
$^{36}_{17}$Cl	0.308×10^6	2.25×10^{-6}
$^{39}_{18}$Ar	269	0.257×10^{-2}
$^{35}_{25}$Mn	3.7×10^6	0.187×10^{-6}
$^{59}_{28}$Ni	8×10^4	0.086×10^{-4}
$^{81}_{36}$Kr	0.213×10^6	3.25×10^{-6}

tion detectors. In recent years, ultrasensitive high-energy tandem mass spectrometers, which literally count ions, have greatly improved the sensitivity with which cosmogenic radionuclides can be detected.

The atmospheric inventories of the cosmogenic radionuclides depend on the rates at which they are produced and on the rates at which they decay or are removed from the atmosphere by meteoric precipitation or by other processes. Once a particular radionuclide has been removed from its atmospheric reservoir, decay becomes the dominant process and its activity declines with time in accordance with equation 16.15. The measured disintegration rate of a particular cosmogenic radionuclide in a geological sample can be used to solve equation 16.15 for t, provided the initial disintegration rate is known or can be derived from a set of data.

a. The Carbon-14 Method

The carbon-14 method of dating was developed in the late 1940s by a group of scientists working at the University of Chicago. The group included W. F. Libby, J. R. Arnold, E. C. Anderson, and A. V. Grosse, who successfully demonstrated the presence of ^{14}C in methane from a sewage-treatment plant. The first age determinations of archaeological objects of known age were made by Arnold and Libby (1949). Two years later, Anderson and Libby (1951) reported that the specific disintegration rate of ^{14}C in modern plants is constant worldwide and has an average value of 15.3 ± 0.5 disintegrations per minute per gram (dpm/g) of carbon. This value was later revised to 13.56 ± 0.07 dpm/g of carbon. These measurements therefore established the feasibility of the ^{14}C method of dating, which Libby described in a book published in 1952. The method was widely acclaimed and earned W. F. Libby the Nobel Prize in chemistry in 1960. In 1959 a journal was founded called *Radiocarbon* in which all age determinations and many research papers dealing with the distribution of ^{14}C continue to be published (Mook, 1980).

The dominant nuclear reaction producing ^{14}C in the atmosphere involves the interaction of neutrons, derived from cosmic rays, with stable $^{14}_7$N:

$$\text{neutron} + {}^{14}_7\text{N} \rightarrow {}^{14}_6\text{C} + \text{proton} \quad (16.48)$$

The radioactive ^{14}C so produced is incorporated into CO_2 molecules and is rapidly mixed throughout the atmosphere. From there it is absorbed by green plants during photosynthesis and enters cellulose and other carbon compounds manufactured by them. It is transmitted further up the food chain by herbivorous and carnivorous animals so that virtually all living organisms contain ^{14}C at about the same specific concentration expressed in terms of disintegrations per minute per gram of carbon.

As long as a plant is alive, it continues to absorb ^{14}C from the atmosphere. However, when it dies, the concentration of ^{14}C in its tissue begins to decline at a rate dependent on the half-life of ^{14}C as shown in Figure 16.8. The decay of ^{14}C takes place by β^- emission to stable $^{14}_7$N:

$$^{14}_6\text{C} \rightarrow {}^{14}_7\text{N} + \beta^- + \bar{\nu} + Q \quad (16.49)$$

where the symbols have the same meaning as in equation 16.1. The half-life of ^{14}C is 5730 years. However, ^{14}C dates are based on a value of 5568 years for the decay constant.

The basic principle and measurement technique of the carbon-14 method of dating are quite simple. The specific activity of ^{14}C in a sample containing biogenic carbon is measured by converting

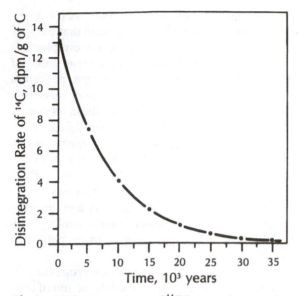

Figure 16.8 Decay curve of ^{14}C in accordance with equation 16.15. The initial decay rate of ^{14}C is 13.56 ± 0.07 dpm/g of C. The ^{14}C date is calculated from equation 16.50 based on the measured disintegration rate of the remaining ^{14}C. This method can be used to date plant remains and calcium carbonate formed in equilibrium with CO_2 of the atmosphere. The limit of the technique based on radiation detectors is about 35,000 years but can be doubled by using ultrasensitive mass spectrometers. Although the carbon-14 dating method is simple in principle, it is complex in reality because the initial activity of ^{14}C has varied with time in the past and because of fractionation of the isotopes of C.

the carbon to CO_2, or another carbon compound, and by counting it in a well-shielded *ionization chamber* or *scintillation counter*. The measured activity (A) is substituted into equation 16.15, which is solved for t using the assumed initial activity of ^{14}C (A_0):

$$t = \frac{1}{\lambda} \ln \frac{A_0}{A} \qquad (16.50)$$

The materials suitable for dating include charcoal and wood, seeds and nuts, grasses, cotton cloth, paper, leather, and peat. In addition, calcium carbonate in mollusk shells and in marine or lacustrine sediment can be dated because such deposits

contain CO_2 derived partly from the atmosphere. The initial activity of ^{14}C in calcium carbonate samples, however, is likely to be different from that observed in plant material. Therefore, it must be evaluated at each locality.

The limit of the carbon-14 method based on standard counting techniques is about 35,000 years. Use of ultrasensitive mass spectrometers can more than double this limit. Dates are sometimes expressed in terms of years *before present* (B.P.), where the present is taken to be 1952. Historical dates are also referred to the Christian calendar by subtracting 1952 years from dates adjusted to that point of reference. Thus a date of 2000 years B.P. becomes 48 B.C. (before Christ), whereas a date of 1500 years B.P. is equivalent to 452 A.D. (*anno domini*).

Although the principles of the ^{14}C method of dating appear to be straightforward, many complications do, in fact, occur. These arise from three main sources:

1. Variations in the production rate of ^{14}C caused by fluctuations in the cosmic-ray flux and by changes in the magnetic field of the Earth, which interacts with the cosmic rays.

2. Variations of the ^{14}C content of plant materials and calcium carbonate caused by isotope fractionation and other local environmental effects.

3. Lowering of the ^{14}C content of the atmosphere by dilution with CO_2 derived by combustion of fossil fuel and increases caused by the explosion of nuclear devices and the operation of nuclear reactors.

The changes in the ^{14}C content of the atmosphere since the start of the industrial revolution around 1860 are so profound that the carbon-14 method of dating is no longer applicable to samples formed since that time.

b. The ^{10}Be and ^{26}Al Methods

Although the presence of ^{10}Be in rainwater has been known since 1955, its use in geological research was retarded by the technical difficulties in measuring its decay rate with radiation detectors. The same is true for ^{26}Al, ^{36}Cl, and most of the other cosmogenic radionuclides listed in Table 16.2. All this changed dramatically in the late 1970s when ultrasensitive mass spectrometers were developed by means of which the abundances of these nuclides can now be measured accurately (Purser et al., 1982).

Both ^{10}Be and ^{26}Al are produced in the atmosphere by highly energetic cosmic rays that fragment stable isotopes of O and N in *spallation reactions*. Both isotopes are rapidly removed from the atmosphere in the form of rain and snow. When they enter the oceans, both radionuclides are scavenged from the water by adsorption on inorganic and organic sediment particles that carry them down to the bottom of the oceans. After deposition, the concentrations (or disintegration rates) of ^{10}Be and ^{26}Al decrease with time as they decay with their characteristic half-lives (Table 16.2). The same is true when ^{10}Be and ^{26}Al are deposited on the continental ice sheets of Greenland and Antarctica. In addition, these and other radionuclides are produced in meteorites and in rocks on other planets exposed to cosmic rays. After meteorites fall to the Earth, the production of cosmogenic radionuclides virtually stops. This makes it possible to determine their *terrestrial residence ages* by measuring the remaining concentrations or activities of ^{10}Be and ^{26}Al.

The decay of ^{10}Be to stable ^{10}B by beta (β^-) decay obeys the law of radioactivity expressed by equation 16.9:

$$^{10}\text{Be} = {}^{10}\text{Be}_0 e^{-\lambda t} \qquad (16.51)$$

This equation can be solved for t based on measurements of ^{10}Be and ^{10}Be$_0$. When samples are taken from a core of sediment or ice, t can be expressed in terms of depth (h) and the sedimentation rate (s) as in equation 16.42. When data points are plotted in coordinates of ln ^{10}Be and h, they define a straight line whose slope yields the sedimentation rate (s) and whose intercept is ln ^{10}Be$_0$, as exemplified by equation 16.44.

In principle, the same suite of samples can be analyzed for both ^{10}Be and ^{26}Al, which act as independent chronometers. On the other hand, the two radionuclides can be combined by taking the ratio of their concentrations or disintegration rates:

$$\frac{^{26}\text{Al}}{^{10}\text{Be}} = \left(\frac{^{26}\text{Al}}{^{10}\text{Be}}\right)_0 \frac{e^{-\lambda_{\text{Al}} t}}{e^{-\lambda_{\text{Be}} t}} \qquad (16.52)$$

Equation 16.52 can be reduced to:

$$\frac{^{26}\text{Al}}{^{10}\text{Be}} = \left(\frac{^{26}\text{Al}}{^{10}\text{Be}}\right)_0 e^{-\lambda' t} \qquad (16.53)$$

where

$$\lambda' = \lambda_{\text{Al}} - \lambda_{\text{Be}}$$

$$= (0.968 - 0.462) \times 10^{-6} \, a^{-1} \qquad (16.54)$$

$$= 0.506 \times 10^{-6} \, a^{-1} \; T_{1/2} = 1.37 \times 10^6 \, a$$

According to equation 16.53, the ^{26}Al/^{10}Be ratio decreases with depth in a sediment or ice core at a rate that depends on the rate of deposition. By substituting $t = h/s$ and by taking natural logarithms of equation 16.53 we obtain:

$$\ln\left(\frac{^{26}\text{Al}}{^{10}\text{B}}\right) = \ln\left(\frac{^{26}\text{Al}}{^{10}\text{Be}}\right)_0 - \frac{\lambda' h}{s} \qquad (16.55)$$

This is the equation of a straight line in coordinates of $\ln(^{26}\text{Al}/^{10}\text{Be})$ and h whose slope $m = -\lambda'/s$ and whose intercept is $\ln(^{26}\text{Al}/^{10}\text{Be})_0$. A limited number of data reviewed by Faure (1986) suggest that the initial ^{26}Al/^{10}Be ratio may be a constant equal to about 0.018 ± 0.010.

c. Exposure Dating of Rock Surfaces

High-energy neutrons associated with cosmic rays produce radionuclides by nuclear spallation reactions involving atoms in rocks and soils exposed at the surface of the Earth (Lal, 1988). The products of these nuclear reactions include $^3_2$He, $^{10}_4$Be, $^{14}_6$C, $^{21}_{10}$Ne, $^{26}_{13}$Al, and $^{36}_{17}$Cl, which accumulate in the target rocks as a function of time. The decay constants and half-lives of the radionuclides are included in Table 16.2, except for $^3_2$He and $^{21}_{10}$Ne, which are stable.

Although nuclear spallation reactions occur in all rocks and minerals that are exposed to cosmic rays, quartz is an especially favorable target for the production of ^{10}Be and ^{26}Al because of its simple chemical composition (SiO_2), its resistance to mechanical and chemical weathering, and its abundance in different kinds of rocks and because cosmogenic radionuclides deposited by meteoric precipitation can be eliminated by a chemical cleaning procedure (Lal and Arnold, 1985; Nishiizumi et al., 1986). In addition, in-situ-produced cosmogenic ^{36}Cl has been used successfully to determine exposure ages of lacustrine sediment and glacial deposits of Pleistocene age (Phillips et al., 1986; Phillips et al., 1990; Jannik et al., 1991).

The accumulation of a radionuclide formed by a nuclear reaction is governed by an equation derivable from the law of radioactivity (Faure, 1986, Section 4.4):

$$P = \frac{R}{\lambda}(1 - e^{-\lambda t}) \qquad (16.56)$$

where P = number of product atoms per gram; R = production rate of the product nuclide in atoms per gram per year; λ = decay constant of the product in reciprocal years; and t = duration of the irradiation in years. The production rates of ^{10}Be and ^{26}Al in quartz at sea level and at geomagnetic latitude $>50°$ have been determined by Nishiizumi et al. (1989) who reported rates of 6.0 atoms $g^{-1} y^{-1}$ for ^{10}Be and 36.8 atoms $g^{-1} y^{-1}$ for ^{26}Al. The production rates rise with increasing altitude above sea level because of decreased atmospheric shielding and also vary with geomagnetic latitude.

When equation 16.56 is used to determine the exposure age of quartz based on the measured concentrations of ^{10}Be or ^{26}Al, several assumptions must be made:

1. The quartz in the target rock did not contain residual cosmogenic radionuclides left over from a previous exposure.
2. The target rock was not shielded from cosmic rays by nearby topographic features.
3. The target rock was irradiated continuously since its initial exposure to cosmic rays.
4. The rate of erosion of the target rock was negligibly low.

These assumptions cannot be evaluated in most cases and therefore introduce uncertainty into the exposure dates calculated from equation 16.56.

The erosion rate of the target rock can be taken into consideration provided it remained constant during the exposure. The relevant equation was derived by Lal (1991):

$$P = \frac{R}{\lambda + \rho\varepsilon/\Lambda}(1 - e^{(\lambda + \rho\varepsilon/\Lambda)t}) \qquad (16.57)$$

where ε = erosion rate of the target in centimeters per year; ρ = density of the target mineral in grams per cubic centimeters; Λ = absorption mean free path of neutrons in the target rock. Nishiizumi et al. (1991) used a value of 165 g/cm^2 for this parameter. Note that the decay constant of the product nuclide in equation 16.57 has been increased by the term $\rho\varepsilon/\Lambda$ and that the apparent half-life has therefore been shortened. In other words, the effect of erosion is to make the product nuclide appear to decay more rapidly than it actually does. Therefore, a longer exposure time is required to achieve the observed concentration of the cosmogenic radionuclide, meaning that the exposure age calculated from equation 16.57 increases with increasing erosion rate.

Equation 16.57 can be used to calculate a minimum exposure date and a maximum erosion rate based on the measured concentrations of ^{10}Be or ^{26}Al in a sample of quartz that was exposed to cosmic rays. If the erosion rate is assumed to be negligibly small, equation 16.57 approaches the form of equation 16.56 and yields a minimum exposure age. However, at higher erosion rates the corresponding exposure ages calculated from equation 16.57 rapidly approach infinity as the erosion rate reaches a maximum. Therefore, the cosmic-ray-exposure date calculated from equation 16.56 is an *underestimate* of the true exposure age.

The discrepancy between the calculated minimum date and the true exposure age rises with increasing values of the erosion rate of the target and with the duration of the exposure. In addition, the sensitivity of this method of dating decreases as the concentration of the product nuclide approaches equilibrium when the rate of production of the radionuclide is equal to its rate of decay. This condition is achieved when the irradiation time is equal to about four half-lives of the product nuclide. Since ^{10}Be has a half-life of 1.5×10^6 years, equilibrium is reached in about 6 million years, assuming that the erosion rate of the target is negligible. This means that a calculated exposure date of that magnitude may actually be older by amounts of time that cannot be specified. However, exposure ages of less than about one million years, calculated from in-situ-produced ^{10}Be in quartz, are reliable, provided that the erosion rates are low (i.e. less than about 10^{-5} cm/year) and that the other assumptions listed above are satisfied. Therefore, the most desirable material for exposure dating are quartz crystals, vein quartz, chert, quartz concretions or agate geodes, quartzite, or well-cemented quartz arenites.

Age determinations based on in-situ-produced ^{10}Be in quartz have been reported by Nishiizumi (1991a,b), Brook et al. (1993), Gosse et al. (1995).

16.5   Summary

The discovery of the radioactivity of the salts of U in 1896 has had a profound effect on geology because radioactivity liberates heat and because it provides clocks that tell geologic time. These geochronometers are based on the slow rate of decay of long-lived radioactive isotopes of K, Rb,

Sm, La, Lu, Re, U, and Th. The interpretation of the measurements is based on the law of radioactivity, which yields information not only about time but also about the past histories of the daughter elements within the Earth. Therefore, isotopic methods of dating based on long-lived radioactive isotopes provide information not only about the history of the Earth but also about the origin of igneous rocks and the chemical differentiation of the crust and mantle.

The short-lived radioactive daughters of ^{238}U and ^{235}U enable us to determine sedimentation rates and ages of minerals that formed within 1 million years of the present. The study of geologic processes of the recent past is especially important because it contributes to our understanding of the terrestrial environment, including variations in climate.

Cosmogenic radionuclides are becoming increasingly important for dating events on the surface of the Earth within the past million years. The nuclides include not only the well-known ^{14}C but also ^{10}Be, ^{26}Al, and ^{36}Cl produced by exposure of rocks and minerals to cosmic rays. The increasing emphasis on dating events that occurred in the recent past reflects the growing conviction that we must understand the recent history of the Earth in order to predict its future.

Problems*

*Problems 2–10 are based on data from Baadsgaard et al. (1976), and Problems 11 and 12 are from Baadsgaard (1973) concerning the geologic history of the Amitsoq gneiss of Greenland, which is one of the oldest rocks known on Earth.

1. Calculate the atomic ^{87}Rb/^{86}Sr ratio of a rock sample containing Rb = 125 μg/g and Sr = 275 μg/g. Atomic weights: Rb = 85.46, Sr = 87.62; isotopic abundances: ^{87}Rb = 27.83%, ^{86}Sr = 9.90%.

(Answer: 1.31)

2. Calculate a K–Ar date for a sample of biotite from the Amitsoq gneiss containing K_2O = 9.00%, ^{40}Ar = 2.424 μg/g. Decay constants: $\lambda_e = 0.581 \times 10^{-10}$ a^{-1}, $\lambda = 5.543 \times 10^{-10}$ a^{-1}; isotopic abundance of ^{40}K = 0.01167%; atomic weights: K = 39.098, O =

15.999, ^{40}Ar = 39.96.

(Answer: 2308 $\times$ 10^6 a)

3. Calculate a Rb–Sr date for the same sample of biotite from the Amitsoq gneiss given that ^{87}Rb/^{86}Sr = 107.1, ^{87}Sr/^{86}Sr = 3.093. Decay constant: $\lambda = 1.42 \times 10^{-11}$ a^{-1}; (^{87}Sr/^{86}Sr)$_0$ = 0.7030.

(Answer: 1554 $\times$ 10^6 a)

4. Calculate the slope and intercept of a Rb–Sr isochron by the method of least squares using data for micas from the Amitsoq gneiss.

Samples	$^{87}Rb/^{86}Sr$	$^{87}Sr/^{86}Sr$
la	763	17.337
1b	41.52	1.739
2	107.12	3.093
3	166.7	4.543
4	138.7	3.858
5	330.7	8.057
6	82.7	2.741

Use the slope to calculate a date for these samples $(\lambda = 1.42 \times 10^{-11}\ a^{-1})$.

(Answer: intercept = 0.8806, slope = 0.021 59,
$t = 1504 \times 10^6\ a$)

5. Calculate the Rb–Sr isochron age and initial $^{87}Sr/^{86}Sr$ ratio of a suite of whole-rock samples from the Amitsoq gneiss near Godthaab, Greenland, based on the following data $(\lambda = 1.42 \times 10^{-11}\ a^{-1})$.

Sample	$^{87}Rb/^{86}Sr$	$^{87}Sr/^{86}Sr$
1	2.098	0.8245
2	0.198	0.7096
3	1.173	0.7668
4	2.033	0.8191
5	1.364	0.7791
6	0.319	0.7163

[Answer: $(^{87}Sr/^{86}Sr)_0 = 0.6971$, slope = 0.0060 22,
$t = 4119 \times 10^6\ a$. For a discussion of these results consult the paper by Baadsgaard et al. (1976)]

6. Calculate a Rb–Sr date from the following analyses of a whole rock and its constituent minerals $(\lambda = 1.42 \times 10^{-11}\ a)$.

Sample	$^{87}Rb/^{86}Sr$	$^{87}Sr/^{86}Sr$
Whole rock	1.173	0.7668
Biotite	166.7	4.543
K-feldspar	0.6437	0.7822
Plagioclase	0.0633	0.7344

Note that the whole rock is sample 3 in Problem 5 and the biotite is sample 3 in Problem 4.

[Answer: $(^{87}Sr/^{86}Sr) = 0.7468$, slope = 0.022 77,
$t = 1586 \times 10^6\ a)]$

7. Plot the whole-rock isochron from Problem 5 and the mineral isochron from Problem 6 and use the diagram to interpret the dates obtained in Problem 5 and 6.

8. Calculate U–Pb and Th–Pb dates for a sample of sphene from the Amitsoq gneiss in Greenland. $^{206}Pb/^{204}Pb = 53.90$, $^{207}Pb/^{204}Pb = 20.76$, $^{208}Pb/^{204}Pb = 36.94$, $^{238}U/^{204}Pb = 88.746$, $^{238}U/^{235}U = 137.88$, $^{232}Th/^{204}Pb = 55.38$, $(^{206}Pb/^{204}Pb)_0 = 12.97$, $(^{207}Pb/^{204}Pb)_0 = 14.17$, $(^{208}Pb/^{204}Pb)_0 = 33.90$, $\lambda_1 = 1.551\,25 \times 10^{-10}\ a^{-1}$, $\lambda_2 = 9.8485 \times 10^{-10}\ a^{-1}$, $\lambda_3 = 4.9475 \times 10^{-11}\ a^{-1}$.

(Answer: $t_{206} = 2445 \times 10^6\ a$, $t_{207} = 2456 \times 10^6\ a$,
$t_{208} = 1080 \times 10^6\ a$)

9. Calculate the U/Th concentration ratio of the sphene in Problem 8 given that $^{238}U/^{204}Pb = 88.746$ and $^{232}Th/^{204}Pb = 55.38$. Atomic weights: U = 238.029, Th = 232.038. Abundance of $^{238}U = 99.28\%$.

(Answer: U/Th = 1.655)

10. Calculate the 207/206 date of sphene from the Amitsoq gneiss given that $^{206}Pb/^{204}Pb = 140.8$, $^{207}Pb/^{204}Pb = 35.36$, $(^{206}Pb/^{204}Pb)_0 = 12.97$, $(^{207}Pb/^{204}Pb)_0 = 14.17$.

(Answer: $t_{207/206} = 2512 \times 10^6\ a$)

11. Plot a concordia curve for values of t from 3.0×10^9 to $3.7 \times 10^9\ a$.

12. Determine the age of zircons from the Amitsoq gneiss by plotting the data given below on the diagram constructed above (Baadsgaard, 1973).

Sample	$^{206}Pb*/^{238}U$	$^{207}Pb*/^{235}U$
1	0.732	32.25
2	0.715	32.25
3	0.714	31.75
4	0.673	29.75
5	0.643	27.50
6	0.625	27.00
7	0.600	25.50
8	0.595	25.60
9	0.500	20.00

$\lambda_1 = 1.551\,25 \times 10^{-10}\ a^{-1}$, $\lambda_2 = 9.8485 \times 10^{-10}\ a^{-1}$.

(Answer: discordia intercept = 0.1324,
slope = 0.018 30, intercept with concordia:
$^{207}Pb*/^{235}U = 33.74$, $^{206}Pb*/^{238}U = 0.754$,
$t = 3622 \times 10^6\ a$)

13. Review the answers to Problems 2–6, 8, 10, and 12, all of which pertain to the Amitsoq gneisses.

(a) Why is the K–Ar date of biotite (Problem 2) older than the Rb–Sr date of the same biotite (Problem 3)?

(b) What is the significance of the fact that micas from different whole-rock samples of the Amitsoq gneisses all fit the same isochron (Problem 4)?

(c) What is the meaning of the whole-rock isochron date of the Amitsoq gneisses (Problem 5)?

(d) Why does the whole-rock sample (3) in Problem 5 also fit the isochron in Problem 6?

(e) What is the significance of the U–Pb dates of sphene from the Amitsoq gneisses in Problems 8 and 10?

(f) Suggest a reason for the low Th–Pb date (Problem 8) and the low U/Th ratio (Problem 9) of the sphenes from the Amitsoq gneisses.

(g) What is the significance of the concordia date of the zircons from the Amitsoq gneisses?

(h) What is the significance of the date of the lower point of intersection of discordia with concordia at $\sim$950 $\times$ 10^6 a? (Baadsgaard, 1973.)

(i) Why is the whole-rock Rb–Sr isochron date of the Amitsoq gneisses older than the concordia U–Pb date of the zircons?

14. Calculate the average growth rate of a manganese nodule from the following information.

Depth, mm	$(^{230}\text{Th}/^{232}\text{Th})_A$, dpm/g
0–0.3	43.7
0.3–0.5	26.2
0.5–0.85	10.4

$\lambda = 9.217 \times 10^{-6}\,\text{a}^{-1}$.

(Answer: 3.2 mm/10^6 yr)

15. A sample of wood from an archeological site was found to be 6367 years old when it was analyzed in 1986. What is its age (a) in years B.P.? (b) in reference to the Christian calendar?

[Answer: (a) 6333 years B.P.; (b) 4381 B.C.]

References

General

BAADSGAARD, H., 1973. U–Th–Pb dates on zircons from the early Precambrian Amitsoq gneisses, Godthaab District, West Greenland. *Earth Planet. Sci. Lett.,* **19**:22–28.

BAADSGAARD, H., R. St. J. LAMBERT, and J. KRUPICKA, 1976. Mineral isotopic age relationships in the polymetamorphic Amitsoq gneisses, Godthaab District, West Greenland. *Geochim. Cosmochim. Acta,* **40**:513–527.

BATEMAN, H., 1910. Solution of a system of differential equations occurring in the theory of radioactive transformations. *Proc. Cambridge Phil. Soc.,* **15**:423.

BECQUEREL, H., 1896. Sur les radiations invisibles emises par les sels d'uranium. *Compt. Rend. Acad. Sci. Paris,* **122**:689.

BROECKER, W. S., D. L. THURBER, J. GODDARD, T.-L. KU, R. K. MATHEWS, and K. J. MESOLELLA, 1968. Milankovitch hypothesis supported by precise dating of coral reefs and deep-sea sediment. *Science,* **159**:287–300.

BURCHFIELD, J. D., 1975. *Lord Kelvin and the Age of the Earth.* Science History Publications, New York, 260 pp.

CREASER, R. A., D. A. PAPANASTASSIOU, and G. J. WASSERBURG, 1991. Negative thermal ion mass spectrometry of osmium, rhenium, and iridium. *Geochim. Cosmochim. Acta,* **55**:397–401.

CURIE, P., and A. LABORDE, 1903. Sur la chaleur degagee spontanement par les sels de radium. *Compt. Rend. Acad. Sci. Paris,* **136**:673–675.

DALRYMPLE, G. B., and M. A. LANPHERE, 1969. *Potassium–Argon Dating.* W. H. Freeman, New York, 258 pp.

DICKIN, A. P., 1995. *Radiogenic Isotope Geology.* Cambridge University Press, New York, 425 pp.

FAURE, G., and J. L. POWELL, 1972. *Strontium Isotope Geology.* Springer-Verlag, New York, 188 pp.

FAURE, G., 1986. *Principles of Isotope Geology,* 2nd ed. Wiley, New York, 589 pp.

HOLMES, A., 1911. The association of lead with uranium in rock-minerals, its application to the measurement of geologic time. *Proc. Roy. Soc. London, Ser. A,* **85**:248–256.

HOLMES, A., 1913. *The Age of the Earth.* Harper and Brothers, London, 194 pp.

JOLY, J. J., 1909. *Radioactivity and Geology.* Archibald Constable, London, 287 pp.

KROGH, T. E., 1982a. Improved accuracy of U–Pb zircon dating by selection of more concordant fractions using a high gradient magnetic separation technique. *Geochim. Cosmochim. Acta,* **46**:631–635.

KROGH, T. E., 1982b. Improved accuracy of U–Pb zircon ages by the creation of more concordant systems using an air abrasion technique. *Geochim. Cosmochim. Acta,* **46**:637–649.

ROMER, A., 1971. *Radiochemistry and the Discovery of Isotopes.* Dover, New York, 261 pp.

RUTHERFORD, E., and F. SODDY, 1902. The cause and nature of radioactivity. Part I. *Phil. Mag., Ser. 6,* **4**:370–396.

RUTHERFORD, E., 1906. *Radioactive Transformations.* Charles Scribner's Sons, New York, 287 pp.

STACEY, J. S., and J. D. KRAMERS, 1975. Approximation of terrestrial lead isotope evolution by a two-stage model. *Earth Planet. Sci. Lett.,* **26**:207–221.

STEIGER, R. H., and E. JÄGER, 1977. Subcommission on geochronology: Convention on the use of decay constants in geochronology and cosmochronology. *Earth Planet. Sci. Lett.,* **36**:359–362.

THOMSON, W. (Lord Kelvin), 1899. The age of the Earth as an abode fitted for life. *Phil. Mag.,* (5), **47**:66–90.

WETHERILL, G. W., 1956. Discordant uranium–lead ages. *Trans. Amer. Geophys. Union,* **37**:320–326.

Cosmogenic

ANDERSON, E. C., and W. F. LIBBY, 1951. Worldwide distribution of natural radiocarbon. *Phys. Rev.,* **81**:64–69.

ARNOLD, J. R., and W. F. LIBBY, 1949. Age determinations by radiocarbon content:Checks with samples of known age. *Science,* **110**:678–680.

BROOK, E. J., M. D. KURZ, R. P. ACKERT, JR., G. H. DENTON, E. T. BROWN, G. M. RAISBECK, and F. YIOU, 1993. Chronology of Taylor Glacier advances in Arena Valley, Antarctica, using in situ cosmogenic ^{3}He and ^{10}Be. *Quaternary Research,* **39**:11–23.

BROOK, E. J., E. T. BROWN, M. D. KURZ, R. P. ACKERT JR., G. M. RAISBECK, and F. YIOU, 1995. Constraints on age, erosion, and uplift of Neogene glacial deposits in the Transantarctic Mountains determined from in situ cosmogenic ^{10}Be and ^{26}Al. *Geology,* **23**(12):1063–1066.

GOSSE, J. C., E. B. EVENSON, J. KLEIN, B. LAWN, R. MIDDLETON, 1995. Precise cosmogenic ^{10}Be measurements in western North America:Support for a global Younger Dryas cooling event. *Geology,* **23**(10):877–880.

IVY-OCHS, S., C. SCHLÜCHTER, P. W. KUBIK, B. DITTRICH-HANNEN, and J. BEER, 1995. Minimum ^{10}Be exposure ages of early Pliocene for the Table Mountain plateau and the Sirius Group at Mount Fleming, Dry Valleys, Antarctica. *Geology,* **23**(11):1007–1010.

JANNIK, N. O., F. M. PHILLIPS, G. I. SMITH, D. ELMORE, 1991. A ^{36}Cl chronology of lacustrine sedimentation in the Pleistocene Owens River system. *Geol. Soc. Amer. Bull.* **103**(9):1146–1159.

LAL, D., 1988. In situ-produced cosmogenic isotopes in terrestrial rocks. *Ann. Rev. Earth Planet. Sci.,* **16**:355–388.

LAL, D., 1991. Cosmic ray labeling of erosion surfaces:in situ nuclide production rates and erosion models. *Earth Planet. Sci. Letters,* **104**:424–439.

LAL, D., and J. R. ARNOLD, 1985. Tracing quartz through the environment. *Proc. Indian Acad. Sci. (Earth Planet. Sci.),* **94**(1):1–5.

LIBBY, W. F., 1952. *Radiocarbon Dating.* University of Chicago Press, 175 pp.

MOOK, W. G. (Ed.), 1980. *The Principles and Applications of Radiocarbon Dating.* Elsevier, Amsterdam.

NISHIIZUMI, K., D. LAL, J. KLEIN, R. MIDDLETON, and J. R. ARNOLD, 1986. Production of ^{10}Be and ^{26}Al by cosmic rays in terrestrial quartz in situ and implications for erosion rates. *Nature,* **319**:134–136.

NISHIIZUMI, K., E. L. WINTERER, C. P. KOHL, and J. KLEIN, 1989. Cosmic ray production rates of ^{10}Be and ^{26}Al in quartz from glacially polished rocks. *J. Geophys. Res.,* **94**(B12):17,907–17,915.

NISHIIZUMI, K., C. P. KOHL, J. R. ARNOLDS, J. KLEIN, D. FINK and R. MIDDLETON, 1991a. Cosmic ray produced ^{10}Be and ^{26}Al in Antarctic rocks:exposure and erosion history. *Earth Planet. Sci. Letters,* **104**:440–454.

NISHIIZUMI, K., C. P. KOHL, E. M. SHOEMAKER, J. R. ARNOLD, J. KLEIN, D. FINK, and R. MIDDLETON, 1991b. In situ ^{10}Be–^{26}Al exposure ages at Meteor Crater, Arizona. *Geochim. Cosmochim. Acta,* **55**:2699–2703.

PHILLIPS, F. M., B. D. LEAVY, N. O. JANNIK, D. ELMORE, and P. W. KUBIK, 1986. The accumulation of cosmogenic chlorine-36 in rocks:A method for surface exposure dating. *Science,* **231**:41–42.

PHILLIPS, F. M., M. G. ZREDA, S. S. SMITH, D. ELMORE, P. W. KUBIK, and P. SHARMA, 1990. A cosmogenic chlorine-36 chronology of glacial deposits at Bloody Canyon, eastern Sierra Nevada, California. *Science,* **248**:1529–1532.

PURSER, K. H., C. J. RUSSO, R. B. LIEBERT, H. E. GOVE, D. ELMORE, R. FERRARO, A. E. LITHERLAND, R. P. BEUKENS, K. H. CHANG, L. R. KILIUS, and H. W. LEE, 1982. The application of electrostatic tandems to ultrasensitive mass spectrometry and nuclear dating. In L. A. Currie (Ed.), Nuclear and Chemical Dating Techniques, 45–74. *Amer. Chem. Soc., Symp. Ser.,* vol. 176, 516 pp.

17

Isotope Fractionation

Several elements of low atomic number are composed of two or more stable isotopes. Therefore, when these elements react to form compounds, the molecules that are produced differ from each other by the particular isotopes they contain. For example, there are nine different isotopic varieties of H_2O and ten isotopic varieties of CO_2 whose molecular weights differ significantly from each other. The differences in the molecular weights affect the way these molecules respond to certain kinds of physical processes that are mass dependent. Moreover, the strengths of covalent bonds in molecules or ions depend on the masses of the atoms. Therefore, certain isotopic varieties of a particular molecule or complex ion are more stable than others because the bonds they contain are stronger than those of other isotopic varieties of the same molecule. The study of these subtle effects has become very important in the Earth Sciences because they enhance our understanding of geochemical processes occurring on the surface of the Earth.

17.1 Principles of Isotope Fractionation

The isotope compositions of certain elements of low atomic number are variable and depend on the compound in which the element resides as well as on the origin of that compound. For example, the isotopic composition of oxygen in rainwater varies widely depending on the latitude and other geographic and climatic factors of the collecting site. Similarly, seawater, marine mollusk shells, and quartz grains on a beach all contain oxygen of different isotope composi-

tions. The observed differences in the isotope compositions of oxygen and other elements of low atomic number are caused by processes referred to as *isotope fractionation*.

The magnitude of isotope fractionation depends on differences in the masses of the isotopes, on the temperature of formation of the compounds in which they occur, on the character of the chemical bonds they form, and on other factors. The elements whose isotopes are commonly fractionated include H, C, N, O, and S (see Table 17.1). In addition, isotope fractionation has been observed in Li, B, and Si. However, isotope fractionation is not possible for Be, F, Na, Al, and P because these elements have only one stable isotope.

Variations in the isotopic compositions of H, C, N, O, and S are important in the study of natural processes because they are among the most abundant elements in the crust of the Earth, because they form a wide variety of solid, liquid, and gaseous compounds, and because they occur in the crust and mantle of the Earth, as well as in its biosphere, hydrosphere, and atmosphere. Therefore, these elements participate in virtually all geochemical processes that take place on the surface of the Earth. As a result of the fractionation of their isotopes by these processes, the elements preserve a record of the environmental conditions under which their compounds formed. The growing importance of isotope fractionation in the study of sedimentary rocks and natural environments is reflected in books by Faure (1986), Hoefs (1996), Arthur et al. (1983), Fritz and Fontes (1980a, 1986), Rundel el al. (1988), and Ding et al. (1996).

Isotope fractionation is caused by isotope exchange reactions and by mass-dependent

Table 17.1 Abundances and Masses of Stable Isotopes of Elements Subject to Isotope Fractionation

Element	Stable isotopes	Mass, amu	Average abundance, %	Mass[a] difference, %
Hydrogen	$_{1}^{1}\text{H}$	1.007 825	99.985	99.8
	$_{1}^{2}\text{H (D)}$[b]	2.0140	0.015	
Carbon	$_{6}^{12}\text{C}$	12.000 000	98.90	8.36
	$_{6}^{13}\text{C}$	13.003 355	1.10	
Nitrogen	$_{7}^{14}\text{N}$	14.003 074	99.63	7.12
	$_{7}^{15}\text{N}$	15.000 108	0.37	
Oxygen	$_{8}^{16}\text{O}$	15.994 915	99.762	12.5
	$_{8}^{17}\text{O}$	16.999 131	0.038	
	$_{8}^{18}\text{O}$	17.999 160	0.200	
Sulfur	$_{16}^{32}\text{S}$	31.972 070	95.02	6.24 ($^{34}\text{S} - {}^{32}\text{S}$)
	$_{16}^{33}\text{S}$	32.971 456	0.75	
	$_{16}^{34}\text{S}$	33.967 866	4.21	
	$_{16}^{35}\text{S}$	35.967 080	0.02	

[a]The mass difference is expressed as $(m_H - m_L)/m_L \times 10^2$, where m_H and m_L are the masses of molecules containing the heavy and light isotopes, respectively.
[b]Deuterium symbolized by D.
SOURCE: Weast et al. (1986).

differences in the rates of certain chemical reactions and physical processes. For example, an important exchange reaction involving two isotopic molecules of CO_2 and H_2O is:

$$\tfrac{1}{2} \text{C}^{16}\text{O}_2 + \text{H}_2{}^{18}\text{O} \rightleftharpoons \tfrac{1}{2} \text{C}^{18}\text{O}_2 + \text{H}_2{}^{16}\text{O} \quad (17.1)$$

In this reaction ^{16}O and ^{18}O are exchanged between the two molecules and the factor of $\tfrac{1}{2}$ is needed to reduce the number of O atoms in CO_2 to one. Isotope exchange reactions progress from some initial state toward equilibrium much like ordinary chemical reactions. At equilibrium the Law of Mass Action applies:

$$\frac{(\text{C}^{18}\text{O}_2)^{1/2}(\text{H}_2{}^{16}\text{O})}{(\text{C}^{16}\text{O}_2)^{1/2}(\text{H}_2{}^{18}\text{O})} = K \quad (17.2)$$

where K is the equilibrium constant. The amounts of reactants and products are expressed in terms of molar concentrations rather than activities because the activity coefficients of isotopic varieties of a molecule cancel in equations like 17.2. If

CO_2 and H_2O molecules accept ^{16}O and ^{18}O without discrimination, then $K = 1.00$ and no isotope fractionation takes place. In fact, the equilibrium constant of reaction 17.1 at 25°C is $K = 1.0412$, which implies that the molecules do favor one of the O isotopes over the other.

The standard free energy change of isotope exchange reactions is of the order of -0.020 kcal, which is quite low compared to free energy changes of chemical reactions (Valley et al., 1986). Nevertheless, isotope exchange reactions can achieve equilibrium in spite of the low driving force, and the existence of isotopic equilibrium generally implies that the chemical reactions are also in equilibrium.

The equilibrium constant of the isotope exchange reaction represented by equation 17.1 differs from unity at 25°C because ^{18}O forms a *stronger* covalent bond with C than ^{16}O does. The explanation for this fact arises from the quantum mechanical treatment of the energy of a diatomic molecule. The strength of the covalent bond in such a molecule arises from the *reduction* of the energy of the molecule as the two atoms approach each other. At room temperature, most molecules of a gas are in the ground vibrational state (quantum number $n = 1$) and their vibrational energy (E) is:

$$E = \tfrac{1}{2}h\nu \qquad (17.3)$$

where h is Planck's constant and ν is the frequency of vibration. When the heavy isotope of an element replaces the light isotope, the energy of the molecule *decreases* because the increase in the mass of the molecule decreases its vibrational frequency. Consequently, the heavy isotope forms a stronger bond than the light isotope and molecules containing the heavy isotope of an element are more stable than those containing the light isotope. The *general rule* regarding isotope exchange reactions stated by Bigeleisen (1965) is "The heavy isotope goes preferentially to the chemical compound in which the element is bound most strongly."

The second cause for isotope fractionation is related to the differences in the masses of isotopes and of the molecules in which they occur. For example, the data in Table 17.1 indicate that deuterium ($^{2}_{1}H$) is 99.8% heavier than protium ($^{1}_{1}H$). The mass differences for the isotopes of the other elements are less than that, but still average $\sim 8.6\%$ and decrease with increasing atomic number of the elements. The molecular weights of *compounds* also vary depending on the isotopes they contain. For example, the molecular weight of $^{2}H_2^{16}O$ (or $D_2^{16}O$) is 11.1% greater than that of $^{1}H_2^{16}O$ (or $H_2^{16}O$). Similarly, $H_2^{18}O$ is also $\sim 11.1\%$ heavier than $H_2^{16}O$. Such differences in the molecular weights of compounds affect the rates of certain physical and chemical processes like diffusion, evaporation, and dissociation.

The fractionation of isotopes during diffusion and evaporation is caused by differences in the velocities of isotopic molecules of the same compound. According to statistical mechanics, all molecules of an ideal gas have the same kinetic energy depending on the temperature. Since the kinetic energy is equal to $\tfrac{1}{2}mv^2$, the ratio of the velocities (v) of two isotopic molecules of mass (m) of the same compound is:

$$\frac{v_L}{v_H} = \left(\frac{m_H}{m_L}\right)^{1/2} \qquad (17.4)$$

where the subscripts L and H identify the molecule containing the *light* and *heavy* isotopes, respectively. In the case of the carbon monoxide molecules $^{12}C^{16}O$ and $^{13}C^{16}O$, the ratio of the velocities is:

$$\frac{v_L}{v_H} = \left(\frac{28.998\ 27}{27.994\ 915}\right)^{1/2} = 1.0177 \qquad (17.5)$$

This means that the velocity of $^{12}C^{16}O$ is 1.0177 times greater than that of $^{13}C^{16}O$ regardless of the temperature.

Velocity differences among isotopic molecules of the same compound contribute to isotope fractionation in a number of ways. For example, molecules containing the light isotope of an element may diffuse out of a system more rapidly than the molecules containing the heavy isotope, thereby enriching the system in the heavy isotope. Therefore, during evaporation of water, $H_2^{16}O$ molecules escape into the vapor phase more rapidly than $H_2^{18}O$ and thus enrich the vapor in ^{16}O leaving the remaining water enriched in ^{18}O.

Differences in bond strengths among isotopic molecules of the same compound affect the kinetics of reactions in which they participate. Since the lighter of two isotopes forms *weaker* bonds than the heavier isotope, molecules containing the lighter isotope react more rapidly and are concentrated preferentially into the products of the reaction. Therefore, unidirectional chemical reactions can be very effective in fractionating isotopes. Such reactions occur in open systems where the products can escape and equilibrium is unattainable. For example, certain bacteria (*Desulfovibrio desulfuricans*) reduce sulfate ions to H_2S^0 or HS^-, which may be removed from the reaction by precipitation of metal sulfides. The sulfides produced by this process are enriched in ^{32}S relative to ^{34}S

because the ^{32}S—O bonds of the sulfate ions are more easily broken by the bacteria than the ^{34}S—O bonds.

The equilibrium constants of isotope exchange reactions in the gas phase can be calculated from so-called partition functions, which are defined in statistical mechanics. However, the fractionation of isotopes by reactions involving solids or liquids must be studied under controlled conditions in the laboratory or by analysis of natural systems believed to have achieved isotopic equilibrium. The magnitude of equilibrium constants of isotope exchange reactions depends primarily on the temperature and approaches 1.00 as the temperature rises to high values because at high temperature the small differences in the strengths of bonds formed by different isotopes of an element in a molecule are not as effective in causing isotope fractionation as they are at low temperature.

17.2   Mathematical Relations

The equilibrium constant of an isotope exchange reaction involving the replacement of only *one* atom in a molecule by two different isotopes of an element is identical to the *isotope fractionation factor*, which is defined by:

$$\alpha_b^a = \frac{R_a}{R_b} \qquad (17.6)$$

where R_a is the ratio of the heavy to the light isotope of an element in a molecule or phase *a*, R_b is the same in molecule or phase *b*, and the reaction is in equilibrium at a specific temperature. For example, when water is in equilibrium with its vapor at 25 °C, the fractionation of O isotopes in water molecules is expressed by:

$$\alpha_v^l = \frac{R_l}{R_v} = 1.0092 \qquad (17.7)$$

where $R_l = {}^{18}O/{}^{16}O$ in the liquid phase and $R_v = {}^{18}O/{}^{16}O$ in the vapor phase. The corresponding fractionation factor for H between water and

its vapor is defined in terms of the D/H ratios and has a value of 1.074 at 25 °C.

The data in Table 17.1 indicate that the $^{18}O/^{16}O$ ratio of *average* terrestrial O is $0.200/99.762 = 0.00200$ and that the D/H ratio is 0.00015. Because it is difficult to measure small differences in such small ratios, the isotope compositions of H, C, N, and S are expressed as the per mille difference of the isotope ratios of a sample (spl) and a standard (std). For O we have:

$$\delta^{18}O = \left(\frac{R_{spl} - R_{std}}{R_{std}}\right) \times 10^3 \, ‰ \qquad (17.8)$$

where R is the ratio of the heavy to the light isotope, that is, $R = {}^{18}O/{}^{16}O$. Similarly for C:

$$\delta^{13}C = \left(\frac{R_{spl} - R_{std}}{R_{std}}\right) \times 10^3 \, ‰ \qquad (17.9)$$

where $R = {}^{13}C/{}^{12}C$. The standard for O is *standard mean ocean water* (*SMOW*), whereas that for C is the calcite of a belemnite from the Peedee Formation in South Carolina known as PDB. SMOW also serves as the standard for H, whereas the isotope composition of O in carbonate rocks and minerals is commonly expressed relative to PDB. The relationship between $\delta^{18}O$ (SMOW) and $\delta^{18}O$ (PDB) is:

$$\delta^{18}O \text{ (SMOW)} = 1.030\,86\delta^{18}O \text{ (PDB)} + 30.86 \qquad (17.10)$$

The standard for N is N_2 of the atmosphere, whereas the isotopic composition of S is expressed relative to the S in troilite (FeS) of the Canyon Diablo meteorite from Meteor Crater, Arizona. In all cases, a positive δ-value indicates that the sample is enriched in the heavy isotope relative to the standard, whereas a negative δ-value indicates depletion in the heavy isotope. The δ-values are determined on specially designed mass spectrometers using gases prepared from the samples. The practical problems associated with the determination of δ-values are discussed by Hoefs (1996).

The relationship between the fractionation factor α_b^a for an element whose isotopes are distributed between molecules or phases a and b and the corresponding δ-values can be derived from their definitions. Since $\alpha_b^a = R_a/R_b$, we use equation 17.8 to express R_a and R_b in terms of δ-values:

$$R_a = \frac{\delta_a R_{std}}{10^3} + R_{std} = \frac{R_{std}(\delta_a + 10^3)}{10^3} \quad (17.11)$$

Therefore,

$$\alpha_b^a = \frac{R_{std}(\delta_a + 10^3)10^3}{10^3 R_{std}(\delta_b + 10^3)} = \frac{\delta_a + 10^3}{\delta_b + 10^3} \quad (17.12)$$

Equation 17.12 enables us to calculate either α_b^a, δ_a, or δ_b when any two of the three variables are known. For example, we can use equation 17.12 to calculate the $\delta^{18}O$ value of water vapor in the air over the oceans at 25°C, assuming that isotope equilibrium is maintained. Since $\alpha_v^l(O) = 1.0092$ and $\delta^{18}O_l = 0.0‰$ relative to SMOW:

$$\delta^{18}O_v = \frac{\delta^{18}O_l + 10^3}{\alpha_v^l} - 10^3 \quad (17.13)$$

$$\delta^{18}O_v = \frac{0 + 10^3}{1.0092} - 10^3 = -9.1‰ \quad (17.14)$$

The minus sign indicates that water vapor is depleted in ^{18}O relative to SMOW. We can also calculate δD of water vapor in equilibrium with seawater at 25°C. Since $\alpha_v^l(D) = 1.074$ and $\delta D_l = 0.0‰$, $\delta D_v = -68.9‰$, which means that the vapor is depleted in deuterium by 68.9‰ or 6.89%. In reality, water vapor over the oceans is even *more* depleted in ^{18}O and D than predicted by these calculations because the air over the oceans is undersaturated with respect to water vapor, which means that seawater is not in equilibrium with its vapor. As a result, the rate of evaporation exceeds the rate of condensation and kinetic isotope fractionation takes place.

The temperature variation of isotope fractionation factors for specific exchange reactions is usually determined experimentally and is expressed by equations of the form:

$$10^3 \ln \alpha_b^a = \frac{A \times 10^6}{T^2} + B \quad (17.15)$$

where A and B are constants and T is the temperature in kelvins. Values of A and B for many mineral–water systems were compiled by Friedman and O'Neil (1977) and Faure (1986).

The natural logarithms of small numbers like $1.00x$ have the interesting property that:

$$10^3 \ln 1.00x \approx x \quad (17.16)$$

For example, $10^3 \ln 1.005 = 4.987$; but the approximation deteriorates when $x > 10$, since $10^3 \ln 1.010 = 9.950$. Nevertheless, the approximation does apply to the fractionation of C, N, O, and S in many natural systems, and we can therefore use it to simplify equation 17.15. We start with the equation:

$$\alpha_b^a - 1 = \frac{R_a}{R_b} - 1 = \frac{R_a - R_b}{R_b} \quad (17.17)$$

Next, we use equation 17.11 to relate R_a and R_b to their respective δ-values and substitute them into equation 17.17. After simplifying the resulting expression we obtain:

$$\alpha_b^a - 1 = \frac{R_a - R_b}{R_b} = \frac{(\delta_a + 10^3) - (\delta_b + 10^3)}{(\delta_b + 10^3)} \quad (17.18)$$

which reduces to:

$$\alpha_b^a - 1 = \frac{\delta_a - \delta_b}{\delta_b + 10^3} \quad (17.19)$$

If α_b^a is less than 1.010, $\delta_b + 10^3 \approx 10^3$ and therefore:

$$\alpha_b^a - 1 \approx \frac{\delta_a - \delta_b}{10^3} \quad (17.20)$$

and

$$10^3(\alpha_b^a - 1) \approx \delta_a - \delta_b \quad (17.21)$$

Therefore, we can state that:

$$10^3(\alpha_b^a - 1) \approx 10^3 \ln \alpha_b^a \approx \delta_a - \delta_b \quad (17.22)$$

If we let $\delta_a - \delta_b = \Delta_b^a$, then:

$$10^3 \ln \alpha_b^a \approx \Delta_b^a \approx \frac{A \times 10^6}{T^2} + B \quad (17.23)$$

This equation indicates that the *difference* in the δ-values of C, N, O, and S in coexisting compounds in isotopic equilibrium *decreases* with *increasing* temperature. Ultimately, as T becomes very large, Δ approaches zero because the two coexisting compounds have virtually the *same* isotope composition. However, in the normal temperature range at or near the surface of the Earth, coexisting compounds of H, C, N, O, and S have *different* isotope compositions when the respective exchange reactions are in equilibrium.

17.3   Isotope Fractionation in the Hydrosphere

Evaporation of water from the surface of the oceans in the equatorial region of the Earth causes the formation of air masses containing water vapor that is depleted in ^{18}O and D compared to seawater. In the northern hemisphere the moist air from the tropics is displaced in a northeasterly direction by the circulation of the atmosphere and therefore enters cooler climatic zones. As a result of the decrease in temperature, the air becomes saturated with water vapor, which causes condensation and precipitation of water as rain, snow, or hail.

The condensate is enriched in the heavy isotopes of O and H because condensation is the opposite of evaporation. However, whereas evaporation may take place under disequilibrium conditions, the condensation of water from its vapor occurs at equilibrium. Therefore, we can calculate $\delta^{18}O$ and δD of the first condensate that forms from water vapor of known isotopic composition at a specified temperature.

According to equation 17.12, the $\delta^{18}O$ value of the condensate is:

$$\delta^{18}O_l = \alpha_v^l(\delta^{18}O_v + 10^3) - 10^3 \quad (17.24)$$

If $\delta^{18}O_v = -13.1‰$ relative to SMOW and $\alpha_v^l(O) = 1.0092$ at 25 °C, then:

$$\delta^{18}O_l = 1.0092(-13.1 + 10^3) - 10^3 = -4.0‰$$
$$(17.25)$$

Similarly, the δD value of the condensate that forms at 25 °C from vapor having $\delta D_v = -94.8‰$ is:

$$\delta D_l = 1.074(-94.8 + 10^3) - 10^3 = -27.8‰$$
$$(17.26)$$

Note that the condensate is enriched in ^{18}O and D relative to the vapor in the air. However, the isotopic composition of the first condensate is not the same as that of the original seawater because we assumed that the isotope fractionation during evaporation was enhanced by kinetic effects.

When the first droplets of condensate fall out of the air mass as rain or snow, the water vapor remaining in the cloud is further depleted in ^{18}O and D. As a result of continuing condensation and loss of water or snow from an air mass, the remaining water vapor becomes progressively more depleted in ^{18}O and D as air masses in the northern hemisphere move in a northeasterly direction. Consequently, the $\delta^{18}O$ and δD values of *meteoric precipitation* become progressively more negative with increasing geographic latitude.

The resulting *latitude effect* on the isotope composition of meteoric water is caused by (1) progressive isotope fractionation associated with condensation of water vapor and the removal of water droplets from the air mass, (2) the increase of the isotope fractionation factors caused by a decrease in temperature, (3) the reevaporation of meteoric water from the surface of the Earth, and (4) evapotranspiration of water by plants. Since H and O occur together in water molecules and since both experience the same sequence of events during the migration of air masses, the δD and $\delta^{18}O$ values of meteoric water are strongly correlated and satisfy an empirical equation known as the *meteoric-water line* (Figure 17.1):

$$\delta D = 8\delta^{18}O + 10 \quad (17.27)$$

The slope of the meteoric-water line is related to the isotopic fractionation factors of H and O during evaporation or condensation of water:

$$\frac{\alpha_v^l(H) - 1}{\alpha_v^l(O) - 1} = \frac{0.074}{0.0092} = 8.0 \quad (17.28)$$

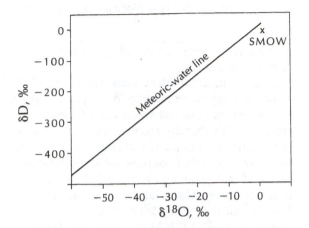

Figure 17.1 Relationship between δD and $\delta^{18}O$ of meteoric precipitation expressed by equation 17.27. Snow in the polar regions is strongly depleted in ^{18}O and D as a result of progressive isotope fractionation of water vapor in air masses moving toward the polar regions.

The intercept at $\delta D = +10‰$ for $\delta^{18}O = 0‰$ is called the *deuterium excess*. Both the slope and the intercept vary depending on local climatic conditions (Hoefs, 1996). Nevertheless, meteoric precipitation is progressively depleted in ^{18}O and D with increasing geographic latitude such that winter snow at the South Pole is characterized by $\delta^{18}O = -55‰$ and $\delta D = -430‰$.

The isotope fractionation of water vapor in an air mass described above can be modeled by *Rayleigh distillation*, provided that the isotopic composition of the water vapor in the air mass is not modified by evaporation of water from the surface of the Earth, or by evapotranspiration, or by partial evaporation of raindrops falling through the cloud. In accordance with Rayleigh distillation, the isotope ratio (R_v) of the remaining vapor is related to its original isotope ratio (R_v^0) by the equation:

$$R_v = R_v^0 f^{\alpha - 1} \qquad (17.29)$$

where $\alpha_v^l = R_l/R_v$ and f is the fraction of vapor remaining. We can use equation 17.11 to convert equation 17.29 to the δ-notation:

$$\delta^{18}O_v = (\delta^{18}O_v^0 + 10^3)f^{\alpha - 1} - 10^3 \quad (17.30)$$

Similarly, for H:

$$\delta D_v = (\delta D_v^0 + 10^3)f^{\alpha - 1} - 10^3 \quad (17.31)$$

When $f = 1.0$ in equation 17.30, $\delta^{18}O_v = \delta^{18}O_v^0$. However, when f varies from 1.0 to 0.0, $\delta^{18}O$ of the *remaining vapor* decreases as shown in Figure 17.2. As the process continues, the isotopic abundances of ^{18}O and D in the *condensate* must also decrease because the isotope ratios of O and H in the liquid are bound to those in the remaining vapor by equation 17.6, that is, $\alpha_v^l = R_l/R_v$. Since $R_v = R_l/\alpha$, we obtain from equation 17.29:

$$R_l = \alpha R_v^0 f^{\alpha - 1} \qquad (17.32)$$

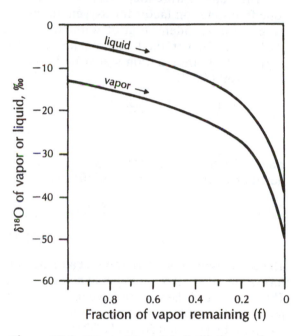

Figure 17.2 Effect of Rayleigh distillation on the $\delta^{18}O$ value of water vapor remaining in an air mass and of meteoric precipitation falling from it at a constant temperature of $25°C$ and $\alpha_v^l = 1.0092$. Note that meteoric precipitation becomes strongly depleted in ^{18}O as the fraction of vapor remaining in the air mass decreases. Such ^{18}O-depleted snow occurs in the polar regions and at high elevations in mountain ranges, but the effect is probably caused more by condensation at decreasing temperature than by Rayleigh distillation.

and by analogy to equation 17.30:

$$\delta^{18}O_l = \alpha(\delta^{18}O_v^0 + 10^3)f^{\alpha-1} - 10^3 \quad (17.33)$$

This equation yields the $\delta^{18}O$ value of meteoric precipitation in Figure 17.2 that forms by progressive condensation of vapor at constant temperature.

The results help to explain why meteoric water is depleted in ^{18}O and D compared to seawater. However, in reality the process is complicated because of the addition of vapor to the air mass by reevaporation of fresh water and because the isotope fractionation factors of O and H in water are not constant but *increase* with decreasing temperature.

The temperature dependence of the isotope fractionation factor for oxygen during the evaporation/condensation of water from 0 to 100 °C has been studied experimentally and is expressed by an equation stated by Friedman and O'Neil (1977):

$$10^3 \ln \alpha = \frac{1.534 \times 10^6}{T^2} - \frac{3.206 \times 10^3}{T} + 2.644$$
$$(17.34)$$

where T is the temperature in kelvins. The fractionation of H in water from 0 to 50 °C is:

$$10^3 \ln \alpha = \frac{24.844 \times 10^6}{T^2} - \frac{76.248 \times 10^3}{T} + 52.612$$
$$(17.35)$$

The condensation of water vapor in air masses to form snow takes place at temperatures below 0 °C. The variation of the isotope fractionation factors for ice and its vapor in this temperature range was discussed by Fritz and Fontes (1980b).

The strong dependence of the isotope fractionation factors of O and H in meteoric water on the temperature of evaporation/condensation gives rise to a relationship between the $\delta^{18}O$ value of meteoric precipitation and average monthly temperatures. The relationship was expressed by Yurtsever (1975) in the form of an empirical equation based on data from northern Europe and Greenland:

$$\delta^{18}O \text{ (SMOW)} = (0.521 \pm 0.014)t - 14.96 \pm 0.21$$
$$(17.36)$$

where t is the average monthly temperature in degrees Celsius.

The temperature dependence of $\delta^{18}O$ and δD values of snow accumulating in the polar regions is reflected in seasonal variations of these parameters, which thereby record long-term changes in climatic conditions. For example, the climatic warming at the Pleistocene–Holocene boundary about 15,000 years ago is recorded in ice cores recovered from Greenland and Antarctica by a 5–10‰ shift in the $\delta^{18}O$ values of the ice (Johnsen et al., 1972; Moser and Stichler, 1981; Lorius et al., 1985).

17.4   Oxygen Isotope Composition of Calcite

After the end of World War II, Harold C. Urey, the discoverer of deuterium, set up a laboratory at the University of Chicago where he began to study the fractionation of O isotopes between calcite and water. This work soon led to the development of a *paleothermometer* based on the isotope composition of O in marine calcite and aragonite. In addition, Urey assembled a group of gifted students and postdoctoral collaborators whose research has contributed significantly to the systematic study of the isotope compositions of H, C, N, O, and S in terrestrial and extraterrestrial samples. Harold Urey's former colleagues and their students are still at the forefront of the study of isotope fractionation and its applications to geochemistry and cosmochemistry (Valley et al., 1986).

Calcite and aragonite are enriched in ^{18}O compared to the water from which they precipitated. When the reaction takes place in isotopic equilibrium in the oceans, the $\delta^{18}O$ value of the calcite (δ_c) is related to the $\delta^{18}O$ of seawater (δ_w) and the temperature (t) in degrees Celsius by the equation:

$$t = 16.9 - 4.2(\delta_c - \delta_w) + 0.13(\delta_c - \delta_w)^2 \quad (17.37)$$

where δ_c is $\delta^{18}O$ of CO_2 prepared by reacting calcium carbonate with 100% H_3PO_4 at 25 °C and δ_w is $\delta^{18}O$ of CO_2 in equilibrium with the water at 25 °C and both δ-values are based on the same standard. Anderson and Arthur (1983) recast equation 17.37 by expressing δ_w on the SMOW scale, whereas δ_c is on the PDB scale:

$$t(°C) = 16.0 - 4.14(\delta_c - \delta_w) + 0.13(\delta_c - \delta_w)^2$$
$$(17.38)$$

It is customary to express the $\delta^{18}O$ values of carbonate minerals relative to the PDB standard (Peedee Formation, belemnite) originally chosen by Urey. Consequently, the $\delta^{18}O$ values of marine carbonates are close to 0‰ on the PDB scale, whereas those of nonmarine carbonates are negative because fresh water is depleted in ^{18}O.

The $\delta^{18}O$ value of seawater is included in equations 17.37 and 17.38 because the isotope composition of water in the oceans is variable on a regional scale, depending on the rate of evaporation from the surface, and because of mixing of seawater with fresh water discharged by rivers or by melting icebergs. Moreover, the $\delta^{18}O$ value of the oceans has varied in the past depending on the extent of continental glaciation. At the height of the Pleistocene glaciation, the oceans became enriched in ^{18}O because the continental ice sheets were strongly *depleted* in ^{18}O. Therefore, the $\delta^{18}O$ of seawater has varied by up to 1.5‰ from +0.09‰ on the SMOW scale (full-scale continental glaciation) to ~0‰ (present state) to −0.60‰ (no continental ice sheets).

The oxygen-isotope paleothermometer is also based on the assumption that biogenic calcites or aragonites were precipitated in isotopic equilibrium with seawater. This requirement is met *only* by a few carbonate-secreting organisms including mollusks and the foraminifera. The latter include both benthic (bottom dwelling) and pelagic (floating) species whose habitats have quite different temperatures. Other complications to be considered are the seasonality of shell growth of mollusks and the mineralogy of the shells because calcite and aragonite have slightly different O isotope fractionation factors.

The complications described above make it impractical to use equation 17.37 or 17.38 to determine the numerical values of paleotemperatures in the oceans. Nevertheless, *variations* in the $\delta^{18}O$ (PDB) values of selected foraminiferal tests from different depths in a sediment core can reveal *changes* of the water temperature in the past at the site where the core was taken. An *increase* in ^{18}O enrichment of calcite may be caused either by a *decrease* in the temperature, or by an *increase* in the ^{18}O content of the water, or both. A *decrease* in the ^{18}O enrichment of calcite may be caused by the opposite changes in the temperature and the $\delta^{18}O$ value of water. The temperature of the oceans and the $\delta^{18}O$ value of seawater may be related because a decrease in the temperature of seawater may be caused by the onset of continental glaciation, which also causes the oceans to become enriched in ^{18}O. Unfortunately, equations 17.37 or 17.38 cannot be used to determine both the temperature and the $\delta^{18}O$ value of seawater because they contain three variables, only one of which can be measured. This difficulty could be overcome by analyzing other authigenic O-bearing minerals such as Ca phosphate or chert. However, the temperature dependence of the isotope fractionation factors for these materials is very similar to that of the carbonates. Therefore, it is not possible to determine both the temperature and isotopic composition of seawater by presently available isotopic methods (Faure, 1986; Hoefs, 1996).

The $\delta^{18}O$ values of recently formed marine limestones are close to zero on the PDB scale but tend to be slightly negative. The small ^{18}O depletion of calcite in modern coral reef deposits occurs during the recrystallization of the carbonate particles in the presence of fresh water that may accumulate in the subsurface in coral atolls and in coastal reef complexes.

The average $\delta^{18}O$ values of marine carbonate rocks of Phanerozoic age become increasingly more negative with age (Veizer and Hoefs, 1976). The apparent ^{18}O depletion of these rocks has been attributed to (1) the long-term effects of oxygen-isotope exchange between carbonate rocks and meteoric water, (2) the existence of *higher* temperatures in the oceans of the Paleozoic Era than at present, and (3) a *lower* ^{18}O content of seawater in Paleozoic time compared to the modern oceans. The explanation for the observed secular changes in the $\delta^{18}O$ values of marine carbonate rocks may have far-reaching consequences to our understanding of the history of the Earth, but there is as yet no consensus on this important question.

Nonmarine carbonate rocks are variably *depleted* in ^{18}O compared to marine carbonates and commonly have negative $\delta^{18}O$ values on the PDB scale because they formed from meteoric water. Some nonmarine carbonate rocks, such as those of the Flagstaff and Green River formations in the western United States, formed by precipitation of calcite in large lakes that became saline as a result of evaporative concentration under arid or semiarid climatic conditions. If the precipitates formed in isotopic equilibrium, the fractionation of O isotopes between calcite and water varies with temperature between 0 and 500°C in accordance with the equation (Friedman and O'Neil, 1977):

$$10^3 \ln \alpha_w^c = \frac{2.78 \times 10^6}{T^2} - 2.89 \quad (17.39)$$

where letters c and w refer to calcite and water, respectively, and T is the temperature in kelvins. This equation can be used to determine the $\delta^{18}O$ value of the water from which calcite precipitated, as illustrated in the following example.

Calcite concretions in the Pagoda Tillite of Permian age in the Queen Alexandra Range of the Transantarctic Mountains have an average $\delta^{18}O$ value of about -20‰ on the PDB scale (Lord et al., 1988). In order to calculate $\delta^{18}O$ of the water from which the concretion was deposited, we convert $\delta^{18}O$ of the concretion from the PDB to the SMOW scale by means of equation (17.10).

$$\delta^{18}O_c \text{ (SMOW)} = 1.03086(-20) + 30.86$$

$$= +10.2‰ \quad (17.40)$$

The temperature at which the calcite precipitated was probably just above the freezing point

because tillite is a glacial deposit. If the temperature was 0°C, equation 17.39 yields the fractionation factor for oxygen in calcite and water:

$$10^3 \ln \alpha_w^c(O) = \frac{2.78 \times 10^6}{(273.15)^2} - 2.89 = 34.36$$

$$(17.41)$$

Therefore, $\ln \alpha_w^c(O) = 0.03436$ and $\alpha_w^c(O) = 1.03495$. Next, we use equation 17.12 to calculate $\delta^{18}O_w$:

$$1.03495 = \frac{10.2 + 10^3}{\delta^{18}O_w + 10^3} \quad (17.42)$$

which yields $\delta^{18}O_w = -23.9$‰ on the SMOW scale. The approximation represented by equation 17.22:

$$\delta_c - \delta_w \approx 10^3 \ln \alpha_w^c \quad (17.43)$$

leads to $\delta_w = 10.2 - 34.36 = -24.16$‰, which differs from the precise value (-23.9‰) by only about 1%.

The results of this calculation indicate that the concretions in the Pagoda Tillite of Antarctica were deposited by water that was strongly depleted in ^{18}O. This result implies that the water formed by melting of the Gondwana ice sheet, which also deposited the sediment of the Pagoda Tillite. Evidently, the $\delta^{18}O$ value of the Gondwana ice sheet, which existed more than 245 million years ago, has been preserved in the calcite concretion. The isotopic composition of H in the glacial meltwater is given by equation 17.27:

$$\delta D = 8(-23.9) + 10 = -181.2‰ \quad (17.44)$$

The O and H isotope compositions of the Gondwana ice sheet are similar to those of snow that is presently forming in the coastal areas of Antarctica and in Greenland. The average monthly temperature of the area of accumulation of the Gondwana ice sheet can be estimated from equation 17.36:

$$-23.9 = 0.521t - 14.96 \quad (17.45)$$

which yields $t = -17.2$°C. This temperature applies to the area of *accumulation* of the Gondwana ice sheet and not to its margin in Antarctica where the ice was melting during the Permian Period.

17.5 Oxygen and Hydrogen in Clay Minerals

Oxygen in the silicate and aluminosilicate minerals of igneous and metamorphic rocks and their sedimentary derivatives is enriched in ^{18}O relative to SMOW. When these aluminosilicate minerals react with acidified water to form kaolinite or other clay minerals (Chapter 16), the isotopic compositions of O and H in the clay minerals depend on several factors:

1. The isotope composition of the water.
2. The isotopic composition of the reactant minerals.
3. The amount of water compared to the amount of reactant minerals (water/rock ratio).
4. The temperature.
5. The isotopic fractionation factors of O and H.

The reactions between feldspar and acidified *meteoric* water to form clay minerals take place in the presence of an excess of water such that the water/rock ratio is large. Therefore, the isotope composition of O and H in clay minerals depends on the *water* rather than on the reactant minerals. Since $\delta^{18}O$ and δD of meteoric water vary depending on geographic latitude and climatic temperatures, the clay minerals formed on the continents inherit the relationship between δD and $\delta^{18}O$ expressed by the meteoric-water line. However, clay minerals are enriched in ^{18}O and depleted in D relative to the water at the site of the reaction. Therefore, the equations relating $\delta^{18}O$ and δD of clay minerals have similar slopes but different intercept values than the meteoric-water line. According to Savin and Epstein (1970), the equations are:

$$\delta D = 7.3\delta^{18}O - 260 \quad \text{(montmorillonite)} \quad (17.46)$$

$$\delta D = 7.5\delta^{18}O - 220 \quad \text{(kaolinite)} \quad (17.47)$$

These equations have been plotted in Figure 17.3 in order to illustrate the relationship between the isotopic composition of clay minerals and meteoric water.

When clay minerals that formed on land are transported and redeposited in the oceans, the

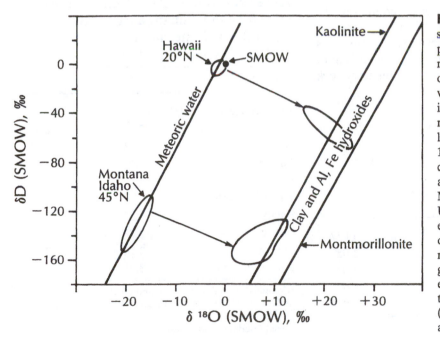

Figure 17.3 Schematic representation of the isotope compositions of O and H in clay minerals and Al, Fe hydroxides of soils and the local meteoric water. The meteoric-water line is equation 17.27, whereas the montmorillonite and kaolinite lines are equations 17.46 and 17.47, respectively. The isotope compositions of meteoric water and soil from Hawaii and the Montana–Idaho area of the United States illustrate the ^{18}O enrichments and D depletion of clay minerals relative to local meteoric precipitation. The diagram also depicts the latitude effect on the isotopic composition of meteoric precipitation (based on data by Lawrence and Taylor, 1971).

isotope compositions of H and O remain unchanged. Other detrital minerals associated with clay minerals in marine muds likewise do not equilibrate isotopically at the time of deposition. The oxygen isotope fractionation factors of clay minerals at Earth-surface temperatures listed in Table 17.2 range from 1.0236 for illite to 1.027 for kaolinite and montmorillonite. Therefore, the $\delta^{18}O$ value of kaolinite that forms from microcline in the presence of meteoric water having $\delta^{18}O = -10‰$ is given by equation 17.12:

$$\delta^{18}O \text{ (kaolinite)} = 1.027(-10 + 10^3) - 10^3$$

$$= +16.7‰ \qquad (17.48)$$

The corresponding δD value is obtained from equation 17.46:

$$\delta D = 7.3(+16.7) - 260 = -138.1‰ \quad (17.49)$$

These values are typical of clay-rich soils as indicated in Figure 17.3 based on measurements by Lawrence and Taylor (1971).

The mineral particles of *clastic* sedimentary rocks approach isotopic equilibrium *only* at elevated temperatures when the rates of the respective isotope exchange reactions become significant. According to O'Neil and Kharaka (1976), temperatures of ~100°C are required to equilibrate H, whereas a temperature of 300°C or more is required to equilibrate O between clay minerals and water. However, note that the achievement of isotopic equilibrium by a clastic sedimentary rock

does *not* mean that all minerals then contain O or H of the same isotope composition.

In order to illustrate this point we use the data in Table 17.3 to calculate O-isotope fractionation factors and $\delta^{18}O$ values of detrital illite, smectite, quartz, and calcite in contact with seawater ($\delta^{18}O = 0.0‰$) at 100°C, assuming that sufficient time is available to achieve isotope equilibrium.

Mineral	α_i^c	$\delta^{18}O$, ‰
Illite	1.0127	+12.7
Smectite	1.0144	+14.4
Quartz	1.0210	+21.1
Calcite	1.0167	+16.7

Evidently, each mineral has a different $\delta^{18}O$ value even though all were assumed to have equilibrated O with water of the same isotopic composition at the same temperature.

Clastic sedimentary rocks are generally composed of mineral particles whose H and O isotope compositions may have been only partly reset

Table 17.2 Isotope Fractionation Factors for Hydrogen and Oxygen Between Clay Minerals (Including Gibbsite) and Water at Earth-Surface Temperatures

Mineral	Hydrogen	Oxygen
Montmorillonite	0.94	1.027
Kaolinite	0.97	1.027
Glauconite	0.93	1.026
Illite	—	1.0236
Gibbsite	0.984	1.018

SOURCE: Faure (1986), Table 25.3.

Table 17.3 Temperature Dependence of the O-Isotope Fractionation Factor Between Minerals and Water in the Form $10^3 \ln \alpha = A \times 10^6/T^2 + B$[a]

Mineral	A	B
Clay Minerals		
Kaolinite	2.5	−2.87
Illite	2.43	−4.82
Smectite	2.67	−4.82
Chlorite	1.56	−4.70
Other Minerals		
Quartz	3.38	−3.40
Calcite	2.78	−2.89[b]
Dolomite	3.14	−2.0
Anhydrite	3.21	−4.72

[a]Compiled from the literature and recalculated by Anderson and Arthur (1983) to be consistent with Friedman and O'Neil (1977).

[b]From Friedman and O'Neil (1977).

during burial and diagenesis depending on the temperature and available time. The $\delta^{18}O$ values of marine shales range from $+14$ to $+19‰$ (Savin and Epstein, 1970), whereas their δD values lie between -33 and $-73‰$ (Yeh, 1980). The $\delta^{18}O$ values of marine limestones decrease with their depositional age from about $+30‰$ (SMOW) for Quaternary rocks to $+21‰$ (SMOW) in specimens of Cambrian age (Keith and Weber, 1964; Veizer and Hoefs, 1976). Since Ca carbonate does not contain H, the δD values of limestones are not defined.

Note that this section deals with systems having *large* water/rock ratios. Under such conditions the water contains a virtually infinite reservoir of H and O whose isotope compositions are not affected by exchange reactions with minerals in contact with them. Such environments occur primarily on the surface of the Earth where mineral particles are exposed to large quantities of water.

The situation *below* the surface of the Earth is commonly quite different because in this environment relatively small quantities of water are in contact with large volumes of rock for long periods of time and at elevated temperatures. In such cases the water/rock ratio is *small* and therefore the isotope compositions of O and H in the water change significantly, whereas those of the rocks change much less.

17.6   Groundwater and Geothermal Brines

The isotope compositions of O and H in groundwater are altered by isotope exchange reactions with the rocks in which the water occurs. The magnitude of this effect increases primarily with temperature and with the subsurface residence age of the water, but depends also on the mineral composition of the rocks and on the presence of other substances such as hydrocarbons and $H_2S(g)$. Several of these environmental properties, including temperature and residence age of the water, may increase with depth below the surface.

Water occurring in the pore spaces and fractures of rocks in the subsurface may originate from several different sources defined by Kharaka and Carothers (1986):

Meteoric water was derived from rain, snow, water courses, and other bodies of surface water that percolates through pore spaces in the rocks and displaces the interstitial water they contain regardless of its origin.

Connate water was deposited with sediments and has been out of contact with the atmosphere since the time of deposition. In the case of sedimentary rocks of marine origin the connate water was derived from seawater trapped in the pore spaces of the sediment.

Diagenetic water is released from solid phases as a result of reactions among minerals during diagenesis, such as the transformation of gypsum to anhydrite and smectite to illite both of which release water.

Formation water is the water that exists in layers of sedimentary rocks prior to drilling regardless of its origin.

Brine is water whose salinity, expressed in terms of total dissolved solids (TDS), exceeds that of seawater, which is 35,000 mg/L.

When meteoric water moves through the pore spaces of subsurface rocks, isotope exchange reactions take place that alter the isotope composition of this water in a systematic way. Figure 17.4 illustrates the case where meteoric water having $\delta^{18}O = -10‰$, $\delta D = -70‰$ on the SMOW scale interacts with shale having $\delta^{18}O = +14‰$ and δD between -33 and $-73‰$ (SMOW). If the water/rock ratio is small, the $\delta^{18}O$ and δD values of the water change along linear trajectories toward the isotope composition of the shale. The location of a particular water sample on such a trajectory depends primarily on the temperature, which controls the values of the isotope fractionation factors. Since the fractionation factors *decrease* with increasing temperature, $\delta^{18}O$ and δD of the water move closer *toward* the isotope composition of the rocks as the equilibrium temperature *rises*.

In order to illustrate this relationship we consider two points, *A* and *B,* in Figure 17.4, which

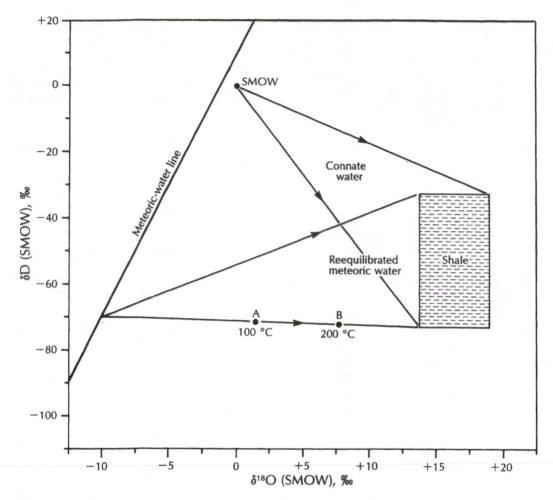

Figure 17.4 Schematic view of the changes in the isotopic composition of groundwater caused by isotope reequilibration with rocks at increasing temperature in systems having low water/rock ratios. Points *A* and *B* are samples of meteoric water ($\delta^{18}O = -10‰$) that reequilibrated with illite at 100 °C (point *A*) and 200 °C (point *B*) assuming that $\delta^{18}O$ of illite is +14‰ (SMOW) and remains constant. The diagram shows that both meteoric water and seawater may be altered isotopically by this process. As a result of isotope reequilibration groundwater may be displaced from the meteoric-water line to form linear trajectories directed toward the isotope composition of the host rocks. Alternatively, such trajectories may also be the result of mixing between old isotopically reequilibrated brines and meteoric water from the surface of the Earth.

represent meteoric water that has equilibrated O with illite at 100 and 200 °C, respectively. The $\delta^{18}O$ of the water was calculated from the appropriate equation in Table 17.3:

$$10^3 \ln \alpha_w^i = \frac{2.43 \times 10^6}{T^2} - 4.82 \quad (17.50)$$

where α_w^i is the fractionation factor for O between illite (i) and water (w). We assume that the water/rock ratio is small, that isotopic equilibrium between illite and water has been established, and that $\delta^{18}O$ of illite is +14‰ (SMOW). The $\delta^{18}O$ of the water can be calculated by solving equation 17.50 for α_w^i at the desired temperatures and by substituting these values into equation 17.12. The results are $\delta^{18}O_w = +1.3‰$ at 100°C (point A in Figure 17.4) and $O^{18}O_w = +7.9‰$ at 200°C (point B). Evidently, $\delta^{18}O$ of the meteoric water is shifted from $-10‰$ at the surface toward $\delta^{18}O$ of the rocks by an amount that increases with increasing temperature.

In the case illustrated in Figure 17.4 the isotope composition of H of the meteoric water changes very little because the water and the rock were assumed to have similar δD values. In general, the δD values of reequilibrated groundwaters do change with increasing temperature depending on the presence of H-bearing compounds, including not only clay minerals but also hydrocarbons and H_2S. However, the temperature dependence of the applicable fractionation factors for H is not yet known.

We conclude that the isotopic compositions of groundwater derived from meteoric precipitation may be displaced from the meteoric-water line by enrichment in ^{18}O and D. Water samples collected at increasing depth below the surface tend to lie along linear trajectories directed toward the $\delta^{18}O$ and δD values of the subsurface rocks with which the water is equilibrating at varying temperatures depending on the depth. Such linear trajectories are commonly observed by isotope analyses of oilfield brines (Clayton et al., 1966; Hitchon and Friedman, 1969).

Seawater originally trapped in the pore spaces of sediment is likewise altered isotopically by isotope exchange reactions with the host rocks. Figure 17.4 outlines the area in the $\delta^{18}O$–δD plane where such waters are located, provided they equilibrated with shale over a range of elevated temperatures in environments having low water/rocks ratios. Therefore, connate water in rock formations of marine origin is ancient seawater whose isotopic and chemical composition has changed as a result of water–rock interactions. Formation waters may be *mixtures* of connate water and meteoric water, both of which had previously reequilibrated H and O depending on the composition of the rocks, the temperature, and residence age of the water.

Evidently, the isotope composition of groundwater can be understood also as a result of mixing of waters of differing compositions and origins. Meteoric water percolating downward from the Earth's surface may mix at depth with brines that may themselves be mixtures of connate water and highly reequilibrated old meteoric water. The resulting water compositions also lie in the area of reequilibrated meteoric water in Figure 17.4. Therefore, mixing is an important process affecting the isotopic and chemical composition of groundwater in ways that may be indistinguishable from the results of progressive chemical and isotopic equilibration between rocks and water in the subsurface (Lowry et al., 1988).

17.7   Isotope Fractionation of Carbon

Carbon is the principal element in the biosphere, but it occurs also in the lithosphere, hydrosphere, and atmosphere of the Earth. It has two stable isotopes (^{12}C and ^{13}C, Table 17.1) whose abundances in terrestrial samples vary by ~10% as a result of isotope fractionation by exchange reactions and kinetic effects. The isotopic composition of carbon is expressed in terms of the $\delta^{13}C$ parameter defined by equation 17.9 relative to the PDB standard.

The fractionation of C isotopes occurs primarily during *photosynthesis* reactions in green plants. As a result, biogenic C compounds are strongly *depleted* in ^{13}C and have negative $\delta^{13}C$ values. However, $CaCO_3$ precipitated in equilibrium with the CO_2 of the atmosphere is *enriched* in ^{13}C relative to CO_2 and has $\delta^{13}C$ values near 0‰ because the PDB standard is itself a marine carbonate. However, under certain circumstances, biogenic C may be oxidized to bicarbonate ions which may be incorporated into $CaCO_3$, giving it negative $\delta^{13}C$ values whose magnitude depends on the proportions of mixing of biogenic and inorganic C.

The photosynthesis process starts with diffusion of CO_2 gas from the atmosphere into the cells of green plants and causes the internal CO_2 to be *depleted* in ^{13}C. The internal CO_2 is subsequently converted into phosphoglyceric acid containing three C atoms (Calvin cycle) or into dicarboxylic acid with four C atoms (Hatch–Slack cycle). These transformations are irreversible and cause further depletion in ^{13}C amounting to 17–40‰ or more in C_3 plants but only 2–3‰ in C_4 plants. The subsequent synthesis of organic compounds causes additional isotope fractionation with maximum ^{13}C-depletion in lipids (oils, fats, and waxes). Consequently, those land plants which use the C_3 metabolic process are strongly depleted in ^{13}C and have $\delta^{13}C$ values between −23 and −34‰ (PDB), whereas C_4 plants (corn and tropical grasses) are less depleted in ^{13}C than C_3 plants, and their $\delta^{13}C$ values range from −6 to −23‰ (PDB). Note that the $\delta^{13}C$ values of C_3 and C_4 plants do not overlap (Deines, 1980).

Fossil fuels (coal, petroleum, and natural gas) are strongly depleted in ^{13}C because they originated primarily from the remains of C_3 plants. The $\delta^{13}C$ values of coal range from about −20 to −30‰ (PDB) and average about −25‰. Petroleum is slightly more depleted in ^{13}C than coal, and its $\delta^{13}C$ values cluster between −27 and −30‰ (PDB)

The isotopic composition of hydrocarbon gases depends on their molecular composition and origin. In general, methane (CH_4) is the most abundant constituent of natural gas and tends to be more depleted in ^{13}C than ethane (C_2H_6), propane (C_3H_8), or butane (C_4H_{10}). Methane formed by bacterial decomposition of cellulose under reducing conditions in soil and recently deposited sediment has strongly negative $\delta^{13}C$ values ranging to about −80‰ (PDB). Methane produced by cracking of larger hydrocarbon molecules or during the formation of coal or petroleum is depleted in ^{13}C relative to its source material, such that its $\delta^{13}C$ values vary between −25 and −50‰ (PDB). Deines (1980) showed that methane recovered from relatively shallow depths between 100 and ~2000 m is more depleted in ^{13}C than methane that originated from greater depths. The greater ^{13}C-depletion of methane originating from near-surface sources may be caused by the presence of varying proportions of bacteriogenic methane.

Carbon dioxide of the atmosphere has a $\delta^{13}C$ value of −7.0‰, but its isotopic composition is changing because of the addition of biogenic C released by the combustion of fossil fuel. When CO_2 dissolves in water to form carbonic acid and its ions, isotope exchange reactions occur that enrich HCO_3^- in ^{13}C relative to atmospheric CO_2 (Table 17.4). When $CaCO_3$ precipitates from such solutions the solid phase is further enriched in ^{13}C. According to Deines et al. (1974), the resulting fractionation of C-isotopes in $CaCO_3$ relative to CO_2 varies with the temperature such that:

$$10^3 \ln \alpha_g^s = \frac{1.194 \times 10^6}{T^2} - 3.63 \quad (17.51)$$

where s and g refer to solid $CaCO_3$ and gaseous CO_2, respectively. Therefore, the isotope fractionation factor for C in the $CaCO_3$–CO_2 system at 20°C is 1.0103. If $\delta^{13}C$ of the CO_2 = −7.0‰, then $CaCO_3$ formed in isotopic equilibrium at 20°C has $\delta^{13}C$ = +3.2‰ (PDB) based on equation 17.12.

The $\delta^{13}C$ values of *marine* carbonate rocks of Phanerozoic age are actually close to 0‰. For example, Keith and Weber (1964) reported an average value of +0.56 ± 1.55‰ (PDB) for 321 selected samples of marine limestones of Phanerozoic age. The C in *nonmarine* carbonate rocks tends to be depleted in ^{13}C and has more variable isotope compositions than marine carbonates. Keith and Weber (1964) reported an average $\delta^{13}C$ = −4.93 ± 2.75‰ (PDB) for 183

Table 17.4 Fractionation of Carbon Isotopes Between Aqueous Carbonate Ions and Calcium Carbonate Relative to $CO_2(g)$

Carbonate species	A^a	B^a
$H_2CO_3^0$	0.0063	−0.91
HCO_3^-	1.099	−4.54
CO_3^{2-}	0.87 (approx.)	−3.4 (approx.)
$CaCo_3(s)$	1.194	−3.63

$^a A$ and B are constants in the equation

$$10^3 \ln \alpha = \frac{A \times 10^6}{T^2} + B$$

where α is defined as $R(\text{carbonate species})/R(CO_2(g))$ and R is the atomic $^{13}C/^{12}C$ ratio.
SOURCE: Deines et al. (1974).

selected samples of nonmarine limestones. The ^{13}C-depletion of nonmarine carbonate rocks is attributable to the presence of biogenic carbon formed by oxidation of plant debris to HCO_3^- in the environment of deposition. However, nonmarine carbonates deposited where no plant debris was present have $\delta^{13}C$ values that are indistinguishable from those of marine carbonates (Lord et al., 1988).

Calcite cements in sandstones may also be depleted in ^{13}C where CH_4 escapes from petroleum pools or coal seams and is oxidized to HCO_3^- (Donovan et al., 1974). Similarly, the calcite in the cap rocks of salt domes has negative $\delta^{13}C$ values because it precipitated from HCO_3^- ions that formed by the bacterial oxidation of petroleum in pools along the flanks of the salt domes. Cheney and Jensen (1967) reported an average $\delta^{13}C = -36.2 \pm 6.2‰$ (PDB) for calcites associated with salt domes in Louisiana and Texas. They also found ^{13}C-depleted calcites in the sandstone-hosted U deposits in the Wind River Formation (Lower Eocene), Gas Hills, Wyoming.

Isotope analyses of O and C in calcite can now be made by releasing CO_2 from "zap pits" or microscopic traverses using a laser. Hence, it is possible to study mineral cements or carbonate concretions in great detail on a scale of about 10 μm by means of laser-released CO_2 gas.

17.8   Isotope Compositions of Strontium in Carbonate Rocks

Carbonate rocks contain not only O and C but also include trace amounts of Sr whose isotope composition is variable because of the decay of naturally occurring radioactive ^{87}Rb to stable ^{87}Sr (Section 16.3b) Therefore, we digress briefly to discuss the isotope composition of Sr in carbonate rocks even though the isotopes of Sr are *not* fractionated by exchange reactions or by kinetic effects (Faure, 1986).

Calcium carbonate permits Sr^{2+} to replace Ca^{2+} but excludes Rb^+. As a result, calcite has a very low Rb/Sr ratio, and its ^{87}Sr/^{86}Sr ratio is not significantly altered by radioactive decay of ^{87}Rb to ^{87}Sr after deposition (Section 16.3b). Instead, the isotope composition of Sr in calcite deposited in the oceans results from mixing of different isotopic varieties of Sr that enter the oceans. The principal sources of marine Sr having distinctive ^{87}Sr/^{86}Sr ratios are (1) old granitic basement rocks of the continental crust (high Rb/Sr, high ^{87}Sr/^{86}Sr); (2) young volcanic rocks along midocean ridges, in oceanic islands, and along continental margins (low Rb/Sr, low ^{87}Sr/^{86}Sr); and (3) marine carbonate rocks on the continents (low Rb/Sr, intermediate ^{87}Sr/^{86}Sr).

The isotopes of Sr are not fractionated during precipitation of calcite or aragonite from aqueous solutions because the mass difference between ^{87}Sr and ^{86}Sr (defined in Table 17.1) is only 1.2%. Because of the absence of isotope fractionation effects and the negligibly small production of radiogenic ^{87}Sr by decay of ^{87}Rb in carbonate rocks, we conclude that marine and nonmarine carbonate rocks, as well as calcite cements and concretions, record the isotope composition of Sr in the *fluid phase* at the time of deposition.

Isotope analyses of Sr in a large number of marine limestones of Precambrian and Phanerozoic age have revealed that the ^{87}Sr/^{86}Sr ratio of seawater has varied systematically with time (Burke et al., 1982). These variations must have been caused by changes in the isotopic compositions of Sr that entered the oceans from various sources and by changes in the relative proportions of these inputs. The systematic variation of the isotope composition of Sr in the oceans therefore reflects the operation of the "geochemical machine" mentioned first in Section 14.5. However, we still do not have a complete explanation of the specific reasons for the way in which the ^{87}Sr/^{86}Sr ratio in the oceans has varied with time (Faure, 1986).

The steady rise in the ^{87}Sr/^{86}Sr ratios of the oceans during the Tertiary period is shown in Figure 17.5 based on measurements by DePaolo and Ingram (1985). This phenomenon can be used for dating marine carbonate samples of Tertiary age and for stratigraphic correlations (Hurst, 1986). The isotope composition of Sr in the modern oceans is constant throughout the oceans

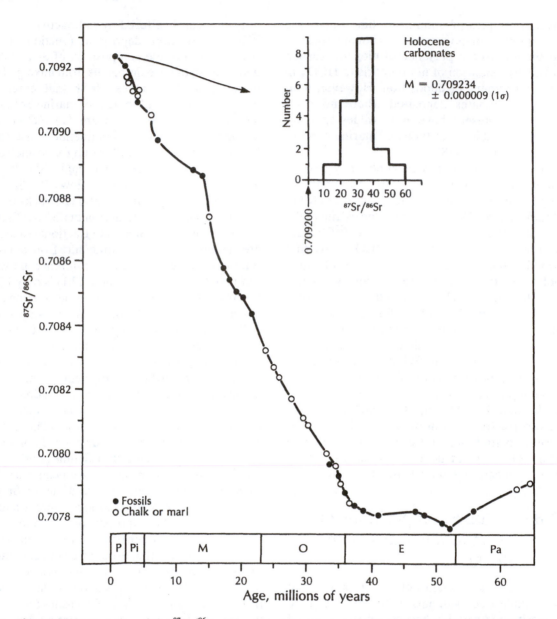

Figure 17.5 Variation of the $^{87}Sr/^{86}Sr$ ratio of marine carbonates with age during the Tertiary Period based on measurements by DePaolo and Ingram (1986). The continuous increase of this ratio since the end of the Eocene epoch can be used to date samples of marine carbonate of Tertiary age. Note that the rate of change of the $^{87}Sr/^{86}Sr$ ratio has varied with time and that the present value of this ratio in the oceans is 0.709234 ± 0.000009. However, the $^{87}Sr/^{86}Sr$ ratio measured by DePaolo and Ingram (1986) are *higher* than those reported by Burke et al. (1982) by 0.00017. The time scale includes P = Pleistocene, Pi = Pliocene, M = Miocene, O = Oligocene, E = Eocene, and Pa = Paleocene.

because its residence time is ~4000 times longer than the time required for mixing. Measurements by DePaolo and Ingram (1986) indicate that the average $^{87}Sr/^{86}Sr$ ratio of modern seawater is $0.709\,234 \pm 0.000\,009$. However, this value must be reduced by $0.000\,17$ to make it compatible with measurements by Burke et al. (1982) and with a compilation of data from the literature by Faure (1982).

The $^{87}Sr/^{86}Sr$ ratios of *nonmarine* limestones reflect the ages and Rb/Sr ratios of the rocks exposed to weathering in the drainage basin in which the carbonates were deposited. Therefore, stratigraphic variations of the $^{87}Sr/^{86}Sr$ ratios of nonmarine carbonate rocks can only be caused by *changes* in the geology or hydrology of the drainage basin (Neat et al., 1979).

The $^{87}Sr/^{86}Sr$ ratios of carbonate cements in arkosic or lithoclastic sandstones and conglomerates are affected by the release of ^{87}Sr from Rb-bearing minerals in the sediment during diagenesis. Therefore, the homogeneity of $^{87}Sr/^{86}Sr$ ratios of calcite cements in a lithologic unit is a reflection of the rate of flow of formation water through the lithologic unit or aquifer. Relatively rapid throughput of water tends to make $^{87}Sr/^{86}Sr$ ratios homogeneous, whereas stagnant conditions caused by low permeability or a low hydrostatic gradient permit local variations to develop in the isotope composition of Sr of calcite or zeolite cements (Stanley and Faure, 1979).

17.9   Isotope Fractionation of Sulfur

The stable isotopes of S listed in Table 17.1 are fractionated during the reduction of sulfate ions by bacteria and as a result of isotope exchange reactions among S-bearing ions and molecules in aqueous solution. The isotope composition is expressed in the δ-notation as:

$$\delta^{34}S = \left(\frac{(^{34}S/^{32}S)_{spl} - (^{34}S/^{32}S)_{std}}{(^{34}S/^{32}S)_{std}} \right) \times 10^3 \quad (17.52)$$

where the standard is the S in troilite (FeS) of the iron meteorite Canyon Diablo collected at

Meteor Crater, Arizona. The isotope composition of S is determined by analyzing SO_2 gas prepared from samples on a multicollector gas-source mass spectrometer (Hoefs, 1996).

The most important process causing isotope fractionation of S is associated with the metabolism of the bacteria *Desulfovibrio* and *Desulfatomaculum*. These bacteria flourish in anoxic environments by oxidizing organic matter using O derived from sulfate ions. In this process, the S is reduced from +6 to −2 and is subsequently expelled as H_2S, which is enriched in ^{32}S compared to the sulfate. We can approximate this transformation by constructing an equation to represent this oxidation–reduction reaction occurring without the aid of bacteria:

$$SO_4^{2-} + CH_4^0 + 2\,H^+ \rightarrow H_2S^0 + CO_2^0 + 2\,H_2O \quad (17.53)$$

The standard free energy change of this reaction is $\Delta G_R^\circ = -26.324$ kcal and its equilibrium constant at 25°C is $K = 10^{19.30}$. Evidently, SO_4^{2-} and aqueous CH_4 react to form H_2S^0, CO_2^0, and H_2O, which dominate at equilibrium. However, the reduction of SO_4^{2-} by bacteria is not at equilibrium and the isotope fractionation of S is caused by kinetic effects (Fry et al., 1986).

The extent of fractionation of S isotopes by bacteria is inversely proportional to the rate of metabolism and depends on the composition and abundance of the food supply, on the size of the sulfate reservoir, on the temperature, and on the rate of removal of the H_2S. In general, bacterial isotope fractionation of S depends on the environmental conditions represented by three alternative scenarios:

1. In *closed* systems, which become anoxic because vertical mixing is inhibited, the concentration of H_2S may rise until it ultimately poisons the bacteria. Such environments exist in the Black Sea, in isolated depressions in the oceans, and in some lakes where H_2S is strongly depleted in ^{34}S.

2. In open systems in which the sulfate reservoir is virtually infinite, but H_2S is continually removed by degassing or by precipitation

as insoluble sulfides, the metabolic S is depleted in ^{34}S but the isotope composition of S remains nearly constant.

3. Systems containing a limited amount of sulfate, but from which the metabolic H_2S can escape, are subject to Rayleigh distillation. Therefore, both the remaining sulfate as well as the H_2S become enriched in ^{34}S because of the systematic and irreversible loss of ^{32}S from the system.

The metabolic H_2S excreted by bacteria in anoxic marine and nonmarine basins may be precipitated as hydrotroilite ($FeS \cdot nH_2O$), which ultimately recrystallizes as pyrite (FeS_2). The S in such sedimentary sulfide minerals is generally depleted in ^{34}S, but the δ^{34}S values may vary stratigraphically, and may even become positive, depending on the extent of Rayleigh distillation. The wide range of variation of δ^{34}S values and the commonly observed depletion in ^{34}S distinguish sedimentary sulfides from those in igneous rocks whose δ^{34}S values cluster close to 0.0‰. However, the isotope composition of S alone is *not* a reliable criterion for distinguishing between sulfides of sedimentary and magmatic origin because:

1. The δ^{34}S values of some sedimentary sulfides may be near 0.0‰ or may even be positive.
2. The δ^{34}S values of some magmatic sulfides may be negative because the S originated from sedimentary rocks in the vicinity of igneous intrusions.
3. The δ^{34}S values of sulfide minerals in metamorphosed igneous or sedimentary rocks may be altered by isotope exchange reactions at elevated temperatures.

When sulfide minerals form by precipitation from a hot aqueous fluid or by crystallization from immiscible sulfide liquids in magmas, the S they contain may have different isotope compositions because of isotope exchange reactions with a common S reservoir. The magnitude of the *differences* in the δ^{34}S values of pairs of coexisting sulfide minerals *increases* with decreasing temperature. If the relevant isotope exchange reac-

tions achieved equilibrium, differences in δ^{34}S values of coexisting sulfide minerals may indicate the isotope equilibration temperature based on the experimentally determined calibration equations listed in Table 17.5.

Fractionation of S isotopes between sphalerite and galena has been investigated several times with somewhat different results. For example, at 200°C the equation of Grootenboer and Schwarcz (1969) yields $\alpha_{gn}^{sp} = 1.00285$, whereas that of Kajiwara and Krouse (1971) yields $\alpha_{gn}^{sp} = 1.00360$, where sp and gn refer to sphalerite and galena, respectively. Since in the example cited above $(\alpha - 1)10^3 < 10$, we can use the approximation:

$$10^3 \ln \alpha_{gn}^{sp} = \delta^{34}S_{sp} - \delta^{34}S_{gn} = \Delta_{gn}^{sp} \quad (17.54)$$

Ohmoto and Rye (1979) used this relationship to express the temperature dependence of S-isotope fractionation between sphalerite and galena:

$$\Delta_{gn}^{sp} = (7.2 \times 10^5)/T^2 \quad (17.55)$$

Evidently, Δ_{gn}^{sp}, which is defined as $\delta^{34}S(\text{sphalerite}) - \delta^{34}S(\text{galena})$, *decreases* with *increasing* equilibrium temperature. When $\Delta_{gn}^{sp} = 4.0$, equation 17.55 yields:

$$T = \left(\frac{7.2 \times 10^5}{4.0}\right)^{1/2} = 424.3 \text{ K} = 151\,°C \quad (17.56)$$

Such isotope equilibration temperatures can be checked by determining the filling temperatures of fluid inclusions in sphalerite.

We now return to S-isotope fractionation in the oceans and to the variation of the δ^{34}S parameter in marine sulfate minerals throughout geological time. Marine sulfate in present-day seawater has δ^{34}S = +20.0‰ throughout the oceans. When gypsum precipitates from sulfate-bearing solutions at Earth-surface temperatures, the solid is enriched in ^{34}S by only ~1.65‰. Therefore, marine gypsum and anhydrite effectively record the δ^{34}S value of sulfate ions in the oceans. Measurements of δ^{34}S values of marine gypsum/anhydrite deposits indicate that the δ^{34}S value of sulfate ions in the oceans has *varied systematically* throughout Phanerozoic and Precambrian time.

Table 17.5 Sulfur Isotope Fractionation Between Sulfide Minerals That Equilibrated Sulfur with the Same Reservoir at Elevated Temperatures in Kelvins

	Reference[a]
Pyrite–galena	
$10^3 \ln \alpha = (1.1 \times 10^6)/T^2$	(1)
Pyrite–sphalerite (or pyrrhotite)	
$10^3 \ln \alpha = (3.0 \times 10^5)/T^2$	(1)
Pyrite–chalcopyrite	
$10^3 \ln \alpha = (4.5 \times 10^5)/T^2$	(1)
Chalcopyrite–galena	
$10^3 \ln \alpha = (6.5 \times 10^5)/T^2$	(1)
Sphalerite–chalcopyrite and pyrrhotite–chalcopyrite	
$10^3 \ln \alpha = (1.5 \times 10^5)/T^2$	(1)
Sphalerite–galena	
$10^3 \ln \alpha = (8.9 \times 10^5)/T^2 - 0.57$	(2)
$10^3 \ln \alpha = (8.0 \times 10^5)/T^2$	(1)
$10^3 \ln \alpha = (7.0 \times 10^5)/T^2$	(3)
$10^3 \ln \alpha = (6.6 \times 10^5)/T^2 - 0.1$	(4)
Molybdenite–galena (experimental)	
$10^3 \ln \alpha = (1.3 \times 10^6)/T^2 - 0.80$	(5)
Molybdenite–sphalerite (experimental)	
$10^3 \ln \alpha = (0.71 \times 10^6)/T^2 - 0.15$	(5)
Molybdenite–pyrite (calculated, 400–750°C)	
$10^3 \ln \alpha = (0.48 \times 10^6)/T^2 - 0.75$	(5)
Molybdenite–chalcopyrite (calculated, 400–750°C)	
$10^3 \ln \alpha = (0.72 \times 10^6)/T^2 - 0.70$	(5)
Molybdenite–pyrrhotite (calculated, 400–750°C)	
$10^3 \ln \alpha = (0.65 \times 10^6)/T^2 - 1.65$	(5)
$SO_2(g)$–$H_2S(g)$ (287–1000°C)	
$10^3 \ln \alpha = (4.54 \times 10^6)/T^2 - 0.30$ (experimental)	(6)
$10^3 \ln \alpha = (3.65 \times 10^6)/T^2$ (calculated)	(6)
$HSO_4^-(aq)$–$H_2S(g)$ (200–320°C)	
$10^3 \ln \alpha = (5.1 \times 10^6)/T^2 + 6.3$	(7)
Sphalerite–$HS^-(aq)$ (50–340°C)	
$10^3 \ln \alpha = (1.11 \times 10^5)/T^2 + 1.36$	(2)
Galena–$HS^-(aq)$ (50–340°C)	
$10^3 \ln \alpha = (7.82 \times 10^5)/T^2 + 1.7$	(2)

[a](1) Kajiwara and Krouse (1971); (2) Kiyosu (1973); (3) Czamanske and Rye (1974); (4) Grootenboer and Schwarcz (1969); (5) Suvorova (1974); (6) Thode et al. (1971); (7) Robinson (1973).
SOURCE: Friedman and O'Neil (1977).

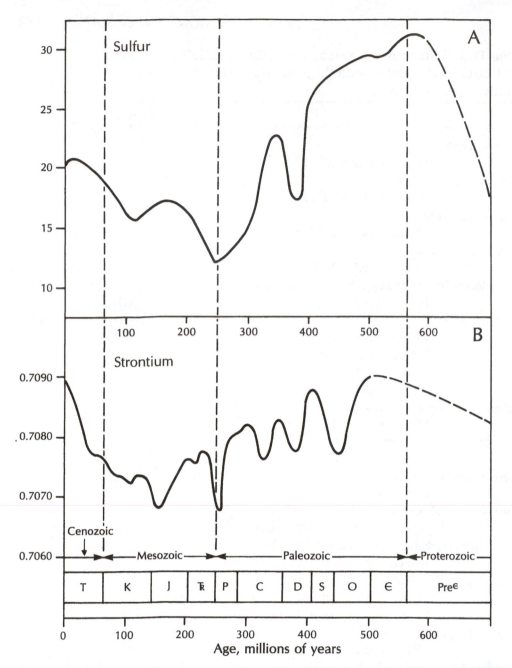

Figure 17.6 Variation of isotope compositions of S and Sr in the oceans in Phanerozoic time. Note that both $\delta^{34}S$ and the $^{87}Sr/^{86}Sr$ ratio declined irregularly during the Paleozoic Era from the Cambrian to the Permian periods. During the Mesozoic Era both parameters had low values but began to rise during the Cretaceous Period and continued to rise through the Tertiary Period. The observed variations of the isotopic compositions of S and Sr in the oceans were caused by global changes in the geochemical cycles of these elements. The curves for S and Sr are based on data by Claypool et al. (1980) and Burke et al. (1982), respectively. These references should be consulted for precise comparisons of the isotopic data pertaining to specific epochs or periods.

The isotope evolution of S in the oceans is remarkably similar to the isotopic evolution of Sr. A comparison of the two curves in Figure 17.6 shows that both the δ^{34}S value and the ^{87}S/^{86}Sr ratio in the oceans declined during the Paleozoic Era from high values during the Cambrian Period to low values during the Permian Period. Subsequently, both parameters remained low during the Mesozoic Era, but began to increase during the Cretaceous Period. The ^{87}Sr/^{86}Sr ratio has continued to increase steadily toward the Present (Figure 17.5), whereas the rise of the δ^{34}S parameter has faltered and may have been reversed in Late Tertiary time.

The isotope compositions of both elements have varied in response to changes in the magnitudes of Sr and S fluxes entering and leaving the oceans as well as in response to changes in the respective isotope compositions of the inputs and outputs. Sulfur enters the oceans primarily as the sulfate ion derived by erosion of sulfate deposits on the continents and by oxidation of sulfide minerals. In addition, S is discharged in hydrothermal fluids along spreading ridges on the ocean floor and by direct interactions between heated seawater and volcanic rocks of the oceanic crust. Sulfur is removed from the oceans by precipitation as insoluble metal sulfides and by direct precipitation of gypsum from seawater.

The isotope composition of S in the oceans is controlled by the δ^{34}S values of the inputs and outputs. For example, sulfate ions formed by oxidation of sedimentary sulfides in shale and limestone are depleted in ^{34}S and have negative δ^{34}S values, whereas sulfate ions released by erosion of marine evaporite rocks have positive δ^{34}S values. Therefore, the isotope composition of S entering the oceans from the continents at a particular time depends on the amounts of S each source contributes to the oceans. The output of S from the oceans by bacterial reduction of sulfate enriches the remaining sulfate in ^{34}S, whereas precipitation of gypsum lowers the S content of the oceans, making the isotope composition of the remaining S more susceptible to change. Evidently, the δ^{34}S value of the oceans as well as its ^{87}Sr/^{86}Sr reflect the operation of the S and Sr cycles, respectively, on the surface of the Earth.

17.10   Summary

Isotope fractionation is a common phenomenon that affects the isotope compositions of H, C, O, N, S, and of a few other elements of low atomic number. It takes place during chemical reactions and in the course of certain physical processes, such as evaporation, condensation, and diffusion, because of differences in the masses of the stable isotopes of these elements. Isotope fractionation during chemical reactions results from the fact that the heavy isotope is concentrated preferentially in a particular compound during isotope exchange reactions that have achieved equilibrium at a specific temperature.

Isotope fractionation also takes place during unidirectional reactions because the masses of isotopes influence the rates of chemical reactions owing to differences in the bond strengths of isotopic molecules of the same compound or because of differences in the diffusion rates or vapor pressures of those isotopic molecules. Kinetic isotope fractionation effects are variable depending on environmental conditions and decrease as the reaction approaches equilibrium.

The isotope compositions of O and H in water molecules may vary widely because of isotope fractionation in the hydrologic cycle. Water vapor in the atmosphere is initially depleted in ^{18}O and D during the evaporation of seawater. Subsequently, the vapor is further depleted in ^{18}O and D because of the preferential removal of these isotopes by condensation of water or snow from moist air masses drifting from the equatorial region toward higher latitudes. Consequently, meteoric water is depleted in ^{18}O and D and has negative δ^{18}O and δD values on the SMOW scale, which are related by the equation for the so-called meteoric-water line.

The extent of isotope fractionation of meteoric water is generally related to latitude and temperature and hence to climatic conditions. The ice sheets of Greenland and Antarctica have preserved a record of climate change that can be recovered by systematic measurements of the isotope compositions of H and O in cores drilled through the ice sheets.

When calcium carbonate precipitates from aqueous solution, either inorganically or by the action of organisms living in the water, the O in the solid phase is *enriched* in ^{18}O compared to the water. When isotopic equilibrium is maintained, the isotope composition of O in the calcite depends on the temperature and on the isotope composition of O in the water. This relationship can be used to detect changes in the temperature of the oceans in the geologic past, provided that the isotope composition of the water has remained constant.

When clay minerals form by decomposition of aluminosilicate minerals, the isotope compositions of O and H in the clay are controlled by the *water* at the site of the reactions. Clay minerals in soil are enriched in ^{18}O but depleted in D compared to the meteoric water with which they are in contact. Their $\delta^{18}O$ and δD parameters are therefore related by linear equations having similar slopes but different intercept values than the meteoric-water line.

The O in groundwater is slowly *enriched* in ^{18}O by exchange with the O of silicate or carbonate minerals. The H-isotope composition of water may also change depending on the H-content of the rocks and on the presence of hydrocarbons or H_2S. The change in the isotope compositions increases with the temperature and salinity of the water and hence with depth below the surface. Seawater, trapped in the pore spaces of marine sediment at the time of deposition, is also altered by isotope exchange reactions with the rocks with which it comes in contact.

Carbon in the biosphere is depleted in ^{13}C during photosynthesis reactions depending on the type of metabolism. Consequently, fossil fuels (coal, petroleum, and natural gas) are characteristically depleted in ^{13}C compared to the PDB standard. In addition, C-isotopes are fractionated by exchange reactions among CO_2 gas, carbonate ions in solution, and solid Ca carbonate. The process tends to enrich calcite or aragonite in ^{13}C relative to CO_2 gas of the atmosphere. However, when part or all of the CO_2 gas in an environment formed by oxidation of biogenic C compounds, the carbonate minerals inherit the biogenic C. Therefore, calcite cements in sandstones associated with petroleum pools and calcites in the caps of salt domes or associated with sedimentary U deposits may have negative $\delta^{13}C$ values on the PDB scale.

The isotope composition of Sr in marine carbonate rocks is *not* affected by isotope fractionation, but has nevertheless varied throughout geologic time as a result of mixing of different isotopic varieties of Sr entering the oceans. The isotope composition of Sr in stratigraphic sections of nonmarine carbonates may also vary in response to changes in the geology or hydrology of the drainage basin. The steady rise of the $^{87}Sr/^{86}Sr$ ratio of marine carbonates since the end of the Eocene epoch can be used to date carbonate rocks or fossil shells of Tertiary age.

The isotope composition of S is changed during the reduction of sulfate to sulfide by anaerobic bacteria. As a result, the remaining sulfate is enriched in ^{34}S, whereas the sulfide is enriched in ^{32}S and depleted in ^{34}S. Therefore, sulfide minerals in unmetamorphosed sedimentary rocks commonly have negative $\delta^{34}S$ values relative to troilite of Canyon Diablo whereas marine sulfate rocks have positive $\delta^{34}S$ values. The isotope composition of S in marine sulfate rocks of Precambrian and Phanerozoic age has varied systematically with time. The time-dependent S-isotope curve in the oceans is similar to the Sr-isotope curve, which suggests that both elements were affected by global geological changes that modified the cycles of Sr and S in similar ways.

Problems

1. Calculate $\delta^{18}O$ of water *vapor* in equilibrium with liquid water at 10°C assuming that $\delta^{18}O_l = -10.0‰$ (SMOW) and $\alpha_v^l = 1.0105$.   (Answer: $-20.3‰$)

2. Calculate $\delta^{18}O$ of liquid water that is condensing from vapor at 10°C having $\delta^{18}O = -25.0‰$ assuming that $\alpha_v^l = 1.0105$.   (Answer: $-14.8‰$)

3. Calculate the value of α_v^l at 35°C from equation 17.34.   (Answer: 1.008 43)

4. Calculate the value of δD of water *vapor* remaining in an air mass based on Rayleigh distillation for the case that $\alpha_v^l(D) = 1.074, f = 0.45, \delta D_v^0 = -94.8‰$
                    (Answer: $-146.7‰$)

5. Calculate the value of δD of *liquid* water condensing from an air mass whose initial $\delta D = -94.8‰$, if only 10% of the original vapor remain and $\alpha_v^l(D) = 1.074$.
                    (Answer: $-180.1‰$)

6. Calculate the $\delta^{18}O$ value of meteoric water whose $\delta D = -70‰$.   (Answer: $-10‰$)

7. What is the $\delta^{18}O$ value of calcite precipitated at 5°C in equilibrium with meteoric water having $\delta D = -85‰$?   (Answer: $+22.7‰$)

8. Given that a fluid inclusion filled with water in calcite that formed in a cave has a measured value of $\delta D = -175‰$ and that $\delta^{18}O$ of the calcite is $+10.0‰$ (SMOW). What was the temperature of deposition of the calcite in °C assuming that the water was of meteoric origin and isotopic equilibrium existed between calcite and water at the time of depositions?
                    (Answer: 3.8°C)

9. What is the $\delta^{18}O$ value of water on the SMOW scale in a fluid inclusion in calcite whose $\delta^{18}O = -2.5‰$ (PDB), if the O in the water reequilibrated with the O of the calcite at 15°C after deposition?
                    (Answer: $-2.70‰$)

10. Compare Problems 8 and 9 and explain why the measured $\delta^{18}O$ values of water in fluid inclusions in calcite or quartz cannot be used to determine the temperature of formation of these minerals.

11. Calculate $\delta^{18}O$ and δD of glauconite in equilibrium with seawater ($\delta^{18}O = 0‰$, $\delta D = 0‰$) using the fractionation factors in Table 17.2.
                    (Answer: $\delta^{18}O = +26‰$, $\delta D = -70‰$)

12. Calculate $\alpha_w^q(O)$ for the quartz–water at temperatures of 20, 40, 60, 80, and 100°C and plot the results in coordinates of α_w^q and $t°C$.
                    (Answer: 1.036 58 at 20°C, 1.031 55 at 40°C, 1.027 42 at 60°C, 1.023 98 at 80°C, and 1.021 09 at 100°C)

13. What is the $\delta^{13}C$ (PDB) value of calcite precipitated at 15°C in equilibrium with CO_2 gas having $\delta^{13}C = -25.0‰$ (PDB)?   (Answer: $-14.5‰$ (PDB))

14. Sulfate in a reservoir of limited size is being reduced by bacteria at 10°C and the resulting H_2S gas escapes from the system. Assume that Rayleigh distillation is occurring and plot $\delta^{34}S$ trajectories both for the remaining sulfate and for the H_2S gas for values of f from 1.0 to 0.01. Assume that $\alpha_{H_2S}^{SO_4} = 1.025$ and that $\delta^{34}S^0$(sulfate) $= +20‰$.

References

ANDERSON, T. F., and M. A. ARTHUR, 1983. Stable isotopes of oxygen and carbon and their application to sedimentologic and paleoenvironmental problems. In M. A. Arthur, T. F. Anderson, I. R. Kaplan, J. Veizer, and L. S. Land (Eds.), *Stable Isotopes in Sedimentary Geology.* SEPM Short Course No. 10, Soc. Econ. Paleont. Mineral., Tulsa, OK.

ARTHUR, M. A., T. F. ANDERSON, I. R. KAPLAN, J. VEIZER, and L. S. LAND (Eds.), 1983. *Stable Isotopes in Sedimentary Geology.* SEPM Short Course No. 10, Soc. Econ. Paleont. Mineral., Tulsa, OK.

BIGELEISEN, J., 1965. Chemistry of isotopes. *Science,* **147**:463–471.

BURKE, W. H., R. E. DENISON, E. A. HETHERINGTON, R. B. KOEPNICK, N. F. NELSON, and J. B. OTTO, 1982. Variation of seawater $^{87}Sr/^{86}Sr$ throughout Phanerozoic time. *Geology,* **10**:516–519.

CHENEY, E. S., and M. L. JENSEN, 1967. Corrections to carbon isotope data of Gulf Coast salt-dome cap rock. *Geochim. Cosmoshim. Acta,* **31**:1345–1346.

CLAYTON, R. N., I. FRIEDMAN, D. L. GRAF, T. K. MAYEDA, W. F. MEENTS, and N. F. SHIMP, 1966. The origin of saline formation waters. 1. Isotopic composition. *J. Geophys. Res.,* **71**(16):3869–3882.

CLAYPOOL, G. E., W. T. HOLSER, I. R. KAPLAN, H. SAKAI, and I. ZAK, 1980. The age curves of sulfur and oxygen isotopes in marine sulfate and their mutual interpretation. *Chem. Geol.,* **28**:199–260.

CZAMANSKE, G. K., and R. O. RYE, 1974. Experimentally determined sulfur isotope fractionations between sphalerite and galena in the temperature range 600° to 275°C. *Econ. Geol.,* **69**:17–25.

DEINES, P., D. LANGMUIR, and R. S. HARMON, 1974. Stable carbon isotope ratios and the existence of a gas phase in the evolution of carbonate groundwater. *Geochim. Cosmochim. Acta,* **38**:1147–1164.

DEINES, P., 1980. The isotopic composition of reduced organic carbon. In P. Fritz and J. Ch. Fontes (Eds.), *Handbook of Environmental Isotope Geochemistry,* vol. 1, 329–406. Elsevier, Amsterdam, 545 pp.

DEPAOLO, D. J., and B. L. INGRAM, 1985. High-resolution stratigraphy with strontium isotopes. *Science,* **227**:938–941.

DING, T., S. JIANG, D. WAN, Y. LI, J. LI, H. SONG, Z. LIU, and X. YAO, 1996. *Silicon Isotope Geochemistry.* Geol. Publ. House, Beijing, P. R. C., 125 pp.

DONOVAN, T. J., I. FRIEDMAN, and J. D. GLEASON, 1974. Recognition of petroleum-bearing traps by unusual isotopic compositions of carbonate-cemented surface rocks. *Geology,* **2**:351–354.

FAURE, G. 1982. The marine-strontium geochronometer. In G. S. Odin (Ed.), *Numerical Dating in Stratigraphy,* vol. 1, 73–79. Wiley, Chichester, England, 630 pp.

FAURE, G., 1986. *Principles of Isotope Geology,* 2nd ed. Wiley, New York, 589 pp.

FRIEDMAN, I., and J. R. O'NEIL, 1977. Compilation of stable isotope fractionation factors of geochemical interest. In M. Fleischer (Ed.), *Data of Geochemistry,* 6th ed., ch. kk. U. S. Geol. Surv. Prof. Paper 440.

FRITZ, P., and J. CH. FONTES (Eds.), 1980a. *Handbook of Environmental Isotope Geochemistry,* vol. 1. Elsevier, Amsterdam, 545 pp.

FRITZ, P., and J. CH. FONTES, 1980b. Introduction. In P. Fritz and J. Ch. Fontes (Eds.), *Handbook of Environmental Isotope Geochemistry,* vol. 1, 1–18. Elsevier, Amsterdam, 545 pp.

FRITZ, P., and J. CH. FONTES (Eds.), 1986. *Handbook of Environmental Isotope Geochemistry,* vol. 2. Elsevier, Amsterdam, 557 pp.

FRY, B., J. COX, H. GEST, and J. M. HAYES, 1986. Discrimination between ^{34}S and ^{32}S during bacterial metabolism of inorganic sulfur compounds. *J. Bacteriol.,* **165**:328–330.

GROOTENBOER, J., and H. P. SCHWARCZ, 1969. Experimentally determined sulfur isotope fractionations between sulfide minerals. *Earth Planet. Sci. Lett.,* **7**:162–166.

HITCHON, B., and I. FRIEDMAN, 1969. Geochemistry and origin of formation waters in the western Canada sedimentary basin—I. Stable isotopes of hydrogen and oxygen. *Geochim. Cosmochim. Acta,* **33**:1321–1349.

HOEFS, J., 1996. *Stable Isotope Geochemistry,* 4th ed. Springer-Verlag, Berlin, 241 pp.

HURST, R. W., 1986. Strontium isotope chronostratigraphy and correlation of the Miocene Monterey Formation in the Ventura and Santa Maria basins of California. *Geology,* **14**:459–462.

KAJIWARA, Y., and H. R. KROUSE, 1971. Sulfur isotope partitioning in metallic sulfide systems. *Can. J. Earth Sci.,* **8**:1397–1408.

KEITH, M. L., and J. N. WEBER, 1964. Carbon and oxygen isotopic composition of selected limestones and fossils. *Geochim. Cosmochim. Acta,* **28**:1787–1816.

KHARAKA, Y. K., and W. W. CAROTHERS, 1986. Oxygen and hydrogen isotope geochemistry of deep basin brines. In P. Fritz and J. Ch. Fontes (Eds.), *Handbook of Environmental Isotope Geochemistry,* vol. 2, 305–360. Elsevier, Amsterdam, 557 pp.

KIYOSU, Y., 1973. Sulfur isotope fractionation among sphalerite, galena and sulfide ions. *Geochem. J.,* **7**:191–199.

JOHNSEN, S., J. W. DANSGAARD, H. B. CLAUSEN, and C. C. LANGWAY, JR., 1972. Oxygen isotope profiles through the Antarctic and Greenland ice sheets. *Nature,* **235**:429–434.

LAWRENCE, J. R., and H. P. TAYLOR, 1971. Deuterium and oxygen-18 correlation: clay minerals and hydroxides in Quaternary soils compared to meteoric waters. *Geochim. Cosmochim. Acta,* **35**:993–1003.

LORD, B. K., L. M. JONES, and G. FAURE, 1988. Evidence for the existence of the Gondwana ice sheet in the ^{18}O depletion of carbonate rocks in the Permian formations of the Transantarctic Mountains. *Chem. Geol., Isotope Geosci. Sect.,* **8**:163–172.

LORIUS, C., J. JOUZEL, C. RITZ, L. MERLIVAT, N. I. BARKOV, Y. S. KOROTKEVICH, and V. M. KOTLYAKOV, 1985. A 150,000-year climatic record from Antarctic ice. *Nature,* **316**:591–596.

LOWRY, R. M., G. FAURE, D. I. MULLET, and L. M. JONES, 1988. Interpretation of chemical and isotopic compositions of brines based on mixing and dilution, "Clinton" sandstones, eastern Ohio, U.S.A. *Appl. Geochem.,* **3**:177–184.

MOSER, H., and W. STICHLER, 1981. Environmental isotopes in ice and snow. In P. Fritz and J. Ch. Fontes (Eds.), *Handbook of Environmental Isotope Geochemistry,* vol. 1, 141–178. Elsevier, Amsterdam, 545 pp.

NEAT, P. L., G. FAURE, and W. J. PEGRAM, 1979. The isotopic composition of strontium in non-marine carbonate rocks: The Flagstaff Formation of Utah. *Sedimentology,* **26**:271–282.

OHMOTO, H., and R. O. RYE, 1979. Isotopes of sulfur and carbon. In H. L. Barnes (Ed.), *Geochemistry of Hydrothermal Ore Deposits,* 2nd ed., 509–567. Wiley, New York, 798 pp.

O'NEIL, J. R., and Y. K. KHARAKA, 1976. Hydrogen and oxygen isotope exchange reactions between clay minerals and water. *Geochim. Cosmochim. Acta,* **40**:241–246.

ROBINSON, R., 1973. Sulfur isotope equilibrium during sulfur hydrolysis at high temperature. *Earth Planet. Sci. Lett.,* **18**:443–450.

RUNDEL, P. W., J. R. EHLERINGER, and K. A. NAGY (Eds.), 1988. *Stable Isotopes in Ecological Research.* Springer-Verlag, New York, 570 pp.

SAVIN, S. M., and S. EPSTEIN, 1970. The oxygen and hydrogen geochemistry of clay minerals. *Geochim. Cosmochim. Acta,* **34**:25–42.

STANLEY, K. O., and G. FAURE, 1979. Isotopic composition and source of strontium in sandstone cements: The High Plains Sequence of Wyoming and Nebraska. *J. Sed. Pet.,* **49**:45–54.

SUVOROVA, V. A., 1974. Temperature dependence of the distribution coefficient of sulfur isotopes between equilibrium sulfides. Fifth National Symp. on Stable Isotope Geochemistry, Moscow, Pt. 1, 128 pp.

THODE, H. G., C. B. CRAGG, J. B. HULSTON, and C. E. REES, 1971. Sulfur isotope exchange between sulfur dioxide and hydrogen sulfide. *Geochim. Cosmochim. Acta,* **35**:35–45.

VALLEY, J. W., H. P. TAYLOR, JR., and J. R. O'NEIL (eds.), 1986. Stable isotopes in high temperature geological processes. *Rev. Mineral.,* vol. 16. Mineral. Soc. Amer., Washington, DC, 570 pp.

VEIZER, J., and J. HOEFS, 1967. The nature of $^{18}O/^{16}O$ and $^{13}C/^{12}C$ secular trends in sedimentary carbonate rocks. *Geochim. Cosmochim. Acta,* **40**:1387–1395.

WEAST, R. C., M. J. ASTLE, and W. H. BEYER (Eds.), 1986. *CRC Handbook of Chemistry and Physics.* CRC Press, Boca Raton, FL.

YEH, H. W., 1980. D/H ratios and late stage hydration of shales during burial. *Geochim. Cosmochim. Acta,* **45**:341–352.

YURTSEVER, Y., 1975. Worldwide survey of stable isotopes in precipitation. Rept. Section Isotope Hydrology, International Atomic Energy Agency, Vienna, 40 pp.

18

Mixing and Dilution

Many natural processes on the surface of the Earth cause mixing of two or more components having different chemical and/or isotopic compositions. This phenomenon applies especially to compositionally distinct water masses that come in contact with each other in estuaries, in streams and lakes, and below the surface of the Earth. Mixing also takes place in depositional basins where detrital mineral particles of different compositions are deposited. The concept even applies to blending of magmas and to the assimilation of country rock by cooling magmas in the continental crust.

The chemical and isotopic compositions of two-component mixtures are related to their end members in simple ways expressed by characteristic linear or hyperbolic patterns formed by their chemical and isotopic compositions. These relations are useful because they help us to recognize mixtures when they occur in nature and because the composition of the pure end members can be deduced from them under favorable circumstances.

18.1 Binary Mixtures

When two components (A and B) of different chemical composition mix in varying proportions, the chemical compositions of the resulting mixtures (M) vary systematically depending on the relative abundances of the end members. We therefore define the mixing parameter:

$$f_A = \frac{W_A}{W_A + W_B} \qquad (18.1)$$

where W_A and W_B are the weights (or volumes) of the components A and B in a given mixture.

Evidently, f_A is a dimensionless number that in reality is constrained to values between zero and unity. When f_A is defined as in equation 18.1, then $f_A = 1.0$ means that the mixture is composed of component A only. When $f_A = 0$, then $W_A = 0$ and only component B is present. The mixing parameter defined above is the "weight fraction" or "volume fraction" of component A in two-component mixtures. The complementary weight fraction of B (f_B) is equal to $1 - f_A$ because:

$$1 - f_A = 1 - \frac{W_A}{W_A + W_B} = \frac{W_A + W_B - W_A}{W_A + W_B}$$

$$= \frac{W_B}{W_A + W_B} = f_B \qquad (18.2)$$

The concentration of any conservative element (X) in a binary mixture of A and B depends on the concentration of that element in components A and B and on the abundances of components A and B in the mixture. Therefore, the concentration of element X in a mixture (M) of components A and B is:

$$(X)_M = (X)_A f_A + (X)_B (1 - f_A) \quad (18.3)$$

where the parentheses mean *concentration*. Equation 18.3 yields the concentrations of element X in mixtures of A and B for different values of f_A between 1.0 and 0 where X_A and X_B are known constants. By rearranging equation 18.3 we obtain:

$$(X)_M = f_A[(X)_A - (X)_B] + (X)_B \quad (18.4)$$

Equation 18.4 is a straight line in coordinates of (X) and f_A having a slope $m = (X)_A - (X)_B$ and an intercept $b = (X)_B$. Evidently, when $f_A = 1.0$, equation 18.4 yields $(X)_M = (X)_A$ (the mixture is composed of A only), and when $f_A = 0.0$, $(X)_M = (X)_B$ (the mixture is composed of B only). If the concentration of element X is expressed in units of weight of X per unit *volume* of A and B, then f_A must also be defined in terms of volumes.

All conservative elements in two-component mixtures satisfy equations like 18.3. Therefore, for another element Y we have:

$$(Y)_M = (Y)_A f_A + (Y)_B (1 - f_A) \quad (18.5)$$

In any given mixture of two components the mixing parameter f_A must have the same value for all conservative elements. This statement allows us to combine equations 18.4 and 18.5 by solving both for f_A and by equating the results:

$$f_A = \frac{(X)_M - (X)_B}{(X)_A - (X)_B} = \frac{(Y)_M - (Y)_B}{(Y)_A - (Y)_B} \quad (18.6)$$

By cross multiplying and solving for $(Y)_M$ we obtain:

$$(Y)_M = (X)_M \frac{(Y)_A - (Y)_B}{(X)_A - (X)_B}$$
$$+ \frac{(X)_A (Y)_B - (X)_B (Y)_A}{(X)_A - (X)_B} \quad (18.7)$$

Equation 18.7 is a straight line in coordinates of (X) and (Y) having a slope:

$$m = \frac{(Y)_A - (Y)_B}{(X)_A - (X)_B} \quad (18.8)$$

and an intercept on the (Y)-axis:

$$b = \frac{(X)_A (Y)_B - (X)_B (Y)_A}{(X)_A - (X)_B} \quad (18.9)$$

Therefore, the slope and intercept of the mixing line depend only on the concentration of X and Y in components A and B. Equation 18.7 indicates that the concentrations of any two conservative elements X and Y in a series of binary mixtures lie on straight lines known as "mixing lines," which include both end members.

A linear relationship between the concentrations of two conservative elements in a suite of *natural* samples is evidence that they may be the products of mixing. In such a case, the slope and intercept of the mixing equation can be determined by least-squares regression of the data array. The composition of the endmembers is constrained by the requirement that they must lie on the mixing line and that their concentrations of elements X and Y must be positive numbers. In case the concentration of either X or Y in the end members is known (e.g., $(X)_A$ and $(X)_B$), then the concentrations of the other element in the two components $(Y)_A$ and $(Y)_B$ can be calculated from the equation of the mixing line. In addition, if $(X)_A$ and $(X)_B$ (or $(Y)_A$ and $(Y)_B$) are known, then either element can be used to calculate f_A from equation 18.3 using the measured concentrations $(X)_M$ or $(Y)_M$ of any mixture in the suite.

An example of mixing occurs in the waters of the North Channel of Lake Huron between Manitoulin Island and the mainland of Canada where water from Lake Superior mingles with water in Lake Huron (Faure et al., 1967). The concentrations of Ca^{2+} and Sr^{2+} in the waters form a linear array of data points in Figure 18.1. The mixing equation fitted to these data by least-squares regression is:

$$(Ca^{2+})_M = 0.178(Sr^{2+})_M + 9.81 \quad (r^2 = 0.989)$$
$$(18.10)$$

where $(Ca^{2+})_M$ is the concentration in mg/L, $(Sr^{2+})_M$ is expressed in $\mu g/L$, and r^2 is the linear correlation coefficient. Next, we use equation 18.9 to calculate the concentrations of Sr^{2+} in the water of Lake Huron and Lake Superior based on $(Ca^{2+})_S = 14.4 \pm 0.2$ mg/L and $(Ca^{2+})_H = 26.9 \pm 0.3$ mg/L, where the subscripts S and H identify Superior and Huron, respectively, and the errors are one standard deviation of the mean. The results are $(Sr^{2+})_S = 25.8 \pm 1.1$ $\mu g/L$ and $(Sr^{2+})_H = 96.0 \pm 1.7$ $\mu g/L$.

This information now enables us to calculate the volume fraction of Lake Superior water in any water sample in the North Channel based on its measured concentration of either $(Ca^{2+})_M$ or $(Sr^{2+})_M$. If we select a water sample having $(Ca^{2+})_M = 22.0$ mg/L, we find from equation 18.3 that $f_S = 0.39$, which means that 39% by volume of this water originated from Lake Superior. Note

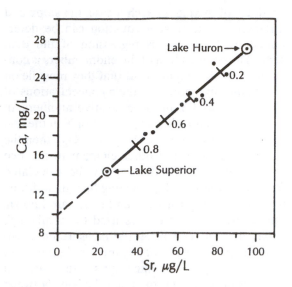

Figure 18.1 Binary mixing of water in the North Channel of Lake Huron between Manitoulin Island and the mainland of Canada. The concentrations of Ca^{2+} and Sr^{2+} define a mixing line whose equation was fitted to the data points by least-squares regression (equation 18.10). The average concentrations of Sr^{2+} in Lake Superior and Lake Huron were calculated by substituting the average measured Ca^{2+} concentrations into equation 18.10. The mixing line was subdivided in 0.2 increments of f_s, which is the volume fraction of Lake Superior water in the North Channel in Lake Huron (data from Faure et al., 1967).

that the mixing line in Figure 18.1 has been subdivided in increments of f_2 to show how the concentrations of (Ca^{2+}) and (Sr^{2+}) vary in relation to the mixing parameter.

Real data sets representing binary mixtures commonly scatter above and below the mixing line, indicating that the assumptions of the mixing model are not strictly satisfied in nature. These assumptions are:

1. The elements under consideration are conservative; that is, their concentrations are not altered by biological activity or by adsorption on particles.

2. Only two components are contributing to the chemical composition of the mixtures.

3. The compositions of the end members are constant.

The scatter of data points about the mixing line formed by real data derived from surface water may be caused by dilution with rain or by the addition of a third component of natural or anthropogenic origin. Moreover, the chemical compositions of surface waters also vary seasonally, reflecting changes in the provenance of the water.

The chemical compositions of *sediment* formed by mixing of mineral grains derived from different sources also tend to scatter because they too may be diluted or because the chemical composition of sediment derived from different sources may vary randomly within certain limits.

18.2 Dilution

When a subsurface brine is progressively mixed with meteoric water, a special kind of linear array is produced, which is directed toward the origin. The resulting *dilution line* is a special case of equation 18.7, where the coordinates of one of the components are virtually equal to zero, excluding exceptional circumstances. If component A is the dilute end member $((X)_A = 0, (Y)_A = 0)$, then equation 18.7 reduces to:

$$(Y)_M = (X)_M \left(\frac{(Y)_B}{(X)_B} \right) \qquad (18.11)$$

Therefore, data arrays that fit a straight line through the origin may have been generated by dilution.

Although dilution of *brines* with meteoric water is a common phenomenon in the subsurface, in estuaries, and in the open ocean, dilution is by no means restricted to the hydrosphere. For example, the concentration of K_2O and Na_2O in arkosic sandstones or in shaley limestones both vary in response to the presence of quartz and calcite, respectively, which dilute the minerals containing the alkali metals.

Figure 18.2 contains the concentrations of K_2O and Na_2O of 29 selected quartz *sandstones* taken from a compilation by Pettijohn (1963). The

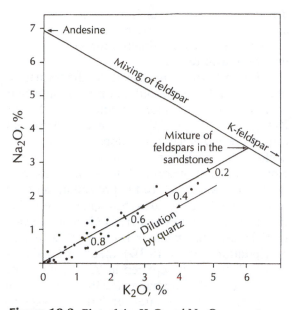

Figure 18.2 Plot of the K_2O and Na_2O concentrations of 29 selected sandstones from a compilation by Pettijohn (1963). The data points scatter about a straight line whose equation was derived by least-squares regression (equation 18.12). The alkali metals in these sandstones are assumed to reside in a mixture of K-feldspar and the plagioclase (Ab = 60% andesine) that lie on a mixing line between K-feldspar and andesine. The coordinates of the point of intersection of equation 18.12 with the feldspar mixing line represent the concentrations of K_2O and Na_2O of the feldspar in the sandstones. The addition of increasing amounts of quartz progressively dilutes the feldspar component and lowers the concentrations of K_2O and Na_2O of the resulting sandstones. The treatment of data is conjectural, but serves to illustrate the effect of dilution on the chemical composition of sandstone.

data points form a linear array directed toward the origin, which indicates that in this case the mineral constituents containing K and Na are progressively diluted with quartz. The equation of the dilution line, based on a least-squares regression, is:

$$(Na_2O) = 0.55(K_2O) + 0.05 \quad (18.12)$$

where (Na_2O) and (K_2O) are concentrations in weight percent. The linear correlation coefficient

is $r^2 = 0.9335$, which indicates a high degree of positive correlation between the two elements.

For the sake of this demonstration we assume that the minerals containing K_2O and Na_2O in these sandstones are K-feldspar and plagioclase having the composition of andesine (Ab = 60%, Section 8.3), respectively. Mixtures of these feldspars lie on a straight line between two points on the K_2O and Na_2O axes in Figure 18.2 that represent K-feldspar $[(K_2O) = 16.9\%, \ (Na_2O) = 0.0\%]$ and andesine $[(K_2O) = 0.0\%, \ (Na_2O)=6.9\%]$. The quartz dilution line of the sandstones intersects the feldspar mixing line at $(K_2O) = 6.0\%$ and $(Na_2O) = 3.5\%$. As increasing amounts of quartz are added to this mixture of feldspars, the concentrations of K_2O and Na_2O decline, but the $(Na_2O)/(K_2O)$ ratio remains constant. When the sediment consists of nearly pure quartz (orthoquartzite), the concentrations of K_2O and Na_2O approach zero.

The abundance of quartz in these feldspar-bearing sandstones can be calculated by modifying equation 18.3. Let f_Q be defined as:

$$f_Q = \frac{W_Q}{W_Q + W_F} \quad (18.13)$$

Where Q and F identify quartz and the mixture of feldspars, respectively. Then, from equation 18.4:

$$(K_2O)_M = f_Q[(K_2O)_Q - (K_2O)_F] + (K_2O)_F \quad (18.14)$$

where $(K_2O)_M$ is the K-concentration of a mixture (i.e., a sandstone). Equation 18.14 was used to subdivide the dilution line in Figure 18.2 in increments of 0.2 of the mixing parameter, assuming that $(K_2O)_Q = 0.0\%$ and $(K_2O)_F = 6.0\%$. Although the treatment of data is based on assumptions that cannot be confirmed in this case, this example demonstrates that the chemical compositions of mixtures of detrital mineral grains may be treated in terms of mixing and dilution.

18.3   Evaporative Concentration

A linear array of concentrations of conservative elements may also be generated in waters undergoing *evaporative concentration* (see Section 10.1). As water is lost, the concentrations of conservative ions or molecules in the solution rise in such a way that their ratios remain constant. Therefore, the concentrations of any two conservative elements (X and Y) vary along straight lines illustrated in Figure 18.3A having slopes $m = (Y)/(X)$ and whose intercepts are $b = 0$.

The concentration of a conservative element X in a solution undergoing evaporative concentration, such that X is *not* involved in the formation of a precipitate, obeys equations of the form:

$$(X)_f = g(X)_i \qquad (18.15)$$

where subscripts f and i signify *final* and *initial* concentrations of X, respectively, and g is the concentration factor. Another conservative element Y in the same solution obeys a similar equation, and the concentrations of both elements are subject to the same concentration factor. Therefore, we can combine the equations for X and Y by equating the concentration factors:

$$g = \frac{(X)_f}{(X)_i} = \frac{(Y)_f}{(Y)_i} \qquad (18.16)$$

from which it follows that:

$$(Y)_f = (X)_f \left(\frac{(Y)_i}{(X)_i} \right) \qquad (18.17)$$

where $((Y)/(X))_i$ is the slope of a straight line that passes through the origin and the concentrations of elements X and Y may be expressed either in terms of weights or numbers of moles of X and Y.

When one of the ions participates in the formation of an insoluble compound, the increase in its concentration is limited by the requirement that the relevant ion activity product (IAP) must remain less than the solubility product constant (K_{sp}) of the compound. In case progressive evaporative concentration of water ultimately forces an insoluble compound containing element X or Y to precipitate, the chemical *composition* of the solution is changed because the concentration of that element (X, for example) declines rapidly as progressive evaporation forces the compound out of the solution, whereas the concentration of the other element (Y) continues to rise. This process is illustrated in Figure 18.3B. As a result, waters from which a compound has precipitated become *depleted* in the ions that formed the compound, and therefore such waters *deviate* from the straight line that was initially generated by evaporative concentration.

If both X and Y participate in the formation of an insoluble compound and if they enter that compound in equal *molar* amounts (i.e., XY or X_2Y_2), then the precipitation of the compound causes the *molar* concentrations of X and Y to decline along a linear trajectory whose slope is equal to unity. If the ratio of the *molar* concentrations in the water before the precipitation of the compound is $Y/X > 1.0$ as shown in Figure 18.3C, the solution becomes depleted in X and enriched in Y. In case the *molar* ratio $Y/X < 1.0$ before precipitation (Figure 18.3D), then the solution is enriched in X and depleted in Y. Therefore, the precipitation of a compound involving the ions of elements X and Y in waters undergoing evaporative concentration functions as a *geochemical divide,* which alters the *chemical composition* of waters that are subjected to this process.

18.4 Ternary Mixtures

When *three* components are mixed in varying proportions, the resulting mixtures lie within a *mixing triangle* whose corners are occupied by the three components. Mixtures of three or more components form distribution patterns in X–Y diagrams, colloquially known as "shotgun patterns," that are difficult to interpret quantitatively. However, when the compositions of the three components are known, even scattered data points representing mixtures of three components can be resolved into their end members.

Ternary mixtures in which one of the components is located at the origin can be treated as though they resulted from a combination of mixing of two components followed by dilution of the resulting mixtures by the third component. In the general case, the binary mixtures of components A and B may be distributed along a mixing line *AB* as shown schematically in Figure 18.4. Dilution by component C subsequently displaces them from the mixing line into a triangle of mixing that has one of its corners at the origin.

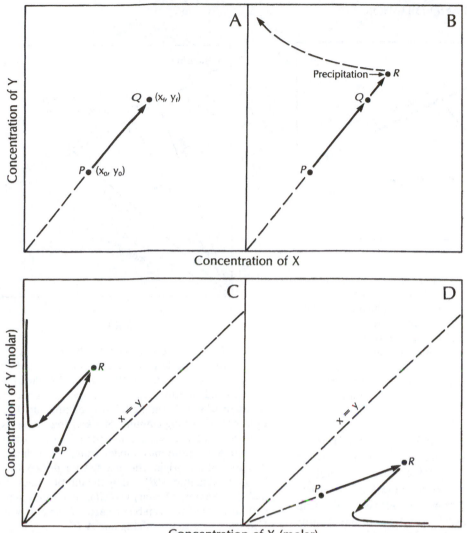

Figure 18.3 Effect of evaporative concentration of aqueous electrolyte solutions of the concentrations of two conservative elements X and Y.

A. Evaporation increases the concentrations of both elements from their initial values at point P to final values at point Q. The evaporation trajectory is a straight line represented by equation 18.17.

B. Continued evaporation increases the concentrations of X and Y to point R, when a compound begins to precipitate that removes element X from the solution but does not contain element Y. Continued evaporation depletes the solution in X but allows the concentration of Y to increase. Hence, the solution moves off the linear trajectory.

C. Evaporative concentration of a solution drives the *molar* concentrations from P to R, when a compound XY begins to precipitate. As a result, the molar concentrations of both X and Y decline along a line of slope $m = 1.0$. Since the ratio Y/X in the original solution is greater than unity, the concentration of X is reduced to near zero, whereas that of Y increases as evaporative concentration continues. The final brine is enriched in Y and is depleted in X

D. Similar to **C** above, except that the solution is depleted in Y and enriched in X because initially at point P the ratio of *molar* concentration Y/X was less than unity. Cases **C** and **D** illustrate the concept of a geochemical divide, which was first mentioned in Section 10.1.

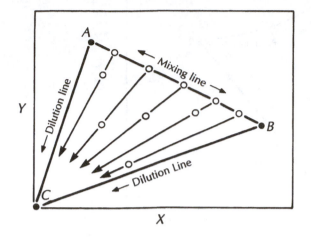

Figure 18.4 Schematic representation of mixing of two components A and B followed by variable dilution with component C. The resulting mixtures initially lie on the mixing line *AB*, but are subsequently drawn into a triangle of mixing by additions of component C whose concentrations of elements X and Y are both zero. The sandstone data in Figure 18.2 are a special case of ternary mixing in which the feldspar mixtures have a narrowly defined range of compositions.

The concentrations of K_2O and Na_2O of the sandstones in Figure 18.2 are a special case of mixing and dilution in which the hypothetical feldspar mixtures appear to have fairly uniform compositions. As a result, progressive dilution with quartz generates a linear array of data points directed toward the origin. However, *greywackes* from Pettijohn (1963) have more variable K_2O and Na_2O concentrations than quartz sandstones and form a cluster of data points in Figure 18.5. The alkali metals in greywackes probably reside not only in the feldspars but also in clay minerals, micas, and even in amphiboles and pyroxenes. Therefore, the quantitative interpretation based on Figure 18.5 is intended *only* as an illustration of the procedure and should *not* be taken literally.

The mixing triangle in Figure 18.5 was formed by drawing dilution lines from the origin to the most widely scattered samples and by extending these lines to the K-feldspar–andesine mixing line. The points A' and B' encompass the

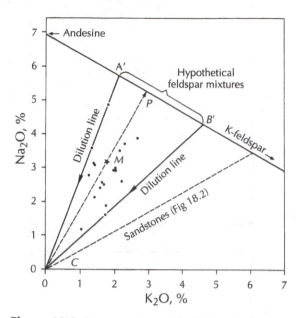

Figure 18.5 Interpretation of the K_2O and Na_2O concentrations of 20 greywackes from Pettijohn (1963) in terms of ternary mixing. We assume that the alkali metals reside in mixtures of K-feldspar and andesine between A' and B' on the feldspar mixing line. The presence of varying amounts of *alkali-free* minerals (component C) causes the feldspar mixtures to scatter within a hypothetical mixing triangle, which has one apex at the origin. The mixing triangle is defined by the mixing line $A'B'$ and by the dilution lines $A'C$ and $B'C$. Sample M (star) was chosen at random to demonstrate how it can be projected from component C at the origin to the mixing line $A'B'$.

entire compositional range of the hypothetical feldspar mixtures in this suite of greywackes. The starred sample (M) in Figure 18.5 was returned to the hypothetical feldspar mixing line by projecting it from the origin to point P. The coordinates of point P are $(K_2O) = 2.9\%$, $(Na_2O) = 5.25\%$, which indicates that the concentration of K-feldspar in this mixture is:

$$\frac{2.9 \times 2 \times 278.33}{94.20} = 17.1\% \qquad (18.18)$$

where 94.20 is the formula weight of K_2O and 278.33 is the formula weight of K-feldspar.

Consequently, the abundance of andesine is $100 - 17.1 = 82.9\%$. The abundance of the *akali-free* phase (f_C) in sample M is ~40% based on equation 18.14, given that $(K_2O)_M = 1.75\%$ and $(K_2O)_F = 2.9\%$. Therefore, the abundances of the hypothetical mineral components in sample M are:

K-feldspar:	$17.1 \times 0.6 = 10.3\%$
plagioclase:	$82.9 \times 0.6 = 49.7\%$
alkali-free minerals	$= 40\%$

Mixing and dilution may also take place in subsurface where two connate brines may mix in varying proportion and where these mixed brines are subsequently diluted with meteoric water from the surface. In such cases, one may wish to eliminate the effect of dilution from the concentrations of conservative elements in the brine mixtures. This can be accomplished by projecting data points from the origin to the mixing line, as already illustrated in Figure 18.5.

Mixing and dilution were demonstrated by Lowry et al. (1988) for brines from the Clinton sandstones of Early Silurian age in eastern Ohio. Figure 18.6 contains a plot of the concentrations of Na^+ and Sr^{2+} in these brines that scatter widely within a mixing triangle. The coordinates of the brine components A and B were derived by Lowry et al. (1988) by pairing Na and Sr with several other elements in X–Y plots. The data points in Figure 18.6 were projected from the origin to the mixing line AB. The coordinates of the projection points can be precisely calculated by simultaneous solution of the equations of the mixing line and the projection (dilution) lines using equations 18.7 and 18.11, respectively. Alternatively, the problem can be solved graphically using diagrams drawn to scale.

18.5 Isotopic Mixtures of One Element

The isotope compositions of certain elements are variable because some of their isotopes are the products of naturally occurring radioactive par-

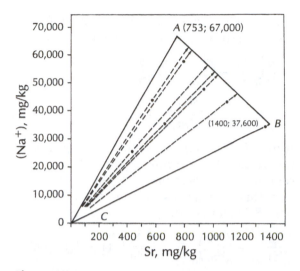

Figure 18.6 Restoration of oilfield brines from the Clinton sandstones of eastern Ohio that were contaminated with varying amounts of meteoric water from the surface. The compositions of the brine components A and B were determined by Lowry et al. (1988) from X–Y plots in which Na and Sr were paired with several other elements. The concentrations of Na^+ and Sr^{2+} in the *undiluted* brine mixtures are the coordinates of the points of intersection of the projection/dilution lines with the mixing line of brines A and B.

ents (Chapter 16) or because of isotope fractionation (Chapter 17). Therefore, we can consider cases where mixing occurs between two materials that contain the same element with different isotope compositions. For example, water masses having different $^{87}Sr/^{86}Sr$ ratios or $\delta^{18}O$ values may mix in lakes, in estuaries, or in the subsurface.

We consider first the formation of mixtures of two end members A and B having different $^{87}Sr/^{86}Sr$ ratios and Sr concentrations. Using the mixing parameter f_A defined in equation 18.1, we express the Sr concentration of any mixture M as:

$$(Sr)_M = (Sr)_A f_A + (Sr)_B(1 - f_A) \qquad (18.19)$$

If the end members have different $^{87}Sr/^{86}Sr$ ratios, then the isotopic composition of any mixture must be the weighted sum of the $^{87}Sr/^{86}Sr$ ratios where the weighting factors are $(Sr)_A/(Sr)_M$ and $(Sr)_B/(Sr)_M$ (a detailed derivation was given by

Faure, 1986). Therefore, the equation for the $^{87}Sr/^{86}Sr$ ratio of a binary mixture is:

$$\left(\frac{^{87}Sr}{^{86}Sr}\right)_M = \left(\frac{^{87}Sr}{^{86}Sr}\right)_A f_A \frac{(Sr)_A}{(Sr)_M}$$

$$+ \left(\frac{^{87}Sr}{^{86}Sr}\right)_B (1 - f_A) \frac{(Sr)_B}{(Sr)_M} \quad (18.20)$$

If the $^{87}Sr/^{86}Sr$ ratio of A and B are equal, equation 18.20 reduces to equation 18.19. If the Sr concentrations of components A and B are equal, then $(Sr)_A/(Sr)_M = (Sr)_B/(Sr)_M = 1.0$ and equation 18.20 reduces to:

$$\left(\frac{^{87}Sr}{^{86}Sr}\right)_M = \left(\frac{^{87}Sr}{^{86}Sr}\right)_A f_A + \left(\frac{^{87}Sr}{^{86}Sr}\right)_B (1 - f_A) \quad (18.21)$$

Equations 18.19 and 18.20 can be used to predict the Sr concentrations and $^{87}Sr/^{86}Sr$ ratio of any mixture of two components having different concentrations and isotope compositions of Sr for different values of the mixing parameter f_A. The same equations apply to all other elements that have radiogenic isotopes such as Nd, Os, Hf, and Pb.

Equation 18.21 applies also to mixtures of the same compound, such as water or carbon dioxide, containing equal concentrations of H and O, or C and O, respectively. In order to adapt equation 18.21 to mixtures of water containing O and H of different isotope compositions, we replace the $^{87}Sr/^{86}Sr$ ratios with $\delta^{18}O$ or δD values:

$$\delta^{18}O_M = \delta^{18}O_A f_A + \delta^{18}O_B(1 - f_A) \quad (18.22)$$

$$\delta D_M = \delta D_A f_A + \delta D_B(1 - f_A) \quad (18.23)$$

Note, however, that the concentrations of O and H in highly concentrated brines are *less* than those of pure water and that equation 18.20 may have to be used to predict the $\delta^{18}O$ and δD values of mixtures of such brines with other brines or with fresh water.

When two components having different concentrations and isotope compositions of Sr (or of another element) are mixed in varying proportions, the resulting mixtures differ from the end members both in terms of the Sr concentration and the $^{87}Sr/^{86}Sr$ ratio. The Sr concentrations and $^{87}Sr/^{86}Sr$ ratios of a series of mixtures can be calcu-

lated from equations 18.19 and 18.20 for selected values of the mixing parameter f_A based on end members of known composition. However, in reality we are usually confronted by the opposite situation, that is, we measure the $^{87}Sr/^{86}Sr$ ratios and Sr concentrations of a suite of samples and then wish to test the hypothesis that they are mixtures of two components. In fact, we want to *derive* the isotopic mixing equation from an array of data points as we were able to do for binary mixtures of different chemical compositions based on equation 18.7.

The desired isotopic mixing equation can be derived from equations 18.19 and 18.20 by solving both for f_A and by equating the results. The outcome of this algebraic manipulation is:

$$\left(\frac{^{87}Sr}{^{86}Sr}\right)_M = \frac{a}{(Sr)_M} + b \quad (18.24)$$

where:

$$a = \frac{(Sr)_A(Sr)_B \left[\left(\frac{^{87}Sr}{^{86}Sr}\right)_B - \left(\frac{^{87}Sr}{^{86}Sr}\right)_A \right]}{(Sr)_A - (Sr)_B} \quad (18.25)$$

$$b = \frac{(Sr)_A \left(\frac{^{87}Sr}{^{86}Sr}\right)_A - (Sr)_B \left(\frac{^{87}Sr}{^{86}Sr}\right)_B}{(Sr)_A - (Sr)_B} \quad (18.26)$$

Equation 18.24 is a hyperbola in coordinates of $(Sr)_M$ (x-coordinate) and $(^{87}Sr/^{86}Sr)_M$ (y-coordinate) whose position and curvature are defined by the numerical values of a and b, which depend entirely on the compositions of the end members. The special virtue of equation 18.24 is that it is transformable into a straight line by defining the x-coordinate as $1/(Sr)_M$. If $x = 1/(Sr)_M$, then equation 18.24 becomes:

$$\left(\frac{^{87}Sr}{^{86}Sr}\right)_M = ax + b \quad (18.27)$$

which is a straight line whose slope $m = a$ and whose intercept on the y-axis is equal to b. Therefore, the isotopic mixing equation for binary mixtures of components that differ in the isotope compositions of one element (such as Sr, Nd, Hf, Os, and Pb) can be derived by fitting a straight line to data points in coordinates of the appropriate isotopic parameter and the reciprocals of the concentration.

This procedure was used successfully by Boger and Faure (1974, 1976) and Boger et al. (1980) to treat sediment samples from the Red Sea and by Nardone and Faure (1978) from the Black Sea. In addition, Shaffer and Faure (1976), Kovach and Faure (1977), and Faure and Taylor (1983) used it to study mixing of detrital sediment in the Ross Sea and under the Ross Ice Shelf adjacent to Antarctica.

Isotopic mixing arrays may consist of three or more components in which case the data points representing such samples scatter on the mixing diagram. If the third component in sediment samples consists of authigenic calcite, or some other acid-soluble mineral, it can be removed from the samples prior to analysis by treating them with dilute solutions of a weak acid such as acetic acid. The tests of diatoms and radiolarians, which are composed of opaline silica, can be removed by flotation in a mixture of bromoform and acetone. The presence of a component of amorphous silica in sediment samples that are binary mixtures causes a *decrease* of the Sr concentrations *without* affecting the $^{87}Sr/^{86}Sr$ ratios, whereas the presence of calcite and aragonite changes *both* parameters of binary sediment mixtures.

Mixing of *water* on or below the surface of the Earth commonly affects the isotope compositions of O and H as well as those of dissolved elements such as Sr and Nd. The isotope compositions of O and H in binary mixtures of water are expressed by equations 18.22 and 18.23. These equations can be combined by eliminating f_A to yield an equation in coordinates of δD and $\delta^{18}O$ equivalent to equation 18.7. However, when a third component is present, a mixing triangle results, which can be reduced to a mixing line by projecting the data points from one of the corners of the triangle to the line connecting the other two components. Changes in the isotope composition of O and H in groundwater may *also* result from exchange reactions with minerals, by the release of water of hydration during dehydration of minerals, and by the *uptake* of water during mineral transformations that may alter the isotope composition of the residual water.

Consequently, in systems where the water/rock ratio is small, the isotope composition of water can be altered in a variety of ways that do not necessarily involve mixing.

The isotopic compositions of Sr and of other elements dissolved in groundwater are likewise changed by the preferential solution of minerals containing Sr whose isotope composition depends on the ages and Rb/Sr ratios of these minerals. However, as the water approaches chemical equilibrium with the minerals with which it is in contact, its $^{87}Sr/^{86}Sr$ ratio is stabilized and any further changes can then be attributed to mixing of water having different $^{87}Sr/^{86}Sr$ ratios and Sr concentrations.

An example of the effect of mixing of subsurface brines on the isotope composition of Sr is presented in Figure 18.7 based on the work of Lowry et al. (1988) on brines from the Clinton sandstones of Ohio. These brines were diluted with varying amounts of meteoric water that lowered their Sr concentrations but did not appreciably alter their $^{87}Sr/^{86}Sr$ ratios. Therefore, the Sr concentrations were first restored to their original values of projecting the data points in Figure 18.6 from the origin to the mixing line AB. The reciprocals of these restored Sr concentrations and the $^{87}Sr/^{86}Sr$ ratios form a linear array, as expected of binary mixtures. The equation of the isotopic mixing line, determined by least-squares regression, is:

$$\frac{^{87}Sr}{^{86}Sr} = 7.786 \frac{1}{(Sr)} + 0.7028 \qquad (18.28)$$

This equation was plotted as a hyperbola in Figure 18.7 in coordinates of the Sr concentration and the $^{87}Sr/^{86}Sr$ ratio. Because of dilution with meteoric water, the measured Sr concentrations were up to 54% *lower* than the values derived from Figure 18.6.

18.6   Isotopic Mixtures of Two Elements

Most natural materials contain not just one but several elements having different isotope compositions. For example, sediment or water derived

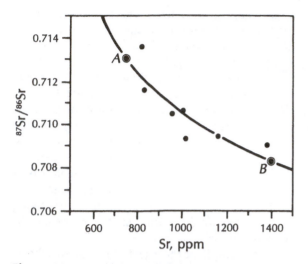

Figure 18.7 Mixing hyperbola formed by the
$^{87}Sr/^{86}Sr$ ratios and Sr concentrations of brines from
the Clinton sandstones of Ohio. The Sr concentrations
were corrected for dilution with meteoric water in
Figure 18.6. The equation of the hyperbola was
derived by fitting a straight line to the data points in
coordinates of $^{87}Sr/^{86}Sr$ and $1/(Sr)$. The isotope compositions of the end members were determined from
the mixing equation based on their Sr concentrations,
which were derived by fitting mixing triangles to pairs
of elements including Sr. Brine A is Na-rich, has
$^{87}Sr/^{86}Sr = 0.7131 \pm 0.0006$, and $\delta^{18}O = -4.1 \pm 0.4$‰;
whereas brine B is Ca-rich, has $^{87}Sr/^{86}Sr = 0.70837$
± 0.00006, and $\delta^{18}O = -1.0 \pm 0.3$‰ (Lowry et al.
1988).

from different sources may contain measurable
concentrations of Sr and Nd or O with distinctive
isotope compositions. The isotope compositions
of these elements in a suite of binary mixtures
vary progressively depending on the isotope compositions of the end members and on the proportions of mixing. The isotope compositions of any
two elements in binary mixtures can be calculated
from equation 18.20 for different values of f_A,
provided the concentrations and isotope compositions of the elements in the end members are
known or assumed. When the results of such systematic calculations are plotted in coordinates of
the appropriate isotope ratios, isotopic mixtures
of two elements are found to form hyperbolas

that are *not* transformable into straight lines
because the mixing equation contains a cross
product term (Faure, 1986). Therefore, such mixtures are commonly modeled by constructing
mixing hyperbolas, as described above, based on
end members whose isotopic and chemical compositions are assumed.

18.7   Summary

Mixing is a common phenomenon in nature that
takes place in the course of geological processes.
The chemical compositions of binary mixtures of
water or sediment vary linearly in coordinates of
the concentrations of any two conservative elements. The end members must also satisfy the
mixing equation, which helps to define them and
permits us to calculate their abundance in any
given mixture.

Three-component mixtures scatter within a
triangle of mixing that can be reduced to two
components by projecting data points from one of
the corners to the opposite side of the triangle.
Mixing of two saline waters followed, or accompanied, by dilution with meteoric water is a special case of three-component mixing that may
occur in some oilfield brines. Ternary mixing may
also explain the chemical compositions of some
detrital sedimentary rocks such as arkosic sandstones and greywackes.

Evaporative concentration of surface water
may initially cause linear relationships among the
concentrations of conservative elements. However,
when the ion activities rise to critical values such
that a compound can precipitate, the ions that
participate in the formation of the compound
deviate from the linear pattern and respond to
the operation of geochemical divides. As a result,
both the concentrations of the ions and the composition of the dissolved salts change.

Mixtures of two components that differ both
in the concentration and in the isotope compositions of one element form data arrays in the form
of hyperbolas in coordinates of an isotope ratio
and concentration of the element. Such mixing
hyperbolas are transformable into straight lines

by inverting the concentration. Ternary mixtures of components having different isotope compositions of *one* element can be reduced to binary mixtures by removing one of the components using appropriate treatment of the samples prior to analysis.

Binary mixtures containing *two* elements with different isotope compositions form hyperbolic arrays in coordinates of the two isotopic ratios (or a related parameter). These hyperbolas are *not* transformable into straight lines because the mixing equation contains additional terms.

Problems

1. Calculate the concentration of Mg^{2+} in a binary mixture of two components, given that $(Mg^{2+})_A = 2.4$ mg/L and $(Mg^{2+})_B = 6.7$ mg/L and $f_A = 0.75$.

(Answer: 3.47 mg/L)

2. Given two end members A and B whose concentrations of K and Mg are $(K)_A = 100$ ppm, $(Mg)_A = 25$ ppm; $(K)_B = 35$ ppm, $(Mg)_B = 250$ ppm. Calculate the slope and intercept of the mixing equation.

(Answer: Slope $= -0.289$, intercept $= 107.2$)

3. Plot the mixing equation and end members of Problem 2 on graph paper and add a data point M containing $(K)_M = 62.0$ ppm and $(Mg)_M = 89.3$ ppm. Calculate the abundances of components A and B in that sample.

(Answer: A $= 41.5\%$, B $= 58.5\%$)

4. A sample P contains $(K)_P = 40$ ppm and $(Mg)_P = 125$ ppm. Assume that this sample formed by mixing of components A and B defined in Problem 2 and was subsequently diluted with a third component located at the origin.

(a) Derive the equation of the dilution line.

(b) Determine the coordinates of the point of intersection of the dilution line with the mixing line AB.

(c) Determine the weight fraction of the dilute component in sample P.

5. A series of brine samples have the following concentrations of Na and Ca.

Na^+, mg/kg	Ca^{2+}, mg/kg
58,800	30,200
51,500	36,100
48,800	38,600
43,200	42,200
35,000	47,600

Derive the equation of the mixing line by least-squares regression.

6. If the end members in Problem 5 contain $(Na^+)_A = 67,000$ mg/kg and $(Na^+)_B = 37,600$ mg/kg, calculate the corresponding concentrations of $(Ca^{2+})_A$ and $(Ca^{2+})_B$.

7. Additional brine samples belonging to the suite of Problem 5 have the following concentrations.

Na^+, mg/kg	Ca^{2+}, mg/kg
11,600	9,700
44,000	16,400
25,500	15,500
35,500	22,500
37,400	26,200

Plot these data points together with the mixing line from Problem 5 and the end members from Problem 6. Estimate the concentrations of Na^+ and Ca^{2+} in the brine samples prior to the dilution by a graphical procedure.

8. Calculate the $^{87}Sr/^{86}Sr$ ratio and Sr concentration of a binary mixture of A and B for $f_A = 0.40$ given that $(^{87}Sr/^{86}Sr)_A = 0.725$, $(Sr)_A = 100$ ppm, $(^{87}Sr/^{86}Sr)_B = 0.704$, $(Sr)_B = 450$ ppm.

9. Plot a mixing hyperbola for mixtures of A and B defined in Problem 8 by calculating the coordinates of mixtures characterized by $f_A = 0.8, 0.6, 0.4,$ and 0.2.

10. Calculate the $\delta^{18}O$ value of water formed by mixing two components such that $\delta^{18}O_A = 10.0‰$, $\delta^{18}O_B = -2.0‰$, and $f_B = 0.65$.

11. Derive the equation for two-component mixtures of water in coordinates of $\delta^{18}O$ and δD.

12. Calculate the $^{87}Sr/^{86}Sr$ ratios and δD values of a series of two-component mixtures defined by $(^{87}Sr/^{86}Sr)_A = 0.750$, $(Sr)_A = 5.0$ ppm, $\delta D_A = -150‰$ and $(^{87}Sr/^{86}Sr)_B = 0.709$, $(Sr)_B = 75$ ppm, $\delta D_B = -30‰$ for $f_A = 0.2, 0.4, 0.6$, and 0.8. Plot the results in coordinates of $^{87}Sr/^{86}Sr$ and δD and draw a smooth curve through the points.

References

BOGER, P. D., and G. FAURE, 1974. Strontium-isotope stratigraphy of a Red Sea core. *Geology,* **2**:181–183.

BOGER, P. D., and G. FAURE, 1976. Systematic variations of sialic and volcanic detritus in piston cores from the Red Sea. *Geochim. Cosmochim. Acta,* **40**:731–742.

BOGER, P. D., J. L. BOGER, and G. FAURE, 1980. Systematic variations of $^{87}Sr/^{86}S$ ratios, Sr compositions, selected major oxide concentrations and mineral abundances in piston cores from the Red Sea. *Chem. Geol.,* **32**:13–38.

FAURE, G., L. M. JONES, R. EASTIN, and M. CHRISTNER, 1967. Strontium isotope composition and trace element concentrations in Lake Huron and its principal tributaries. Rept. for Office of Water Resources Research, U.S. Dept. Interior, Project No B-004-OHIO, 109 pp.

FAURE, G., and K. S. TAYLOR, 1983. Sedimentation in the Ross Embayment: Evidence from RISP core 8 (1977/78). In R. L. Oliver, P. R. James, and J. B. Jago (Eds.), *Antarctic Earth Sci.,* 546–550. Australian Acad. Sci., Canberra.

FAURE, G., 1986. *Principles of Isotope Geology,* 2nd ed. Wiley, New York, 589 pp.

KOVACH, J., and G. FAURE, 1977. Sources and abundances of volcanogenic sediment in piston cores from the Ross Sea, Antarctica. *N.Z. J. Geol. Geophys.,* **20**:1017–1026.

LOWRY, R. M., G. FAURE, D. I. MULLET, and L. M. JONES, 1988. Interpretation of chemical and isotopic compositions of brines based on mixing and dilution, "Clinton" sandstones, eastern Ohio, U.S.A. *Appl. Geochem.,* **3**:177–184.

NARDONE, C. D., and G. FAURE, 1978. A study of sedimentation at DSDP Hole 379A, Black Sea, based on the isotopic compositions of strontium. *Initial Reports of the Deep Sea Drilling Project,* vol. 42, Part 2, 607–615.

PETTIJOHN, F. J., 1963. Chemical composition of sandstones-excluding carbonate and volcanic sands. In M. Fleischer (Ed.), *Data of Geochemistry,* 6th ed. U. S. Geol. Surv. Prof. Paper, 440-S.

SHAFFER, N. R., and G. FAURE, 1976. Regional variation of $^{87}Sr/^{86}Sr$ ratios and mineral compositions of sediment from the Ross Sea, Antarctica. *Geol. Soc. Amer. Bull.,* **87**:1491–1500.

V

APPLICATIONS OF GEOCHEMISTRY TO THE SOLUTION OF GLOBAL PROBLEMS

The chapters of Part V apply the principles of geochemistry to important environmental phenomena, such as chemical weathering, variations in the quality of surface water, oxidation of sulfide-bearing mineral deposits, the transfer of chemical elements among its natural reservoirs, consequences of the contamination of the atmosphere, disposal of radioactive waste, and the effects of environmental lead on human health. The selected topics demonstrate that geochemistry is central to our efforts to understand the environment in which we live.

19

Consequences of Chemical Weathering

When igneous and metamorphic rocks are exposed to water and atmospheric gases at or near the surface of the Earth, the minerals they contain decompose selectively and at varying rates. The chemical reactions that cause the decomposition of minerals were discussed in Chapters 9–15 and constitute the process we call *chemical weathering*. The products of chemical weathering consist of (1) the compounds formed during incongruent solution of the primary minerals, (2) the ions and molecules that dissolve in the water, and (3) the resistant minerals in the rocks that are liberated by the decomposition of their more susceptible neighbors. In addition, the noble gases (He, Ne, Ar, Kr, Xe, and Rn) and certain other gases (H_2, hydrocarbons, N_2, etc.) are released into the atmosphere. The newly formed compounds (oxides, hydroxides, clay minerals, etc.) may accumulate at the site of weathering together with unreacted mineral grains (quartz, feldspars, micas) and undecomposed pieces of rock (pebbles, cobbles, boulders) as a residual mantle or *regolith*, also called *saprolite*. Further modification of the saprolite leads to the formation of *soils,* which sustain rooted plants and the fauna that feeds on these plants. However, ultimately, saprolite is removed by *erosion* through transport of the material downslope under the influence of gravity or by the actions of running water or wind.

Therefore, from a geological perspective, chemical weathering is the start of the mass migration of sediment from the continents into basins of deposition in the oceans. Chemical weathering is also the start of the geochemical cycles of the elements, which are released from their rock reservoirs into solution for a trip to the oceans where they reside for varying lengths of time before they reenter the crust of the Earth in sedimentary rocks (Section 4.4).

Much of the subject matter to be discussed in the fifth part of this book is concerned with the practical consequences of chemical weathering, which not only consists of the decomposition of rocks but also affects the chemical composition of water and causes dispersion of ions from metallic and nonmetallic mineral deposits. Chemical weathering therefore contributes to the ability of the environment on the surface of the Earth to sustain life.

The natural environment is increasingly affected in unforeseen ways by the activities of the growing human population. The changes in the chemical composition of the atmosphere and hydrosphere caused by our activities may alter our habitat in ways that are difficult to predict with certainty, but may in fact be harmful to us. For this reason, geochemists are increasingly called upon to study how natural geochemical processes on the surface of the Earth are perturbed by anthropogenic activities, to predict the effect of these changes on the quality of the environment, and to sound the alarm when the resulting changes in the global environment are potentially harmful to us.

19.1   Changes in Chemical Composition of Rocks

The effects of chemical weathering on the chemical composition of rocks are readily apparent by comparing the concentrations of major-element oxides in fresh and weathered samples of a rock. However, such direct comparisons may be misleading because of the "closed table effect," which is the property of chemical analyses, expressed in weight percent, of having to add up to 100. A serious consequence of this property of chemical analyses is that a *real* change in one constituent causes *apparent* changes in all other constituents. For example, if a rock *loses* a significant amount of one major element as a result of chemical weathering, the concentrations of all other elements must *increase*, thereby erroneously suggesting additions of these elements to the rock. Similarly, if a rock actually *gains* one element, the concentrations of all others must *decrease*. As a result, differences in the concentrations of major elements expressed in percent between weathered and unweathered specimens of a homogeneous rock are *not* reliable indicators of actual gains and losses caused by chemical weathering.

The procedure for identifying real gains and losses of elements as a result of chemical weathering is based on the assumption that one of the major-element oxides has remained constant in *amount* even though its concentration may appear to have changed. The constituent chosen most often for this purpose is Al_2O_3, consistent with the limited solubility of $Al(OH)_3$ at pH values between 6 and 8 (Figure 12.6). This choice is also consistent with the convention regarding the conservation of Al during incongruent solution of aluminosilicate minerals (Section 10.6). Alternatively, Fe_2O_3, TiO_2, or ZrO_2 may be selected in cases where Al may have been mobile in highly acid or basic environments.

Detailed chemical and mineralogical studies of chemical weathering of igneous rocks of basaltic composition have been published by Eggleston et al. (1987), Claridge and Campbell (1984), Colman (1982), and Hendricks and Whittig (1968). Loughnan (1969), Carroll (1970), Lindsay (1979), whereas Colman and Dethier (1986) discussed weathering of silicate minerals in general.

In order to demonstrate the effects of weathering on the chemical composition of a rock we examine the data in Table 19.1 for a basalt from the state of Paraná, Brazil, and for a sample of saprolite that formed from it.

We observe that the saprolite has significantly *lower* concentrations of SiO_2, CaO, MgO, Na_2O, K_2O, MnO, and P_2O_5 than the basalt. However, it has *higher* concentrations of Al_2O_3, Fe_2O_3, TiO_2, and LOI (loss on ignition), which includes adsorbed water, water in crystal lattices and fluid inclusions, CO_2 of carbonates, and SO_2 of sulfides. Note that the decreases and increases of the concentrations of the major-element oxides in the saprolite balance in such a way that the sum of the concentrations is equal to 100. Therefore, the real losses of some of the oxide components that may have occurred during weathering of the basalt are partly compensated by apparent increases in the concentrations of the other components. Similarly, real gains in some components of the basalt have caused apparent losses in others. When both gains and losses have occurred, the resulting chemical composition of the weathering products can only be interpreted by recalculating the analysis based on the assumption that one of the oxides has remained constant in amount.

In the case at hand, we select Al_2O_3 as the constant oxide and note that its concentration apparently increased from 12.3% to 26.3% as a result of weathering of the basalt. Note that the *concentration* of a constituent expressed in weight percent is equivalent to an *amount* in grams per 100 g of rock. Therefore, 100 g of basalt originally contained 12.3 g Al_2O_3. If we assume that the *amount* of Al_2O_3 remained *constant*, the apparent *increase* in the *concentration* of Al_2O_3 must be caused by a *reduction* in the weight of the rock from 100 g to some smaller amount derivable from the relation

$$\frac{\text{weight of constituent}}{\text{weight of rock}} \times 100$$

$$= \text{percent concentration} \quad (19.1)$$

Table 19.1 Comparison of Chemical Analyses of Basalt from Paraná, Brazil, and Its Weathering Products

	Basalt, %	Saprolite, %	Amount remaining, g	Gain + or loss −, g	Gain + or loss −, %
SiO_2	50.7	35.1	16.4	−34.3	−67.6
Al_2O_3	12.3	26.3	12.3	0	0
CaO	7.83	0.06	0.028	−7.80	−99.6
MgO	4.18	0.26	0.12	−4.06	−97.1
Na_2O	2.53	0.02	0.009	−2.52	−99.6
K_2O	1.71	0.13	0.061	−1.65	−96.5
Fe_2O_3	15.1	21.5	10.04	−5.06	−33.5
MnO	0.20	0.10	0.046	−0.15	−75.0
TiO_2	3.19	3.96	1.85	−1.34	−42.0
P_2O_5	0.46	0.11	0.051	−0.41	−89.1
LOI[a]	1.70	12.4	5.79	+4.09	+241
Sum	100.0	100.0	46.7	−53.3	

[a]Loss on ignition.

SOURCE: Faure, unpublished data.

In the case under consideration, the weight of the rock must be $(12.3/26.3) \times 100 = 46.7$ g, which implies that 53.3 g of rock was removed by weathering from each 100 g of basalt.

The *amounts* of the other oxide constituents remaining in the altered basalt can now be calculated by multiplying their percent concentrations in the saprolite by the *weight-loss factor* (w) derived from the ratio of the concentrations of the constant oxide in the fresh and weathered rock. If Al_2O_3 is chosen as the constant oxide:

$$w = \frac{(Al_2O_3)_{fresh}}{(Al_2O_3)_{weath.}} = \frac{12.3}{26.3} = 0.467 \quad (19.2)$$

where the parentheses signify concentrations in weight percent of the selected oxide in the fresh and weathered rock, respectively. For example, the amount of SiO_2 remaining after weathering is $35.1 \times 0.467 = 16.4$ g. All of the concentrations of the oxide components of the saprolite in Table 19.1 were multiplied similarly by the weight-loss factor and the remaining amounts are listed in column 3. Note that these amounts sum to 46.7 g, confirming that $100 - 46.7 = 53.3$ g of material was

lost as a result of weathering per 100 g of original basalt.

The actual gains and losses of each component can now be determined by comparing the data in column 3 (amounts remaining) to the analysis of the basalt in column 1 of Table 19.1. For SiO_2 we have a real loss of $16.4 - 50.7 = -34.3$ g, whereas for Fe_2O_3 we have a real loss of $10.04 - 15.1 = -5.06$ g. Note that Fe_2O_3 was lost even though its concentration in the saprolite is higher (21.5%) than in the basalt (15.1%). The same is true for TiO_2, which was diminished by 1.34 g even though its concentration in the saprolite is higher than in the basalt. The data in column 4 indicate that weathering caused real losses of all constituents, except H_2O, CO_2, and SO_2, relative to constant Al_2O_3.

The final step in the evaluation of the data in Table 19.1 is to rank the oxide components in terms of the magnitude of the losses that have occurred. For this purpose we express the losses in terms of percent of the amount originally present in the basalt. For example, the basalt lost $(34.3/50.7) \times 100 = 67.6\%$ of its SiO_2. The greatest losses occurred for CaO and Na_2O (99.6%),

MgO (97.1%), K_2O (96.5%), and P_2O_5 (89.1%) followed by MnO (75.0%), SiO_2 (67.6%), TiO_2 (42.0%), and Fe_2O_3 (33.5%). Evidently, chemical weathering of basalt in Paraná has led to the formation of saprolite that is composed primarily of Al_2O_3, SiO_2, Fe_2O_3, and H_2O.

19.2 Normative Mineral Composition of Weathering Products

The mineral composition of the saprolite in Table 19.1 can be studied by means of x-ray diffraction or by electron-transmission microscopy. However, when weathering products are poorly crystallized or amorphous, as is often the case, identification of minerals is difficult or impossible. A similar problem arises in the study of volcanic igneous rocks, which may be cryptocrystalline or glassy. In igneous petrology this problem is overcome by calculating the abundances of selected minerals from chemical analyses of the rocks. The resulting mineral composition is called its *norm,* and the minerals that make up the norm are the *normative* minerals. The procedures used to calculate mineral norms of volcanic rocks can be adapted to estimate the abundances of selected minerals in partly amorphous weathering products.

For this purpose we recognize that saprolite contains varying amounts of certain minerals derived from the original rock as well as amorphous compounds and minerals formed during weathering. The most common primary minerals are:

quartz	biotite
K-feldspar	hornblende, pyroxene, olivine
plagioclase	rutile
muscovite	magnetite, ilmenite

Among the secondary minerals we expect to find:

kaolinite, smectite	hematite
gibbsite	goethite
amorphous silica	pyrolusite

The choice of minerals to be included in the norm must be made by the investigator and depends on the composition of the bedrock, the climatic conditions at the site, and on any available information regarding the mineral composition of the bedrock or its weathering products. However, we recall that the mineral norm is only an *approximation* to reality and therefore the selection of minerals and their composition is arbitrary.

We have seen that the saprolite in Table 19.1 is composed primarily of Al_2O_3, SiO_2, Fe_2O_3, and H_2O, which presumably occur in the secondary minerals kaolinite and goethite. In addition, we assign K_2O to K-feldspar, Na_2O and CaO to plagioclase, MgO to the pyroxene enstatite, MnO to pyrolusite, TiO_2 to ilmenite, and P_2O_5 to apatite and berlinite. Other selections are certainly possible but seem less likely. For example, SiO_2 is not likely to occur as quartz grains in this case because basalt does not contain quartz. However, where the bedrock is composed of granite or granite–gneiss, quartz should be present in the saprolite. Similarly, K_2O could be assigned to muscovite or illite, MgO to smectite, CaO to calcite, TiO_2 to sphene or rutile, etc. In most cases such alternative selections have little effect on the conclusions derivable from the norm.

The calculation of the mineral norm of the saprolite in Table 19.1 is carried out in Table 19.2. Note that the formulas of the minerals are expressed in oxide form and that the percent concentrations of the oxide components in the analysis are recalculated to millimoles per 100 g of saprolite. Note also that LOI was *assumed* to be H_2O. The primary minerals derived from the basalt are satisfied first and the secondary weathering products (goethite, pyrolusite, kaolinite, and gibbsite) are then made from the remaining oxides. TiO_2 probably occurred in the basalt as ilmenite ($FeTiO_3$), which is resistant to weathering but may alter to TiO_2 and/or to a variety of sphene ($CaTiSiO_5$) called leucoxene (Correns, 1978). However, in this case the saprolite contains insufficient CaO for leucoxene to form, and TiO_2 is therefore expressed as ilmenite in the norm.

We begin the norm calculation with the primary minerals and convert all of the available K_2O, 1.38 millimoles (mmol), into K-feldspar. In accordance with the oxide formula for K-feldspar we reduce the available Al_2O_3 by 1.38 mmol and

Table 19.2 Calculation of a Mineral Norm from the Chemical Analysis of the Saprolite in Table 19.1

	SiO_2	Al_2O_3	CaO	MgO	Na_2O	K_2O	Fe_2O_3	MnO	TiO_2	P_2O_5	LOI	
In saprolite												
%	35.1	26.3	0.06	0.26	0.02	0.13	21.5	0.10	3.96	0.11	12.4	
mmol/100 g	584	258	1.07	6.44	0.322	1.38	134	1.41	49.5	0.775	689	

Minerals	Millimoles/100 g											Normative minerals, %
$K_2O \cdot Al_2O_3 \cdot 6SiO_2$ K-feldspar	575.7	256.6	1.07	6.44	0.322	0	134	1.41	49.5	0.775	689	0.77
$Na_2O \cdot Al_2O_3 \cdot 6SiO_2$ albite	573.7	256.3	1.07	6.44	0	0	134	1.41	49.5	0.775	689	0.17
$3CaO \cdot P_2O_5$ apatite	573.7	255.9	0	6.44	0	0	134	1.41	49.5	0.418	689	0.11
$Al_2O_3 \cdot P_2O_5$ berlinite	573.7	255.5	0	6.44	0	0	134	1.41	49.5	0	689	0.10
$CaO \cdot Al_2O_3 \cdot 2SiO_2$ anorthite	573.7	255.5	0	6.44	0	0	134	1.41	49.5	0	689	0.00
$MgO \cdot SiO_2$ enstatite	567.3	255.5	0	0	0	0	134	1.41	49.5	0	689	0.65
$FeO \cdot TiO_2$ ilmenite	567.3	255.5	0	0	0	0	109.3	1.41	0	0	689	7.51
$Fe_2O_3 \cdot H_2O$ goethite	567.3	255.5	0	0	0	0	0	1.41	0	0	579.7	19.42
MnO_2 pyrolusite	567.3	255.5	0	0	0	0	0	0	0	0	579.7	0.12
$Al_2O_3 \cdot 2SiO_2 \cdot 2H_2O$ kaolinite	56.3	0	0	0	0	0	0	0	0	0	68.3	66.0
$Al_2O_3 \cdot 3H_2O$ gibbsite	56.3	0	0	0	0	0	0	0	0	0	68.3	0.00
SiO_2 amorph. SiO_2	0	0	0	0	0	0	0	0	0	0	68.3	3.38
H_2O water	0	0	0	0	0	0	0	0	0	0	0	1.22
											Total	99.45

SiO_2 by $6 \times 1.38 = 8.28$ mmol. The amount of K-feldspar in grams per 100 g of saprolite is:

$$\frac{1.38 \times 556.67}{1000} = 0.77 \text{ g/100 g} \quad (19.3)$$

Next, all available Na_2O is used to make albite, and the remaining molar amounts of Al_2O_3 and SiO_2 are reduced proportionately. The millimolar amount of albite is then converted to a weight in grams, as illustrated in equation 19.3, and the amount is listed in Table 19.2. Note that the amounts of feldspar remaining in the saprolite are very small, but that the residual K-feldspar (0.77 g/100 g) is more abundant than albite (0.17 g/100 g). The amount of CaO is used to make 0.35 mmol of apatite ($3CaO \cdot P_2O_5$), and the remaining 0.418 mmol of P_2O_5 is combined with Al_2O_3 to make secondary berlinite $(Al_2O_3P_2O_5)$ leaving 255.5 mmol of Al_2O_3. Since all of the available CaO was used to make the phosphate apatite, none is left for anorthite, which presumably does not occur in the saprolite. MgO is converted into 0.65 g of enstatite $(MgOSiO_2)$ without difficulty. However, in order to convert TiO_2 into ilmenite $(FeOTiO_2)$, the Fe_2O_3 in the analysis must be recast as FeO. The amount of Fe in the saprolite is $134 \times 2 = 268$ mmol/100 g, which makes an equal number of moles of FeO. Therefore, 49.5 mmol of FeO is consumed in making 7.51 g of ilmenite from 49.5 mmol of TiO_2, leaving $268 - 49.5 = 218.5$ mmol of FeO, which is equivalent to 109.3 mmol of Fe_2O_3.

The remaining Fe_2O_3 is now converted into 19.4 g of the secondary mineral goethite $(Fe_2O_3 \cdot H_2O)$ by combining Fe_2O_3 with an equal millimolar amount of H_2O. MnO in the analysis is assigned to 0.12 g of pyrolusite (MnO_2), and all of the remaining Al_2O_3 is made into 66.0 g of kaolinite $(Al_2O_3 \cdot 2SiO_2 \cdot 2H_2O)$. We are left with a residue of 56.3 mmol of SiO_2 and 68.3 mmol of H_2O. The former is converted into 3.38 g of amorphous silica, or opal, or chalcedony, depending on its crystallinity. The remaining H_2O is converted into 1.22 g of water, which coats the solid particles of the saprolite. Note that gibbsite does not occur in the norm because SiO_2 is more abundant than Al_2O_3.

The calculation carried out in Table 19.2 indicates that the normative mineral composition of the saprolite in units of grams per 100 g or weight percent is:

	%		%
kaolinite	66.0	goethite	19.4
ilmenite	7.5	amorph. SiO_2	3.38
misc. traces	1.9	water	1.2

The significant conclusions to be derived from these results are that chemical weathering has converted the basalt into a mixture of kaolinite and goethite with lesser amounts of Ti-bearing minerals and amorphous silica. Only trace amounts of the primary minerals K-feldspar, albite, pyroxene, and apatite remain. The absence of gibbsite suggests that the activity of silicic acid in the water at 25 °C was greater than $10^{-4.68}$ mol/L in accordance with the kaolinite–gibbsite boundary in Figure 12.1.

19.3   Susceptibility of Minerals to Weathering

The minerals of igneous and metamorphic rocks differ widely in their susceptibility to chemical weathering. Another way of expressing this fact is to say that the dissolution reactions of minerals take place at different rates. In fact, the kinetics of many dissolution reactions have been studied, and much has been learned not only about the *rates* of these reactions but also about the *mechanisms* by which minerals dissolve in water (Colman and Dethier, 1986). Such studies were initiated by soil chemists who are interested in knowing the rates at which plant nutrients (Na^+, K^+, Ca^{2+}, Fe^{2+}, Mg^{2+}, and P) are released by minerals in soils under differing environmental conditions (McLelland, 1950). Later, this subject began to attract the attention of geochemists who have been increasingly interested in the processes that control the rates of congruent and incongruent dissolution. Some important contributions to this subject include papers by Wollast (1967), Helgeson (1971, 1972), Busenberg and Clemency (1976), Holdren and Berner (1979), Berner and Holdren (1979) and Tole et al. (1986).

Experimental studies with powdered mineral samples have commonly indicated a *decrease* in the rate of dissolution with time, suggesting that several different processes may occur simultaneously. For example, Garrels and Howard (1957) demonstrated that cations are released and that the pH rises within one minute after powdered feldspar is placed in water, suggesting that cations from grain surfaces are exchanged for H^+ to form a layer of "hydrogen-feldspar." This phenomenon is equivalent to *hydrolysis* (Section 10.2) if we regard feldspar as a salt of the weak hypothetical acid $HAlSi_3O_8$. Subsequently, cations and silica are lost from feldspar grains in nonstoichiometric proportions such that the feldspar grains *presumably* become coated with a layer of kaolinite or gibbsite whose presence retards further dissolution.

However, it was discovered that the surfaces of grains of powdered minerals used in dissolution experiments are coated with small adhering particles having large surface areas (Grandstaff, 1977). The presence of these particles undoubtedly increases the initial dissolution rates. In addition, grinding may damage the crystal lattice of surface layers of grains, and this phenomenon also increases initial dissolution rates of mineral powders. Moreover, when mineral grains from weathering environments were examined with a scanning electron microscope, the expected layer of alteration products was *not* found (Berner and Schott, 1982).

Apparently, mineral grains dissolve preferentially where crystallographic dislocations exist on the grain surface. As a result, etch pits are produced that expand and deepen as the process continues. The rate of dissolution of a mineral therefore depends in part on the density of etch pits per unit area of surface and hence on the abundance of dislocations within the mineral grains. The layer of weathering products that may adhere to grain surfaces is *permeable* and *nonprotective* and does *not* affect the kinetics of dissolution.

The relative susceptibility of the common rock-forming silicate and oxide minerals can also be determined directly from the observed abundances of primary minerals in saprolites formed at different stages of weathering of a particular type of rock. Such a study was carried out by Goldich (1938) on the Archean gneisses in the valley of the Minnesota River in southern Minnesota. The degree of weathering experienced by a series of saprolites that formed from the same kind of rock is indicated by their concentrations of Al_2O_3. The reason is that Al_2O_3 is very nearly immobile during chemical weathering, whereas other oxide components are lost, so that the concentration of Al_2O_3 rises as weathering progresses. Goldich (1938) used this criterion to classify samples of saprolite that had formed from granitic gneiss in the Morton–Redwood Falls area of Minnesota whose chemical compositions are shown graphically in Figure 19.1A.

The data of Goldich (1938) indicate that chemical weathering of granitic gneiss in southern Minnesota causes rapid apparent decreases of the concentrations of Na_2O, CaO, and MgO, whereas K_2O and SiO_2 concentrations decline more slowly as weathering progresses. However, the concentrations of H_2O, total Fe as Fe_2O_3, TiO_2, and Al_2O_3 actually increase. A constant-oxide calculation for the most highly weathered saprolite from Minnesota confirms that real gains in Fe_2O_3, TiO_2, and H_2O occurred relative to constant Al_2O_3, whereas all other constituents were lost in varying amounts.

The changes in the chemical compositions of progressively weathered saprolites must reflect what is happening to the primary minerals as a result of chemical weathering. Goldich (1938) actually determined the mineral compositions of the granitic gneiss and of two saprolites formed from it. His data are presented in Figure 19.1B in the form of smooth curves that show the rapid loss of plagioclase and biotite and the much slower removal of K-feldspar and quartz. Whereas K-feldspar is almost completely removed from the most highly weathered saprolite, about 20% of quartz and 6% of magnetite and ilmenite remain. The principal weathering product in this case is kaolinite, which accumulates in the saprolite and makes up 66% by volume of the most highly weathered rock.

The changing abundances of the common rock-forming minerals in saprolites reflect their susceptibilities to chemical weathering. Goldich's data in Figure 19.1B clearly show that plagioclase is much more susceptible to weathering than K-feldspar and that biotite is likewise destroyed during the early stages of weathering. The abundances of quartz and K-feldspar actually increase initially because of the preferential loss of plagioclase and biotite. However, ultimately K-feldspar is completely removed, whereas the abundance of quartz decreases only by about 30% in the most highly weathered saprolite in Minnesota. The concentrations of magnetite and ilmenite in the saprolite also rise, indicating their resistance to weathering. However, both eventually do

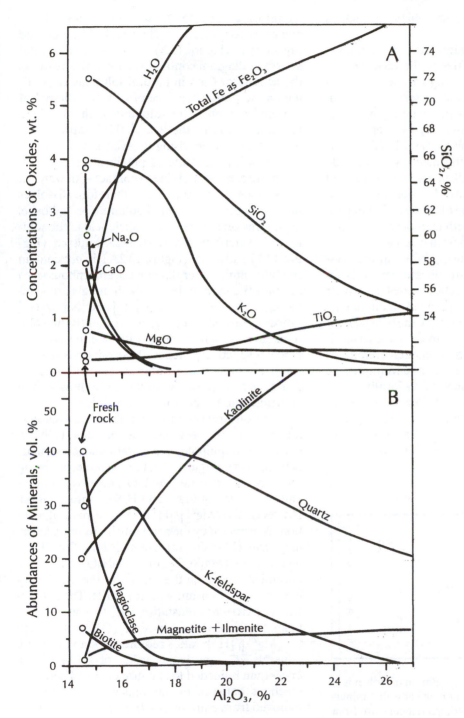

Figure 19.1 A. Variation of the chemical compositions of saprolites representing increasing intensities of chemical weathering of granitic gneisses in the Morton–Redwood Falls area of Minnesota (Goldich, 1938). Note that the concentrations of Na_2O, CaO, MgO, and SiO_2 decline continuously, whereas K_2O declines more slowly and the concentrations of total Fe as Fe_2O_3, Al_2O_3, and H_2O increase with increasing degree of weathering.

B. Variation of the measured abundances of minerals in the saprolites shown above. The rapid decrease in the abundance of plagioclase accounts for the loss of Na_2O and CaO shown in A. K-feldspar actually increases in abundance, as does quartz, but both ultimately decline. However, magnetite and ilmenite resist weathering and persist in the saprolite. Kaolinite is the principal weathering product and accumulates in the saprolite as the primary aluminosilicate minerals are decomposed (data from Goldich, 1938).

decompose to form secondary minerals such as goethite and leucoxene.

Goldich (1938) used information of this kind to show that the susceptibilities of the rock-forming silicate minerals to weathering can be related to their position in Bowen's reaction series depicted in Figure 19.2. Accordingly, olivine and Ca-plagioclase are most susceptible followed by pyroxene, hornblende, biotite, and Na-plagioclase. K-feldspar is more resistant than any of the minerals that precede it in the sequence of crystallization and is surpassed only by muscovite and quartz. In a general way, the susceptibility to weathering of the rock-forming silicate minerals decreases with their decreasing temperatures of crystallization from magma.

The apparent relationship between the crystallization temperature of silicate minerals, as expressed in Bowen's reaction series, and their susceptibility to weathering is a useful mnemonic device. However, we know that the progress of reactions that constitute chemical weathering is strongly dependent on environmental conditions and therefore recognize that exceptions to Goldich's generalization may occur locally.

An alternative explanation for the observed differences in the susceptibility to weathering of rock-forming silicate minerals arises from a consideration of activity diagrams. For this purpose we return to Chapter 12, which contains activity diagrams for anorthite (Figure 12.2B) and the Mg silicates (Figure 12.5). We recall first that the anorthite diagram contains calcite, which restricts the activity of Ca^{2+} in natural solutions, depending on the partial pressure of CO_2 and the pH of the environment. The solubility limit of calcite therefore prevents the $[Ca^{2+}]/[H^+]^2$ ratio of natural solutions from rising to levels at which equilibrium with anorthite is possible. As a result, anorthite cannot be stable in contact with water at the surface of the Earth and therefore dissolves incongruently to form kaolinite or gibbsite, depending on the activity of silicic acid in the solution. The activity diagrams of K-feldspar (Figure 12.1) and albite (Figure 12.2A) do *not* contain solubility limits other than that of amorphous silica because the carbonates of alkali metals are soluble. Consequently, the $[K^+]/[H^+]$ and $[Na^+]/[H^+]$ ratios of natural solutions *can* rise until the solutions are in equilibrium with K-feldspar and albite, respectively. Hence, these minerals may be stable in contact with water at the surface of the Earth and therefore weather more slowly than anorthite and Ca-bearing plagioclase.

The differences in susceptibility to weathering among the Mg-bearing silicates (olivine, pyroxene, amphibole) can be rationalized similarly by means of Figure 12.5. For example, olivine in the form of forsterite (Mg_2SiO_4) dissolves congruently by releasing Mg^{2+} and H_4SiO_4 into solution. However, the $[Mg^{2+}]/[H^+]^2$ ratio of natural solutions is *restricted* by the solubility limits imposed by magnesite ($MgCO_3$), serpentine ($Mg_3Si_2O_5(OH)_4$), and sepiolite ($Mg_4Si_6O_{15}(OH)_2 \cdot 6H_2O$). Therefore, natural solutions on the surface of the Earth cannot be in equilibrium with forsterite. The same is true of pyroxenes (enstatite, $MgSiO_3$) and amphiboles (anthophyllite, $Mg_7Si_8O_{22}(OH)_2$), although the $[Mg^{2+}]/[H^+]^2$ ratios required to stabilize these minerals at a particular value of $[H_4SiO_4]$ are *lower* than is needed to stabilize olivine. Therefore, reactions between olivine and water are further removed from equilibrium than reactions between pyroxene and water and between amphibole and water. Consequently, olivine is more susceptible to weathering than pyroxene, which is more suscepti-

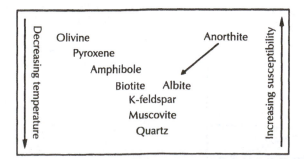

Figure 19.2 Schematic representation of Bowen's reaction series, which is used here to show that minerals that crystallize early and at high temperature from magma (olivine, Ca-plagioclase) are more susceptible to weathering than those that form later at somewhat lower temperatures (muscovite, quartz).

ble than amphibole, as implied also by Bowen's reaction series.

In conclusion, we consider briefly the susceptibility to weathering of other kinds of minerals such as the oxides, carbonates, and sulfides. Minerals consisting of these compounds may coexist with silicate minerals in certain kinds of rocks that may be classified as ore deposits. The effects of chemical weathering on the chemical and mineralogical compositions of such rocks are therefore of great interest to geologists in the mining industry.

The relative susceptibilities to weathering of sulfide minerals will be considered in Chapter 21 by means of Eh–pH diagrams for the sulfides of different metals. The dispersion of metals that may result from the chemical weathering of sulfide-bearing rocks causes the formation of local geochemical anomalies in the saprolite and soil derived from such deposits. These geochemical anomalies are used to locate potential ore deposits by means of systematic geochemical exploration surveys. In addition, the presence of anomalous concentrations of metals and certain nonmetals in soils associated with so-called *mineralized* rocks may pose a natural geochemical health hazard to humans living in the area or consuming food grown on such soils.

In comparison with the silicate minerals taken as a group, sulfides and carbonates (calcite, aragonite, dolomite) are *more* susceptible to weathering than silicate minerals (olivine, plagioclase, pyroxene, etc.), whereas the oxides (magnetite, ilmenite, chromite, rutile, cassiterite, etc.) are *less* susceptible. The resistance to chemical weathering of certain accessory minerals occurring in ordinary igneous and metamorphic rocks and in certain kinds of ore deposits has interesting consequences, which we consider in the next section.

19.4   Formation of Placer Deposits

All igneous, sedimentary, and metamorphic rocks may contain accessory minerals that resist both chemical and mechanical weathering. These minerals are liberated by the decomposition of the rock-forming minerals with which they are associated, and subsequently they may enter the sediment-transport system operating on the surface of the Earth. As a result of transport by mass wasting (downslope movement) and by water in motion or by wind action, the grains of accessory minerals are concentrated on the basis of size, specific gravity, and shape into deposits of sand and gravel. All deposits of sand and gravel contain some of the common accessory minerals of igneous and metamorphic rocks. In cases where these minerals have economic value and when their concentration is sufficient to make their extraction profitable, such sand and gravel deposits are called *placers* (the *a* is pronounced as in plaza).

The accessory minerals that may be concentrated in sand and gravel include the native metals or nonmetals (Au, Pt group elements, diamonds), oxides (cassiterite, magnetite, ilmenite, chromite, rutile, corundum, etc.), silicates (zircon, garnet, sphene, staurolite, tourmaline, etc.), and phosphates (apatite, monazite, etc.). Some of these are *ore minerals* from which one or several metals can be extracted at a profit (Au, Pt group metals, cassiterite, magnetite, chromite, monazite), whereas others have intrinsic value as gems, including diamond and corundum (sapphire or ruby), may be semiprecious stones (zircon, garnet), or are used as abrasives (zircon, garnet, staurolite, etc.). Placer deposits may be buried and cemented to become special kinds of sandstones or conglomerates that may be classified as ore deposits because of the presence of valuable minerals. Many such fossil placers are known, but their occurrence, origin, and mineral composition are properly the subject of Economic Geology (Park and MacDiarmid, 1970).

The resistant minerals listed above have higher densities than quartz, feldspar, calcite, etc., and therefore are grouped with the *heavy minerals,* which include also olivine, pyroxenes, amphiboles, biotite, and others whose densities are all greater than 2.90 g/cm^3. Placer deposits form when grains of minerals that are resistant to weathering and have high densities settle out of water in which they were suspended.

The movement of spherical particles through water under the influence of gravity is described by equations derived by Stokes (1851, 1901). Particles having other shapes can be treated using relations summarized by Lerman (1979). A spherical particle of radius r and density ρ suspended in water (density $= 1.0$ g/cm^3) is acted upon by a gravitational force (F_g):

$$F_g = \tfrac{4}{3}\pi r^3 g(\rho - 1) \tag{19.4}$$

where g is the acceleration due to gravity. As a result, the particle acquires a velocity (v_s) that generates an opposing force known as *viscous drag* (F_d):

$$F_d = 6\pi\eta r v_s \tag{19.5}$$

Note that the magnitude of the viscous drag on the spherical particle increases with its velocity. As a result, the velocity of a sphere sinking through water increases until the magnitude of the viscous drag equals the gravitational force. When $F_g = F_d$,

$$6\pi\eta r v_s = \tfrac{4}{3}\pi r^3 g(\rho - 1) \tag{19.6}$$

Therefore, the velocity of a spherical particle sinking through still water can increase only to a value given by the equation:

$$v_s = \frac{2}{9}\left(\frac{r^2 g}{\eta}\right)(\rho - 1) \tag{19.7}$$

Evidently, the terminal velocity of a spherical particle increases with its density (ρ) and with the square of its radius (r). Since the viscosity of water is $\eta = 0.01$ g/cm/sec and $g = 980$ cm/sec^2, equation 19.7 reduces to the formula (Lerman, 1979):

$$v_s = \frac{2 \times 980}{9 \times 0.01} r^2(\rho - 1)$$

$$= (2.18 \times 10^4) r^2(\rho - 1) \text{ cm/sec} \tag{19.8}$$

where r is expressed in centimeters. This formula was used to construct the settling curves in Figure 19.3 for spherical particles of native gold ($\rho = 19.0$ g/cm^3) and quartz ($\rho = 2.65$ g/cm^3) in water. The diagram illustrates the point that gold particles are approximately one-third smaller in diameter than quartz grains with which they are deposited, i.e., that have the same settling velocity.

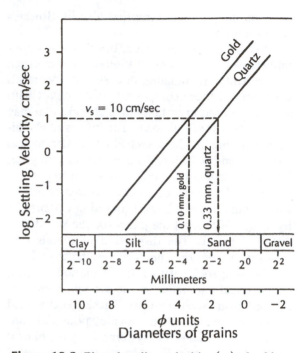

Figure 19.3 Plot of settling velocities (v_s) of gold and quartz particles in water versus their grain diameters expressed in millimeters and ϕ units in accordance with equation 19.7. The densities of gold and quartz were taken to be 19.0 and 2.65 g/cm^3, respectively. In order for particles of two different densities to be deposited together from still water they must have identical settling velocities. Therefore, gold particles settling with a velocity of 10 cm/sec are about one third smaller in diameter than quartz grains.

Placer deposits of native gold have played an important role in the history of the United States, starting in 1848 when gold nuggets were discovered in present-day stream gravels at Sutter's Fort near Coloma, California. In the following year, a gold rush ensued as the so-called forty-niners moved into the area to pan for gold in the streams emerging from the Sierra Nevada Mountains near Sacramento, California. Eventually, fossil placers were discovered in the area in gravel terraces of an earlier river of Tertiary age, and the source of all of the placer gold was subsequently traced to gold-bearing hydrothermal quartz veins in the Mariposa slate of the Mother Lode mining district of California.

19.5   Provenance Determination by Isotopic Dating

The heavy minerals in detrital sedimentary rocks have been studied extensively by sedimentary petrologists in order to identify their sources or *provenance*. Certain heavy minerals originate from specific kinds of igneous or metamorphic rocks. For example, Pt group metals and chromite occur in peridotites and pyroxenites, diamonds occur in kimberlites, and cassiterite is found in certain granites, whereas garnet, staurolite, sillimanite, epidote, and glaucophane are associated most often with metamorphic rocks of various kinds. In addition, a purple variety of zircon called *hyacinth* occurs only in granitic gneisses of Archean age (>2500 Ma) because the diagnostic color is caused by radiation damage resulting from the emission of energetic alpha particles by the isotopes of U, Th, and their unstable daughters (Tyler et al., 1940; Poldervaart, 1955). However, ordinary zircon, magnetite, ilmenite, rutile, sphene, monazite, tourmaline, pyroxene, etc., occur in both igneous and metamorphic rocks, so that their presence in sand or sandstones is not necessarily diagnostic. Therefore, studies of the mineral compositions of heavy-mineral assemblages in clastic sedimentary rocks are effective *only* under special circumstances.

The provenance of certain minerals that resist weathering can also be determined by isotopic dating. For example, muscovite and K-feldspar both contain K and Rb and are therefore dateable by the K–Ar and Rb–Sr methods described in Section 16.3 and used in Section 13.6 to date certain clay minerals. In addition, zircon, monazite, and apatite, which contain U and Th, can be dated by the U–Pb and Th–Pb methods. The results reflect the age of the rocks from which the minerals were derived and thereby identify their provenance. For example, Faure and Taylor (1981) and Faure et al. (1983) dated detrital feldspar in samples of till from the Transantarctic Mountains by the Rb–Sr method and concluded that the feldspar had originated primarily from Cambro-Ordovician granites of the local basement complex. However, some samples appeared to be mixtures of grains derived from sources of different ages. Moreover, efforts to date detrital feldspar from Paleozoic sandstones had only limited success because the feldspar grains disintegrated into small fragments during crushing of the rock in the laboratory (Szabo and Faure, 1987).

The difficulties encountered with provenance dating of feldspar and muscovite concentrates can be avoided by dating hand-picked suites or even single grains of zircon by the U–Pb method. Zircon crystals are more resistant to chemical weathering and have greater hardness than either feldspar or muscovite. As a result, zircon crystals tolerate weathering, transport, and deposition quite well and are preserved virtually unaltered in detrital sedimentary rocks. The problem of dating multicomponent mixtures of zircons derived from sources having different ages can be avoided by sorting them on the basis of color, shape, and magnetic susceptibility. Each variety can then be analyzed separately to identify its source from its measured U–Pb dates.

An excellent example of this technique is the study by Gaudette et al. (1981) of zircons extracted from the Potsdam sandstone of Cambrian age in upper New York State. The authors collected a rock sample weighing 35 kg from an outcrop located just east of the Adirondak Mountains, which consist of igneous and metamorphic rocks of the Grenville structural province of the Canadian Precambrian Shield. The zircons recovered from this specimen were sorted into four varieties: (1) brown, elongate, (2) brown, rounded, (3) clear, rounded, and (4) clear, elongate. The U–Pb dates of these *combined* zircon varieties range from 968 Ma to 1466 Ma and do not permit a meaningful interpretation. However, each variety yields distinctive dates on concordia diagrams, as shown in Table 19.3 and illustrated in Figure 19.4.

The upper concordia intercept date of 1170 ± 80 Ma of the brown, elongate zircons in Table 19.3 is similar to the age of charnockitic gneiss of the Adirondack Highlands. Therefore, these least-abraded zircons were probably derived from sources in the Adirondacks a short distance west of the collecting site. The brown, rounded zircons are older (1320 ± 80) than the brown,

Table 19.3 Summary of U–Pb Dates Derived from Different Varieties of Zircon Extracted from a Sample of the Potsdam Sandstone (Cambrian) from Upper New York State

Variety of zircon	Upper concordia intercept, Ma	Lower concordia intercept, Ma
Brown, elongate	1170 ± 100	325
Brown, rounded	1320 ± 80	380
Clear, rounded	2160 ± 500	985 ± 40
Clear, elongate	2700 ± 250	985 ± 15

SOURCE: Gaudette et al. (1981).

elongate ones and may have originated from the Grenville structural province of southern Quebec where rocks of this age have been recognized. The clear zircons yield upper intercept dates of 2160 ± 500 Ma (clear, rounded) and 2700 ± 250 (clear, elongate). Rocks in this age range occur in the Superior structural province about 600 km northwest of the collecting site. However, the well-defined lower intercept dates of 985 Ma indicate that both clear zircon fractions experienced severe Pb losses during the Grenville orogeny. Therefore, these zircons were initially incorporated into pre-Grenville sediment, were subsequently metamorphosed during the Grenville orogeny, and were then re-eroded and deposited in the Potsdam sandstone during the Cambrian period. The lower-intercept dates of the brown zircons (325 and 380 Ma) appear to coincide with orogenies in the Appalachian Mountains during the Paleozoic era. However, in the absence of other supporting evidence, these dates may be artifacts created by the linear extrapolations of discordia lines (Faure, 1986).

Evidently, the zircon grains in the Potsdam sandstone have participated in the major geological events on the North American continent since Archean time.

The technique of U–Pb dating has been developed to the point where single zircon grains (Schärer and Allègre, 1982a) and even fragments of one zircon grain have been dated (Schärer and Allègre, 1982b). In addition, the ion microprobe called SHRIMP at the Australian National University was used by Froude et al. (1983) to analyze spots about 25 μm in diameter on four zircon grains from the Mt. Narryer quartzite in the Yilgarn Block of Western Australia. The results indicate that these zircons crystallized between 4100 and 4200 Ma, which makes them the oldest known terrestrial objects and indicates that magma of granitic composition formed only about 300–400 Ma years after the accretion of the Earth about 4.5×10^9 years ago.

19.6 Formation of Soils

Perhaps the most important consequence of chemical weathering is that it permits the development of soils (Bohn et al., 1979; Sposito, 1981). Although chemical weathering is *destructive* of rocks and minerals, its products promote the existence of life on Earth and thus are certainly a very positive attribute of chemical weathering. Soils are the *interface* between the lithosphere (in the geochemical sense) and the biosphere because they permit the growth of rooted plants and other life forms, which themselves affect the chemical processes that characterize soils. We have so far presented the weathering of rock-forming minerals entirely in terms of inorganic chemical reactions. This may be a necessary oversimplification, but in reality *bacteria* and other life forms not only modify the natural environment but are active contributors to the process of chemical weathering.

Soils are characteristically layered as a result of the redistribution of material into *horizons,* which are identified by the letters O, A, E, B, C, and R. These layers form a *soil profile* described briefly in Table 19.4. The O, A, E, and B horizons constitute the actual soil or *solum.* The soil horizons are the products of soil-forming processes that are affected by environmental parameters, including temperature and meteoric precipitation (climate), vegetation, pH and O_2 content, time, and the composition of bedrock. Consequently, the principal soil hori-

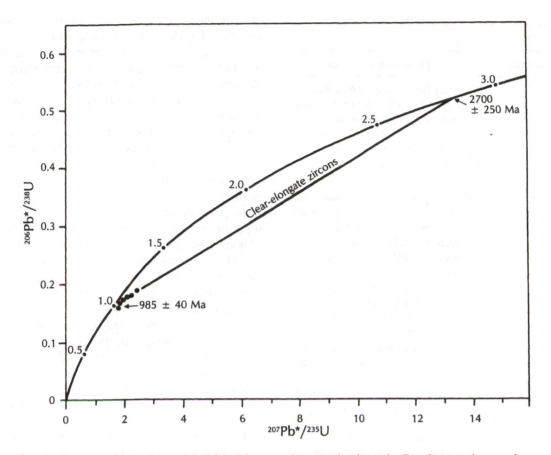

Figure 19.4 Concordia diagram for clear, elongate zircon grains from the Potsdam sandstone of Cambrian age from upper New York State. The zircons form a discordia line that intersects concordia at 2700 ± 250 and 985 ± 40 Ma. The upper intercept is the crystallization age of these zircons in Archean time. The lower intercept indicates that the zircons lost a large proportion of their radiogenic Pb during the Grenville orogeny, presumably because they were liberated by weathering and were incorporated into a sedimentary rock in pre-Grenville time. Subsequently, they were liberated a second time from the resulting metamorphic rock and were redeposited in the Potsdam sandstone during the Cambrian period. (Data and interpretation from Gaudette et al., 1981.)

zons are not equally developed under the wide range of environmental conditions that exist on the surface of the Earth. For example, an O horizon forms only where the rate of production of plant debris exceeds the rate at which it is decomposed. This condition arises in tropical rainforests where plant debris accumulates rapidly and in regions where decomposition of plant debris is retarded either by low temperatures or because the soil is waterlogged and therefore becomes depleted in O_2.

The amount of time required for a soil profile to develop from freshly deposited volcanic ash, alluvium, till, or bedrock is of the order of hundreds to thousands of years or more, depending on climatic conditions and the character of the starting material. Volcanic ash composed of small particles of silicate glass weathers relatively rapidly, whereas igneous or metamorphic rocks resist weathering and prolong the time required to develop soil. For example, lava flows extruded

Table 19.4 Description of Soil Horizons Starting at the Top

Soil horizon	Properties and origin
O	surface layer composed primarily of organic matter with only a small content of mineral material; black or dark brown in color
A	mineral particles mixed with finely divided organic matter that produces a dark grey color
E	primarily mineral grains that resist dissolution and are too large to be translocated in colloidal suspension, low content of organic matter causes light grey color
B	enriched in clay minerals, oxides, and hydroxides of Fe, Al, Mn, etc., removed from overlying A and E horizons; may also contain calcite or gypsum that can precipitate from aqueous solution within this layer
C	saprolite or sediment that is largely unaffected by soil-forming processes; may be the product of chemical weathering of the underlying bedrock (saprolite) or have been transported and deposited by water (alluvium) or ice (till) or by volcanic activity (tuff)
R	unweathered bedrock underlying the C horizon

SOURCE: Foth (1984).

through fissure eruptions on Mt. Kilauea on Hawaii can support rooted plants in only a few years, whereas outcrops of granitic gneiss north of Lake Superior in Canada show little evidence of weathering after exposure for about 14,000 years. In general, the thickness of a soil profile depends on the rate of the soil-forming processes and on the rate of erosion by running water, gravitational sliding, and wind action. In places where the rate of erosion exceeds the rate of soil formation, the top layer of the parent material is removed before soil horizons can develop within it.

The most important factor in the development of soil profiles is the *climate,* represented by the temperature and by the amount and seasonality of meteoric precipitation. As a result, soils within a particular climatic zone tend to have similar profiles regardless of the composition of the bedrock from which they formed. However, soils *within* a climatic zone may vary locally depending primarily on their location within the topography of the landscape, which affects the movement of water, the rate of erosion, and other factors. Such topographically controlled variations in soil profiles give rise to a *catena* or a *toposequence* of soils.

The classification of soils used in Europe and the United States was formerly based on the premise that soils reflect the environmental conditions under which they formed. However, this classification was replaced in 1960 by a scheme that gives more weight to the morphology of soils than to their genesis. The new soil classification was devised by the Soil Survey Staff of the U.S. Department of Agriculture and was reissued in revised form in 1975 as USDA Handbook 436. The new classification contains ten *soil orders,* listed in Table 19.5, whose names are intended to convey important properties of the soils to which they apply. Some of these soil orders evolve sequentially as a function of time starting with the weathering of bedrock or with the deposition of volcanic ash, alluvium, or till that become the parent materials from which the soils develop. The sequential development of soil orders is indicated diagramatically in Figure 19.5. However, the sequence of soil development may vary depending on local conditions or because of climate change.

The chemical and physical properties of soils are greatly affected by the small grain size of their constituent particles. As a result, chemical reactions in soils are strongly influenced by ion exchange phenomena discussed in Section 13.5. The cation exchange capacity (CEC) of soils depends primarily on the presence of organic matter, clay minerals, and oxides of Al and Fe in the different soil horizons. Consequently, the CEC varies with depth depending on the compo-

Table 19.5 Soil Orders and Their Properties

Order	Description
Entisols	early stage of development composed of A and C or A and R horizons but lacking a B horizon
Inseptisols	young soil with some leaching in the A horizon but only weakly developed B horizon
Alfisols	strongly leached acidic A horizon containing decomposing organic matter; B horizon well developed and clay-rich
Spodosols	composed of A and E horizons with accumulation of oxides/hydroxides of Al and Fe that form "hardpan" in the B horizon
Ultisols	highly weathered with well-developed clay-rich B horizon underlying well-leached and acidic A and E horizons
Oxisols	like Ultisols, but clay in B horizon is decomposed to oxides of Al and Fe, making this soil lateritic in character
Mollisols	thick, dark-colored A horizon that is soft when dry; clay-rich B horizon; high content of Ca^{2+} and Mg^{2+}
Aridisols	little organic material, essentially unleached mineral soil with deposits of calcite (caliche)
Vertisols	rich in swelling clay, soil profile poorly developed because of vertical mixing caused by seasonal cracking
Histosols	thick O horizon formed by accumulation of plant debris that fails to decompose

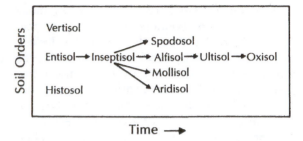

Figure 19.5 Approximate sequence of development of different soil orders. Entisols and Inseptisols are early stages of soil development leading to Spodosols, Alfisols, Mollisols, and Aridisols depending on the mineral composition of the parent material and climatic conditions. Ultisols and Oxisols occur in humid tropical regions as end stages of soil development enriched in oxides of Al and Fe. Vertisols develop only from parent material composed of swelling clay in regions with alternating wet and dry seasons. Histosols consist of partly decomposed plant material in places that are deprived of O_2 because they are waterlogged (after Foth, 1984).

sition of the soil horizons. Moreover, the CEC of soils depends on the pH because the charges on the surfaces of both organic and inorganic particles are determined by the abundance of H^+ in the aqueous phase. The adsorbed ions are available to plants as nutrients and contribute to the *fertility* of soils. Therefore, the phenomenon of ion exchange is *vital* to the ability of soils to sustain rooted plants and the other organisms that are associated with them.

The development of soil and its ability to sustain rooted plants are strongly affected by the life forms that coexist within or on it. The community of organisms that coexist in a particular environment constitutes an *ecosystem*. The organisms of an ecosystem can be classified into the *producers* (autotrophs) and *consumers* (heterotrophs). The producers are able to use C in the form of CO_2 and chemical nutrients from the environment to feed themselves, whereas the consumers sustain themselves by *decomposing* organic matter generated by the producers. The energy required to produce or decompose organic matter is derived either from sunlight or from the oxidation or reduction of chemical elements including S, Fe, and Mn. Accordingly, all organisms can be classified into the *photoautotrophs* (plants and many algae), *chemoautotrophs* (nitrifying and S-oxidizing bacteria), and *chemoheterotrophs* (animals, protozoa, fungi, and most bacteria). In a viable ecosystem the primary producers provide food for the consumers who recycle nutrients for use by the producers.

The ecosystem of soils contains rooted plants, bacteria, fungi, actinomycetes, algae, and protozoans. In addition, the animal kingdom is represented by vertebrates (mice, voles, moles, rabbits, squirrels, gophers, foxes, badgers, bear, deer, and birds), arthropods (ants, millipedes, centipedes, mites, springtails, harvestmen, diplopoda, diptera, and crustaceans), as well as nonarthropod animals such as earthworms, nematodes, and snails. Although bacteria constitute only about 1.0% of the organic matter of a fertile soil in a temperate region, they are by far the most abundant life form with a concentration of more than 1×10^9 individuals per gram of soil (Foth, 1984).

The nutrients required by plants and the other members of the soil ecosystem can be divided into the *macronutrients* and the *micronutrients*. The former occur in plants at concentrations of more than 500 ppm, whereas the latter are usually present at concentrations of less than 50 ppm. The macronutrients required by plants include N, P, K, Ca, Mg, and S; the micronutrients are B, Fe, Mn, Cu, Zn, Mo, Co, and Cl. The nutrient elements therefore tend to be concentrated in plant material relative to their abundance in the soil water and compared to the crust of the Earth as a whole. For example, corn plants are enriched in Ca, Mg, K, N, P, and S compared to the water in the soil by factors ranging from 2.6 to 67,000 (Foth, 1984, Table 11.3).

The chemical composition of plants varies widely depending on the chemical composition of the soil on which they grow and on the additions of fertilizers and other soil conditioners used in agriculture. Native plants tend to favor soils in which certain nutrients are abundant, and their distribution may therefore reflect regional variations in the chemical composition of the soil and of the bedrock from which it formed. This principle is used in geobotanical surveys to locate geological contacts and potential ore deposits (Brooks, 1972).

The wide variation in the chemical composition of cultivated and native plants in the United States is reflected by the data in Table 19.6 based on the work of Connor and Shacklette (1975). Cultivated plants tend to have more uniform compositions than native plants, but they are also enriched in some elements (primarily P and K) that are added to soil in the form of fertilizer.

19.7 Geomicrobiology

Microscopic organisms including bacteria, algae, protozoans, and fungi are widely distributed on the Earth and occur in large numbers. The *bacteria* (Class Schizomycetes), which play an especially large role in geochemistry, are divided into ten orders of about 1500 species. They occur in virtually all environments on the Earth, including the polar regions, the oceans, the atmosphere, and the continents.

Bacteria are a highly diversified group of organisms. Most species prefer temperatures between 5 and 37 °C, but some tolerate temperatures as low as 0 °C, and others survive heat up to about 70 °C. Most bacteria favor neutral environments, but some species thrive under acidic conditions, and others prefer basic or saline living conditions. Some bacteria become dormant when living conditions are unfavorable and can remain in this state for long periods of time. Bacteria reproduce asexually by binary fission and grow best in the dark, although photosynthetic bacteria do require ultraviolet radiation. Some bacterial colonies may grow in the absence of oxygen (anaerobes), others require oxygen (aerobes), and some can tolerate oxygen even though they do not require it (facultative anaerobes). Bacteria, which range from 1 to 5 μm in diameter, may be spherical in shape (coccus), rodlike (bacillus), or spiral (spirillum). All three types inhabit the human body. Although some cause disease (pathogens), most are benign.

The important role played by bacteria and other microorganisms in geological processes is receiving increasing attention from geochemists (Ehrlich, 1981; Krumbein, 1978a–c). In addition, the subject is represented by several scientific journals (e.g., *Geomicrobiology Journal, Global Biogeochemical Cycles, Applied Environmental Microbiology*). These publications contain a rapidly growing body of knowledge about the geochemical transformations that result from the activities of microorganisms in nature. They affect the weathering of silicate, carbonate, and sulfide minerals and participate directly in the geochemical cycles of many elements, including Fe, Mn, S, P, Se,

Table 19.6 Average Chemical Composition of Plants and Plant Ash (in parts per million)

Element	Plants[1] (dry weight)	Ash[2] Cultivated[2] plants	Ash[2] Native[2] plants
Li	0.1	—	4.0–15
Be	—	—	0.64–2.0
B	5	37–540	140–600
C	49.65[a]	—	—
N	0.92[a]	—	—
O	43.20[a]	—	—
F	—	0.43–0.49	0.50–1.6
Na	200	25–39	200–3100
Mg	700	$(1.5–13) \times 10^4$	$(1.6–10) \times 10^4$
Al	20	200–4000	1000–39,000
Ga	—	—	1.5–2.8
Si	1500	—	—
P	700	$(1.2–22) \times 10^4$	$(0.71–3.1) \times 10^4$
S	500	—	—
K	3000	$(18–41) \times 10^4$	$(2.9–23) \times 10^4$
Ca	5000	$(0.29–20) \times 10^4$	$(13–35) \times 10^4$
Ti	2	4.7–250	69–1200
V	1	—	2.6–2.3
Mn	400	96–810	470–14,000
Cr	2.40	0.42–6.6	2.2–22
Fe	500	$(0.06–0.27) \times 10^4$	$(0.08–0.93) \times 10^4$
Co	0.40	0.50–6.2	0.65–400
Ni	3	2.7–130	0.81–130
Cu	9	21–230	50–270
Zn	70	180–1900	170–1800
Se	0.1	0.04–0.17	0.01–0.42
Rb	2	—	—
Sr	20	14–880	320–5300
Y	—	—	2.1–47
Zr	—	—	2.4–85
Mo	0.65	2.5–20	0.76–7.6
Ag	0.05	—	—
Cd	—	0.37–2.3	0.95–20
I	—	4.6–13	2.8–5.4
Ba	30	15–450	270–11,000
La	—	—	14–270
Cl	0.40	—	350
Yb	—	—	1.1–1.8
Pb	—	7.1–87	24–480
U	0.05	—	—

[a]"Wood" according to Foth (1984) expressed in percent by weight.
SOURCES: (1) Brooks (1972), Table 11.2; (2) Connor and Shacklette (1975).

Te, As, Sb, and Hg. In addition, they contribute to the transformation of organic matter into peat, coal, petroleum, and natural gas.

Although geologists and geochemists have been slow to accept the importance of microorganisms in geological processes, biologists have been aware of their abilities for more than a century. For example, Ehrenberg (1838) reported that the stalked bacterium *Gallionella ferruginea* participates in the deposition of Fe hydroxides in bogs and Winogradsky in 1887 discovered that *Beggiatoa* can oxidize H_2S to elemental S. This work was later extended by V. I. Vernadsky, whose important contributions to geochemistry were mentioned in Chapter 1 of this book, and by Stutzer (1911). The importance of acid-producing microorganisms to the weathering of silicate minerals was suggested by A. Muentz in 1890 and by G. P. Merrill in 1895. It was later elaborated in a textbook by Waksman (1932). The role of bacteria in the deposition of the Precambrian iron formation of the Lake Superior district of the United States was discussed in an important paper by Harder (1919).

Microorganisms, including bacteria and algae, were the first life forms to appear on the Earth during the Archean Eon more than three billion years ago (Schopf, 1983). In fact, the oldest known fossil is a filamentous bacterium found in stromatolitic black chert of the Warrawoona Group (Pilbara Supergroup) of Western Australia (Awramik et al., 1983). The age of the rocks of the Warrawoona Group is 3.556 ± 0.032 Ga. These fossilized bacteria resemble several existing varieties, including green sulfur bacteria (*Chloroflexus aurantiacus*), methanogenic bacteria (*Methanospirillum hungatti*), and the cyanobacteria (*Oscillatoria angustissima, Phormidium angustissium*). The metabolism of most of the *existing* bacteria that resemble their Archean ancestor requires the absence of oxygen, although some can tolerate it. However, the cyanobacteria are photosynthetic and actually release oxygen into the environment.

Bacteria are primarily heterotrophic organisms that *decompose* complex organic molecules into simpler ones. Some species also decompose inorganic compounds (carbonates, sulfides, silicates) and even corrode metals. In fact, any natu-

rally occurring substance can be degraded by some species of bacteria. The ability of bacteria to carry out biochemical transformations has been used by man to make cheese, to leaven bread dough, and to ferment carbohydrates to make alcohol. Modern sewage-treatment facilities include holding tanks in which bacteria decompose organic solids in the presence of oxygen in order to avoid the severe oxygen depletion that follows when untreated sewage is discharged into the environment.

Bacteria in soil decompose plant debris and animal waste and thereby *recycle* nutrients. Without the recycling of nutrient elements, soils could not sustain rooted plants and life on Earth as we know it would be impossible. In addition, bacteria also assist plants to obtain nitrogen, which is the most important of all the nutrients required by plants.

The atmosphere of the Earth contains about 79% N_2, which acts like an inert gas and cannot be used by plants. Therefore, the nitrogen of the atmosphere must be processed into a form that plants can absorb through their roots. Evidently, plants have *not* evolved a N-fixing process similar to photosynthesis that enables them to absorb CO_2 from the atmosphere. Some N_2 from the atmosphere is converted to NO_3^- by lightning and enters the soil in meteoric precipitation. However, the principal suppliers of soluble N in soil are bacteria of the genus *Rhizobium* that live in the roots of leguminous plants such as soybeans, alfalfa, clover, and peas in a symbiotic arrangement. However, some N-fixing bacteria in soil are nonsymbiotic (*Azotobacter* and *Clostridium*), and some nonlegumes also harbor symbiotic N-fixing bacteria. The amount of N in soil is maintained at a steady state by another group of denitrifying bacteria that convert the soluble N compounds in soil back to N_2 gas. In addition, between 2 and 4% of N in soils is converted to NH_4^+ during the decomposition of organic matter by heterotrophic bacteria. The ammonium ion can substitute for K^+ in clay minerals and mica and is thus made unavailable to plants. Consequently, the formation of NH_4^+ is called *mineralization* by soil scientists.

The diversity of bacteria and their very large numbers may be exploited in the mining industry to recover metals and fossil fuels (Murr et al., 1978).

For example, the Society of Mining Engineers organized a symposium on "Bioleaching" during their annual meeting in 1988 in Las Vegas, Nevada. The participants in this symposium discussed the feasibility of using bacteria to dissolve sulfide minerals in ore deposits, to recover Mn from MnO_2 ores, to precipitate selenates from contaminated water, and to purify water discharged by manufacturing industries. These applications include manipulation of the genetic material of native species of bacteria in order to develop strains that are selective in their metabolic processing of chemical compounds. Significant contributions to this important and rapidly growing subject are by Trudinger (1971), Meyer and Yen (1976), Lundgren and Silver (1980), Monticello and Finnerty (1985), Hutchins et al. (1986), and Eligwe (1988). These and other references to microorganisms are listed separately in the References to emphasize the importance of this subject area to geochemistry.

19.8 Food Production and Population Growth

Increases in the population of the Earth are historically related to improvements in the methods of food production and increases in the area of land being cultivated. The onset of the Industrial Revolution around 1650 marked the start of a rapid rise in the world's population from about 545 million to a projected level of 6000 million by the year 2000. This *increase* in the population has been accelerated by a *decrease* in the death rate and now threatens to exceed the existing capacity for food production. Although Figure 19.6 indicates that the problem is *global* in scope, the balance between the rate of production and the rate of consumption of food shows wide *regional* variation.

The average annual *increase* in the population of the world is about 1.7%, with a population-doubling time of 40 years. However, the growth rate ranges from 2.9% in Africa to 0.2% in Europe, with corresponding doubling times of 24 years and 423 years, respectively. Several European countries have actually achieved zero population growth (Denmark, East Germany, Italy, United

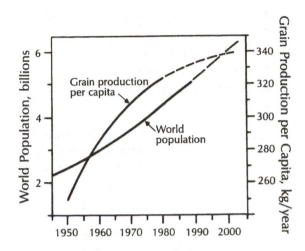

Figure 19.6 Population growth and per capita grain production worldwide. The graph indicates that the population of the world is growing and will probably exceed 6 billion by the year 2000. However, per capita grain production appears to be leveling off. The data for the world as a whole obscure regional differences in both food production and population growth. As a result, malnutrition and episodic starvation occur in some regions, whereas others have surplus food (after Foth, 1984).

Kingdom) and others even have declining populations (Austria, Hungary, West Germany).

The average per capita grain production, which has been rising worldwide, appears to be leveling off and is actually *declining* in some parts of Africa, Asia, and South America that have above-average rates of population growth. As a result, widespread malnutrition and episodes of starvation occur in these areas. The regional differences in the rates of per capita grain production are large and appear to be increasing with time, as shown in Figure 19.7.

The annual rate of grain production varies regionally for many different reasons and is not solely attributable to population growth. The contributory factors include:

1. The amount of arable land under cultivation.
2. The availability of fertilizers.
3. Effective weed and pest control.

4. Use of improved strains of cereal grains.
5. Farming practices.
6. Character of the soil.
7. Climatic conditions.
8. Economic incentives.
9. Civil wars.

In Japan, intensive farming practices result in an average yield of 5.8 metric tons of grain per hectare (ha; $= 10^4$ m^2). However, other nations do less well (Brown et al., 1987).

	Metric tons/ha
United States	4.8
China	3.9
Indonesia	3.7
Bangladesh	2.2
Mexico	2.1
Brazil	1.8
India	1.6
Pakistan	1.6
Soviet Union	1.6
Nigeria	0.8

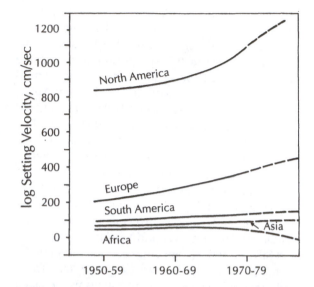

Figure 19.7 Average per capita grain production in different parts of the world. The grain production of North America and Europe is higher than those of South America, Asia, and Africa and is increasing more rapidly. The areas in the world having low rates of grain production have above-average rates of population growth. Note that "Africa" excludes South Africa and "Asia" excludes China and Japan. (Adapted from Foth, 1984, Figure 15.2.)

The high yields achieved in Japan cannot be realized elsewhere because of drought, mineral stress, shallow depth of soil development, excess water, or presence of permafrost. *Mineral stress* is caused by nutritional deficiencies in or actual toxicity of some soils depending on their chemical composition or origin.

Another way to increase food production in the world would be to increase the amount of land under cultivation. The data in Table 19.7 demonstrate that only 44% of the arable land in the world is presently under cultivation. Europe is most heavily farmed with 88% of all arable land in actual use, whereas South America and Africa still have large amounts of unused arable land. However, the productivity of the unused arable land is limited by unfavourable soil properties and climatic conditions. When these factors are taken into consideration, only a small fraction of the arable land is

without limitations, ranging from 36% in Europe to 22% in North America, about 16% in Africa and South America, and 10% in north and central Asia. Consequently, it is unlikely that the present rate of grain production in the world could be doubled merely by farming all of the available arable land. Nevertheless, given the present rate of population growth, the global output of cereal grains must be doubled every 40 years to maintain the average level of nutrition. Because of the uneven distributions of population growth and grain production, some regions of the world will experience shortages, whereas others will have a surplus.

Available agricultural technology (including the use of chemical fertilizers, effective weed and pest control, improved strains of food crops, and soil-management practices) is capable of increasing grain production significantly where yields are presently low. However, the cost of fertilizers

Table 19.7 Summary of Arable Land and Its Present Use (in billions of acres)

Region	Area	Potentially arable[a]	Under cultivation[b]	Unused arable land[c]
Africa	7.46	1.81 (24)	0.39 (22)	1.42 (78)
Asia	6.67	1.55 (23)	1.28 (83)	0.27 (17)
U.S.S.R.	5.52	0.88 (16)	0.56 (64)	0.32 (36)
North America	5.21	1.15 (22)	0.59 (51)	0.56 (49)
South America	4.33	1.68 (39)	0.19 (11)	1.49 (89)
Australia and New Zealand	2.03	0.38 (19)	0.08 (21)	0.30 (79)
Europe	1.18	0.43 (36)	0.38 (88)	0.05 (12)
Totals	32.49	7.88 (24)	3.47 (44)	4.41 (56)

[a]Number in parentheses is percent of total area in each region.
[b]In parentheses: percent of arable land presently under cultivation.
[c]In parentheses: percent of arable land not under cultivation.
SOURCE: Foth (1984), Table 15.2.

and pesticides is high and the introduction of more efficient farming practices may disrupt the established fabric of society. On the other hand, climate changes resulting from the greenhouse effect may be beneficial in some areas where present climatic conditions are unfavorable to agriculture.

19.9   Summary

Chemical weathering initiates the mass migration of sediment and chemical compounds in solution from the continents to the oceans and thereby starts the transport cycles of matter that together constitute the geochemical machine first mentioned in Section 4.5. In addition, chemical weathering contributes to formation of soil that sustains rooted plants whose presence on the Earth has enabled other life forms to evolve. Therefore, chemical weathering plays a vital role in making the Earth "an abode fitted for life," to use a phrase coined by Lord Kelvin in 1897.

The most obvious effect of chemical weathering is the decomposition of solid rocks into loose sediment whose chemical and mineral compositions differ greatly from those of the parent rocks. However, the apparent changes in the concentra- tions of the major elements must be evaluated carefully in order to identify which constituents were actually lost and to determine the amounts removed from the rock.

The newly formed products of weathering are commonly amorphous or only poorly crystallized. It is therefore not easy to determine the mineral composition of weathering products just as it is difficult or impossible to identify the minerals in aphanitic or glassy volcanic rocks. This difficulty can be overcome by recasting chemical analyses of weathering products in the form of mineral norms adapted to this purpose from igneous petrology.

The changes in chemical and mineral composition caused by weathering indicate that certain rock-forming silicate minerals are more suscepti- ble to decomposition than others. This information can be represented in the form of Bowen's reac- tion series, thereby implying that minerals that crystallize from magma at high temperatures are more susceptible to weathering than minerals that form at somewhat lower temperatures. However, the differences in susceptibility to weathering of minerals may also be understood in the context of the stabilities of minerals with respect to ions in aqueous solution. Certain common or accessory minerals in igneous and metamorphic rocks are strongly resistant to weathering and enter the sediment-transport system intact.

The resistant minerals tend to become concentrated in sand and gravel deposits, which may acquire economic value if they contain nuggets of gold or Pt group elements, or particles of diamond, cassiterite, or some other valuable minerals. Such sand and gravel deposits, called placers, are a special type of ore deposit.

Some of the minerals that survive weathering and are removed from their place of origin are suitable for dating by isotopic methods. The resulting dates reflect the time elapsed since their crystallization and can therefore be helpful in identifying the rocks from which they originated. Zircons are especially good for this purpose because they strongly resist weathering and because the art of dating them by the U–Pb method has been raised to a high level.

The most important consequence of chemical weathering is the development of soil, which sustains life on the land surface of the Earth. Soils provide nutrients to rooted plants and other organisms that constitute the soil ecosystem. Microorganisms play a vital role in soils by recycling nutrients through decomposition of organic matter and by converting N_2 from the atmosphere into water-soluble forms.

We can relate these geochemical processes to our own welfare by contemplating the fact that the annual per capita grain production in some parts of the world is low and is not keeping pace with the growth of the population. The problem is aggravated by the realization that existing farming technology cannot be introduced rapidly into areas that need it the most because of social, economic, and political barriers.

Problems

1. Carry out a constant-oxide calculation to determine the gains or losses in terms of weights and percent values for a granitic gneiss from the valley of the Minnesota River (Goldich, 1938), assuming that Al_2O_3 remained constant in amount.

	Gneiss, %	Saprolite, %
SiO_2	71.45	55.25
Al_2O_3	14.60	26.23
CaO	2.08	0.16
MgO	0.77	0.33
Na_2O	3.84	0.05
K_2O	3.92	0.14
Fe_2O_3	2.51	6.55
MnO	0.04	0.03
TiO_2	0.26	1.03
P_2O_5	0.10	0.11
CO_2	0.14	0.36
H_2O^+	0.30	9.76
Sum	100.00	100.00

2. Determine whether Fe was lost or gained as a result of weathering of granitic gneiss from the Minnesota River valley assuming that Al_2O_3 remained constant in amount.

	Gneiss, %	Saprolite, %
Al_2O_3	14.62	18.34
Fe_2O_3	0.69	1.55
FeO	1.64	0.22

3. Calculate normative K-feldspar and albite for a suite of saprolites ordered in terms of increasing intensity of weathering. Plot the results versus the concentration of Al_2O_3 and draw smooth curves through the data points. Rank the minerals in terms of their susceptibility to weathering.

	SiO_2, %	Al_2O_3, %	Na_2O, %	K_2O, %
1	69.89	16.54	0.43	5.34
2	68.09	17.31	0.12	3.48
3	61.75	18.54	0.10	3.54
4	70.30	18.34	0.09	2.47
5	57.53	23.57	0.06	0.35
6	55.07	26.14	0.05	0.14

4. A detrital muscovite from the Tyee Formation of Oregon (Eocene) has $^{87}Rb/^{86}Sr = 34.34$ and $^{87}Sr/^{86}Sr = 0.7840$ (Heller et al., 1985). What is the age of the rocks from which this muscovite originated? [Assume $(^{87}Sr/^{86}Sr)_0 = 0.7060$, $\lambda(^{87}Rb) = 1.42 \times 10^{-11}$ a^{-1}].

(Answer: 163 Ma)

5. Muscovite concentrates separated from the Tyee Formation of Oregon yielded the following results (Heller et al., 1985).

	$^{87}Rb/^{86}Sr$	$^{87}Sr/^{86}Sr$
1	34.34	0.7840
2	20.63	0.7446
3	27.96	0.7646
4	35.87	0.7917
5	14.46	0.7392
6	27.44	0.7671
7	17.27	0.7484
8	33.61	0.7874
9	30.16	0.7717

Fit a straight line to these data, and calculate the age of the muscovites from its slope.

(Answer: $t = 170$ Ma)

6. Calculate a K–Ar date for the detrital muscovite from the Tyee Formation in Problem 4 based on the following data (Heller et al., 1985): K = 8.157%, $^{40}Ar = 973.1 \times 10^{-12}$ mol/g. (Let $\lambda_\beta = 4.963 \times 10^{-10}$ a^{-1}, $\lambda_e = 0.581 \times 10^{-10}$ a^{-1}, $^{40}K/K = 1.167 \times 10^{-4}$ atom/atom, see equation 16.27.) (Answer: 67.5 Ma)

7. A concentrate of detrital K-feldspar from the same rock (Problems 4–6) yielded the following results: K = 10.294%, $^{40}Ar = 1129.4 \times 10^{-12}$ mol/g. Using the same constants as in Problem 6, calculate the K–Ar date of the feldspar. [See Heller et al. (1985) for an interpretation of these dates.] (Answer: 62.2 Ma)

8. A sample of brown, round zircons from the Potsdam sandstone yielded the following results (Gaudette et al., 1981): $^{206}Pb/^{204}Pb = 448$, $^{238}U/^{204}Pb = 3473$, $(^{206}Pb/^{204}Pb)_0 = 18.02$. Calculate a date based on the decay of ^{238}U to ^{206}Pb. ($\lambda_1 = 1.551\,25 \times 10^{-10}$ a^{-1}.)

(Answer: 752 Ma)

9. Plot a concordia curve for values of t between 0 and 1400 Ma in coordinates of $^{207}Pb*/^{235}U$ (x-coordinate) and $^{206}Pb*/^{238}U$ (y-coordinate) based on the equations (Section 8.3a): $^{207}Pb*/^{235}U = e^{\lambda_2 t} - 1$ and $^{206}Pb*/^{238}U = e^{\lambda_1 t} - 1$ where $\lambda_1 = 1.55125 \times 10^{-10}$ a^{-1}, $\lambda_2 = 9.8485 \times 10^{-10}$ a^{-1}.

10. Plot the $^{206}Pb*/^{238}U$ and $^{207}Pb*/^{235}U$ of the brown, round zircons on the concordia diagram, fit a straight line to the data points, and determine approximate dates from the intersections of the line with concordia.

	$^{206}Pb*/^{238}U$	$^{207}Pb*/^{235}U$
1	0.1238	1.279
2	0.1431	1.552
3	0.1025	1.0196
4	0.1446	1.564
5	0.1584	1.776
6	0.1354	1.445
7	0.1404	1.476
8	0.1174	1.236

(Answer: 1320 Ma, 380 Ma (Gaudette et al., 1981))

11. Calculate the terminal velocity of spherical quartz grains ($\rho = 2.65$ g/cm^3) having diameters of 500 μm settling in still water. (Answer: 22.5 cm/sec)

12. How long does it take quartz particles of 10 μm diameter to settle in a water column that is 25 cm long?

(Answer: 46.3 min)

References

BERNER, R. A., and G. R. HOLDREN, 1979. Mechanism of feldspar weathering—II. Observations of feldspar from soils. *Geochim. Cosmochim. Acta,* **43**:1173–1186.

BERNER, R. A., and J. SCHOTT, 1982. Mechanism of pyroxene and amphibole weathering II: Observations of soil grains. *Amer. J. Sci.,* **282**:1214–1231.

BOHN, L., B. L. McNEAL, and G. A. O'CONNOR, 1979. *Soil Chemistry.* Wiley, New York.

BROOKS, R. R., 1972. *Geobotany and Biogeochemistry in Mineral Exploration.* Harper & Row, New York, 290 pp.

BROWN, L. R., et al., 1987. *State of the World.* Norton, New York, 268 pp.

BUSENBERG, E., and C. V. CLEMENCY, 1976. The dissolution kinetics of feldspars at 25°C and 1 atm CO_2 partial pressure. *Geochim. Cosmochim. Acta,* **40**:41–50.

CARROLL, D., 1970. *Rock Weathering.* Plenum Press, New York, 203 pp.

CLARIDGE, G. G. C., and I. B. CAMPBELL, 1984. Mineral transformations during weathering of dolerite under cold arid conditions in Antarctica. *New Zealand J. Geophys.,* **27**:537–546.

COLMAN, S. M., 1982. Chemical weathering of basalts and andesites: Evidence from weathering rinds. *U.S. Geol. Surv. Prof. Paper* 1246, 51 pp.

COLMAN, S. M., and D. P. DETHIER (Eds.), 1986. *Rates of Chemical Weathering of Rocks and Minerals.* Academic Press, New York, 603 pp.

CONNOR, J. J., and H. T. SHACKLETTE, 1975. Background geochemistry of some rocks, soils, plants and vegetables in the conterminous United States. *U.S. Geol. Surv. Prof. Paper* 547-F, 168 pp.

CORRENS, C. W., 1978. Abundance (of Ti) in rock-forming minerals, phase equilibria, Ti-minerals. In K. H. Wedepohl (Ed.), *Handbook of Geochemistry,* II-2: 22D1–22. Springer-Verlag, Berlin.

EGGLESTON, R. A., C. FOUDOULIS, and D. VARKEVISSER, 1987. Weathering of basalt: changes in rock chemistry and mineralogy. *Clays and Clay Minerals,* **35**:161–169.

FAURE, G., 1986. *Principles of Isotope Geology,* 2nd ed. Wiley, New York, 589 pp.

FAURE, G., and K. S. TAYLOR, 1981. Provenance of glacial deposits in the Transantarctic Mountains based on Rb–Sr dating of feldspars. *Chem. Geol.,* **32**:271–290.

FAURE, G., K. S. TAYLOR, and J. H. MERCER, 1983. Rb–Sr provenance dates of feldspar in glacial deposits of the Wisconsin Range, Transantarctic Mountains. *Geol. Soc. Amer. Bull.,* **94**:1275–1280.

FOTH, H. D., 1984. *Fundamentals of Soil Science,* 7th ed. Wiley, New York, 435 pp.

FROUDE, D. O., T. R. IRELAND, P. D. KINNY, I. S. WILLIAMS, W. COMPSTON, I. R. WILLIAMS, and J. S. MYERS, 1983. Ion microprobe identification of 4,100–4,200-Myr-old terrestrial zircons. *Nature,* **304**:616–618.

GARRELS, R. M., and P. HOWARD, 1957. Reactions of feldspar and mica with water at low temperature and pressure. *Clays and Clay Minerals,* **6**:68–88.

GAUDETTE, H. E., A. VITRAC-MICHARD, and C. J. ALLÈGRE, 1981. North American Precambrian history recorded in a single sample: high-resolution U–Pb systematics of the Potsdam sandstone detrital zircons, New York State. *Earth Planet. Sci. Letters,* **54**:248–260.

GOLDICH, S. S., 1938. A study in rock weathering. *J. Geol.,* **46**:17–58.

GRANDSTAFF, D. E., 1977. Some kinetics of bronzite orthopyroxene dissolution. *Geochim. Cosmochim. Acta,* **41**:1097–1103.

HELGESON, H. C., 1971. Kinetics of mass transfer among silicates and aqueous solutions. *Geochim. Cosmochim. Acta,* **35**:421–469.

HELGESON, H. C., 1972. Kinetics of mass transfer among silicates and aqueous solutions: Correction and Clarification. *Geochim. Cosmochim. Acta,* **35**:1067–1070.

HELLER, P. L., Z. E. PETERMAN, J. R. O'NEIL, and M. SHAFIQULLAH, 1985. Isotopic provenance of sandstone from the Tyee Formation, Oregon Coast Range. *Geol. Soc. Amer. Bull.,* **96**:770–780.

HENDRICKS, D. M., and L. D. WHITTIG, 1968. Andesite weathering, Part II. Geochemical changes from andesite to saprolite. *Soil Sci.,* **19**:147–153.

HOLDREN, G. R., and R. A. BERNER, 1979. Mechanism of feldspar weathering—I. Experimental studies. *Geochim. Cosmochim. Acta,* **43**:1161–1171.

LERMAN, A., 1979. *Geochemical Processes; Water and Sediment Environments.* Wiley, New York, 481 pp.

LINDSAY, W. L., 1979. *Chemical Equilibria in Soils.* Wiley, New York, 449 pp.

LOUGHNAN, F. C., 1969. *Chemical Weathering of the Silicate Minerals.* Elsevier, New York, 154 pp.

McLELLAND, J. E., 1950. The effect of time, temperature and particle size on the release of bases from some common soil-forming minerals of different crystal structure. *Soil Sci. Soc. Amer. Proc.,* **15**:35–50.

PARK, C. F., JR., and R. A. MACDIARMID, 1970. *Ore Deposits,* 2nd ed. W. H. Freeman, New York, 522 pp.

POLDERVAART, A., 1955. Zircons in rocks. 1. Sedimentary rocks. *Amer. J. Sci.,* **253**:433–461.

SCHÄRER, U., and C. J. ALLÈGRE, 1982a. Investigation of the Archaean crust by single-grain dating of detrital zircon: a greywacke of the Slave Province, Canada. *Canadian J. Earth Sci.,* **19**:1910–1918.

SCHÄRER, U., and C. J. ALLÈGRE, 1982b. Uranium–lead system in fragments of a single zircon grain. *Nature,* **295**:585–587.

SOIL SURVEY STAFF, 1975. *Soil Taxonomy.* USDA Handbook 436, Washington, DC.

SPOSITO, G., 1981. *The Thermodynamics of Soil Solutions.* Clarendon Press, Oxford.

STOKES, G. G., 1851. On the effect of the internal friction of fluids on the motion of pendulums. *Trans. Cambridge Phil. Soc.,* **9**:8–106.

STOKES, G. G., 1901. *Mathematical and Physical Papers,* vol. 3. Cambridge University Press, Cambridge, England, pp. 1–141.

SZABO, Z., and G. FAURE, 1987. Isotopic studies of carbonate cements and provenance dating of feldspar in basin analysis:

The Berea sandstones of Ohio, U.S.A. In R. Rodriguez-Clemente and Y. Tardy (Eds.), *Geochemistry and Mineral Formation in the Earth Surface,* 51–66. C.S.I.C. and C.N.R.S., Madrid, Spain, 893 pp.

TOLE, M. P., A. C. LASAGA, C. PANTANO, and W. B. WHITE, 1986. The kinetics of dissolution of nepheline ($NaAlSiO_4$). *Geochim. Cosmochim. Acta,* **50**:379–392.

TYLER, S. A., R. W. MARSDEN, F. F. GROUT, AND G. A. THIEL, 1940. Studies of the Lake Superior pre-Cambrian by accessory mineral methods. *Geol. Soc. Amer. Bull.,* **51**:1429–1538.

WOLLAST, R., 1967. Kinetics of K-feldspar in buffered solutions at low temperature. *Geochim. Cosmochim. Acta,* **31**:635–648.

Geomicrobiology

AWRAMIK, S. M., J. W. SCHOPF, and M. F. WALTER, 1983. Filamentous fossil bacteria 3.5×10^9 years old from the Archean of Western Australia. *Precamb. Res.,* **20**:357–374.

EHRENBERG, C. G., 1838. *Die Infusionsthierchen als vollkommene Organismen* (The infusorians as complete organisms). L. Voss, Leipzig, Germany.

EHRLICH, H. L., 1981. *Geomicrobiology.* Marcel Dekker, New York, 393 pp.

ELIGWE, C. A., 1988. Microbial desulphurization of coal. *Fuel,* **67**:451–458.

HARDER, E. C., 1919. Iron-depositing bacteria and their geological relations. *U.S. Geol. Surv. Prof. Paper* 113, 89 pp.

HUTCHINS, S. R., M. S. DAVIDSON, J. A. BRIERLEY, and C. L. BRIERLEY, 1986. Microorganisms in reclamation of metals. *Ann. Rev. Microbiol.,* **40**:311–336.

KRUMBEIN, W. E. (Ed.), 1978. *Environmental Biogeochemistry and Geomicrobiology.* Ann Arbor Science Pub., Ann Arbor, MI. (a) Vol. 1, *The Aquatic Environment,* 394 pp. (b) Vol. 2, *The Terrestrial Environment,* 397–711. (c) Vol. 3, *Methods, Metals and Assessment,* 717–1055.

LUNDGREN, D. G., and M. SILVER, 1980. Ore leaching by bacteria. *Ann. Rev. Microbiol.,* **34**:263–283.

MEYER, W. C., and T. F. YEN, 1976. Enhanced dissolution of oil shale by bioleaching with thiobacilli. *Appl. Environ. Microbiol.,* **34**(4): 610–616.

MONTICELLO, D. J., and W. R. FINNERTY, 1985. Microbial desulfurization of fossil fuels. *Ann. Rev. Microbiol.,* **39**:371–389.

MURR, L. E., A. E. TORMA, and J. A. BRIERLEY (Eds.), 1978. *Metallurgical Applications of Bacterial Leaching and Related Microbiological Phenomena.* Academic Press, New York.

SCHOPF, J. W. (Ed.), 1983. *Earth's Earliest Biosphere: Its Origin and Evolution.* Princeton Univ. Press, Princeton, NJ, 543 pp.

STUTZER, O., 1911. *Die wichtigsten Lagerstaetten der Nicht-Erze* (The most important nonmetallic mineral deposits). Part 1. Berlin.

TRUDINGER, P. A., 1971. Microbes, metals and minerals. *Minerals Sci. Eng.,* **3**(4): 13–26.

WAKSMAN, S. A., 1932. *Principles of Soil Microbiology,* 2nd ed. Williams and Wilkins, Baltimore, MD.

20

Chemical Composition of Surface Water

The chemical compositon of water depends primarily on the minerals which have dissolved in it. In addition, the chemical composition of water is modified by ion-exchange equilibria, by precipitation of compounds during evaporative concentration, by mixing and dilution, by the uptake and recycling of nutrient elements, by exchange with the gases of the atmosphere, and by discharge of municipal and industrial waste.

We have already discussed some of these processes in previous chapters. For example, the geochemistry of water in streams and in the oceans was reviewed briefly in Section 4.4. In Section 12.3 we pointed out how the solubilities of certain compounds *limit* the chemical composition of natural water. In Section 13.5 we took up the properties of colloidal suspensions and the phenomenon of ion exchange. Chapter 14 was devoted to oxidation–reduction reactions, which affect the abundances of the large number of elements (including Fe, Mn, and S) that can have different oxidation states. Finally, we discussed the fractionation of the isotopes of oxygen and hydrogen in the hydrologic cycle in Section 17.3 and devoted most of Chapter 18 to the study of mixing, dilution, and evaporative concentration of water.

In this chapter we consider the chemical composition of surface water and attempt to relate it to the minerals that have reacted with it. In addition we take up again the phenomenon of evaporative concentration and the resulting evolution of brine lakes. Finally, we treat water as a natural resource whose quality determines its suitability for domestic, agricultural, and industrial use.

20.1 Chemical Analysis of Water in Streams

The chemical composition of streams varies between wide limits depending on climatic and hydrologic conditions in the drainage basin, on the mineral composition of the regolith and underlying bedrock, and on anthropogenic inputs, including municipal sewage effluent, industrial and agricultural wastes, and leachate from municipal landfills and mine dumps.

The total chemical load carried by a stream consisits of:

1. Ions and molecules in solution.
2. Ions adsorbed on colloidal particles and on sediment carried in suspension.
3. The chemical composition of the *colloidal* particles.
4. Microorganisms in the water, such as bacteria and algae.
5. Dissolved gases whose concentrations depend on the temperature of the water and their partial pressures in the atmosphere.

Evidently, stream water is a complex multicomponent system whose bulk chemical composition varies as a function of time and location along the stream channel because of continuous adjustments in the abundances of the constituents listed above.

The chemical composition of water in a stream is also affected by seasonal changes in the *source* of the water. Water originating as surface runoff and derived from groundwater reservoirs is mixed in streams in continuously changing proportions. During periods of low precipitation, streams carry primarily groundwater whose chemical composition depends on the geology of the drainage basin. During rainy seasons or after storms, surface runoff dominates and may cause flooding. As a result, the level of the water table along a stream may rise temporarily. When flow returns to normal, this "bank storage" water is drained off until the water table has returned to its former level. When the level of the water table drops *below* the surface of the water in the channel, the stream loses water and may eventually "dry up." The balance between gain and loss of water varies seasonally and may change along the course of a given stream. Under temperate climatic conditions most streams gain or lose water depending on the season, whereas in arid regions streams may lose water continuously along most of their lengths and may therefore be dry some or most of the time.

In view fo the complexity of stream water and its origin, it is by no means easy to determine the average chemical composition or water in a stream (Brown et al., 1970). In general, the results depend on the time and place where the water was collected, how the water was treated and stored prior to analysis, and the methods that were used to analyze it. The chemical composition of water in a stream may vary across the channel because of the existence of *plumes* of chemically distinct water discharged by tributaries, from springs, or by waste disposal. Moreover, the concentrations of nutrients may vary diurnally in response to the activities of organisms in the water. Therefore, in order to determine the average chemical composition of a stream, samples must be collected systematically across the channel at regular time intervals throughout at least one year. The concentrations of the ions in these samples must be weighted by the discharge in order to obtain a representative average chemical composition. Consequently, the discharge of the stream must also be recorded during the sampling period so that its *hydrograph* can be constructed (Fetter, 1988).

Certain parameters such as pH, Eh, and electrical conductivity should be measured *in situ*. In addition, the concentrations of HCO_3^- and dis-

solved Fe should also be determined immediately to avoid changes that may occur after collection. The samples should be *filtered* through $0.45\text{-}\mu m$ membranes to remove most of the suspended solids, microorganisms, and large colloidal particles. However, filtration also removes a significant fraction of the adsorbed ions from the water. After filtration, the water is acidified to pH < 3.0 to suppress further biological activity and to counteract the adsorption of ions on the walls of the containers in which the samples are stored. In addition, water samples should be refrigerated at $0.2\,°C$ to prevent further growth of microorganisms.

Several different analytical instruments are now available for the analysis of water samples. Some of these are based on the absorption or emission of eletromagnetic radiation by atoms in the vapor state (at elevated temperatures), whereas others operate at room temperature. For example, inductively coupled plasma spectrometers (ICPs) operate at temperatures in excess of $10,000\,°C$ that are sufficient to break chemical bonds and to disintegrate solid particles suspended in the water. In contrast, colorimetric and spectrophotometric methods respond to the presence of specific molecular or ionic species in the water and do no necessarily record elements that occur in colloidal particles or are adsorbed on their surfaces.

When water samples are collected from a stream at different times throughout a year, the concentrations of elements are commonly found to vary by factors of two or more (Eastin and Faure, 1970). Seasonal variations in the concentrations of nitrogen and phosphorus compounds and of molecular oxygen reflect the biological activity in the stream. However, the seasonal variations in the concentrations of *conservative* elements are caused by *mixing* of two or more components of water, by ion-exchange equilibria, and by evaporative concentration.

In cases where *mixing* is the dominant process, the concentrations of the conservative elements may be controlled by the proportions of groundwater, surface runoff, and anthropogenic wastewater that are present. Therefore, the concentrations of any two conservative elements in water samples collected in the course of a year form a triangle of mixing (Section 18.4). Surface runoff that originated as rain or meltwater is likely to be dilute and therefore lies close to the origin on such a plot, whereas the chemical compositions of groundwater and wastewater depend on the geology of the drainage basin and on the population density and industrial activities, respectively. After the mixing triangle has been defined by direct analysis of the pure components or by enclosing the data points within mixing or dilution lines, the proportions of each of the three components in any sample in the array can be determined either graphically or algebraically using equations 18.3 and 18.7 derived in Chapter 18.

This information can be used to delineate the abundances of each component in a stream as a function of time. Moreover, when the fractional abundances of the components are multiplied by the discharge at the time of collection, the *hydrograph* of the stream can be resolved into its components. In general, one expects that meteoric precipitiation is the dominant component when a stream is in flood and that the groundwater component dominates at low flow, provided the stream is not actually losing water. The abundance of the anthropogenic component may vary irregularly depending on the distribution of its sources along the channel. Moreover, the anthropogenic component may be associated with the surface runoff, as, for example, in wastewater derived from agriculture.

20.2   Chemical Composition of Streams

Few streams have been studied as thoroughly and over a sufficient number of years as may be necessary to reliably determine their average annual compositions. Therefore, inadequate data may have to be included in estimates of the continental or global fluxes of elements transported by streams into the oceans. The problem is compounded by the fact that a significant fraction of the present solution load of streams is derived from anthropogenic sources, which have contaminated the rivers of the world.

A recent reevalutation by Berner and Berner (1987) of data compiled by Meybeck (1979), Livingstone (1963), and others indicates that the concentrations of most major ions in average river water have been significantly *increased* by anthropogenic contamination as shown below:

	Increase, %		Increase, %
Na^+	28	SO_4^{2-}	43
Ca^{2+}	9	Cl^-	30
Mg^{2+}	8	HCO_3^-	2
K^+	7	SiO_2	0

The increases in concentrations of these ions in the rivers of the world constitute a significant perturbation of the geochemical cycles of these elements.

A listing of the concentrations of the major ions in the largest rivers of the world in Table 20.1 indicates that the water of most rivers is dominated by Ca^{2+} and HCO_3^-. These ions originate primarily by dissolution of calcite and plagioclase in rocks exposed to weathering on the continents. The only exceptions in Table 20.1 are the Colorado and Rio Grande rivers in North America, both of which contain more SO_4^{2-} than HCO_3^-. In addition, all rivers have $Na^+ > K^+$, $Ca^{2+} > Mg^{2+}$, and most have $SO_4^{2-} > Cl$. The concentrations of SiO_2 (actually H_4SiO_4) range from 2.4 mg/L in the St. Lawrence River to 30.4 mg/L in the rivers of the Philippines.

The sum of the concentrations of the major ions in solution, known as *total dissolved solids* (TDS), varies from 33 (Congo) to 881 (Rio Grande at Laredo, Texas). Other rivers having TDS > 300 include the Colorado (703), the Danube (307), and the Rhine (307). The major rivers of South America and Africa, on the other hand, have conspicuously *low* TDS values: Amazon (38), Parana (69), Guyana (36), Orinoco (34), Zambeze (58), Congo (Zaire) (33), and the Niger (66). The average TDS of the rivers of the world is about 110, which reduces to 100 when anthropogenic pollution is removed (Berner and Berner, 1987). The solute content, expressed as TDS, is used to classify waters as follows (Gorrell, 1953):

Table 20.1 Average Chemical Composition of Some of the Major Rivers of the World (in mg/L)[a]

River	Ca^{2+}	Mg^{2+}	Na^+	K^+	Cl^-	SO_4^{2-}	HCO_3^-	SiO_2	Discharge km^3/y
North America									
Mississippi	34	8.9	11.0	2.8	10.3	25.5	116	7.6	580
St. Lawrence	25	3.5	5.3	1.0	6.6	14.2	75	2.4	337
Mackenzie	33	10.4	7.0	1.1	8.9	36.1	111	3.0	304
Columbia	19	5.1	6.2	1.6	3.5	17.1	76	10.5	250
Yukon	31	5.5	2.7	1.4	0.7	22	104	6.4	195
Nelson	33	13.6	24	2.4	30.2	31.4	144	2.6	110
Frazer	16	2.2	1.6	0.8	0.1	8.0	60	4.9	100
Colorado	83	24	95	5.0	82	270	135	9.3	20
Rio Grande	109	24	117	6.7	171	238	183	30	2.4
Europe									
Danube	49	9	(9)	(1)	19.5	24	190	5	203
Rhine									
(upper)	41	7.2	1.4	1.2	1.1	36	114	3.7	—
Azov rivers	43	8.6	17.1	1.3	16.5	42	136	—	158
South America									
Amazon									
(lower)	5.2	1.0	1.5	0.8	1.1	1.7	20	7.2	7245
Orinoco	3.3	1.0	(1.5)	(0.65)	2.9	3.4	11	11.5	946
Parana	5.4	2.4	5.5	1.8	5.9	3.2	31	14.3	567
Guyana									
rivers	2.6	1.1	2.6	0.8	3.9	2.0	12	10.9	240
Magdalena	15.0	3.3	8.3	1.9	(13.4)	14.4	49	12.6	235
Africa									
Congo									
(Zaire)	2.4	1.3	1.7	1.1	2.9	3	11	9.8	1230
Zambezi	9.7	2.2	4.0	1.2	1	3	25	12	224
Niger	4.1	2.6	3.5	2.4	1.3	(1)	36	15	190
Nile	25	7.0	17	4.0	7.7	9	134	21	83
Orange	18	7.8	13.4	2.3	10.6	7.2	107	16.3	10
Asia									
Yangtze	45	6.4	4.1	1.2	4.1	17.9	148	5.8	1063
Mekong	14.2	3.2	3.6	2.0	(5.3)	3.8	58	8.9	577
Ganges	24.5	5.0	4.9	3.1	3.4	8.5	105	12.8	450
New Zealand									
rivers	8.2	4.6	5.6	0.7	5.8	6.2	50	7	400
Philippines	31	6.6	10.4	1.7	3.9	13.6	131	30.4	332
World Average									
Actual									
values	14.7	3.7	7.2	1.4	8.3	11.5	53.0	10.4	37,400
Unpolluted	13.4	3.4	5.2	1.3	5.8	6.6	52.0	10.4	37,400

[a]Values in parentheses are estimates.

SOURCE: A compilation by Berner and Berner (1987), based in part on data from Meybeck (1979) and Livingstone (1963).

Water type	TDS, mg/L
Fresh	<1000
Brackish	1000–10,000
Saline	10,000–100,000
Brine	>100,000

The total average annual discharge of all streams is 37,400 km^3/yr (Table 20.1), about 20% of which is contributed by the Amazon River of South America. The Congo (Zaire) River in Africa and the Yangtze River in China contribute 3.3% and 2.8% respectively. Some other well-known rivers, including the Mississippi in North America, the Mekong in Vietnam, the Danube in Europe, and the Nile in Africa together contribute only 3.8% of the total annual discharge. In fact, the Nile has a discharge of only 83 km^3/yr (0.22% of the world's total), which makes it one of the smallest major rivers of the world in contrast to its historical and present importance in Egypt. Moreover, data compiled by Livingstone (1963) indicate a total discharge of only 0.32 km^3/yr for the entire continent of Australia.

The chemical composition of surface waters was used by Gibbs (1970) to classify them based on their TDS values and their $Na^+/(Na^+ + Ca^{2+})$ ratios. The resulting distribution of data points for the world's major rivers can be circumscribed by a "boomerang" shown in Figure 20.1. The rivers that define the lower end of the boomerang are dominated by meteoric precipitation, those in the upper end are affected primarily by evaporation and precipitation of calcite, whereas the rivers in the center of the boomerang are influenced by rock weathering.

The $Na^+/(Na^+ + Ca^{2+})$ ratio of water in both ends of the boomerang tends toward a value of 1.0. At the lower end, the concentration of Na^+ in *rainwater* is commonly greater than that of Ca^{2+}, especially in coastal areas, so that the $Na^+/(Na^+ + Ca^{2+})$ ratio approaches unity. At the other end of the boomerang, evaporative concentration causes precipitation of calcite, which depletes the water in Ca^{2+}, but enriches it in Na^+ and causes the $Na^+/(Na^+ + Ca^{2+})$ ratio to increase toward

unity. Therefore, rivers in arid regions, saline lakes, and the oceans have high $Na^+/(Na^+ + Ca^{2+})$ ratios. Some representative examples of each type of water are shown in Figure 20.1.

The chemical composition of river water does not necessarily reflect unambiguously the geochemical processes that affected it. For example, the elevated TDS and $Na^+/(Na^+ + Ca^{2+})$ ratios of the Pecos and Rio Grande rivers in New Mexico and Texas (Figure 20.1) are caused primarily by the inflow of NaCl brines derived from halite deposits rather than by evaporative concentration and precipitation of calcite (Feth, 1971; Stallard and Edmond, 1983). In addition, Ca^{2+} is more abundant than Na^+ in rain that falls in the interiors of continents, and the $Na^+/(Na^+ + Ca^{2+})$ ratio of rainwater in continental areas is therefore less than one. For example, the $Na^+/(Na^+ + Ca^{2+})$ ratio of rain in the interior of the United States is only 0.22 (Berner and Berner, 1987). Therefore, dilution by rain in continental interiors does *not* increase the $Na^+/(Na^+ + Ca^{2+})$ ratio of stream water.

The difficulties of genetic classifications of natural waters are avoided by a graphical method proposed by Piper (1944) and used frequently in hydrogeology (Fetter, 1988). The so-called Piper diagram consists of two equilateral triangles and a diamond-shaped field onto which points are projected from the cation and anion triangles. The two triangles are used to represent the chemical compositions of waters expressed in percent of the milliequivalents per liter of the cations Ca^{2+}, Mg^{2+}, and $Na^+ + K^+$ and of the anions Cl^-, SO_4^{2-}, and $HCO_3^- + CO_3^{2-}$. The triangles also serve to classify water based on the dominant cations and anions as shown in Figure 20.2.

The Piper diagram is a graphical representation of the chemical composition of water in terms of the *eight* major cations and anions, whereas the Gibbs diagram involves only *two* cations directly. Moreover, the Piper diagram serves to classify waters without genetic implications, whereas the Gibbs diagram presupposes the geochemical processes that affect waters in different parts of the boomerang.

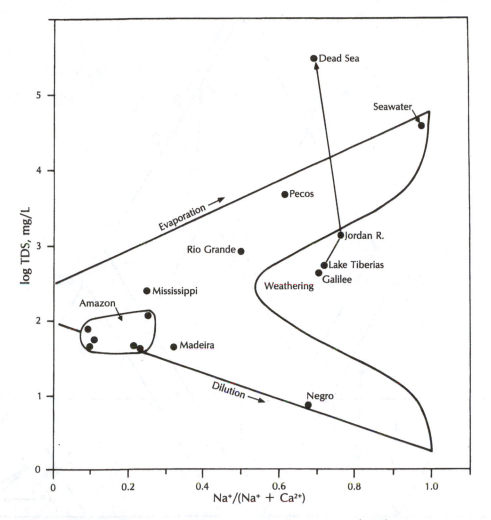

Figure 20.1 Classification of surface waters according to Gibbs (1970). Water in the central part of the "boomerang" is dominated by weathering of silicate minerals. Progressive evaporation and precipitation of calcite increase TDS and enrich the water in Na^+ while depleting it in Ca^{2+}. Hence such waters reside in the *upper* part of the boomerang. Waters in the *lower* part are dominated by rainwater of coastal areas where $Na^+ > Ca^+$. Waters from the Jordan River–Dead Sea area of Israel (Livingstone, 1963) illustrate the effect of progressive evaporation but plot outside the boomerang. The Amazon River and its tributaries (Livingstone, 1963; Stallard, 1985) exemplify water with large rainwater components. The chemical composition of the Mississippi River (Meybeck, 1979) is controlled by chemical weathering of rocks, whereas the Rio Grande (Livingstone, 1963) reflects the effects of evaporative concentration or the influx of NaCl brines. Note the location of seawater on the diagram (Berner and Berner, 1987).

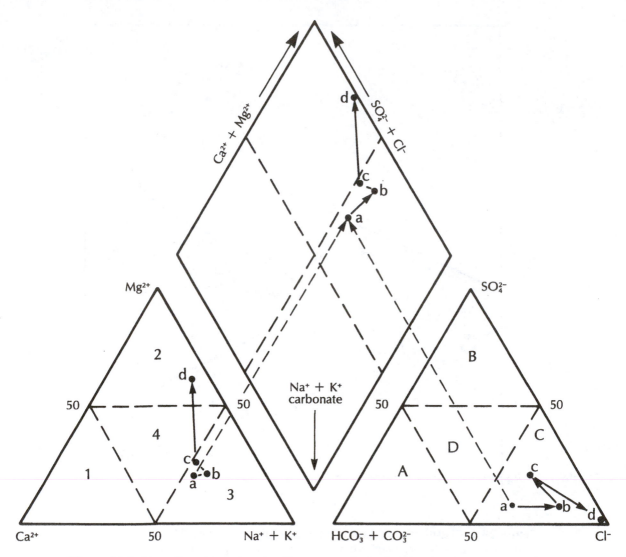

Figure 20.2 Piper diagram used to classify water on the basis of its chemical composition.

1. Ca-rich   A. bicarbonate-rich
2. Mg-rich   B. sulfate-rich
3. alkali-rich   C. chloride-rich
4. mixed cations   D. mixed anions

Data points are from the compositional triangle onto the diamond as shown by the dashed arrows. The waters are from the Dead Sea area: (a) Sea of Galilee; (b) Lake Tiberias; (c) Jordan River at Jericho; (d) Dead Sea, surface water (Livingstone, 1963, Table 51). The composition of the water in this drainage system changes from alkali chloride water (TDS ~500 mg/L) to a Mg-chloride brine (TDS ~270,000 mg/L). The compositions of water samples are plotted as percent milliequivalent per liter.

20.3   Chemical Composition of Meteoric Precipitation

The chemical composition of rain and snow is affected by processes that are characteristic of the atmosphere:

1. Dissolution of solid particles that serve as centers of nucleation for raindrops or snowflakes.
2. Dissolution of the gases of the atmosphere.
3. Discharge of anthropogenic contaminants into the atmosphere.

The solid particles (0.1–20 μm is diameter) include mineral grains derived from soil, volcanic ash, various kinds of combustion products, and salts derived from the surface of the ocean. *Hygroscopic* compounds such as NaCl, H_2SO_4, and HNO_3, which absorb water from the air and are very soluble in it, are especially effective in promoting condensation. The gases that dissolve in rainwater include CO_2, SO_2, and oxides of nitrogen, all of which not only occur naturally but also are released into the atmosphere as a result of combustion of fossil fuel. In the unpolluted atmosphere these gases react with water and molecular oxygen to form acids that cause rainwater to be naturally acidic.

For example, CO_2 gas reacts with water to form carbonic acid, which dissociates to form bicarbonate and hydrogen ions:

$$CO_2 + H_2O \rightarrow HCO_3^- + H^+ \quad (20.1)$$

$$\frac{[H^+][HCO_3^-]}{[CO_2]} = 10^{-7.83} \quad (20.2)$$

where the brackets represent the *activities* of ions and *fugacities* of gases (Section 9.2). If $[CO_2] = 3 \times 10^{-4}$ atm, the pH of rainwater is 5.68. However, where large amounts of SO_2 are released into the atmosphere by combustion of sulfur-bearing coal or by roasting of sulfide ore, the pH of rainwater may be significantly less than 5.68. Such *acid rain* promotes the dissolution of solid particles in the atmosphere, which affects the chemical composition of rain or snow. In addition, the acidification of surface water and soil is harmful to plants and animals and therefore causes deterioration of the environment.

The chemical composition of rainwater varies between wide limits. The data in Table 20.2 and Figure 20.3 indicate that Na^+ and Cl^- are the dominant ions in coastal areas and in rain over the oceans and oceanic islands. However, the concentrations of these ions *decrease* rapidly with increasing distance from the coast, whereas the concentrations of Ca^{2+} and SO_4^{2-} may rise somewhat, as illustrated in Figure 20.3 for the Amazon basin of South America. Nevertheless, the average total dissolved solids of rainwater in the Amazon basin is only 1.3 mg/L compared to 11.8 mg/L in the coastal areas worldwide and 21.2 mg/L in rain over the oceans. Rain in the coastal areas of the United States has TDS = 11.8 mg/L, whereas in the interior rain has TDS = 7.0 mg/L on the average.

The chloride ion in rainwater originates primarily from seawater in the form of NaCl crystals. Therefore, concentration ratios of the elements relative to Cl^- can be used to determine the presence of "excess" amounts of these elements in rain that are attributed to the dissolution of dust particles in the air and/or to anthropogenic pollution. Junge and Werby (1958) studied the distribution of Cl^-, Na^+, K^+, Ca^{2+}, and SO_4^{2-} in rainwater in the United States and calculated excess amounts of these elements relative to seawater based on equations of the form:

$$\left(\frac{Na}{Cl}\right)_r = \frac{Na_s + Na_x}{Cl_s} \quad (20.3)$$

where the subscripts refer to the rain (r), seawater (s), and excess (x).

A compilation of concentration ratios of ions in rainwater relative to Cl^- in Table 20.3 indicates that rain over the oceans and in coastal areas of the world resembles seawater in its chemical composition. For example, the Na^+/Cl^- and K^+/Cl^- ratios of rain over the oceans and in coastal areas differ from seawater values of these ratios by a factor of 1.3 or less. However, rain in these areas is somewhat *enriched* in Ca^{2+} and SO_4^{2-} compared to seawater, presumably because of the dissolution of mineral particles

Table 20.2 Chemical Composition of Meteoric Precipitation (in mg/L)

Region	Na^+	K^+	Mg^{2+}	Ca^{2+}	Cl^-	SO_4^{2-}	NO_3^-	NH_4^+	pH
United States									
Coastal	3.68	0.24	—	0.58	4.83	2.45	—	—	—
Inland	0.40	0.20	0.10	1.4	0.41	3.0	1.20	0.30	—
Europe									
Coastal	11.4	0.60	1.51	1.09	20.1	6.35	1.80	0.45	4.87
Inland	0.64	0.20	0.22	0.78	1.27	3.02	1.42	0.28	4.74
U.S.SR.	1.4	0.6	0.6	1.3	1.6	4.9	0.6	0.7	5.6
Austrailia									
southeast	2.46	0.37	0.50	1.2	4.43	trace	—	—	—
Japan	1.1	0.26	0.36	0.97	1.2	4.5	—	—	—
Oceans and									
islands of the world	5.79	0.31	0.47	0.89	11.0	2.44	0.21	0.075	5.0
World Average,									
coastal (<100 km)	3.45	0.17	0.45	0.29	6.0	1.45	—	—	—
Amazon River									
670 km inland	0.50	0.020	0.036	0.028	0.87	0.64	0.19	0.002	4.71
1700 km inland	0.23	0.039	0.024	0.056	0.30	0.55	0.25	0.007	5.32
1930 km inland	0.21	0.035	0.034	0.060	0.41	0.70	0.18	0.00	4.97
2050 km inland	0.12	0.094	0.012	0.056	0.24	0.56	—	—	5.04
2230 km inland	0.23	0.012	0.012	0.008	0.39	0.28	0.056	—	5.31
Peru 3000 km from									
Atlantic Ocean	0.039	0.039	0.020	0.184	0.12	0.18	—	—	5.67

Adapted from a compilation by Berner and Berner (1987).

derived from soil on islands and coastal areas. The abundance of Mg^{2+} in rain over the oceans and in coastal areas world-wide varies irregularly.

The data in Table 20.3 pertaining to rain along the coast and in the *interior* of the United States indicate significant *enrichments* in all five ions relative to seawater. This is especially true of inland rain, which is enriched in SO_4^{2-} by a factor of 52 with respect to seawater, followed by Ca^{2+} (34), K^+ (23), Mg^{2+} (3.6), and Na^+ (1.8). Rain in the coastal areas is also enriched in these ions, but the enrichment factors range only from 5.7 (Ca^{2+}) to 1.4 (Na^+). The regional variation of excess concentrations of these elements in rain over the United States was also documented by Munger and Eisenreich (1983).

20.4 Normative Minerals from Water Compositions

We now return to the discussion of river water and restate the obvious fact that its chemical composition is largely the result of the minerals that have dissolved in it. Consequently, it should be possible to use the chemical composition of water to identify these minerals and to calculate their relative abundances. Such a calculation was demonstrated by Garrels (1967) based in part on preceding work of Feth et al. (1964).

Before we proceed with the quantitative evaluation of the chemical composition of water, we recall some of the conclusions presented in earlier

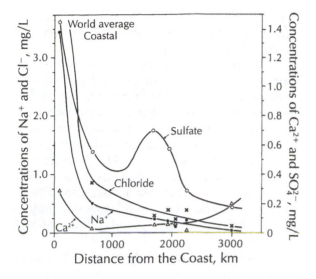

Figure 20.3 Variation of the major cations and anions in rainwater of the Amazon basin with distance from the Atlantic Ocean. The data for the Amazon basin in Table 20.2 are from Stallard and Edmond (1981) as compiled by Berner and Berner (1987).

kinds of environments. In Chapter 14 we discussed the oxidation of Fe and showed the Fe^{2+} oxidizes to Fe^{3+} in the presence of molecular oxygen and subsequently precipitates as $Fe(OH)_3$, which is ultimately converted to Fe_2O_3 (hematite). When we constructed equations to represent weathering reactions, we assumed that the necessary H^+ was provided by the environment, but we did not consider what reactions actually contribute the H^+ ions in most natural environments.

In order to interpret the chemical composition of water quantitatively we now refine our understanding of chemical weathering reactions:

1. Any H^+ required by the reaction is derived by dissociation of carbonic acid, which is in equilibrium with CO_2 whose partial pressure may exceed that of the atmosphere by a factor of 10 or more. Where pyrite/marcasite are present, H^+ may also originate by oxidation of these minerals and by the hydrolysis of Fe^{3+} to form $Fe(OH)_3$.

2. Na^+ and Ca^{2+} ions in water originate by the incongruent dissolution of plagioclase to form kaolinite. Some Na^+ may be contributed by discharge of subsurface brines that previously interacted with evaporite rocks. Ca^{2+} may also be derived by dissolution of calcite, gypsum/anhydrite, or dolomite.

chapters. For example, in Section 10.6 and in Chapter 12 we showed that the incongruent dissolution of aluminosilicate minerals commonly results in the formation of kaolinite, although other clay minerals or gibbsite may form in some

Table 20.3 Concentration *Ratios* of the Major Ions in Rainwater Relative to the Chloride Ion

| Ion | Seawater[1] | Rainwater[2]a | | | |
		Oceans	Coastal Worldwide	Coastal U.S	Inland U.S.
Na^+	0.56	0.53 (0.95)	0.58 (1.04)	0.76 (1.4)	0.98 (1.8)
K^+	0.021	0.028 (1.3)	0.028 (1.3)	0.050 (2.4)	0.49 (23)
Ca^{2+}	0.021	0.081 (3.9)	0.048 (2.3)	0.12 (5.7)	0.71 (34)
Mg^{2+}	0.067	0.043 (0.64)	0.075 (1.1)	—	0.24 (3.6)
SO_4^{2-}	0.14	0.22 (1.6)	0.24 (1.7)	0.51 (3.6)	7.3 (52)

aNumbers in parentheses are enrichment factors relative to seawater.
SOURCES: (1) Berner and Berner (1987), Table 3.6; (2) Table 20.2, this book.

3. Mg^{2+} and K^+ are released by weathering of biotite. Additional Mg^{2+} comes from the solution of ferromagnesian minerals (olivine, pyroxenes, and amphiboles) or dolomite, and some K^+ is released by K-feldspar.

4. Silicic acid enters the water during weathering of silicate minerals, whereas quartz contributes almost none.

5. Sulfate ion is produced primarily by oxidation of sulfide minerals and by dissolution of gypsum/anhydrite or other sulfate minerals, or it originates from meteoric precipitation.

6. Chloride is likewise contributed by meteoric precipitation, by discharge fo subsurface brines, and by fluid inclusions in minerals.

7. Bicarbonate forms from CO_2 gas and by dissolution of calcite and other carbonate minerals.

8. The concentration of Fe^{2+} in water is kept low by the formation of $Fe(OH)_3$ except under strongly acid and reducing conditions.

The proportions of ions released by the incongruent dissolution of different aluminosilicate minerals can be determined from the appropriate equations, assuming that kaolinite is the stable phase. For *albite* we obtain:

$$2\,NaAlSi_3O_8 + 2\,CO_2 + 3\,H_2O \rightarrow$$
$$Al_2Si_2O_5(OH)_4 + 2\,Na^+ + 2\,HCO_3^- + 4\,SiO_2 \tag{20.4}$$

We balance this equation by assuming that Al is conserved and that H^+ is provided by the dissociation of carbonic acid, which produces both H^+ and HCO_3^-. The number of moles of HCO_3^- is determined from the requirement for *electrical neutrality*. In this case, the release of 2 moles of Na^+ requires that 2 moles of HCO_3^- appear as products of the reaction. Note that the HCO_3^- ions are derived entirely from CO_2 in this case, so that the ^{14}C content and the isotopic compo-

sitions of C and O of the bicarbonate ions depend on the source of the CO_2 (Section 16.4, 17.4, and 17.7).

The possible sources of CO_2 include the atmosphere, oxidation of biogenic carbon compounds in soil, oxidation of methane released by bacteria or by hydrocarbons in subsurface reservoirs, or decarbonation reactions of carbonate rocks being heated at depth. Note also that we represent silicic acid by "SiO_2" because it is commonly expressed this way in chemical analyses of water. Therefore, the incongruent dissolution of one mole of albite releases one mole of Na^+, one mole of HCO_3^-, and 2 moles of SiO_2 into the water.

Similarly, the weathering of *anorthite* is represented by the equation:

$$CaAl_2Si_2O_8 + 2\,CO_2 + 3\,H_2O \rightarrow$$
$$Al_2Si_2O_5(OH)_4 + Ca^{2+} + 2\,HCO_3^- \tag{20.5}$$

which indicates that dissolution of 1 mole of anorthite releases 1 mole of Ca^{2+} and 2 moles of HCO_3^- (but no SiO_2) into solution. *Biotite* is a bit more complicated because of its variable composition (Section 13.2b). We assume that any Fe^{2+} in the octahedral layer is oxidized to Fe^{3+} and subsequently precipitates as $Fe(OH)_3$. Therefore, the Fe released into soluton by biotite does not appear in the chemical analysis of the water. In the case of the Mg-mica *phlogopite*, the reaction is:

$$2\,K(Mg_3)(AlSi_3O_{10})(OH)_2 + 14\,CO_2 + 7\,H_2O \rightarrow$$
$$Al_2Si_2O_5(OH)_4 + 2\,K^+ + 6\,Mg^{2+}$$
$$+ 14\,HCO_3^- + 4\,SiO_2 \tag{20.6}$$

Evidently, 1 mole of phlogopite releases 1 mole of K^+, 3 moles of Mg^{2+}, 7 moles of HCO_3^-, and 2 moles of SiO_2 into solution.

In *biotite*, some of the K^+ may be replaced by Na^+, and the octahedral layer may contain Al^{3+}, Ti^{4+}, Fe^{3+}, Fe^{2+}, Mn^{2+} as well as Mg^{2+}. Moreover, more than one Si^{4+} in the tetrahedral layers may be replaced by Al^{3+} (Garrels, 1967). However, for the sake of simplicity, we simulate the incongruent dissolution of *biotite* by the equation:

$$2 \, K(Mg_2Fe^{2+})AlSi_3O_{10}(OH)_2$$
$$+ \, 10 \, CO_2 + \tfrac{1}{2}O_2 + 8 \, H_2O \rightarrow$$
$$Al_2Si_2O_5(OH)_4 + 2 \, Fe(OH)_3$$
$$+ \, 2 \, K^+ + 4 \, Mg^{2+} + 10 \, HCO_3^- + 4 \, SiO_2 \quad (20.7)$$

In this example, 1 mole of biotite reacts to form 1 mole of K^+, 2 moles of Mg^{2+}, 5 moles of HCO_3^-, and 2 moles of SiO_2. Note that $\tfrac{1}{2}$ mole of molecular oxygen is required to act as the acceptor of the electrons (oxidizing agent) released by Fe^{2+}.

Magnesium is also released by the incongruent dissolution of other ferromagnesian minerals such as the pyroxenes and amphiboles. The weathering of *proxene* is represented by the equation:

$$2 \, CaMgFeAl_2Si_3O_{12} + \tfrac{1}{2}O_2 + 11 \, H_2O + 8 \, CO_2 \rightarrow$$
$$2 \, Al_2Si_2O_5(OH)_4 + 2 \, Fe(OH)_3 + 2 \, Ca^{2+}$$
$$+ \, 2 \, Mg^{2+} + 2 \, SiO_2 + 8 \, HCO_3^- \quad (20.8)$$

Each mole of this pyroxene releases 1 mole of Ca^{2+}, 1 mole of Mg^{2+}, 1 mole of SiO_2, and 4 moles of HCO_3^-. The reactions of other minerals can be written in the same way, and the number of moles of ions released per mole of mineral are listed in Table 20.4.

We are now ready to examine the chemical composition of a sample of spring water from the Sierra Nevada Mountains in California (Garrels and Mackenzie, 1967), which has reacted with granitic igneous rocks composed primarily of feldspars, biotite, and quartz. The analysis of this water in Table 20.5 indicates that Ca^{2+} and HCO_3^- are the dominant ions, that the concentration of $Na^+ > K^+$ and $Ca^{2+} > Mg^{2+}$, TDS = 45.83 mg/L, and SiO_2 = 16.4 mg/L, which is unusually high compared to river water in Table 20.1,

We begin by converting the concentrations from milligrams per liter to micromoles per liter, followed by rounding to whole numbers. Next, we subtract the micromolar quantities of ions in meltwater derived from local snow (Garrels and Mackenzie, 1967). The concentrations of SO_4^{2-} and

Cl^- in the groundwater are thereby accounted for as well as 29% of K^+, 24% of Mg^{2+}, 18% of Na^+, and 13% of Ca^{2+}. Before we can distribute the remaining amounts of the ions among the minerals from which they originated, we check the solution for electrical neutrality. We find a small excess of negative charge (Table 20.5), which we correct by reducing HCO_3^- by 2 μmol/L, leaving 308 μmol/L.

Now we convert the 108 μmol of Na^+ in one liter of water into 108 μmol of albite. Therefore, we reduce SiO_2 in Table 20.5 by 216 μmol and HCO_3^- by 108 μmol, in accordance with the data in Table 20.4. The resulting amount of albite dissolved in one liter of water is:

$$\frac{108 \times 262.15}{1000} = 28.3 \text{ mg} \quad (20.9)$$

Next, we assign all of the Mg^{2+} to biotite and combine it with 11 μmol of K^+, 22 μmol of SiO_2, and 55 μmol of HCO_3^- consistent with the formula of biotite we adopted in equation 20.7. The Fe^{2+} of this biotite was oxidized to Fe^{3+} and formed solid $Fe(OH)_3$ (equation 20.7). The amount of biotite, calculated as illustrated by equation 20.9 is 4.9 mg per liter of solution. The remainder of the K^+ is converted into K-feldspar, with proportional reductions of SiO_2 and HCO_3^-, yielding 2.5 mg of K-feldspar per liter of solution. Finally, we convert Ca^{2+} into 68 μmol of anorthite, which consumes all of the available HCO_3^- and is equivalent to 18.9 mg of anorthite dissolved in one liter of water.

We are left with a residue of 14 μmol of SiO_2, which we attribute to the dissolution of 0.8 mg of amorphous SiO_2 and speculate that it may have been present in the soil or saprolite overlying the granitic bedrock.

The total amount of mineral matter of the granitic bedrock that dissolved per liter of water is 54.6 mg (excluding the amorphous silica). Therefore, the abundances of the minerals we have considered are albite, 51.8%; anorthite, 34.6%; K-feldspar, 4.6%; and biotite, 9.0%. The composition of the plagioclase that dissolved expressed by the abundance of albite (Ab), is:

$$Ab = \frac{108 \times 10^2}{108 + 68} = 61 \text{ mol \%} \quad (20.10)$$

Table 20.4 Number of Moles of Ions Released into Solution per Mole of Mineral[a] in the Presence of CO_2 Gas

Mineral	Na^+	K^+	Mg^{2+}	Ca^{2+}	SiO_2	HCO_3^-	SO_4^{2-}
Albite, $NaAlSi_3O_8$	1	0	0	0	2	1	0
Anorthite, $CaAl_2Si_2O_8$	0	0	0	1	0	2	0
Orthoclase, $KAlSi_3O_8$	0	1	0	0	2	1	0
Biotite, $K(Mg_2Fe)AlSi_3O_{10}(OH)_2$	0	1	2	0	2	5	0
Pyroxene, $CaMgFeAl_2Si_3O_{12}$	0	0	1	1	1	4	0
Enstatite, $MgSiO_3$	0	0	1	0	1	2	0
Diopside, $CaMg(SiO_3)_2$	0	0	1	1	2	4	0
Calcite, $CaCO_3$	0	0	0	1	0	2	0
Dolomite, $CaMg(CO_3)_2$	0	0	1	1	0	4	0
Gypsum, $CaSO_4 \cdot 2H_2O$	0	0	0	1	0	0	1
Pyrite, FeS_2	0	0	0	0	0	0	2

Formula Weights

albite	262.2	pyroxene (eq. 20.8)	450.3
biotite (eq. 21.7)	448.7	calcite	100.1
K-feldspar	278.3	dolomite	184.4
anorthite	278.1	anhydrite	136.1
diopside	216.5	pyrite	120.0

[a]Note that aluminosilicates are assumed to dissolve incongruently to kaolinite and that Fe-bearing minerals form $Fe(OH)_3$.

Recalling the fictitious geochemist A. O. Alba from Section 8.3 we identify the plagioclase that dissolved in the water as *andesine* (Ab_{50} to Ab_{70}). However, the plagioclase in the rocks probably was more Na-rich than andesine because anorthite is *more* susceptible to dissolution than albite (Section 19.3). In addition, the rock may well have contained quartz, magnetite, zircon, and other accessory minerals that resist chemical weathering.

The reconstruction of the minerals dissolved in a sample of water provides a clue to the kinds of rocks with which the water has interacted. In the water from the Sierra Nevada, the composition of the solute is consistent with the expected weathering products of granitic igneous or metamorphic rocks when allowance is made for differences in the relative susceptibilities of rock-forming silicate minerals.

Table 20.5 Reconstruction of the Minerals That Contributed Ions to a Sample of Groundwater from the Sierra Nevada Mountains, California

	Na^+	K^+	Mg^{2+}	Ca^{2+}	SiO_2	HCO_3^-	SO_4^{2-}	Cl^-	mg	$Wt.\%$
Analysis										
mg/L	3.03	1.09	0.70	3.11	16.4	20.0	1.0	0.5		
μmol/L	132	28	29	78	273	328	10	14		
Snow, μmol/L	24	8	7	10	3	18	10	14		
Net μmol/L	108	20	22	68	270	310	0	0		
Electric neutrality, μeq/L			308+			310−				
Adjusted μmol/L	108	20	22	68	270	308	0	0		
Albite, μmol/L	0	20	22	68	54	200	0	0	28.3	51.8
Biotite, μmol/L	0	9	0	68	32	145	0	0	4.9	9.0
K-feldspar, μmol/L	0	0	0	68	14	136	0	0	2.5	4.6
Anorthite, μmol/L	0	0	0	0	14	0	0	0	18.9	34.6
Amorph. silica, μmol/L	0	0	0	0	0	0	0	0	0.8	—
									54.6[a]	100

[a]Excluding the amorphous silica.
SOURCE: Garrels and Mackenzie (1967).

Bedrock containing significant amounts of sulfide minerals causes water that interacts with it to become enriched in sulfate and in certain *chalcophile* and siderophile elements identified in Table 8.2, such as Cu, Co, Ni, Zn, Mo, Ag, As, Sb, Se, and Te, which form soluble salts with the most abundant anions in groundwater. The presence of unusual concentrations of these elements in water samples is used in *geochemical prospecting* surveys (discussed in Chapter 22) to locate deposits of "mineralized" rocks below the surface of the Earth.

In order to demonstrate the relationship between the mineral composition of the rocks in a watershed and the resulting chemical composition of water, we consider a water sample of as yet unidentified origin in Table 20.6. The water conforms to the normal pattern in the sense that Ca^{2+} and HCO_3^- are dominant, $Na^+ > K^+$, and $Ca^{2+} > Mg^{2+}$; however, TDS = 430 mg/L compared to only 46 mg/L in the spring water from the Sierra Nevada, and the concentration of SiO_2 is low at 6.0 mg/L. In addition, the analysis includes NO_3^- and PO_4^{3-} and the pH = 7.7.

We correct the analysis for meteoric precipitation using the composition of average inland rain

in the United States in Table 20.2. The concentration of HCO_3^- in average inland rain was estimated from equation 20.2 for $[CO_2] = 3 \times 10^{-4}$ atm to be 2 μmol/L. A check of the electrical balance among the remaining ions reveals an excess of 15 μeq of negative charge, which we correct by arbitrarily decreasing HCO_3^- by that amount, thereby decreasing its concentration by only 0.3%.

The small amount of SiO_2 in this water is assigned to 50 μmol of albite, and the chloride is combined with Na^+ to make 1060 μmol of NaCl. The large amount of Mg^{2+} in the water suggests the presence of dolomite, but its amount is limited by HCO_3^- to 1118.5 μmol, which leaves a residue of 871.5 μmol of Ca^{2+} and 152.5 μmol of Mg^{2+}. We combine the remaining Ca^{2+} with SO_4^{2-} to make 871.5 μmol of anhydrite ($CaSO_4$) and use up the K^+ as 41 μmol K_2SO_4, which is a common ingredient in mixed fertilizers. Similarly, the 58 μmol of PO_4^{3-} are attributed to fertilizer in the form of $NH_4H_2PO_4$, where the NH_4^+ is assumed to have oxidized to NO_3^- after applicaton. The small amounts of Na^+, Mg^{2+}, SO_4^{2-}, and NO_3^- are now arbitrarily assigned to 29 μmol $NaNO_3$ and 152.5 μmol of $MgSO_4$, although other choices are certainly possible. These

Table 20.6 Reconstruction of Minerals Dissolved in the Water of an Unidentified Stream

	Na^+	K^+	Mg^{2+}	Ca^{2+}	SiO_2	HCO_3^-	SO_4^{2-}	Cl^-	NO_3^-	PO_4^{3-}
Analysis										
mg/L	31	3.4	31	81	6.0	274	101	38	6.6	5.5
μmol/L	1348	87	1275	2021	100	4491	1097	1072	106	58
Rain, μmol/L	17	5	4	35	0	2	31	12	19	0
Net μmol/L	1331	82	1271	1986	100	4489	1066	1060	87	58
Electrical neutrality,										
μeq/L			7927+				7942−			
Adjusted μmol/L	1331	82	1271	1990	100	4474	1066	1060	87	58
Albite, μmol/L	1281	82	1271	1990	0	4474	1066	1060	87	58
NaCl, μmol/L	221	82	1271	1990	0	4474	1066	0	87	58
Dolomite, μmol/L	221	82	152.5	871.5	0	0	1066	0	87	58
Anhydrite, μmol/L	221	82	152.5	0	0	0	194.5	0	87	58
K_2SO_4, μmol/L	221	0	152.5	0	0	0	153.5	0	87	58
$NH_4H_2PO_4$, μmol/L	221	0	152.5	0	0	0	153.5	0	29	0
$NaNO_3$, μmol/L	192	0	152.5	0	0	0	153.5	0	0	0
$MgSO_4$, μmol/L	192[a]	0	0	0	0	0	1	0	0	0

[a]The apparent excess of positively charged ions was caused in part by our choice to treat NO_3^- as NH_4^+.

alternatives do not affect the clear evidence that the water in Table 20.6 originated by dissolution of very different minerals than that in Table 20.5.

This point is emphasized in Table 20.7 in which the micromolar amounts of the minerals are expressed in terms of weight percent. The results indicate that dolomite, anhydrite, and halite contributed 89% of the solute and an additional 3% was derived from silicate minerals represented by albite. The remaining 8% can be attributed to various kinds of fertilizers, implying that the watershed of this stream is being farmed. The dominant minerals reflected in the chemical composition of the water are characteristic of carbonate-evaporite rocks of marine origin. Soluble minerals like halite and anhydrite may be present in subsurface, and their ions may enter the stream as effluent of saline groundwater. The SiO_2 may have been contributed by weathering of detrital silicate minerals in the carbonate rocks or in shale, by weathering of volcanic ash or igneous rocks, or by dissolution of silicate minerals in transported regolith such as till.

In fact, the water sample we have been discussing was collected from the Great Miami River at West Carrollton is southwestern Ohio. The river drains marine carbonate–evaporite rocks of early to middle Paleozoic age that are overlain by Wisconsinan and Illinoian till and

Table 20.7 Abundances of Minerals That Dissolved to Form the Water Presented in Table 20.6.

Mineral	μmol/L	mg/L	Weight %
albite	50	13.1	3.0
halite	1060	61.9	14.2
dolomite	1118.5	206.2	47.5
anhydrite	871.5	118.6	27.3
K_2SO_4	41	7.1	1.6
$NH_4H_2PO_4$	58	6.7	1.5
$NaNO_3$	29	2.5	0.6
$MgSO_4$	152.5	18.4	4.2
Totals		434.5	99.9

outwash deposits derived from the Precambrian Shield of Canada north of Ohio. Evidently, the chemical composition of the water in the Great Miami River *does* reflect the geology of its watershed.

Reconstructions of minerals are most appropriate for surface water whose chemical composition has *not* been altered by ion exchange and/or precipitation of insoluble compounds. However, mixed waters that have interacted with different kinds of rocks also provide meaningful information about the geology of the watershed from which they originated.

20.5 Evaporative Concentration

Since the chemical composition of water in streams depends on the minerals that have dissolved in it, we now consider whether these minerals can *precipitate* if the water is subjected to evaporative concentration. We already know that chemical weathering of most minerals is an *irreversible* process, and therefore we do *not* expect that evaporative concentration of water can reverse weathering. We recall that silicate minerals containing Al and Fe dissolve *incongruently*. Therefore, the ions that go into solution represent only a fraction of the original minerals. Although we can use these ions and molecules to *calculate* the amounts of the minerals that dissolved, the process is not reversible unless the ions are able to react with the appropriate secondary minerals that formed during incongruent dissolution. The weathering of sulfide minerals is commonly accompanied by the oxidation of S^{2-} to SO_4^{2-} and by the precipitation of some of the metals, so that these minerals also weather irreversibly. Even the carbonate mineral *dolomite* does not precipitate from aqueous solutions, except under unusual conditions defined in terms of the salinity and Mg/Ca ratio of the solution (Folk and Land, 1975). Other examples of the irreversibility of mineral dissolution reactions occur in the geochemistry of SiO_2 (Section 9.6) and $CaSO_4$ (Section 10.1).

Nevertheless, evaporative concentration is an important process that can change not only the *concentrations* of the dissolved ions and molecules in natural water but also the *composition* of the solute (see Section 18.3). As a result, evaporative concentration of surface water may result in the formation of brines of widely differing chemical compositions and can lead to the deposition of minerals that may form nonmarine *evaporite rocks*. These rocks may be composed of calcite, dolomite, anhydrite, and a wide variety of other evaporative minerals of potential economic value. Some lacustrine carbonate rocks in the Green River Formation of Wyoming and its correlatives in Colorado and Utah contain *kerogen* in sufficient concentrations to warrant calling them *oil shale* (Eugster, 1985).

The carbonate ions of groundwater discharged by springs are not in equilibrium with CO_2 of the atmosphere. Therefore, the activities of the carbonate ions and the pH of spring water change rapidly as equilibrium is established after the water is discharged by the spring. Subsequently, the activities of all ions may increase as a result of the progressive evaporation of water, and ultimately compounds may precipitate when their ion activity products (IAP) exceed the solubility product constants (K_{sp}). Garrels and Mackenzie (1967) carried out the necessary calculations to model the evaporative concentration of spring water from the Sierra Nevada Mountains of California and concluded that the compounds most likely to form are calcite, gypsum, sepiolite, and amorphous silica. As a result, the water is depleted in Ca^{2+}, Mg^{2+}, HCO_3^-, SO_4^{2-}, and SiO_2 (H_4SiO_4). In addition, it is enriched in Na^+, K^+, and Cl^-, and its pH rises to 10 or higher.

The changes in the chemical composition of spring water from the Sierra Nevada Mountains as a result of evaporative concentration are displayed graphically in Figure 20.4 according to Garrels and Mackenzie (1967). *Calcite* precipitates first and, as water continues to evaporate, the concentration of Ca^{2+} declines, whereas that of HCO_3^- increases because the original water contained more HCO_3^- than Ca^{2+}. *Sepiolite* precipitates next after the concentrations of Mg^{2+} and SiO_2 have tripled. Dolomite does not form under these conditions because of unfavorable

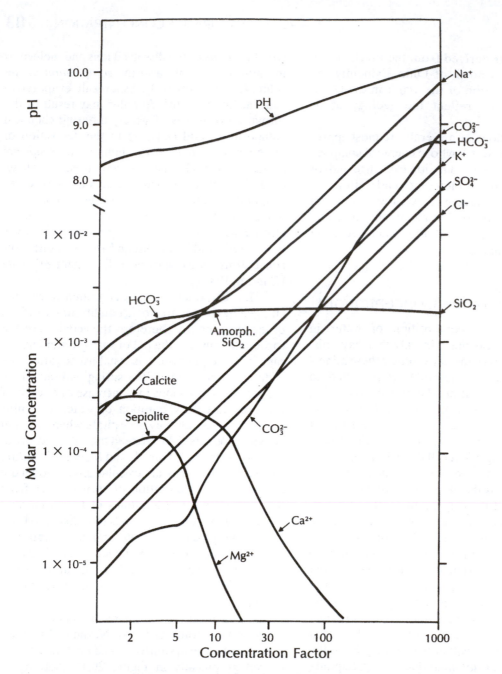

Figure 20.4 Evolution of spring water from the Sierra Nevada Mountains, California, as a result of evaporative concentration and the precipitation of calcite, sepiolite, and amorphous silica. Note that calcite precipitation caused by continuing loss of water depletes the water in Ca^{2+} but enriches it in carbonate ions because the mole ratio $Ca^{2+}/(CO_3^{2-} + HCO_3^-)$ of the spring water is less than one. Similarly, the water is depleted in Mg^{2+} but enriched in SiO_2 by sepiolite precipitation because the Mg^{2+}/SiO_2 ratio is less than one. The concentration of SiO_2 in the brine is ultimately held constant by the precipitation of solid amorphous SiO_2. The result of a 1000-fold increase in the concentration of dissolved species caused by loss of water is a Na + K + carbonate + sulfate + chloride brine (adapted from Garrels and Mackenzie, 1967).

kinetics, even though the solution becomes super-saturated with respect to dolomite. Progressive evaporation and continuing precipitation of sepiolite force Mg^{2+} out of the solution but cause the concentration of SiO_2 to rise. The solution ultimately becomes supersaturated with respect to amorphous SiO_2 (concentration factor = 10). Subsequently, the concentration of SiO_2 in the solution remains constant in equilibrium with solid amorphous SiO_2 (Section 9.6).

In this case the concentrations of Na^+, K^+, Cl^-, and SO_4^{2-} increase continuously with progressive evaporation because the chlorides, sulfates, and carbonates of the alkali metals are very soluble in water. Gypsum cannot form in this case because the water was depleted in Ca^{2+} by the precipitation of calcite. The pH of the water in the spring rises initially from 6.8 to 8.2 as the water equilibrates with CO_2 of the atmosphere ($P_{CO_2} = 10^{-3.5}$ atm) at 25 °C and stabilizes temporarily at about 8.5 while sepiolite precipitates because that reaction releases H^+:

$$2\,Mg^{2+} + 3\,H_4SiO_4 + nH_2O \rightarrow$$

$$Mg_2Si_3O_8 \cdot nH_2O + 4\,H_2O + 4\,H^+ \quad (20.11)$$
$$\text{sepiolite}$$

After sepiolite precipitation ceases, the pH rises and reaches a value of about 10.1 at a concentration factor of 1000. The increase in the pH causes the concentration of CO_3^{2-} to rise even more rapidly than the progress of the evaporation because of the dissociation of bicarbonate ion.

The calculation of Garrels and Mackenzie (1967) are based on familiar concepts of aqueous geochemistry and lead to plausible conclusions regarding the evolution of the water they treated. Compounds of Na^+ and K^+ precipitate only after the brines have become so highly concentrated that the Debye–Hückel theory no longer applies and the more complete treatment of ion interactions by Pitzer (1973) must be used. Alternatively, the crystallization of minerals from highly concentrated brines may be predicted by means of the appropriate ternary phase diagrams used by Hardie and Eugster (1970).

The results of these initial attempts to model the evolution of saline lakes demonstrate that brines of widely diverging chemical compositions can form by progressive evaporation of water and precipitation of compounds. The evolutionary paths depend on the operation of several *geochemical divides* (Section 10.1) associated with the precipitation of calcite, gypsum, and sepiolite. The subject was summarized by Eugster and Hardie (1978), Eugster and Jones (1979), Al-Droubi et al. (1980), and Drever (1982). The principles and resulting computational schemes have been used to explain the chemical compositions of many saline lakes, including Great Salt Lake in Utah (Spencer et al., 1985a,b), Lake Magadi in Kenya (Jones et al., 1977), Lake Chad in west central Africa (Al-Droubi et al., 1976; Eugster and Maglione, 1979), and many others included in the study of Hardie and Eugster (1970).

The concept of a *geochemical divide* was first discussed in Section 10.1 in connection with the precipitation of gypsum. We now return to this topic to consider more specifically how such a geochemical divide works based on an illustration by Drever (1982). We assume that the *concentrations* of Ca^{2+} and SO_4^{2-} in a *saturated* solution of gypsum are such that:

$$(Ca^{2+}) = 2(SO_4^{2-}) \quad (20.12)$$

and that the applicable activity coefficients are both equal to unity. Since the solubility product constant of gypsum is $10^{-4.6}$ (Table 10.1), the application of the Law of Mass Action yields:

$$2(SO_4^{2-})(SO_4^{2-}) = 10^{-4.6} \quad (20.13)$$

from which it follows that:

$$(SO_4^{2-}) = 3.54 \times 10^{-3}\ \text{mol/L}$$
$$(Ca^{2+}) = 7.09 \times 10^{-3}\ \text{mol/L}$$

When one liter of this solution is evaporated until the volume of the remaining water is $1/n$, where n is the concentration factor, x moles of gypsum are forced to precipitate and the concentrations of the ions in equilibrium with gypsum are modified accordingly:

$$(SO_4^{2-}) = n[(3.54 \times 10^{-3}) - x] \quad (20.14)$$
$$(Ca^{2+}) = n[(7.09 \times 10^{-3}) - x] \quad (20.15)$$

If the ions remain in equilibrium with gypsum,

$$n[(3.54 \times 10^{-3}) - x]n[(7.09 \times 10^{-3}) - x]$$
$$= 10^{-4.6} \quad (20.16)$$

which yields the quadratic equation:

$$n^2[x^2 - (10.63 \times 10^{-3})x + (25.1 \times 10^{-6})]$$
$$= 10^{-4.6} \quad (20.17)$$

This equation was solved for x for preselected values of n ranging from 1 to 100. The resulting values of x were then used to calculate the concentrations of Ca^{2+} and SO_4^{2-} from equations 20.14 and 20.15.

The results in Figure 20.5 show that the molar concentration of Ca^{2+} *increases*, whereas that of SO_4^{2-} decreases, with increasing values of n. In other words, as gypsum continues to precipitate in the course of evaporative concentration, the brine is enriched in Ca^{2+} and depleted in SO_4^{2-}. The decisive criterion is the cation/anion ratio of the solution at the initial point of saturation. The ion that dominates initially is ultimately concentrated in the brine, whereas the less abundant ion is depleted. Hence we can justifiably describe this phenomenon as a *geochemical divide*.

The computations required to predict the chemical evolution of spring water to form brines begin with the equilibrium of the carbonate species in water with CO_2 gas of the atmosphere. For this purpose we assume that $T = 25\,^\circ C$, $P = 1$ atm, $[CO_2] = 10^{-3.5}$ atm, and $[H_2O] = 1.0$, following Hardie and Eugster (1970). Therefore, the bicarbonate ion is related to carbon dioxide by:

$$CO_2 + H_2O \rightleftharpoons HCO_3^- + H^+ \quad K_1 = 10^{-7.82} \quad (20.18)$$

Since $[CO_2] = 10^{-3.5}$ atm,

$$[HCO_3^-] = \frac{10^{-11.32}}{[H^+]} \quad (20.19)$$

and

$$(HCO_3^-) = \frac{10^{-11.32}}{\gamma_{\pm 1}[H^+]} \quad (20.20)$$

Similarly,

$$HCO_3^- \rightleftharpoons CO_3^{2-} + H^+ \quad K_2 = 10^{-10.3} \quad (20.21)$$

and therefore:

$$[CO_3^{2-}] = \frac{10^{-10.3}[HCO_3^-]}{[H^+]} = \frac{10^{-21.62}}{[H^+]^2} \quad (20.22)$$

$$(CO_3^{2-}) = \frac{10^{-21.62}}{\gamma_{\pm 2}[H^+]^2} \quad (20.23)$$

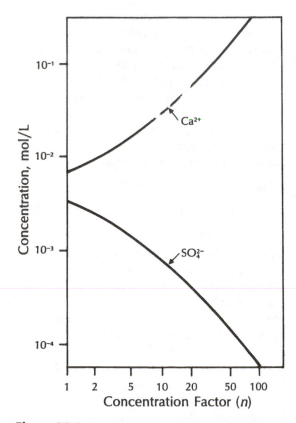

Figure 20.5 Operation of a geochemical divide caused by the precipitation of gypsum from a saturated solution as a result of progressive loss of water by evaporation. The concentration of Ca^{2+} increases and that of SO_4^{2-} decreases because the concentration of Ca^{2+} in the saturated solution before any gypsum precipitated ($n = 1$) was *twice* that of the SO_4^{2-} concentration. The concentrations of the ions were calculated by solving equation 20.17 for selected values of n and substituting the resulting values of x into equations 20.14 and 20.15.

In addition, the dissociation of water leads to the relation:

$$(OH^-) = \frac{10^{-14}}{\gamma_{\pm 1}[H^+]} \qquad (20.24)$$

Since the water must maintain electrical neutrality as the carbonate ions equilibriate with CO_2 of the atmosphere, we have the requirement that the major ions must always satisfy the equation:

$$2(CO_3^{2-}) + (HCO_3^-) + (OH^-)$$
$$= 2(Ca^{2+}) + 2(Mg^{2+}) + (Na^+) + (K^+)$$
$$-2(SO_4^{2-}) - (Cl^-) \qquad (20.25)$$

Note that all of the pH-dependent ions are assembled on the *left* side and that the molar concentrations of SO_4^{2-}, Cl^-, and any other anions not related to the carbonate equilibria are moved to the *right* side of the electrical-neutrality equation. The concentrations of all of the ions on the right side of equation 20.25 are known from the chemical analysis of the water. By substituting equations 20.20, 20.23, and 20.24 into equation 20.25, we obtain:

$$\frac{2 \times 10^{-21.62}}{\gamma_{\pm 2}[H^+]^2} + \frac{10^{-11.32}}{\gamma_{\pm 1}[H^+]} + \frac{10^{-14}}{\gamma_{\pm 1}[H^+]} = A \quad (20.26)$$

where

$$A = 2(Ca^{2+}) + 2(Mg^{2+}) + (Na^+) + (K^+)$$
$$- 2(SO_4^{2-}) - (Cl^-) \qquad (20.27)$$

In order to solve equation 20.26 for $[H^+]$ we must first determine the activity coefficients for CO_3^{2-}, HCO_3^-, and OH^- from the Debye–Hückel theory (Section 10.3) based on the ionic strength (I) of the solution. However, we cannot calculate the ionic strength accurately because the concentrations of the ions related by the carbonate equilibria change as a result of the equilibration of the water with CO_2 of the atmosphere. We are therefore confronted by a dilemma: in order to calculate the concentrations of the carbonate ions and the pH of the water we must *know* their concentrations. The impasse is circumvented by making a series of successive approximations. We first calculate I from the analysis of the water and hence determine the activity coefficients, which then allow us to solve equation 20.26 for $[H^+]$. We then use that value to calculate (HCO_3^-), (CO_3^{2-}), and (OH^-) from equations 20.20, 20.23, and 20.24, respectively. These values are then used to recalculate I and a new set of activity coefficients until the $[H^+]$ and the resulting concentrations of HCO_3^-, CO_3^{2-}, and OH^-, determined by successive iterations, become constant. Only after this computation has been carried out are we able to proceed with the calculations of the effects of evaporative concentration.

We begin this process by doubling the concentrations of all ions. Next, we repeat the calculations required to maintain the carbonate ions and pH in equilibrium with CO_2 of the atmosphere. Then we calculate the IAPs of compounds that may precipitate and compare them to the respective K_{sp} values.

When *calcite* begins to precipitate from the water, the ions must maintain equilibrium with it:

$$CaCO_3 \rightleftharpoons Ca^{2+} + CO_3^{2-} \quad K_{sp} = 10^{-8.34} \quad (20.28)$$

and therefore:

$$[Ca^{2+}] = \frac{10^{-8.34}}{[CO_3^{2-}]} \qquad (20.29)$$

Substituting equation 20.22 yields:

$$[Ca^{2+}] = \frac{10^{-8.34}[H^+]^2}{10^{-21.62}} = 10^{13.28}[H^+]^2 \quad (20.30)$$

and

$$(Ca^{2+}) = \frac{10^{13.28}[H^+]^2}{\gamma_{\pm 2}} \qquad (20.31)$$

Evidently, Ca^{2+} is now a pH-dependent species and must be transferred to the left side of the electrical neutrality equation (20.25):

$$2(CO_3^{2-}) + (HCO_3^-) + (OH^-) - 2(Ca^{2+})$$
$$= 2(Mg^{2+}) + (K^+) + (Na^+) - 2(SO_4^{2-}) - (Cl^-)$$
$$(20.32)$$

Once again we must determine activity coefficients from the Debye–Hückel theory by recalculating the ionic strength of the solution until the H^+ ion activity indicated by equation 20.32 reaches a constant value.

The entire calculation must be repeated each time the concentrations are doubled until a second salt begins to precipitate. If that salt is *gypsum,* we must include it in the equation for electrical neutrality:

$$CaSO_4 \cdot 2H_2O \rightleftharpoons Ca^{2+} + SO_4^{2-} + 2\,H_2O$$

$$K_{sp} = 10^{-4.6} \qquad (20.33)$$

$$[SO_4^{2-}] = \frac{10^{-4.6}}{[Ca^{2+}]} \qquad (20.34)$$

Since Ca^{2+} is also in equilibrium with calcite, it must satisfy equation 20.30:

$$[SO_4^{2-}] = \frac{10^{-4.6}}{10^{13.28}[H^+]^2} = \frac{10^{-17.88}}{[H^+]^2} \quad (20.35)$$

Therefore, equation 20.35 expresses the activity of SO_4^{2-} in equilibrium with both *calcite* and *gypsum* in a solution that is also in equilibrium with CO_2 of the atmosphere. The equation for electrical neutrality is now written:

$$2(CO_3^{2-}) + (HCO_3^-) + (OH^-) - 2(Ca^{2+})$$
$$+ 2(SO_4^{2-}) = 2(Mg^{2+}) + (Na^+) + (K^+) - (Cl^-)$$
$$(20.36)$$

which is then solved by iteration for the $[H^+]$ of the solution.

If *sepiolite* begins to precipitate after the onset of calcite formation, it is included in the calculation. According to Hardie and Eugster (1970),

$$Mg_2Si_3O_8 \cdot nH_2O + 4\,H^+ \rightleftharpoons$$
$$\qquad (20.37)$$
$$2\,Mg^{2+} + 3SiO_2(aq) + (n + 2)\,H_2O$$

where $K_{sp} = 10^{18.8}$. Therefore:

$$[Mg^{2+}] = \left(\frac{10^{18.8}[H^+]^4}{[SiO_2]^3} \right)^{1/2}$$

$$= \frac{10^{9.4}[H^+]^2}{[SiO_2]^{3/2}} \qquad (20.38)$$

and

$$(Mg^{2+}) = \frac{10^{9.4}[H^+]^2}{\gamma_{\pm 2}[SiO_2]^{3/2}} \qquad (20.39)$$

If calcite and sepiolite precipitate together, the equation for electrical neutrality becomes:

$$2(CO_3^{2-}) + (HCO_3^-) + (OH^-) - 2(Ca^{2+})$$
$$- 2(Mg^{2+}) = (Na^+) + (K^+) - 2(SO_4^{2-}) - (Cl^-)$$
$$(20.40)$$

If calcite, sepiolite, and gypsum precipitate together, (SO_4^{2-}) expressed by equation 20.35 must be included among the pH-dependent species of the left side of equation 20.40.

This procedure requires knowledge of activity coefficients of the ions in brines formed by evaporative concentration. In addition, it is clear that the computations are *unmanageable* without the aid of a *computer.* In fact, Hardie and Eugster (1970) wrote a computer program to carry out the calculations outlined above and applied it to more than 60 water samples.

The chemical evolution predicted by Hardie and Eugster (1970) for average North American river water (Livingstone, 1963) is illustrated in Figure 20.6 by means of two equilateral triangles similar to those devised by Piper (1944). The diagram shows that *calcite* precipitates first, causing the water to become depleted in Ca^{2+} and $CO_3^{2-} + HCO_3^-$. As a result, the composition of the water initially moves away from the Ca^{2+} and carbonate corners or the cation and anion triangles, respectively. Since the molar $Ca^{2+}/(HCO_3^- + CO_3^{2-})$ ratio of this water is greater than one, calcite precipitation virtually eliminates carbonate ions from the solution *before* all of the Ca^{2+} has been removed. The cation path then turns away from the Mg^{2+} corner because *sepiolite* precipitates next. As evaporation continues, the concentration of Ca^{2+} rises until *gypsum* begins to precipitate. Continued precipitation of gypsum and sepiolite eventually depletes the water in Ca^{2+} and Mg^{2+}, leaving Na^+ and K^+ as the principal cations. When gypsum precipitation started, the mole ratio of Ca^{2+}/SO_4^{2-} was *less* than one. Therefore, an excess of SO_4^{2-} remains after Ca^{2+} has been virtually eliminated

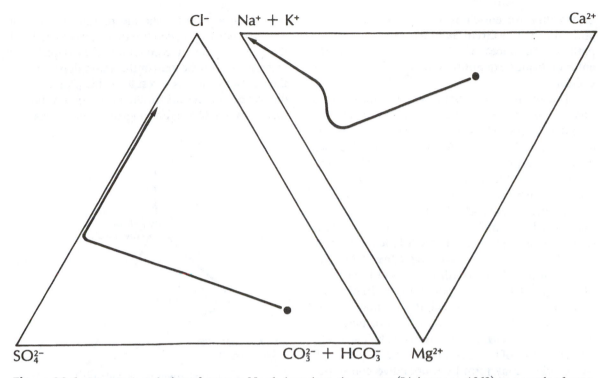

Figure 20.6 Chemical evolution of average North American river water (Livingstone, 1963) as a result of evaporative concentration according to Hardie and Eugster (1970). Calcite precipitates first, depleting the water in Ca^{2+} and carbonate ions. Since the molar $Ca^{2+}/(CO_3^{2-} + HCO_3^-)$ ratio of the water was initially greater than one, some Ca^{2+} remains when the carbonate ions are exhausted. Sepiolite precipitates next, causing the composition of the water to move away from the Mg^{2+} corner but not altering the anion composition. The abundance of Ca^{2+} rises because of the continuing loss of water until gypsum begins to precipitate. Continuing precipitation of sepiolite and gypsum depletes the water in Ca^{2+} and Mg^{2+} and enriches it in Na^+ and K^+. At the end (ionic strength = 5.0) some SO_4^{2-} remains because the molar SO_4^{2-}/Ca^{2+} ratio at the onset of gypsum precipitation was greater than one. The final result is Na + K + Cl + SO₄ brine depleted in Ca^{2+}, Mg^{2+}, and carbonate ions (adapted from Hardie and Eugster, 1970).

from the solution. The end result is a Na + K + Cl + SO₄ brine depleted in Ca^{2+}, Mg^{2+}, and carbonate ions.

The chemical evolution of surface water as a result of evaporative concentration was represented by Hardie and Eugster (1970) by a flowsheet, as shown in Figure 20.7A. *Calcite* precipitates first, either *depleting* the brine in Ca^{2+} and enriching it in carbonate ions (path I) or *enriching* the brine in Ca^{2+} and depleting it in carbonate (path II), depending on the initial molar $Ca^{2+}/(CO_3^{2-} + HCO_3^-)$ ratio. If the water takes path I, it may become supersaturated with respect to *sepiolite* as evaporation continues. In that case, all Mg^{2+} may be removed

(path IA), or all silica may be removed (path IIA), as more and more water evaporates. Water that follows paths I and IA becomes depleted in alkaline earths and forms an alkali metal–carbonate–sulfate–chloride brine exemplified by Lake Magadi in Kenya. Note that water that becomes enriched in Ca^{2+} (path II) may also precipitate sepiolite, but this is not shown explicitly in the diagram. If the water takes either path II or I plus IIA, continued evaporative concentration may enrich it in alkaline earths to the point of *gypsum* precipitation. As a result, the water may evolve into a sulfate-rich and Ca-poor brine (path III) or into a Ca + Mg-rich but sulfate-poor brine (path IV). Path III leads to brines that are enriched in alkali metals, sulfate, and chloride like Great Salt Lake in Utah, whereas path IV produces alkali metal–alkaline earth–chloride brines exemplified by Bristol Dry Lake in California.

The evolutionary pathways in Figure 20.7A do not explain the origin of certain Mg-rich brines found in Little Manitou Lake in Saskatchewan and Poison Lake in Washington. For Mg-rich brines to be produced, the formation of *sepiolite* must be inhibited either by a process that removes silicic acid from the water or by a failure of sepiolite to nucleate. If Mg^{2+} is *not* removed, then two kinds of Mg-rich brines may form, as suggested in Figure 20.7B. Path II leads to gypsum precipitation and produces either a Mg sulfate (Ca-poor) brine (path III, Poison Lake) or a Mg + Ca + Cl (sulfate-poor) brine (path IV). Alternatively, gypsum precipitation may be bypassed and a Mg + Na + SO_4 + Cl brine may form directly after calcite precipitation (broken arrow).

The point of this discussion is that a wide variety of brines may form by evaporative concentration of surface water. The mechanism depends on the operation of geochemical divides associated with the precipitation of compounds that deplete the water in either the cation or the anion, depending on their molar ratio. As a result of this process, small differences in the *initial* chemical composition of surface water in neighboring drainage basins may

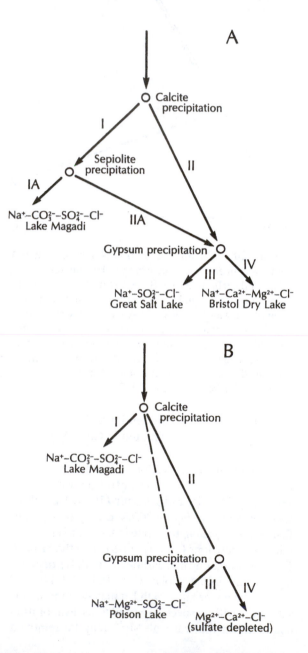

Figure 20.7 A. Chemical evolution of surface water as a result of precipitation of calcite, sepiolite, and gypsum during evaporative concentration. Calcite precipitates first, and the water becomes depleted in Ca^{2+} (path I) or enriched in Ca^{2+} (path II) as evaporation of water continues. Sepiolite may precipitate next, causing the water to become depleted in Mg^{2+} (path IA) or enriched in Mg^{2+} (path IIA). The diagram implies, but does not show explicitly, that water enriched in Ca^{2+} (path II) may also precipitate sepiolite. Ultimately such waters (paths II and IIA) become supersaturated with respect to gypsum and are then either enriched in sulfate and depleted in Ca^{2+} (path III) or depleted in sulfate and enriched in Ca^{2+} (path IV).

B. Flowsheet devised to explain the formation of Mg-rich brines. If sepiolite does not form because it does not nucleate or because the concentration of H_4SiO_4 is too low, then Mg-rich brines can form by precipitation of calcite and gypsum (path II). After gypsum has precipitated, path III leads to Mg sulfate brines, whereas path IV produces Mg + Ca + Cl brines. A Mg sulfate brine may also form directly after removal of Ca^{2+} by calcite precipitation as indicated by the dashed arrow (adapted from Hardie and Eugster, 1970).

be amplified to produce brines of quite different chemical compositions.

The treatment of this process by Hardie and Eugster (1970) was limited by the failure of the Debye–Hückel theory to account for ion interactions in concentrated electrolyte solutions. Recent advances in the quantitative treatment of ion interactions by Pitzer (1973, 1979, and 1987) have made it possible to predict the solubilities of compounds in electrolyte solutions having ionic strengths of up to 20. Such calculations are exemplified by the work of Harvie et al. (1982, 1984), Weare (1987), and Møller (1988), which permit the study of brine evolution under a wide range of chemical and geological conditions.

The computations required to make these kinds of calculations are quite time-consuming and therefore require the use of computers. In fact, the study of the geochemistry of natural waters is so complex that computer-based models are required to carry out the necessary computations. Several such programs are listed in Table 20.8.

Table 20.8 Computer Programs for Modeling the Chemical Compositions of Water

Name	Purpose	Reference
WATEQ	chemical equilibria of natural waters	Truesdell and Jones (1974)
WATEQF	chemical equilibria of natural waters (Fortran IV)	Plummer et al. (1976)
EVAPOR	chemical composition of water and amount of salt precipitated by progressive concentration	Al-Droubi et al. (1976)
EQUIL	ionic speciation and degree of saturation with respect to selected minerals and salts	Al-Droubi et al. (1976)
DISSOL	incongruent dissolution of minerals in water and formation of secondary minerals	Al-Droubi et al. (1976)
MINEQL	chemical equilibria in aqueous solution	Westall et al. (1976)
ISOTOP	modeling of ^{13}C and ^{14}C in groundwater	Reardon and Fritz (1978)
GEOCHEM	ionic speciation of soil water	Sposito and Mattigod (1980)
PHREEQ	chemical equilibria in aqueous solution	Parkhurst et al. (1980)
WATEQ2	major and trace-element speciation in aqueous solution	Ball et al. (1980)
EQCALC	thermodynamic properties of reactions at elevated P and T	Flowers (1985)
SALTNORM	calculation of normative salts from chemical composition of natural water	Bodine and Jones (1986)

The mineral deposits that form by the evaporation of water in saline lakes and in isolated marine basins are important sources of certain industrial raw material including compounds of Li, B, Na, Mg, S, Cl, K, and Ca. In addition, such deposits may contain kerogen, zeolites, and certain clay minerals that have desirable chemical or physical properties. The occurrence, composition, and origin of evaporite rocks have been described by Stewart (1963), Braitsch (1971), Kirkland and Evans (1973), and Sonnenfeld (1984).

20.6   Water Quality

The chemical composition of water determines its suitability for human consumption, for use in the operation of household appliances, as drinking water for domestic animals, for irrigation of agricultural land, and for its many uses in the manufacturing industry. Each of these uses requires the water to meet certain standards that define its *quality*. The chemical composition is one of the principal criteria of the quality of water. Also considered are biological oxygen demand, chemical oxygen demand, specific conductance, hardness, alkalinity, presence of harmful trace metals and organic compounds, turbidity, temperature, and other properties as required for specific uses (Hem, 1985).

Biological oxygen demand (BOD) is the weight of oxygen consumed per unit volume of sample by the oxidation of organic matter in the water when oxygenated water is added.

Total organic carbon (TOC) includes all forms of organic matter such as live microorganisms, suspended particles, and dissolved molecules. It is determined by measuring the amount of CO_2 produced by burning of the dry residue of the water sample.

Chemical oxygen demand (COD) is a measure of the amount of oxidizable organic matter in a water sample that reacts with a strong oxidizing agent. The amount of oxidizing agent consumed by the reaction, expressed in equivalents of oxygen, is the COD of the sample.

Total dissolved solids (TDS) is determined either by weighing the residue after evaporating a known weight of the water or by summing the concentrations of ions and molecules based on a complete chemical analysis.

Specific conductance reflects the concentration of ions in a water sample and is therefore related to TDS. However, the relationship is only empirical and does not hold for complex solutions like seawater.

Alkalinity is defined as the ability of a water sample to react with H^+ ions and is based primarily on the presence of carbonate and bicarbonate ions. The determination is made by titration and depends on the pH of the solution because of its effect on the distribution of carbonate species (Section 10.4).

Hardness is expressed in terms of the concentrations of Ca^{2+} and Mg^{2+} because these ions react with soap to form insoluble compounds and precipitate as carbonates and sulfates when the water is heated. The hardness is the sum of the milliequivalent weights of Ca^{2+} and Mg^{2+} per liter, which is then converted to milligrams of $CaCO_3$ per liter:

$$\text{hardness} = \left(\frac{2(Ca^{2+})}{\text{at. wt. Ca}} + \frac{2(Mg^{2+})}{\text{at. wt. Mg}} \right) \times \frac{\text{mol. wt. } CaCO_3}{2} \quad (20.41)$$

where (Ca^{2+}) and (Mg^{2+}) are the concentrations in mg/L. The hardness has no geochemical significance but helps to evaluate the quality of water for domestic use.

Acidity is the capacity of water to react with hydroxyl ions and reflects the presence of H^+, Fe^{2+}, and HSO_4^- ions. It may be expressed as the equivalent concentration of H^+ or H_2SO_4.

The quality of water for human consumption not only is affected by the concentrations of the *major* ions but depends also on the presence of *trace* amounts of other metals, nonmetals, and organic compounds, many of which become toxic above certain levels. These trace elements and compounds may dissolve in the water as a result of chemical weathering of rocks and minerals (Salomons and Förstner, 1984). They may also be released into the hydrologic cycle by human activities, in which case they are called *contaminants*. When the concentration of a contaminant is large enough to make the water unfit for use, the contaminant is classified as a *pollutant*.

The contamination of surface water in lakes and streams takes many forms, most of which degrade its quality. For example, Beeton (1965) documented a 36% increase of TDS in Lake Ontario and Lake Erie from 1900 to 1960. The increase was caused by discharges of various kinds of wastewaters, dumping of solid waste into the lakes, and deposition of particles from the air. In addition deforestation, farming, and construction projects increase the rate of erosion, which in turn increases the amount of sediment in suspension and accelerates chemical weathering.

The introduction of nutrients such as P, N, and K by agricultural and urban runoff stimulates growth of plants, especially algae, and other unicellular organisms. The subsequent decay of organic matter in the water consumes dissolved oxygen and the lack of sufficient oxygen may ultimately kill fish and other organisms in the water. The process leading to this condition is called *eutrophication*. Lakes and streams affected by this process are degraded because of the unpleasant odor that emanates from the water, because of the accumulation of dead fish and algal slime on the beaches, and because these conditions detract from the recreational use of the water bodies.

The consequences of the contamination of the environment were brought to our attention by Rachel L. Carson through her popular books (Carson, 1951, 1962). Although her warnings initially went unheeded, others eventually took up the cause of *environmental science* and helped us to become aware that our activities are changing the world in ways we had not anticipated (Turk et al., 1972). The growing public concern about the possible deterioration of the environment is

expressed in many forms including the annual publication of the *State of the World* by the World Watch Institute of Washington, D.C. (Brown et al., 1988). The Environmental Protection Agency (EPA) of the Unites States government now sets and enforces standards for the purity of *drinking water* that cover not only metals and nonmetals and their ions but also organic compounds used as pesticides, industrial solvents, fuels, dyes, etc. (U.S. Environmental Protection Agency, 1975, 1976).

The authority of the EPA to set standards for the quality of drinking water is based on the Safe Drinking Water Act (U.S. Public Law 93-523). The *maximum contaminant levels* (mcl) established by the EPA are based on the known toxicity to humans of elements and compounds at a daily rate of consumption of 2 L/day per person and include appropriate safety factors. In addition, the EPA has specified analytical procedures for the determination of each contaminant. The Congress has further directed each state to set standards for the quality of water in lakes, streams, and bodies of seawater (Public Law 92-500, Section 302). These standards are intended to protect recreational users of such water bodies, the animals that live in or on them, and the persons who hunt and fish these waters. The consumers of fish, shellfish, and waterfowl may be "at risk" because of the phenomenon of *bioaccumulation* of toxic elements and compounds in the food chain. The water-quality standards for surface water are supposed to take this phenomenon into consideration, based on what is called the "latest scientific knowledge."

Virtually all elements become toxic or detrimental to human health at some level of intake, especially in cases of prolonged exposure. Therefore, the permissible concentrations of the most toxic elements in drinking water listed in Table 20.9 have been set at conservative levels. Among the most toxic elements and compounds in drinking water are arsenic, cadmium, chromium, lead, manganese, mercury, silver, selenium; the chlorinated hydrocarbons endrin, lindane, and toxaphene; the chlorophenoxys 2,4-D and

Table 20.9 Primary and Secondary Water-Quality Standards and Sources of Contamination for Selected Elements and Compounds in Drinking Water

Element or compound	Acceptable concentration, mg/L	Sources of contamination
Cations and anions		
Arsenic	0.05	herbicide used on land and water
Barium	1.0	barite ($BaSO_4$) and witherite ($BaCO_3$)
Cadmium	0.01	mine tailings and industrial effluents
Chromium	0.05	industrial effluents
Copper	1.0	aquatic herbicide
Iron	0.3	minerals of Fe
Lead	0.05	industrial effluents and uses
Manganese	0.05	minerals of Mn
Mercury	0.002	minerals of Hg, antifungal agent, combustion of coal and petroleum, mining and smelting of Hg
Silver	0.05	bactericide natural occurrence
Zinc	5	natural occurrence, industrial effluent
Fluoride	2.0	industrial effluent natural occurrence
Chloride	250	NaCl brines
pH	6.5–8.5	acid rain, mine drainage, industrial effluent
Nitrate (as N)	10	farm runoff
Selenium	0.01	natural occurrence, fertilizer, combustion of coal and paper
Sulfide	0.002	oxidation of organic matter
Sulfate	250	oxidation of sulfides, oilfield brines, brine lakes
TDS	500	evaporative concentration, discharge of brines

Table 20.9 (continued)

Element or compound	Acceptable concentration, mg/L	Sources of contamination
Chlorinated hydrocarbons		pesticides
Endrin	0.0002	
Lindane	0.004	
Methoxychlor	0.1	
Toxaphene	0.005	
Chlorophenoxys		herbicides
2,4-D	0.1	
2,4,5-TP (Silvex)	0.01	
Radioactivity		
Ra	5 pCi/L	
Gross alpha	15 pCi/L	
Gross beta	4 millirem/yr	

SOURCE: U.S. Environmental Protection Agency (1986).

2,4,5-TP (Silvex); and the radioactivity in terms of Ra and emitters of α and β radiation. The dispersion of Pb is discussed in Chapter 25.

These elements and compounds are present in water partly or entirely because they are released into the environment by the activities of the human population and are therefore classified as *anthropogenic contaminants.* However, virtually all elements and many compounds also occur naturally as a consequence of chemical weathering of rocks or as a result of chemical reactions in nature. These *natural contaminants* may constitute a natural hazard to humans and other life forms living in areas where they are present.

20.7   Summary

In order to assess the rate of movement of elements and compounds from the continents into the oceans, we need to know the average chemical compositions of the world's major rivers and their total annual discharge. This information is not easy to obtain because streams are complex systems whose chemical compositions vary as a function of both time and space. Moreover, the chemical load of streams takes many forms that need to be considered in selecting appropriate methods of sampling and analysis. The information available for many of the major rivers is not as good as we would like it to be.

The chemical composition of surface water varies because of geological factors and because of the discharge of domestic, agricultural, and industrial wastewaters. Several different schemes have been devised to classify surface waters based on their chemical composition. However, genetic classifications are troubled by the problem that the chemical composition of surface water cannot be attributed to a unique source.

The chemical composition of rainwater on the continents depends on the abundance of salts derived from seawater, which decreases with distance from the coast; the dissolution of mineral particles in the atmosphere; and the presence of anthropogenic contaminants. In general, rain in the interior of the United States is strongly enriched in K^+, Ca^{2+}, and SO_4^{2-} relative to Cl^- in seawater and slightly enriched in Na^+ and Mg^{2+}.

The chemical composition of surface water can be used to calculate the abundances of the rock-forming minerals that dissolved in it. The results are instructive because they provide information about the geology of the watershed and about the agricultural and industrial activity on it.

When water at the surface of the Earth evaporates, the concentrations of ions increase until salts begin to precipitate. The removal of ions by precipitation causes major changes in the chemical composition of the evolving brines. In most cases, calcium carbonate is the first compound to precipitate, followed by sepiolite and sometimes by gypsum, depending on the chemical composition of the brine and the extent of evaporative concentration.

The precipitation of a compound as a result of progressive evaporative concentration constitutes a geochemical divide. If the initial molar ratio of the cation and anion of a particular salt in the water was *greater* than the ratio in that compound, then the water ultimately becomes enriched in the cation and is depleted in the anion, and vice versa. The occurrence of geochemical divides during evaporative concentration explains why brines of widely differing chemical compositions can form from seemingly similar surface waters. The minerals precipitated in saline lakes may accumulate to form valuable nonmarine evaporite deposits containing salts of Li, B, Na, K, Ca, Mg, and other elements.

The computations necessary to predict the chemical evolution of brines during evaporative concentration are virtually unmanageable without the aid of computers. The same is true of calculations to determine the ionic speciation of natural waters, the composition of salts they contain, and the ion activity coefficients for electrolyte solutions whose ionic strength exceeds the range at which the Debye–Hückel theory applies.

The quality of water is expressed in terms of certain defined parameters and by the concentrations of toxic elements or compounds whose presence may constitute a health hazard to humans, domestic animals, and wildlife. The toxic elements and compounds may have natural and/or anthropogenic sources. The natural sources of many trace elements are rocks that contain minerals of these elements, some of which have economic value. Therefore, in the next chapter we discuss the presence of these metals in water, soil, vegetation, and atmospheric or soil gases and how they may be used to discover mineral deposits.

Problems

1. Calculate the total amount of Ca^{2+} transported annually to the oceans from North America based on the data in Table 20.1. Express the result in grams per year.

2. Write an equation for the chemical weathering of hornblende to kaolinite and ions, including HCO_3^-. The formula for hornblende is:

$$(Ca_{1.78}Na_{0.28}K_{0.12})(Ti_{0.16}Mg_{3.32}Fe^{3+}_{0.60}Fe^{2+}_{0.96}Al_{0.14})$$
$$\times Si_6(Si_{1.04}Al_{0.96})O_{22.24}(OH)_{1.57}$$

Assume that Ti forms solid $Ti(OH)_4$ and that Fe forms $Fe(OH)_3$ (Garrels, 1967).

3. Calculate the abundances of minerals that reacted with water from the Tsurukabato area of Japan to give it the following chemical composition (Takamatsu et al., 1981): $Na^+ = 7.2$ mg/L, $K^+ = 0.77$ mg/L, $Ca^{2+} = 13.2$ mg/L, $Mg^{2+} = 0.79$ mg/L, $SiO_2 = 15.5$ mg/L, $Cl^- = 6.5$ mg/L, $SO_4^{2-} = 10.4$ mg/L, $HCO_3^- = 33.6$ mg/L. Use an appropriate analysis of rainwater from Table 20.2. What kinds of rocks interacted with this water?

4. Calculate the variations of the concentrations of Ca^{2+} and SO_4^{2-} as a result of progressive precipitation of gypsum during evaporative concentration of a solution saturated with respect to gypsum in which initially $(Ca^{2+}) = 3(SO_4^{2-})$. Plot the results for a range of concentration factors from 1 to 100.

5. What should be the pH of the water in Problem 3 after it has equilibrated with $[CO_2] = 3 \times 10^{-4}$ atm?

6. At what value of the enrichment factor n does the water in Problem 3 first reach saturation with respect to calcite as a result of evaporative concentration?

7. Calculate the "hardness" of the water in Problem 3.

References

AL-DROUBI, A., C. CHEVERRY, B. FRITZ, and Y. TARDY, 1976. Geochemie des eaux et des sels dans les sols des polders du Lac Tchad: application d'un modéle thermodynamique de simulation de l'évaporation. *Chem. Geol.,* **17**:165–177.

AL-DROUBI, A., B. FRITZ, J.-Y. GAC, and Y. TARDY, 1980. Generalized residual alkalinity concept; application to prediction of natural waters by evaporation. *Amer. J. Sci.,* **280**:560–572.

BALL, J. W., D. K. NORDSTROM, and E. A. JENNE, 1980. Additional and revised thermochemical data and computer code for WATEQ2. A computerized chemical model for trace and major element speciation and mineral equilibria of natural waters. *U.S. Geol. Surv., Water Res. Invest.,* **78**(116):109 pp.

BEETON, A. M., 1965. Eutrophication of the St. Lawrence Great Lakes. *Limnol. Oceanog.,* **10**:240–254.

BERNER, E. K., and R. A. BERNER, 1987. *The Global Water Cycle.* Prentice-Hall, Upper Saddle River, NJ, 397 pp.

BODINE, M. W., JR., and B. F. JONES, 1986. The SALT-NORM: A quantitative chemical-mineralogical characterization of natural waters. *U.S. Geol. Survey, Water Resources Investigations Rept.* 86-4086, 130 pp.

BRAITSCH, O., 1971. *Salt Deposits: Their Origin and Composition.* Springer-Verlag, New York, 297 pp.

BROWN, E., M. W. SKOUGSTAD, and M. S. FISHMAN, 1970. Methods for collection and analysis of water samples for dissolved minerals and gases. In *Techniques of Water Resources Investigations,* U.S. Geol. Survey.

BROWN, L. R., W. U. CHANDLER, C. FLAVIN, J. JACOBSON, C. POLLOCK, S. POSTEL, L. STARKE, and E. C. WOLFE, 1988. *State of the World 1988.* Norton, New York, 268 pp.

CARSON, R. L., 1951. *The Sea Around Us.* Oxford Univ. Press, New York, 230 pp.

CARSON, R. L., 1962. *Silent Spring.* Houghton Mifflin, Boston, 368 pp.

DREVER, J. I., 1982. *The Geochemistry of Natural Waters.* Prentice-Hall, Upper Saddle River, NJ, 388 pp.

EASTIN, R., and G. FAURE, 1970. Seasonal variation of the solute content and the $^{87}Sr/^{86}Sr$ ratio of the Olentangy and Scioto rivers at Columbus, Ohio. *Ohio J. Sci.,* **70**: 170–179.

EUGSTER, H. P., 1985. Oil shales, evaporites and ore deposits. *Geochim. Cosmochim. Acta,* **49**:619–636.

EUGSTER, H. P., and L. A. HARDIE, 1978. Saline lakes, 237–293. In A. Lerman (Ed.), *Lakes—Chemistry, Geology, Physics.* Springer-Verlag, New York, 363 pp.

EUGSTER, H. P., and B. F. JONES, 1979. Behavior of major solutes during closed-basin brine evolution. *Amer. J. Sci.,* **279**:609–631.

EUGSTER, H. P., and G. MAGLIONE, 1979. Brines and evaporites of the Lake Chad basin, Africa. *Geochim. Cosmochim. Acta,* **43**:973–982.

FETH, H. J., 1971. Mechanisms controlling world water chemistry: Evaporation-crystallization process. *Science,* **172**:870–871.

FETH, J. H., C. E. ROBERSON, W. L. POLZER, 1964. Sources of mineral constituents in water from granitic rocks, Sierra Nevada, California and Nevada. *U.S. Geol. Surv. Water-Supply Paper,* 1535-I.

FETTER, C. W., Jr., 1988. *Applied Hydrogeology,* 2nd ed. Merrill Publ. Co., Columbus, OH, 488 pp.

FLOWERS, G. C., 1985. EQCALC: a BASIC program designed to calculate the thermodynamic properties of reactions at elevated temperatures and pressures. Tierra Consulting, Inc., P. O. Box 15585, New Orleans, LA. 70175 (Tel. 504-861-0494).

FOLK, R. L., and L. S. LAND, 1975. Mg/Ca ratio and salinity: Two controls over crystallization of dolomite. *Amer. Assoc. Petrol. Geol. Bull.,* **59**:60–68.

GARRELS, R. M., 1967. Genesis of some ground waters from igneous rocks. In P. H. Abelson (Ed.), *Researches in Geochemistry*, 405–420. Wiley, New York, 663 pp.

GARRELS, R. M., and F. T. MACKENZIE, 1967. Origin of the chemical composition of some springs and lakes, 222–242. In W. Stumm (Ed.), *Equilibrium Concepts in Natural Water Systems*. Advances in Chemistry Series No. 67, Amer. Chem. Soc., Washington, DC.

GIBBS, R. J., 1970. Mechanisms controlling world water chemistry. *Science,* **170**:1088–1090.

GORRELL, H. A., 1953. Classification of formation waters based on sodium chloride content. *Amer. Assoc. Petrol. Geol. Bull.,* **42**:2513.

HARDIE, L. A., H. P. EUGSTER, 1970. The evolution of closed-basin brines. *Mineral. Soc. Amer. Spec. Paper,* **3**:273–290.

HARVIE, C. E., and H. P. EUGSTER, and J. H. WEARE, 1982. Mineral equilibria in the six-component seawater system, Na-K-Mg-SO_4-Cl-H_2O at 25°C. II. Compositions of saturated solutions. *Geochim. Cosmochim. Acta,* **46**:1603–1618.

HARVIE, C. E., N. MØLLER, and J. W. WEARE, 1984. The prediction of mineral solubilities in natural waters: the Na-K-Mg-Ca-H-Cl-SO_4-OH-HCO_3-CO_3-CO_2-H_2O system to high ionic strengths at 25°C. *Geochim. Cosmochim. Acta,* **48**:723–751.

HEM, J. D., 1985. Study and interpretation of the chemical characteristics of natural waters, 3rd ed. *U.S. Geol. Survey Water-Supply Paper* No. 1473, 363 pp.

JONES, B. F., H. P. EUGSTER, and S. L. RETTIG, 1977. Hydrogeochemistry of the Lake Magadi basin, Kenya. *Geochim. Cosmochim. Acta,* **41**:53–72.

JUNGE, C. E., and R. T. WERBY, 1958. The concentrations of chloride, sodium, potassium, calcium and sulfate in rainwater over the United States. *J. Meteorol.,* **15**:417–425.

KIRKLAND, D. W., and R. EVANS (Eds.), 1973. Marine evaporites: Origin, diagenesis and geochemistry. *Benchmark Papers in Geology*. Dowden, Hutchinson and Ross, Inc., Stroudsburg, PA, 426 pp.

LIVINGSTONE, D. A., 1963. Chemical composition of rivers and lakes. In M. Fleischer (Ed.), Data of geochemistry, 6th ed. *U.S. Geol. Surv. Prof. Paper* 440.

MEYBECK, M., 1979. Concentrations des eaux fluviales en éléments majeurs et apports en solution aux océans. *Rev. Géol. Dyn. Geogr. Phys.,* **21**(3):215–246.

MØLLER, N., 1988. The prediction of mineral solubilities in natural waters: A chemical equilibrium model for the Na-Ca-Cl-SO_4-H_2O system, to high temperature and concentration. *Geochim. Cosmochim. Acta,* **52**:821–837.

MUNGER, J. W., and S. J. EISENREICH, 1983. Continental-scale variations in precipitation chemistry. *Environ. Sci. Technol.,* **17**:32A–42A.

PARKHURST, D. L., C. T. DONALD, and L. N. PLUMMER, 1980. PHREEQ: a computer program for geochemical calcula-

tions. *U.S. Geol. Survey, Water Resources Investigations,* Rept. 80.

PIPER, A. M., 1944. A graphic procedure in the geochemical interpretation of water analyses. *Trans. Amer. Geophys. Union,* **25**:914–923.

PITZER, K. S., 1973. Thermodynamics of electrolytes, I. Theoretical basis and general equations. *J. Phys. Chem.,* **77**:268–277.

PITZER, K. S., 1979. Theory: ion interaction approach. In R. Pytkowitcz (Ed.), *Activity Coefficients in Electrolyte Solutions*, pp. 157–208. CRC Press, Inc., Boca Raton, FL.

PITZER, K. S., 1987. Thermodynamic model for aqueous solutions of liquid-like density. In I. S. E. Carmichael and H. P. Eugster (Eds.), *Thermodynamic Modeling of Geological Materials: Minerals, Fluids and Melts*. Rev. Mineral., **17**:97–142. Mineral. Soc. Amer., Washington, DC.

PLUMMER, L. N., B. F. JONES, and A. H. TRUESDELL, 1976. WATEQF—a Fortran IV version of WATEQ, a computer program for calculating chemical equilibrium of natural waters. *U.S. Geol. Survey, Water Resources Investigations, Rept.* 76-13, 61 pp.

REARDON, E., and P. FRITZ, 1978. Computer modelling of groundwater ^{13}C and ^{14}C isotope compositions. *J. Hydrol.,* **36**:201–224.

SALOMONS, W., and U. FÖRSTNER, 1984. *Metals in the Hydrocycle*. Springer-Verlag, Berlin, 349 pp.

SONNENFELD, P., 1984. *Brines and Evaporites*. Academic Press, New York, 613 pp.

SPENCER, R. J., H. P. EUGSTER, and B. F. JONES, 1985a. Geochemistry of Great Salt Lake, Utah II. Pleistocene-Holocene evolution. *Geochim. Cosmochim. Acta,* **49**:739–748.

SPENCER, R. J., H. P. EUGSTER, B. F. JONES, and S. L. RETTIG, 1985b. Geochemistry of Great Salt Lake, Utah, I. Hydrochemistry since 1850. *Geochim. Cosmochim. Acta,* **49**:727–738.

SPOSITO, G., and S. V. MATTIGOD, 1980. *GEOCHEM: A Computer Program for the Calculation of Chemical Equilibria in Soil Solutions and Other Natural Water Systems*. Kearney Foundation of Soil Science, University of California, Riverside.

STALLARD, R. F., 1985. River chemistry, geology, geomorphology and soils in the Amazon and Orinoco basins. In J. I. Drever (Ed.), *The Chemistry of Weathering*, 293–316. Reidel, Boston.

STALLARD, R. F., and J. M. EDMOND, 1981. Chemistry of the Amazon, precipitation chemistry and the marine contribution to the dissolved load at the time of peak discharge. *J. Geophys. Res.,* **86**(C10):9844–9858.

STALLARD, R. F., and J. M. EDMOND, 1983. Geochemistry of the Amazon 2: The influence of the geology and weathering environment on the dissolved load. *J. Geophys. Rev.,* **88**:9671–9688.

STEWART, F. J., 1963. Marine evaporites. In M. Fleischer (Ed.), Data of geochemistry (6th ed.). *U.S. Geol. Survey, Prof. Paper* 440-Y, 53 pp.

TAKAMATSU, N., K. SHIMODAIRA, M. IMAHASHI, and R. YOSHIOKA, 1981. A study on the chemical composition of groundwaters from granitic areas. *Geochemistry,* **15**:69–76.

TRUESDELL, A. H., and B. F. JONES, 1974. WATEQ: a computer program for calculating chemical equilibria of natural waters. *U.S. Geol. Survey, J. Research,* **2**:233–248.

TURK, A., J. TURK, and J. T. WITTES, 1972. *Ecology, Pollution, Environment.* Saunders, Philadelphia, PA, 217 pp.

UNITED STATES ENVIRONMENTAL PROTECTION AGENCY, 1975. National interim primary drinking water regulations. *Federal Register,* **40**(248):59,566–88.

UNITED STATES ENVIRONMENTAL PROTECTION AGENCY, 1976. *Quality Criteria for Water.* U.S. Government Printing Office, Washington, DC, 256 pp.

UNITED STATES ENVIRONMENTAL PROTECTION AGENCY, 1986. Environmental Protection Agency national drinking water regulations. *Code of Federal Regulations,* **40**(143), Appendix V, p. 187, and **40**(265), Appendix V, p. 621.

WEARE, J. H., 1987. Models of mineral solubility in concentrated brines with application to field observations. Amer. Mineral Soc. Short Course on Thermodynamic Modeling of Geologic Systems, Oct. 26–29, 1987, Phoenix, Arizona.

WESTALL, J. C., J. L. ZACHARY, and F. M. M. MOREL, 1976. MINEQL, a computer program for the calculation of chemical equilibrium composition of aqueous systems. Dept. Civil Eng., M.I.T., *Technical Note* No. 18, 91 pp.

21

Chemical Weathering
of Mineral Deposits

Mineral deposits of all kinds are distinguished from ordinary rocks by the fact that they contain certain valuable minerals in uncommon concentrations. When these kinds of rocks are exposed to chemical weathering, the elements they contain can be dispersed by geological and geochemical processes. The presence of some of these elements in surface water or soil can constitute a health hazard. On the other hand, their dispersal patterns in soil or surface water can also be used to locate the rocks from which they originated. Many geochemists are engaged in the search for mineral deposits by means of geochemical surveys because the metals or compounds that can be recovered from them are of vital importance to society (Chapter 1).

In this chapter, we want to apply all of the relevant information available to us to consider the dispersal of a few selected metals and to outline how such knowledge is used to search for mineral deposits. The principles of geochemical exploration will then lead us to consider questions regarding the present and future availability of mineral resources.

21.1 Metallic Mineral Deposits

The term *mineral deposit* is applied to rocks that contain minerals that either have useful intrinsic physical or chemical properties or from which metals can be extracted by appropriate metallurgical procedures. Such rocks may be igneous, sedimentary, or metamorphic in origin. In addition, some mineral deposits form by filling of fractures

or cavities within the host rock, by replacement of the original minerals, or by the accumulation of chemical weathering products (Park and MacDiarmid, 1970; Cox and Singer, 1986; Barnes, 1988; Dennen, 1989).

The dispersal of elements from mineral deposits depends primarily on the solubility of their minerals under the conditions that can occur at the surface of the Earth and at shallow depth. Since we have already considered the solubility of the common rock-forming silicate and oxide minerals in Chapters 12 and 19, we concentrate our attention in this chapter on the *sulfides*. Consequently, we will discuss the geochemical properties of the *chalcophile* elements identified in Figure 8.3 (including Cu, Zn, Ag, Hg, Pb, As, and Sb) as well as some *siderophile* elements (Fe, Co, Ni, and Mo), which also form sulfide minerals. The elements we have just listed are important industrial metals that are recovered by mining their *ore deposits* and by smelting their ore minerals.

In this connection, an *ore deposit* is defined as a volume of rock containing one or more ore minerals in *sufficient* concentration that one or several metals can be recovered from it *at a profit*. Evidently, the distinction between a *metallic mineral deposit* and an *ore deposit* is based partly on *economic considerations*, such as the price of the metals to be recovered, the cost of the mining operation, and the initial investment of funds needed to bring the mine into production. Each of these factors depends on a multitude of other geological, geographical, economic, legal, and even political circumstances that determine whether a mineral deposit can be mined profitably (Harris, 1984).

The geologic aspects that must be considered in the evaluation of a mineral deposit include:

1. The amount of ore in the deposit.
2. The concentrations of recoverable metals.
3. The depth of the orebody below the surface.
4. The shape of the orebody.
5. The hardness of the surrounding rocks.
6. The availability of water at the site.
7. The presence of elements whose release may be harmful to the environment.
8. The disposal of waste rock and necessary treatment of water.

These geological factors affect the choice of the mining method to be used and hence the investment of capital required to bring a mine into production. The mining method that is chosen also plays a role in determining the cost of operating the mine and hence the profit margin.

The chemical weathering of sulfide minerals is accompanied by the oxidation of the sulfide ions to sulfate or to native sulfur. In addition, some metals (Fe, Cu, Hg, As, etc.) also change oxidation state. Hence, we are concerned here with oxidation–reduction reactions presented in Chapter 14. In Section 14.1 we considered the oxidation of pyrite or marcasite to solid ferric hydroxide and sulfate ions, whereas in Section 14.5 we constructed stability diagrams for the oxides, carbonate, and sulfides of Fe in coordinates of Eh and pH (Figures 14.5–14.11).

21.2 Oxidation of Iron Sulfides and the Role of Bacteria

We now resume the discussion of the oxidation of pyrite because it is commonly associated with sulfides of other metals in mineral deposits hosted in igneous, sedimentary, and metamorphic rocks. In addition, pyrite occurs in many hydrothermal vein deposits formed over a wide range of temperatures, it occurs in coal and associated carbonaceous shale, and it is a common accessory mineral in carbonate rocks (limestone and dolomite), which may be quarried as building stone, for road construction, or for metallurgical applications.

The oxidation of pyrite to solid ferric hydroxide can be subdivided into a series of steps (Stumm and Morgan, 1981):

$$FeS_2 + \tfrac{7}{2} O_2 + H_2O \rightarrow Fe^{2+} + 2 SO_4^{2-} + 2 H^+ \tag{21.1}$$

$$Fe^{2+} + \tfrac{1}{4} O_2 + H^+ \rightarrow Fe^{3+} + \tfrac{1}{2} H_2O \tag{21.2}$$

$$Fe^{3+} + 3 H_2O \rightarrow Fe(OH)_3 + 3 H^+ \tag{21.3}$$

$$FeS_2 + \tfrac{15}{4} O_2 + \tfrac{7}{2} H_2O \rightarrow Fe(OH)_3 + 2 SO_4^{2-} + 4 H^+ \tag{21.4}$$

Equation 21.4, which is equivalent to equation 14.13, illustrates several important points regarding the oxidation of pyrite:

1. Ferric hydroxide is an insoluble base, which means that pyrite is ultimately converted to *limonite* by the transformation of ferric hydroxide:

$$2 Fe(OH)_3 \rightarrow Fe_2O_3 \cdot 3H_2O \tag{21.5}$$
$$\text{limonite}$$

The limonite may accumulate to form a special type of saprolite known as *gossan*.

2. The oxidation of Fe^{2+} to Fe^{3+} is a slow process having a half-time of ~1000 days at pH = 3 (Stumm and Morgan, 1981). Therefore, Fe^{2+} may be *transported* by the movement of water before precipitation of ferric hydroxide takes place. This means that gossans may be *displaced* from the mineralized rocks in which the pyrite resides. However, in spite of the slowness of the oxidation of Fe^{2+}, iron is generally *not* a mobile element except in acidic environments with low O_2 fugacities (Eh < +0.6 V, Figure 14.6).

3. The oxidation of pyrite releases H^+ into the environment and causes acidification of surface waters.

4. The sulfate ion goes into solution and serves as a *messenger* that the water in which it occurs has interacted with sulfide minerals.

The oxidation of pyrite can also take place in the *absence* of O_2 by the action of Fe^{3+}, which serves as the electron acceptor (oxidizing agent):

$$FeS_2 + 14\,Fe^{3+} + 8\,H_2O \rightarrow$$

$$15\,Fe^{2+} + 2\,SO_4^{2-} + 16\,H^+ \quad (21.6)$$

This reaction is fast and releases additional H^+ into the environment (Stumm and Morgan, 1981). The Fe^{3+} required for this reaction is provided by the oxidation of Fe^{2+} and by the dissolution of solid $Fe(OH)_3$.

The decomposition of pyrite is also accelerated by the activity of the bacterium *Thiobacillus ferrooxidans* and others that oxidize Fe^{2+} to Fe^{3+} and thereby promote reaction 21.6 (Ehrlich, 1981, ch. 12; also Section 19.7). *Thiobacillus* is a rod-shaped microorganism that feeds on Fe^{2+} and CO_2 in acid environments. It is the best known of a group of bacteria that interact with Fe in a variety of ways: (1) enzymatic *oxidation* of Fe^{2+} to Fe^{3+}; (2) nonenzymatic *oxidation* of Fe^{2+} to Fe^{3+} by localized changes in Eh, or pH, or both; (3) adsorption of $Fe(OH)_3$ on cell walls; (4) synthesis of organic chelators for Fe^{3+} that prevent its precipitation and increase its mobility; and (5) enzymatic and nonenzymatic *reduction* of Fe^{3+} by *Bacillus polymyxa* and others discussed by Ehrlich (1981).

Bacteria contribute to the deposition of *ochre* and other varieties of Fe oxides and hydroxides that form in pipes and channels draining waterlogged soils as well as in swamps and marshes. Such places appear to be the habitat of *Gallionella ferruginea* (Section 19.7), which thrives in neutral environments (pH = 6–7) containing Fe^{2+} and CO_2. This bacterium has also been identified in a bay of the Greek island Palaea Kameni, where Fe oxide is presently accumulat-

ing (Puchelt et al., 1973). Another example of bacterial Fe oxidation is the deposition of Fe oxides in the Pine Barrens on the coastal plain of southern New Jersey, which Crerar et al. (1979) attributed to the oxidation of Fe^{2+} by *Thiobacillus ferrooxidans* and other species. One of these is *Metallogenium,* which occurs abundantly in acid mine waters at pH > 3.5. In more acidic waters *Thiobacillus ferrooxidans* and others become dominant, including *Sulfolobus ferrooxidans,* which can attack sulfide minerals like *chalcopyrite* ($CuFe_2$) and *molybdenite* (MoS_2) by oxidizing the sulfide ions (Brierley and Murr, 1973).

When pyrite and other sulfide minerals are in contact with sulfate-rich solutions as a result of evaporation of water or for other reasons, minerals of the *alunite* group may precipitate. Iron-bearing species of that group are called *jarosite* and have the general formula $M^+Fe_3^{3+}(OH)_6(SO_4)_2$ where the place of the metal ion M^+ may be taken by K^+, Na^+, NH_4^+, or Ag^+. Jarosite can also accommodate Pb^{2+} (plumbojarosite) in the form $PbFe_6(OH)_{12}(SO_4)_4$. The solubility product constant of potassium jarosite can be calculated from the dissolution reaction:

$$KFe_3(OH)_6(SO_4)_2 \rightarrow$$

$$K^+ + 3\,Fe^{3+} + 6\,OH^- + 2\,SO_4^{2-} \quad (21.7)$$

$$\Delta G_R^\circ = +140.596 \text{ kcal (Appendix B)}$$

$$K_{sp} = 10^{-103.07}$$

The mineral ranges in color from bright yellow to brown and occurs near hot springs and fumaroles around volcanic craters and as an alteration product of sulfide minerals where it may form from $Fe(OH)_3$:

$$K^+ + 3\,Fe(OH)_3 + 2\,SO_4^{2-} + 3\,H^+ \rightarrow$$

$$KFe_3(OH)_6(SO_4)_2 + 3\,H_2O \quad (21.8)$$

Therefore, jarosite may occur in *gossans* together with the insoluble weathering products of other metals that may be associated with Fe in a particular mineral deposit.

21.3 Eh–pH Diagram for Copper Minerals

The oxidation of sulfide minerals at or near the surface of the Earth depends on the Eh and the pH of the environment (Chapter 14). The pH is controlled largely by the presence of pyrite or other iron-bearing sulfides because the formation of insoluble $Fe(OH)_3$ causes the release of H^+. The Eh is effectively controlled by the availability of molecular O_2, which is the principal, but not the only, oxidizing agent at the surface of the Earth. The oxidation of sulfide minerals ultimately depends on the availability of water at the site of weathering and hence on climatic and hydrologic conditions.

Copper is commonly associated with Fe in mineral deposits where it occurs either as the pure sulfides *chalcocite* (Cu_2S) or *covellite* (CuS) and as mixed sulfides: *chalcopyrite* ($CuFeS_2$), *bornite* (Cu_5FeS_4), *cubanite* ($Cu_2Fe_4S_6$), and other rare minerals. In addition, Cu sulfides may coexist with sulfides of other metals, including Ni, Co, Zn, Pb, Mo, Ag, As, and Sb and may also be combined with them in sulfide minerals: stromeyrite ($Ag, Cu)_2S$, carrollite ($CuS \cdot Co_2S_3$), enargite ($3Cu_2S \cdot As_2S_5$), bournonite ($2PbS \cdot Cu_2S \cdot Sb_2S_3$), and many others.

The dissolution of *chalcocite* can be expressed by the equations:

$$Cu_2S \rightarrow 2\,Cu^+ + S^{2-} \quad (21.9)$$

$$S^{2-} + 2\,O_2 \rightarrow SO_4^{2-} \quad (21.10)$$

$$2\,Cu^+ + \tfrac{1}{2}\,O_2 + 2\,H^+ \rightarrow 2\,Cu^{2+} + H_2O \quad (21.11)$$

$$\overline{Cu_2S + \tfrac{5}{2}\,O_2 + 2\,H^+ \rightarrow 2\,Cu^{2+} + SO_4^{2-} + H_2O}$$
$$(21.12)$$

The products of these reactions are Cu^{2+} and SO_4^{2-}, both of which go into solution and are therefore mobile in the presence of water. The oxidation of Cu^+ to Cu^{2+} by molecular oxygen consumes H^+ and is favored in acid environments, which can result from the oxidation of coexisting pyrite.

The oxidation of *chalcocite* and *covellite* can be depicted in coordinates of Eh and pH by the procedures presented in Chapter 14. We consider first a very simple system composed of only Cu, O_2, and H_2O in which Cu appears as the *native metal*, as *cuprite* (Cu_2O), and as *tenorite* (CuO). We begin with the most reduced form of copper, which is the native metal, and write an electrode half-reaction for the oxidation of metallic Cu to tenorite (Cu_2O), which contains Cu^+:

$$2\,Cu + H_2O \rightarrow Cu_2O + 2\,H^+ + 2\,e^- \quad (21.13)$$

From Appendix B,

$$\Delta G_R^\circ = -35.1 - (-56.687) = +21.587 \text{ kcal}$$

and therefore,

$$E^\circ = \frac{21.587}{2 \times 23.06} = +0.468 \text{ V} \quad (21.14)$$

Hence:

$$Eh = 0.468 + \frac{0.0591}{2} \log [H^+]^2 \quad (21.15)$$

and

$$Eh = 0.468 - 0.0591\,pH \quad (21.16)$$

This line is the locus of points whose coordinates are the Eh and pH generated by the electrode in equation 21.13. When the Eh and pH are *imposed* on the electrode by the environment, equation 21.13 can be used to predict which phase is stable. If the point representing an environment in the Eh–pH plane lies to the *left* of the line represented by equation 21.16, an excess of H^+ for the particular Eh is present in the environment and therefore metallic Cu is stable. Therefore, equation 21.16 is a boundary between metallic Cu and cuprite on the Eh–pH diagram in Figure 21.1A. The Eh–pH equation for the cuprite–tenorite boundary is derived similarly:

$$Cu_2O + H_2O \rightarrow 2\,CuO + 2\,H^+ + 2\,e^- \quad (21.17)$$

$$\Delta G_R^\circ = +30.387 \text{ kcal} \quad E^\circ = +0.659 \text{ V}$$

$$Eh = 0.659 - 0.0591\,pH \quad (21.18)$$

Next, we consider the solubility of native Cu to form Cu^+:

$$Cu \rightarrow Cu^+ + e^- \quad (21.19)$$

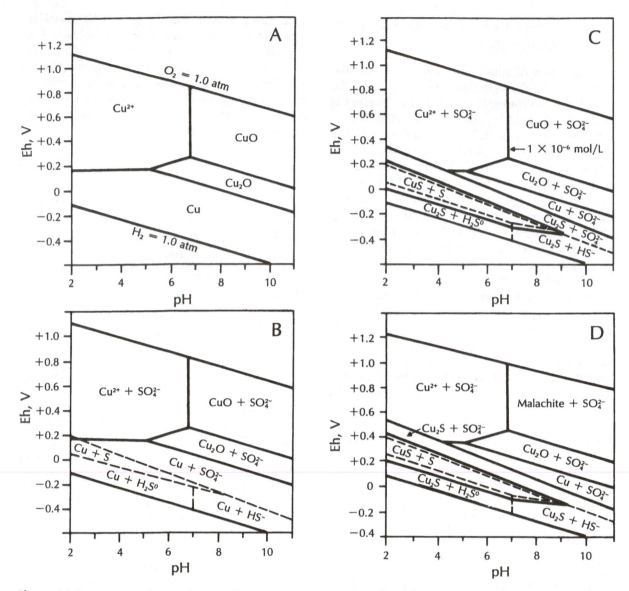

Figure 21.1 **A.** Eh–pH diagram for metallic Cu and its oxides cuprite (Cu_2O) and tenorite (CuO). The solids dissolve in acid solutions to form Cu^{2+}, which is more abundant than Cu^+. The solubility boundaries are at $[Cu^{2+}] = 1 \times 10^{-6}$ mol/L. The diagram is based on equations 21.16 ($Cu–Cu_2O$), 21.18 ($Cu_2O–CuO$), and 21.24 ($Cu–Cu^{2+}$). The solubilities of Cu_2O and CuO are the subject of Problems 1 and 2. The standard free energies are compiled in Appendix B.

B. Eh–pH diagram for Cu superimposed on the Eh–pH diagram for S. The dashed lines indicate the distribution of S species taken from Figure 14.10. The sulfides of Cu that can form in this system are shown in C.

C. The stability fields of chalcocite (Cu_2S) and covellite (CuS) in a system containing $\Sigma S = 1 \times 10^{-1}$ mol/L. The boundaries are defined by equations 21.27, 21.29, 21.31, and 21.33 for chalcocite and by equations 21.35, 21.37, and 21.39 for covellite. The sulfates of Cu are soluble and occur naturally only under arid climatic conditions.

D. When the fugacity of $CO_2 > 1 \times 10^{-3.00}$ atm, tenorite becomes unstable with respect to malachite (equation 21.40), which takes its place on the Eh–pH diagram. Malachite dissolves to form Cu^{2+} in accordance with equation 21.42.

$$\Delta G_R^\circ = +11.95 \text{ kcal} \qquad E^\circ = 0.518 \text{ V}$$

$$\text{Eh} = 0.518 + \frac{0.0591}{1} \log [\text{Cu}^+] \quad (21.20)$$

If $[\text{Cu}^+] = 10^{-6}$ mol/L is taken as the limiting solubility of metallic Cu and its compounds in geologic settings (Section 14.4),

$$\text{Eh} = 0.518 + 0.0591(-6) = 0.163 \quad (21.21)$$

The oxidation of metallic Cu to Cu^{2+} is represented by the electrode half-reaction:

$$\text{Cu} \rightarrow \text{Cu}^{2+} + 2\,e^- \quad (21.22)$$

$$\Delta G_R^\circ = +15.66 \text{ kcal} \qquad E^\circ = +0.340 \text{ V}$$

and hence,

$$\text{Eh} = 0.340 + \frac{0.0591}{2} \log [\text{Cu}^{2+}] \quad (21.23)$$

If $[\text{Cu}^{2+}] = 10^{-6}$ mol/L,

$$\text{Eh} = 0.340 + \frac{0.0591}{2}(-6) = +0.163 \quad (21.24)$$

Evidently, metallic Cu dissolves to form almost equal concentrations of Cu^+ and Cu^{2+} at Eh = 0.163 V. In fact, $[\text{Cu}^{2+}]/[\text{Cu}^+] = 1.0$ at Eh = 0.161 V. Therefore, at Eh = 0.163 V the concentration of Cu^{2+} is greater than that of Cu^+. Consequently, we conclude that metallic Cu as well as cuprite and tenorite dissolve to form Cu^{2+} rather than Cu^+ as shown in Figure 21.1A (see also Problems 1–3). The same diagram was derived by Garrels and Christ (1965) and by Brookins (1988).

The stability limits of the sulfides chalcocite (Cu_2S) and covellite (CuS) can be added to Figure 21.1A by adding S to the system such that $\Sigma \text{S} = 10^{-1}$ mol/L, which is sufficient for native S to precipitate (Section 14.5d). In order to determine the speciation of S in the Eh–pH plane, the lines governing the distribution of different forms of S (Table 14.5) have been superimposed in Figure 21.1B on the $\text{Cu–O}_2\text{–H}_2\text{O}$ system. It becomes apparent that native copper is associated with native S, H_2S^0, HS^-, and SO_4^{2-} in different parts of its Eh–pH stability field, whereas cuprite, tenorite and Cu^{2+} lie in the SO_4^{2-} field. Therefore, the Eh–pH boundaries for the Cu sulfides must be based on reactions involving native Cu

because it is the only form of Cu considered in Figure 21.1A that shares its stability field with reduced forms of S.

In the $\text{Cu} + \text{H}_2\text{S}^0$ field a reaction should occur to form chalcocite (Cu_2S). The applicable electrode reaction is:

$$2\,\text{Cu} + \text{H}_2\text{S}^0 \rightarrow \text{Cu}_2\text{S} + 2\,\text{H}^+ + 2\,e^- \quad (21.25)$$

$$\Delta G_R^\circ = -13.94 \text{ kcal} \qquad E^\circ = -0.302 \text{ V}$$

$$\text{Eh} = -0.302 - 0.0591\,\text{pH} - \frac{0.0591}{2} \log [\text{H}_2\text{S}^0] \quad (21.26)$$

When $[\text{H}_2\text{S}^0] = 10^{-1}$ mol/L, we obtain:

$$\text{Eh} = -0.272 - 0.0591\text{pH} \quad (21.27)$$

Because this equation plots *below* the lower stability limit of water $\text{H}_2 = 1$ atm), Cu and H_2S^0 cannot coexist with water and Cu_2S is the stable phase in the field labeled $\text{Cu} + \text{H}_2\text{S}^0$.

The reaction between metallic Cu and HS^- in aqueous solution yields the electrode:

$$2\,\text{Cu} + \text{HS}^- \rightarrow \text{Cu}_2\text{S} + \text{H}^+ + 2\,e^- \quad (21.28)$$

$$\Delta G_R^\circ = -23.53 \text{ kcal} \qquad E^\circ = -0.510 \text{ V}$$

$$\text{Eh} = -0.480 - 0.0296\text{pH}$$

$$([\text{HS}^-] = 1 \times 10^{-1} \text{ mol/L}) \quad (21.29)$$

This line also lies below the lower stability limit of water; thus chalcocite is stable in the $\text{Cu} + \text{HS}^-$ field as well.

The boundary between chalcocite and native Cu may lie in the $\text{Cu} + \text{SO}_4^{2-}$ field, in which case the electrode half-reaction is:

$$\text{Cu}_2\text{S} + 4\,\text{H}_2\text{O} \rightarrow 2\,\text{Cu} + \text{SO}_4^{2-} + 8\,\text{H}^+ + 6\,e^- \quad (21.30)$$

$$\Delta G_R^\circ = +69.648 \text{ kcal} \qquad E^\circ = +0.503 \text{ V}$$

$$\text{Eh} = 0.493 - 0.0788\text{pH}$$

$$([\text{SO}_4^{2-}] = 1 \times 10^{-1} \text{ mol/L}) \quad (21.31)$$

The solubility limit of Cu_2S with respect to $[\text{Cu}^{2+}] = 10^{-6}$ mol/L is given by:

$$\text{Cu}_2\text{S} + 4\,\text{H}_2\text{O} \rightarrow$$
$$2\,\text{Cu}^{2+} + \text{SO}_4^{2-} + 8\,\text{H}^+ + 10\,e^- \quad (21.32)$$

$$\Delta G_R^\circ = +100.968 \text{ kcal} \qquad E^\circ = +0.438 \text{ V}$$

$$Eh = 0.361 - 0.0473pH$$

$$([SO_4^{2-}] = 1 \times 10^- \text{ mol/L}),$$

$$([Cu^{2+}] = 1 \times 10^{-6} \text{mol/L}) \qquad (21.33)$$

The stability field of chalcocite delineated by these equations is outlined in Figure 21.1C.

Finally, we consider electrode half-reactions based on the conversion of chalcocite (Cu_2S) to covellite (CuS) and H_2S^0, HS^-, and SO_4^{2-}. In the H_2S field we obtain:

$$Cu_2S + H_2S \rightarrow 2 \text{ CuS} + 2 H^+ + 2 e^- \qquad (21.34)$$

$$\Delta G_R^\circ = +1.66 \text{ kcal} \qquad E^\circ + 0.0360 \text{ V}$$

$$Eh = 0.00645 - 0.0591pH$$

$$([H_2S^0] = 1 \times 10^{-1} \text{ mol/L}) \qquad (21.35)$$

At pH > 7.0 where HS^- is the dominant S species:

$$Cu_2S + HS^- \rightarrow 2 \text{ CuS} + H^+ + 2 e^- \qquad (21.36)$$

$$\Delta G_R^\circ = -7.93 \text{ kcal} \qquad E^\circ = -0.172 \text{ V}$$

$$Eh = -0.142 - 0.0296pH$$

$$([HS^-] = 1 \times 10^{-1} \text{mol/L}) \qquad (21.37)$$

When SO_4^{2-} is the dominant ion, the boundary between chalcocite and covellite is defined by the equation:

$$2 \text{ CuS} + 4 H_2O \rightarrow Cu_2S + SO_4^{2-} + 8 H^+ + 6 e^- \qquad (21.38)$$

$$\Delta G_R^\circ = +54.048 \text{ kcal} \qquad E^\circ = +0.391 \text{ V}$$

$$Eh = 0.381 - 0.0788pH$$

$$([SO_4^{2-}] = 1 \times 10^{-1} \text{ mol/L}) \qquad (21.39)$$

Equations 21.35, 21.37, and 21.39 outline the stability field of covellite in Figure 21.1C, which includes the Eh–pH region in which native S is dominant in systems having a total S content of 1×10^{-1} mol/L.

The *sulfates* of Cu are quite soluble and occur naturally only under arid climatic conditions. Therefore, the *oxides* of Cu are stable in the presence of water under oxidizing conditions in systems containing insufficient carbonate ions to precipitate the Cu carbonates *malachite* ($CuCO_3 \cdot Cu(OH)_2$) and *azurite* ($2CuCO_3 \cdot Cu(OH)_2$). The largest deposit of Cu sulfates occurs in the orebody of the *Chuquicamata Mine* in the Atacama Desert of Chile, which formed by oxidation of enargite, bornite, chalcopyrite, and other sulfide minerals. The Cu-sulfate minerals in the oxidized ore include antlerite ($CuSO_4 \cdot 2Cu(OH)_2$), brochantite ($CuSO_4 \cdot 3Cu(OH)_2$), kröhnkite ($Na_2Cu(SO_4)_2 \cdot 2H_2O$), chalcanthite ($CuSO_4 \cdot 5H_2O$), atacamite ($Cu_2Cl(OH)_3$), and several others described by Jarrell (1944). An Eh–pH diagram for some Cu minerals including brochantite was constructed by Hem (1977).

We complete the development of the Eh–pH diagram for Cu by adding CO_2 to the system. The carbonates *malachite* and *azurite* should now be stable depending on the fugacity of CO_2, as shown in Figure 12.10. Since *malachite* contains divalent Cu and requires a lower fugacity of CO_2 than azurite, it should displace tenorite (CuO) from its stability field in Figure 21.1A by the reaction:

$$2 \text{ CuO} + CO_2 + H_2O \rightarrow CuCO_3 \cdot Cu(OH)_2 \qquad (21.40)$$

$$\Delta G_R^\circ = -4.099 \text{ kcal} \qquad K = 10^{3.00}$$

Therefore, tenorite can coexist in equilibrium with malachite only when:

$$[CO_2] = 10^{-3.00} \text{ atm} \qquad (21.41)$$

Partial pressures of CO_2 of this magnitude do occur in soil gases, indicating that malachite is stable with respect to tenorite in such environments. The solubility of malachite with respect to Cu^{2+} is given by:

$$CuCO_3 \cdot Cu(OH)_2 + 4 H^+ \rightarrow$$
$$2 Cu^{2+} + CO_2 + 3 H_2O \qquad (21.42)$$

$$\Delta G_R^\circ = -16.555 \text{ kcal} \qquad K = 10^{12.137}$$

Therefore, the activity of $Cu^{2+} = 10^{-6}$ mol/L at pH = 6.78, if $[CO_2] = 1 \times 10^{-3.00}$ atm.

The Eh–pH diagrams in Figure 21.1 indicate that the sulfides of Cu, as well as the native element, are *unstable* under oxidizing conditions at the surface of the Earth and form insoluble oxides or carbonates at pH > 6.8. In addition, the oxides and carbonates become increasingly *soluble* with increasing acidity of the environment, and Cu becomes mobile in environments where pH < 6.8. Since the oxidation of pyrite, which is commonly associated with Cu sulfide minerals, acidifies the environment, the *presence of pyrite enhances the mobility of Cu²⁺ ions at the site of weathering*. Therefore, Cu^{2+} ions may be *removed* from the site of weathering by the movement of acidic groundwaters, but may precipitate elsewhere to form oxides or carbonates as the pH of the solution increases. The pH of acid groundwater rises because of *hydrolysis* reactions with silicate and carbonate minerals that consume H^+ to form silicic and carbonic acid, respectively.

21.4   Supergene Enrichment of Fe–Cu Sulfide Deposits

Some of the largest ore deposits of Cu and Ag were significantly enriched as a result of chemical weathering by a process called *supergene enrichment* (Garrels, 1954). Such deposits typically include mineralized breccia pipes, granitic stocks, or irregular veinlets forming stockworks. Igneous rocks of granitic composition in the form of small plutons or stocks, which contain disseminated Cu sulfide minerals, are known as *porphyry* copper deposits. Such deposits are the source of most of the Cu (and Mo) mined in the southwestern United States. In addition, Ag was recovered from supergene enrichment deposits in a belt extending from the western United States through Mexico to southern Chile (Park and MacDiarmid, 1970).

The supergene enrichment of disseminated Cu deposits depends on the common association of Cu sulfides with *pyrite* and other Fe-bearing sulfides such as chalcopyrite or bornite. The oxidation of pyrite and other Fe-bearing sulfides creates a localized acid environment (equations 21.1–21.4). Consequently, Cu^{2+} remains in solution and is carried downward to the water table. The Cu^{2+} ions may precipitate as oxides or carbonates if the pH increases sufficiently as a result of hydrolysis of carbonate or silicate ions in the country rock. However, under ideal conditions, the Cu^{2+} is transported to the water table where it is reduced to Cu^+ and comes in contact with Fe sulfides of the unweathered rock called the *protore*. When pyrite is exposed to a solution having a large Cu^+/Fe^{2+} ratio, equilibrium is reestablished by the *precipitaiton* of Cu sulfide and *dissolution* of Fe sulfide. In other words, pyrite is replaced by, or converted into, *chalcocite*. As a result, the rock below the water table is significantly enriched in Cu derived from the Cu sulfide minerals *above* the water table.

This process has resulted in the enrichment of Cu and Ag deposits but rarely those of Pb and Zn because the latter elements form insoluble carbonate and sulfate minerals that precipitate in the zone of oxidation. Therefore, supergene enrichment of an Fe–Cu–Pb–Zn sulfide deposit may cause *differentiation* of the deposit into an Fe-rich gossan, Pb–Zn carbonates or sulfates in the zone of oxidation, and Cu sulfides below the present or former water table. In addition, the Cu carbonates malachite and azurite may occur with *cerussite* ($PbCO_3$) and *smithsonite* ($ZnCO_3$) and attract the attention of *prospectors* by their brilliant color.

The effectiveness of supergene enrichment depends on the occurrence of favorable geologic, hydrologic, and climatic conditions:

1. The presence of pyrite or other Fe sulfides in the protore.
2. Sufficient porosity and permeability of the protore to allow movement of solutions.
3. A slow rate of erosion to provide the time required for the process to run to completion.
4. A deep water table that determines the volume of protore exposed to oxidation.

5. A low rate of annual precipitation, which provides the water required for the reactions and for transport of ions, but keeps the water table low and prevents flushing of ions released by weathering.

6. Low topographic relief, which affects the rate of erosion and controls the hydraulic gradient of the groundwater.

The ideal conditions include a semiarid climate, a deep or decreasing water table, low topographic relief, and low hydraulic gradients in rocks having good porosity and permeability. Such conditions exist in the southwestern United States and have caused significant supergene enrichment of porphyry copper deposits at Miami, Arizona (Peterson, 1962); Ely, Nevada (Bauer et al., 1966); and Bingham Canyon, Utah (Stringham, 1953). Figure 21.2 is a schematic and idealized cross section of a porphyry-Cu deposit that was modified by oxidation of sulfides and supergene enrichment.

21.5 Replacement of Pyrite by Chalcocite

The reactions of pyrite and Cu–Fe sulfides in the oxidized zone *above* the water table are embodied in the Eh–pH diagrams for Fe (Figure 14.11) and Cu (Figure 21.1C). Therefore, we consider next reactions that occur when water containing Cu^{2+} in solution percolates from the acid and oxidizing environment close to the surface to the water table where conditions are less acid and frequently reducing. The migration of the solution is marked on the combined Fe–Cu Eh–pH diagram in Figure 21.3.

At point P in Figure 21.3, pyrite and chalcocite are unstable. Pyrite is transformed into hematite and $SO_4^{2-} + H^+$ ions (equation 21.4), whereas chalcocite is oxidized to $Cu^{2+} + SO_4^{2-}$ in solution:

$$2\,Cu_2S + 5\,O_2 + 4\,H^+ \rightarrow$$
$$4\,Cu^{2+} + 2\,SO_4^{2-} + 2\,H_2O \quad (21.43)$$

Note that the dissolution and oxidation of chalcocite are favored in acid environments created

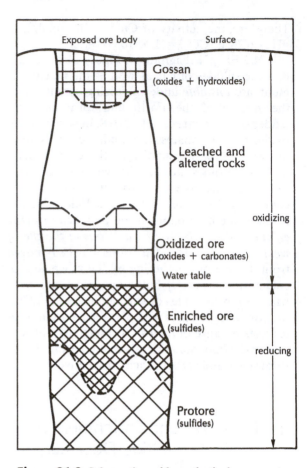

Figure 21.2 Schematic and hypothetical representation of oxidation and supergene enrichment of a low-grade mineral deposit containing disseminated sulfides of Fe, Cu, and other metals. As a result of oxidation of Cu sulfide in an acid environment caused by the oxidation of pyrite to solid $Fe(OH)_3$, Cu^{2+} ions are transported from the zone of oxidation to the water table. Secondary oxides and carbonates may be precipitated on the way. The Cu^{2+} ions that reach the water table "displace" Fe^{2+} from pyrite, chalcopyrite or bornite, thereby converting them to chalcocite (Cu_2S). As a result, the low-grade protore is significantly enriched in Cu and may be converted into Cu ore that is minable at a profit (adapted from Batemen, 1950, p. 245).

by the oxidation of pyrite. Therefore, the reactions at point P form a solution containing Cu^{2+}, SO_4^{2-}, H^+, and a small amount of Fe^{2+} that is in equilibrium with hematite (or ferric hydroxide).

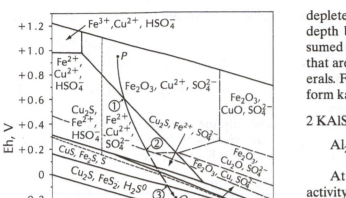

Figure 21.3 Combined Eh–pH diagram for Fe and Cu with $\Sigma S = 1 \times 10^{-1}$ mol/L, based on Figures 14.11 and 21.1D, and simplified in the area where native S precipitates. Point P represents a near-surface environment in which both pyrite and chalcocite are unstable. Iron is precipitated as Fe_2O_3, whereas Cu^{2+} and SO_4^{2-} are transported in solution to the water table. As a result, the ions enter an environment represented by point Q where both chalcocite and pyrite are stable. Equilibrium is established by reactions that cause pyrite to dissolve and chalcocite to precipitate, thereby enriching the rock in Cu. The water table is assumed to be a major discontinuity in the Eh and pH conditions such that the environmental conditions change rapidly between points 2 and 3 on the migration path.

The activity of Fe^{2+} in equilibrium with hematite depends on the Eh and pH of the environment at point P in accordance with equation 14.89:

$$Eh = 0.65 - 0.177pH - 0.05916 \log [Fe^{2+}] \quad (21.44)$$

Since at P we have pH = 2.75 and Eh = +0.95 V, we obtain $[Fe^{2+}] = 10^{-13.30}$ mol/L.

The solution percolates downward, thereby entering environments that are progressively depleted in O_2. The pH begins to rise slowly with depth because H^+ ions in the solution are consumed by hydrolysis of aluminosilicate minerals that are converted to kaolinite or other clay minerals. For example, the reaction of microcline to form kaolinite consumes H^+ ions by the reaction:

$$2\ KAlSi_3O_8 + 9\ H_2O + 2\ H^+ \rightarrow$$
$$Al_2Si_2O_5(OH)_4 + 2\ K^+ + 4\ H_4SiO_4 \quad (21.45)$$

At point 1 along the path in Figure 21.3, the activity of Fe^{2+} in equilibrium with hematite has increased from $10^{-13.30}$ mol/L at point P to 1×10^{-6} mol/L. Consequently, the solution is able to dissolve a small amount of hematite as it moves from point P to point 1 and to transport Fe^{2+} ions downward. The activities of Cu^{2+} and SO_4^{2-} are not directly affected by the dissolution of hematite, which continues as the solution percolates into environments having decreasing Eh and increasing pH values.

At point 2 along the path, Cu^{2+} is converted into Cu^+ by the electrode reaction:

$$Cu^+ \rightarrow Cu^{2+} + e^- \quad (21.46)$$

$$\Delta G_R^\circ = +3.71 \text{ kcal} \qquad E^\circ = +0.161 \text{ V}$$

$$Eh = 0.161 + \frac{0.0591}{1} \log \frac{[Cu^{2+}]}{[Cu^+]} \quad (21.47)$$

Therefore, $[Cu^{2+}] = [Cu^+]$ when the Eh = +0.161 V. At point 2 on the solubility boundary of chalcocite, Cu^{2+} is still the dominant Cu ion, but its activity is now limited by equation 21.33. If $[SO_4^{2-}] = 1 \times 10^{-1}$ mol/L, then $[Cu^{2+}] = 1 \times 10^{-6}$ mol/L. The activity of Fe^{2+} at point 2 (Eh = +0.190 V, pH = 3.8) in equilibrium with hematite has increased by about 10 orders of magnitude to $[Fe^{2+}] = 10^{-3.59}$ mol/L since the solution left the original site of weathering at point P. Therefore, the amount of *hematite* dissolved between point P and point 2 in Figure 21.3 is approximately:

$$\frac{(10^{-3.59} - 10^{-13.30}) \times 159.69}{2} \approx 2 \times 10^{-2} \text{ g/L} \quad (21.48)$$

provided that equilibrium was maintained. This explains why the Fe-rich *gossan* in Figure 21.2 is underlain by a *leached zone,* which is depleted in Fe and in which aluminosilicate minerals are altered to kaolinite and other clay minerals.

We assume that point 2 is close to the water table and that the Eh–pH conditions now change so rapidly with depth that chemical equilibria between ions and solids are *not* maintained. Consequently, the solution approaches the environment represented by point Q in Figure 21.3 after it enters the zone of saturation below the water table.

At point 3 along the route, SO_4^{2-} is converted to H_2S^0, which is now the dominant S species. In addition, pyrite and chalcocite in the protore are stable in this environment, but may be out of equilibrium with Fe^{2+} and Cu^+ in the solution. Point Q represents the Eh–pH conditions of the environment in the zone of saturation where pyrite and chalcocite of the protore react to achieve equilibrium with Fe^{2+}, Cu^+, and H_2S^0.

The activity of Fe^{2+} in equilibrium with pyrite at Eh = -0.2 V and pH = 5.0 (point Q, Figure 21.3) in a system in which $H_2S^0 = 1 \times 10^{-1}$ mol/L is derivable from equation 14.134:

$$Eh = -0.148 + \frac{0.0591}{2}\log\frac{[H^+]^4}{[Fe^{2+}][H_2S^0]^2} \quad (21.49)$$

If $[H_2S^0] = 1 \times 10^{-1}$ mol/L,

$$Eh = -0.0889 - 0.1182pH - 0.029\,55\log[Fe^{2+}]$$
$$(21.50)$$

Therefore, at point Q in equilibrium with pyrite:

$$[Fe^{2+}] = 10^{-16.24}\text{ mol/L} \quad (21.51)$$

The solubility of chalcocite with respect to Cu^+ is given by:

$$Cu_2S + 2\,H^+ \rightleftharpoons 2\,Cu^+ + H_2S^0 \quad (21.52)$$

$$\Delta G_R^\circ = +37.84\text{ kcal} \qquad K = 10^{-27.742}$$

Therefore, at equilibrium:

$$\frac{[Cu^+]^2[H_2S^0]}{[H^+]^2} = 10^{-27.742} \quad (21.53)$$

and at pH = 5.0 and $[H_2S^0] = 1 \times 10^{-1}$ mol/L:

$$[Cu^+] = \left(\frac{10^{-27.742} \times 10^{-10.0}}{10^{-1.0}}\right)^{1/2}$$

$$= 10^{-18.37}\text{ mol/L} \quad (21.54)$$

Evidently, the $[Cu^+]/[Fe^{2+}]$ ratio of a solution in equilibrium with both chalcocite and pyrite at point Q in Figure 21.3 is:

$$\frac{[Cu^+]}{[Fe^{2+}]} = \frac{10^{-18.37}}{10^{-16.24}} = 10^{-2.13} = 7.41 \times 10^{-3} \quad (21.55)$$

In other words, pyrite is about 270 times as soluble as chalcocite, if we include the fact that each mole of chalcocite produces two moles of Cu^+.

Equation 21.55 explains why pyrite can be replaced by chalcocite during supergene enrichment. If the $[Cu^+]/[Fe^{2+}]$ ratio of a solution at point Q in Figure 21.3 is *greater* than 7.41×10^{-3}, chalcocite must precipitate and pyrite must dissolve in order to establish equilibrium between the solids and their ions. Hence, pyrite is *replaced* by chalcocite and the Cu ions, derived by oxidation of chalcocite at point P, are thus precipitated at point Q. The Cu concentration of the protore is thereby *increased* significantly.

The process we have outlined does not include the formation of carbonate or oxide minerals of Cu because the path PQ does not enter their stability fields in Figure 21.3. In reality, the physical and chemical conditions to which a solution, formed by oxidation of sulfides of Fe and Cu, is exposed as it percolates downward may vary widely. As a result, tenorite, cuprite, native Cu, malachite, azurite, and the Cu silicate chrysocolla ($CuSiO_3 \cdot 2H_2O$) may precipitate under appropriate conditions, thereby preventing some of the Cu ions from reaching the water table. In addition, the solution may precipitate hematite (or limonite), jarosite, or pyrite, as well as covellite and native S, depending on environmental conditions. Evidently, the outcome of this process depends on the geologic, hydrologic, and climatic factors (listed in Section 21.3), which may be highly site-specific.

21.6   Oxidation of Ore Minerals of Other Metals

We conclude from the foregoing discussion that the supergene enrichment of low-grade Cu deposits occurs because chalcocite is less soluble than pyrite under the reducing conditions that prevail in the zone of saturation below the water table. We can calculate the solubilities of the sulfides of other metals that may coexist with pyrite and thereby predict whether supergene enrichment is possible for them. The results in Table 21.1 are ordered in terms of decreasing activities of metal ions in equilibrium with their sulfides at the conditions represented by point Q in Figure 21.3, namely, Eh = -0.2 V, pH = 5.0, $\Sigma S = [H_2S] = 1 \times 10^{-1}$ mol/L, $T = 25$ °C, and $P = 1.0$ atm. The solubilities of the sulfides included in Table 21.1 range over 44 orders of magnitude from 7.46×10^{-47} mol/L for cinnabar (HgS) to 3.97×10^{-3} mol/L for alabandite (MnS). The sulfides of Hg, Ag, Cu, and Cd are *less* soluble than pyrite and therefore could replace it under favorable conditions. The sulfides of Pb, Zn, Ni, Co, and Mn are *more* soluble than pyrite in the specified environment and therefore could replace it *only* when the [metal ion]/[Fe^{2+}] activity ratio in the solution exceeds the equilibrium value. The listing of metals in Table 21.1 resembles the

Schürmann series, which was derived experimentally more than 100 years ago on the basis of the affinity of metals for sulfur (Schürmann, 1888).

The geochemical properties of Ag, Hg, Mo, and U are illustrated by means of their Eh–pH diagrams in Figure 21.4. Silver sulfide deposits containing argentite (Ag_2S), are oxidized under near-surface conditions, and the resulting Ag^+ ions may be reduced to native Ag or may replace metals whose sulfides are *more* soluble than argentite. Therefore, supergene enrichment has significantly increased the value of the many Ag deposits in the Cordilleran ranges of North, Central, and South America (Park and MacDiarmid, 1970).

In contrast to the sulfides of Cu and Ag, ore deposits of Hg containing HgS (cinnabar) are not significantly enriched as a result of oxidation. For example, at the Mt. Jackson mine in Sonoma County, California, secondary cinnabar apparently did not form by replacement of pyrite and is not economically important (Tunnel, 1970). The aqueous geochemistry of Hg is affected by the stability of complex ions such as $HgCl_4^{2-}$ and HgS_2^{2-}, that significantly increase the amount of this element that can occur in solution (Krauskopf, 1951). Consequently, cinnabar deposits form by precipitation from aqueous solutions at low temperature

Table 21.1 Listing of the Activities of Metal Ions in Equilibrium with Their Sulfides (Eh = -0.20 V, pH = 5.0, $\Sigma S = [H_2S^0] = 1 \times 10^{-1}$ mol/L, $T = 25$ °C, and $P = 1.0$ atm.)

Metal	Mineral	Activity of the ion, mol/L
Hg	cinnabar (HgS)	$[Hg_2^{2+}] = 7.46 \times 10^{-47}$
Ag	argentite (Ag_2S)	$[Ag^+] = 9.53 \times 10^{-20}$
Cu	chalcocite (Cu_2S)	$[Cu^+] = 4.27 \times 10^{-19}$
Cd	greenockite (CdS)	$[Cd^+] = 1.14 \times 10^{-18}$
Fe	pyrite (FeS_2)	$[Fe^{2+}] = 5.75 \times 10^{-17}$
Pb	galena (PbS)	$[Pb^{2+}] = 2.42 \times 10^{-17}$
Zn	sphalerite (ZnS)	$[Zn^{2+}] = 2.37 \times 10^{-14}$
Ni	millerite (NiS)	$[Ni^{2+}] = 8.80 \times 10^{-11}$
Co	cobaltite (CoS)	$[Co^{2+}] = 7.80 \times 10^{-10}$
Mn	alabandite (MnS)	$[Mn^{2+}] = 3.97 \times 10^{-3}$

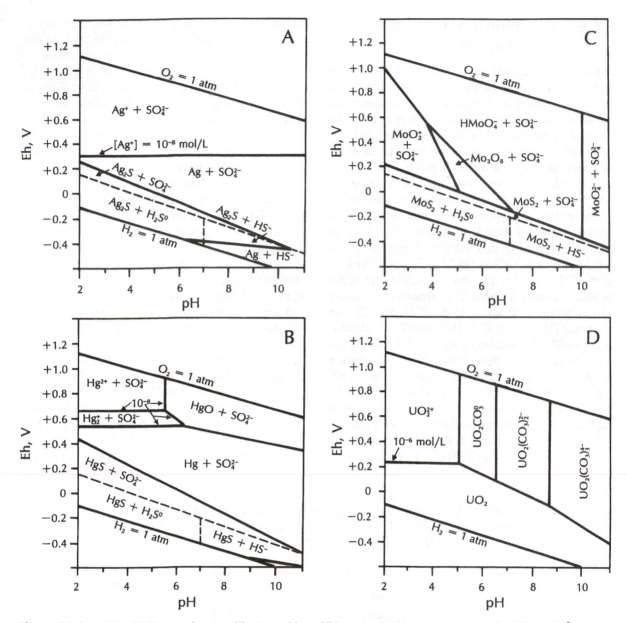

Figure 21.4 A. Eh–pH diagram for metallic Ag and its sulfide argentite in a system containing $\Sigma S = 10^{-3}$ mol/L at 25 °C and 1 atm. The solubility boundary is drawn at $[Ag^+] = 10^{-8}$ mol/L.

B. Eh–pH diagram for metallic Hg, the sulfide cinnabar and the oxide montroydite in a system containing $\Sigma S = 10^{-3}$ mol/L at 25 °C and 1 atm. The solubility boundaries are drawn at $[Hg^{2+}]$ and $[Hg_2^+] = 10^{-8}$ mol/L.

C. Eh–pH diagram for molybdenite (MoS_2) and ilsemannite (Mo_3O_8) in a system containing $\Sigma S = 10^{-3}$ mol/L at 25 °C and 1 atm. The solubility boundaries are drawn at $[MoO_2^{2+}]$, $[HMoO_4^-]$, and $[MoO_4^{2-}] = 10^{-8}$ mol/L.

D. Eh–pH diagram for uraninite (UO_2), the uranyl ion (UO_2^{2-}), and its carbonate complexes in a system containing $\Sigma C = 10^{-3}$ mol/L and $\Sigma S = 10^{-3}$ mol/L at 25 °C and 1 atm. The solubility boundaries are at 10^{-6} mol/L of the respective ions and molecules (after Brookins, 1988).

(epithermal) and thereby become separated from the sulfides of Fe, Cu, Mo, and other metals that are deposited at higher temperatures (hypothermal–mesothermal).

Molybdenum is an important element that occurs as the sulfide molybdenite (MoS_2) and is commonly associated with Cu sulfides. Molybdenum is a member of the Cr group (group VIB, Figure 8.3), which also includes W and U. These elements may be oxidized to the +6 state and then combine with O to form oxy ions. Molybdenite dissolves under oxidizing conditions to form the bimolybdate ($HMoO_4^-$) ion, which gives way to the molybdate (MoO_4^{2-}) ion at a pH above ~10 (Brookins, 1988). In addition, Mo forms MoO_2^+ ions under strongly acidic conditions in which Mo has a valence of +5. Consequently, Mo ions released by the oxidation of molybdenite tend to be dispersed, but may be reprecipitated under reducing conditions with secondary U minerals. The mobility of Mo under oxidizing conditions makes it useful in geochemical exploration surveys as an indicator of both Cu–Mo and secondary U deposits. The bimolybdate ion reacts with Pb^{2+} to form the mineral wulfenite ($PbMoO_4$), which is much less soluble than the molybdates of other metals (Landergren and Manheim, 1978).

The aqueous geochemistry of U and its deposits has been studied intensively because of the importance of this element as a source of energy and as an explosive based on induced fission of ^{235}U (Rich et al., 1977). All aspects of the geochemistry and technology of U are the subject matter of the scientific journal *Uranium*. The element forms complex ions with O, hydroxyl, carbonate, phosphate, and other ions under oxidizing conditions where U assumes a valence of +6 (Langmuir, 1978). Therefore, U is *mobile* under oxidizing conditions but can be precipitated in the tetravalent state as *uraninite* (UO_2), *coffinite* ($USiO_4$), and a few other compounds not all of which have mineral names. In addition, the uranyl ion (UO_2^+) forms a very large number of insoluble oxides, carbonates, sulfates, phosphates, arsenates, vanadates, silicates, and molybdates, including *rutherfordine* (UO_2CO_3), *autunite* ($Ca(UO_2)_2(PO_4)_2 \cdot 8–12H_2O$),

and *carnotite* ($K_2(UO_2)_2(VO_4)_2 \cdot 1–3H_2O$) (Rogers and Adams, 1969).

21.7 Geochemical Exploration

When certain kinds of mineral deposits weather, ions may be released into the ambient groundwater or saprolite in unusual concentrations. The resulting geochemical anomalies can be detected by analyzing systematically collected samples of water, soil, colluvium, or plants. Once geochemical anomalies have been located, they become the targets for more detailed exploration based on geological, geophysical, and additional geochemical methods. The ultimate objective is to discover the sources of the geochemical anomalies and to determine the economic potential of the mineral deposits. Therefore, geochemical exploration is one of the techniques geologists use in their quest to discover new ore deposits (Chapter 1).

Geochemical prospecting requires an understanding of the geological processes that apply to the formation of mineral deposits and that come into play during their eventual alteration at the surface of the Earth. In addition, information is required regarding the chemical properties of the elements and how these properties affect the migration and dispersion of selected elements under a given set of environmental circumstances.

Geochemical exploration is the principal occupation of a large number of geologists and geochemists working in the mineral industry. They have organized several professional societies, hold national and international symposia, and publish the results of their research in specialized journals, including the *Journal of Geochemical Exploration, Mineralium Deposita,* and *Economic Geology.* The principles of geochemical exploration have been summarized in many textbooks and monographs, e.g. Hawkes and Webb (1962), Brooks (1972), Levinson (1974), Fletcher (1981), Boyle (1982), Howarth (1982), Govett (1983), Davy and Mazzucchelli (1984), Thornton and Howarth (1986), and Wignall and de Geoffrey (1987). In addition, the *Proceedings of the International*

Geochemical Exploration Symposia have been reprinted from the *Journal of Geochemical Exploration* under the general title: *Geochemical Exploration . . .* published annually by Elsevier Science Publishers in Amsterdam.

Geochemical methods of exploration are appropriate in areas where the bedrock is not exposed because of the accumulation of weathering products or because of the presence of alluvial sediment or till. In addition, these methods can be used in areas that are inaccessible because of rugged topography, heavy vegetation, or the absence of adequate roads. Such conditions prevail in many areas of the world, including some parts of Africa, South America, Asia, Australia, and Antarctica. Geochemical surveys are relatively inexpensive and can be adapted to achieve specific objectives under a given set of circumstances.

Before any work is done in the field, available sources of information should be thoroughly studied to determine the bedrock geology and overburden of the area to be investigated, as well as its vegetation and climate, roads and settlements, existing mines and smelters, legal requirements to obtain permits, and the availability of housing, food, and fuel. Such preliminary work is necessary not only because the study area may be in a remote location in a foreign country, but also because the most appropriate method of collecting and analyzing samples must be selected.

The preliminary study of the geology of the area to be surveyed will suggest what kinds of deposits may be present and what minerals they may contain. For example, low-Ca granites are enriched in Sn, Hg, Li, and Be, whereas Fe, Co, Ni, and Cr are associated with ultramafic rocks. Sedimentary rocks may contain fossil fuels as well as deposits of Fe, Mn, P, U, and more rarely Cu, Pb, or Zn. All of the information gathered by geochemists about the chemical compositions of different kinds of rocks finds direct practical application when it becomes necessary to predict the composition of potential mineral deposits in an area.

After the metals to be searched for have been identified, the solubility of the minerals in which they occur and the properties of their ions and molecules that form during dissolution of these minerals need to be reviewed. According to information in Chapter 19, some minerals resist chemical weathering and are ultimately concentrated in deposits of sand and gravel. The valuable minerals in this category include Au, Pt group metals, diamonds, rutile, various phosphate minerals, magnetite, corundum, and others. The mobility of some elements is limited by the formation of secondary minerals and the extent to which their ions may be adsorbed on clay minerals, oxide and hydroxide particles or grain coatings, and organic matter.

The geochemical properties of the elements to be included in a survey must also be considered in choosing appropriate methods of sampling and analysis. In addition, the geochemist planning a survey must consider the objectives and budgetary limitations when deciding what kinds of samples to collect. In general, a wide variety of materials can be collected for study:

1. Rock samples from outcrops or boulders embedded in the regolith.

2. Soil in bulk or from specific horizons in which the elements of interest may be concentrated.

3. Stream or lake water.

4. Sediment carried in suspension in the water or from the bottom of streams or lakes.

5. Organic matter in streams or lakes including molluscs, fish, larvae, and aquatic plants in bulk, or selectively.

6. Water from springs and water wells.

7. Land plants, including leaves, pine needles, twigs, seeds, or soil humus.

8. Selected land animals, their feces, and solid body parts such as antlers, bones, or hair.

9. Soil gases and air.

Each of these is appropriate for certain elements, represents a different part of the environment, and can be collected in ways that reflect the scope of the survey to be undertaken.

Geochemical methods of exploration may be used to evaluate the resource potential of an area in a *preliminary manner*. In this case, water and/or sediment samples can be collected from streams in existing drainage basins. Metal ions may enter a stream through its tributaries or as effluent from the groundwater. Therefore, samples should be taken at the confluence of each tributary and along the main stream. Alternatively, preliminary surveys may be carried out by sampling soil along the existing network of roads. In both cases, the objective is to identify areas that warrant more detailed investigation.

For a *detailed survey*, samples are collected at regular intervals either based on surveyed lines or by reference to topographic maps or aerial photographs of the area. The spacing of sampling sites must reflect the size of the deposit being searched for. Samples may consist of soil, vegetation, or soil gases, as appropriate for the elements being sought. If the bedrock is well exposed, rock samples can also be taken. The samples collected in the field should be analyzed while the survey is still in progress so that the sampling can be extended based on the initial results.

The analysis of the samples should be done by experts, but the pretreatment of the samples requires careful consideration by the geochemists in charge of the survey. For example, soil samples may be collected selectively from the A, B, or C horizons, they may be analyzed by complete dissolution of the bulk sample or of a selected grain-size fraction, or they may be leached with acids or other solvents to selectively extract elements that are adsorbed or exist in secondary acid-soluble compounds. In some cases, certain minerals may be selectively concentrated prior to analysis based on differences in specific gravity, magnetic susceptibility, or electrical conductivity. Whatever pretreatment procedure and analytical method is chosen, must be applied uniformly to all samples in order to generate an internally consistent data set.

The interpretation of the results of a systematic geochemical survey usually *starts* with the preparation of a contoured map for each of the elements included in the survey. In the ideal case, such a map

describes a *geochemical landscape* within which may occur *geochemical anomalies*. An anomaly is an area on the contoured geochemical map in which the concentration of a selected element is consistently and significantly above the background based on systematically collected and analyzed samples of a certain kind. Therefore, the *background* must be defined for each element included in the survey.

The geochemical background depends on the lithologic composition of the bedrock in the study area and therefore may change at geological contacts. This principle has been used in geologic mapping of areas where bedrock is covered by overburden that formed in place. Certain kinds of igneous or sedimentary rocks may give rise to high geochemical backgrounds in surficial deposits because the rocks are enriched in certain elements. For example, carbonaceous shales and phosphate rocks commonly produce high geochemical backgrounds in saprolite or soil formed from them. Ultramafic igneous rocks and certain syenites also give rise to high background values of elements that are concentrated in them (Table 4.5). Therefore, such elevated local backgrounds must be distinguished from the somewhat lower regional background. In addition, the *range* of random variations of the concentrations of elements in the background must be known.

Large geochemical anomalies have resulted from the dispersal of contaminants released as a consequence of human activity. Some common sources of such anthropogenic contamination of the environment include:

1. Effluent from piles of waste rock or tailings ponds associated with mining operations.
2. Particles deposited from plumes of smoke and gases emitted by smelters.
3. Effluent from chemical or manufacturing plants.
4. Leachate from municipal landfills and effluent from sewage-treatment facilities.

The geochemical anomalies that can be attributed to anthropogenic causes must be identified and removed from further consideration. The remaining *natural* anomalies become targets for further evaluation.

Natural geochemical anomalies can be classified on the basis of their formation into *clastic, hydromorphic,* and *biogenic* dispersion patterns. However, all three processes of dispersal may contribute to the formation of a particular geochemical anomaly (Hawkes and Webb, 1962). *Clastic* dispersion patterns result primarily from the movement of solid particles under the influence of gravity or by entrainment in moving water or ice. *Hydromorphic* dispersal is caused by the transport of ions and molecules in solution and their ultimate deposition by adsorption on solid particles or by precipitation of compounds. *Biogenic* patterns result from the absorption of ions by plant roots, the accumulation of metals in leaf litter and soil humus, or from the dispersal of elements in the metabolic products of animals.

The *contrast* between an anomaly in the concentration of a particular element in soil samples and the local background or *threshold* depends on the geochemical properties of the element and the environmental conditions in the soil. *High* contrast between anomalies and the background result from the following conditions occurring singly or in combination:

1. Shallow depth of chemical weathering.
2. Incomplete soil development.
3. Basic conditions (pH > 7).
4. Presence of carbonates and/or Fe oxides.
5. Presence of clay minerals and other adsorbers of ions.
6. Low rainfall.
7. Poor drainage.
8. Rapid erosion.

Conversely, the opposite conditions promote the *mobility* of ions and result in *low* contrast between anomalies and the local background.

Differences in the geochemical properties of metals occurring together in a deposit may affect the contrast of the anomalies they cause in soil. In general, metals whose compounds are soluble are *more mobile* under a given set of environmental conditions than metals whose compounds are insoluble. Such metals therefore form large, but low-contrast, anomalies compared to the more insoluble metals with whom they may be associated in the deposit. For example, when Cu, Zn, and Pb occur together in a mineral deposit, the contrast of the resulting anomalies in soil may be Pb > Zn > Cu, regardless of their abundances in the deposit. Similarly, when Cu and Mo occur together, the contrast of the Cu anomaly is commonly greater than that of Mo because the latter is more mobile under oxidizing conditions than Cu. However, Mo is dispersed *more widely* in a low-contrast anomaly that can be found more easily by a geochemical survey than the high-contrast more localized Cu anomaly. Therefore, Mo acts as a *pathfinder* for deposits that contain both Cu and Mo.

In general, pathfinders are elements that are more easily detected in geochemical surveys than the valuable metals with which they are commonly associated in different kinds of mineral deposits. The list of pathfinders in Table 21.2 includes not only Mo, but also SO_4, As, Hg, Se, and Ag.

One of the most important aspects of the evaluation of geochemical anomalies is to discover their *sources,* which, in many cases, are not exposed to direct observation. The first step toward that objective is to recognize the process by which an anomaly was produced. In general, the sources of *clastic* dispersal patterns are located upslope of the anomaly, whereas the sources of hydromorphic patterns lie upstream or in the up-gradient direction of the water table. Consequently, the sources of geochemical anomalies are commonly *not* located in the rocks that *underlie* them.

This insight is helpful in planning the evaluation of geochemical anomalies by means of more detailed geochemical surveys, by geophysical methods, and ultimately by core drilling. If the further exploration of a deposit yields positive results, a thorough drilling program is undertaken to determine its volume and tenor, which must be known in order to determine whether the deposit can be mined at a profit. Most natural geochemical anomalies do *not* result in the discovery of a

Table 21.2 Pathfinders of the Ore Deposits from Which They Originate

Pathfinder	Samples taken	Type of mineral deposit
SO_4^{2-}	water	all sulfide deposits
Mo	water, sediment, soil	porphyry Cu deposits
As	soil, sediment	Au-bearing veins
Hg	soil	Pb–Zn–Ag deposits
Se	soil	hydrothermal sulfide deposits (epithermal)
Ag	soil	Au–Ag deposits

SOURCE: Hawkes and Webb (1962).

minable deposit because the volume of mineralized rock, or the concentrations of metals, or both, are insufficient to warrant the expense of bringing the deposit into production. However, some large but low-grade deposits may ultimately become minable if the market value of the metals it contains rises or if large-scale surface mining methods are adopted.

21.8 Production and Consumption of Mineral Resources

At the *start* of the 20th century, when the population of the Earth was 1.6 billion, the natural resources of the world appeared to be inexhaustible. As we approach the end of this century, the situation has changed dramatically. The world population is rapidly approaching 6.0 billion, and serious questions have been raised about the future availability of adequate natural resources. Alarming predictions have been made about the decline of living standards based on extrapolations of present rates of consumption of natural resources. These concerns are based in part on our instinctive urge to assure the survival of the human species on Earth, not just for another century, but for all time.

Ore deposits, in the present sense of the term, are rocks whose chemical composition differs greatly from that of the average continental crust.

Consequently, such rocks are relatively *rare* and their total volume, in that part of the crust that is accessible to us, is *small*. In addition, metallic and nonmetallic deposits are *nonrenewable*. These facts combine to confront geologists with the challenging task of finding new deposits to replace those that have been mined out and thereby to assure an adequate supply of all kinds of natural resources for the future.

The limited volume and the nonrenewable character of metallic and nonmetallic mineral deposits affect the way in which the rate of production can vary as a function of time. The rates of production of all mineral commodities have increased with time since the middle of the 19th century. When these rates are projected into the future, the total amount produced increases and approaches infinity, which is obviously inconsistent with reality. Therefore, the *rates* of production of mineral commodities cannot continue to increase indefinitely, but eventually must reach a peak and must then *decline*. In spite of short-term fluctuations, a plot of annual production versus time must form a bell-shaped curve, sometimes irreverently called Hubbert's pimple after M. King Hubbert, who first drew attention to these essential facts of mineral economics (Hubbert, 1969; 1973).

Hubbert was concerned primarily with the production of fossil fuels at a time when the energy supply of the United States had become a matter of growing concern. He prefaced one of his papers with a quotation from James B. Cabell (1926, p. 129).

> The optimist proclaims that we live in the best of all possible worlds; and the pessimist fears that this is true.

Hubbert (1969) estimated that the total amount of petroleum that will ultimately be produced in the United States (excluding Alaska) is about 165×10^9 barrels and used this result to predict the complete cycle of petroleum production in the United States. Figure 21.5 shows that 80% of all recoverable petroleum will be consumed in 65 years between 1934 and 1999 and that the highest production rate occurred around 1970. The decline of the rate of petroleum production in the United States was slowed temporarily by the discovery of oil on the North Slope of Alaska in 1968 and may be further slowed by improvements in the efficiency of oil recovery, which presently is less than 50%. However, Hubbert's

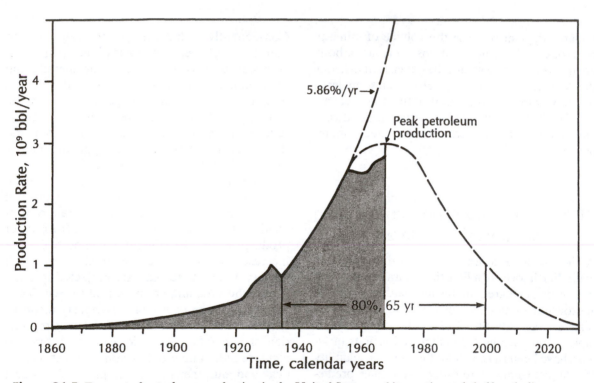

Figure 21.5 Forecast of petroleum production in the United States and its continental shelf excluding Alaska based on the estimate that the total amount of recoverable petroleum in the United States is 165×10^9 barrels (bbl). Therefore, the annual production rate must reach a peak and subsequently decline so that the area under the curve equals the total amount of available petroleum. Adding the expected production in Alaska raises the total amount of recoverable petroleum to about 190×10^9 bbl, which slows the rate of decline, but does not alter the forecast significantly. The shape of the annual production curve indicates that the United States will consume 80% of the available petroleum in 65 years and that by the year 2000 only 10% of the original amount will remain. The dashed line rising from the bell-shaped curve indicates the increase of the production rate based on an annual growth rate of 5.68%, which occurred between 1935 and 1955. This rate of growth cannot be sustained because it allows the total amount of petroleum to approach infinity (based on Hubbert, 1969).

model predicts the ultimate exhaustion of this and other nonrenewable natural resources.

The implications of the requirement that the rates of production of all nonrenewable mineral commodities must ultimately decline has given rise to a new genre of books on mineral economics. Some of these have quite alarming titles, as, for example, *Affluence in Jeopardy* (Park, 1968), *Earthbound* (Park, 1975), and *At the Crossroads* (Cameron, 1986). However, the basic facts regarding production and consumption of metals, nonmetallic minerals, and fossil fuels can be found in the *Minerals Yearbook* published annually in three volumes by the U.S. Bureau of Mines. In addition, the Bureau issues annual summaries of production and consumption of nonfuel mineral commodities (Bureau of Mines, 1988).

An important aspect of mineral economics is the extent to which the demand for mineral commodities can be met by *domestic* sources. Since no country is entirely self-sufficient, trade with other nations is required. Consequently, mineral economics affects foreign policy as nations develop alliances with trading partners who can provide the raw materials needed for industrial development. The position of the United States in this regard is shown in Table 21.3, which indicates that in 1987 more than 50% of the demand for 24 commodities was met by *imports* from other countries. In addition to the 37 metals and industrial minerals listed in Table 21.3, the United States also imported significant amounts of andalusite, Sb, Bi, Ga, Ge, I, ilmenite, Hg, Re, rutile, and V. The imports originated from 50 countries, 11 of which contributed 5 or more commodities. The list is led by Canada (24 commodities), Mexico (16), Republic of South Africa (13), Australia (9), Brazil (9), China (9), United Kingdom (6), Spain (6), Japan (5), Belgium (5), and Peru (5). The value of imports of mineral commodities in 1987 was $4 billion for ores and raw materials and $33 billion for metals and processed compounds. *Exports* from the United States in 1987 of raw materials, metals, and processed compounds were valued at $24 billion, which resulted in a *trade imbalance* of about $13 billion (Bureau of Mines, 1988).

The consumption of metals and industrial minerals has increased throughout most of the 20th century following the rise of the world's population and increases in per capita consumption (Park, 1968). However, between 1964 and 1985 the annual rates of production of most metals and industrial minerals in Table 21.4 *declined* in spite of the continuing growth of the population (Figure 19.6). The changes in production rates take the form of progressively smaller gains (Al, Cr, Fe, Ni, cement, phosphate, and potash), as well as actual reductions in the amounts produced, especially in the period 1981–1985 (Co, Mo, Sn, Hg, W, barite, fluorite, ilmenite/rutile, bauxite, and salt). Such declines in the amounts of these commodities may represent short-term fluctuations that will reverse in the future. Some evidence of such fluctuations can be seen in the production rates of Cu, Pb, Mn, Hg, W, barite, and feldspar. Nevertheless, fluctuations in the production of metals and industrial minerals are partly driven by prices and reflect the state of the balance between supply and demand. Such short-term fluctuations are a property of the world mineral economy and do not provide a reliable basis for predicting the availability of mineral resources in the distant future.

Even the rational analysis of the *petroleum production* in the United States by Hubbert (1969) *cannot* be used to predict the future availability of *metals* because the definition of what constitutes an *ore deposit* has changed and is continuing to do so. In the past, the nonferrous metals were typically recovered from high-grade deposits mined by underground methods. Most of the richest deposits have been mined out or are becoming unprofitable because the cost of underground mining has increased, whereas the prices of metals fluctuate and were depressed in the 1980s because of overproduction. Consequently, the industry has concentrated on mining low-grade deposits at shallow depth that can be recovered by less expensive open-pit methods. The trend toward large-scale mining of low-grade deposits began in 1907 when D. C. Jackling and R. C. Gemmell began open-pit mining of the porphyry-copper deposit in Bingham Canyon in Utah (Skinner, 1987).

As a result of the decrease in the minimum tenor of ore deposits, the *volume* of ore available

Table 21.3 Imports of Mineral Commodities to the United States in 1987 Expressed as Percent of Apparent Consumption

Commodity	Imports, %	Sources
Arsenic	100	Sweden, Canada, Mexico
Niobium	100	Brazil, Canada, Thailand, Nigeria
Graphite	100	Mexico, China, Brazil, Madagaskar
Manganese	100	S. Africa, France, Gabon, Brazil
Mica	100	India, Belgium, Japan, France
Celestite	100	Mexico, Spain, China
Yttrium	100	Australia
Gem stones	99	Belgium, Luxembourg, Israel, India, S. Africa
Bauxite	97	Australia, Guinea, Jamaica, Suriname
Tantalum	92	Thailand, Brazil, Australia, Canada
Diamond (industrial)	89	S. Africa, United Kingdom, Ireland, Belgium, Luxembourg
Fluorite	88	Mexico, S. Africa, Spain, Italy, China
Platinum group metals	88	S. Africa, United Kingdom, Russia
Cobalt	86	Zaire, Zambia, Canada, Norway
Tungsten	80	China, Canada, Bolivia, Portugal
Chromium	75	S. Africa, Zimbabwe, Turkey, Yugoslavia
Nickel	74	Canada, Australia, Norway, Botswana
Tin	73	Brazil, Thailand, Indonesia, Bolivia
Potassium salts	72	Canada, Israel, Germany, Russia
Zinc	69	Canada, Mexico, Peru, Australia
Cadmium	66	Canada, Australia, Mexico, Germany
Barite	63	China, Morocco, India
Silver	57	Canada, Mexico, United Kingdom, Peru
Asbestos	51	Canada, S. Africa
Gypsum	37	Canada, Mexico, Spain
Silica	33	Brazil, Canada, Norway, Venezuela
Fe ore	28	Canada, Brazil, Venezuela, Liberia
Copper	25	Canada, Chile, Peru, Zaire, Zambia, Mexico
Aluminum	24	Canada, Japan, Venezuela, Brazil
Cement	20	Canada, Mexico, Spain
Iron + steel	19	Europe, Japan, Canada, Rep. Korea
Lead	15	Canada, Mexico, Peru, Australia, Honduras
Salt (NaCl)	12	Canada, Mexico, Bahamas, Chile, Spain
Beryllium	11	Brazil, China, Switzerland, S. Africa
Titanium	8	Japan, Russia
Nitrogen	7	Canada, Russia, Trinidad and Tobago, Mexico
Sulfur	6	Canada, Mexico

SOURCE: Bureau of Mines (1988).

Table 21.4 Annual Growth Rates of Production of Selected Mineral Commodities in Percent per Year

Commodity	1964–1973	1973–1980	1981–1985
Metals			
Al	8.12	3.45	0.28
Cr	5.21	5.17	1.86
Fe (steel)	5.37	0.24	0.24
Co	5.96	0.87	−0.14
Cu	4.87	1.01	4.33
Pb	3.87	0.13	0.53
Mn	3.36	−1.85	1.00
Mo	7.74	4.2	−2.07
Ni	6.5	2.36	1.42
Sn	2.21	0.48	−3.58
Hg	0.88	−5.13	−1.39
W	3.43	5.23	−0.29
Industrial Minerals			
Barite	3.96	7.33	−5.27
Cement	5.99	3.44	1.92
Feldspar	4.26	3.12	4.11
Fluorite	6.55	0.29	−1.24
Ilmenite/rutile	4.27	2.91	−1.64
Bauxite	9.73	4.25	−0.05
Phosphate	6.24	4.88	1.64
Salt	5.13	0.91	−0.13
Potash (K_2O eq.)	6.26	4.01	1.13

SOURCE: Cameron (1986), Table 6.2; Bureau of Mines (1985).

for mining has increased and continues to grow. Therefore, predictions based on the ore reserves available at one time must be revised later when large amounts of previously unprofitable deposits are reclassified as ore. Mining technology has changed greatly during the 20th century and is likely to continue to develop methods that are relatively inexpensive, can recover *groups* of metals from low-grade deposits, and do not alter the environment irreversibly. Such methods may include *solution mining* of artificially fractured rocks aided by *bacteria* that decompose sulfide minerals and thereby increase their solubility in aqueous solutions.

The long-range outlook for the availability of metals and industrial minerals depends more on the *cost* and *availability* of the *energy* required for mining and processing of ore than on the volume of minable rocks. The mineral resources of the Earth become very large if metals and industrial compounds can be recovered from certain kinds of sedimentary rocks such as carbonaceous shale, phosphorite, or carbonate rocks of marine and nonmarine origin.

21.9 Summary

The principles of geochemistry and the accumulated information about the properties of the elements and their compounds are applied in the search for ore deposits from which metals can be extracted at a profit. The applicable principles concern the oxidation–reduction reactions that occur during the chemical weathering of sulfide minerals of the transition metals, including those of Fe, Cu, Pb, Zn, Hg, Ag, and Mo.

The oxidation of pyrite and the subsequent formation of solid $Fe(OH_3)$ causes acidification that enhances the solubility of the compounds of other metals. The process is greatly accelerated by the metabolic activities of certain bacteria, some of which can also attack sulfide minerals.

Copper is commonly associated with Fe either in the form of discrete sulfide minerals, such as chalcocite and pyrite, or in mixed compounds, such as chalcopyrite, bornite, or cubanite. Under acidic and oxidizing conditions, these minerals dissolve to form solutions of Cu^{2+} and SO_4^{2-}, which percolate downward towards the water table. If the solutions encounter basic environments, they may precipitate Cu oxides (cuprite or tenorite) or the carbonates (malachite or azurite). However, if the Cu^{2+} ions reach the reducing environment below the water table, they are reduced to Cu^+ and may replace Fe^{2+} in pyrite. As a result of the formation of secondary chalcocite by replacement of pyrite, the concentration of Cu in the rock is increased.

This process of supergene enrichment has played an important role in the formation of minable deposits of Cu and Ag. In principle, one can rank the sulfides of different metals in terms of *increasing* solubility and predict, on that basis, that the ions of one metal can replace the ions of another in a mineral that is *more* soluble. The effectiveness of supergene enrichment is enhanced when the water table is low, the rocks are permeable, hydrologic gradients and erosion rates are low, and carbonate rocks are absent. Supergene enrichment is ineffective for Pb and Zn, because these metals form insoluble sulfates and carbonates, and for Mo because its ions remain in solution over a wide range of environmental conditions.

The mobility of ions released by oxidation of sulfide deposits determines their dispersion and hence the contrast of the geochemical anomaly they form in surficial deposits. Geochemical exploration, which is based on this phenomenon, is an effective and widely practiced technique for locating mineral deposits. However, some of the anomalies identified by such surveys may be of anthropogenic origin and most of the natural anomalies do not actually result in the discovery of minable deposits because either the tenor or the volume of the deposit, or both, are insufficient to warrant the expense of bringing it into production.

The consumption of mineral resources has increased historically in step with the growing population of the world. However, the average annual rates of production of some important metals and industrial minerals decreased in the second half of the 20th century and some have actually become negative. The shortages that may occur in the future may spur production, thereby contributing to the fluctuations in supply and demand that are characteristic of the mining economy of the world. The trend appears to be toward mining of low-grade near-surface deposits that can be worked by open-pit methods and by leaching metals from rocks either on the surface or below it. As a result of the changing definition of what constitutes a minable deposit, the volume of ore reserves is expanding. The ultimate limitation on the production of natural mineral resources is the cost of the necessary energy.

Problems

1. Derive the Eh–pH equation for the dissolution of cuprite (Cu_2O) to Cu^{2+}.

2. Derive the Eh–pH equation for the dissolution of tenorite (CuO) to Cu^{2+}.

3. Calculate the ration of $[Cu^{2+}]/[Cu^+]$ in an aqueous solution at Eh = +0.80 V.

4. Derive an Eh–pH diagram for sphalerite (ZnS) and zincite (ZnO) and their ions Zn^{2+} and ZnO_2^{2-} in a system at 25°C and 1 atm, where $\Sigma S = 10^{-3}$ mol/L (i.e., native S does not occur). Draw the solubility limits at activities of 10^{-8} mol/L (see Brookins, 1988, p. 55).

5. Calculate the solubility of molybdenite (MoS_2) dissolving to form $HMoO_4^-$ at Eh = −0.20 V, pH = 5.0, $\Sigma S = 10^{-1}$ mol/L, 25°C at 1 atm.

6. Calculate the solubility of wulfenite ($PbMoO_4$) in pure water at pH = 7 and express it in terms of $[Pb^{2+}]$ at equilibrium (check Figure 21.4C).

7. If the system is open to the atmosphere ($[CO_2] = 3 \times 10^{-4}$ atm), would cerussite ($PbCO_3$) replace wulfenite assuming pH = 7.0 and $[HMoO_4^-]$ as calculated in Problem 6?

8. If the activity ratio $[Ag^+]/[Pb^{2+}] = 10^{-2}$ in contact with galena, how many moles of Ag_2S will form per liter of solution before equilibrium is restored? (Assume that the data in Table 21.1 apply.)

9. Consider whether supergene enrichment of Cd can occur by replacement of Zn in sphalerite in systems having $\Sigma S = 10^{-3}$ mol/L, if pH = 5.0, Eh = −0.20 V, 25°C and 1 atm (consult Brookins, 1988).

References

BARNES, J. W., 1988. *Ores and Minerals.* Open University Press, New York, 192 pp.

BATEMAN, A. M., 1950. *Economic Mineral Deposits,* 2nd ed. Wiley, New York, 916 pp.

BAUER, H. L., Jr., R. A. BREITRICK, J. J. COOPER, and J. A. ANDERSON, 1966. Porphyry copper deposits in the Robinson mining district, Nevada, 233–244. In S. R. Titley and C. L. Hicks (Eds.), *Geology of the Porphyry Copper Deposits.* University of Arizona Press, Tucson.

BOYLE, R. W., 1982. *Geochemical Prospecting for Thorium and Uranium Deposits.* Elsevier, Amsterdam, 498 pp.

BRIERLEY, C. L., and L. E. MURR, 1973. Leaching: Use of a thermophilic and chemoautotrophic microbe. *Science,* **179**:488–499.

BROOKINS, D. G., 1988. *Eh–pH Diagrams for Geochemistry.* Springer-Verlag, New York, 176 pp.

BROOKS, R. R., 1972. *Geobotany and Biogeochemistry in Mineral Exploration.* Harper & Row, New York, 290 pp.

BUREAU OF MINES, 1985. *Minerals Yearbook; Metals and Minerals,* vol. 1. U.S. Government Printing Office, Washington, DC, 1104 pp.

BUREAU OF MINES, 1988. *Mineral Commodity Summaries.* U.S. Government Printing Office, Washington, DC, 193 pp.

CABELL, J. B., 1926. *The Silver Stallion.* McBride, New York.

CAMERON, E. N., 1986. *At the Crossroads: The Mineral Problems of the United States.* Wiley, New York, 320 pp.

COX, D. P., and D. A. SINGER (Eds.), 1986. *Mineral Deposit Models.* U.S. Geol. Survey Bulletin, 1693, 379 pp.

CRERAR, D. A., G. W. KNOX, and J. L. MEANS, 1979. Biogeochemistry of bog iron in the New Jersey Pine Barrens. *Chem. Geol.,* **24**:111–135.

DAVY, R., and R. H. MAZZUCCHELLI (Eds.), 1984. *Geochemical Exploration in Arid and Deeply Weathered Environments.* Elsevier, Amsterdam, 376 pp.

DENNEN, W. H., 1989. *Mineral Resources.* Taylor and Francis, New York, 300 pp.

EHRLICH, H. L., 1981. *Geomicrobiology.* Marcel Dekker, New York, 393 pp.

FLETCHER, W. K., 1981. *Analytical Methods in Geochemical Prospecting.* Elsevier, Amsterdam, 256 pp.

GARRELS, R. M., 1954. Mineral species as functions of pH and oxidation–reduction potentials, with special reference to the zone of oxidation and secondary enrichment of sulphide ore deposits. *Geochim. Cosmochim. Acta,* **5**:153–168.

GARRELS, R. M., and C. L. CHRIST, 1965. *Solutions, Minerals and Equilibria.* Harper & Row, New York, 450 pp.

GOVETT, G. J. S., 1983. *Rock Geochemistry in Mineral Exploration.* Elsevier, Amsterdam, 462.

HARRIS, D. P., 1984. *Mineral Resources Appraisal.* Oxford University Press, New York, 448 pp.

HAWKES, H. E., and J. S. WEBB, 1962. *Geochemistry in Mineral Exploration.* Harper & Row, New York, 415 pp.

HEM, J. D., 1977. Reactions of metal ions at surfaces of hydrous iron oxide. *Geochim. Cosmochim. Acta,* **41**:527–538.

HOWARTH, R. J. (Ed.), 1982. *Statistics and Data Analysis in Geochemical Prospecting.* Elsevier, Amsterdam, 438 pp.

HUBBERT, M. K., 1969. Energy resources. In P. Cloud (Ed.), *Resources and Man,* 157–242. Committee on Resources and Man, National Academy of Sciences–National Research Council. W. H. Freeman, San Francisco, 259 pp.

HUBBERT, M. K., 1973. Survey of world energy resources. *Can. Min. Metall. Bull.,* **66**(753):37–54.

JARRELL, O. W., 1944. Oxidation at Chuquicamata, Chile. *Econ. Geol.,* **39**:215–286.

KRAUSKOPF, K. B., 1951. Physical chemistry of quicksilver transportation in vein fluids. *Econ. Geol.,* **46**:498–523.

LANDERGREN, S., and F. T. MANHEIM, 1978. Molybdenum. In K. H. Wedepohl (Ed.), *Handbook of Geochemistry,* vol. II (4). Springer-Verlag, Berlin, ch. 42.

LANGMUIR, D., 1978. Uranium solution—mineral equilibria at low temperatures with applications to sedimentary ore deposits. *Geochim. Cosmochim. Acta,* **4**:547–569.

LEVINSON, A. A., 1974. *Introduction to Exploration Geochemistry.* Applied Publ. Calgary, 612 pp.

PARK, C. F., Jr., 1968. *Affluence in Jeopardy: Minerals and the Political Economy.* Freeman, Cooper and Co., San Francisco, 368 pp.

PARK, C. F., Jr., and R. A. MACDIARMID, 1970. *Ore Deposits,* 2nd ed. W. H. Freeman, New York, 522 pp.

PARK, C. F., 1975. *Earthbound: Minerals, Energy, and Man's Future.* W. H. Freeman, San Francisco, 270 pp.

PETERSON, N. P., 1962. Geology and ore deposits of the Globe–Miami district, Arizona. *U. S. Geol. Surv. Prof. Paper* 342.

PUCHELT, H., H. H. SCHOCK, and H. HANERT, 1973. Rezente marine Eisenerze auf Santorin, Griechenland. *Geol. Rundschau,* **62**:786–812.

RICH, R. A., H. D. HOLLAND, and U. PETERSON, 1977. *Hydrothermal Uranium Deposits.* Elsevier, Amsterdam, 264 pp.

ROGERS, J. J. W., and J. A. S. ADAMS, 1969. Uranium. In K. H. Wedepohl (Ed.), *Handbook of Geochemistry,* vol. II(5). Springer-Verlag, Berlin, ch. 92 (D-O).

SCHÜRMANN, E., 1888. Über die Verwandtschaft der Schwermetalle zum Schwefel. *Justus Liebigs Ann. Chem.,* **249**:326–350.

SKINNER, B. J., 1987. Changing thoughts about mineral deposits: The 5th to 28th IGC. *Episodes,* **10**(4):297–301.

STRINGHAM, B., 1953. Granitization and hydrothermal alteration at Bingham, Utah. *Geol. Soc. Amer. Bull.,* **64**:945–991.

STUMM, W., AND J. J. Morgan, 1981. *Aquatic Chemistry,* 2nd ed. Wiley, New York, 780 pp.

THORNTON, I., and R. J. HOWARTH, 1986. *Applied Geochemistry in the 1980s.* Wiley, New York, 347 pp.

TUNNEL, G., 1970. Mercury. In K. H. Wedepohl (Ed.), *Handbook of Geochemistry,* B to O, vol. II(5). Springer-Verlag, Berlin, ch. 80.

WIGNALL, T. K., and J. DE GEOFFREY, 1987. *Statistical Models for Optimizing Mineral Exploration.* Plenum Pub. Corp., New York, 444 pp.

22

Geochemical Cycles

The geochemical processes on the surface of the Earth work like a machine through which matter is cycled continuously (Section 4.3). The plumbing system of the machine consists of several reservoirs of different volumes connected by pipes through which matter moves from one reservoir to another. All chemical elements are continually being processed by this geochemical machine. However, the volumes of the reservoirs and the period of time a particular element resides in each reservoir depend on the geochemical properties of the element. Therefore, each element has a characteristic geochemical cycle that reflects its particular set of geochemical properties. The geochemical cycles of the elements have influenced the environment on the surface of the Earth and have, in turn, been affected by changes that have occurred on that surface. The most recent perturbation of geochemical cycles has been caused by human activities (Stumm, 1977).

The rock cycle in Figure 22.1 illustrates how the major rock types are related to each other by geological and geochemical processes. For example, chemical and mechanical weathering convert igneous rocks into sediment and ions in solution that are transported to the oceans where they are deposited (Sections 4.3 and 4.4). The resulting layers of sediment are lithified into sedimentary rocks that may be altered to metamorphic rocks or may be partially melted to form magma from which new igneous rocks crystallize. Since the rock cycle has operated throughout the history of the Earth, we can assume that it has achieved a *steady state*. Consequently, the chemical compositions of the conduits through which the ele-

ments pass and the reservoirs in which they reside for varying periods of time should have remained constant. This generalization is helpful because it implies that the chemical composition of sedimentary rocks, averaged over sufficiently long intervals of geologic time, is constant and that the chemical composition of seawater has likewise been constant.

The chemical composition of seawater and sedimentary rocks is, however, affected by *long-term* changes such as the evolution of life and the related changes in the chemical composition of the atmosphere (Holland, 1984). The reservoirs also respond to orogenies on continents, volcanic activity in subduction zones and along midocean ridges, and to the resulting chemical as well as isotopic exchange between seawater and basalt (Wolery and Sleep, 1988). Moreover, the geochemical differentiation of the mantle and continental crust in the course of geologic time has allowed differences to develop in the isotopic abundances of stable, *radiogenic* daughters of long-lived radioactive parents (Section 16.3). Consequently, the isotopic compositions of the nonvolatile elements Ca, Sr, Nd, Hf, Os, and Pb in the oceans and in clastic sedimentary rocks have changed over geologic time in response to variations in the inputs from the continental crust and the mantle of the Earth (Faure, 1986).

Some important contributions to the literature on geochemical cycles include the books by Garrels and Mackenzie (1971), Garrels et al. (1975), Salomons and Förstner (1984), Berner and Berner (1987), Gregor et al. (1988), and Lerman and Meybeck (1988).

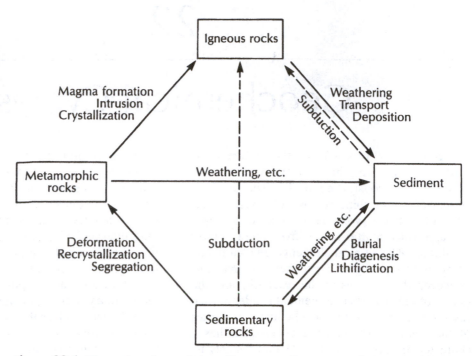

Figure 22.1 The rock cycle consisting of igneous, sedimentary, and metamorphic rocks. The arrows indicate the geological and geochemical processes that cause the required transformations of matter.

22.1 The Principle of Mass Balance

If the geochemical cycles of the elements are in a steady state, the total mass of a nonvolatile element released by weathering of igneous rocks must be equal to its mass in sedimentary rocks and in the oceans. This postulate is expressed by the equation (Drever et al., 1988):

$$M_{ig} C_{ig}^i = M_{sed} \sum a_j C_{sed}^i + M_{sw} C_{sw}^i \quad (22.1)$$

where
M_{ig} = mass of all igneous rocks in the continental crust that have weathered

C_{ig}^i = concentration of element i in igneous rocks of the crust

M_{sed} = mass of all sedimentary rocks in existence at the present time, including sediment in the oceans

C_{sed}^i = concentration of element i in sedimentary rocks and sediment in the oceans

M_{sw} = mass of water in the oceans

C_{sw}^i = concentration of element i in seawater

a_j = mass fractions of different kinds of sedimentary rocks and sediment in the oceans

Equation 22.1 neglects groundwater, surface water, and ice sheets because together they constitute only 2.75% of the water in the hydrosphere (Table 4.6) and because the concentrations of most elements they contain are low compared to seawater (Table 4.7). The equation *does* include the clay minerals, oxides, and hydroxides that form during incongruent dissolution of aluminosilicate minerals because they are ultimately incorporated into sedimentary rocks. The equation can be expanded to include saprolite, till, and soil on the continents and sediment in suspension in rivers, lakes, and in the oceans, if desired.

The masses of the various reservoirs have been estimated as follows:

M_{sed} = 2.5 $\pm$ 0.4 $\times$ 10²⁴g (Li, 1972; Ronov and Yaroshevsky, 1976)

M_{ig} = 0.88 M_{sed}, where M_{sed} includes volatile elements and compounds degassed from the interior of the Earth, such as H_2O, CO_2, and the noble gases (Li, 1972)

M_{sw} = 1.4 $\times$ 10²⁴ g (Sverdrup et al., 1942)

The mass of sediment in the oceans is about 0.12 $\times$ 10²⁴ g (Drever et al., 1988) distributed in approximately equal proportions between pelagic clay (deep-sea sediment) and calcareous ooze. The abundances of the different kinds of sedimentary rocks have been estimated in various ways with somewhat different results (Sections 4.2 and 4.3). Garrels and Mackenzie (1971) estimated the proportions of shale/sandstone/carbonates/evaporites to be 74 : 11 : 15 : 2. These proportions indicate that the mass fractions (a_j) are:

a_{sh} (shale) = 0.725 a_{cc} (carbonate) = 0.147

a_{ss} (sandstone) = 0.108 a_{ev} (evaporite) = 0.020

Drever et al. (1988) treated oceanic pelagic (op) and oceanic carbonate (oc) separately and adjusted the mass fractions in equation 22.1 accordingly:

$$M_{ig} C_{ig}^i = M_{sed}(0.702C_{sh}^i + 0.108C_{ss}^i$$
$$+ 0.122C_{cc}^i + 0.02C_{ev}^i + 0.024C_{op}^i$$
$$+ 0.024C_{oc}^i) + M_{sw}C_{sw}^i \qquad (22.2)$$

Since $M_{sed} = M_{ig}/0.88 = 2.5 \pm 0.4 \times 10^{24}$ g, we divide equation 22.2 by M_{sed}:

$$0.88C_{ig}^i = 0.702C_{sh}^i + 0.108C_{ss}^i$$
$$+ 0.122C_{cc}^i + 0.02C_{ev}^i + 0.024C_{op}^i$$
$$+ 0.24C_{oc}^i + 0.56C_{sw}^i \qquad (22.3)$$

This equation can be used to calculate the chemical composition of seawater or to assess the geochemical balance of individual elements by substituting appropriate estimates of the concentrations in the different kinds of rocks and in seawater. In doing so, we make use of all of the accumulated information about the chemical composition of terrestrial rocks summarized in Chapter 4.

In order to assess the state of the mass balance of the elements, it is convenient to evaluate each side of equation 22.3 separately and then to form the ratio of the results. If the right side of equation 22.3 is represented by $\overline{C}_s$, a perfect balance is indicated when the ratio:

$$\frac{\overline{C}_s}{0.88C_{ig}} = 1.00 \qquad (22.4)$$

However, because of the uncertainties in the magnitudes of the concentrations and mass fractions in equation 22.3, we can be satisfied with values between 0.8 and 1.30, as suggested by Drever et al. (1988).

The state of the mass balance for Mg can be evaluated by using data from Table 4.5. For this purpose we estimate the average Mg concentration of igneous rocks by combining low-Ca and high-Ca granites in equal proportions to obtain an average value for *granite* and by defining C_{ig}^i as follows:

$$C_{ig}^i = 0.35C_{basalt}^i + 0.65C_{granite}^i \qquad (22.5)$$

The necessary data for the assessment of the Mg balance based on Table 4.5 (rocks) and Table 4.7 (seawater) are:

	%		%
C_{ig}^{Mg}	1.95	C_{ev}^{Mg}	4.70 (?)
C_{sh}^{Mg}	1.50	C_{op}^{Mg}	2.10
C_{ss}^{Mg}	0.70	C_{oc}^{Mg}	4.70 (?)
C_{cc}^{Mg}	4.70		

Based on these data, $C_s^{Mg} = 1.645\%$, $0.88C_{ig}^{Mg} = 1.716\%$, and the ratio:

$$\frac{C_s^{Mg}}{0.88C_{ig}^{Mg}} = \frac{1.645}{1.716} = 0.96 \qquad (22.6)$$

which confirms the mass balance for this element.

The result for Ca does not work out this well. Using the data listed below from Tables 4.5 and 4.7 we find *excess* Ca in sedimentary rocks and seawater:

	%		%
C_{ig}^{Ca}	3.51	C_{ev}^{Ca}	30.23
C_{sh}^{Ca}	2.21	C_{op}^{Ca}	2.90
C_{ss}^{Ca}	3.91	C_{oc}^{Ca}	30.23
C_{cc}^{Ca}	30.23		

$$\frac{C_s^{Ca}}{0.88 C_{ig}^{Ca}} = \frac{7.061}{3.088} = 2.3 \qquad (22.7)$$

The Ca imbalance can be resolved either by questioning the validity of the data used in the calculation or by postulating the existence of additional sources for Ca that are not represented in equation 22.3, such as:

1. Leaching of Ca from volcanogenic sediment (Garrels and Mackenzie, 1971).
2. Preferential exchange of Mg^{2+} and Na^+ in seawater with Ca^{2+} in basalt in convecting hydrothermal systems along midocean ridges (Wolery and Sleep, 1976; Edmond et al., 1979), followed by precipitation of the Ca as calcite.

These and other suggestions were evaluated by Drever et al. (1988) without resolving the dilemma. Nevertheless, 43 elements listed in Table 22.1 have acceptable mass balances, whereas 18 elements do not. The balance ratios of these elements range from ~1.5 for U to 151 for Cl. Evidently, there is much more Cl in the oceans and in sedimentary rocks than can be accounted for by weathering of igneous rocks in the crust.

22.2 Mass Balance for Major Elements in the Ocean

The ocean plays an important role in the operation of geochemical cycles because both solid particles and ions/molecules pass through it. In Section 4.4, we calculated the mean oceanic residence times (MORT, Table 4.7) of the elements and discussed the importance of solid particles in selectively removing the ions of some elements

Table 22.1 State of the Geochemical Mass Balance for the Chemical Elements (Based on the requirement that the amount supplied by weathering of igneous rocks is equal to the amounts in sedimentary rocks and in the oceans.)

Element	Balance ratio[a]	Element	Balance ratio
A. Satisfactory balance			
Be	1.26	Y	1.05
F	1.24	Zr	1.07
Na	0.84	Nb	0.81
Mg	1.02	Pd	~1
Al	0.91	Ag	0.92
Si	0.94	Cd	1.29
P	0.81	In	0.91
K	0.96	Ba	1.02
Sc	0.81	La	1.01
Ti	0.87	Ce	1.13
V	0.88	Nd	1.24
Cr	1.18	Sm	0.97
Mn	1.27	Eu	1.25
Fe	0.94	Tb	1.00
Co	0.90	Yb	1.11
Ni	1.24	Hf	1.02
Cu	1.02	Ta	0.99
Zn	1.25	W	1.26
Ga	1.06	Au	0.98
Ge	1.13	Tl	1.17
Rb	1.20	Th	1.05
Sr	1.15		
B. Unsatisfactory balance			
Li	2.12	Sn	2.44
B	11.9	Sb	6.28
S	16.5	Te	10.4
Cl	151	I	7.52
Ca	2.1	Cs	1.79
As	6.1	Hg	15.1
Se	10.1	Pb	1.55
Br	12.7	Bi	8.5
Mo	2.34	U	1.47

[a]The balance ratio is defined in equation 22.4. A value between 0.80 and 1.30 implies a satisfactory mass balance, whereas a value >1.30 implies an imbalance between the mass of an element in sedimentary rocks (including the oceans) and crustal igneous rocks that have weathered.
SOURCE: Drever et al. (1988).

from seawater. We now return to this subject in order to assess the mass balance of elements entering and leaving the oceans by considering the relevant inputs and outputs of both major and trace elements.

The *major elements* enter the oceans primarily by discharge of river water and to some extent by ion exchange reactions between seawater and suspended sediment, and between seawater and hot basalt along midocean ridges. The input of major elements to the ocean by rivers has been evaluated by Meybeck (1979) based on estimates of the average chemical compositions and discharges of the rivers of the world (Sections 20.1 and 20.2 and Table 20.1). The amounts of each element discharged into the oceans as a result of chemical weathering of rocks on the continents must be reduced by subtracting cyclic marine salts and anthropogenic contaminants as shown in Table 22.2. Note that anthropogenic contamination contributes ~30% of the Na^+, Cl^-, and SO_4^{2-} and from 7 to 10% of the Mg^{2+}, Ca^{2+}, and K^+ dissolved in rivers at the present time (Section 20.2). In addition, marine salts represent an additional 32% of the Cl^- and 20% of Na^+ dissolved in rivers, which means that only ~40–50% of the Cl^- and Na^+ in solution in river water was released by chemical weathering.

If the ocean is in a *steady state,* the inputs must be equal to the outputs for each element. The principal outputs of major elements from the ocean include:

1. Burial of pore water.
2. Ion exchange reactions and water–rock interactions.
3. Diagenesis.
4. Chemical precipitates of carbonates, sulfides, oxides, phosphates, silica, sulfates, and chlorides.

Each year a certain amount of seawater is removed from the oceans because it occupies pore spaces in sediment. The porosity of the sediment decreases as a result of compaction and therefore cannot be stated precisely. Drever et al. (1988) adopted a value of 30%, but considered

Table 22.2 Inputs of Major Elements in Solution in River Water That Enter the Oceans Each Year in Units of 10^{12} mol/yr Based on a Total Discharge of 37.4×10^{15} L/yr

Ion or compound	Total flux	Marine salt	Anthropogenic contaminant	Net flux
Na^+	11.7	2.39	3.39	5.91
K^+	1.36	0.05	0.13	1.17
Mg^{2+}	5.59	0.29	0.41	4.85
Ca^{2+}	13.7	0.05	1.18	12.36
Cl^-	8.69	2.82	2.63	3.27
SO_4^{2-}	4.49	0.16	1.29	3.07
HCO_3^-	33.4	—	1.64	32.09
H_4SiO_4	6.47	—	—	6.47

SOURCE: Drever et al. (1988); Meybeck (1979).

that it could range up to 50%. Ion exchange of suspended clay minerals entering the oceans is relatively unimportant for the major elements and involves primarily Na^+ replacing Ca^{2+} on adsorption sites. However, water–rock interaction may be a source for some elements and a sink for others. Chemical reactions between mineral particles in the sediment and trapped seawater (diagenesis) also release some ions while consuming others. An evaluation by Sayles (1979) indicated that K^+ and Mg^{2+} are consumed, whereas Na^+, Ca^{2+}, and HCO_3^- are released by diagenetic reactions. Seawater is close to being saturated with respect to several types of compounds (listed above), which may precipitate from it under favorable physical/chemical conditions or as a result of biological activity.

The quantitative estimates of outputs from the oceans in Table 22.3 (Drever et al., 1988) can be compared to the inputs by rivers to test the balance for each major element. For example, Na^+ is removed from seawater by burial of pore water and by ion exchange reactions, but is added by diagenetic reactions in the sediment. The sum of the inputs and outputs for Na^+ is $+5.9 \pm 2.4 \times 10^{12}$ mol/yr, where the uncertainty is the sum of the errors of the estimated fluxes in

Table 22.3 Inputs and Outputs of Major Elements in Solution in the Ocean[a] in Units of 10^{12} mol/yr

Ion or compound	River[b] input	Burial, pore water	Ion exchange	Diagenetic reactions	Precipitates	Net balance
Na^+	+5.91	-0.96 ± 0.64	-1.53 ± 0.06	$+2.5 \pm 1.7$	—	$+5.9 \pm 2.4$
K^+	+1.17	-0.02 ± 0.01	-0.20 ± 0.08	-0.9 ± 0.5	—	$+0.05 \pm 6$
Mg^{2+}	+4.85	-0.09 ± 0.04	-0.32 ± 0.08	-2.3 ± 1.6	-0.26 ± 0.02	$+1.9 \pm 1.7$
Ca^{2+}	+12.36	-0.02 ± 0.01	-0.96 ± 0.10	$+3.6 \pm 3.0$	-12.5 ± 1.8	$+4.5 \pm 4.9$
Cl^-	+3.27	-1.07 ± 0.71	—	—	—	$+2.2 \pm 0.7$
SO_4^{2-}	+3.07	-0.06 ± 0.04	—	—	$-0.60 \pm ?$	$+2.4$
HCO_3^-	+32.09	—	-0.42 ± 0.16	$+4.7 \pm 4.0$	-24.2 ± 3.6	$+11.6 \pm 7.8$
H_4SiO_4	+6.47	—	—	—	-7.0 ± 0.09	-0.5 ± 0.09

[a]A plus sign indicates gain by and a minus sign means loss from the ocean.
[b]From Table 22.2.
SOURCE: Drever et al. (1988).

Table 22.3. Evidently, more Na^+ appears to be entering the ocean than is leaving it, which may indicate either that the estimates of the input or output fluxes are incorrect or that the ocean is not in a steady state with respect to Na^+; that is, the Na^+ concentration of seawater is increasing. Similar positive imbalances exist for Ca^{2+}, Cl^-, HCO_3^-, SO_4^{2-}, and Mg^{2+}, whereas K^+ and H_4SiO_4 appear to be in balance.

The excess Ca^{2+} in the ocean is partly matched by an excess of SO_4^{2-}, whereas the Na^+ excess correlates with the excess in Cl^-. Therefore, we may conclude that the concentrations of these elements in seawater are rising in preparation for the next episode of deposition of $CaSO_4$ and $NaCl$ in marine evaporite basins. The last such episode occurred ~5.5×10^6 yr ago when the Mediterranean basin was isolated from the Atlantic Ocean (Hsü, 1983). However, the imbalances for Na^+ and Ca^{2+} persist even after allowances are made for the future precipitation of gypsum and halite.

This difficulty serves as a reminder that the estimates of inputs and outputs in Table 22.3 do not include the interactions of seawater with hot basalt along midocean ridges. According to studies cited by Drever et al. (1988), these reactions involve the conversion of anorthite ($CaAl_2Si_2O_8$) to talc ($Mg_3Si_4O_{10}(OH)_2$) by the uptake of Mg^{2+} from seawater:

$$6 Mg^{2+} + 4 CaAl_2Si_2O_8 + 4 H_2O \rightarrow$$
$$2 Mg_3Si_4O_{10}(OH)_2 + 4 Al_2O_3$$
$$+ 4 Ca^{2+} + 4 H^+ \quad (22.8)$$

Complex interactions of fayalite (Fe_2SiO_4) with SO_4^{2-} and Mg^{2+} form sepiolite ($Mg_2Si_3O_6(OH)_4$), pyrite, amorphous silica, and magnetite:

$$2Mg^{2+} + 2 SO_4^{2-} + 11 Fe_2SiO_4 + 2 H_2O \rightarrow$$
$$Mg_2Si_3O_6(OH)_4 + FeS_2$$
$$+ 8 SiO_2 + 7 Fe_3O_4 \quad (22.9)$$

In addition, Na^+ replaces Ca^{2+} and K^+ as a result of the albitization of Ca-plagioclase and K-feldspar, respectively:

$$2 Na^+ + CaAl_2Si_2O_8 + 4 SiO_2 \rightarrow$$
$$Ca^{2+} + 2 NaAlSi_3O_8 \quad (22.10)$$

and:

$$Na^+ + KAlSi_3O_8 \rightarrow K^+ + NaAlSi_3O_8 \quad (22.11)$$

These reactions *reduce* the concentrations of Mg^{2+}, SO_4^{2-}, and Na^+ in seawater and *increase* the concentrations of Ca^{2+} and K^+. However, the

Ca^{2+} released by this process precipitates as calcite in fractures and cavities in the basalt and does not remain in seawater. Moreover, the fluxes of the major elements leaving and entering the ocean as a result of water–rock interactions have not been quantified. Therefore, we can say only that these reactions probably remove Na^+, Mg^{2+}, and SO_4^{2-} from the oceans and that some K^+ may be released. However, water–rock interaction does *not* help to explain the apparent excess of Ca^{2+} in the oceans.

The attempt to balance inputs and outputs of major elements in the oceans is only partly successful. This could mean that the ocean is not in a steady state and that the composition of seawater is changing due to natural causes and probably has done so in the geologic past. The imbalances discussed above become *larger* when anthropogenic contamination of rivers is included. For example, the riverine Na^+ flux to the ocean at the *present time* is not 5.91×10^{12} mol/yr (Table 22.2) but is actually 9.30×10^{12} mol/yr when the anthropogenic input of 3.39×10^{12} mol/yr is added to the natural flux. However, anthropogenic contamination has a *negligible* effect on the concentrations of the *major elements* in the ocean. For example, Drever et al (1988) reported that it would take 300×10^6 yr before the anthropogenic input of SO_4^{2-} (1.29×10^{12} mol/yr) would saturate seawater with respect to gypsum, assuming that the concentration of Ca^{2+} in the ocean remains constant. The ocean resists change because it is a very large reservoir of the major elements compared to the annual inputs by rivers.

22.3   Mass Balance for Trace Elements in the Ocean

Trace elements in solution in river water and adsorbed on suspended particles are generally deposited in estuaries and on the continental shelves or accumulate in deep-sea clay (Sections 4.3 and 4.4). Therefore, the annual riverine input of a trace element to the oceans must be equal to the output of that element associated with marine muds. The annual input of trace element i by rivers can be represented by:

$$\text{input}_i = M_{sp}C_{sp}^i + M_{rw}C_{rw}^i \quad (22.12)$$

where M_{sp} = mass of suspended sediment transported annually by rivers into the ocean

M_{rw} = mass of river water entering the ocean annually

C_{sp}^i = concentration of element i in suspended particles

C_{rw}^i = concentration of element i in river water

The output from the oceans is given by:

$$\text{output}_i = M_{ds}C_{ds}^i + M_{sh}C_{sh}^i \quad (22.13)$$

where M_{ds} = mass of deep-sea clay deposited annually

M_{sh} = mass of mud deposited annually in estuaries and on continental shelves

C_{ds}^i = concentration of element i in deep-sea clay

C_{sh}^i = concentration of element i in shelf sediment (shale)

The mass of sediment deposited near shore each year is equal to the input by rivers in the form of suspended particles and dissolved ions minus the mass of deep-sea clay:

$$M_{sh} = M_{sp} + M_{rw} \sum C_{rw}^i - M_{ds} \quad (22.14)$$

The term $\sum C_{rw}^i$ in equation 22.14 is the sum of the concentrations of all ions in river water, excluding Ca^{2+} and HCO_3^-, after reducing the concentrations by the amount of recycled marine chloride. Drever et al. (1988) excluded Ca^{2+} and HCO_3^- because they precipitate as calcite in the ocean and because calcite concentrates only a few of the trace elements.

By substituting equation 22.14 in equation 22.13 and by letting $\text{input}_i = \text{output}_i$, we obtain the complete mass balance equation (Drever et al., 1988):

$$M_{sp}C_{sp}^i + M_{rw}C_{rw}^i = M_{ds}C_{ds}^i$$
$$+ \left(M_{sp} + M_{rw} \sum C_{rw}^i - M_{ds} \right) C_{sh}^i \quad (22.15)$$

The masses of sediment entering and leaving the ocean annually have been estimated and can be specified as follows:

$$M_{sp} = 18 \pm 3 \times 10^{15} \text{ g/yr}$$
(Holeman, 1968; Meybeck, 1979)

$$M_{ds} = 1.1 \pm 0.4 \times 10^{15} \text{ g/yr}$$
(Ku et al., 1968)

$$M_{rw} = 33 \pm 3 \times 10^{18} \text{ g/yr}$$
(Garrels and Mackenzie, 1971)

$$M_{rw} \sum C_{rw}^i = 0.92 \times 10^{15} \text{ g/yr}$$
(Drever et al., 1988)

Substituting these values into equation 22.15 yields:

$$(18 \times 10^{15})C_{sp}^i + (33 \times 10^{18})C_{rw}^i$$
$$= (1.1 \times 10^{15})C_{ds}^i + [(18 \times 10^{15})$$
$$+ (0.92 \times 10^{15}) - (1.1 \times 10^{15})]C_{sh}^i \quad (22.16)$$

which reduces to:

$$C_{sp}^i + (1.8 \times 10^3)C_{rw}^i = 0.06C_{ds}^i + 0.99C_{sh}^i \quad (22.17)$$

where the concentrations are expressed in parts per million. We now define the input (M_{in}^i) as:

$$M_{in}^i = C_{sp}^i + (1.8 \times 10^3)C_{rw}^i \quad (22.18)$$

and the output M_{out}^i as:

$$M_{out}^i = 0.06C_{ds}^i + 0.99C_{sh}^i \quad (22.19)$$

If the ocean is in a steady state with respect to a particular element, the ratio of its output and input functions must be:

$$\frac{M_{out}^i}{M_{in}^i} = 1.00 \quad (22.20)$$

As before, we accept values of this ratio between 0.80 and 1.30 as evidence of a satisfactory balance between output and input and use this criterion to classify the elements.

Data assembled from the literature by Drever et al. (1988) indicate that 30 elements identified in Table 22.4A have satisfactory input–output balances. Lithium, As, and Rb, whose output/input

Table 22.4 Mass Balance of Elements (Equation 22.20) Entering the Ocean by Discharge of Rivers (in Solution and in Suspended Particles) and Leaving the Ocean by Deposition of Sediment (as Shale and Deep-Sea Clay)

Element	Output/input	Element	Output/input
A. Satisfactory Balance (Output $\approx$ Input)			
Li	2.26	Ga	0.80
B	1.3	As	1.58
Mg	0.83	Rb	1.43
Al	0.89	Sr	1.07
Si	0.97	Mo	1.0
K	1.14	Cs	0.89
Ca[a]	1.02	Ba	1.1
Sc	0.78	La	0.96
Ti	0.86	Ce	0.93
V	0.79	Nd	1.14
Cr	0.93	Sm	1.13
Mn	1.2	Eu	1.13
Fe	1.05	Lu	1.27
Co	1.2	Th	0.90
Ni	0.89	U	1.02
B. Unsatisfactory Balance (Input > Output)			
Na	0.65	Ag	0.11
P	0.66	Cd	0.29
S	0.37	Sb	0.25
Cl	0.10	I	0.30
Cu	0.53	Au	0.026
Zn	0.27	Hg	0.52
Br	0.20	Pb	0.16

C. Not Classified (Nonvolatile)			
Be	Pd	Ho	Re
F	In	Er	Os
Ge	Te	Tm	Ir
Zr	Pr	Yb	Pr
Nb	Gd	Hf	Tl
Ru	Tb	Ta	
Rh	Dy	W	

[a]Oceanic carbonate included by adding 0.08 C_{oc}^i to the right side of equation 22.17 (Drever et al., 1988).
SOURCE: Drever et al. (1988).

Na	2.0×10^8
S	5.0×10^8
Cl	6.3×10^8
Br	7.8×10^8
I	$<4.0 \times 10^6$

ratios (equation 22.20) are actually greater than 1.30 are, nevertheless, included in this group. A second group of 14 elements are out of balance because their *inputs* to the ocean are significantly *greater* than their outputs. A large number of elements are not listed at all, either because they are volatile or because the information necessary for an evaluation is lacking. Table 22.4 contains 26 unclassified nonvolatile elements without including the noble gases, C, H, N, O, and the radioactive elements in the decay chains of U and Th: Po, At, Fr, Ra, Ac, and Pa.

The apparent imbalance of the inputs and outputs of elements in Table 22.4B may be caused by *inaccurate* analytical data or by the existence of outputs that are not included in equation 22.17. Drever et al. (1988) considered the possible removal of these elements from the ocean by *water–rock interactions* in the hydrothermal systems along midocean ridges. However, they concluded that only a small fraction of the excess can be removed in this way.

Another possibility is that the apparent *excess* in the riverine *input* may be caused by the recycling of marine aerosols. In other words, elements are removed from the ocean in aerosols, are washed out of the atmosphere by rain or snow deposited on the continents, and are then returned to the oceans in solution in river water. Additional sources of the excess riverine input of some of the Table 22.4B elements may be volcanic gases, exhalations by plants, and anthropogenic contamination. Drever et al. (1988) concluded that the apparent excess of Cl, Br, I, Na, Cd, and Hg *can* be accounted for by recycling of marine aerosols.

However, the apparent excess in the riverine input of Na and Cl may also be an indication, already suggested by the data in Table 22.3, that these elements as well as Br, I, and S are actually *accumulating* in the ocean. Therefore, the mass balance on the basis of particulate fluxes through the ocean (equation 22.17) confirms conclusions regarding these elements based on the rock cycle (equation 22.3; i.e., weathered igneous rocks = sedimentary rocks + seawater). Moreover, these elements all have long oceanic residence times expressed in years (Table 4.7):

as expected, if they are accumulating in the ocean.

Although Hg and Cd may be supplied to the rivers by marine aerosols, the excess may also originate from exhalations by volcanoes and plants on the continents. In this case, they should have been accumulating in the ocean because the known outputs are unable to remove them completely. However, their short oceanic residence times (Cd, 7.9×10^4 yr; Hg, 7.9×10^3 yr, Table 4.7), indicate that the apparent excess input is probably caused by anthropogenic contamination or by bad data.

The other elements of Table 22.4B (P, Cu, Zn, Ag, Sb, Au, and Pb) are not explainable by recycling of marine aerosols or by natural emissions by volcanoes and plants. Moreover, only Sb and Au have long oceanic residence times of 1.3×10^5 yr and 1.6×10^6 yr, respectively. Drever et al. (1988) concluded that the apparent excess of the riverine inputs of all of these elements is caused either by anthropogenic contamination of streams or by inaccurate data.

22.4   The Cycles of C-H-O-N

The geochemical cycles of the volatile elements, including C-H-O-N and the noble gases, cannot be represented adequately in terms of solids and ions in solution because these elements occur in the gas phase and together constitute the atmosphere, the biosphere, and the hydrosphere of the Earth. The geochemical cycles of these elements are therefore of paramount importance because they create the environment in which we live and of which we are an integral part.

The C-H-O-N elements dominate the *exogenic cycle*, which operates on the surface of the Earth. Their principal reservoirs and the processes that move them from one reservoir to another are depicted in Figure 22.2 by boxes and arrows, respectively. The resulting "plumbing system" shows that the exogenic cycle is connected to the *endogenic cycle,* which operates in the interior of the Earth and involves magma, igneous rocks, sedimentary rocks, and metamorphic rocks (Wolery and Sleep, 1976). The principal connection between the cycles occurs in sediment and soil because both are the result of weathering of endogenic rocks and also interact with the biosphere, the hydrosphere, and the atmosphere. Additional connections take the form of water–rock interactions along midocean ridges and the exhalation of volcanic gases.

The nonvolatile elements tend to reside preferentially in the reservoirs of the *endogenic* cycle, whereas the volatile and the water-soluble elements occupy reservoirs in the *exogenic* cycle. However, all elements occur in all of the available reservoirs even though their residence times may be short and their concentrations may be low.

We have already considered parts of this geochemical machine by evaluating the rates at which nonvolatile elements enter and leave the ocean and by calculating their mean oceanic residence times (Table 4.7). We have also specified the volumes of the reservoirs of the hydrologic cycle (Table 4.6) and made an inventory of the composition of the atmosphere (Table 4.8).

The geochemical cycles of C-H-O-N are interconnected because these elements form compounds involving two or more of the elements in

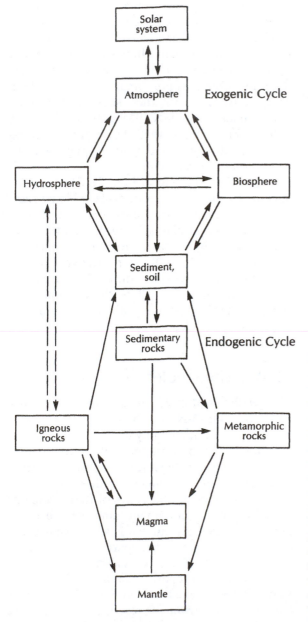

Figure 22.2 Interactions between the exogenic and endogenic cycles of the elements. The boxes are reservoirs in which elements reside for varying periods of time depending on their geochemical properties. The arrows are processes that move elements from one reservoir to another. The two cycles are connected primarily through sediment and soil, which interact directly with the atmosphere, the biosphere, and the hydrosphere. Other connections take the form of subduction and partial melting of sedimentary rocks or their metamorphic descendents and the exchange of elements between hot basalt and seawater. The mantle of the Earth is the ultimate source of all matter in the two cycles except for a small amount derived from the solar system as meteorites, interplanetary dust, and the solar wind. Some of the processes that make the geochemical cycles work are *reversible* as indicated by double arrows, whereas others are unidirectional and are *irreversible*.

this group, such as H_2O, CO_2, CH_4, CH_2O, and NO_2. These elements also occur in minerals and therefore participate in the *endogenic* cycle. For example, O is the most abundant chemical element in the rocks of the continental crust (Table 4.3), whereas C occurs in mantle-derived kimberlites as diamond.

a. The Carbon Cycle

The geochemistry of C is dominated by the reduction of C in CO_2 during photosynthesis and by its eventual reoxidation to CO_2, some of which may react with water to form bicarbonate ions and hence $CaCO_3$. The movement of C through the *endogenic* and *exogenic* cycles can be traced in Figure 22.3. The amounts of C in the reservoirs and the magnitudes of the fluxes are listed in Table 22.5.

Reduced C enters the crust from the mantle by the intrusion of magma and resides in igneous rocks or is exhaled through volcanoes directly into the atmosphere primarily as CO_2. The CO_2 of the atmosphere dissolves in the water of the hydrosphere and also enters the biosphere by photosynthesis and by absorption of carbonate ions through plant roots. The hydrosphere and biosphere in Figure 22.3 are each divided into oceanic and continental reservoirs. The bicarbonate ions of the hydrosphere are incorporated into sediment in the form of $CaCO_3$, whereas a fraction of the biospheric C is deposited as organic matter in sediment and soil. Sediment composed primarily of $CaCO_3$ forms *limestone,* whereas sediment composed of plant debris forms *coal.* In addition, the organic matter in shale may alter to *kerogen* from which *petroleum* and *natural gas* may form.

Once it has entered the *endogenic* cycle, the C in sedimentary rocks may be transformed into anthracite or graphite by thermal metamorphism of its sedimentary host, whereas limestones may recrystallize to form marble. Some C in sedimentary rocks may be incorporated into magma during partial melting and thus enters igneous rocks of the crust. In addition, C may return to the mantle by subduction of sedimentary and volcanic rocks.

The pathways of C in the exogenic cycle are *reversible,* whereas those of the endogenic cycle are generally *unidirectional* and *irreversible.* Once C has become enclosed in sedimentary, igneous, or metamorphic rocks, it can return to the exogenic cycle only as a result of weathering of its host, by venting of magma through volcanoes, or by decarbonation reactions of carbonate rocks at elevated temperatures.

If the geochemical machine in Figure 22.3 is in a steady state, the amounts of C in the various reservoirs and the fluxes between them are invariant with respect to time, with the exception of the long-term changes discussed before. Therefore, the present magnitudes of these reservoirs and fluxes can provide a valid description of how the C-cycle works at the present time and how it has worked in the not-too-distant geologic past. The data listed in Table 22.5 indicate that sediment in the oceans and on the continents, and the sedimentary rocks derived from them, together form the largest reservoir with 7643×10^{18} moles of C. The ocean contains only 3.3×10^{18} moles of C, whereas the oceanic and continental biospheres together contain even less at $\sim 0.39 \times 10^{18}$ moles of C. The atmosphere, with only 0.058×10^{18} moles of C, nevertheless plays an important role in the C-cycle because it allows C to pass between the hydrosphere and the biosphere. Therefore, it acts more as a conduit of C than as a reservoir and is the point of entry of C released into the atmosphere by volcanoes and by the combustion of fossil fuel by humans.

The fluxes of C between the atmosphere and the ocean are very large ($\sim 8000 \times 10^{12}$ mol/yr), which favors the establishment of a steady state. Similarly, the fluxes between the atmosphere and the continental biosphere ($\sim 4800 \times 10^{12}$ mol/yr) and between the ocean and the marine biomass ($\sim 3100 \times 10^{12}$ mol/yr) are quite large and likewise permit C to move freely among these reservoirs. As a result, changes in the C-content of one reservoir are rapidly transmitted to the others. For example, a reduction of the continental biomass releases C into the atmosphere as a result of bacterial degradation of organic matter. Consequently, the C-content of the atmosphere rises temporarily

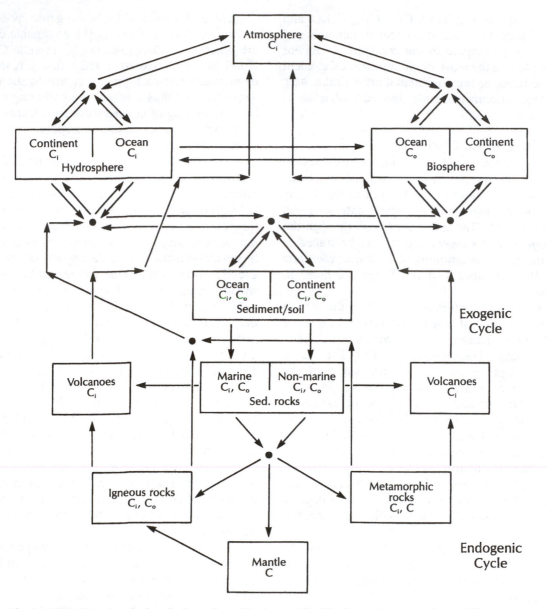

Figure 22.3 Geochemical cycle for carbon. The boxes identify the major reservoirs, and the arrows are the fluxes of C moving between them annually. The magnitudes of the C-reservoirs and fluxes are stated in Table 22.5. C_i = inorganic C (carbonate and CO_2), C_o = organic C (hydrocarbons, carbohydrates, etc.), C = native C (diamond and graphite). All other elements circulate through the exogenic and endogenic cycles in a similar way as C, except that the amounts in the reservoirs and the fluxes of other elements may be very different from those of C (adapted from Holser et al., 1988).

Table 22.5 Reservoirs and Fluxes of the Natural Carbon Cycle[a]

Reservoir		Flux		
Name	Amount, mol	From	To	Amount, 10^{12} mol/yr
Atmosphere	58×10^{15} C_i	atmosphere	ocean	8000 C_i
		atmosphere	biosphere (cont.)	4800 C_i
		atmosphere	sed. and soil (cont.)	35 C_i
Hydrosphere	3.3×10^{18} C_i	ocean	atmosphere	8042 C_i
(ocean)		ocean	biosphere (ocean)	3100 C_i
		ocean	sediment (ocean)	5 C_i
		ocean	sediment (cont.)	46 C_i
				8 C_o
Hydrosphere	2×10^{12} C_o	hydrosphere (cont.)	ocean	86 C_i
(continent)				18 C_o
Biosphere	130×10^{15} C_o	biosphere (ocean)	ocean	3090 C_o
(ocean)		biosphere (ocean)	sediment (ocean)	0.2 C_o
		biosphere (ocean)	sediment (cont.)	10 C_o
Biosphere	260×10^{15} C_o	biosphere (cont.)	hydrosphere (cont.)	10 C_o
(continent)		biosphere (cont.)	atmosphere	4790 C_i
Sediment/soil	1290×10^{18} C_i	sediment (ocean)	mantle	5 C_i
(ocean)	63×10^{18} C_o	sediment (ocean)	volcanoes	0.2 C_o
		sediment (ocean)	met. and igneous rocks	0.02 C_i
Sediment/soil	5170×10^{18} C_i	sediment (cont.)	hydrosphere (cont.)	86 C_i
(continent)	1120×10^{18} C_o			8 C_o
		sediment (cont.)	volcanoes	3 C_i
		sediment (cont.)	met. and igneous rocks	?
Metamorphic and igneous rocks	?	met. and igneous rocks	hydrosphere (cont.)	?
Volcanoes	?	volcanoes	atmosphere	3 C_i
Mantle	?	mantle	hydrosphere (ocean)	6 C_i

[a] C_i means inorganic C in carbonate or carbon dioxide; C_o means organic carbon.
SOURCE: Holser et al. (1988).

until some of the excess is transferred to the ocean. Carbon dioxide entering the ocean forms bicarbonate ions, which may cause $CaCO_3$ to precipitate. Therefore, a fraction of the C released by the continental biosphere is put in storage as $CaCO_3$ in marine sediment. This C could be retrieved by dissolution of carbonate sediment in the ocean in response to a decrease in the atmospheric C-content.

The rapid exchange of C among the exogenic reservoirs stabilizes the C-cycle by preventing excessive concentrations from building up in any one of the major reservoirs. This fact may moderate the release of CO_2 into the atmosphere by combustion of fossil fuel and may reduce the severity of the resulting changes in the atmospheric circulation and climatic conditions of the world.

b. The Water Cycle

Hydrogen and O are combined in the molecules of water in the hydrologic cycle, which operates primarily on the surface of the Earth (Section 4.4). In addition, O is the most abundant element in the continental crust (45.5%) and mantle of the Earth (44.1% in pyrolite, Table 4.2). Water not only makes up the hydrosphere but also occurs in common minerals, including amphiboles, micas, clay minerals, and in the oxides, sulfates, and carbonates of some of the abundant elements in the crust (Na, Mg, Al, Ca, Mn, Fe, etc.). In this section we consider only the water in the exogenic cycle of the hydrosphere. The presence of water on the surface of the Earth is an essential prerequisite to life, and its movement permits transport of sediment and ions produced by chemical reactions that can occur only in its presence.

The history of the atmosphere and oceans of the Earth has been discussed by many authors including Rubey (1951) whose work was a milestone in the solution of the problem of their origin. Most of the proposed models for the formation of the atmosphere and oceans have been based on the abundances of the noble gases in the atmosphere, including particularly Ar, which has a radiogenic isotope (^{40}Ar) that forms by decay of ^{40}K (Section 16.3a). Important contributions to the history of outgassing of the Earth are by Damon and Kulp (1958), Turekian (1959, 1974,) Fanale (1971), Hamano and Ozima (1978), Fisher (1978), Hart and Hogan (1978), Schwartzman (1978), and Holland (1984). The continuing quest for understanding indicates that the history of the oceans and atmosphere of the Earth is one of the most fundamental questions in the Earth Sciences.

Water was an original constituent of the planetesimals as water of hydration of certain minerals and as ice. Water was probably also contributed to the Earth by impacting comets or the so-called cometesimals (Section 2.2). Most of the water and other volatiles were initially in the interior of the Earth and may have been dissolved in a magma ocean during the early molten stage of the Earth. Most evidence supports the conclusion that the volatile elements and compounds were outgassed very early in the history of the Earth. A model by Holland (1984) indicates that about two thirds of the atmosphere was released at the time of accretion or very soon thereafter. About 30% of the water was subsequently consumed by chemical weathering and now resides in the continental crust. Nevertheless, Holland (1984) concluded that large bodies of water existed on the surface of the Earth more than ~4 billion years ago. This conclusion is strongly supported by the presence of (3.7×10^9)-year-old sedimentary rocks, including chemically precipitated carbonate rocks, at Isua in Greenland (Schidlowski et al., 1979; Appel, 1990; Baadsgaard et al., 1986).

The geochemical cycle of water therefore started more than 4×10^9 years ago with the release of water during cooling and crystallization of the mantle and later by the formation and subsequent upward movement of magma. Some water was trapped in igneous rocks, but most of it was initially discharged into the atmosphere by volcanic activity. Outgassing of the Earth has continued to the Present, but at a greatly diminished rate, as indicated by the discharge of primordial ^{3}He by submarine volcanic activity (Kurz et al., 1982).

The movement of water through the exogenic and endogenic cycle at the present time is outlined in Figure 22.4. Volcanoes still discharge water vapor, but almost all of it now originates from groundwater and by dehydration of crustal rocks. A small fraction of water in the upper atmosphere (4.8×10^{-4} km^3/yr, Walker, 1977) dissociates into its constituent gases and the H$_2$ subsequently escapes into space. The biosphere occupies a central place in the hydrologic cycle because it depends on the presence of water for its existence and because it receives and discharges water to the other reservoirs on the surface of the Earth. Figure 22.4 does *not* show that water contained in crustal rocks may be released during chemical weathering and that water may also return to the mantle by the subduction of wet sediment and hydrated volcanic rocks.

The ocean is by far the largest reservoir, containing 95.25% of all the water in the hydrosphere (Tables 4.3 and 22.6). Glaciers and continental ice sheets make up 2.05%, whereas groundwater

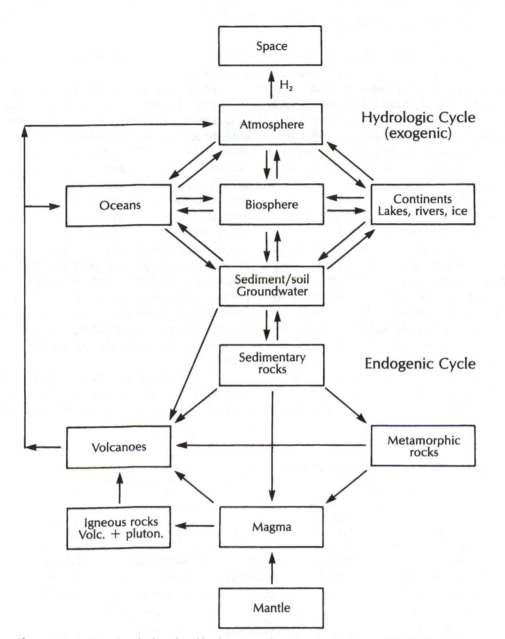

Figure 22.4 Geochemical cycle of hydrogen and oxygen in the form of H_2O. The water on the surface of the Earth was outgassed from the mantle by volcanic activity in Early Archean time. Since then, most of it has been circulating through the reservoirs of the *hydrologic cycle*. The biosphere occupies a central position in the cycle of water because it communicates with all other surface reservoirs and because it depends on water for its existence. Water enters the *endogenic* cycle via sedimentary rocks and may be stored in hydrous minerals of metamorphic and igneous rocks. Volcanoes continue to vent groundwater and water derived from sedimentary, igneous, and metamorphic rocks back into the atmosphere. The diagram does not show that water may return to the mantle by subduction of crustal rocks and that water in these rocks may be released into the hydrologic cycle by chemical weathering.

(shallow and deep), lakes, and rivers together contain only ~0.70% of the water on the surface of the Earth. The actual volume of fresh water that is available for domestic, agricultural, and industrial use is about 9.86×10^6 km^3.

Data cited by Berner and Berner (1987) indicate that the hydrologic cycle is driven largely by solar energy (0.5 cal cm^{-2} min^{-1}, 99.98%), with much smaller contributions by heat flow from the interior of the Earth (0.9×10^{-4} cal cm^{-2} min^{-1}, 0.018%), and from tides (0.9×10^{-5} cal cm^{-2} min^{-1}, 0.0012%). The amount of heat received by the Earth from the Sun must be equal to heat losses in order to maintain the thermal balance. According to Berner and Berner (1987), ~30% of the solar energy received by the Earth is reflected back into space as short-wave radiation ($\lambda < 4$ μm). The remainder is absorbed by the surface (51%) and by the atmosphere (water vapor, dust, ozone, CO$_2$, 16%; clouds, 3%). The amount of heat absorbed by the atmosphere depends on its chemical composition, including primarily the concentrations of CO$_2$, H$_2$O, CH$_4$, and certain fluorocarbon gases used as refrigerants and propellants in aerosol cans. (See Chapter 23.)

The input of energy from the Sun causes water to evaporate from the surface of the ocean. Some of that water is returned to the ocean by direct precipitation. However, the data in Table 22.6 indicate a net flux of 37,400 km^3/yr of water from the oceans to the continents in the form of atmospheric water vapor. This amount of water must return to the oceans annually in order to balance the cycle. The atmosphere over the continents also receives 72,900 km^3/yr of water by evaporation from the continents, so that the total amount of precipitation over the continents is 110,300 km^3/yr. Some of this water is stored temporarily in the polar ice sheets or glaciers and as groundwater. However, ~6% of the return flow to the oceans is by discharge of groundwater directly into the ocean (Berner and Berner, 1987, Table 2.2).

The ocean acts as a giant heat reservoir that moderates temperature fluctuations of the atmosphere. Therefore, the climate of the Earth, both

Table 22.6 Volumes and Fluxes of Water Within the Hydrologic Cycle

Reservoir		Flux		
Name	*Amount,* 10^6 km^3	*From*	*To*	*Amount,* km^3/yr
Ocean	1370	ocean	atmosphere	423,100
Ice sheets, glaciers	29	atmosphere	ocean	385,700
Groundwater (750–4000 m)	5.3	ocean	continents (water vapor)	37,400
Groundwater (<750 m)	4.2	continents	atmosphere	72,900
Lakes	0.125	atmosphere	continents	110,300
Soil moisture	0.065	continents	ocean	37,400
Atmosphere (liquid equivalent)	0.013	whole Earth	atmosphere	496,000
Rivers	0.0017	atmosphere	whole Earth	496,000
Biosphere	0.0006			
Total	1408.7			

SOURCE: Berner and Berner (1987).

globally and locally, depends on complex interactions between the atmosphere and the oceans. These interactions include feedback mechanisms, which may amplify small random perturbations into major events that affect global weather patterns. These events are virtually impossible to predict because they result from the *chaotic* behavior of the ocean–atmosphere system.

c. The Nitrogen Cycle

Nitrogen is strongly concentrated in the atmosphere, which contains 78.084 volume % (Table 4.8) or 2.82×10^{20} moles of nitrogen as $N_2(g)$ and much smaller amounts ($\sim 10^{12}$ moles) of six other molecular species: NH_3, NH, N_2O, NO, NO_2, and HNO_3 (Table 22.7). The biosphere in the oceans and on the continents contains 6.40×10^{16} and 5.52×10^{16} moles of N, respectively. In addition, the oceans contain 2.40×10^{16} moles of N in dissolved organic

Table 22.7 Inventory of Nitrogen Compounds in the Major Exogenic Reservoirs

Reservoir	Amount, mol N	Residence time
Atmosphere		
N_2	2.82×10^{20}	44×10^6 yr
NH_3	1.83×10^{12}	3–4 mo
NH	0.29×10^{12}	2–3 wk
N_2O	91×10^{12}	12–13 yr (?)
NO	0.22×10^{12}	1 mo (?)
NO_2	0.22×10^{12}	1–2 mo
HNO_3	small	2–3 wk
Biosphere		
Continents	5.52×10^{16}	
Ocean	6.40×10^{16}	
Hydrosphere (oceans)		
Dissolved N_2	1.57×10^6 (N)	
Dissolved organic	2.40×10^{16} (N)	
Sediment		
Inorganic N	14.2×10^{18}	
Organic N	57.2×10^{18}	400×10^6 yr

SOURCE: Garrels et al. (1975).

compounds but only 1.57×10^6 moles of N as dissolved N_2. Sediment and sedimentary rocks are a major sink for N containing 14.2×10^{18} moles of inorganic N and 57.2×10^{18} moles of N in organic compounds.

Nitrogen is an important element in the biosphere and its geochemical cycle (Delwiche, 1970) reflects the metabolic activity of bacteria (Section 19.7). In addition, the combustion of fossil fuel at high temperatures causes atmospheric N_2 to be converted to oxides, including NO_2, which is a toxic constituent of urban smog. Atmospheric N_2 is also used to make fertilizer containing NO_2^- or NO_3^-. However, the removal of N_2 from the atmosphere by this process does not threaten to reduce the atmospheric reservoir appreciably.

The geochemical cycle, outlined in Figure 22.5, starts with the "fixation" of atmospheric N_2 by bacteria, including *Azotobacter*:

$$2 N_2 + 6 H_2O \rightarrow 4 NH_3 \text{ (organism)} + 3 O_2 \tag{22.21}$$

The NH_3 is released following the death of the organism. Subsequent hydrolysis of NH_3 produces NH_4^+:

$$4 NH_3 + 4 H_2O \rightarrow 4 NH_4^+ + 4 OH^- \tag{22.22}$$

The NH_4^+ is oxidized to NO_2^- and NO_3^- by nitrifying bacteria (*Nitrosomonas* and *Nitrobacter*):

$$4 NH_4^+ + 6 O_2 \rightarrow 4 NO_2^- + 8 H^+ + 4 H_2O \tag{22.23}$$

$$4 NO_2^- + 2 O_2 \rightarrow 4 NO_3^- \tag{22.24}$$

The NO_2^- and NO_3^- as well as NH_4^+ are assimilated by plants through their roots and are incorporated into cell material as the amino radical NH_2. However, NO_3^- may also be reduced to N_2 by *denitrifying* bacteria (e.g., *Pseudomonas*):

$$4 NO_3^- + 2 H_2O \rightarrow 2 N_2 + 5 O_2 + 4 OH^- \tag{22.25}$$

This process is a form of anaerobic respiration from which bacteria derive O_2.

The biological cycling of N_2 is augmented by *abiologic* processes in the troposphere (<17 km

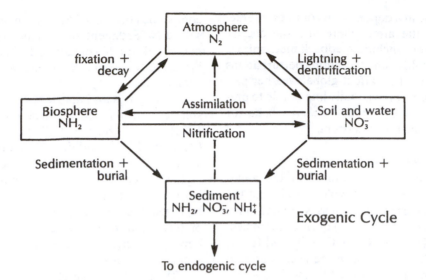

Figure 22.5 Exogenic cycle of nitrogen. The N_2 of the atmosphere is *fixed* by bacteria as NH_3, which hydrolyzes to NH_4^+, and is converted to NO_2^- and NO_3^- by *nitrification*. Nitrite and nitrate ions are *assimilated* by plants through their roots and are incorporated into plant tissue as NH_2. Nitrogen is also fixed by the conversion of N_2 to NO_3^- during lightning discharge. Some of the nitrogen is incorporated into sediment in the form of organic matter and as NH_4^+, which substitutes for K^+ in clay minerals and mica. Nitrogen returns to the atmosphere by decay of organic matter, which releases NH_3 and NH_4^+, and by the action of denitrifying bacteria in the soil, which convert NO_3^- to N_2 (adapted from Walker and Drever, 1988).

above the equator), which are initiated by the formation of nitric oxide (NO) during lightning discharge. The NO is subsequently converted to NO_3^- by photochemical reactions and is removed from the atmosphere in rain or snow. The nitrate produced by lightning in the atmosphere is assimilated by plants and thus enters the biosphere.

A small fraction of N in plant tissue enters the *endogenic cycle* when organic material is deposited in sedimentary basins and is ultimately buried and incorporated into sedimentary rocks by diagenesis and lithification. In addition, N may be present in sedimentary rocks as NH_4^+, which replaces K^+ in clay minerals and mica in accordance with Goldschmidt's rules (Section 8.1). The concentration of N in sedimentary rocks (200–4000 ppm), results primarily from the pres-

ence of organic matter. Igneous and metamorphic rocks contain even less N (1–20 ppm), presumably in mica minerals (Wlotzka, 1972).

The return flow of N from the biosphere to the atmosphere takes place during decay of organic matter, which releases NH_3. The fluxes of NH_3 to the atmosphere are about 5.4×10^{12} mol/yr from the continental biosphere and 2.9×10^{12} mol/yr from the oceans (Garrels et al., 1975). The concentration of NH_3 in the atmosphere is about 6 ppb, which gives a total NH_3 inventory of 1.83×10^{12} moles. The NH_3 is hydrolyzed to NH_4^+, which is rapidly removed from the atmosphere. Once the atmospheric NH_4^+ has entered the soil, it may be converted to NO_2^- and NO_3^- by nitrifying bacteria (equations 22.23 and 22.24), or it is assimilated by plants.

The release of nitric oxide (NO) into the atmosphere by high-temperature combustion of fossil fuel has undesirable environmental consequences because NO reduces the *ozone* concentration of the stratosphere located between 17 and 50 km above the surface of the Earth at the equator (Chapter 23).

22.5   The Sulfur Cycle

The geochemistry of S on the surface of the Earth is dominated by the activity of the anaerobic bacteria *Desulfovibrio desulfuricans,* which reduce sulfate to sulfide or, more rarely, to native S (Section 17.9). These bacteria occur widely in marine as well as lacustrine sediment and may also occur in the overlying water column, provided conditions are sufficiently anoxic (Eh < +0.100 V), as in the Black Sea and in the water of the Carioca Trench (Holser et al., 1988). In order to survive, the bacteria require a food supply, which they derive from the organic matter that is deposited with the sediment. The bacteria prefer simple hydrocarbons and short-chain carboxylic acids and may enter into symbiotic relationships with fermentative bacteria that break down complex organic molecules into edible form.

The S-reducing bacteria can tolerate a wide range of pH conditions and salt concentrations and are not poisoned by the sulfide they excrete. The reduction of sulfate takes place primarily a few centimeters below the sediment–water interface, but may continue to a depth of several meters. The sulfate is supplied by seawater trapped in the pores of marine sediment and by diffusion from the overlying water column. In lacustrine environments the sulfate originates by dissolution of marine evaporite deposits and by oxidation of sulfide minerals in the drainage basin.

The sulfide ions produced by the bacteria precipitate as iron sulfide, which recrystallizes to pyrite. The necessary Fe is derived from Fe-oxide coatings of detrital mineral particles, which release Fe^{3+} that must be reduced to Fe^{2+} by oxidation of organic carbon or sulfide. Therefore, the accumulation of pyrite in the sediment depends not only on the availability of HS^- but also on the availability of Fe^{2+}. When the supply of Fe^{2+} is inadequate to precipitate all of the sulfide that is produced, the bacteria may manufacture native S, as they seem to do in the caprock of salt domes.

The ratio of organic C to pyrite S in marine shales of different ages has varied with time, but shales from a particular period of geologic time tend to have similar C/S ratios (Raiswell and Berner, 1986). This relationship presumably arises because the rate of pyrite production depends on the supply of edible C-compounds, which make up a certain fraction of the total organic matter in the sediment (Goldhaber and Kaplan, 1974). As a result, the amount of S in pyrite and the amount of residual inedible C in marine shale are correlated. Unaltered and unmetamorphosed marine shales of Cambrian and Ordovician age have significantly lower average C/S ratios (0.5 ± 0.1) than shales of Devonian to Tertiary age (1.8 ± 0.5), presumably because of the absence of vascular land plants in early Paleozoic time. Vascular plants produce lignin and therefore are a major source of refractory C in shale of post-Ordovician age. Marine sediment of Holocene age has a mean C/S ratio of 2.8 ± 0.2.

The preanthropogenic S cycle depicted in Figure 22.6 has many reservoirs whose S contents are listed in Table 22.8. We start the cycle with the bacterial reduction of sulfate in the ocean and in bodies of fresh water. The reduced S is deposited as pyrite in marine and nonmarine shales. In addition, sulfate salts may be precipitated episodically in marine and nonmarine evaporite basins. The S sequestered in sedimentary rocks returns to the hydrosphere in the form of dissolved sulfate when the sedimentary rocks are exposed to chemical weathering.

Sulfur from the mantle may enter the hydrosphere as a result of alteration of basalt in the ocean, as a result of chemical weathering of mafic igneous rocks, and by the emission of SO_2 by volcanoes. However, the S discharged by volcanoes is partly derived from sedimentary rocks as well as from igneous and metamorphic rocks of the crust.

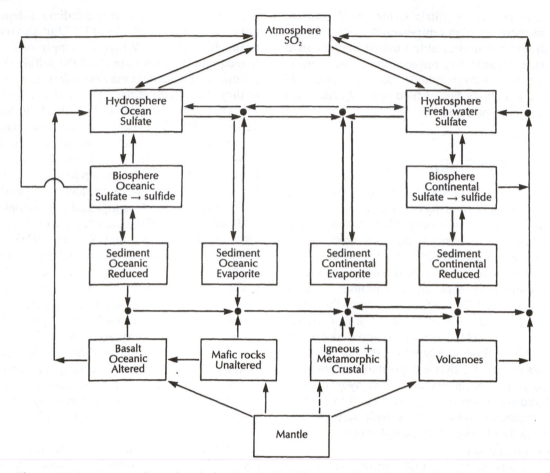

Figure 22.6 The natural geochemical cycle of sulfur. The exogenic cycle of S is dominated by the reduction of dissolved sulfate to sulfide by anaerobic bacteria (*Desulfovibrio*) living just below the sediment–water interface in the oceans and in bodies of fresh water. The sulfide is precipitated as pyrite in the presence of an adequate supply of Fe^{2+}. In addition, sulfate salts may form in marine and continental evaporite basins. The return flow of the S from sedimentary rocks to the hydrosphere takes place by chemical weathering. In the endogenic cycle, S from the mantle is brought into the crust by basalt and other mafic rocks from which it may be released by weathering and by alteration of basalt on the floor of the ocean. A third outlet for S from the interior of the Earth is by volcanic activity that may release a mixture of S derived partly from the mantle and partly from igneous, sedimentary, and metamorphic rocks of the continental and oceanic crust. The S released by volcanic activity may enter the hydrosphere directly or may enter the atmosphere first and be subsequently removed as sulfate in meteoric precipitation. The igneous and metamorphic rocks contain S derived primarily from oceanic or continental sediment with contributions from the mantle via volcanic rocks of basaltic composition. Anthropogenic effects on the S cycle are not shown here.

Table 22.8 Principal Reservoirs of Sulfur in the Endogenic and Exogenic Cycles

Reservoir	Moles of S	Principal form
Ocean	40×10^{18}	sulfate
Atmosphere	56×10^{9}	SO_2
Fresh water	40×10^{15}	sulfate
Sediment		
Oceanic	8×10^{18}	sulfide
Continental, reduced	145×10^{18}	sulfide
Continental, oxidized, evaporite	192×10^{18}	sulfate
Oceanic basalt, altered	25×10^{18}	sulfide
Mafic rocks, unaltered	106×10^{18}	sulfide
Igneous and metamorphic rocks	430×10^{18}	sulfide

SOURCE: Holser et al. (1988).

Figure 22.6 does not show that even seawater or continental brines may contribute part of the S that is recycled by volcanic activity. Volcanic emissions of S may contain H_2S, but it is rapidly oxidized during the eruption and enters the atmosphere as SO_2.

The atmosphere contains S derived primarily by volcanic activity, by exhalation from the biosphere, and by recycling of sulfates from the surface water of the ocean, from surficial salt deposits, and from brine lakes. Sulfur compounds of volcanic or biogenic origin are oxidized by reactions with O_2 and H_2O in the atmosphere to form sulfuric acid. The resulting acidification of rain and snow is a natural phenomenon that commonly follows large volcanic eruptions. Evidence for this process is preserved in the ice sheets of Antarctica and Greenland in the form of layers of ice containing tephra overlain by ice containing excess sulfate. Evidently, the S emitted by major volcanic eruptions remains in the atmosphere somewhat longer than the tephra, which are deposited downwind from their source (Palais and Kyle, 1988).

The S cycle has been significantly disturbed by the emission of SO_2 into the atmosphere from coal-burning electric power plants and by roasting of sulfide ores. In addition, mining operations and construction projects cause exposure of sulfide-bearing rocks, which accelerates chemical weathering and causes localized release of H_2SO_4 into streams, lakes, and groundwater (Sections 14.5d and 21.2).

The isotope composition of S processed by bacteria is altered because of the preferential enrichment of the metabolic H_2S in ^{32}S (Section 17.9). As a result, sedimentary sulfide deposits may be variably enriched in ^{32}S, whereas marine sulfate deposits are enriched in ^{34}S and depleted in ^{32}S. The $\delta^{34}S$ values of marine sulfate deposits have varied systematically throughout geologic time (Figure 17.6) in response to secular variations in the isotope compositions of the inputs and outputs of S to and from the ocean. The models that have been proposed to account for the observed variation of the $\delta^{34}S$ value of marine sulfate deposits were reviewed by Holser et al. (1988). We focus attention here on the possibility that the alteration of the isotope composition of S by bacteria can be used to detect the first appearance of S-reducing bacteria on the Earth in Early Archean time.

Measurement by Monster et al. (1979) indicate that sulfide minerals in different facies of the banded iron formation at Isua in Greenland have an average $\delta^{34}S$ value of $+4.5 \pm 0.90‰$ relative to S in the Canyon Diablo troilite. Associated tuffaceous amphibolites have $\delta^{34}S = +0.27 \pm 0.91‰$, and sulfide veins in the iron formation and amphibolite yielded $\delta^{34}S = +0.78 \pm 0.63‰$. These values are *indistinguishable* from each other and from meteoritic S, which represents S in igneous rocks derived from the mantle. Therefore, Monster et al. (1979) concluded that sulfate-reducing bacteria had *not* yet evolved 3.7×10^9 years ago. The reason may be that these bacteria require an ample supply of sulfate, which was not available until photosynthetic organisms had begun to release O_2 into solution in standing bodies of water. The first appearance of S-reducing bacteria occurred prior to about 2.0 billion years ago, but the precise timing of this event is still controversial (Schidlowski, 1979; Cameron, 1983).

22.6 Summary

Geological and geochemical processes have been operating on the surface of the Earth throughout geologic time. Therefore, we may assume that these processes are in a steady state such that the chemical compositions of the major reservoirs have remained constant except for long-term changes caused by the evolution of life, convection in the mantle, and the decay of long-lived radioactive nuclides to stable daughters.

The assumptions can be tested by equating the mass of an element released by chemical weathering of igneous rocks with the sum of its masses in sedimentary rocks and in the ocean at the present time. The resulting mass-balance equation can be evaluated for each element using its average concentrations in the three principal reservoirs. The results yield satisfactory balances for many elements with some notable exceptions, including S, Cl, Ca, Hg, and 14 other elements. All of these elements have *higher* concentrations in the oceans and in sedimentary rocks than can be accounted for by the mass of igneous rocks that has weathered.

Another kind of balance involves a comparison of the mass of an element entering the ocean annually and its rate of removal from the ocean. The annual inputs by rivers must be corrected for recycling of marine salts, which are transported to the continents as atmospheric aerosols, and for anthropogenic contamination of river water. Outputs from the ocean take the form of burial of pore water, ion exchange reactions, diagenesis and precipitation of salts all of which are difficult to evaluate quantitatively. Nevertheless, satisfactory input–output balances can be demonstrated for 30 elements, whereas 14 elements have *larger inputs* than outputs.

The group of elements with unsatisfactory oceanic input–output balances includes some of the same elements that have unsatisfactory mass balances (weathered granite = sedimentary rocks + oceans): S, Cl, Hg, as well as Na, P, Cd, and Pb. These results suggest that the concentrations of Na, Cl, and S in seawater are actually *increasing* at the present time. The excess inputs of other elements may only be apparent because they may be caused by recycling of marine salt (Hg and Cd), by anthropogenic contamination of river water, or by bad data.

The geochemical balances of the volatile elements (C-H-O-N and noble gases) cannot be evaluated in terms of their concentrations in rocks or water alone because significant proportions of these elements reside in the atmosphere and in the biosphere. Their complete geochemical cycles include both an *endogenic* (magma, igneous, sedimentary, and metamorphic rocks) and an *exogenic* (atmosphere, biosphere, ocean, sediment/soil) subcycle. If the amounts of an element in all of the available reservoirs and the fluxes between them are known, then it is possible to describe the complete geochemical cycle of that element and to calculate residence times in each reservoir. Such complete descriptions of geochemical cycles are difficult to achieve partly because of the complex interactions among reservoirs and partly because of anthropogenic intervention in the natural operation of the geochemical cycles of many elements. Nevertheless, even qualitative descriptions of the circulation of an element among its exogenic and endogenic reservoirs promote *understanding* of the interactions among the reservoirs.

The discussion of the exogenic cycles of the C-H-O-N group of elements leads to important questions about the CO_2 content of the atmosphere, the history of outgassing of the Earth, and the role of bacteria in the global circulation of N. The exogenic cycle of S is likewise dominated by the bacterial reduction of sulfate to sulfide, which is accompanied by significant fractionation of the stable isotopes of S. A study of the isotope composition of S in pyrite in rocks of Archean age indicates that bacterial reduction of sulfate *postdates* photosynthesis because the bacteria require sulfate ions that could not form until after green plants had begun to release O_2 into solution in standing bodies of water on the surface of the Earth.

Problems

1. Check equation 22.9 for proper exchange of electrons and material balance.

2. Determine whether this reaction can be in equilibrium in seawater at 25°C and 1 atm pressure (use data from Table 4.7 and Appendix B).

3. Calculate the total amount of Na^+ in the ocean (Table 4.7), mass of seawater = 1.4×10^{24} g).

4. How long would it take to increase the concentration of Na^+ in seawater by 10% based on the present anthropogenic flux (Table 22.2), assuming that the increase was not counteracted by increases in the output?

5. Calculate the average residence time of water (liquid equivalent) in the atmosphere based on data in Table 22.6.

6. Calculate a mass balance for K using data from Tables 4.5 and 4.7. Assume that average granitic rocks consist of 35% basalt and 65% granite and that granite is the average of high-Ca and low-Ca granite.

7. Calculate the output/input ratio for Fe entering and leaving the ocean using equations 22.18 and 22.19 and the following data (Drever et al., 1988):

	ppm
Fe_{sp}	48,000
Fe_{rw}	0.0388
Fe_{ds}	65,000
Fe_{sh}	47,272

8. Draw the endogenic and exogenic cycles of P including its major reservoirs and pathways.

References

APPEL, P. W. U., 1980. On the Early Archean Isua iron-formation. *Precamb. Res.,* **11**:73–87.

BAADSGAARD, H., A. P. NUTMAN, and D. BRIDGWATER, 1986. Geochronology and isotopic variation of the Early Archaean Amitsoq gneisses of the Isukasis area, southern West Greenland. *Geochim. Cosmochim. Acta,* **50**:2173–2184.

BERNER, E. K., and R. A. BERNER, 1987. *The Global Water Cycle.* Prentice-Hall, Upper Saddle River, NJ, 397 pp.

CAMERON, E. M., 1983. Genesis of Proterozoic iron formation: Sulphur isotopic evidence. *Geochim. Cosmochim. Acta,* **47**:1069–1074.

DAMON, P. E., and J. L. KULP, 1958. Inert gases and the evolution of the atmosphere. *Geochim. Cosmochim. Acta,* **13**:280–292.

DELWICHE, C. C., 1970. The nitrogen cycle. *Sci. Amer.,* Sept., 137–146.

DREVER, J. I., Y.-H. LI, and J. B. MAYNARD, 1988. Geochemical cycles: the continental crust and the oceans. In C. B. Gregor, R. M. Garrels, F. T. Mackenzie, and J. B. Maynard (Eds.), *Chemical Cycles in the Evolution of the Earth,* 17–53. Wiley, New York, 276 pp.

EDMOND, J. M., J. B. CORLISS, and L. I. GORDON, 1979. Ridge-crest hydrothermal metamorphism at the Galápagos spreading center and reverse weathering. In M. Talwani, C. G. Harrison, and D. E. Hayes (Eds.), *Deep Drilling Results in the Atlantic Ocean: Ocean Crust,* 383–390. American Geophysical Union, Washington, DC.

FANALE, F. P., 1971. A case for catastrophic early degassing of the Earth. *Chem. Geol.,* **8**:79–105.

FAURE, G., 1986. *Principles of Isotope Geology,* 2nd ed. Wiley, New York, 589 pp.

FISHER, D. E., 1978. Terrestrial potassium and argon abundances as limits to models of atmospheric evolution. In E. C. Alexander, Jr., and M. Ozima (Eds.), *Terrestrial Rare Gases,* 173–183. Japanese Scientific Societies Press, Tokyo.

GARRELS, R. M., and F. T. MACKENZIE, 1971. *Evolution of Sedimentary Rocks. A New Approach to the Study of Material Transfer and Change Through Time.* Norton, New York, 397 pp.

GARRELS, R. M., F. T. MACKENZIE, and C. HUNT, 1975. *Chemical Cycles and the Global Environment: Assessing Human Influences.* Kaufmann, Los Altos, CA, 206 pp.

GOLDHABER, M. B., and I. R. KAPLAN, 1974. Mechanisms of sulfur incorporation and isotope fractionation during early diagenesis in sediments of the Gulf of California. *Marine Chem.,* **9**:95–143.

GREGOR, C. B., R. M. GARRELS, F. T. MACKENZIE, and J. B. MAYNARD (Eds.), 1988. *Chemical Cycles in the Evolution of the Earth.* Wiley, New York, 276 pp.

HAMANO, Y., and M. OZIMA, 1978. Earth-atmosphere evolution model based on Ar isotopic data. In E. C. Alexander, Jr., and M. Ozima (Eds.), *Terrestrial Rare Gases,* 155–171. Japanese Scientific Societies Press, Tokyo.

HART, R., and L. HOGAN, 1978. Earth degassing models and the heterogeneous vs. homogeneous mantle. In E. C. Alexander, Jr., and M. Ozima (Eds.), *Terrestrial Rare Gases,* 193–206. Japanese Scientific Societies Press, Tokyo.

HOLEMAN, J. N., 1968. The sediment yield of major rivers of the world. *Water Resources Res.,* **4**:737–747.

HOLLAND, H. D., 1984. *The Chemical Evolution of the Atmosphere and Oceans.* Princeton University Press, Princeton, NJ, 568 pp.

HOLSER, W. T., M. SCHIDLOWSKI, F. T. MACKENZIE, and J. B. MAYNARD, 1988. Geochemical cycles of carbon and sulfur. In C. B. Gregor, R. M. GARRELS, F. T. Mackenzie, and J. B. Maynard (Eds.), *Chemical Cycles in the Evolution of the Earth,* 105–173. Wiley, New York, 276 pp.

HSÜ, K. J., 1983. *The Mediterranean Was a Desert.* Princeton University Press, Princeton, NJ, 197 pp.

KU, T. L., W. B. BROECKER, and N. OPDYKE, 1968. Comparison of sedimentation rates measured by paleomagnetic and the ionium methods of age determination. *Earth Planet. Sci. Lett.,* **4**:1–16.

KURZ, M. D., W. J. JENKINS, J.-G. SCHILLING, and S. R. HART, 1982. Helium isotopic variations in the mantle beneath the central North Atlantic Ocean. *Earth Planet. Sci. Lett.,* **58**:1–14.

LERMAN, A., and M. MEYBECK (Eds.), 1988. *Physical and Chemical Weathering in Geochemical Cycles.* NATO ASI Series, Mathematical and Physical Sciences, vol. 251. Kluwer Acad. Publishers, Dordrecht, The Netherlands, 375 pp.

LI, Y.-H., 1972. Geochemical mass balance among lithosphere, hydrosphere and atmosphere. *Amer. J. Sci.,* **272**:119–137.

MEYBECK, M., 1979. Concentrations des eaux fluviales en éléments majeurs et apports en solution aux oceans. *Rev. Geol. Dyn. Geogr. Phys.,* **21**:215–246.

MONSTER, J., P. W. U. APPEL, H. G. THODE, M. SCHIDLOWSKI, C. M. CARMICHAEL, and D. BRIDGWATER, 1979. Sulfur isotope studies in early Archean sediments from Isua, West Greenland: Implications to the antiquity of bacterial sulfate reduction. *Geochim. Cosmochim. Acta,* **43**:405–413.

PALAIS, J. M., and P. R. KYLE, 1988. Chemical composition of ice containing tephra layers in the Byrd Station ice core, Antarctica. *Quat. Res.,* **30**:315–330.

RAISWELL, R., and R. A. BERNER, 1986. Pyrite and organic matter in Phanerozoic normal marine shales. *Geochim. Cosmochim. Acta,* **50**:1967–1976.

RONOV, A. B., and A. A. YAROSHEVSKY, 1976. A new model for the chemical structure of the Earth's crust. *Geochem. Int.,* **13**(6):89–121.

RUBEY, W. W., 1951. Geologic history of seawater: An attempt to state the problem. *Geol. Soc. Amer. Bull.,* **62**:1111–1147.

SALOMONS, W., and U. FÖRSTNER, 1984. *Metals in the Hydrocycle.* Springer-Verlag, Berlin, 349 pp.

SAYLES, F. L., 1979. The composition and diagenesis of interstitial solutions: I. Fluxes across the seawater–sediment interface in the Atlantic Ocean. *Geochim. Cosmochim. Acta,* **43**:527–546.

SCHIDLOWSKI, M., 1979. Antiquity and evolutionary status of bacterial sulfate reduction: Sulfur isotope evidence. In *Origins of Life,* **9**:299–311. Reidel, Dordrecht and Boston.

SCHIDLOWSKI, M., P. W. U. APPEL, R. EICHMANN, and C. E. JUNGE, 1979. Carbon isotope geochemistry of the 3.7×10^9-yr old Isua sediments, West Greenland. Implications for the Archaean carbon and oxygen cycles. *Geochim. Cosmochim. Acta,* **43**:189–200.

SCHWARTZMAN, D. W., 1978. On the ambient mantle $^4He/^{40}Ar$ ratio and the coherent model of degassing of the Earth. In E. C. Alexander, Jr., and M. Ozima (Eds.), *Terrestrial Rare Gases,* 185–191. Japanese Scientific Societies Press, Tokyo.

STUMM, W. (Ed.), 1977. *Global Geochemical Cycles and Their Alterations by Man.* Heyden & Son, Inc., Philadelphia, 347 pp.

SVERDRUP, H. N., M. W. JOHNSON, and R. H. FLEMING, 1942. *The Oceans.* Prentice-Hall, Upper Saddle River, NJ, 1087 pp.

TUREKIAN, K. K., 1959. The terrestrial economy of helium and argon. *Geochim. Cosmochim. Acta,* **17**:37–43.

TUREKIAN, K. K., 1964. Degassing of argon and helium from the Earth. In P. J. Brancazio and A. G. W. Cameron (Eds.), *The Origin and Evolution of Atmospheres and Oceans,* 74–82. Wiley, New York.

WALKER, J. C. G., 1977. *Evolution of the Atmosphere.* Macmillan, New York, 318 pp.

WALKER, J. C. G., and J. I DREVER, 1988. Geochemical cycles of atmospheric gases. In C. B. Gregor, R. M. Garrels, F. T. Mackenzie, and J. B. Maynard (Eds.), *Chemical Cycles in the Evolution of the Earth,* 55–76. Wiley, New York, 276 pp.

WLOTZKA, F., 1972. Nitrogen. In K. H. Wedepohl (Ed.), *Handbook of Geochemistry,* Sections E and K. Springer-Verlag, Berlin.

WOLERY, T. J., and N. H. SLEEP, 1976. Hydrothermal circulation and geochemical flux at mid-ocean ridges. *J. Geol.,* **84**:249–275.

WOLERY, T. J., and N. H. SLEEP, 1988. Interactions of geochemical cycles with the mantle. In C. B. Gregor, R. M. Garrels, F. T. Mackenzie, and J. B. Maynard (Eds.), *Chemical Cycles in the Evolution of the Earth,* 77–103. Wiley, New York, 276 pp.

23

Chemistry of the
Atmosphere

Until recently, the atmosphere of the Earth has not received the attention from geochemists it deserves, even though our very lives depend on the availability of oxygen in the air we breathe. Our complacency about the atmosphere was shattered by the publication of a paper by Farman et al. (1985) in which they reported a drastic decline in the ozone (O_3) content of the stratosphere in the austral spring (October to November) at Halley Bay in Antarctica. This discovery raised worldwide concern because ozone contributes significantly to the absorption of ultraviolet light, which can cause skin cancer in humans and may have other harmful biological effects.

Another danger signal is the steady increase of the carbon dioxide (CO_2) concentration of the atmosphere recorded continuously since 1958 at the Mauna Loa observatory in Hawaii (Keeling et al., 1982). Ice cores collected at Siple Station in Antarctica (Friedli et al., 1986) extended this geochronological record to about 1750 A.D. and demonstrate that the concentration of carbon dioxide in the atmosphere has been rising at an accelerating rate from about 275 ppmv (parts per million by volume) in 1750 A.D. to about 350 ppmv in 1988. Increases have also been recorded in the concentrations of methane (CH_4) (Khalil and Rasmussen, 1987), carbon monoxide (CO) (Zander et al., 1989), and nitrous oxide (NO_x) (Pearman et al., 1986). These gases, as well as water vapor and ozone, absorb infrared radiation ($\lambda > 750$ nanometers)

emitted by the surface of the Earth, thus causing the atmosphere to heat up in the so-called greenhouse effect.

A third reason for the rising interest in the atmosphere of the Earth is the exploration of the earthlike planets and the satellites of the gaseous planets of the solar system. The atmospheres of Venus and Mars are different from that of Earth in composition, temperature, and density, whereas Mercury and the Moon lack atmospheres altogether (excluding very small concentrations of He on the former, Lewis and Prinn, 1983, p. 121). The exploration of the solar system has therefore focused attention on the evolution of planetary atmospheres (Lewis and Prinn, 1983; Chamberlain and Hunten, 1987) and on the origin of the atmosphere of Earth (Holland, 1979, 1984; Holland et al., 1986).

Our increased awareness of the place of the Earth in the solar system has added urgency to the study of impacts by meteoroids, asteroids, and comets on the Earth. The energy released by such events cause catastrophic temporary changes in the atmosphere, including an increase in the dust content, global fire storms, and acid rain (Prinn and Fegley, 1987). The traumatic effect on the biosphere of such sudden environmental changes may have been responsible for large-scale extinctions of species that were unable to cope with the environmental consequences of these celestial accidents. (Silver and Schultz, 1982; Alvarez et al., 1984; Fassett and Rigby, 1987).

The growing realization of the importance of atmospheric processes is reflected in the publication of monographs on the chemistry of the atmosphere (Warneck, 1988) and of interdisciplinary textbooks concerned with climate change on the Earth (Graedel and Crutzen, 1993).

23.1   Structure and Composition

The vertical structure of the atmosphere in Figure 23.1 is based on the observed variation of the temperature as a function of altitude above sea level and consists of the troposphere (0–12 km), stratosphere (12–50 km), mesosphere (50–100 km), and the thermosphere (100 to >1000 km). The boundaries between these atmospheric shells are called the tropopause, stratopause, etc. The chemical compositions of the troposphere and the stratosphere affect the climate and hence the living conditions on the surface of the Earth in ways to be discussed in this chapter.

The temperature in the troposphere decreases with altitude from 288.15 K (15.0°C) at the surface to 216.65 K (−56.5°C) at the tropopause (12 km) (Anonymous, 1976). The temperature in the stratosphere actually *increases* with altitude and reaches 270.65 K (−2.5°C) at the stratopause 50 km above the surface of the Earth. In the overlying mesosphere the temperature decreases again to 186.87 K (−86.28°C), but rises in the thermosphere to values in excess of 373.15 K (100°C). However, the thermosphere contains very little heat because of its low density.

The pressure of the atmosphere decreases with altitude from 1010 millibars (mbar) at the surface to 194 mbar at 12 km (tropopause), to 0.800 mbar at 50 km (stratopause), and to 0.00184 mbar at 90 km (mesopause). Therefore, the pressure range of the stratosphere (194 to 0.800 mbar) includes the atmospheric pressure on the surface of Mars of about 8 mbar.

The atmospheric pressure can also be expressed as the "number density" in terms of the number of molecules per cubic centimeter of space. The number density of the atmosphere decreases from 2.55×10^{19} molecules/cm^3 at the surface to 6.49×10^{18} molecules/cm^3 at the tropopause (12 km) and to 2.14×10^{16} molecules/cm^3 at the stratopause (50 km). In other words, the number of molecules in a cubic centimeter of space 50 km above the surface of the Earth is about 1200 times smaller than at the surface of the Earth. The lower number density of the stratosphere reduces the frequency of molecular collisions and hence affects the rates of chemical reactions.

The chemical composition of the atmosphere can be described in terms of only 10 constituents listed in Table 23.1. (See also Table 4.8.) The data indicate that N_2 and O_2 make up 99.0% of the atmosphere followed by the noble gases at

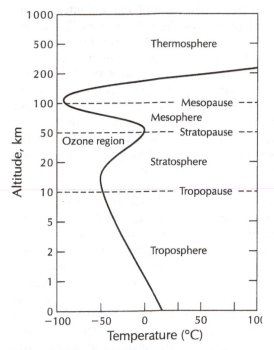

Figure 23.1 The structure of the atmosphere based on the variation of the temperature (°C) with altitude (km) above the surface of the Earth. Note that the boundaries between the major units are placed at or close to reversals in the temperature gradient and that the altitude scale is logarithmic rather than linear. [Modified from Graedel and Crutzen (1993) who adapted it from Chameides and Davis (1982).]

Table 23.1 Chemical Composition of Dry Air at the Surface in Remote Continental Areas

Constituent	Symbol or formula	Concentration, %
Nitrogen	N_2	78.1
Oxygen	O_2	20.9
Argon	Ar	0.93
Carbon dioxide	CO_2	0.035
Neon	Ne	0.0018
Helium	He	0.0005
Methane	CH_4	0.00017
Krypton	Kr	0.00011
Hydrogen	H_2	0.00005
Ozone	O_3	0.000001–0.000004

SOURCE: Graedel and Crutzen (1993).

0.932% and CO_2 with 0.035%. These gases together amount to 99.967% of the atmosphere, leaving only 0.0326% for methane, hydrogen, and ozone. The compounds in the atmosphere participate in chemical reactions when molecules are energized or broken up by ultraviolet radiation from the Sun. Therefore, chemical reactions in the stratosphere reduce the amount of ultraviolet radiation that reaches the surface of the Earth and thus protect the biosphere from the harmful effects of this radiation.

23.2   Ultraviolet Radiation

The Sun emits electromagnetic radiation having a wide range of wavelengths (λ) consistent with its surface temperature of 5780 K. The peak intensity occurs at $\lambda \sim 500$ nanometers (nm; 1nm = 10^{-9}m). The wavelength spectrum of solar radiation consists of three types of radiation:

ultraviolet, $\lambda < 400$ nm

visible light, $\lambda = 400$ to 750 nm

infrared, $\lambda > 750$ nm

The energy (E) transmitted by electromagnetic radiation is related to its wavelength through the equation:

$$E = h\nu \qquad (23.1)$$

where h is Planck's constant (6.626×10^{-34} Js) and ν is the frequency, equal to c/λ, where c is the speed of light (2.998×10^8 m/s). Therefore, the energy of electromagnetic radiation of wavelength λ, stated in Chapter 5 equation 5.16, is:

$$E = \frac{hc}{\lambda} \qquad (23.2)$$

Ultraviolet radiation is further subdivided into UV-A(400–320 nm), UV-B(320–290 nm), and UV-C(<290 nm). The most energetic ultraviolet light (UV-C) is absorbed primarily by N_2O(20–100 nm), O_2 (80–210 nm), and O_3 (ozone) (200–290 nm). The presence of these molecules in the stratosphere therefore prevents UV-C from reaching the surface of the Earth.

The efficiency with which molecules absorb ultraviolet radiation is expressed by the absorption cross section, defined as a quantitative measure of the absorption of ultraviolet light at a specified wavelength by atoms or molecules in the atmosphere. Ozone plays an important role in absorbing UV-C because its absorption cross section for ultraviolet light (Figure 23.2) rises to a maximum at $\lambda = 250$ nm (Hartley band) and then declines to low values above $\lambda = 290$ nm. The decrease of the absorption cross section of ozone for wavelengths between 290 and 320 nm causes the magnitude of the UV-B flux received by the surface of the Earth to increase with increasing wavelength by a factor of about 10^4. Since the least energetic form of ultraviolet light (UV-A) is not significantly absorbed by the atmosphere, most incident UV-A reaches the surface. The biosphere has evolved a tolerance for UV-A, but the sensitivity of DNA increases rapidly with decreasing wavelength of UV-B. Therefore, the reduction of the ozone concentration in the stratosphere increases the exposure of the biosphere to the harmful effects of UV-B.

Research results summarized by Graedel and Crutzen (1993, p. 262) indicate that many plants

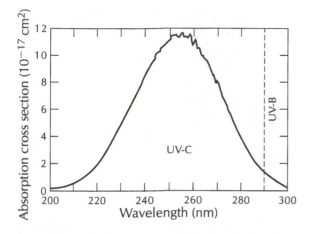

Figure 23.2 Efficiency of absorption of the most energetic ultraviolet light (UV-C) by ozone expressed here as the absorption cross section. The diagram demonstrates that the ozone of the stratosphere helps to shield the surface of the Earth from UV-C. The efficiency of ozone as an absorber of UV-B (290–320 nm) is much less than for the more damaging UV-C. [Modified from Graedel and Crutzen (1993) who reproduced it from Brasseur and Solomon (1986).]

are adversely affected by exposure to UV-B, showing reductions in the size of leaves, in the length of shoots, and in the rate of photosynthesis. For example, a 25% reduction in the concentration of stratospheric ozone is projected to cause a 50% reduction in soybean yields. Since planktonic organisms in the oceans are likewise sensitive to UV-B, they live at a depth below the surface that permits only long-wave UV radiation needed for photosynthesis to penetrate. An increase in the UV-B flux may force plankton to live at greater depth, thereby causing a reduction of the food supply for organisms higher in the food chain, including invertebrates, fish, and whales. Whether the increase in UV-B radiation caused by the present ozone depletion of the stratosphere over Antarctica has affected plankton in the Southern Ocean is not yet known.

Exposure of unprotected humans to UV-B causes cancer, increases the occurrence of cataracts, and weakens the immune system. According to Graedel and Crutzen (1993, p. 262), a 1% reduction in the concentration of stratospheric ozone increases the UV-B flux at the surface of the Earth by 2% which raises the incidence of basal-cell carcinomas by 4% and the occurrence of squamous-cell carcinomas by 6%. In addition, exposure to UV-B also causes the more dangerous melanoma skin cancers.

23.3 Ozone in the Stratosphere

When molecular O_2 in the stratosphere absorbs UV-C, it dissociates into two atoms of O:

$$O_2 + h\nu \rightarrow O + O \qquad (23.3)$$

where $h\nu$ is the energy of the UV radiation having $\lambda < 240$ nm. Each of the two free oxygen atoms then reacts with diatomic oxygen molecules to form ozone (O_3):

$$O_2 + O + M \rightarrow O_3 + M \qquad (23.4)$$

resulting in a net transformation of:

$$3\,O_2 + h\nu \rightarrow 2\,O_3 \qquad (23.5)$$

Reaction 23.4 requires the presence of a molecule (M), such as N_2 or O_2, that absorbs the energy liberated by this reaction and disperses it by collisions with other molecules in the surrounding space.

The ozone produced by this process absorbs UV-C most effectively at $\lambda \simeq 250$ nm (Hartley band). As a result, the ozone dissociates:

$$O_3 + h\nu \rightarrow O_2 + O \qquad (23.6)$$

followed by:

$$O + O_3 \rightarrow 2\,O_2 \qquad (23.7)$$

for a net change of:

$$2\,O_3 \rightarrow 3\,O_2 \qquad (23.8)$$

The process outlined above is modified by the catalytic destruction of ozone in the stratosphere caused by the presence of certain trace constituents, symbolized by $X\cdot$ and $XO\cdot$, where $X = NO$, HO, or Cl. For example, N_2O emitted by soil bacteria passes through the troposphere with a residence time of 150–200 years. Once it reaches the stratosphere, the N_2O molecules are decomposed into NO by several processes, including:

$$N_2O + O(^1D) \rightarrow 2\,NO \qquad (23.9)$$

where $O(^1D)$ is an oxygen atom that has been energized by ultraviolet radiation.

The NO radical then reacts with ozone in a sequence of steps that transform it into diatomic oxygen"

1. $NO\cdot + O_3 \rightarrow NO_2\cdot + O_2$

2. $O_3 + h\nu \rightarrow O + O_2$

3. $O + NO_2 \rightarrow NO\cdot + O_2$

4. $2\,O_3 + h\nu \rightarrow NO\cdot + 3\,O_2$ (23.10)

The net result (step 4) is the same as that of reactions (23.6–23.8). Although the catalysts listed above (NO, HO, and Cl) are present in the stratosphere in very small concentrations under natural conditions, the destruction of ozone in the stratosphere is accelerated when the concentrations of these catalysts are increased as a result of anthropogenic activities. For example, NO may be injected into the stratosphere by high-flying aircraft or by nuclear explosions in the atmosphere. In addition, the catalytic destruction of ozone in the stratosphere is linked to the release of chlorofluorocarbon gases used in refrigerators and as propellants in aerosol spray cans. These compounds, exemplified by $CFCl_3$ and CF_2Cl_2, are very stable and are not soluble in water. The so-called "CFCs" are not removed from the troposphere by meteoric precipitation, but rise to altitudes of about 25 km in the stratosphere where they are decomposed by UV-C in reations such as:

$$2\,CFCl_3 + h\nu + \tfrac{3}{2}O_2 + H_2O \rightarrow$$
$$2\,CO_2 + 2\,HF + 6\,Cl \qquad (23.11)$$

The Cl acts as a catalyst in the destruction of ozone by the sequence of reactions:

1. $Cl\cdot + O_3 \rightarrow ClO\cdot + O_2$

2. $O_3 + h\nu \rightarrow O + O_2$

3. $ClO\cdot + O \rightarrow Cl\cdot + O_2$ (23.12)

thereby causing the conversion of $2\,O_3$ into $3\,O_2$ (Molina and Rowland, 1974).

The catalytic "ozone killers" may be neutralized in reactions such as:

$$HO\cdot + NO_2 + M \rightarrow HNO_3 + M \qquad (23.13)$$
$$ClO\cdot + NO_2 + M \rightarrow ClONO_2 + M \qquad (23.14)$$

where $ClONO_2$ is a molecule of chorine nitrate. However, these compounds are short-lived reservoirs of $HO\cdot$ and $ClO\cdot$ because the HNO_3 and $ClONO_2$ molecules are decomposed by ultraviolet radiation.

Chlorine is supplied to the stratosphere by sea spray, volcanic eruptions, and in the form of methyl chloride gas (CH_3Cl) released by seaweed. According to Graedel and Crutzen (1993, p. 143), the Cl concentration of the stratosphere has increased by a factor of five, from 0.6 to about 3 ppbv (parts per billion by volume), as a consequence of anthropogenic emissions of CFCs and other Cl-bearing industrial gases. The increase in the Cl concentration of the stratosphere caused by the release of CFC gases has contributed to a decrease of the ozone concentration of the atmosphere, especially over Antarctica.

23.4 Ozone Hole Above Antarctica

The abundance of ozone in a column of air having a diameter of 1.0 cm^2 varies seasonally and with latitude. In the equatorial region of the Earth, the ozone content of the atmosphere is about 250 Dobson units (DU), where:

$$1\,DU = 2.69 \times 10^{16} \text{ molecules of } O_3/cm^2$$

The DU is defined by the statement that 100 DU corresponds to a layer of ozone 10 mm thick at a pressure of 1.0 atmosphere and a temperature of 0 °C.

The amount of ozone in the polar regions varies seasonally. In the Arctic the range is from about 300 DU in the early fall up to about 460 DU in the spring. In the Antarctic the ozone levels in the spring reach only about 400 DU under natural conditions. However, the measurements by Farman et al. (1958) shown in Figure 23.3 demonstrated that the average ozone concentration in October over Halley Bay, Antarctica, declined from about 330 DU in 1957 to about 200 Du in 1984. The monthly average ozone content for October has continued to decline since then and fell to just over 120 DU in 1989.

The cause for this alarming phenomenon was ultimately deduced from the following observations: (1) Direct measurements by a high-flying ER-2 plane on September 21, 1987 (Anderson et al., 1989), demonstrated a strong anticorrelation between the concentrations of ClO and O_3 in the stratosphere over Antarctica at latitudes greater than about 70°S; (2) a balloon launched from McMurdo on October 27, 1987, detected a significant decrease in the ozone profile in the lower stratosphere between altitudes of about 12 and 21 km (Hofmann et al., 1989); (3) balloon

launches demonstrated a 20-degree increase in the temperature at 21 km in the stratosphere over McMurdo between August 29 and October 27, 1987. These direct observations indicated that the reduced ozone concentrations in Antarctica occur in the early austral spring (September–October), coincide with elevated concentrations of ClO, and are associated with an increase in the temperature of the lower stratosphere. These are the principal clues that were used to explain the development of the ozone hole over Antarctica.

During the austral winter months (June–September), the atmosphere over Antarctica is isolated from the global circulation by the polar vortex and the stratosphere receives no direct sunlight. As a result, the temperature of the stratosphere over Antarctica declines to −80°C. This permits the formation of ice crystals composed of a mixture of water and nitric acid. The absence of UV radiation caused by the absence of solar radiation retards the reactions that produce and destroy ozone in the stratosphere. Under these conditions, chlorine nitrate ($ClONO_2$) is incorporated into the ice crystals and is adsorbed on their surfaces where it reacts with HCl to form Cl_2 and HNO_3:

$$ClONO_2 + HCl \xrightarrow{ice} Cl_2 + HNO_3 \quad (23.15)$$

The Cl_2 molecules are released into the air of the stratosphere, whereas the HNO_3 remains in the ice crystals. When sunlight returns to Antarctica in the early spring, the ice crystals sublime and the chlorine molecules are dissociated by UV ($\lambda < 450$ nm), thus releasing the Cl catalyst:

$$Cl_2 + h\nu \rightarrow 2\,Cl\cdot \quad (23.16)$$

Consequently, the return of the Sun to Antarctica triggers rapid catalytic destruction of ozone by the Cl· stored in the stratosphere. However, as the stratosphere warms up, the circumpolar circulation breaks down and the ozone-depleted air over Antarctica mixes with ozone-rich air at lower latitudes. This phenomenon causes the "ozone hole" to spread to populated areas in the southern hemisphere until normal transport

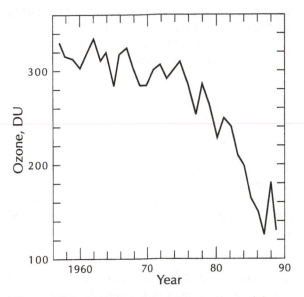

Figure 23.3 Average ozone concentrations of the stratosphere during the month of October over Halley Bay, Antarctica, from 1957 to 1989. The data show that the average ozone concentrations for October at this site declined from 330 DU in 1957 to about 120 DU in 1989, or about 64% in 33 years. [Modified from Graedel and Crutzen (1993) who adapted it from Farman et al. (1985).]

processes in the stratosphere restore the global distribution of ozone.

The potential hazard to the biosphere of increased exposure to UV-B caused by the large-scale destruction of ozone by CFCs was considered in 1987 by an international conference in Montreal (Graedel and Crutzen, 1993, p. 372). The protocol arising from that conference called for a voluntary 50% reduction in the manufacture of CFCs by 1998. This recommendation was strengthened at a subsequent conference in 1990 in London by specifying maximum allowable production of CFCs relative to 1986: 80% by 1993; 50% by 1995; 15% by 1997; and 0% by 2000. In addition, the production of methylchloroform (CH_3CCl_3) and carbon tetrachloride (CCl_4) is to end in 2005. Although the production of CFCs in Europe and North America has been reduced, the targets have not been met by some major industrial nations.

Model calculations by Brühl and Crutzen (1988) predict a significant 28% reduction of stratospheric ozone (relative to 1965) in the southern hemisphere by the year 2030 even if the production targets of London are actually realized. However, if CFC production continues at 1974 levels, the ozone depletion of the southern hemisphere will reach 48% of the 1965 level in 2030. If the recommendations for reduced production of CFCs are not implemented and production rises above 1974 levels, the ozone depletion will be greater than 48% and the resulting health hazard to humans may become quite serious.

For this reason, Cicerone et al. (1991) considered the feasibility of reducing the chlorine content of the stratosphere over Antarctica by injecting ethane (C_2H_6) and propane (C_3H_8). These gases react with $Cl \cdot$ to form HCl, thus sequestering it and preventing the destruction of ozone:

$$Cl \cdot + C_2H_6 \rightarrow HCl + C_2H_5 \quad (23.17)$$

$$Cl \cdot + C_3H_8 \rightarrow HCl + C_3H_7 \quad (23.18)$$

Unfortunately, a fleet of high-flying aircraft would be required to disperse about 50,000 tons of the hydrocarbons annually. In addition, the gas must be dispersed widely within the atmosphere

encompassed by the polar vortex whose position and integrity vary unpredictably. It is probably better to stop production of chlorine-bearing gases worldwide than to attempt to repair the atmosphere. However, even in the unlikely event that production of these gases is actually stopped in 2005, ozone depletion in the southern hemisphere will continue for many decades until the atmosphere gradually cleans itself up.

Ozone depletion by CFCs is, of course, a world-wide phenomenon that is aggravated in the Southern Hemisphere because of the isolation of Antarctica during the winter months by the polar vortex. In the Northern Hemisphere the problem is not as severe because the Arctic region does not develop a polar vortex. Consequently, the ozone content of the stratosphere in the Arctic is reduced by only 10–20% of normal levels when the sun returns at the end of the winter.

23.5 Infrared Radiation and the Greenhouse Effect

Ninety percent of the sunlight reaching the Earth is in the visible (λ 400–750 nm) and infrared ($\lambda > 750$ nm) part of the spectrum. The radiation that reaches the surface is converted into heat and causes the surface to emit infrared radiation consistent with its temperature. This radiation is not sufficiently energetic to break chemical bonds, but it is absorbed by molecules of the air and increases their vibrational and rotational energies. The excess energy is transformed into kinetic energy and results in an increase in the frequency of molecular collisions (heat). The principal constituents of the atmosphere, O_2 and N_2, are not efficient absorbers of infrared radiation. Instead, minor constituents including H_2O, CO_2 and O_3 are primarily responsible for absorption of infrared radiation emitted from the surface of the Earth.

The radiation budget of the Earth indicates that almost 30% of the incoming solar radiation is reflected back into space, approximately 25% is absorbed by the atmosphere, and the rest (about 47%) is absorbed at the surface. Most of that radiation is consumed as latent heat of evaporation of water and is released back into the atmosphere when water vapor condenses in the troposphere. In addition, some of the secondary infrared radiation is absorbed by H_2O and CO_2. The recycling of solar energy between the surface of the Earth and its atmosphere increases the temperature of the surface quite significantly compared to what it would be without this natural greenhouse effect.

Certain trace constituents of the atmosphere (O_2, N_2O, CH_4, $CFCl_3$, etc.) play a sensitive role in the heat budget of the atmosphere because they absorb infrared radiation in the wavelength band 8000–12,000 nm where H_2O and CO_2 are not effective. Therefore, these trace gases trap some of the infrared radiation that would otherwise escape and thereby enhance the greenhouse warming of the atmosphere.

The effective radiation temperature of the Earth (T_E) depends on the amount of solar energy the Earth receives from the Sun (E_S) and the amount of energy emitted by the Earth as infrared radiation. This balance is expressed by the equation:

$$\frac{E_S}{4}(1 - \alpha) = \sigma T_E^4 \qquad (23.19)$$

where $E_s = 1380\,\text{W m}^{-2}$ (solar constant) is the solar energy received by the Earth at the average Sun–Earth distance in units of watts per square meter; $\alpha = 0.28$ (planetary albedo) is the fraction of solar radiation reflected into space; $\sigma = 5.67 \times 10^{-8}\,\text{W m}^{-2}\,\text{K}^{-4}$ (Stefan–Boltzmann constant); and the factor 1/4 is the ratio of the planetary disk to its surface area, required because the solar radiation is intercepted by the area of the disk. Substituting the appropriate values into equation 23.19:

$$\frac{1380}{4}(1 - 0.28) = 5.67 \times 10^{-8}T_E^4$$

yields:

$$T_E = 257\ \text{K}\ or\ -16°C$$

The recycling of infrared radiation emitted by the surface of the Earth can be taken into account by means of the effective atmospheric infrared transmission factor, $f = 0.61$; that is, only about 61% of the secondary infrared radiation actually escapes

from the Earth. Therefore, equation 23.19 is modified to include the infrared transmission factor:

$$\frac{E_S}{4}(1 - \alpha) = f\sigma T_S^4 \qquad (23.20)$$

where T_S is the temperature of the Earth's surface as before. The result is:

$$T_S = 291\ \text{K}\ or\ 18°C$$

which is close to the observed global temperature of about 15°C. This calculation demonstrates the importance of the greenhouse effect of the natural atmosphere, which increases the average global surface temperature of the Earth by 34°C compared to what it would otherwise be.

The beneficial effects of energy recycling in the atmosphere are amplified by increases in the concentrations of the so-called greenhouse gases. These gases have different heating powers depending partly on their concentrations in the natural atmosphere. For example, an increase in the concentration of methane from 0 to 2 ppmv adds about the same amount of heat to the atmosphere as an increase in the CO_2 content from 275 to 375 ppmv. The reason is that CO_2 is already absorbing most of the infrared radiation in its spectral region because of its high natural concentration. Therefore, the heating power of the greenhouse gas (called radiative forcing) can be expressed in terms of the reduction of the amount of infrared radiation leaving the Earth caused by a unit increase in the concentration of the gas in the atmosphere. A comparison by Houghton et al. (1990) of atmospheric heating by greenhouse gases indicates that the effect of CFCs, CH_4, N_2O, and stratospheric water vapor has been rising and nearly equalled the heating caused by CO_2 in the decade of the 1980s.

The climatic consequences of the release of different greenhouse gases into the atmosphere depend not only on their concentration and hence on their radiative forcing but also on their residence time in the atmosphere. Therefore, the global warming potential (GWP) of a greenhouse gas is computed by including its atmospheric lifetime and by comparing the resulting heating function to that of CO_2. Houghton et al. (1990) listed values of the global warming potentials per kg of the principal greenhouse gases relative to CO_2:

Greenhouse gas	GWP
CO_2	1
CH_4	21
N_2O	290
CFCs	3000–8000

The global warming potential of 1 kg of CH_4 causes 21 times as much absorption of infrared radiation as the release of 1 kg of CO_2. The heating potentials of the CFCs are obviously very large compared to that of CO_2 because their concentration in the natural atmosphere is zero, because they absorb infrared radiation in the wavelength band between 8000 and 12,000 nm where CO_2 is ineffective, and because they have long atmospheric residence times.

The global warming potentials of greenhouse gases help to quantify the effect these gases have on the average global temperature. Predictions of the increase in the surface temperature to be expected during the 21st century vary depending on the assumed rate of release of greenhouse gases, on the uptake of the resulting heat by the oceans, and on other factors such as changes in the planetary albedo caused by increased cloudiness, etc. Nevertheless, virtually all models predict an increase in the average global surface temperature during the 21st century (Houghton et al., 1990):

	Temperature increase, °C	
	by 2050 A.D.	by 2100 A.D.
High estimate	3.75	6.25
Best estimate	2.75	4.25
Low estimate	1.75	3.00

Even the lowest estimate predicts an increase of 1.75 °C in the average global temperature by the year 2050.

Changes in the surface temperature of the Earth have certainly occurred in the past. For example, the Little Ice Age (1450–1890) was caused by a *decrease* in the average global temperature of about 1 °C, and the end of the last ice age about 15,000 years ago was accompanied by an *increase* in the average global temperature of about 3 °C. However, the warming at the end of the last ice age occurred gradually over a period of many centuries, whereas the expected global warming caused by the release of greenhouse gases will take only a few decades.

23.6 Prediction of the Future Global Climate

The geological record contains clear evidence that the global climate has changed significantly during the Earth's history as a result of both terrestrial and extraterrestrial factors. The extraterrestrial causes of global climate change include:

1. The variation of the luminosity of the Sun in the past 4.5×10^9 years (Graedel et al., 1991).

2. Periodic changes in the celestial mechanics of the Earth (Milankovitch theory) that affect the amount of sunlight it receives (Berger, 1988).

The terrestrial causes of climate change include:

1. The chemical evolution of the atmosphere.

2. The amount of dust emitted by volcanic eruptions and meteoroid impacts.

3. Changes in the circulation pattern of the oceans.

4. Changes in the albedo caused by the formation of continental ice sheets.

5. Anthropogenic emissions of CO_2 and other gases.

However, even though the average global temperature has undoubtedly changed significantly in the course of geologic time, life on the Earth has continued without interruption since its first recorded presence about 3.5×10^9 years ago.

The extraterrestrial factors affecting the average global temperature (solar luminosity and celestial mechanics) will continue to do so in the future. Volcanic eruptions and infrequent meteoroid impacts will also continue to cause short-term catastrophic climate perturbations. An evaluation by Lamb (1977) of the effects of orbital changes on the amount of solar radiation reaching the top of the atmosphere of the Earth suggests that episodic decreases in the winter temperatures will occur during the next several tens of thousands of years. The effectiveness of celestial mechanics to cause global temperature fluctuations was clearly demonstrated by Prell and Kutzbach (1987). Therefore, the orbital parameters are likely to cause a decrease in the average global temperature with minima at 5 and 22 kyr after the present (AP), leading to continental glaciation at about 50 kyr AP (Berger, 1987).

However, the climate predictions based on Milankovitch theory (celestial mechanics) must be modified to include the anthropogenic emission of greenhouse gases. Continued combustion of fossil fuel, deforestation, and other factors may increase concentrations of CO_2 and other greenhouse gases in the atmosphere by up to a factor of four in the next 1000 years or less. Broecker (1987) in his book *How to Build a Habitable Planet* predicted that in the next 5000 years global temperature could rise by about 4.5°C above normal values, but would then decline as predicted by Milankovitch theory.

The extent of global warming depends significantly on the response of the C-cycle (Section 22.4, Figure 22.3) to the release of greenhouse gases. If the increase in the concentration of atmospheric CO_2 stimulates the growth of marine organisms, the increase in the oceanic biomass will result in the burial of a portion of the carbon in marine sediment, thus taking it out of circulation for long periods of geologic time. Therefore, the anthropogenic greenhouse warming of the Earth may be a short-lived episode on the geological time scale, producing an "anthropogenic superinterglacial" epoch to be followed by the next glaciation mandated by celestial mechanics (Broecker, 1987). However, this long-range prediction provides little comfort to humans who must cope with the effects of global warming and increased sea level in the next several centuries.

23.7 Summary and Prognosis for the Future

The atmosphere of the Earth is composed primarily of N_2 and O_2 with minor concentrations of Ar, CO_2, Ne, He, CH_4, Kr, H_2, and O_3. The structure of the atmosphere consists of the troposphere (0–10 km), stratosphere (10–50 km), mesosphere (50–100 km), and the thermosphere (100–1000 km). The pressure of the atmosphere decreases exponentially from 1.0 atmosphere at sea level to 10^{-4} atmospheres close to 70 km above the surface of the Earth. The temperature in the troposphere decreases with increasing altitude and reaches about −50°C at the tropopause located at an altitude of 10 km at the equator. Subsequently, the temperature rises in the stratosphere, decreases in the mesosphere, but rises again in the thermosphere.

The gases of the atmosphere absorb ultraviolet radiation and thereby protect the surface of the Earth from it. The wavelength spectrum of ultraviolet radiation includes UV-C ($\lambda < 290$ nm), UV-B (290–320 nm), and UV-A (320–400 nm). The most energetic form of ultraviolet radiation (UV-C) is absorbed by N_2O, O_2, and O_3 in the stratosphere. However, UV-B does reach the surface and can cause biological damage unless precautions are taken to limit exposures.

Ozone plays an important role in reducing the fluxes of UV-C and UV-B that reach the surface of Earth. Ozone is produced in the stratosphere when UV-C breaks the bond of O_2 molecules. The resulting oxygen atoms then react with O_2 to form ozone (O_3). Under natural conditions the inventory of O_3 in the stratosphere is maintained in a steady state by the production and destruction of this molecule.

The introduction into the atmosphere of Cl-bearing anthropogenic gases (CFCs and others) has resulted in an acceleration of the destructive reactions causing the ozone content of the stratosphere to decline. The effect is especially severe over Antarctica during the spring (September–October) because of the combination of factors related to the geographic location and atmospheric circulation during the austral winter.

The reduction in the ozone content of the stratosphere over Antarctica (ozone hole) has caused an increase in the UV-B flux to that region, and this increase endangers all life forms that are exposed to it. The increase in the flux of UV-B also raises concerns about the increased risk to life on a global scale.

The release of excess amounts of CO_2, CH_4, and certain other gases (e.g., CFCs) has increased the ability of the atmosphere to absorb infrared radiation ($\lambda > 750$ nm) emitted by rocks and soil on the surface of the Earth. The expected increase in the average annual global temperature will change weather patterns and may increase the frequency and severity of storms. Global warming may also lead to an increase in sea level caused by the recession of glaciers and polar ice sheets and by an increase in the temperature of seawater. Existing models all predict an increase in the average annual global temperature, but differ in the effect that the increase in temperature will have on the albedo of the Earth and the response of the polar ice sheets.

Long-range predictions suggest that global warming will be a relatively short-lived phenomenon on a time scale of 10^4 years because the excess C in the atmosphere will be buried in marine sediment. Subsequently, changes in the celestial mechanics of the Earth will cause cooling, leading to another ice age at about 50,000 years after the present.

Problems

1. Given that 100 Dobson units (DU) correspond to a layer of ozone 1mm thick at atmospheric pressure and 0°C, demonstrate that 1 DU = 2.69×10^{16} molecules of O_3 cm^{-2}.

2. Calculate the number density of air at 1 atm and 0°C assuming that it is composed of $N_2 = 78.15\%$, $O_2 = 20.95\%$. Express the result in molecules per cm^3.
(Answer: 2.66×10^{19} molecules/cm^3)

3. Using equation 23.2, calculate the energy of a photon of ultraviolet light (UV-C) having a wavelength of 250 nm ($= 250 \times 10^{-9}$ m).
(Answer: 7.94×10^{-18} J)

4. Use equation 23.19 to calculate the effective radiation temperature of the Earth if the albedo of the Earth increases by 10% to 0.31, assuming that the solar constant remains 1380 W m^{-2}.
(Answer: -18.6°C)

5. Explain why the average annual global temperature at the surface of the Earth is significantly higher than the radiation temperature.

6. How will the atmosphere infrared transmission factor (f) change when anthropogenic emissions of greenhouse gases cause an increase in global temperatures?

References

ALVAREZ, W., E. G. KAUFFMAN, F. SURLYK, L.W. ALVAREZ, F. ASARO, and H. V. MICHEL, 1984. Impact theory of mass extinctions and the invertebrate fossil record. *Science*, **223**:1135–1141.

ANDERSON, J. G., W. H. BRUNE, AND M. H. PROFFITT, 1989. Ozone destruction by chlorine radicals within the Antarctic vortex: The spatial and temporal evolution of ClO–O_3 anticorrelation based in situ ER-2 data. *J. Geophysics, Res.*, **94**:11,465–11,479.

ANONYMOUS, 1976. U.S. standard atmosphere. Nat. Ocean Atmosph. Admin., Washington, D. C.

BERGER, A., 1987. The Earth's future climate at the astronomical time scale. In C. Goodess and S. Pakstikof (Eds), *Future, Climate Change and Radioactive Waste Disposal.* Norwich, U.K.

BERGER, A., 1988. Milankovitch theory and climate. *Rev. Geophys.*, **26**:624–657.

BROECKER, W. S., 1987. *How to Build a Habitable Planet.* Eldigio Press, Palisades, NY.

BRASSEUR, G., and S. SOLOMON, 1986. *Aeronomy of the Middle Atmosphere,* 2nd ed. Kluwer Acad. Pub.

BRÜHL, C., and P. J. CRUTZEN, 1988. Scenarios for possible changes in atmospheric temperatures and ozone concentrations due to man's activities as estimated with a one-dimensional coupled photochemical climate model. *Climate Dynamics,* **2**:173–203.

CHAMBERLAIN, J. W., and D. M. HUNTEN, 1987. *Theory of Planetary Atmospheres: An Introduction to their Physics and Chemistry,* 2nd ed. *Internat. Geophys. Series,* vol. 36. Academic Press, San Diego, CA.

CHAMEIDES, W. L., and D. D. DAVIS, 1982. Chemistry in the troposphere. *Chem. Eng. News,* **60**(40):38–52.

CICERONE, R. J., S. ELLIOTT, and R. P. TURCO, 1991. Reduced Antarctic ozone depletions in a model with hydrocarbon injections. *Science,* **254**:1191–1193.

FARMAN, J. C., B. G. GARDINER, and J. D. SHANKLIN, 1985. Large losses of total ozone in Antarctica reveal seasonal interaction. *Nature,* **315**:207–210.

FASSETT, J. E., and J. K. RIGBY, JR., 1987. The Cretaceous-Tertiary boundary in the San Juan and Raton basins, New Mexico and Colorado. *Geol. Soc. Amer., Spec. Paper* **209**:200 pp.

FRIEDLI, H., H. LÖTSCHER, H. OESCHGER, U. SIEGENTHALER, and B. STAUFER,1986. Ice core record of the $^{13}C/^{12}C$ ratio of atmospheric CO_2 in the past two centuries. *Nature,* **324**:237–238.

GRAEDEL, T. E., and P. J. CRUTZEN, 1993. *Atmospheric Change: an Earth System Perspective.* W. H. Freeman, New York.

GRAEDEL, T. E., I.-J. SACKMAN, and A. I. BOOTHROYD, 1991. Early solar mass loss: a potential solution to the weak sun paradox. *Geophy. Res. Letters,* **18**:1881–1884.

HOFMANN, D. J., J. W. HARDER, J. M. ROSEN, J. V. HEREFORD, and J. R. CARPENTER, 1989. Ozone profile measurements at McMurdo Station, Antractica, during the spring of 1987. *J. Geophys. Res.,* **94**:16,527–16,536.

HOLLAND, H. D., 1979. *The chemisty of the atmosphere and oceans.* Wiley, New York.

HOLLAND, H. D., 1984. *The chemical evolution of the atmosphere and oceans.* Princeton University Press, 582 pp.

HOLLAND, H. D., B. LAZAR, and M. M. McCAFFREY, 1986. Evolution of the atmosphere and oceans. *Nature,* **320**:27–33.

HOUGHTON, J. T., G. J. JENKINS, and J. J. EPHRAUMS, (Eds.), 1990. *Climate Change: The IPCC Scientific Assessment.* Cambridge Univ. Press, New York.

KEELING, C. D., R. B. BACASTOW, and T. P. WHORF, 1982. Measurements of the concentrations of carbon dioxide at Mauna Loa Observatory, Hawaii. In W. C. Clark (Ed.), *Carbon Dioxide Review* 1982, 377–385. Oxford University Press, New York.

KHALIL, M. A. K., and R. A. RASMUSSEN, 1987. Atmospheric methane: trends over the last 10,000 years. *Atmospheric Environment,* **21**:2445–2452.

LAMB, H. H., 1977. *Climate: present, past, and future,* vol. 2, Methuen Co., London.

LEWIS, J. S., and R. G. PRINN, 1983. Planets and their atmospheres: origin and evolution. *Internat. Geophys. Series,* vol. 33. Academic Press, San Diego, CA.

MOLINA, M. J., and F. S. ROWLAND, 1974. Stratospheric sink for chlorofluoromethanes: chlorine atom catalized destruction of ozone. *Nature,* **249**:810–812.

PEARMAN, G. E., D. ETHERIDGE, F. DE SILVA, and P. J. FRASER, 1986. Evidence of changing concentrations of atmospheric CO_2, N_2O, and CH_4 from air bubbles in the Antarctic ice. *Nature,* **320**:248–250.

PRELL, W. L. and J. E. KUTZBACH, 1987. Monsoon variability over the past 150,000 years. *J. Geophys. Res.,* **92**:8411–8425.

PRINN, R. G., and B. FEGLEY JR., 1987. Bolide impacts, acid rain, and biospheric traumas at the Cretaceous–Tertiary boundary. *Earth Planet. Sci. Letters,* **83**:1–15.

SILVER, L. T., and P. H. SCHULTZ, 1982. Geological implications of impacts of large asteroids and comets on Earth. *Geol. Soc . Amer., Spec. Paper* **190**:528 pp.

WARNECK, P., 1988. Chemistry of the natural atmosphere. *Internat. Geophys. Series.* vol. 41, Academic Press, San Diego, CA.

ZANDER, R., PH. DEMOULIN, D. H. EHHALT, U. SCHMIDT, and C. P. RINSLAND, 1989. Secular increase of the total vertical column abundance of carbon monoxide above central Europe since 1950. *J. Geophys. Res.* **94**:11021–11028.

24

Environmental Geochemistry: Disposal of Radioactive Waste

We now return to the idea that geochemists have a role to play in the struggle of the human race to survive on the Earth. Although we have succeeded in controlling the environment on a local scale, we cannot escape the fact that the capacity of the Earth to support the human population is limited. Moreover, our efforts to improve the quality of life worldwide have unexpectedly caused the environment to deteriorate in ways that threaten the very quality of life we seek to improve. This dilemma can be avoided only by prudent management of all of the Earth's resources based on an understanding of the natural processes that make the Earth "an abode fitted for life."

24.1 The Big Picture

Life on the Earth has always been hazardous because of the occurrence of violent geological events, including volcanic eruptions, earthquakes, floods, and landslides. In addition, we are exposed to *geochemical hazards* associated with the ionizing radiation emitted by naturally occurring long-lived radionuclides, by the unstable daughters of U and Th, and by cosmogenic radionuclides including tritium, [10]Be, and [14]C (Faure, 1986). Other natural geochemical hazards consist of anomalous concentrations of certain elements whose presence or absence in our diet can cause medical problems (Cannon, 1974). In addition, our individual survival is endangered by toxins

generated by bacteria and by certain plants and animals of the biosphere. We even face danger from outer space in the form of meteorite impacts, ultraviolet radiation, and cosmic rays. All of these hazards are a natural property of the environment on the surface of the Earth and ultimately control our life expectancy. Some of these natural hazards can be minimized by a judicious choice of life style and habitat. However, all forms of life are exposed to these natural risks and have adapted themselves to survive as long as possible. The subject of this chapter concerns new anthropogenic hazards that threaten not only the human population but could affect the entire biosphere on the surface of the Earth.

Geologists and geochemists have studied the natural hazards of life for many years and have attempted to inform the public by pointing out that the loss of life and property resulting from natural disasters was predictable and *could* have been and therefore *should* have been avoided in many cases. Inevitably, the subject matter of environmental geology and environmental geochemistry has been presented in the form of case studies of natural disasters in the hope that the horror stories of the past will not be repeated in the future.

In addition to the natural geological and geochemical hazards, we now face threats that are a direct consequence of our own actions. These threats were not anticipated in many cases because we believed that the capacity of the Earth to *provide* the *resources* we require and to *absorb* our *waste products* is virtually unlimited. We have begun to realize only recently that nonrenewable natural resources are limited (Section 21.8) and that the improper disposal of waste products can cause a deterioration of the environment in which we live. Instead of presenting the evidence for these newly recognized threats in another series of horror stories, we attempt here to integrate them into a big picture.

We begin by asserting that all human beings have a right to live and are therefore entitled to have access to the natural resources required to provide the necessities of life: food, water, air, housing, tools and utensils, transportation, and energy. The consumption of natural resources inevitably creates waste products, which must be disposed of in such a way that they do not threaten the quality of life of future generations. Therefore, the problem to be addressed includes both the *supply* of natural resources and the *disposal* of the resulting waste products. The severity of the problem we face is increased by the rapid increase of the population during the 20th century, by the uneven distribution of natural resources, and by the need to transport food products and mineral commodities to population centers. All of these activities require the use of *energy,* which also must be provided from natural sources, resulting in additional waste products that aggravate the potential for environmental degradation. The role of *geochemists* in this formidable task centers increasingly on evaluations of the environmental consequences arising both from the *exploitation* of natural resources and from the *disposal* of waste products.

We have already considered some aspects of the production of food (Section 19.8), water (Section 20.6), and mineral resources (Section 21.8). These can be treated as *global abstractions* but are expressed directly at the *local level* where they affect the quality of life of the people and the environment in which they live. In this chapter we draw attention to problems of *waste disposal,* which can also be treated on a global scale as well as at the local level. We select for discussion the consequences of energy production by induced fission of ^{235}U because the resulting

waste products remain harmful to life for very long periods of time. The uncontrolled dispersal of radioactive isotopes on the surface of the Earth, and the safe disposal of nuclear waste have become important problems that require the attention of geochemists throughout the world.

24.2   High-Level Radioactive Waste

Nuclear power reactors generate heat by means of a controlled chain reaction based on the fission of ^{235}U by thermal neutrons (Section 16.1c). The resulting fission-product nuclides accumulate in the fuel rods of nuclear reactors together with transuranium elements, including plutonium (Pu), neptunium (Np), americium (Am), and curium (Cm), which are produced by neutron-capture reactions starting with ^{238}U and ^{235}U. The fuel rods of nuclear reactors therefore become highly radioactive as a result of the decay of the fission products, of the transuranium elements, and of their unstable daughters. The activity of spent fuel elements decreases *slowly* with time and reaches the level of U ore only after 10,000–100,000 years, depending on whether or not the fuel elements are reprocessed. The radioactive wastes generated by nuclear power reactors not only are highly *radiotoxic* but also release significant amounts of *heat.* For these reasons, such wastes must be isolated in repositories designed to contain them safely until the rate of decay has diminished to levels encountered in U ore.

The rate of decay of a radioactive nuclide is expressed in the *becquerel* (Bq), defined as one disintegration per second (dis/s). Alternatively, the rate of decay may be stated in *curies* (Ci), where 1 Ci = 3.70×10^{10} Bq. The decay rates encountered in nuclear waste are of the order of terabecquerels (1 TBq = 10^{12} Bq) and petabecquerels (1 PBq = 10^{15} Bq).

The exposure to ionizing radiation (α-, β-, γ-, and x-rays) is measured in terms of the energy that is released when the radiation interacts with a unit weight of the absorber such as biological tissues. The basic unit of measurement of the *radiation dose* is called the *gray* (Gy), which is equal to 1 J/kg or to 100 *rad* (radiation absorbed dose). One rad is the amount of radiation required to liberate 1×10^{-5} J of energy per gram of absorbing material (Roxburgh, 1987, p. 18). In order to make the radiation dose relevant to humans, the *rem* (roentgen equivalent man) is defined as one rad of x-rays or γ-rays. The *sievert* (Sv) is defined as 100 rem.

All forms of ionizing radiation are harmful, but the amount of tissue damage depends on the intensity and energy of the radiation, on the distance between the source of the radiation and the object being irradiated, on the shielding provided by the matter between the source and the object, on the duration of the exposure, on the type of radiation, and on the tissue that is irradiated. The natural, whole-body, background radiation from rocks, cosmic rays, and atmospheric gases is ~0.1 rem per year, depending on the composition of the rocks, the elevation of the site, and other circumstances. Anthropogenic sources of radiation, such as medical or dental x-rays, just about double the average exposure to 0.2 rem per year, although wide variations occur here as well, depending on the circumstances. The physiological effects of full-body irradiations described in Table 24.1 require *much larger* doses than are provided by the natural environment. However,

Table 24.1 Physiological Effects of Whole-Body Irradiation

Dose, rem	Effects
0–25	reduction of white blood cells
25–100	nausea, fatigue, changes in blood
100–200	nausea, vomiting, fatigue, susceptible to infection due to low count of white blood cells; death possible
200–400	fatal in 50% of cases, especially without treatment; damage to bone marrow and spleen
>600	fatal, even with treatment

SOURCE: Roxburgh (1987).

high-level radioactive waste from nuclear reactors can administer large doses of ionizing radiation in a short time that severely affect tissues, the digestive system, and the nervous system. As a result, a deficiency of white blood cells (leukopenia) occurs, followed by hemorrhaging, fever, vomiting, and death.

The radioactive waste generated in the course of the fuel cycle is indicated in Figure 24.1. The waste rock and mineral residues produced during the mining and extraction of U from the ore are commonly stored at the mine in tailings ponds and piles of waste rock (Brookins, 1984, ch. 10). The mine tailings are radioactive because of the presence of the unstable daughters of U and Th, including the isotopes of the noble gas radon (Rn). In addition, the mine tailings may contain pyrite and other minerals that create an environmental hazard when they react with water and oxygen (Sections 14.1 and 21.2). Chemical weathering of mine tailings may also release potentially toxic elements into the local groundwater including Mo, Se, and As. The environmental hazards of mine tailings of any kind are magnified when such sites are later used to construct housing developments and shopping malls. The most dangerous high-level waste (HLW) in the U-fuel cycle is produced during the initial fabrication of the fuel elements, during their subsequent reprocessing after use in a reactor, and by discarding the spent fuel elements without reprocessing them.

When the U fuel of a nuclear reactor has been depleted in ^{235}U, the fuel rods are removed from the core of the reactor and are stored at the reactor site for up to five years in order to allow the intensity of the radioactivity to decrease. Subsequently, the fuel elements may be reprocessed to recover the U and Pu they contain. For this purpose, the U fuel is separated from the metal containers into which it is packed and is then dissolved in nitric acid. The U and Pu are separated from this solution by solvent extractions with tributyl phosphate (TBP) and odorless kerosene (OK). The residual nitric acid solution contains 99.9% of the nonvolatile fission products, 0.1% of the U, <1% of the Pu, and most of the other transuranium elements that form in the reactor by neutron capture and

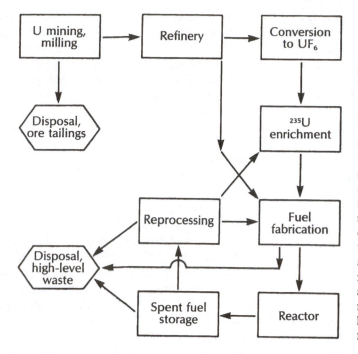

Figure 24.1 Nuclear fuel cycle starting with the mining and milling of U-ore leading to the fabrication of fuel elements and their use in nuclear power reactors. In the United States the spent fuel is stored at the reactor site for up to five years in order to allow the short-lived fission-product nuclides and some of the transuranium elements to decay. Subsequently, the spent fuel may be reprocessed or discarded. Radioactive waste products arise during the initial mining and milling of U-ore, during the reprocessing of spent fuel elements, and by discarding spent fuel without reprocessing. The spent fuel and high-level waste remain radiotoxic for more than 10,000 years.

β-decay (Roxburgh, 1987). The heat liberated by the decay of these radionuclides is used to remove excess water by boiling. After further cooling, the residue is converted into a borosilicate glass, which is poured into steel containers for ultimate disposal in a geological repository. The U recovered in this process can be reused after some additional enrichment of the remaining ^{235}U. The ^{239}Pu produced from ^{238}U during the operation of the reactor is itself fissionable and is used in nuclear weapons (Blasewitz et al., 1983).

The volume of HLW produced by reprocessing one metric ton (1 t = 10^3 kg) of spent fuel is about 5 m^3, according to an estimate by the Nuclear Energy Agency (NEA) of the Organization for Economic Cooperation and Development (OECD) (Anonymous, 1984). To put it another way, the generation of 1 GW (1 GW = 10^9 W) of electric power (GWe) produces 6 m^3 of HLW in vitrified form and ~50 m^3 of additional highly radioactive waste containing high concentrations of alpha emitters. The NEA has estimated that the nuclear generating capacity of the world in

the year 2000 will be 600 GWe, which will result in the discharge of about 15,000 metric tons of spent fuel annually. If half of that is reprocessed, it would result in 750 m^3 of HLW per year worldwide. According to data cited by Roxburgh (1987, Table 1.9), the total amount of HLW in the United States by the year 2000 will have a volume of 10,500 m^3 with an activity of 126,000 megacuries (MCi) or 4662 × 10^{18} Bq. Various other forms of radioactive waste (*radwaste*) will have a volume of 2.28 × 10^6 m^3 and a total activity of 1980 MCi, which is equivalent to 73.26 × 10^{18} Bq and will raise the total accumulated radioactivity in the United States alone to 4735 × 10^{18} Bq.

The radiochemical composition of the fission products in HLW can be predicted from the isotopic abundance of ^{235}U in the fuel, from the energy spectrum of the neutrons in the reactor, from possible neutron-capture reactions of product nuclides, and from the decay time since the fuel was removed from the reactor. The results of such a calculation are listed in Table 24.2 for a reactor that operated for one year at one MWe

Table 24.2 List of Fission-Product Nuclides and Their Activities in Terabecquerels (1 TBq = 10^{12} Bq) at Different Times After Removal from a Reactor That Operated for One Year at One MWe

Radionuclide	t = 0	t = 100 days	t = 5 years
^{85}Kr	7.07	6.92	4.88
^{89}Sr	1413.40	381.10	0
^{90}Sr	52.91	52.54	44.40
^{90}Y[a]	52.91	52.54	44.40
^{91}Y	1809.30	536.50	0
^{95}Zr	1820.40	629.00	0
^{95}Nb[a]	1783.40	1061.90	0
^{103}Ru	1143.30	219.04	0
^{103m}Ru	1143.30	219.04	0
^{106}Ru	80.66	66.60	2.59
^{106}Rh[a]	80.66	66.60	2.59
^{131}I	932.40	0.15	0
^{133}Xe	2046.10	0	0
^{137}Cs	39.96	39.59	35.89
^{137m}Ba[a]	38.11	37.74	34.04
^{140}Ba	1912.90	8.51	0
^{140}La[a]	1912.90	9.99	0
^{141}Ce	1768.60	175.38	0
^{143}Pr	1676.10	10.73	0
^{144}Ce	987.90	769.60	9.99
^{144}Pr[a]	987.90	769.60	9.99
^{147}Nd	806.60	1.48	0
^{147}Pm[a]	181.30	177.60	50.32

[a]Daughter of a fission-product nuclide.
SOURCE: Roxburgh (1987) after Pentreath (1980).

(Roxburgh, 1987, after Pentreath, 1980). The data indicate that the activity and hence the radiotoxicity of the principal fission-product nuclides and their daughters decrease rapidly with time. After a "cooling time" of 100 days, the total activity of the radionuclides decreases to about 23% of its original value and is further reduced to 0.84% of its initial value in 5 years when its activity is about 190 TBq.

The principal radionuclides remaining in HLW after a cooling period of 5 years are isotopes of some of the rare earth elements (Pm, Pr, and Ce), platinum group metals (Ru and Rh), the alkaline earths (Sr and Ba), plus Kr, Y, and Cs. In addition, *spent unreprocessed fuel* (SURF) contains isotopes of the actinide elements, including specifically ^{239}Pu ($T_{1/2} = 2.44 \times 10^4$ yr), ^{240}Pu ($T_{1/2} = 6.54 \times 10^3$ yr), ^{241}Am ($T_{1/2} = 433$ yr), ^{237}Np ($T_{1/2} = 2.14 \times 10^6$ yr), ^{238}U ($T_{1/2} = 4.47 \times 10^9$ yr), and residual ^{235}U ($T_{1/2} = 7.04 \times 10^8$ yr). All of these nuclides have long half-lives and give rise to series of unstable daughters whose decay augments the activity of their parents. As a result, SURF not only contains more radioactivity than HLW but also remains active longer than reprocessed HLW (Roxburgh, 1987). This point is illustrated in Figure 24.2, which shows that the fission-product nuclides of HLW and SURF decay to less than one TBq in 1000 years, whereas the actinides of SURF require more than 100,000 years to decay to the same level of activity.

The radioactivity of HLW, SURF, and of the cladding hulls of spent nuclear fuel elements releases very substantial amounts of heat for about 10,000 years after they were removed from a power reactor. The heat output decreases with time, as shown in Figure 24.3; but even after 100,000 years, HLW produces ~1 W/metric ton of nuclear fuel, whereas SURF gives off about six times that much heat. An even more graphic illustration of the thermal energy produced by SURF is to note that the spent fuel removed *annually* from the Pickering A power reactor in Canada produces the same amount of heat as the combustion of *25 million barrels of oil* (Barnes, 1979).

24.3 Geological Disposal of Radioactive Waste

The radwaste that is accumulating at nuclear reactors is a threat to all life forms for more than 10,000 years after its removal from the reactor. Therefore, this material must be isolated from the biosphere in

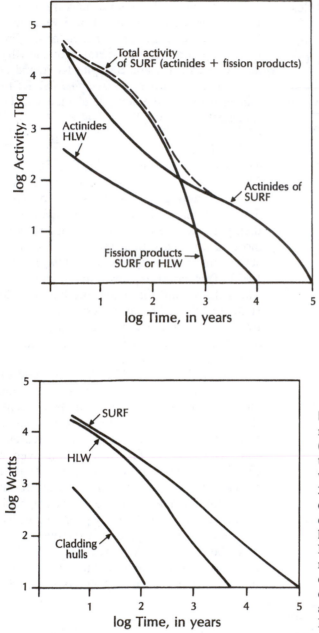

Figure 24.2 Decay rate of reprocessed high-level waste (HLW) and spent unreprocessed fuel (SURF) in units of terabecquerels (1 TBq = 10^{12} dis/s) and time in years. The data pertain to a pressurized water reactor generating 33 GW day tonne^{-1} (1 GW = 10^9 W) after five years of cooling and are normalized to one metric tonne of heavy metal in the original fuel elements. The fission-product nuclides in both HLW and SURF decay to <1 TBq in ~1000 years, whereas U and the transuranium elements (^{238}U, ^{235}U, ^{237}Np, ^{239}Pu, ^{241}Am, and their daughters) in SURF require >100,000 years to decay to that level of activity (plotted from data listed in Roxburgh, 1987).

Figure 24.3 Heat output from spent unreprocessed fuel (SURF), high-level waste (HLW), and cladding hulls as a function of time. The data pertain to a pressurized water reactor with a fuel consumption of 33 GW day tonne^{-1} followed by a 5-year cooling period. The thermal energy is expressed in watts per metric tonne of heavy metal in the original fuel elements. Note that SURF generates significantly more heat and does so for a longer period of time than HLW because of the presence of U and certain transuranium elements and their unstable daughters (data from Roxburgh, 1987).

such a way that all accidental contact with it is prevented (Blomke, 1976; Chapman and McKinley, 1987). In general, this can be achieved by storing the material in appropriately designed underground repositories within certain kinds of rocks that can tolerate the heat generated by the radioactivity and are sufficiently impermeable to prevent the contamination of groundwater by radionuclides.

The design and construction of underground nuclear-waste repositories have been technically feasible for many years. Nevertheless, construction of such facilities has been delayed for a number of reasons:

1. The need to make detailed evaluations of proposed sites to determine the hydrologic conditions as well as the geochemical and structural stability of the rocks at these sites.

2. Reluctance by people living near proposed repositories to permit their construction because of potential hazards resulting from accidents during transport of rad-waste to the site or from unforeseen problems with its containment within the repository.

The social and political resistance to the siting of repositories has given rise to the NIMBY complex (not in my back yard). Few people want their area to become the dumping ground of other people's radwaste. At the present time, each country with a significant nuclear industry is planning to build repositories for waste created within its own borders. These repositories will be constructed within different kinds of rocks, will be based on a variety of designs, and will use different packaging methods. The construction of a large number of radwaste repositories all over the world may increase the probability of accidents but may also reduce the severity of their consequences. Moreover, shorter transport distances from reactors to the national or regional disposal facilities may reduce the risk of accidents en route.

Underground storage of nuclear waste provides a number of important advantages compared to other disposal methods that have been considered, such as deposition in deep-sea trenches, burial in continental ice sheets, and injection into the Sun (Anonymous, 1981, 1983, 1984). Geological repositories of nuclear waste have the following desirable attributes:

1. They are *safe* because the host rock (if properly chosen) can absorb the radiation and dissipate the heat, and because acci-dental or malicious intrusion can be prevented by the depth of the repository below the surface. The migration of certain radionuclides by diffusion or advection by groundwater can be minimized by the use of multiple barriers. Any future movement of radionuclides out of the repository can be predicted from the known properties of the nuclides, the hydrologic conditions in the host rock, and the mineral and chemical composition and ion-exchange properties of the rocks.

2. Underground repositories require *no maintenance* after they are filled and sealed. This is important because the time required to detoxify radioactive waste is longer than the rise and fall of most human civilizations in the past.

3. Geological repositories offer *flexibility* and *convenience* in site selection and construction because repositories can be constructed within different kinds of rocks, including granitic rocks, evaporites, shale, tuff, or basalt. Therefore, most nations wishing to build a repository can do so within their borders, and the repositories can be located so as to minimize transport distances from reactor sites.

4. The construction of underground repositories is *feasible* based on well-established methods of mining engineering. The cost of building such installations does not add significantly to the price of electricity generated by nuclear power reactors.

5. The radioactive waste is *retrievable* even though underground geological repositories are designed to contain the waste products indefinitely.

Several countries have made substantial progress toward the construction of underground repositories of nuclear waste in different kinds of rocks: the Federal Republic of Germany in salt; Canada, Finland, France, Sweden, and United Kingdom, granitic rocks; Belgium and Italy, clay; and the United States, tuff, basalt, and salt deposits.

The safety of underground nuclear repositories during the time interval in the future when the waste material remains toxic must be predicted by modeling. The validity of such predictions depends on factual information about the composition of the waste material and about the local and regional geologic environment of the site. The specific properties of the host rock to be evaluated include (Fried, 1979; Barney et al., 1984; Tsang, 1987):

1. The presence and movement of gases, water, and brine within the host rock at the site and in the surrounding area as well as above and below the repository.
2. The response of the host rock to the input of heat from the radwaste and the effect of the temperature increase on its mechanical and hydrologic properties.
3. The behavior of radionuclides within the host rock and their migration routes.
4. The degradation of the packaging of the radwaste and of the barriers designed to contain radionuclides.
5. Risk of disruption of the repository by geological events such as earthquakes, movement along faults, volcanic eruptions, formation of continental ice sheets, subsidence, or sea-level rise.

The role of *geochemists* in this enterprise is to obtain the information that is needed to predict the future safety of each proposed nuclear waste repository. The necessary information is partly site-specific, but also requires laboratory studies of basic properties and investigations of natural analogs in which the movement of radionuclides in the geologic past can be observed.

The future safety of a particular nuclear-waste repository depends not only on its geologic setting and on geochemical processes but also on its design and on the complementarity of various design features. These include the packaging of the waste material, the depth of the repository and the spacing of waste packages, use of buffers around individual packages, and backfilling of tunnels after the waste packages have been emplaced.

The packaging of nuclear waste must allow it to be transported and stored safely prior to being placed in its final repository. In addition, the form of packaging must contribute to the long-term containment of radionuclides by being mechanically strong and chemically stable. Canisters containing vitrified HLW will probably have overpacks that will enhance safety during transport, will assure their physical integrity for 1000 years or more, and will retard the release of radionuclides in combination with clay (bentonite) buffers placed around each canister.

The estimated release rates of radionuclides from repositories containing several thousand canisters after 10^5 years are listed in Table 24.3 (Anonymous, 1984). Most of the remaining activity is due to isotopes of Pu whose solubility in water is sufficiently low (~2.4 ppm) that only a small fraction (~2×10^{-6}) escapes annually. The *dilution* of the radionuclides released by repositories in the ambient groundwater is expected to reduce their environmental impact. The amount of ^{239}Pu that can be maintained in the human body without significant injury is 0.13 μg whose decay rate is 0.008 μCi or ~300 Bq.

24.4   Geochemistry of Plutonium

Plutonium and many other transuranium elements, some of which are shown in Figure 24.4, originally existed in the solar nebula, but subsequently decayed because all of their isotopes are radioactive and have short half-lives compared to the age of the Earth. Only ^{238}U ($T_{1/2} = 4.47 \times 10^9$ yr) and ^{235}U ($T_{1/2} = 7.04 \times 10^8$ yr) have survived to the present time, whereas ^{244}Pu ($T_{1/2} = 8.3 \times 10^7$ yr) did not and ^{234}U ($T_{1/2} = 2.44 \times 10^5$ yr) occurs only because it is a short-lived daughter of ^{238}U. The *former* presence of ^{244}Pu in meteorites was confirmed by Alexander et al. (1971) by the presence of fissiogenic xenon. However, a search by Wetherill et al. (1964) for ^{247}Cm ($T_{1/2} = 1.54 \times 10^7$ yr) in the rare-earth concentrate of terrestrial *euxenite* (REE niobate and titanate) failed to confirm the pres-

Table 24.3 Estimated Rates of Leaching of Radioactive Isotopes of Actinide Elements from Repositories Containing Several Thousand Canisters of High-Level Waste (HLW) and Spent Unreprocessed Fuel (SURF) After 10^5 Years

Nuclide	Solubility, ppm	Leach rate,[a] fraction/yr		Activity released per year	
		HLW	SURF	HLW, Ci/3000 canister	SURF, Ci/7000 canister
^{238}U	485	1.4×10^{-4}	3.1×10^{-7}	129	948
^{242}Pu[b]	2.4	1.3×10^{-6}	2.3×10^{-6}	3,940	29,500
^{237}Nb[c]	0.007	1.2×10^{-9}	1.45×10^{-9}	2.1	16
^{230}Th[d]	0.024	1.3×10^{-5}	1.68×10^{-7}	196	1,470
^{239}Pu	2.4	1.3×10^{-6}	2.3×10^{-6}	61,900	464,000

[a]Flow of water assumed to be limited by diffusion through a bentonite buffer.
[b]Table 7a of Anonymous (1984) lists ^{237}Pu for HLW, which may be a misprint because this isotope has a half-life of only 45.6 days.
[c]Reducing conditions.
[d]Daughter of ^{238}U.
SOURCE: Anonymous (1984).

ence of this isotope. All of the other isotopes of Np, Pu, Am, and Cm in Figure 24.4 have shorter half-lives than ^{244}Pu and ^{247}Cm. Nevertheless, several of these do occur in small concentrations in terrestrial environments.

The presence of ^{239}Pu ($T_{1/2} = 2.44 \times 10^4$ yr) in U minerals was originally predicted by Goldschmidt (1942) based on the neutron capture of ^{238}U followed by β-decay of the resulting ^{239}U and ^{239}Np. Subsequently, Seaborg and Perlman (1948) and Garner et al. (1948) actually detected ^{239}Pu in pitchblende and carnotite. This and other isotopes of Np, Pu, Am, and Cm form not only in U minerals but also in nuclear reactors and during the explosion of nuclear fission devices. Consequently, the isotopes occupying the shaded squares in Figure 24.4 occur in radwaste and spent unreprocessed U-fuel and have been dispersed all over the Earth by fallout resulting from testing of nuclear weapons in the atmosphere and by the burnup of radioisotope thermoelectric generators (RTGs) fueled with ^{238}Pu and used in satellites and space probes. These anthropogenic sources now contribute more transuranium isotopes to the environment than the natural neutron-capture reactions (Harley, 1980).

The discharge of various radionuclides into the environment by fallout from nuclear weapons tests and by reprocessing plants may be used to date marine and lacustrine sediment (Stanners and Aston, 1984) as well as ice in Greenland and Antarctica (Koide et al., 1985).

The Pu isotopes that have been identified in natural materials include ^{238}Pu ($T_{1/2} = 87.8$ yr), ^{239}Pu ($T_{1/2} = 2.44 \times 10^4$ yr), ^{240}Pu ($T_{1/2} = 6.54 \times 10^3$ yr), ^{241}Pu ($T_{1/2} = 15$ yr), and ^{242}Pu ($T_{1/2} = 3.87 \times 10^5$ yr). The *atomic ratios* of Pu isotopes in fallout from nuclear weapons tests are:

$$^{240}Pu/^{239}Pu = 0.18 \pm 0.01$$

$$^{241}Pu/^{239}Pu = 0.009 \pm 0.002$$

$$^{242}Pu/^{239}Pu = 0.004 \pm 0.001$$

whereas the *activity ratio* (in the radiological sense) of $^{240}Pu/^{239}Pu$ released by weapons tests is ~0.67 (Sholkovitz, 1983).

The concentrations of Pu are usually determined by separating and purifying the element

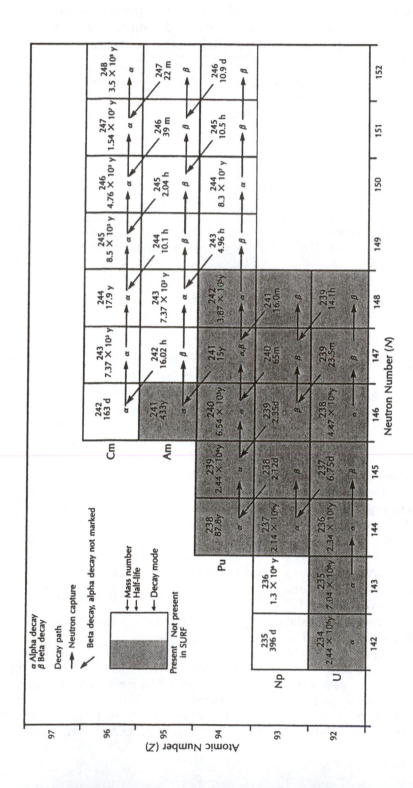

Figure 24.4 Partial chart of the nuclides showing the principal isotopes of U and of the transuranium elements neptunium (Np), plutonium (Pu), americium (Am), and curium (Cm). Although all of the isotopes of these elements are radioactive, those that are shaded occur in spent unreprocessed fuel elements (SURF) as a result of sequential neutron capture and β-decay starting with ^{238}U and ^{235}U. The isotope ^{234}U occurs because it is a daughter of ^{238}U. The neutron-capture reactions take place during the operation of nuclear reactors as well as during the explosion of nuclear fission devices. In addition, isotopes of Np and Pu form in U-rich rocks as a result of the neutron flux maintained by spontaneous fission of ^{238}U, by (α, n) reactions, and by cosmic-ray neutrons.

chemically and by measuring the rate of emission of α-particles by the isotopes of Pu. Unfortunately, the α-particles emitted by ^{240}Pu and ^{239}Pu have similar energies and cannot be resolved by this technique. Therefore, the two isotopes are determined together and the results are reported as $^{240,\,239}$Pu. Alternatively, both the isotope composition and the concentration of Pu in natural samples can be determined by mass spectrometry. However, this technique is time consuming and is therefore not used for studies of Pu concentrations requiring large numbers of analyses.

The occurrence of Pu in nature implies that short-lived ^{238}Np ($T_{1/2} = 2.12$ d), ^{239}Np ($T_{1/2} = 2.35$ d), ^{240}Np ($T_{1/2} = 65$ m), and ^{241}Np ($T_{1/2} = 16.0$ m) must also be present because they are intermediate products during the neutron-capture reactions that produce the Pu isotopes. However, they all decay rapidly as soon as the process stops, whereas ^{237}Np ($T_{1/2} = 2.14 \times 10^6$ yr), ^{239}Pu, ^{240}Pu, and ^{242}Pu persist for tens of thousands of years because of their long half-lives.

The ^{241}Pu, formed by ^{240}Pu(n, γ)^{241}Pu, decays by β-emission to ^{241}Am ($T_{1/2} = 433$ yr), which occurs in fallout and radwaste and is now widely distributed all over the Earth (Livingston and Bowen, 1976; Koide et al., 1980; Day and Cross, 1981). Americium is strongly adsorbed onto the surfaces of particles including crystals of calcite and aragonite (Shanbhag and Morse, 1982). The equilibrium constant ($K_d = $ Am in solid/Am in solution) for calcite is $>10^5$ and the adsorption appears to be irreversible. Hence, ^{241}Am is scavenged rapidly from seawater and is incorporated into the sediment.

^{241}Am decays by α-emission to ^{237}Np ($T_{1/2} = 2.14 \times 10^6$ yr), which grows in and becomes one of the dominant radionuclides in old radwaste. The ^{241}Am ($T_{1/2} = 433$ yr) may also capture a neutron to form ^{242}Am ($T_{1/2} = 16.02$ h), which decays by β-emission to ^{242}Cm ($T_{1/2} = 163$ d). This isotope emits an α-particle as it decays to ^{238}Pu ($T_{1/2} = 87.8$ yr). This sequence of events therefore contributes to the presence of ^{242}Am, ^{243}Cm, ^{237}Np, and ^{238}Pu in fallout and in radwaste from nuclear reactors. Figure 24.4 indicates that a multitude of other radionuclides may form both

by neutron capture and by subsequent decay depending on the magnitude of the neutron flux, the duration of the irradiation, and the time elapsed since the irradiation ended.

The natural neutron flux of U-bearing rocks also causes the formation of other radioactive isotopes including ^{3}H, ^{14}C, ^{36}Cl, ^{37}Ar, and ^{39}Ar, whereas induced fission of ^{235}U in rocks produces ^{81}Kr, ^{85}Kr, and ^{129}I (Andrews et al., 1989). The magnitude of the neutron flux in different kinds of rocks depends on the concentrations of U and Th in the rocks because the neutrons arise primarily by spontaneous fission of ^{238}U and by (α, n) reactions. Cosmic-ray neutrons also contribute to the neutron flux, depending on the depth below the surface, and are most effective in the atmosphere. The neutron flux in rocks is affected by their neutron-absorbing capacity, which arises from the presence of O, F, Na, Al, Si, K, and Mg. The neutron fluxes in different kinds of rocks in Table 24.4 range from 4.07×10^{-4} n cm^{-2} s^{-1} in the U-rich Stripa granite in Sweden to 4.5×10^{-8} n cm^{-2} s^{-1} in the salts formed by evaporating seawater.

The geochemistry of Pu in the oceans and on the continents was reviewed and reinterpreted by Sholkovitz (1983) based on the voluminous literature that has accumulated on this subject. Bowen et al. (1980) estimated that more than 400 kCi of ^{239}Pu and ^{240}Pu have been released by nuclear weapons tests in the form of about 3900 kg of ^{239}Pu and 700 kg of ^{240}Pu. About 70 kCi of the total amount of Pu remained close to the test sites, including the Bikini and Eniwetok atolls in the Pacific Ocean, which contain ~2.7 kCi of Pu in lagoonal sediment (Noshkin and Wong, 1979) in addition to ~2 kCi in sediment in the surrounding ocean (H. D. Livingston in Sholkovitz, 1983).

In addition to its dispersion in fallout, Pu has been injected into the Earth's surface environment as a result of accidents. These include the burnup over the Indian Ocean of the SNAP-9A navigational satellite, which contained an RTG fueled with 1 kg (17 kCi) of ^{238}Pu in April of 1964 (Hardy et al., 1973), the crash of a B-52 bomber at Thule, Greenland, which released weapons-grade Pu (Harley, 1980), and the explosion of the nuclear reactor in Chernobyl, Ukraine, on April 6, 1986.

Table 24.4 Calculated Neutron Flux in Rocks of Various Compositions

Rock type	U, ppm	Th, ppm	Production n cm^{-3} y^{-1}	Absorption cross section, cm^2/g	Neutron flux, n cm^{-2} s^{-1}
Granite, Stripa, Sweden	44.1	33.0	217.1	0.0065	4.07×10^{-4}
Ultramafic	0.001	0.004	0.004	0.0044	1.0×10^{-8}
Basalt	1.0	4.0	10.2	0.0073	1.6×10^{-5}
Granite	3.5	18.0	36.7	0.0069	6.4×10^{-5}
Clay, shale	3.2	11.0	17.6	0.0098	3.0×10^{-5}
Sandstone	0.45	1.7	2.0	0.0061	4.6×10^{-6}
Carbonates	2.2	1.7	5.45	0.0039	2.0×10^{-5}
Evaporite[a]	0.1	$<10^{-5}$	1.02	0.3260	4.5×10^{-8}

[a]Salts formed by evaporation of seawater.
SOURCE: Andrews et al. (1989).

Other point sources of Pu and other radionuclides in the oceans are the reprocessing facilities at Windscale, United Kingdom (Aston and Stanners, 1981; Livingston et al., 1982), at LaHague near Cherbourg, France (Murray et al., 1979), at Savannah, Georgia (Hayes and Sackett, 1987), as well as dumping sites of low-level radwaste, especially in the northeastern Atlantic Ocean (Needler and Templeton, 1981). One reason for concern about the release of Pu into the oceans is the possibility that it can be remobilized from sediment after burial.

Plutonium can exist in four oxidation states (+3, +4, +5, and +6), it forms complex ions with both organic and inorganic ligands, and it can form colloidal suspensions (Cleveland, 1979). As a result, Pu is a *mobile* element under a wide range of environmental conditions and is therefore able to enter the food chain in the oceans. Consequently, fish and other kinds of seafood may become contaminated with Pu.

Noshkin and Wong (1979) reported that the concentration of Pu (^{239}Pu and ^{240}Pu) in the water of the lagoons of the Bikini and Eniwetok atolls ranges between 14×10^{-15} and 114×10^{-15} Ci/L, whereas that of the open ocean lies between 0.3×10^{-15} and 0.5×10^{-15}Ci/L. The Pu in the lagoonal water constitutes only 0.08% of the Pu in the upper 60 cm of sediment in the lagoons. Noshkin and Wong (1979) therefore predicted that it will take between 400 and 500 years for all of the Pu in the sediment to be released into the lagoons and hence into the open ocean. Plutonium is also being released by sediment deposited west of the islands that contains close-in fallout.

When Pu first enters the ocean, it is rapidly scavenged by adsorption on biogenic particles. However, as the particles sink they decompose, thereby releasing the adsorbed Pu, which goes into solution in deep water. The Pu remains in solution until it is readsorbed onto another particle that may carry it all the way to the bottom of the ocean. In addition, Pu may be carried downward rapidly on large particles and may be released from them while still in the water column, at the sediment–water interface, or after burial (Sholkovitz, 1983; Bowen et al., 1980). Evidently, the geochemistry of Pu in the oceans involves complex processes reflecting not only the chemical properties of the element but also the biological productivity of the water and the input of sediment particles, both of which vary regionally within the ocean.

24.5   Eh–pH Diagrams for Neptunium and Plutonium

The geochemical properties of the transuranium elements can be presented in the form of Eh–pH diagrams (Brookins, 1988). The mobility of these elements, indicated by these diagrams, depends on whether the hydroxide or oxide is assumed to be the stable phase. For example, Np may form $Np(OH)_4$, $NpO(OH)_2$, or NpO_2, all of which contain Np^{4+}. Similarly, Pu may exist as $Pu(OH)_4$ or PuO_2. The solid hydroxides transform spontaneously into the oxides when the water/rock ratio is reduced after initial precipitation of the hydroxide. In the case of Pu, the transformation is represented by the reaction:

$$Pu(OH)_4(s) \rightarrow PuO_2(s) + 2\,H_2O(l) \quad (24.1)$$

$$\Delta G_R^\circ = -11.08 \text{ kcal}$$

Hence, PuO_2 is the stable phase in the standard state, although metastable $Pu(OH)_4$ may initially precipitate from aqueous solution. When carbonate ions are present, the transuranium elements can form solid carbonates such as $Np(CO_3)_2$, $Pu_2(CO_3)_3$, and $Am_2(CO_3)_3$, as well as complex carbonate ions $PuO_2(OH)CO_3^{2-}$ and $NpO_2(OH)_2CO_3^{2-}$.

The transformation of $Np(OH)_4$ to $NpO(OH)_2$ and ultimately to NpO_2 is accompanied by a dramatic *decrease* in the solubility and hence in the mobility of this element at 25°C under conditions that prevail at the surface of the Earth or at shallow depth. The Eh–pH diagrams of Brookins (1988) in Figure 24.5 illustrates the progressive expansion of the stability fields of the solid phases $Np(OH)_4$, $NpO(OH)_2$, and NpO_2, respectively. At pH values between 2 and 10 in Figure 24.5A, three ions [NpO_2^+, $Np(OH)_5^-$, and Np^{3+}] have equilibrium activities $>10^{-6}$ mol/L in contact with $Np(OH)_4$. When $NpO(OH)_2$ is the solid, only NpO_2^+ has equilibrium activities of that magnitude (Figure 24.5B). The solubility of NpO_2 to form NpO_2^+ is further restricted to environments having pH <8.0 and Eh >0.753 V (Figure 24.5C). Consequently, Np is *not* expected to be mobile in

groundwater under normal pH conditions (pH >3 but <7).

The mobility of Np in the *oceans* (pH ≈ 8.2) predicted from Eh–pH diagrams depends on the solid phase that actually forms in this environment. If $Np(OH)_4$ is present (Figure 24.5A), Np can have activities $>10^{-6}$ mol/L both as $Np(OH)_5^-$ and as NpO_2^+ depending on the Eh. However, if NpO_2 is present (Figure 24.5C), then NpO_2^+, $Np(OH)_5^-$, and $NpO_2(OH)_2CO_3^{2-}$ all have equilibrium activities $<10^{-6}$ mol/L and Np is expected to remain immobile in the ocean and in marine sediment.

The solubility of the solids $Pu(OH)_4$ and $Pu_2(CO_3)_3$ with respect to Pu^{3+} and $Pu(OH)_5^-$ is depicted in Figure 24.6A, based on Brookins (1988). The ions PuO_2^+ and Pu^{4+} are dominant only at pH < 2 and Eh > +1.0 V and therefore are rarely important in natural environments. When PuO_2 is the stable phase, only Pu^{3+} can have activities $>10^{-6}$ mol/L at pH < 5.5 in systems represented by Figure 24.6B. Consequently, Pu may be *mobile* in acidic groundwaters under reducing as well as oxidizing conditions. As in the case of Np, the mobility of Pu in the oceans depends on whether solid $Pu(OH)_4$ or PuO_2 is actually present.

In reality, the geochemistry of Pu in the oceans is more complicated than it appears to be, based on the Eh–pH diagrams in Figure 24.6, because Pu occurs in four oxidation states: Pu^{3+}, Pu^{4+}, Pu^{5+}, and Pu^{6+}. Nelson and Lovett (1978) reported that the ratio $(Pu^{5+} + Pb^{6+})/(Pu^{3+} + Pu^{4+})$ of dissolved $^{239,\,240}Pu$ in the Irish Sea (derived largely from the Windscale reprocessing plant) ranges from about 2.4 to 12, whereas Pu adsorbed on suspended particles has lower values of this ratio between 0 and 0.4. Evidently, Pu^{5+} and Pu^{6+} are concentrated in the *aqueous phase,* whereas Pu^{3+} and Pu^{4+} dominate on *suspended particles.* However, the opposite appears to apply to some lakes in central North America and Canada where the (+5, +6)/(+3, +4) Pu ratio is <1 (Wahlgren et al., 1977). Evidently, the higher oxidation states dominate in the water of the Irish Sea, whereas the lower oxidation states are

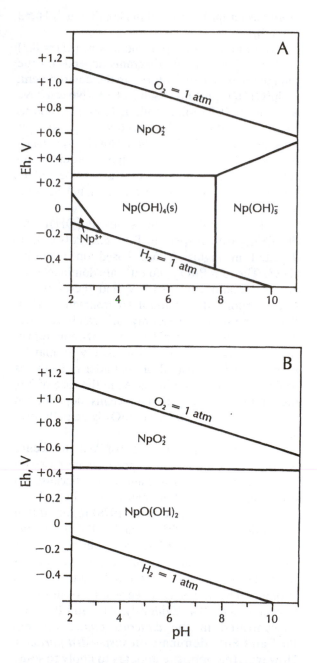

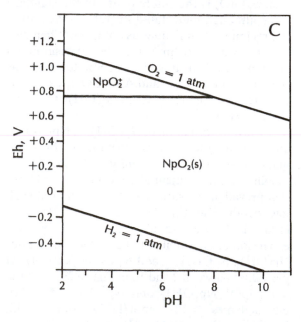

Figure 24.5 A. Eh–pH diagram for Np based on $Np(OH)_4$ as the solid phase in the presence of H_2O and containing $\Sigma(carbonate) = 10^{-3}$ mol/L. The solubility boundaries are drawn at activities of 10^{-6} mol/L of the ionic species, including NpO_2^+, $Np(OH)_5^-$, and Np^{3+}. In addition, Np^{4+} is the dominant ion at pH <2.0 and Eh >0.147 V but is not shown here.

B. Eh–pH diagram for Np based on $NpO(OH)_2$ as the solid phase in the presence of H_2O and containing $\Sigma(carbonate) = 10^{-3}$ mol/L. Evidently, $NpO(OH)_2$ is less soluble than $Np(OH)_4$, and NpO_2^+ is the only ion that can have activities $>10^{-6}$ mol/L in equilibrium with $NpO(OH)_2$ between pH 2 and pH 10.

C. Eh–pH diagram for Np based on NpO_2 as the solid phase in the presence of H_2O and containing Σ (carbonate) $= 10^{-3}$ mol/L. The Eh–pH conditions in which NpO_2^+ can have an activity $>10^{-6}$ mol/L in equilibrium with NpO_2 are further reduced. The equilibrium activity of the oxyhydroxycarbonate complex $(NpO_2(OH)_2CO_3^{2-})$ is $<10^{-6}$ mol/L and does not appear on this diagram. (Adapted from Brookins, 1988.)

dominant in the water of some North American lakes. Actually, Pu^{5+} and Pu^{6+} also dominate in Lake Michigan and Clear Lake, which have $(+5, +6)/(+3, +4)$ Pu ratios between 4 and 6 like the Irish Sea.

Sholkovitz (1983) reviewed evidence that Pu^{4+} forms strong complexes with dissolved organic compounds, whereas Pu^{6+} forms carbonate complexes (Aston, 1980). Therefore, the concentration of dissolved Pu and the distribution of

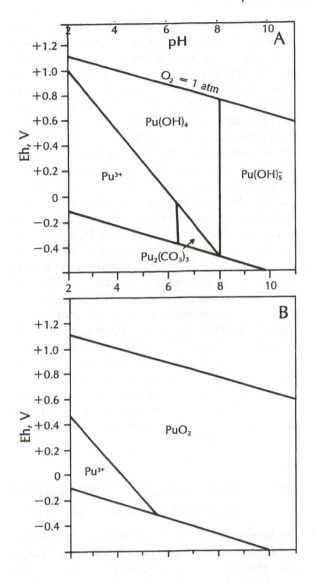

Figure 24.6 A. Eh–pH diagram for Pu based on the presence of $Pu(OH)_4$ and $Pu_2(CO_3)_3$ in a system in which $\Sigma(carbonate) = 10^{-3}$ mol/L. The solubility boundaries are drawn at ion activities of 10^{-6} mol/L.

B. Eh–pH diagram for Pu based on the presence of PuO_2 in a system in which $\Sigma(carbonate) = 10^{-3}$ mol/L. Note that PuO_2 is much less soluble than $Pu(OH)_4$, but that Pu^{3+} can have activities $>10^{-6}$ mol/L in acidic environments. That the geochemistry of Pu is strongly affected by ion exchange equilibria is not apparent from these Eh–pH diagrams.

its oxidation states may depend on the abundances of dissolved organic compounds (fulvic acids) of low molecular weight (5000–10,000), on the carbonate content, and hence on the *alkalinity* of the water.

Some of the Pu in seawater is associated with suspended particles and can be removed from it by 0.45-μm filters. Sholkovitz (1983) estimated that from 5 to 20% of Pu in the open ocean occurs in particulate form. Nelson and Lovett (1978) reported that the distribution coefficients (solid/liquid) of "reduced" (+3, +4) Pu in the Irish Sea are consistently *larger* than those of "oxidized" (+5, +6) Pu. In fact, the values of K_d for reduced particulate Pu are from 21 times to 1000 times larger than the values of K_d for oxidized particulate Pu. Evidently, the particles suspended in the water of the Irish Sea preferentially scavenge reduced Pu from seawater and thereby concentrate the oxidized forms in the aqueous phase. However, the numerical values of K_d are probably variable because they depend on the character of the particles (inorganic, biogenic, and living organisms), on their composition (mineralogy and organic matter content), and on their size distribution. In addition, because the abundance of suspended particles varies regionally within the ocean, the fraction of adsorbed Pu also varies.

Sholkovitz (1983) expressed the fraction of adsorbed Pu as a function of the concentration of suspended particles in a system in which $K_d(Pu^{3+}, Pu^{4+}) = 25 \times 10^5$ kg/L, whereas $K_d(Pu^{5+}, Pu^{6+}) = 0.15 \times 10^5$ kg/L. The resulting graph in Figure 24.7 illustrates that ~95% of the "reduced" Pu is adsorbed when the suspended load is 10 mg/L, whereas only ~10% of the "oxidized" Pu is adsorbed under these conditions. Consequently, the ratio of "oxidized" to "reduced" Pu in solution *does* depend on the concentration of suspended particles and on their characteristic properties, which determine their ability to adsorb Pu ions.

The preceding discussion makes clear that Eh–pH diagrams are *incomplete* representations of the geochemistry of Pu in the ocean as well as in lakes and rivers. Presumably, ion exchange phenomena also affect the geochemical properties of

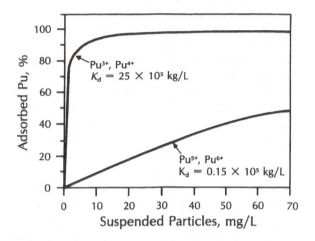

Figure 24.7 Fraction of Pu ions that are adsorbed onto suspended particles depending on their oxidation states, the concentration of suspended particles, and the values of the distribution constants (K_d) of the Pu^{3+}, Pu^{4+} (reduced) and Pu^{5+}, Pu^{6+} (oxidized) Pu. The graph illustrates the phenomenon that reduced Pu is scavenged more effectively than oxidized Pu and that the ratio of oxidized to reduced Pu remaining in solution can vary widely depending on the concentration of suspended particles and their compositions, which determine the magnitude of the distribution constants (adapted from Sholkovitz, 1983, Figure 9).

For example, the thermal gradient around a canister of HLW in a subsurface repository may cause silicate minerals to break down in its vicinity releasing silicic acid into the pore fluid which may transport it away from the canister to areas of lower temperature where the silicic acid precipitates as amorphous silica, thereby reducing the permeability of the rocks. Under the same set of circumstances, carbonates and sulfates, whose solubilities decrease with increasing temperature, may be deposited on the outer walls of the canister, thereby creating an additional barrier to the movement of radionuclides. Therefore, the escape of radionuclides from a canister of HLW is affected by geochemical processes that change the chemical composition and mechanical properties of the rocks surrounding the canister. The coupling of the process that may change the geochemical environment around an aging HLW repository makes it more difficult to predict the release of radionuclides and to evaluate its future safety. Consequently, such evaluations are not totally reliable and need to be confirmed by examining natural repositories of radioactive elements such as ore deposits of U in the form of veins or as mineral cements in sandstone (Brookins, 1984).

Pu ions in solution in groundwater. In addition, the study of dissolution, transport, and deposition of long-lived radionuclides, such as the isotopes of Pu, Np, and Am, in underground repositories must be coupled to other processes occurring at the same time.

The *coupling* of thermal, hydrological, mechanical, and chemical processes in predicting the long-term behavior of nuclear-waste repositories was the topic of an international conference in 1985 in Berkeley, California. The Proceedings of this conference were edited by Tsang (1987). It contains a paper by Langmuir (1987) in which he discussed the difficulties of predicting the outcome of geochemical processes in waste repositories because of the lack of sufficient thermodynamic and kinetic data at elevated temperature and pressure. In addition, even geochemical processes are *coupled*.

24.6 Analog Studies: The Natural Reactors at Oklo, Gabon

On September 25, 1972, the former chairman of the French High Commission for Atomic Energy, Dr. Francis Perrin, reported to the French Academy of Sciences that a uranium deposit in West Africa had undergone a chain reaction based on induced fission of ^{235}U (Section 16.1c). The deposit at Oklo, Gabon, had been put into production in the summer of 1970 and the U concentrate from the open-pit mine was shipped to Pierrelatte, France, for enrichment in ^{235}U prior to its use as reactor fuel. However, isotope analyses revealed that the U from Oklo was variably *depleted* in ^{235}U and that the abundance of ^{235}U in some batches of U concentrate was only 0.440 atom % compared to 0.720 atom % in normal terrestrial U. The results also suggested that the abundance of ^{235}U decreased with increasing U content of the ore. In addition, analyses by Neuilly et al. (1972) at the French atomic center in Cadarach indicated that Nd, Sm, Eu, and Ce in the ore have anomalous isotope compositions. The scientists who had participated in these discoveries therefore concluded that ^{235}U in the orebodies at Ok^{1}o had fissioned in self-sustaining chain reactions with groundwater acting as the *moderator* of neutrons emitted by each fission event. The chain reaction probably operated intermittently because the reaction may have stopped when the rising temperature of the U-rich rocks caused the water to be converted to steam (Sullivan, 1972).

The discovery that natural chain reactions had occurred at Oklo was not entirely unexpected. Sixteen years earlier, Kuroda (1956) had demonstrated that 2100 million years ago a mass of pitchblende a few feet in diameter could have sustained a chain reaction in the presence of water because at that time the isotopic abundance of ^{235}U was 4.0% compared to 0.720% at the present time. Dr. Kuroda concluded that "The effect of such an event could have been a sudden elevation of the temperature, followed by a complete destruction of the critical assemblage." As a historical footnote let it be recorded that Dr. Kuroda obtained the analysis of the pitchblende (from Johanngeorgenstadt, Germany) out of F. W. Clarke's book, *The Data of Geochemistry* (see Chapter 1).

The unique natural reactors at the Oklo mine provide an opportunity to study the migration of fission-product nuclides in a natural setting. Consequently, the deposit has been studied intensively to determine its origin, to ascertain the conditions that permitted the occurrence of chain reactions, and to evaluate the mobility of long-lived radionuclides produced as fission products and by neutron capture (Brookins, 1978a,b; de Laeter et al., 1980; Loss et al., 1984). The initial results were summarized by Naudet (1974) and other French investigators in the June issue (No. 193) of the *Bulletin d'Informations Scientifiques et Techniques du Commissariat à l'Energie Atomique*. Later, the International Atomic Energy Agency in Vienna issued two reports on the Oklo reactors (IAEA, 1975, 1978). In addition, Kuroda (1982) included a description of the natural fission reactors at Oklo in his book on nuclear processes in nature.

The U ore at the Oklo mine occurs in sandstones of the Precambrian Francevillian Series. The U minerals were initially deposited syngenetically, but were subsequently remobilized during structural deformation of the host rocks and redeposited in small bodies containing between 50 and 70% U. The age of the U ore is 2.05×10^9 years based on the U–Pb method (Gancarz, 1978). As a result of the dissolution and reprecipitation, the remobilized U was not only concentrated but was also purified of certain elements (V, Se, Mo, and Fe) that could act as reactor poisons by absorbing neutrons.

The ore at the Oklo mine is composed of pitchblende (complex U-oxides, amorphous or cryptocrystalline), with impurities of chlorite and illite. Quartz is rare in the ore but common in the surrounding rocks. The ore also contains small amounts of pyrite, hematite, barite, chalcopyrite, and other minerals. Brookins (1984) pointed out that the presence of both pyrite and hematite fixes the range of Eh and pH conditions that existed in the ore at the time the chain reactions occurred (see Section 14.5d, Figure 14.11).

At the time of formation of the enriched orebodies at Oklo, the abundance of ^{235}U was 3.2%, which is within the range of modern reactor fuel. However, in order for a chain reaction to start based on *induced* fission of ^{235}U, groundwater must have been present in sufficient amounts to reduce the kinetic energies of neutrons to permit them to be absorbed by nuclei of ^{235}U, causing it to fission into complementary pairs of fission-product nuclides that subsequently decayed to stable daughters. About 17% of the time, the absorption of a slow neutron by ^{235}U leads to the formation of ^{236}U (Figure 24.4), which then decays by α-emission to long-lived ^{232}Th. Some of the ^{236}U atoms formed from ^{235}U may capture another neutron to form ^{237}U, which decays by a long series of α- and β-emissions to stable ^{209}Bi. Therefore, it is significant that Th and Bi occur in the richest U ore at Oklo but not elsewhere in the deposit (Cowan, 1976).

Studies by many investigators have demonstrated that the stable daughters of about 30 fission-product nuclides are present in the natural reactors at Oklo in just about the expected amounts and proportions. This not only confirms the occurrence of chain reactions but also indicates that the fission-product nuclides in most cases did *not* migrate out of the volume of rock in which they were formed. Among the *immobile* elements are the REEs (La, Ce, Pr, Nd, Eu, Sm, and Gd) and the Pt group elements (Ru, Rh, and Pd), as well as Y, Zr, Nb, Th, and Bi. The REEs and Pt-group elements include the long-lived radionuclides that may survive in aging repositories of nuclear waste. Evidently, these potentially dangerous radionuclides did *not* migrate at Oklo. In addition, the transuranium elements Np, Pu, Am, and Cm were *not* mobile in the orebodies while the chain reactions were in progress or afterward (Brookins, 1984).

The elements that appear to have been only partially retained in the reactors at Oklo include Mo, Ag, I, Kr, and Xe. In addition, Rb, Cs, Sr, Cd, and Pb were either lost completely or show evidence of local migration (Cowan, 1976).

The chain reactions within the Oklo orebodies were based primarily on induced fission of the ^{235}U in the U ore. However, the indigenous ^{235}U was augmented by ^{235}U produced by decay of ^{239}Pu. The chain reactions were also based to some extent on induced fission of ^{239}Pu itself and on fission of ^{238}U caused by *fast* neutrons. Altogether, six tons of ^{235}U were consumed and 15,000 megawatt-years of energy were released by the chain reactions that occurred for at least 500,000 years (Cowan, 1976; Brookins, 1984).

During the period the chain reactions were occurring, the temperatures within the reactors probably exceeded 400°C, but the rocks did not melt and the deposit did not explode. The reason is that the chain reactions were controlled by the water content of the ore, which must have ranged from 12 to 15% to permit induced fission of ^{235}U to occur. When the temperature rose, the water content of the rocks was reduced, the flux of slow

neutrons was therefore decreased, and the chain reactions stopped. The evidence for relatively low temperatures in the reactor zones is supported by the fact that Bonhomme et al. (1965) obtained a Rb–Sr whole-rock isochron date of 1.85 ± 0.1 Ga for sedimentary rocks that overlie the U deposits at Oklo. Moreover, they reported a K–Ar date of 1.8 ± 0.1 Ga for concentrates of the $1M_d$ polymorph of illite from these rocks (Sections 16.3a, 16.3b, 13.2b.3, and 13.6). The $1M_d$ polymorph of illite does not retain radiogenic ^{40}Ar above ~125°C and is converted into the 2M polymorph between 150 and 250°C. Therefore, the survival of the $1M_d$ polymorph in the immediate vicinity of the reactors indicates that the chain reactions did not generate high temperatures in the surrounding rocks. In addition, these rocks could not have been buried more than a few hundred meters without destroying the $1M_d$ illite. However, the depth of the U ore at the time the natural reactors were operating is not known precisely.

The relevance of the natural reactors at Oklo to the evaluation of the safety of geological repositories of HLW is that the long-lived radionuclides of the REEs, the Pt-group metals, and the transuranium elements were quantitatively retained in rocks that have widely varying permeabilities and are also strongly jointed and fractured. Because of these properties, the Oklo site would *not* have been selected for the construction of a nuclear waste repository. Therefore, the evidence derived from studies of the natural reactors at Oklo indicates that rocks having low permeability and high thermal stability should retain radwaste even better than the rocks at Oklo.

24.7   Reactor Accidents: Chernobyl, Ukraine

The energy liberated by induced nuclear fission is very large compared to that of conventional energy sources, such as the combustion of fossil fuel or chemical explosives, including trinitrotoluene (TNT). Nevertheless, the process can be controlled by modulating the flux of thermal neutrons in the core of the reactor. However, if for any reason the process goes out of control, the temperature in the reactor core can increase very rapidly, causing an explosion of the core and leading to the dispersal of radionuclides. Although the design of nuclear reactors is supposed to prevent such occurrences, several accidents have taken place since 1942, when the first reactor was built by Enrico Fermi and his associates in a squash court in the basement of the football stadium of The University of Chicago. The most recent reactor accident occurred on April 26, 1986, in Chernobyl, Ukraine.

The events leading up to this accident were described by Baier (1989) and Anspaugh et al. (1988). According to these sources, the Chernobyl Nuclear Power Station near the city of Kiev operated four power reactors of the "RBMK type" in which each fuel rod is contained within a pipe through which water is pumped at high pressure. The fuel assemblies were embedded in graphite, which served as the moderator. Reactor No. 4 at Chernobyl had 1659 fuel assemblies, each of which contained 114.7 kg of U. The fuel elements had been in the reactor for 610 days on average and had therefore accumulated more than 200 different fission-product nuclides including ^{90}Sr ($T_{1/2}$ = 29 yr) and ^{137}Cs ($T_{1/2}$ = 30.1 yr).

The power output of the reactor was controlled by means of 211 rods composed of neutron absorbers that could be inserted into the core to stop the chain reaction. Under normal operating conditions, 30 control rods had to be in place in the core of reactor No. 4 in order to keep the chain reaction under control. However, on the day before the accident the operators had reduced power to determine whether the heat in the graphite core of the reactor was sufficient to run the main electrical generator. While this experiment was in progress, the demand for electricity in the transmission lines fed by the generator suddenly increased. Therefore, the operators attempted to bring the reactor back to full power by withdrawing all but eight of the control rods, in violation of mandatory operating procedures. Consequently, the energy output rose within three seconds from 200 to 530 megawatts. In the next fraction of a second the reactor went out of

control. As the core temperature increased by 3000 °C, Zr in the fuel elements reacted with water to form oxides, thereby releasing H_2 gas, which reacted explosively with O_2 and burst the steel housing containing the 2500 tons of graphite of the reactor core. The initial explosion caused a large-scale release of fission-product nuclides, including parts of the fuel elements, and resulted in the ultimate combustion of 250 tons of graphite.

After the initial explosive release of radionuclides, the temperature of the reactor core continued to rise for several days as a result of decay of radionuclides and because the cooling system had been destroyed in the explosion. The release of radionuclides, which had decreased drastically after the explosion, increased with the rising temperature and almost equaled the initial rate nine days earlier. The secondary release consisted primarily of the isotopes of volatile elements, including all of the noble gases, ^{131}I ($T_{1/2}$ = 8.041 d), ^{134}Cs ($T_{1/2}$ = 2.06 yr), and ^{137}Cs ($T_{1/2}$ = 30.1 yr), but only small amounts of nonvolatile radionuclides such as ^{89}Sr ($T_{1/2}$ = 50.59 d), ^{90}Sr ($T_{1/2}$ = 29 yr), ^{141}Ce ($T_{1/2}$ = 32.53 d), ^{144}Ce ($T_{1/2}$ = 284.44 d) and the Pu isotopes ^{238}Pu ($T_{1/2}$ = 87.8 yr), ^{239}Pu ($T_{1/2}$ = 2.439 × 10^4 yr), and ^{240}Pu ($T_{1/2}$ = 6.54 × 10^3 yr). The release was terminated nine days after the accident by corrective measures.

The plume of radionuclides emitted from Chernobyl moved northwest and reached Sweden and Finland, where it was quickly detected. Subsequently, the plume changed direction several times, resulting in the deposition of fallout over western Russia and most of Europe. The total amount of ^{137}Cs dispersed over the land areas of the northern hemisphere was about 10 × 10^{16} Bq, of which western Russia and Europe received 79% and eastern Russia and Asia 20%, whereas the fallout over Canada and the United States amounted to 2.5 × 10^{14} and 2.8 × 10^{14} Bq, respectively, or about 0.5% of the total. In addition to the former Soviet Union, the countries that were hardest hit by the fallout from the accident are identified in Table 24.5.

As a direct consequence of the accident, 31 persons in Ukraine died, 237 suffered acute radiation sickness, and more than 100,000 persons had

Table 24.5 Fallout of ^{137}Cs and Consequent Individual Radiation Dose of Persons in Selected Countries in Order of the Arrival of the Plume After the Reactor Accident in Chernobyl, Ukraine, on April 26, 1986

Country	Time of plume arrival, hr	Total ^{137}Cs deposition, Bq	50-year external dose, Gray/person
Finland	66	1.9×10^{15}	4.3×10^{-4}
Norway	78	1.1×10^{15}	2.1×10^{-4}
Sweden	78	3.4×10^{15}	5.8×10^{-4}
Poland	78	9.2×10^{15}	2.2×10^{-3}
Romania	90	6.7×10^{15}	2.1×10^{-3}
East Germany	102	5.8×10^{14}	4.0×10^{-4}
Switzerland	114	2.0×10^{14}	3.7×10^{-4}
Austria	114	1.1×10^{15}	9.5×10^{-4}
Bulgaria	132	2.7×10^{15}	1.8×10^{-3}
Czechoslovakia	132	5.9×10^{14}	3.5×10^{-4}
France	132	8.3×10^{14}	1.1×10^{-4}
Luxembourg	132	3.8×10^{12}	1.1×10^{-4}
West Germany	156	1.6×10^{15}	4.9×10^{-4}
Hungary	156	7.9×10^{14}	6.3×10^{-4}
Italy	156	1.9×10^{15}	4.7×10^{-4}
United States	372	2.8×10^{14}	2.4×10^{-6}
Canada	372	2.5×10^{14}	2.0×10^{-6}

SOURCE: Anspaugh et al. (1988).

to be relocated. In addition, between 1988 and 1998 the risk of leukemia may double in the population living within 30 km of Chernobyl (Anspaugh et al., 1988). Although the cloud of radionuclides released by the accident contaminated most of the northern hemisphere inhabited by about 3 billion people, no cases of acute radiation sickness were reported outside of the former Soviet Union. Nevertheless, the lifetime expectation of fatal radiogenic cancer increased from 0 to 0.2% in Europe and from 0 to 0.003% in the northern hemisphere (Anspaugh et al., 1988).

The economic and social damage caused by this accident is very large. By extrapolating initial estimates by the former Soviet Union, the total damage may amount to $15 billion, of which 90% occurred within the borders of the former Soviet Union. In addition, the accident has heightened concern about the safety of nuclear reactors and has reinforced the conviction that *fossil fuels* are a *safer* source of energy than nuclear reactors. Unfortunately, the release of CO_2, SO_2, and oxides of N caused by the combustion of fossil fuel also has far-reaching environmental consequences that manifest themselves in the form of urban smog, acid rain, and climate change on a global scale.

A decade after the explosion of the nuclear reactor in Chernobyl, Ukraine, the damage to the health of the population and to the economy of the region continues to increase. Estimates now place the total amount of radioactive materials released by the explosion at about 90 million Curies.

The economic impact of this catastrophe is beyond reckoning because the expenses for remedial activities, such as the relocation of the inhabitants of the area around Chernobyl and their continuing medical care, as well as the stabilization of the explosion site, are still rising. The failure of the Soviet government to act decisively to evacuate the population and to notify governments of neighboring countries ultimately contributed to the disintegration of the Soviet Union (Shcherbak, 1996).

The fire fighters who were called upon to extinguish the fire were exposed to high doses of gamma radiation ranging from about 100 to more than 100,000 rem. As a result of this exposure, 187 persons had to be hospitalized and 31 of them died. Subsequently, about 400,000 workers participated in the construction of a protective containment building (the "sarcophagus") and in the burial of contaminated materials. In this group about 30,000 persons became ill and 5,000 are too sick to work.

The symptoms of the so-called Chernobyl syndrome include fatigue, apathy, and a decrease in the number of white blood cells. As a result, there is an increase in the occurrence of malignant tumors, leukemia, and severe cardiac conditions, as well as a weakening of the immune system leading to an increase in common infections such as bronchitis, tonsillitis, and pneumonia.

Although ^{131}I has a relatively short half-life of 8 days, this isotope is especially dangerous because it is concentrated in the thyroid gland. According to Shcherbak (1996), about 4000 children received doses up to 2000 rems to their thyroid glands and a larger number received more than 200 rems. These exposures initially caused inflammation of their thyroids that in some cases has led to thyroid cancer. As a result, the incidence of thyroid cancers among children and adolescents in the Ukraine has increased from about five per year before the accident to an average of 43 cases per year in 1995. The only effective treatment is to surgically remove the thyroid gland, thereby making the patients dependent on supplemental thyroid hormones for the rest of their lives. The experience of Hiroshima and Nagasaki suggests that the population affected by the accident at Chernobyl will continue to experience an abnormally high rate of medical problems for many years to come.

The reactor building still contains several tens of thousands of tons of radioactive material with a decay rate of about 20 million curies. In addition, radioactive materials, including trees that absorbed radionuclides from the atmosphere, have been buried in about 800 clay-lined pits within an area of about 190,000 square kilometers centered on Chernobyl. Nevertheless, sediment suspended in the Dnieper River and its tributary the Pripyat River has become contaminated with ^{90}Sr, ^{137}Cs, and Pu.

24.8   Summary

The consumption of natural resources and the disposal of the resulting waste generated by the world's growing population are threatening to alter the environment in ways that may be harmful. The need for increasing amounts of energy puts an added burden on the environment because both the combustion of fossil fuel and controlled fission of U generate waste products that can adversely affect the quality of life on the Earth in the future.

The waste products of nuclear reactors consist of radioactive fission products and of the unstable isotopes of the transuranium elements Np, Pu, Am, and Cm. The decay of these radionuclides, which is accompanied by the emission of α-particles, β-particles, and γ-rays, makes them lethal to all life forms for thousands of years. In fact, high-level waste (HLW) and spent unreprocessed fuel (SURF) remain harmful to life for more than 100,000 years and must therefore be isolated in subsurface repositories.

The construction of such repositories is technically feasible but has been delayed by the need for thorough studies of the proposed sites and by the reluctance of people living near proposed repository sites to permit the storage of nuclear waste in their neighborhood. Nevertheless, countries that have a nuclear-power industry are proceeding toward the construction of nuclear-waste repositories involving different kinds of rocks. The evaluation of the long-term safety of these repositories depends partly on their ability to retain long-lived radioactive isotopes of the transuranium elements, including Np and Pu.

These two elements, as well as Am and Cm, have been dispersed over the surface of the Earth as fallout from nuclear-weapons tests and by controlled discharge from fuel-reprocessing plants. Although the chemical properties of these elements can be predicted from Eh–pH diagrams, their behavior in the natural environment is strongly modified by the formation of complex ions and by the adsorption of their ions on organic and inorganic particles. For example, the distribution of Pu in the oceans and in lakes is controlled by the magnitude of the flux of suspended particles, by the composition and size distribution of the particles, and by their ability to adsorb Pu in the lower oxidation states ($+3$ and $+4$).

Because of the complex interaction of geochemical, hydrologic, thermal, and mechanical processes in the rocks surrounding aging nuclear-waste repositories, model calculations, intended to evaluate their future safety, must be tested by the study of natural analogs. One of the best opportunities to do so is provided by the high-grade U orebodies at Oklo, Gabon, in which chain reactions based on induced fission of ^{235}U occurred ~2×10^9 yr ago. Careful studies of these natural reactors since their discovery in 1972 has indicated that Np and Pu, and many other elements having long-lived radioisotopes, did *not* migrate out of the U ore in which they were formed, even though the properties of the rocks are far from ideal.

The energy released by nuclear fission is very large compared to combustion of fossil fuel. Therefore, nuclear reactors pose an inherent danger because accidents not only may liberate large amounts of energy but also can release radionuclides into the environment. The most recent reactor accident occurred on April 26, 1986, at Chernobyl, Ukraine. As a direct consequence of this accident, 31 persons were killed, 237 suffered acute radiation sickness, more than 100,000 people from the surrounding area had to be relocated, and the risk of leukemia and thyroid cancer in the affected population was increased significantly.

In addition, ^{137}Cs and other fission products were spread over the western region of the former Soviet Union and Europe, ultimately affecting 3 billion people in the entire northern hemisphere. Although no cases of acute radiation poisoning were reported outside the former Soviet Union, severe property loss and social damage did result, especially in Scandinavia, which was the first region to receive fallout from the plume emanating from Chernobyl.

Problems

1. Calculate the activity ratio (i.e., the ratio of the decay rates) of Pu having an atomic ratio $^{240}Pu/^{239}Pu = 0.18$. The half-lives are 6.54×10^3 yr for ^{240}Pu and 2.44×10^4 yr for ^{239}Pu.

2. Calculate the weight in grams of ^{238}Pu ($T_{1/2} = 8.78$ yr) that produces a decay rate of 17 kCi.

3. Calculate the disintegration rate of 1.0 g of ^{238}U ($T_{1/2} = 4.47 \times 10^9$ yr) and express the results in becquerels and in picocuries.

4. Between 1957 and 1978, 14 kCi of $^{239, 240}Pu$ was released from the Windscale reprocessing plant. If the atomic $^{240}Pu/^{239}Pu$ ratio was 0.18, estimate the weight of the released Pu in grams. The masses of the isotopes are ^{239}Pu : 239.052 18 amu, ^{240}Pu : 240.053 83. The half-lives are given in Problem 1.

5. Convert 1.0 pCi (10^{-12} Ci) into the corresponding decay rate in disintegrations per minute (dpm).

6. Calculate the activity in picocuries of 1 kg of brine from Mono Lake, California, whose Pu concentration is 4.4 dpm/kg.

7. Calculate the concentration of Pu in seawater in parts per million by weight, given that the concentration of $^{239}Pu + ^{240}Pu$ is 0.4 fCi/L (1 femtocurie = 10^{-15} Ci). Assume that the density of seawater is 1.03 and that the atomic ratio $^{240}Pu/^{239}Pu = 0.18$.

8. Construct an Eh–pH diagram for Am in a system containing 10^{-3} mol/L of carbonate ions at 25°C. Include stability fields for the solids $Am_2(CO_3)_3, AmO_2$, and $Am(OH)_3$ and for the ions Am^{3+} and $Am(OH)_5^-$. Draw the solubility boundaries at 10^{-8} mol/L.

9. Write a brief essay on the mobility of Am in natural environments *and* suggest a method of retarding the movement of Am ions out of a nuclear-waste repository.

10. Calculate the activity of $^{241}Am^{3+}$ in fCi/L of a solution that is in equilibrium with solid $Am_2(CO_3)_3$ at pH = 7.0. Assume that the activity coefficient (γ) of Am^{3+} in the solution is equal to 0.75.

References

ALEXANDER, E. C., Jr., R. S. LEWIS, J. H. REYNOLDS, and M. C. MICHEL, 1971. Plutonium-244: confirmation as an extinct radioactivity. *Science,* **172**:837.

ANDREWS, J. N., S. N. DAVIS, J. FABRIJKA-MARTIN, J.-CH. FONTES, B. E. LEHMANN, H. H. LOOSLI, J.-L. MICHELOT, H. MOSER, B. SMITH, and M. WOLF, 1989. The in situ production of radioisotopes in rock matrices with particular reference to the Stripa granite. *Geochim. Cosmochim. Acta,* **53**:1803–1815.

ANONYMOUS, 1981. *Siting of Radioactive Waste Repositories in Geological Formations.* Nuclear Energy Agency, Organisation for Economic Cooperation and Development, Paris, France, 259 pp.

ANONYMOUS, 1983. *A Study of the Isolation System for Geologic Disposal of Radioactive Wastes.* Waste Isolation Systems Panel. Board on Radioactive Waste Management, National Academy Press, Washington, DC, 345 pp.

ANONYMOUS, 1984. *Geological Disposal of Radioactive Waste.* Organisation for Economic Cooperation and Development, Paris, France, 116 pp.

ANSPAUGH, L. R., R. J. CATLIN, and M. GOLDMAN, 1988. The global impact of the Chernobyl reactor accident. *Science,* **242**:1513–1519.

ASTON, S. R., 1980. Evaluation of the chemical forms of plutonium in seawater. *Marine Chem.,* **8**:319–325.

ASTON, S. R., and D. A. STANNERS, 1981. Plutonium transport to and deposition and immobility in Irish Sea intertidal sediments. *Nature,* **289**:581–582.

BAIER, W., 1989. Eine Anlage zur kontinuierlichen Gewinnung von Bomben–Plutonium. *Frankfurter Rundschau,* April 25, No. 96:24.

BARNES, R. W., 1979. The management of irradiated fuel in Canada. In C. R. Barnes (Ed.), *Canadian Geological Survey,* Paper 79-10, Ottawa.

BARNEY, G. S., J. D. NAVRATIL, and W. W. SCHULZ (Eds.), 1984. *Geochemical Behavior of Disposed Radioactive Waste.* ACS Symposium Series 246, American Chemical Society, Washington, DC, 413 pp.

BLASEWITZ, A. G., J. M. DAVIS, and M. R. SMITH, 1983. *The Treatment of Radioactive Wastes.* Springer-Verlag, New York, 658 pp.

BLOMKE, J. O., 1976. Management of radioactive wastes. In J. H. Rust and L. W. Weaver (Eds.), *Nuclear Power Safety.* Pergamon, Oxford, England.

BONHOMME, M., F. WEBER, and R. FAUVRE-MERCURET, 1965. Age par la methode rubidium–strontium des sediments du bassin de Franceville, République Gabonaise. *Bull. Serv. Carte Geol. Alsace Lorraine,* **18**:243.

BOWEN, V. T., V. E. NOSHKIN, H. D. LIVINGSTON, and H. L. VOLCHOK, 1980. Fallout radionuclides in the Pacific Ocean: vertical and horizontal distributions, largely from Geosecs stations. *Earth Planet. Sci. Lett.,* **49**:411–434.

BROOKINS, D. G., 1978a. Retention of transuranic and actinide elements and bismuth at the Oklo natural reactor, Gabon: application of Eh–pH diagrams. *Chem. Geol.,* **23**:309–323.

BROOKINS, D. G., 1978b. Eh–pH diagrams for elements from $Z = 40$ to $Z = 52$; application to the Oklo natural reactor. *Chem. Geol.,* **23**:324–342.

BROOKINS, D. G., 1984. *Geochemical Aspects of Radioactive Waste Disposal.* Springer-Verlag, New York, 347 pp.

BROOKINS, D. G., 1988. *Eh–pH Diagrams for Geochemistry.* Springer-Verlag, New York, 176 pp.

CANNON, H. L., 1974. *Geochemistry and the Environment, vol. 1. The Relation of Selected Trace Elements to Health and Disease.* National Academy of Sciences, Washington, DC, 105 pp.

CHAPMAN, N. A., and I. G. MCKINLEY, 1987. *The Geological Disposal of Nuclear Waste.* Wiley, New York, 232 pp.

CLEVELAND, J. M., 1979. Critical review of plutonium equilibria of environmental concern. In E. A. Jenne (Ed.), *Chemical Modeling in Aqueous Systems,* 321–338. ACS Symposium Series 93, American Chemical Society, Washington, DC.

COWAN, G. A., 1976. A natural fission reactor. *Sci. Amer.,* No. 7:376–47.

DAY, J. P., and Cross, J. E., 1981. ^{241}Am from the decay of ^{241}Pu in the Irish Sea. *Nature,* **292**:43–45.

DE LAETER, J. R., K. J. R. ROSMAN, and C. L. SMITH, 1980. The Oklo natural reactor: cumulative fission yields and retentivity of the symmetric mass region fission products. *Earth Planet. Sci. Lett.,* **50**:238–244.

FAURE, G., 1986. *Principles of Isotope Geology,* 2nd ed., Wiley, New York, 466 pp.

FRIED, S. (Ed.), 1979. *Radioactive Waste in Geologic Storage.* ACS Symposium Series 100, American Chemical Society, Washington, DC, 344 pp.

GANCARZ, A., 1978. U–Pb age (2.05×10^9 years) of the Oklo uranium deposit. *Internat. Atomic Energy Agency, Technical Communication* **119**:513–520.

GARNER, C. S., N. A. BONNER, and G. T. SEABORG, 1948. Search for elements 94 and 93 in nature. Presence of ^{239}Pu in carnotite. *J. Amer. Chem. Soc.,* **70**:3453.

GOLDSCHMIDT, V. M., 1942. Om super-uraner, grunnstoffer med større kjerladning enn 92. *Fysik Verden,* **3**:179.

HARDY, E. P., P. W. KREY, and H. L. VOLCHOK, 1973. Global inventory and distribution of fallout plutonium. *Nature,* **241**:444–445.

HARLEY, J. H., 1980. Plutonium in the environment—a review. *J. Radiat. Res.,* **21**:83–104.

HAYES, D. W., and W. M. SACKETT, 1987. Plutonium and cesium radionuclides in the sediments of the Savannah River estuary. *Estuarine, Coastal and Shelf Sci.,* **25**:169–174.

IAEA, 1975. *The Oklo Phenomenon.* IAEA Symposium, vol. 204.

IAEA, 1978. *Natural Fission Reactors.* IAEA Tech. Communication, No. 119.

KOIDE, M., E. D. GOLDBERG, and V. F. HODGE, 1980. ^{241}Pu and ^{241}Am in sediments from coastal basins off California and Mexico. *Earth Planet. Sci. Lett.,* **48**:250–256.

KOIDE, M., K. K. BERTINE, T. J. CHOW, and E. D. GOLDBERG, 1985. The ^{240}Pu/^{239}Pu ratio, a potential geochronometer. *Earth Planet. Sci. Lett.,* **72**:1–8.

KURODA, P. K., 1956. On the nuclear physical stability of the uranium minerals. Letters to the Editor, *J. Chem. Phys.,* **25**:781–782.

KURODA, P. K., 1982. *The Origin of the Chemical Elements and the Oklo Phenomenon.* Springer-Verlag, New York, 165 pp.

LANGMUIR, D., 1987. Overview of coupled processes with emphasis in geochemistry. In C.-F. Tsang (Ed.), *Coupled Processes Associated with Nuclear Waste Repositories,* 67–101. Academic Press, Orlando, FL, 801 pp.

LIVINGSTON, H. D., and V. T. BOWEN, 1976. Americium in the marine environment: relationships to plutonium, 107–130. In M. W. Miller and J. N. Stannard (Eds.), *Environmental Toxicity of Aquatic Radionuclides.* Ann Arbor Science Publishers, Ann Arbor, MI.

LIVINGSTON, H. D., V. T. BOWEN, and S. L. KUPFERMAN, 1982. Radionuclides from Windscale discharges, I. Non-equilibrium tracer experiment in high latitude oceanography. *J. Marine Res.,* **40**:253–272.

LOSS, R. D., K. J. R. ROSMAN, and J. R. DE LAETER, 1984. Transport of symmetric mass region fission products at the Oklo natural reactors. *Earth Planet. Sci. Lett.,* **68**:240–248.

MURRAY, C. N., H. Kautsky, and H. F. EICKE, 1979. Transfer of actinides from the English Channel into the southern North Sea. *Nature,* **276**:225–230.

NAUDET, R., 1974. Le phénomène d'Oklo. *Bull. Informat. Scient. Tech.,* **193**:7–86.

NEEDLER, G. T., and W. C. TEMPLETON, 1981. Radioactive waste: the need to calculate an oceanic capacity. *Oceanus,* **24**:60–67.

NELSON, D. M., and M. B. LOVETT, 1978. Oxidation state of plutonium in the Irish Sea. *Nature,* **276**:599–601.

NEUILLY, M., J. BUSSAC, C. FREJACQUES, G. NIEF, G. VENDRYES, and J. YVON, 1972. Evidence of early spontaneous chain reaction found in Gabon mine. Text presented by F. Perrin at the press Conference, September 25, 1972. Commissariat a l'Energie Atomique, France.

NOSHKIN, V. E., and K. M. WONG, 1979. Plutonium mobilization from sedimentary sources to the marine environment. In Third Nucl. En. Ag. Seminar Marine Radioecology, Tokyo, Japan.

PENTREATH, R. J., 1980. *Nuclear Power, Man and the Environment.* Taylor and Francis, London.

ROXBURGH, I. S., 1987. *Geology of High-Level Nuclear Waste Disposal.* Chapman & Hall, London, 229 pp.

SEABORG, G. T., and M. L. PERLMAN, 1948. Presence of ^{239}Pu in pitchblende. *J. Amer. Chem. Soc.,* **70**:1571.

SHANBHAG, P. M., and J. W. MORSE, 1982. Americium interaction with calcite and aragonite surfaces in seawater. *Geochim. Cosmochim. Acta,* **46**:241–246.

SHCHERBAK, Y. M., 1996. Ten years of the Chornobyl era. *Sci. Amer.,* 274(4): 44–49.

SHOLKOVITZ, E. R., 1983. The geochemistry of plutonium in fresh and marine water environments. *Earth-Sci. Rev.,* **19**:95–161.

STANNERS, D. A., and S. A. ASTON, 1984. The use of reprocessing effluent radionuclides in the geochronology of recent sediments. *Chem. Geol.,* **44**:19–32.

SULLIVAN, W., 1972. Evidence shows a nuclear reaction occurred spontaneously long ago. *The New York Times,* Sept. 26.

TSANG, C.-F. (Ed.), 1987. *Coupled Processes Associated with Nuclear Waste Repositories.* Academic Press, Orlando, FL, 801 pp.

WAHLGREN, M. A., D. M. NELSON, K. A. ORLANDINI, and E. T. KUCERA, 1977. Study of the occurrence of multiple oxidation states of plutonium in natural water systems. *Argonne Natl. Lab. Ann. Rep.,* ANL-77-65(III): 95–98.

WETHERILL, G. W., W. F. LIBBY, and G. W. BARTON, 1964. Search for natural curium 247. *J. Geophys. Res.,* **69**(8):1603–1605.

25

Effect of Environmental Lead on Human Health

Lead occurs naturally in all kinds of rocks in a wide range of concentrations. It is released during chemical weathering of minerals and is present in soil, water, and air. Lead is absorbed by plants from soil and thus enters the food chain. However, since Pb is toxic to most plants, its concentration in plants is generally low. Similarly, surface water and groundwater have low Pb concentrations because most compounds of Pb (e.g., $PbCO_3$ and $PbSO_4$) have low solubilities. Therefore, food, water, and air under natural environmental conditions are not a source of Pb contamination of humans.

However, the mining and smelting of Pb and its use in household implements have contaminated the surface of the Earth and jeopardized the health of humans, domestic animals, and even wildlife. The dispersal of large quantities of lead alkyls by the combustion of leaded gasoline has become the most pervasive form of Pb contamination. The resulting health effects in humans are difficult to recognize because all humans alive today carry an excessive burden of anthropogenic Pb.

The contamination of soil and dust in metropolitan areas and near smelters of Pb-bearing ores was demonstrated by Warren (1972). His data in Table 25.1 reveal that Pb, Cu, and Zn have been strongly enriched in such samples from metropolitan areas (e.g., Vancouver, Liverpool, and New York), from mining or smelting centers in Canada (e.g., Trail, Sudbury, and Noranda), and from industrial sites (e.g., Vancouver and Toronto). Contamination of the soil with Ni was detected only at Sudbury and Noranda where Ni-bearing ore was mined.

Lead is a neurotoxin and associates with the alkaline earths in the human body. Consequently, Pb is stored in bones together with Ca, Sr, and Ba. The biogeochemistry of Pb and its effects on human health were summarized in a two-volume set of books edited by Nriagu (1978). The occurrence of Pb and other trace metals in the environment was also treated by Cannon and Hopps (1971), Thornton (1983), Salomons and Förstner (1984), and Fuge et al. (1996). The biogeochemistry of trace metals in aquatic organisms was the subject of a book edited by Tessier and Turner (1995). The contamination of even the most remote sites on the Earth with Pb since the onset of the industrial revolution in the middle of the 19th century was thoroughly documented by C. C. Patterson and his collaborators at the California Institute of Technology (Patterson, 1981). A set of papers in honor of Dr. Patterson's retirement was published in 1994 in *Geochimica et Cosmochimica Acta*, 58(15).

25.1 Isotope Composition of Environmental Lead

The study of Pb in the environment not only is based on the concentrations of this element in different kinds of samples, but relies also on its isotope composition. Lead in the environment may have a range of isotopic compositions because, in many cases, it is derived from different sources that have characteristic isotope compositions depending on the type of ore deposits from which the Pb originated. The isotope composition

Table 25.1 Enrichment Factors of Selected Metals in Contaminated Soils and Industrial Dust Compared to Those of Normal Soil[a] (Warren, 1972)

Locality	Pb	Cu	Zn	Ni
Soil[a]				
Vancouver, industrial	20.5	13.0	6.6	0.43
Vancouver, nonindustrial	24.0	2.7	6.0	0.63
Toronto, industrial	19.5	4.1	6.6	0.65
Liverpool, allotment gardens	27.5	3.1	3.7	0.68
Trail, British Columbia	65.0	2.1	10.6	0.48
Sudbury, Ontario	2.9	12.5	1.34	10.4
Noranda, Quebec	12.5	22.5	6.2	1.15
Dust				
Richmond, B.C. (contam.)	4000	47.5	23.2	—
Vancouver, industrial	1200	16.0	108	—
Vancouver, west end	600	31.3	33.6	—
New York, heavy traffic	2000	—	—	—
Toronto, industrial	6780	6.1	6.4	—

[a]Average soil in ppm: Pb = 10, Cu = 20, Zn = 50, Ni = 40 (Vinogradov, 1959).
(—, not determined.)

of Pb in ore deposits depends on the U/Pb and Th/Pb ratios and on the geologic ages of the source rocks from which Pb was removed during the ore-forming process (Faure, 1986). The ultimate cause for the variations of the isotope composition of Pb is that three of its four stable isotopes are decay products of isotopes of U and Th (Section 16.3e, Table 16.1):

$$^{238}U \rightarrow {}^{206}Pb$$

$$^{235}U \rightarrow {}^{207}Pb$$

$$^{232}Th \rightarrow {}^{208}Pb$$

Since ^{204}Pb is the only nonradiogenic stable isotope, the isotopic composition of Pb can be expressed in the form of atomic ratios: $^{206}Pb/^{204}Pb$, $^{207}Pb/^{204}Pb$, and $^{208}Pb/^{204}Pb$, which can be combined in various ways for specific purposes. For example, the isotopic composition of environmental Pb is commonly expressed in terms of the $^{206}Pb/^{204}Pb$, $^{206}Pb/^{207}Pb$, and $^{206}Pb/^{208}Pb$ ratios. The $^{206}Pb/^{207}Pb$ ratio can be calculated from:

$$\frac{\frac{^{206}Pb}{^{204}Pb}}{\frac{^{207}Pb}{^{204}Pb}} = \frac{^{206}Pb}{^{204}Pb} \times \frac{^{204}Pb}{^{207}Pb} = \frac{^{206}Pb}{^{207}Pb}$$

Automobile exhaust fumes became the most important source of environmental Pb after 1923 when tetraethyl Pb was added to gasoline. Chow and Johnstone (1965) measured the isotope composition of Pb in gasoline sold in May of 1964 in San Diego, California, in the hope that the results would help to identify the sources of Pb in the atmosphere. The results in Figure 25.1 indicate that the $^{206}Pb/^{204}Pb$ and $^{206}Pb/^{207}Pb$ ratios of gasoline have characteristic values, except for gasoline sold by the Shell Oil Co.

The data in Figure 25.1 demonstrate that the Pb in aerosol collected by an electrostatic air filter

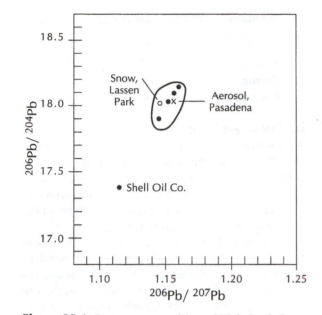

Figure 25.1 Isotope composition of Pb in leaded gasoline (solid circles) purchased in May 1964 in San Diego, California. The distribution of data points indicates that the Pb in aerosol particles on the Caltech campus in Pasadena (cross) and in snow of Lassen National Park (open circle) has virtually the same isotope composition as Pb in gasoline. This result reveals that automobile exhaust fumes were the dominant source of environmental Pb in 1963 and 1964 in California. The difference in the isotope composition of Pb in the gasoline sold by the Shell Oil Co. presumably indicates that the Pb originated from different ore deposits (from Chow and Johnstone, 1965).

at the geochemistry laboratory of the California Institute of Technology in the spring of 1964 originated from leaded gasoline. The concentration of Pb in the ash of the dust collected on the air filter was 5000 parts per million (ppm). In addition, the data show that the isotope composition of Pb in snow in Lassen National Park is also similar to that of Pb in gasoline. Therefore, the Pb in snow of Lassen Park could have originated by combustion of leaded gasoline like that sold in San Diego. The deviation of the Pb in gasoline sold by the Shell Oil Co. from the others in Figure 25.1 presumably indicates that the Pb originated from differ-

ent ore deposits. Nevertheless, these results reveal that in 1963 and 1964 automobile exhaust fumes were the dominant source of atmospheric Pb in California.

25.2   Lead in the Environment

The environmental Pb derived from natural sources has been overwhelmed by anthropogenic Pb, most of which is discharged into the atmosphere and is widely dispersed over the surface of the Earth. Therefore, the atmosphere plays an important role in the geochemistry of environmental Pb.

a.   Sources of Atmospheric Lead

Natural sources of atmospheric Pb include (Nriagu, 1978):

1. Windblown dust (85%).
2. Plant exudates (10%).
3. Forest fires, volcanic eruptions, and meteorite impacts (3%).
4. Radioactive decay and seasalt spray (2%).

The total amount of Pb released annually into the atmosphere from these sources is about 18,600 tons or 1.86×10^{10} g, compared to about 44×10^{10} g released in 1974/75 from anthropogenic sources. In other words, natural sources contributed only about 4% of the total amount of Pb that was discharged annually into the atmosphere in the 1970s.

The principal sources of anthropogenic Pb are (Nriagu, 1978):

1. Combustion of leaded gasoline (61%).
2. Production of steel and base metals (23%).
3. Mining and smelting of Pb (8%).
4. Combustion of coal (5%) (Chow and Earl, 1972).

The amounts of anthropogenic Pb discharged into the atmosphere from these sources vary regionally around the world and change with time. For example, the use of leaded gasoline has been discontinued in the USA, Canada, and many European countries. Consequently, there has been a welcome reduction in the amount of anthropogenic Pb deposited on the surface of the Earth from the atmosphere. In fact, Wu and Boyle (1997) reported that anthropogenic Pb has been virtually eliminated from the surface water of the North Atlantic Ocean.

Both natural and anthropogenic Pb in the atmosphere occur in the form of aerosol particles whose median diameters range from 0.12 μm (Los Angeles freeway) to 0.7 μm (Toronto). The particles have a variety of shapes including spheres, single crystals, and aggregates of smaller particles. In rural air, the particles consist of a large number of Pb compounds, such as $PbCO_3$ (30%), $(PbO)_2 PbCO_3$ (28%), PbO (21%), $PbCl_2$ (5.4%), $PbOPbSO_4$ (4%), etc. (Ter Haar and Bayard, 1971). The compositions of aerosol particles in automobile exhaust are dominated by $PbBrCl$, but the particles also contain oxides, oxyhalides, sulfate, phosphate, and oxysulfate. The residence time of Pb-bearing aerosol particles in the troposphere ranges from 2 to 10 days.

Patterson (1965) estimated that the Pb concentration of pristine (preindustrial) air was about 0.6 nanogram per cubic meter (ng/m^3); 1 nanogram $= 10^{-9}$). Even lower values of 0.046 ng/m^3 have been reported for air over the North Atlantic (Nriagu, 1978). However, at most other sites on the Earth today the Pb concentrations of air are much higher than Patterson's value for pristine air, especially in metropolitan areas and near Pb smelters. The data tabulated by Nriagu (1978) include 0.63 ng/m^3 at the South Pole, 5.5 ng/m^3 at Mauna Loa, Hawaii, and 8.0 ng/m^3 at White Mountain, California.

The concentration of Pb in the air of large cities worldwide is up to three orders of magnitude greater than that of air at the remote locations cited above. A sampling of the data compiled by Nriagu (1978) indicates the following Pb concentrations in air on busy streets of European capitals:

		$\mu g/m^3$
Berlin, Germany	1966–67	3.8
Rome, Italy	1972–73	4.5
Paris, France	1964	4.6
London, England	1972	5.1

The situation in the USA was not much better:

New York	1960–65	4.1
Boston	1972	4.5
Detroit	1960–65	4.8
Los Angeles	1954–55	6.6

b. Snow and Ice

Atmospheric transport of anthropogenic Pb has caused the contamination of meteoric precipitation, soil, and surface water in the oceans. Records of the history of contamination of the surface of the Earth have been preserved in ice sheets of the polar regions and in lacustrine sediment.

The extent of contamination of remote sites by airborne Pb was originally demonstrated by Tatsumoto and Patterson (1963). Subsequently Chow and Johnstone (1965) reported that the isotope composition of Pb in snow collected in 1963 in Lassen State Park in California is indistinguishable from that of gasoline sold in San Diego and of aerosols in Los Angeles (Figure 25.1). In addition, they cited an estimate by Patterson that 2.7×10^{11} g of Pb was released annually by combustion of leaded gasoline in the northern hemisphere and used it to calculate that the Pb concentration of snow in Lassen Park from this source alone is 1.1 μg/kg. The predicted value is similar to the concentration of 1.6 μg/kg measured by Tatsumoto and Patterson (1963) and thereby confirmed that most of the Pb in the snow at Lassen Park originated from automobile exhaust fumes.

The definitive study of Pb in snow and ice in Greenland and Antarctica was carried out by Murozumi et al. (1969). They took extreme precautions in the collection and subsequent analysis of snow and ice samples and reported much lower concentrations of Pb than had been observed in snow and ice at temperate regions. Their measured Pb and Ca concentrations in Figure 25.2 in ice at Camp Century (77°10'N, 61°08'W) in northern West Greenland reveal the history of anthropogenic emission of Pb into the atmosphere. The Pb concentrations generally increased from 0.011 μg/kg in 1753 A.D. to about 0.068 μg/kg at about 1945. Subsequently, the Pb concentrations

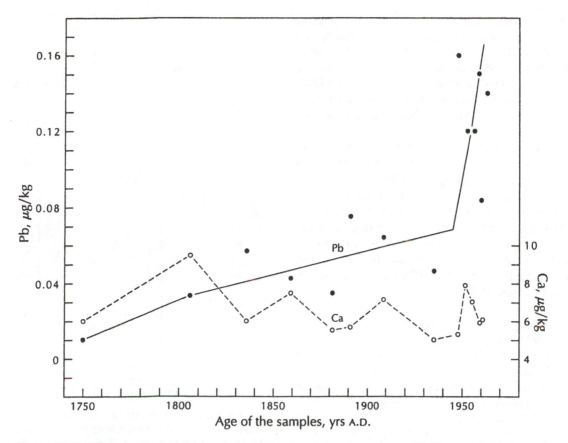

Figure 25.2 Evidence for anthropogenic contamination of the atmosphere with Pb recorded in ice and snow deposited between 1750 and 1963 A.D. at Camp Century (77°10'N, 61°08'W) about 200 km east of Thule located on the west coast of North Greenland. The concentrations of Pb in ice and snow at this site rise from 0.011 $\mu g/kg$ in 1753 A.D. to about 0.068 $\mu g/kg$ in 1945 A.D. Subsequently, the Pb concentrations rise steeply to about 0.16 $\mu g/kg$ in 1960. In contrast to the increasing Pb concentrations, the Ca concentrations of the ice at Camp Century remained constant at an average value of 6.4 ± 0.7 $\mu g/kg$, (data from Murozumi et al., 1969).

rose steeply to about 0.16 $\mu g/kg$ by 1960. In contrast to Pb, the Ca concentrations did not change systematically in the same time interval and have an average value of 6.4 ± 0.7 μ /kg.

The increasing concentrations of Pb in ice and snow at Camp Century between 1750 and 1920 are attributable primarily to emissions from Pb smelters, with a minor contribution from combustion of coal. Starting in 1923, when tetraethyl Pb was introduced as an additive to gasoline, automobile exhaust fumes became a growing source of Pb contamination, whereas emission of

smelters declined starting about 1880 because of efforts to prevent the loss of Pb for economic reasons. The rapid increase in the concentrations of Pb after 1945 in the Camp Century ice core was caused almost entirely by automobile exhaust fumes (Murozumi et al., 1969).

In a recent review of the concentrations of Pb and other metals in snow and ice, Boutron et al. (1994) reported that the concentration of Pb in snow in Antarctica increased by a factor of about 10 from 1900 to 1980, whereas the Pb content of Greenland snow rose by a factor of more than 200 between 1750 and 1989 as a result of anthropogenic inputs of Pb to the troposphere. However, they also concluded that the Pb concentration of snow in Greenland has decreased significantly after the use of tetraethyl lead in gasoline declined in the USA and Canada. The decrease in Pb concentrations in the western North Atlantic Ocean was also reported by Wu and Boyle (1997).

The concentrations of Pb in snow in the interior of Antarctica are very low and range from 0.5 picogram/g (1 picogram = 10^{-12}/g) for ice at Dome C about 7500 years B.P. to about 5 pg/g for snow in 1980. In order to accurately determine such low concentrations of Pb, extreme precautions are required to prevent contamination of snow or ice during collection, transport, and analysis in the laboratory (Patterson and Settle, 1976; Boutron, 1990).

c. Lacustrine Sediment

The contamination of the Earth's surface with anthropogenic Pb was also demonstrated by Shirahata et al. (1980) who determined concentrations of Pb in sediment deposited in a remote pond in Thompson Canyon in Yosemite National Park, California. The data in Figure 25.3 reveal that the concentration of Pb increased from 9.6 μg/g at about 1700 A.D. to 14.8 ± 0.8 μg/g between 1853 and 1894 A.D. Subsequently, the Pb concentration increased to 42.4 μg/g in sediment deposited between 1962 and 1968. The most recent sample, deposited between 1968 and 1976, contains about 10% less Pb than the sediment in the preceding interval.

The foregoing example suggests that the distribution of Pb in lacustrine sediment is a record of the industrial development of the surrounding area and therefore has archaeological value. This aspect of the environmental geochemistry of Pb is illustrated by the work of Ritson et al. (1994) on the concentrations and isotope composition of Pb in the sediment of Lake Erie. Their paper is the latest contribution to the study of the contamination of the Great Lakes of North America with heavy metals of anthropogenic origin (Flegal et al., 1989; Nriagu et al., 1979; Edgington and Robbins, 1976).

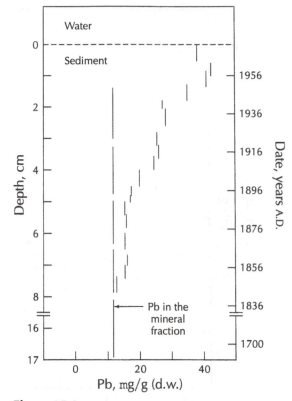

Figure 25.3 Concentrations of Pb in dry sediment in a core recovered from a remote pond in Thompson Canyon, Yosemite National Park, California. The sediment contains both anthropogenic Pb adsorbed on humus and Pb contained within the mineral grains. The average concentration of Pb in the mineral fraction of the sediment is 11.8 ± 1.1 μg/g ($2\overline{\sigma}$, $N = 7$). The diagram shows that the amount of anthropogenic Pb increased with time after about 1850 A.D. and reached a maximum between 1962 and 1968 A.D. (from Shirahata et al., 1980).

The data for the western basin of Lake Erie in Figure 25.4 reveal a sharp increase in the total concentration of Pb in sediment at a depth of 15 cm below the sediment–water interface. The sediment at that depth was deposited in 1963. Ritson et al. (1994) reported that the isotope ratios of Pb also changed at that time from $^{206}Pb/^{207}Pb = 1.243$ before 1963 to 1.202 after that year. The simultaneous change of both the concentration and the iso-

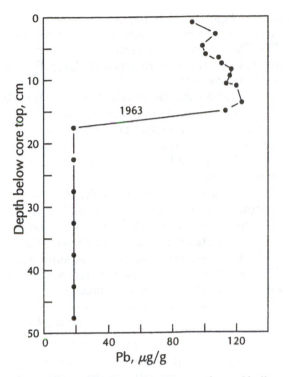

Figure 25.4 Profile of Pb concentrations of bulk sediment in a short core (0–50 cm) taken in the western basin of Lake Erie, USA and Canada. The increase of the Pb concentration from an average of 17.3 ± 0.3 $\mu g/g$ to 124.4 ± 0.5 $\mu g/g$ occurred at 1963 A.D. The $^{206}Pb/^{207}Pb$ ratios also changed from 1.243 ± 0.0006 to 1.202 ± 0.0001. The sediment deposited prior to 1963 contains Pb derived from natural sources. The anthropogenic Pb in the more recent sediment is attributable to municipal wastewater as well as to the combustion of coal and leaded gasoline (data and interpretation from Ritson et al., 1994).

tope composition of Pb in the western basin of Lake Erie indicates a change in the sources of Pb.

According to Ritson et al. (1994), the dominant *natural* source of Pb in the western basin of Lake Erie is sediment derived from the Detroit River, the Maumee River, and from the erosion of shoreline bluffs. *Anthropogenic* Pb is contributed by combustion of coal and leaded gasoline as well as by discharge of municipal wastewater. In the

western basin, municipal wastewater is the dominant source of anthropogenic Pb because of the large populations of the cities of Detroit, Toledo, and Cleveland. The Pb in the central and eastern basins originated primarily by atmospheric deposition of Pb derived from combustion of coal and leaded gasoline, as well as by transport of sediment from the western basin. Anthropogenic Pb is detectable in sediment deposited in the western basin starting in 1963 and in the eastern basin starting in 1926. Prior to these dates, the average concentrations of Pb in sediment throughout the lake were fairly uniform, ranging from about 17.3 ± 0.3 $\mu g/g$ (western basin) to 17.6 ± 0.5 $\mu g/g$ (central basin) to 18.4 ± 0.5 $\mu g/g$ (eastern basin). The $^{206}Pb/^{207}Pb$ ratios were also similar, ranging from 1.242 ± 0.0006 (west) to 1.228 ± 0.0013 (central) to 1.228 ± 0.0020 (east). In addition, the $^{206}Pb/^{207}Pb$ ratio of sediment from long cores (1.80–10.25 m) in Lake Erie have an average value of 1.229 ± 0.004, which is indistinguishable from the Pb in the short cores. For these and other reasons Ritson et al. (1994) concluded that these concentrations and isotope ratios represent Pb derived from natural sources.

d. Mixing of Lead in Lake Erie Sediment

The data of Ritson et al. (1994) indicate that the recently deposited sediment in Lake Erie contains mixtures of natural and anthropogenic Pb. Therefore, the concentrations and isotope ratios of the Pb are related to the end members by equations 18.19 and 18.20 (Section 18.5), restated here for convenience in terms of Pb concentrations and $^{206}Pb/^{204}Pb$ ratios:

$$Pb_M = Pb_A f_A + Pb_B (1 - f_A) \qquad (25.1)$$

$$\left(\frac{^{206}Pb}{^{204}Pb}\right)_M = \left(\frac{^{206}Pb}{^{204}Pb}\right)_A \frac{Pb_A}{Pb_M} f_A$$

$$+ \left(\frac{^{206}Pb}{^{204}Pb}\right)_B \frac{Pb_B}{Pb_M} (1 - f_A) \qquad (25.2)$$

where Pb is the concentration, f is the weight fraction, and the subscripts A, B, and M refer to the components A and B and the mixture M, respectively. These equations can be used to construct mixing hyperbolas in coordinates of the Pb concentration (x-coordinate) and the $^{206}Pb/^{204}Pb$ ratio (y-coordinate) for any two end-member components A and B. Equations 25.1 and 25.2 can be combined to yield the equation of the mixing hyperbola:

$$\left(\frac{^{206}Pb}{^{204}Pb}\right)_M = \frac{a}{Pb_M} + b \qquad (25.3)$$

where a and b are constants whose values are derived from the isotope ratios and concentrations of Pb in the end-member components (Section 18.5). The significance of equation 25.3 arises from the fact that it is a straight line when the isotope ratios are plotted versus the reciprocal concentrations of Pb.

For this reason, the data reported by Ritson et al. (1994) for short cores from the western, central, and eastern basins of Lake Erie form a linear array of points in Figure 25.5. The fit of the data points to a straight line is not perfect because the isotope ratios and concentrations of Pb in sediment containing primarily anthropogenic or natural Pb vary in terms of time and space. Nevertheless, the sediment samples deposited after 1948 in the eastern and western basins cluster within narrow limits and have average values approaching those of sediment containing only the anthropogenic component:

	Western basin	Eastern Basin
Pb, $\mu g/g$	110.5 ± 5.5	68.5 ± 6.9
$\dfrac{^{206}Pb}{^{204}Pb}$	18.79 ± 0.04	18.81 ± 0.03

The isotope compositions and concentrations do not represent Pb of purely anthropogenic origin because even the sediment deposited after 1948 contains some Pb derived from natural sources. The $^{206}Pb/^{204}Pb$ ratio of anthropogenic Pb lies on the extension of the mixing line and must have a value greater than about 18.6 when the concentration of Pb in the sediment becomes very large.

Sediment samples deposited in the eastern basin between 1926 and 1946 form a separate array in Figure 25.5 and approach the cluster of data points representing sediment deposited after 1948. The source of Pb deposited between 1926 and 1948 in the eastern basin of Lake Erie is presently unknown.

The sediment in the central basin contains Pb whose isotope ratios encompass the entire range of values from nearly pure anthropogenic to natural Pb. However, anthropogenic Pb is less abundant in the central basin than in the eastern and western basins.

The sediment deposited in Lake Erie prior to the introduction of large amounts of anthropogenic Pb is characterized by high $^{206}Pb/^{204}Pb$ ratios and low concentrations. The sediment deposited prior to 1968 in the western basin forms a separate array. Nevertheless, the average isotope ratios and concentrations of Pb from *natural* sources are:

	Western basin	Eastern basin
Pb, $\mu g/g$	17.3 ± 0.3	18.4 ± 0.5
$^{206}Pb/^{204}Pb$	19.45 ± 0.05	19.21 ± 0.11

The combination of isotope ratios and concentrations of Pb in the sediment of Lake Erie in Figure 25.5 reveals that the Pb deposited in the lake is a mixture derived primarily from two sources in varying proportions. Therefore, mixing theory applied to closely spaced samples of lacustrine sediment of known age can recover information about the history of industrial activity in the area.

25.3   Lead in Plants

The growth of plants depends on the availability of certain essential nutrients:

C, H, O, N, P, S, K, Ca, Mg, Fe, as well as B, Cl, Cu, Mn, Mo, Zn.

In addition, small concentrations of other elements appear to be beneficial to plants, including Se, Co, Si, and V. However, plant tissue may contain many other elements that do not necessarily have a useful function.

Chemical elements enter plants either from soil through the roots or by absorption from the

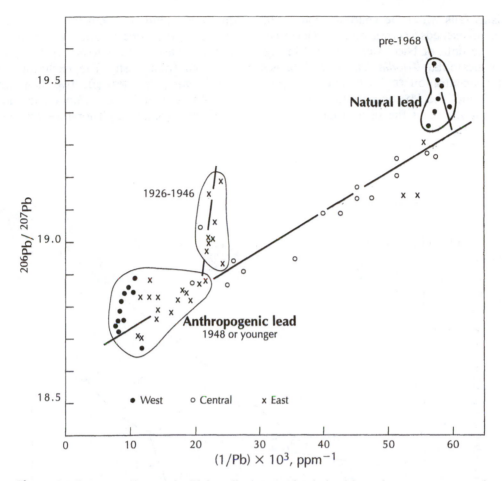

Figure 25.5 Mixing diagram for Pb in sediment samples derived from short cores recovered from the western, central, and eastern basins of Lake Erie. Sediment deposited after 1968 in the western and, after 1946, in the eastern basins contains primarily anthropogenic Pb, whereas sediment deposited before 1968 in the western basin and before 1926 in the eastern basin is dominated by Pb derived from natural sources. The sediment in the central basin contains mixtures of anthropogenic and natural Pb in varying proportions. The isotope compositions of the anthropogenic and natural Pb changed significantly in the time intervals represented by sediment samples that form separate arrays of data points (data from Ritson et al., 1994).

atmosphere by leaves. Plants can absorb only those elements from soil that occur in aqueous solution or are adsorbed on the surfaces of solid particles composed of minerals, amorphous oxides and hydroxides, or organic matter. Ions in solution can diffuse directly into the cells of root tips. Cations that are adsorbed on the surfaces of soil particles must be replaced by H^+ released by the roots as a result of dissociation of carbonic acid

formed from CO_2. The H^+ ions release nutrients (Ca^{2+}, Mg^{2+}, K^+, etc.) from exchangeable sites on the surfaces of clay minerals and other types of soil particles. The cations then diffuse to the root tip where they are exchanged for H^+ ions that move into the soil to release additional nutrients.

Consequently, plants are selective in the absorption of elements from soil and exclude those elements that are toxic. However, the exclusion mechanisms fails when the concentration of the toxic element exceeds a certain threshold level. For example, the data of Nicolls et al. (1965) in Figure 25.6 indicate that *Triodia pungens* maintains a constant concentration of Pb of about 10 ppm (plant ash) until the Pb concentration of the soil within about 15 cm of the plant approaches the

threshold value of about 580 ppm. Above this threshold, the concentration of Pb in the plant increases rapidly to a toxic level of more than 150 ppm (plant ash). The exclusion of Cu by *Triodia pungens* is less effective than that for Pb, and Zn is assimilated without exclusion up to 10,000 ppm (plant ash). Therefore, *Triodia pungens*

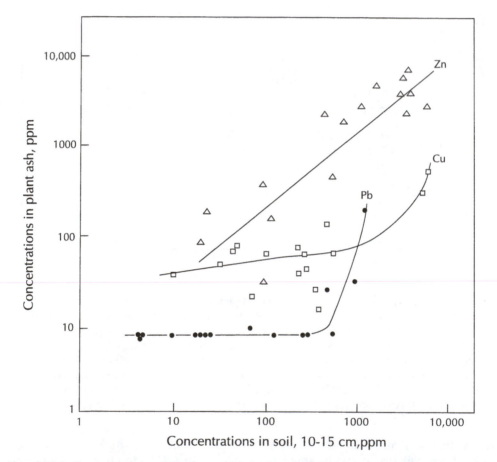

Figure 25.6 Concentrations of Pb, Cu, and Zn in the ash of *Triodia pungens* growing in soil containing a range of concentrations of these metals. The plant maintains a constant Pb concentration of <10 ppm until the Pb concentration of the soil within 10–15 cm of the plant increases to the threshold value of about 580 ppm. At higher soil concentrations, the Pb concentration of the plant rises to toxic levels at about 170 ppm. The data show that the threshold for Cu is higher than that for Pb and that the plant admits Zn without discrimination up to almost 10,000 ppm (from Brooks, 1972; after Nicolls et al., 1965).

tends to exclude Pb and Cu, but admits Zn up to high levels of concentration. In general, different types of plants growing in the same soil may contain different concentrations of metals depending on the nutritional requirements of each species and on the depth of penetration of their roots (Peterson, 1978).

Average concentrations of Pb in plants growing in uncontaminated soil range from about 26 to 85 $\mu g/g$ (ash). Plants growing in Pb-rich soil may contain up to 100 times more Pb. Mosses, liverworts, and lichens are especially efficient in collecting Pb from the atmosphere and from dilute aqueous solutions. Peterson (1978) cited data from the literature indicating extremely high Pb concentrations in these kinds of plants (e.g. 11,611 $\mu g/g$ in mosses; 15,900 $\mu g/g$ in liverworts; and 12,045 $\mu g/g$ in lichens).

The contamination of plants with Pb occurs naturally in areas where soils have formed from Pb-rich rocks. However, in most cases plants are contaminated with Pb released by mine dumps, landfills, and aerosol particles deposited by smelter plumes, and by combustion of leaded gasoline. As a result, wildlife as well as domestic animals feeding on contaminated vegetation may be adversely affected.

25.4   Lead Poisoning of Cows and Horses

Plants discriminate against Pb such that the Pb/Ca ratios of plants are lower than those of the soil in which they grow. Similarly, herbivorous and carnivorous animals preferentially excrete Pb in relation to Ca. This process of *biopurification* causes the Pb/Ca ratios of successive members of the food chain to decline.

Lead poisoning of domestic animals generally results from the ingestion of plants containing excessive concentrations of anthropogenic Pb. Cattle are especially vulnerable because of their indiscriminate eating habits. Forbes and Sanderson (1978) listed the symptoms of acute lead poisoning

in cattle: diarrhea, colic, disorders of the central nervous system, excitement or stupor or depression, loss of coordination, and blindness. However, not all of the symptoms may be present and some cattle have died of Pb poisoning without displaying any of the symptoms. The internal organs of cattle that died from Pb poisoning generally have high concentrations of this element (e.g., livers: 43 ppm, from 0 to 1300; kidneys: 137 ppm, from 2 to 2355). The livers and kidneys of healthy cattle contain only 0.3 to 1.5 ppm of Pb.

A case of Pb poisoning of horses on two pastures near Benicia, California, was investigated by Rabinowitz and Wetherill (1972). They reported that the principal source of anthropogenic Pb in this area was a Pb smelter at Benicia that daily discharged about 45 kg of Pb into the atmosphere. Automobile exhaust contributed an additional 15 kg of Pb per day. Grass collected about 500 meters from the smelter had a Pb concentration of 9645 $\mu g/g$. Grass along the local freeways also had high lead concentrations, ranging from 275 to 6141 $\mu g/g$.

The data in Figure 25.7 demonstrate that the concentrations of Pb in grass near Benicia, California, decreased with increasing distance from the smelter, but remained anomalously high (268 $\mu g/g$) for at least 50 km downwind from the smelter. Grass at a remote location about 100 km downwind of the smelter was found to have a Pb concentration of 36 $\mu g/g$. However, even that value exceeds the average concentrations of Pb in different vegetables reported by Warren and Delavault (1971). The grass on the horse pastures had very high Pb concentrations of 5975 $\mu g/g$ at 3.7 km and 680 $\mu g/g$ at 8.5 km downwind from the smelter.

The symptoms displayed by the horses in the contaminated area included anemia, nucleated red blood cells, peripheral neuritis, and poor health. The cause of death was identified as respiratory complications of chronic Pb poisoning. The Pb concentrations of the kidneys of two horses that had been pastured at 3.7 and 8.5 km downwind from the smelter were found to be 16 and 9.4 $\mu g/g$, respectively, compared to concentrations

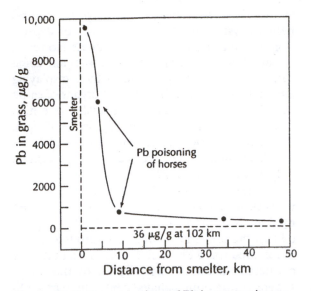

Figure 25.7 Concentrations of Pb in grass at increasing distances downwind from a smelter at Benicia, north of San Francisco, California. The Pb concentrations decrease with increasing distance downwind from the smelter. However, horses pastured in the smelter plume up to 10 km downwind died of chronic Pb poisoning (from Rabinowitz and Wetherill, 1972).

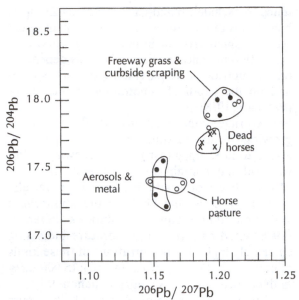

Figure 25.8 Isotope composition of Pb in grass (open circles), aerosols and curbside scrapings (solid circles), and internal organs of horses that died of Pb poisoning (crosses) near Benicia, north of San Francisco, California. The isotope composition of Pb in grass in the horse pastures is similar to that of aerosols derived from a nearby Pb smelter. The Pb in grass along local freeways and curbside scrapings has a distinctly different isotope composition than the smelter Pb in the aerosols. The isotope composition of Pb in kidneys, livers, and bones of horses that died of Pb poisoning is intermediate between those of the pasture grass and in gasoline emissions. Evidently, the horses assimilated not only smelter Pb in the grass they ate but also Pb from automobile exhaust (from Rabinowitz and Wetherill, 1972).

of 1–5 μg/g in the kidneys of horses in the Los Angeles basin that died of other causes. Rabinowitz and Wetherill (1972) concluded that the smelter was the dominant source of Pb in grass growing up to 35 km downwind of it. At greater distances, gasoline became a significant source, reflected in a different isotopic composition of the Pb.

The isotope ratios of Pb in the grass of the horse pastures in Figure 25.8 is similar to that of aerosols collected at the pastures and elsewhere within the plume emitted by the smelter. The Pb in grass growing along the local freeways and in curbside scrapings and aerosols in Berkeley originated primarily from automobile emissions and has a distinctly different isotope composition than the smelter Pb. The kidneys, livers, and bones of horses that died of Pb poisoning on the pastures in the smelter plume contain Pb of intermediate isotope composition. Evidently, the horses

ingested not only smelter Pb in the grass but also freeway Pb in the air they breathed.

Ironically, Hippocrates (460–377 B.C.) reported 2400 years ago that cows and horses could not be pastured near the mines of Laurion near Athens in classical Greece because they became sick and died (Lessler, 1988). It is now known that farm animals have a low tolerance for Pb compared to humans. For example, Zmudski et al. (1983) reported that horses receiving only 3–4 mg of Pb per kg of feed per day died in a few weeks.

25.5   Human Bones and Tissues

In 1965 Patterson published an invited paper in the Archives of Environmental Health in which he estimated that the natural concentration of Pb in human bones is about 0.85 μg/g and that the present Pb content of humans is about 200 times *higher* than expected on geochemical grounds. He concluded that the present Pb content of most humans in developed countries is a toxic burden that jeopardizes their health. Patterson and his research group later demonstrated that 1600-year-old human bones from South America actually do contain less Pb by several orders of magnitude than the bones of present-day humans (Ericson et al., 1979).

The outward symptoms of Pb poisoning in humans were described by Hippocrates (460–377 B.C.) as loss of appetite, colic, palor, weight loss, fatigue, and irritability (Lessler, 1988). A more recent review of the effects of Pb on human health by Posner et al. (1983) included the same symptoms and discussed the specific effects on the digestive, renal, muscular, nervous, and vascular systems.

Lead poisoning of humans generally results from the use of this metal in the home and from occupational exposures. Sources of Pb in the home include paint (white and red), water pipes made from Pb, litharge (glazing compound), solder, leaded gasoline, ammunition, storage batteries used in automobiles, lead-bearing glaze in pottery, etc. Occupational exposure to Pb is associated with mining and smelting of ores of Pb and related chalcophile metals, printing with Pb type, ship and bridge repair releasing flakes of Pb paint, manufacture of Pb compounds, etc.

The use of Pb in the home has been greatly reduced in many countries following the recognition of the harmful effects of this metal, especially for children. In the USA and Canada the manufacture and sale of leaded gasoline, Pb-based paint, and Pb-bearing glazes on pottery have been discontinued. As a result, the incidence of acute Pb poisoning has declined, but the effect of low-level Pb exposures on the mental development of prenatal children is still a subject of concern.

a.   Bones

Most of the Pb in the human body is stored in bones where it accumulates throughout life starting before birth. Soft tissues (e.g., liver, kidney, aorta, pancreas) have different but lower concentrations of Pb than bones. In general, the bones of adult males in the UK contain between about 91.6 and 96.0% of the total body burden of Pb. The amount of Pb in the bones of adult males in the UK is larger than that of females, but both increase with age. This conclusion is illustrated in Figure 25.9 based on the data of Barry (1975, 1976).

The soft tissues contain less Pb than bones and the amount is dependent of age (Figure 25.9). The amount of Pb in the soft tissue is balanced between the uptake of this element in food and water and from the air and its excretion or storage in the bones. The Pb in the vital organs, in blood, and in muscle tissue interferes with metabolic processes and therefore gives rise to the symptoms of Pb poisoning.

The concentration of Pb in bones of modern residents of the USA varies with age, according to data by Schroeder and Tipton (1968). The bones of adults between the ages of 30 and 69 have an average Pb concentration of 39.0 μg/g (ash) based on a sample of 50 individuals. The increase in the concentration of bone Pb with age in Figure 25.10 indicates that the intake of this element exceeded the ability of the body to excrete it. The low concentrations of Pb in the bones of individuals more than 70 years of age was attributed by Schroeder and Tipton (1968) to the survival of those person who had the lowest concentrations of Pb. Alternatively, these person may have had low Pb-exposures in their youth prior to the introduction of leaded gasoline in 1923.

b.   Hair

Lead and other trace elements also occur in human hair, which can be preserved for hundreds of years. Weis et al. (1972) reported that the hair of children living in the USA between 1871 and 1923 A.D. contained almost twice as much Pb as the hair of adults, and that the Pb concentrations of both

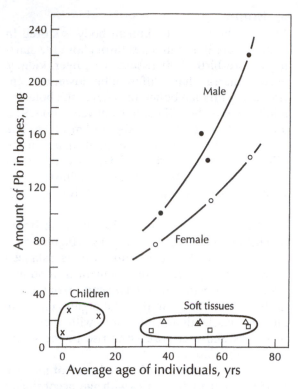

Figure 25.9 Accumulation of Pb in bones of adult humans (male and female) and children with average age of the individuals represented by each data point in the UK. The evidence indicates that Pb accumulates in the bones with increasing age, starting before birth, and that males carry more Pb in their bones than females. However, the amount of Pb in soft tissues is only about 6% of the total amount and is independent of age. The Pb in soft tissues represents a balance between intake and excretion. All amounts are relative to a standard body weight of 70 kg including the data for children (from Barry, 1975, 1976).

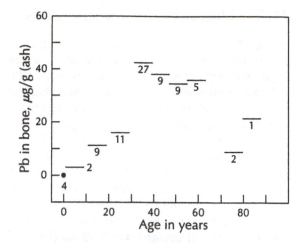

Figure 25.10 Average concentrations of Pb in bones of Americans of different ages, 1952–1957. The number of individuals included in each age group is indicated. The data reveal that Pb accumulated in the bones of Americans with age because the intake exceeded their ability to excrete it. The relatively low concentration of bone Pb of humans more than 70 years of age is attributable to the survival of individuals having low body burdens of Pb, perhaps because they grew up before leaded gasoline was introduced in 1923. However, individuals with high Pb levels did not display symptoms of Pb intoxication, nor were they chronically ill (data and interpretation by Schroeder and Tipton, 1968).

children and adults living in 1971 were significantly *lower* than those of the antique population. Weis et al. (1972) attributed the high Pb levels in the hair of the antique population to contamination with Pb in the homes of that period (water collected from leaded roofs, use of Pb-bearing glaze on pottery, Pb-based paint, etc.). They also suggested that the reduction of the Pb content of hair in the modern population (1971) resulted from increased awareness of the danger of Pb contamination in the home.

A worldwide survey by Barry (1978) of human hair of individuals who were not exposed to high levels of environmental Pb indicated average values between 39.6 ppm (39 individuals, India) to 6.6 ppm (28 individuals, USA; 33 individuals, UK). The weighted mean Pb concentration of hair of 1055 individuals in the 20th century worldwide is about 24 ppm.

c. Effect of Lead on Children

The evidence that children are especially susceptible to Pb is cause for concern because Davis and Svendsgaard (1987) reported that even low concentrations of this element affect their mental development. Although the maximum permissible Pb concentration of whole blood has been lowered from 600 to 250 $\mu g/L$ (Anonymous, 1986), the present limit may still be too high. The effect of Pb on young children is difficult to ascertain because of the large number of variables that affect their physical health and mental development (e.g., nutrition, heredity, home environment, use of tobacco and alcohol by the mother). The Pb concentrations of the blood of prenatal and postnatal children in the studies cited by Davis and Svendsgaard (1987) were less than 150 $\mu g/L$ and therefore considerably *less* than the maximum permissible concentration. Nevertheless, the data demonstrated several adverse effects.

1. The risk of premature birth increased significantly for mothers whose blood contained between 80 and 140 $\mu g/L$ of Pb.
2. Children whose blood in the umbilical cord contained an average of 140 $\mu g/L$ of Pb were measurably impaired in their mental development at the age of six months.
3. High prenatal blood levels in children are associated with low birth weight and a decrease in their length.
4. The adverse effects of high Pb content in the blood of prenatal children are detectable up to at least 12 months after birth.

The authors noted that in 1980 alone nearly one million children were born with sufficiently high Pb concentrations in their blood to adversely affect their mental and physical development.

25.6   Lead in the Bones of Ancient Peoples

The use of Pb among the peoples of Europe probably increased with time and therefore may be reflected by increasing Pb concentrations of human bones of decreasing historical age. The data of Grandjean and Holma (1973) in Figure 25.11 actually suggest such a relationship.

a. Denmark

The bones of humans who lived in what is now Denmark from 1500 B.C. to about 1400 A.D. contain less than $1\mu g/g$ of Pb (Grandjean and Holma, 1973). Subsequently, the concentrations increased significantly and reached levels of about 7 $\mu g/g$ in the 18th century A.D. In marked contrast, Romans who lived in York (UK) up to about 320 A.D. had high concentrations of Pb in their bones because the food they ate was contaminated by kitchen utensils containing high concentrations of Pb. The concentrations of Pb in bones of the Anasazi people in Utah and of ancient Peruvians (Ghazi, 1994) are consistent with the pattern of increasing concentrations of Pb in the former inhabitants of Denmark in Figure 25.11.

The lowest concentrations of Pb, ranging from 0.11 to 2.7 $\mu g/g$, were reported by Ericson et al. (1979) for bones of ancient Peruvians who died between 4500 and 1400 years ago. The average Pb concentration of these bones is 1.0 ± 0.9 $\mu g/g$ ($2\overline{\sigma}$, $N = 5$). The same authors also measured a Pb concentration of 1.1 $\mu g/g$ in the vertebra of a 2200-year-old Egyptian mummy. These values are similar to the Pb concentrations of ancient bones from Denmark (4000 B.C. to 1000 A.D.) analyzed by Grandjean and Holma (1973).

Another example of historical changes of Pb in human bones was reported by Jaworowski (1968). His data demonstrate that the concentration of Pb in human bones in Poland increased from an average of 2.3 ± 0.6 $\mu g/g$ in the 3rd century A.D. to about 60 ± 25 $\mu g/g$ in the 19th century. He attributed the increase in Pb concentrations to the use of Pb-bearing kitchen utensils and noted that the concentration of Pb in the bones of modern residents of Poland has declined in spite of the evidence of a rise in atmospheric Pb contamination recorded in the glaciers of the Tatra Mountains.

In summary, the data suggest that the bones of humans who lived in uncontaminated environments have Pb concentrations of no more than 1 or 2 $\mu g/g$. Higher concentrations indicate either that the individuals ingested excess amounts of Pb or that the bones were contaminated after burial

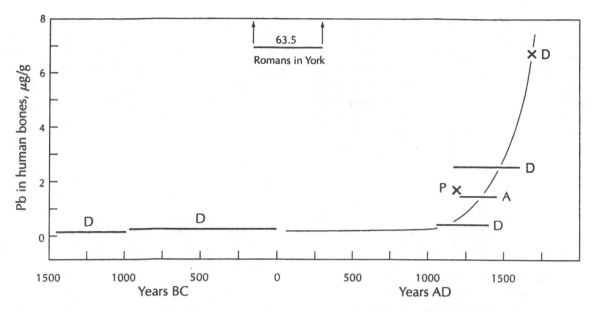

Figure 25.11 Average Pb concentrations of human bones (μg/g dry weight) as a function of historical time (D = Denmark, A = Anasazi, Utah, P = Chiribaya, Peru). The data suggest that the Pb concentration of human bones increased as this metal became more widely used in household articles. The Romans are known to have used Pb both in metallic form and as a food additive (sapa). Consequently, they became significantly contaminated with Pb and experienced ill health as a result of its toxicity (from Grandjean and Holma, 1975, Denmark; Ghazi, 1994, Anasazi and Peruvians; Mackie et al., 1975, Romans, York, UK).

by diagenetic alteration or after subsequent exposure and transport to the analytical laboratory.

b. Lead Poisoning Among the Romans

The contamination of humans with Pb started long before the industrial revolution and the introduction of tetraethyl lead as an additive to gasoline. Lead was used in water pipes by the Romans, in drinking cups, and in other household implements, as well as in cosmetics. A significant cause of Pb poisoning in the Roman Empire (Figure 25.11) was the custom of adding "sapa" containing Pb acetate to wine in order to sweeten it and to prevent spoilage. Sapa was prepared by boiling acidic wine in lead-lined pots and therefore contained approximately 240 mg of Pb per liter (Lessler, 1988).

Mackie et al. (1975) reported that the bones of Romans who lived in York (UK) have excessively high concentrations of Pb averaging 60.2 μg/g for males and 85.2 μg/g for females (dry weight). The higher concentration of Pb in

the bones of females is remarkable and is consistent with the assumption that the Pb contamination occurred during the preparation of food.

The use of Pb in household articles and of sapa by the Roman aristocrats may have adversely affected their health and thereby contributed to the disintegration of the Roman Empire (Lessler, 1988; Nriagu, 1983a,b; Gilfillan, 1965).

c. The Omaha Indians of Nebraska

High concentrations of Pb ranging from 4.8 to 2570 μg/g were reported in bones of the Omaha tribe of American Indians who lived in northeastern Nebraska near the town of Homer between 1780 and 1820 A.D. (Ghazi, 1994). The data indicate that the bones of children under the age of eight had an average Pb concentration of 663 ± 518 μg/g (range: 4.8–2570 μg/g), whereas the bones of adults more than 18 years of age contained 97.5 ± 80 μg/g of Pb (range: 13.7–271). The data of Ghazi (1994) strongly suggest that the highest Pb concentrations occur in the bones of children, even though not all of the children were significantly contaminated with Pb.

The chemical composition of bones buried in soil may be altered during diagenesis by the transfer of elements from the soil to the bone. For this reason, Ghazi (1994) analyzed soil and animal bones from the burial site of the Omaha Indians. The results demonstrate that the soil (total dissolution) has an average Pb concentration of $19.9 \pm 2.1\ \mu g/g$ and that the animal bones recovered at the burial site contain only $1.4 \pm 1.2\ \mu g/g$. Ghazi (1994) concluded that the Pb in the human bones is not of diagenetic origin because:

1. The total Pb concentration of the soil is less than that of most of the human bones that were buried in it.
2. The animal bones that were in contact with the same soil have much lower Pb concentrations than the human bones.
3. The human bones have a wide range of Pb concentrations, indicating that additions of Pb from the soil were insignificant.

Therefore, the high concentration of Pb in the human bones, compared to the animal bones at the same site, indicates that the individuals of the Omaha tribe buried near Homer, Nebraska, suffered from Pb poisoning and that some of the children may have died of this condition.

The excess Pb in the bones of the Omaha Indians may have originated from musket balls, ornamental coils of metallic Pb, cosmetic pigments containing Pb and Hg, and Pb-bearing alloys used in belt buckles and other household implements. In order to relate these artifacts to the Pb in the human bones, Ghazi (1994) determined the isotope compositions of the Pb. The resulting $^{206}Pb/^{204}Pb$ and $^{207}Pb/^{204}Pb$ ratios of the artifacts used by the Omaha Indians and the Pb in their bones form a straight line in Figure 25.12. This relationship not only confirms that the excess Pb in the bones originated from the artifacts and pigments used by the Omahas, but also suggests that the Pb originated from deposits of galena (PbS) in Missouri, Oklahoma, and Arkansas (Heyl et al., 1974). Isotope compositions of Pb in ancient

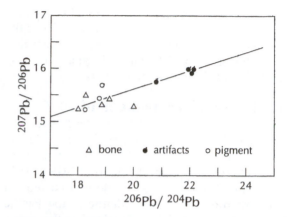

Figure 25.12 Isotope ratios of Pb in the bones of Omaha Indians at a burial site near Homer, Nebraska, and in pigment and Pb artifacts they used. The linear relationship of the isotope ratios suggests that the skeletal Pb originated primarily from Pb artifacts and pigments used by the Omaha tribe and that the Pb originated from deposits in the Tri-State District of Missouri, Oklahoma, and Arkansas (adapted from Ghazi, 1989).

objects have also been reported by Brill (1970), Brill and Wampler (1967), Gale (1989), and Gale and Stos-Gale (1989).

In marked contrast to the Omaha Indians of Nebraska, the bones of the Anasazi people from the Poly Secrest burial site in Moab, Utah, (1225–1405 A.D.) have low Pb concentrations with an average of $1.52 \pm 0.45\ \mu g/g$ and a range from 0.6 to $2.7\ \mu g/g$. Similarly, the bones of Alta people who lived at about 1200 A.D. near Chiribaya, Peru, have a low average Pb concentration of $1.70 \pm 0.70\ \mu g/g$.

Since neither the Anasazi nor the Alta people used Pb artifacts or pigments like those of the Omaha Indians, the Pb of their bones is presumably derived from the natural environment. The bones of the herbivorous animals recovered from the burial site of the Omaha Indians presumably also contain only natural environmental Pb derived from the soil by ingestion of plants and water. If that is true, then the Pb content of bones of herbivorous animals at Tonwantonga village is lower than that of the soil by a factor of about

0.07. In addition, the bones of adult Omaha Indians contain about 70 times as much Pb on the average as herbivorous animals, whereas children's bones are enriched in Pb by factors ranging from 3.4 to 1835 with a mean of about 475.

d. The Franklin Expedition of 1845

One of the worst cases of Pb poisoning occurred between 1845 and 1847 in the Canadian Arctic. Sir John Franklin, who had led two previous expeditions to the Canadian Arctic from 1819 to 1822 and from 1825 to 1827, was ordered to explore a northern route from the Atlantic to the Pacific Ocean. He departed from England in the spring of 1845 in two ships (Erebus and Terror) and a crew of 129 officers and men. They were last seen by a Scottish whaler in the waters north of Baffin Island and then disappeared forever. After twelve years of fruitless searches by several relief expeditions, the skeletons of some of the crew and a diary with a final entry dated April 25, 1848, were found on King William Island.

The expedition had spent the winter of 1845–1846 on Beechey Island off the southern coast of Devon Island. During the following summer the ships moved south through Peel Sound and Franklin Strait, but became trapped in the ice near King William Island. During the following winter Franklin and 24 of his men died. On April 22, 1848, the survivors abandoned the ships and attempted to reach the Canadian mainland by walking south across King William Island. All of them died during that attempt. The skull of one of Franklin's men was discovered 133 years later on June 29, 1981, on the south coast of King William Island along Simpson Strait, which separates it from the Canadian mainland (Beattie and Geiger, 1987).

Subsequent research by anthropologists from the University of Alberta eventually led to the discovery of three graves on Beechey Island (Beattie and Savelle, 1983). The bodies in these graves were exhumed and autopsied. Samples of hair, bone, and internal organs were found to have anomalously high concentrations of Pb (Kowal et al., 1989).

Hair samples collected from the bodies of John Torrington, John Hartnell, and William Braine contained high concentrations of Pb (Beattie and Geiger, 1987):

Torrington: 413–657 ppm

Hartnell: 138–313 ppm

Braine: 145–280 ppm

Because of the rapid growth of hair, the Pb must have entered the bodies of these men after the start of the expedition. It could have originated from several sources, including Pb foil used to wrap tea, pewterware, pottery covered with Pb-bearing glaze, and the solder of the tin cans in which most of the food was stored.

The solder used to seal the metal cans contained more than 90% Pb. The cans were soldered not only on the outside but also on the inside. Kowal et al. (1989) demonstrated that the isotope composition of the Pb in the solder is identical to that in the soft tissues and bones of Franklin's men. Therefore, the solder was probably the principal source of Pb ingested by all members of the expedition.

Franklin's expedition ended tragically not only because of the harsh climatic conditions and bad luck but also because acute Pb intoxication of the officers and crewmen caused anorexia, fatigue, and colic as well as neurotic behavior leading to inappropriate decisions. As a result, all 129 men perished.

Much later, men like Fridtjof Nansen (1861–1930) and Roald Amundsen (1872–1928) adopted the life style of the Inuits by hunting and eating game and wearing fur clothing. Their success in polar exploration demonstrates the wisdom of living in harmony with the polar environment instead of attempting to maintain the life style appropriate to a much more temperate climate.

25.7   Summary

Lead and many other industrial metals have been widely dispersed over the surface of the Earth. The sources of anthropogenic Pb include combustion of leaded gasoline, mining and smelting of Pb-bearing ores, and the use of Pb in household articles such as solder, paint, glazing compound, storage batteries, pottery covered with Pb-bearing glaze, etc. Many of these sources of Pb have been eliminated and acute Pb poisoning among humans is not as common as it used to be. However, Pb continues to be a hazard in certain occupations related to mining and smelting of Pb-bearing ores and to the use of Pb in the manufacturing and chemical industries.

In spite of precautions against the ingestion of Pb, most humans alive today contain significant amounts of Pb in their bones and soft tissues. The amounts increase with age because the rate of intake exceeds the body's ability to excrete it. The health effects caused by the accumulation of Pb are difficult to evaluate because everybody is contaminated.

The lead enters the human body primarily in contaminated food, water, and air. For example, vegetables grown in the vicinity of smelters or other industries that use Pb contain significantly more Pb than vegetables grown elsewhere. Even air con-tains Pb-bearing aerosol particles that may release Pb into the bloodstream after they enter the lungs.

The growing burden of Pb and other metals may adversely affect the life expectancy of humans by making them more vulnerable to other kinds of illnesses. The evidence that the bones of humans over 70 years of age contain less Pb than those of persons who died in their prime is an ominous sign that the chances for survival to old age are reduced by the accumulation of lead and other metals in bones and internal organs.

Children appear to be especially at risk because Pb affects not only their physical growth but also their mental development. Consequently, children who are exposed to anthropogenic Pb in their environment may never reach their full potential. The tragic deaths of the officers and crew of the Franklin Expedition are a reminder that even adults can suffer loss of good judgment in stressful situations as a result of Pb poisoning.

Problems

1. Calculate the concentration of Pb in gasoline sold in the USA in 1964 given that it contained 1.82 mL of tetraethyl lead (TEL) per gallon and that 1 mL of tetraethyl lead contained 1.057 g of Pb (1 gallon, US = 3.785 L).

2. Calculate the concentration of Pb in sediment that is a mixture of two components labeled A and B. Assume that the sediment mixture contains 35% of component B and that the Pb concentrations of the components are $Pb_A = 110$ ppm, $Pb_B = 18$ ppm.

3. Given that the Pb concentration of a mixture of two components A and B is 75 ppm, calculate the abundance of component A assuming that $Pb_A = 125$ ppm, and $Pb_B = 30$ ppm.

4. Calculate the $^{206}Pb/^{204}Pb$ ratio of Pb in a mixture of two components in which the abundance of component A is 35%, given that $(^{206}Pb/^{204}Pb)_A = 18.60$, $Pb_A = 175$ ppm, $(^{206}Pb/^{204}Pb)_B = 19.50$, and $Pb_B = 28$ ppm.

5. Plot a mixing hyperbola for a series of two-component mixtures such that the abundance of component A has values of 0.1, 0.2, 0.4, 0.6, 0.8, and 0.9.

Assume that $(^{206}Pb/^{204}Pb)_A = 18.60$, $Pb_A = 175$ ppm, $(^{206}Pb/^{204}Pb)_B = 19.50$, $Pb_B = 28$ ppm.

6. Use the results of problem 5 to convert the mixing hyperbola into a straight line by plotting the $^{206}Pb/^{204}Pb$ ratios versus the reciprocal concentrations of Pb.

7. Calculate the $^{207}Pb/^{206}Pb$ ratio of a mixture of aerosol particles derived from two sources A and B having different isotope ratios of Pb but identical concentrations. The abundance of component B is 65% and the isotope ratios are $(^{207}Pb/^{206}Pb)_A = 0.859$, $(^{207}Pb/^{206}Pb)_B = 0.832$.

8. The *average* $^{206}Pb/^{204}Pb$ ratio of Pb in the internal organs and bones of horses that died of Pb poisoning on the pastures near Benicia, California, is 17.62 ± 0.18. Assume that the Pb was derived from the nearby Pb smelter $(^{206}Pb/^{204}Pb = 17.33)$ and by combustion of leaded gasoline $(^{206}Pb/^{204}Pb = 18.09)$ and that it was transported in aerosol particles having equal concentrations of Pb. What fraction of the Pb in the internal organs of horses originated from the Pb smelter? State the error limits of the result based on the error of the average $^{206}Pb/^{204}Pb$ ratio of the Pb in the horses.

References

ANONYMOUS, 1986. Preventing lead poisoning in young children. U.S. Centers for Disease Control, U.S. Dept. of Health and Human Services, No. 99–2230, Atlanta.

BARRY, P. S. I., 1975. A comparison of concentrations of lead in human tissues. *British J. Industrial Med., 32*:119–139.

BARRY, P. S. I., 1976. Concentrations of lead in human tissues. Conference Roy. Soc. Health on Lead in the Environment, Birmingham, March 18, 1976.

BARRY, P. S. I., 1978. Distribution of lead in human tissues. In J. O. Nriagu (Ed.), *The Biogeochemistry of Lead in the Environment,* Part B:97–150. Elsevier and North Holland, Amsterdam, 397 pp.

BEATTIE, O., and J. GEIGER, 1987. *Frozen in Time: Unlocking the Secrets of the Franklin Expedition.* E. P. Dutton, New York, 180 pp.

BEATTIE, O. B., and J. M. SAVELLE, 1983. Discovery of human remains from Sir John Franklin's last expedition. *Historical Archaeology, 17*:100–105.

BOUTRON, C. F., 1990. A clean laboratory for ultralow concentration heavy metal analysis. *Fresenius J. Anal. Chem. 337*:482–491.

BOUTRON, C. F., J.-P. CANDELONE, and S. HONG, 1994. Past and recent changes in the large-scale tropospheric cycles of lead and other heavy metals as documented in Antarctic and Greenland snow and ice: a review. *Geochim. Cosmochim. Acta, 58*(15):3217–3225.

BRILL, R. H., 1970. Lead and oxygen isotopes in ancient objects. *Phil. Trans. Roy. Soc. London,* 269A:143–164.

BRILL, R. H., and J. M. WAMPLER, 1967. Isotope studies of ancient lead. *Amer. J. Archaeol., 71*:63–77.

BROOKS, R. R., 1972. *Geobotany and Biogeochemistry in Mineral Exploration.* Harper & Row, New York, 290 pp.

CANNON, H. L., and H. C. HOPPS (Eds.), 1971. *Environmental Geochemistry in Health and Disease.* Geol. Soc. Amer., Mem. 123:230 pp.

CHOW, T. J., and J. L. EARL, 1972. Lead isotopes in North American coals. *Science. 176*:510–511.

CHOW, T. J., and M. S. JOHNSTONE, 1965. Lead isotopes in gasoline and aerosols of Los Angeles Basin, California. *Science, 147*:502–503.

DAVIS, J. J., and D. J. SVENDSGAARD, 1987. Lead and child development. *Nature. 329*:297–300.

EDGINGTON, D. N., and J. A. ROBBINS, 1976. Records of lead deposition in Lake Michigan sediments since 1800. *Environ. Sci. Technol., 10*:266–274.

ERICSON, J. E., H. SHIRAHATA, and C. C. PATTERSON, 1979. Skeletal concentrations of lead in ancient Peruvians. *New England J. Med., 300*:946–951.

FAURE, G., 1986. *Principles of Isotope Geology,* 2nd ed. Wiley, New York, 589 pp.

FLEGAL, A. R., J. O. NRIAGU, S. NIEMEYER, and K. COALES, 1989. Isotopic tracers of lead contamination in the Great Lakes. *Nature. 339*:455–458.

FORBES, R. M., and G. C. SANDERSON, 1978. Lead toxicity in domestic animals and wildlife. In J. O. Nriagu (Ed.), *The Biogeochemistry of Lead in the Environment,* Part B:225–278. Elsevier–North Holland, Amsterdam, 397 pp.

FUGE, R., M. BILLET, and O. SELINUS, 1996. Environmental geochemistry, *App. Geochem., 11*(1/2):385 pp.

GALE, N. H., 1989. Lead isotope studies applied to provenance studies—a brief review. In Y. Maniatis (Ed.), *Archaeometry,* Proc. 25th Internat. Symp., 469–502. Elsevier, Amsterdam, *The Netherlands.*

GALE, N. H., and Z. A. STOS-GALE, 1989. Bronze Age archaeometallurgy of the Mediterranean: the impact of lead isotope studies. In R. O. Allen (Ed.), *Archaeological Chemistry IV,* 165–169. Amer. Chem. Soc., Adv. Chem. Ser., vol. 220.

GHAZI, A. M., 1994. Lead in archaeological samples: an isotopic study by ICP-MS. *Appl. Geochem., 9*:627–636.

GILFILLAN, S. C., 1965. Lead poisoning and the fall of Rome. *J. Occup. Med. 7*:53–60.

GRANDJEAN, P., and B. HOLMA, 1973. A history of lead retention in the Danish population. *Environ. Physiol. Biochem., 3*:268–273.

HEYL, A. V., G. P. LANDIS, and R. E. ZARTMAN, 1974. Isotopic evidence for the origin of the Mississippi Valley–type mineral deposits: a review. *Econ. Geol., 69*:992–1106.

JAWOROWSKI, I., 1968. Stable lead in fossil ice and bones. *Nature, 217*:152–153.

KOWAL, W., P. KRAHN, and O. B. BEATTIE, 1989. Lead levels in human tissues from the Franklin forensic project. *Internat. J. Environ. Anal. Chem., 35*:119–126.

LESSLER, M. A., 1988. Lead and lead poisoning from antiquity to modern times. *Ohio J. Sci., 88*:78–84.

MACKIE, A. C., A. TOWNSEND, and H. A. WALDRON, 1975. Lead concentrations in the bones of Roman York. *J. Archaeol. Soc., 2*:235–237.

MUROZUMI, M., T. J. CHOW, and C. C. PATTERSON, 1969. Chemical concentrations of pollutant lead aerosols, terrestrial dusts and sea salts in Greenland and Antarctic snow strata. *Geochim. Cosmochim. Acta, 33*:1247–1294.

NICOLLS, O. W., D. M. J. PROVAN, M. M. COLE, and J. S. TOOMS, 1965. Geobotany and biogeochemistry in mineral exploration in the Dugald River area, Cloncurry district, Australia. *Inst. Mining Metallurgy Trans., 74*:696–799.

NRIAGU, J. O. (Ed.), 1978. *The Biogeochemistry of Lead in the Environment,* Parts A and B. Elsevier–North Holland, Amsterdam.

NRIAGU, J. O., 1978. Lead in the atmosphere. In J. O. Nriagu (Ed.), *The Biogeochemistry of Lead in the Environment,* Part A, 137–184. Elsevier, Amsterdam, The Netherlands, 422 pp.

NRIAGU, J., A. L. KEMP, H. K. T. WONG, and N. HARP, 1979. Sedimentary record of heavy metal pollution in Lake Erie. *Geochim. Cosmochim. Acta,* **43**:247–258.

NRIAGU, J. O., 1983a. *Lead and Lead Poisoning in Antiquity.* Wiley, New York.

NRIAGU, J. O. 1983b. Saturnine gout among Roman aristocrats. *New England J. Med.,* **308**:660–663.

PATTERSON, C. C., 1965. Contaminated and natural lead environments of man. *Arch. Environ. Health,* **11**:344–360.

PATTERSON, C. C., 1981. Acceptance speech for the V. M. Goldschmidt Medal. *Geochim. Cosmochim. Acta,* **45**:1385–1388.

PATTERSON, C. C., and D. M. SETTLE, 1976. The reduction of orders of magnitude errors in lead analysis of biological materials and natural waters by evaluating and controlling the extent and sources of industrial lead contamination introduced during sample collection and analysis. In P. La Fleur (Ed.), *Accuracy in Trace Analysis.* Natl. Bur. Stand. Spec. Pub. 422, 321–351.

PETERSON, P. J., 1978. Lead in vegetation. In J. O. Nriagu (Ed.), *The Biogeochemistry of Lead in the Environment,* Part B, 355–384. Elsevier–North Holland, Amsterdam, 397 pp.

POSNER, H. S., T. DAMSTRA, and J. O. NRIAGU, 1978. Human health effects of lead. In J. O. Nriagu (Ed.), *The Biochemistry of Lead in the Environment,* Part B, 173–224. Elsevier–North Holland, Amsterdam, 397 pp.

RABINOWITZ, M. B., and G. W. WETHERILL, 1972. Identifying sources of lead contamination by stable isotope techniques. *Environ. Sci. Technol.* (Current Research), **6**(8):705–709.

RITSON, P. I., B. K. ESSER, S. NIEMEYER, and A. R. FLEGAL, 1994. Lead isotope determination of historical sources of lead in Lake Erie, North America. *Geochim. Cosmochim. Acta,* **58**:3297–3305.

SALOMONS, W., and U. FÖRSTNER, 1984. *Metals in the Hydrocycle.* Springer-Verlag, Heidelberg, 349 pp.

SCHROEDER, H. A., and I. H. TIPTON, 1968. The human body burden of lead. *Arch. Environ. Health.* **17**:965–978.

SHAPIRO, I. M., G. MITCHEL, I. DAVIDSON, and S. H. KATZ, 1975. The lead content of teeth. *Arch. Enrivon. Health,* **30**:438–486.

SHIRAHATA, H., R. W. ELIAS, and C. C. PATTERSON, 1980. Chronological variations in concentrations and isotopic compositions of anthropogenic lead in sediments of a remote subalpine pond. *Geochim. Cosmochim. Acta,* **44**:149–162.

TATSUMOTO, M., and C. C. PATTERSON, 1963. Concentrations of common lead in some Atlantic and Mediterranean waters and in snow. *Nature.* **199**:350.

TER HAAR, G. L., and M. A. BAYARD, 1971. Composition of airborne lead particles. *Nature,* **232**:553–554.

TESSIER, A., and D. R. TURNER, 1995. *Metal Speciation and Bioavailability in Aquatic Systems.* Wiley and Sons, Chichester, UK, 679 pp.

THORNTON, I., (Ed), 1983. *Applied Environmental Geochemistry.* Academic Press, London, 501 pp.

VINOGRADOV, A. P., 1959. *The Geochemistry of Rare and Dispersed Chemical Elements in Soils,* 2nd ed. Consultants Bureau, New York, 209 pp.

WARREN. H. V., 1972. Geology and medicine. *Western Miner,* vol. 45.

WARREN, H. V., and R. E. DELAVAULT, 1960. Observations on the biogeochemistry of lead in Canada. *Trans. Roy. Soc. Canada,* **54**:11–20.

WARREN, H. V., and R. E. DELAVAULT, 1971. Variations in the copper, zinc, lead, and molybdenum contents of some vegetables and their supporting soils. In H. L. Cannon and H. C. Hopps (Eds.), *Environmental Geochemistry in Health and Disease,* 97–108. Amer. Geol. Soc. Mem. 123, 230 p.

WEIS, D., B. WHITTEN, and D. LEDDY, 1972. Lead content of human hair (1871–1971). *Science,* **178**:69–70.

WILKINSON, D. R., and W. PALMER, 1975. Lead in teeth as a function of age. *Internat. Lab.,* **67**:41–46.

WU, J.-F, and E. A. BOYLE, 1997. Lead in the western North Atlantic Ocean; completed response to leaded gasoline phaseout. *Geochim. Cosmochim. Acta,* 61(15):3279–3283.

ZMUDSKI, J., G. R. BRATTON, C. WOMAC, and L. ROWE, 1983. Lead poisoning in cattle: reassessment of the minimum toxic oral doses. *Bull. Environ. Contam. Toxicol,* **30**:435–441.

Appendix A

COMPILATIONS OF GEOCHEMICAL DATA

Table A.1 Abundances of the Chemical Elements in the Sun in Units of Atoms per 10^6 Si Atoms

Z	Symbol	Abundance[a]		Z	Symbol	Abundance[a]	
1	H	2.2×10^{10}		41	Nb	1.8	(D⁻)
2	He	1.4×10^9	(C)	42	Mo	3.2	(D⁻)
3	Li	2.2×10^{-1}	(B)	44	Ru	1.5	(D⁻)
4	Be	3.1×10^{-1}	(B)	45	Rh	5.6×10^{-1}	(D⁻)
5	B	<2.8		46	Pd	7.1×10^{-1}	(D⁻)
6	C	9.3×10^6	(B)	47	Ag	1.6×10^{-1}	(C)
7	N	1.9×10^6	(B)	48	Cd	1.58	(B)
8	O	1.5×10^7	(A)	49	In	1.0	(C)
9	F	8.1×10^2	(C)	50	Sn	2.2	(C⁻)
10	Ne	8.3×10^5	(B)	51	Sb	2.2×10^{-1}	(D)
11	Na	4.2×10^4	(A)	55	Cs	<1.8	
12	Mg	8.9×10^5	(C)	56	Ba	2.75	(B)
13	Al	7.4×10^4	(C)	57	La	3.0×10^{-1}	(C)
14	Si	1.0×10^6	(C)	58	Ce	7.9×10^{-1}	(C)
15	P	7.1×10^3	(C)	59	Pr	1.0×10^{-1}	(C)
16	S	3.5×10^5	(C)	60	Nd	3.8×10^{-1}	(C)
17	Cl	7.1×10^3	(D)	62	Sm	1.2×10^{-1}	(C)
18	Ar	2.2×10^4	(C)	63	Eu	1.1×10^{-1}	(D)
19	K	3.2×10^3	(B)	64	Gd	2.9×10^{-1}	(D)
20	Ca	5.0×10^4	(B)	66	Dy	2.6×10^{-1}	(D)
21	Sc	2.4×10^1	(B)	68	Er	1.3×10^{-1}	(D)
22	Ti	2.5×10^4	(B)	69	Tm	4.1×10^{-2}	(C)
23	V	2.3×10^2	(C)	70	Yb	1.8×10^{-1}	(D)
24	Cr	1.1×10^4	(C)	71	Lu	1.3×10^{-1}	(C⁻)
25	Mn	5.9×10^3	(C)	72	Hf	1.4×10^{-1}	(C)
26	Fe	7.1×10^5	(B?)	74	W	1.1	(D⁻)
27	Co	1.8×10^3	(C⁻)	75	Re	$<1.1 \times 10^{-2}$	
28	Ni	4.3×10^4	(B)	76	Os	1.1×10^{-1}	(D)
29	Cu	2.6×10^2	(B)	77	Ir	1.6×10^{-1}	(C)
30	Zn	6.3×10^2	(C)	78	Pt	1.25	(C)
31	Ga	1.4×10^1	(C)	79	Au	1.3×10^{-1}	(C⁻)
32	Ge	7.1×10^1	(C)	80	Hg	<2.8	
37	Rb	8.9	(B)	81	Tl	1.8×10^{-1}	(C)
38	Sr	1.8×10^1	(B)	82	Pb	1.90	(B)
39	Y	2.8	(D)	83	Bi	<1.8	
40	Zr	1.3×10^1	(D)	90	Th	3.5×10^{-2}	
				92	U	$<8.9 \times 10^{-2}$	

[a]Rating of the quality of the determination. A = abundance known to an accuracy of about 10%; B = reasonably good determination; C = fair determination; and D = poor determination.

Recalculated from Ross and Aller (1976).

Table A.2 Representative Chemical Compositions of Meteorites in Weight Percent[a]

	Carbonaceous chondrite[1]	Enstatite chondrite[2]	Ca-poor achondrites[3]	Ca-rich achondrites[4]	Average iron meteorites[5]
Fe	—	20.04	2.92	0.80	90.6
Ni	—	1.96	0.17	—	7.9
Co	—	0.07	—	—	0.5
P	—	—	—	—	0.2
S	—	—	—	—	0.7
FeS	15.07	7.27	1.25	0.41	—
SiO_2	22.56	41.53	54.01	48.17	—
TiO_2	0.07	—	0.06	0.51	—
Al_2O_3	1.65	1.55	0.67	13.91	—
MnO	0.19	—	0.14	0.46	—
FeO	11.39	0.34	0.97	15.99	—
MgO	15.81	23.23	35.92	7.10	—
CaO	1.22	0.74	0.91	10.94	—
Na_2O	0.74	1.26	1.32	0.67	—
K_2O	0.07	0.32	0.10	0.13	—
P_2O_5	0.28	0.8	0.22	0.11	—
H_2O	19.89	—	1.14	0.44	—
Cr_2O_3	0.36	0.56	0.06	0.39	—
NiO	1.23	—	0.26	—	—
CoO	0.06	—	—	—	—
C	3.10	—	—	—	0.04
LOI[b]	6.96	0.86(CaS)	0.51(CaS)	—	—
Sum	100.65	99.91	100.00	100.3	99.94

[a]A dash (—) means "not reported and probably zero," although in some cases the element in question was reported in different form.

[b]Loss on ignition.

SOURCES: (1)Orgueil, type I, from Henderson (1982, Table 1.3); (2) Hvittis, from Henderson (1982, Table 1.3); (3) average aubrite, from Henderson (1982, Table 1.3); (4) average eucrites, from Henderson (1982, Table 1.3); (5) average iron meteorite from Glass (1982, Table 4.3).

Table A.3 Abundance of the Chemical Elements in Chondrite Meteorites in Number of Atoms per 10^6 Si Atoms

Z	Element	Abundance	Z	Element	Abundance
1	H	—	44	Ru	1.5
2	He	1.1×10^{-1}	45	Rh	2.5×10^{-1}
3	Li	$\sim 5.0 \times 10$	46	Pd	$\sim 9 \times 10^{-1}$
4	Be	6.9×10^{-1}	47	Ag	$\sim 9 \times 10^{-2}$
5	B	6.2	48	Cd	7.9×10^{-2}
6	C	2.0×10^3	49	In	1.0×10^{-2}
7	N	9.0×10	50	Sn	5.6×10^{-1}
8	O	3.7×10^6	51	Sb	1.1×10^{-1}
9	F	$\sim 7.0 \times 10^2$	52	Te	6.0×10^{-1}
10	Ne	1.5×10^{-3}	53	I	5.1×10^{-2}
11	Na	4.6×10^4	54	Xe	7×10^{-6}
12	Mg	9.4×10^5	55	Cs	1.4×10^{-1}
13	Al	6.0×10^4	56	Ba	5.0
14	Si	1.0×10^6	57	La	3.9×10^{-1}
15	P	5.3×10^3	58	Ce	1.2
16	S	1.1×10^5	59	Pr	1.4×10^{-1}
17	Cl	7.0×10^2	60	Nd	6.4×10^{-1}
18	Ar	4.0×10^{-1}	62	Sm	2.3×10^{-1}
19	K	3.5×10^3	63	Eu	8.2×10^{-1}
20	Ca	4.9×10^4	64	Gd	3.4×10^{-1}
21	Sc	2.9×10	65	Tb	5.1×10^{-2}
22	Ti	$\sim 2.6 \times 10^3$	66	Dy	3.3×10^{-1}
23	V	2.0×10^2	67	Ho	7.6×10^{-2}
24	Cr	6.6×10^3	68	Er	2.3×10^{-1}
25	Mn	7.4×10^3	69	Tm	3.1×10^{-2}
26	Fe	6.9×10^5	70	Yb	1.8×10^{-1}
27	Co	1.3×10^3	71	Lu	3.1×10^{-2}
28	Ni	4×10^4	72	Hf	1.7×10^{-1}
29	Cu	2.5×10^2	73	Ta	1.6×10^{-2}
30	Zn	1.3×10^2	74	W	1.2×10^{-1}
31	Ga	1.2×10	75	Re	4.6×10^{-2}
32	Ge	2.0×10	76	Os	5.4×10^{-1}
33	As	4.6	77	Ir	3.5×10^{-1}
34	Se	1.9×10	78	Pt	1.30
35	Br	5	79	Au	1.8×10^{-1}
36	Kr	—	80	Hg	1.0×10^{-1}
37	Rb	~ 5	81	Tl	1.0×10^{-3}
38	Sr	2.0×10	82	Pb	1.4×10^{-1}
39	Y	3.4	83	Bi	2.0×10^{-3}
40	Zr	$\sim 3.7 \times 10$	90	Th	2.7×10^{-2}
41	Nb	1.0	92	U	9.0×10^{-3}
42	Mo	2.5			

SOURCE: Brownlow (1979)

Table A.4 Average Chemical Composition of Selected Lunar Rocks

	1[a]	2[b]	3[c]	4[d]	5[e]	6[f]	7[g]
SiO_2	40.67	40.37	44.32	47.88	44.08	45.40	47.04
TiO_2	10.18	11.77	2.65	1.82	0.02	0.32	1.21
Al_2O_3	10.40	8.84	8.03	9.46	35.49	28.63	18.69
FeO	18.68	19.28	21.11	20.13	0.23	4.25	9.45
MnO	0.27	0.24	0.28	0.28	0.00	0.06	0.11
MgO	6.92	7.56	14.07	8.74	0.09	4.38	10.14
CaO	11.70	10.59	8.61	10.54	19.68	16.39	11.52
Na_2O	0.41	0.52	0.22	0.31	0.34	0.41	0.48
K_2O	0.07	0.31	0.06	0.06	<0.01	0.06	0.34
Cr_2O_3	0.29	0.36	0.63	0.47	—	—	0.20
Total	99.59	99.52	99.98	99.69	99.94	99.9	99.18

[a]Low K basalt, Mare Tranquillitatis, Apollo 11, average of eight analyses.

[b]High K basalt, Mare Tranquillitatis, Apollo 11, average of six analyses.

[c]Olivine basalt, Oceanus Procellarum, Apollo 12, average of nine analyses.

[d]Pigeonite basalt, Hadley-Apennine, Apollo 15 Mare Imbrium, average of 13 analyses.

[e]Anorthosite, lunar highlands, Hadley-Apennine, Apollo 15, 15415, Lunar Sample Preliminary Examination Team (1972).

[f]Gabbroic anorthosite, lunar highlands, Descartes Crater, Southern Highlands, Apollo 16, 68415, Lunar Sample Preliminary Examination Team (1972).

[g]KREEP basalt, Fra Mauro Crater, Apollo 14, Philpotts et al. (1972).

SOURCE: Glass (1982).

Table A.5 Abundance of Chemical Elements in the Moon Based on 10^6 Si Atoms

Z	Element	Abundance	Z	Element	Abundance
3	Li[a]	1.7×10^2	27	Co[a]	5.6×10^2
6	C[a]	1.1×10^2	28	Ni	1.2
8	O[a]	3.5×10^6	37	Rb	4.6×10^{-1}
9	F[a]	2.1×10^2	38	Sr	6.5×10^1
11	Na[a]	5.3×10^3	39	Y	1.2×10^1
12	Mg	1.0×10^6	40	Zr	4.3×10^1
13	Al	2.2×10^5	41	Nb	3.1
14	Si	1.0×10^6	55	Cs	1.3×10^1
15	P[a]	2.3×10^3	56	Ba	1.7×10^1
16	S[a]	1.6×10^4	57	La	1.1
19	K	3.5×10^2	62	Sm	9.1×10^{-1}
20	Ca	1.5×10^5	63	Eu	2.4×10^{-1}
21	Sc	6.1×10^1	65	Tb	1.5×10^{-1}
22	Ti	5.1×10^3	70	Yb	5.6×10^{-1}
23	V	1.3×10^2	71	Lu	8.3×10^{-2}
24	Cr	3.5×10^3	72	Hf	4.9×10^{-1}
25	Mn	2.5×10^3	90	Th	1.3×10^{-1}
26	Fe	2.0×10^5	92	U	3.4×10^{-2}

[a]From Ganapathy and Anders (1974).

Recalculated from Taylor (1975) except as noted.

Table A.6 Chemical Compositions of the Atmospheres of Mars and Earth

Component	Mars %	Mars ppm	Earth %	Earth ppm
CO_2	95.32		0.033	
N_2	2.7		78.084	
Ar	1.6		0.934	
O_2	0.13		20.946	
CO	0.07			
H_2O[a]	0.03			
Ne		2.5		18.2
Kr		0.3		1.2
Xe		0.08		0.9
O_3[a]		0.03		

[a]Variable.

SOURCE: Owen et al. (1977); Glass (1982).

Table A.7 Chemical Composition of Martian Soil at the Landing Sites of Viking 1 and 2 in Weight Percent

	Viking 1 1	Viking 1 2	Viking 1 3	Viking 2 1
SiO_2	44.7	44.5	43.9	42.8
Al_2O_3	5.7	—	5.5	—
Fe_2O_3	18.2	18.0	18.7	20.3
MgO	8.3	—	8.6	—
CaO	5.6	5.3	5.6	5.0
K_2O	<0.3	<0.3	<0.3	<0.3
TiO_2	0.9	0.9	0.9	1.0
SO_3	7.7	9.5	9.5	6.5
Cl	0.7	0.8	0.9	0.6
Sum	92.1	incomplete	93.9	incomplete

SOURCE: Toulmin et al. (1977).

Table A.8 Chemical Composition of Surface Material at the Landing Sites of Venera 13 and 14 on Venus in Weight Percent

	Venera 13	Venera 14	Average terrestrial oceanic basalt
MgO	10 ± 6	8 ± 4	7.56
Al_2O_3	16 ± 4	18 ± 4	16.5
SiO_2	45 ± 3	49 ± 4	51.4
K_2O	4 ± 0.8	0.2 ± 0.1	1.0
CaO	7 ± 1.5	10 ± 1.5	9.4
TiO_2	1.5 ± 0.6	1.2 ± 0.4	1.5
MnO	0.2 ± 0.1	0.16 ± 0.08	0.26
FeO	9 ± 3	9 ± 2	12.24
Total	92.7	95.56	99.86

SOURCE: Data from Barsukov quoted by Carr et al. (1984).

Table A.9 Chemical Composition of the Atmosphere of Venus

Component	Concentration[a]	
	%	ppm
CO_2	96.4 ± 1.03	
N_2	3.4 ± 0.02	
H_2O	0.14 ± 0.01	
O_2		69.3 ± 1.3
Ar		18.6 ± 2.4
Ne		$4.3^{+5.5}_{-3.9}$
SO_2		186^{+349}_{-156}

[a]Analysis by gas chromatograph on Pioneer Venus.
SOURCE: Oyama et al. (1979).

Table A.10 Average Chemical Composition of the Continental Crust of the Earth

Element	Units		Elements	Units	
	$\mu g/g$	atoms/10^6 Si atoms		$\mu g/g$	atoms/10^6 Si atoms
Li	13	2.0×10^2	Cd	9.8×10^{-2}	9.1×10^{-2}
Be	1.5	1.7×10^1	In	5.0×10^{-2}	4.6×10^{-2}
B	10	9.7×10^1	Sn	2.5	2.2
O	45.5×10^4	3.08×10^6	Sb	0.2	2×10^{-1}
Na	2.3×10^4	1.05×10^5	Cs	1.0	8×10^{-1}
Mg	3.2×10^4	1.38×10^5	Ba	250	1.9×10^2
Al	8.40×10^4	3.27×10^5	La	16	1.2×10^1
Si	26.8×10^4	1.00×10^6	Ce	33	2.5×10^1
K	0.90×10^4	2.44×10^4	Pr	3.9	2.9
Ca	5.3×10^4	1.38×10^5	Nd	16	1.2×10^1
Sc	30	7.0×10^1	Sm	3.5	2.4
Ti	5.3×10^3	1.18×10^4	Eu	1.1	7.6×10^{-1}
V	2.3×10^2	4.74×10^2	Gd	3.3	2.2
Cr	1.85×10^2	3.73×10^2	Tb	0.60	4.0×10^{-1}
Mn	1.40×10^3	2.67×10^3	Dy	3.7	2.4
Fe	7.06×10^3	1.32×10^5	Ho	0.78	5.0×10^{-1}
Co	29	5.2×10^1	Er	2.2	1.4
Ni	105	1.88×10^2	Tm	0.32	2.0×10^{-1}
Cu	75	1.2×10^2	Yb	2.2	1.3
Zn	80	1.3×10^2	Lu	0.30	1.8×10^{-1}
Ga	18	2.7×10^1	Hf	3.0	1.8
Ge	1.6	2.3	Ta	1.0	5.8×10^{-1}
As	1.0	1.4	W	1.0	5.7×10^{-1}
Se	5×10^{-2}	7×10^{-2}	Re	5×10^{-4}	3×10^{-4}
Rb	32	3.9×10^1	Ir	1×10^{-4}	5×10^{-5}
Sr	260	3.1×10^2	Au	3×10^{-3}	2×10^{-3}
Y	20	2.4×10^1	Tl	0.36	1.8×10^{-1}
Zr	100	1.15×10^2	Pb	8.0	4.1
Nb	11	1.2×10^1	Bi	6×10^{-2}	3×10^{-2}
Mo	1.0	1.1	Th	3.5	1.6
Pd	1×10^{-3}	1×10^{-3}	U	0.91	4.0×10^{-1}
Ag	8.0×10^{-2}	8×10^{-2}			

SOURCE: Taylor and McLennan (1985).

Table A.11 Chemical Composition of Igneous and Sedimentary Rocks (in parts per million unless otherwise indicated)

Element, Z	Ultramafic[a]	Basalt[a]	High-Ca granites[b]	Low-Ca granites[b]	Shale[b]	Sandstone[b]	Carbonate rocks[b]	Deep-sea clay[b]
3 Li	0.5	16	24	40	66	15	5	57
4 Be	0.2	0.7	2	3	3	—	—	2.6
5 B	2	5	9	10	100	35	20	230
9 F	100	385	520	850	740	270	330	1300
11 Na (%)	0.49	1.87	2.84	2.58	0.96	0.33	0.04	4
12 Mg (%)	23.2	4.55	0.94	0.16	1.50	0.70	4.70	2.10
13 Al (%)	1.2	8.28	8.20	7.20	8	2.50	0.42	8.40
14 Si (%)	19.8	23.5	31.40	34.70	7.30	36.80	2.40	25
15 P	195	1130	920	600	700	170	400	1500
16 S	200	300	300	300	2400	240	1200	1300
17 Cl	45	55	130	200	180	10	150	21000
19 K (%)	0.017	0.83	2.52	4.20	2.66	1.07	0.27	2.50
20 Ca (%)	1.6	7.2	2.53	0.51	2.21	3.91	30.23	2.90
21 Sc	10	27	14	7	13	1	1	19
22 Ti	300	11400	3400	1200	4600	1500	400	4600
23 V	40	225	88	44	130	20	20	120
24 Cr	1800	185	22	4.1	90	35	11	90
25 Mn	1560	1750	540	390	850	—	1100	6700
26 Fe (%)	9.64	8.60	2.96	1.42	4.72	0.98	0.33	6.50
27 Co	175	47	7	1	19	0.3	0.1	74
28 Ni	2000	145	15	4.5	68	2	20	225
29 Cu	15	94	30	10	45	—	4	250
30 Zn	40	118	60	39	95	16	20	165
31 Ga	1.8	18	17	17	19	12	4	20
32 Ge	1.3	1.4	1.3	1.3	1.6	0.8	0.2	2
33 As	0.8	2.2	1.9	1.5	13	1	1	13
34 Se	0.05	0.05	0.05	0.05	0.6	0.05	0.08	0.17
35 Br	0.8	3.3	4.5	1.3	4	1	6.2	70
37 Rb	1.1	38	110	170	140	60	3	100
38 Sr	5.5	452	440	100	300	20	610	180
39 Y	—	21	35	40	26	40	30	90
40 Zr	38	120	140	175	160	220	19	150
41 Nb	9	20	20	21	11	—	0.3	14
42 Mo	0.3	1.5	1	1.3	2.6	0.2	0.4	27
47 Ag	0.05	0.11	0.051	0.037	0.07	0.01	0.01	0.11
48 Cd	0.05	0.21	0.13	0.13	0.3	—	0.035	0.42
49 In	0.012	0.22	—	0.26	0.1	—	—	0.08
50 Sn	0.5	1.5	1.5	3	6	0.1	0.1	1.5
51 Sb	0.1	0.6	0.2	0.2	1.5	0.01	0.2	1
53 I	0.3	0.5	0.5	0.5	2.2	1.7	1.2	0.05
55 Cs	0.1	1.1	2	4	5	0.1	0.1	6
56 Ba	0.7	315	420	840	580	10	10	2300
57 La	1.3[c]	6.1[d]	45	55	92	30	1	115
58 Ce	3.5[c]	16[d]	81	92	59	92	11.5	345

Table A.11 (continued)

Element, Z	Ultramafic[a]	Basalt[a]	High-Ca granites[b]	Low-Ca granites[b]	Shale[b]	Sandstone[b]	Carbonate rocks[b]	Deep-sea clay[b]
59 Pr	0.49[c]	2.7[d]	7.7	8.8	5.6	8.8	1.1	33
60 Nd	1.9[c]	14[d]	33	37	24	37	4.7	140
62 Sm	0.42[c]	4.3[d]	8.8	10	6.4	10	1.3	38
63 Eu	0.14[c]	1.5[d]	1.4	1.6	1	1.6	0.2	6
64 Gd	0.54[c]	6.2[d]	8.8	10	6.4	10	1.3	38
65 Tb	0.12[c]	1.1[d]	1.4	1.6	1	1.6	0.2	6
66 Dy	0.77[c]	5.9[d]	6.3	7.2	4.6	7.2	0.9	27
67 Ho	0.12[c]	1.4[d]	1.8	2	1.2	2	0.3	7.5
68 Er	0.30[c]	3.6[d]	3.5	4	2.5	4	0.5	15
69 Tm	0.041[c]	0.60[d]	0.3	0.3	0.2	0.3	0.04	1.2
70 Yb	0.38[c]	3.2[d]	3.5	4	2.6	4	0.5	15
71 Lu	0.036[c]	0.55[d]	1.1	1.2	0.7	1.2	0.2	4.5
72 Hf	0.4	1.5	2.3	3.9	2.8	3.9	0.3	4.1
73 Ta	0.5	0.8	3.6	4.2	0.8	0.01	0.01	0.1
74 W	0.5	0.9	1.3	2.2	1.8	1.5	0.6	1
79 Au	0.006	0.004	0.004	0.004	—	—	—	—
80 Hg	0.01	0.09	0.08	0.08	0.4	0.03	0.04	0.1
81 Tl	0.04	0.21	0.72	2.3	1.4	0.82	0.01	0.8
82 Pb	0.5	7	15	19	20	7	9	80
90 Th	0.0045	3.5	8.5	17	12	1.7	1.7	7
92 U	0.002	0.75	3	3	3.7	0.45	2.2	1.3

[a]Average of Turekian and Wedepohl (1961) and Vinogradov (1962).
[b]Turekian and Wedepohl (1961).
[c]Calculated from data listed by Herrmann (1970).
[d]Average values calculated by Herrmann (1970).

Table A.12 Average Composition of Water in Streams and in the Oceans in micrograms per gram

Element	Classification[a]	Stream water	Seawater	Seawater enrichment	MORT[b]
Li	I	3×10^{-3}	1.7×10^{-1}	56.7	2.5×10^{6}
Be	IV	1×10^{-5}	2×10^{-7}	0.02	6.3×10^{1}
B	I	1×10^{-2}	4.5	450	1.6×10^{7}
F	I	1×10^{-3}	1.3	1300	7.9×10^{5}
Na	I	6.3	1.08×10^{4}	1714	2.0×10^{8}
Mg	I	4.1	1.29×10^{3}	315	5.0×10^{7}
Al	IV	5×10^{-2}	8×10^{-4}	0.016	7.0
Si	II	6.5	2.8	0.43	7.9×10^{3}
P	II	2×10^{-2}	7.1×10^{-2}	3.6	4.0×10^{4}
S	I	3.7	9.0×10^{2}	243	5.0×10^{8}
Cl	I	7.8	1.95×10^{4}	2500	6.3×10^{8}
K	I	2.3	3.99×10^{2}	173	1.3×10^{7}
Ca	I	15	4.13×10^{2}	27.5	1.3×10^{6}
Sc	IV	4×10^{-6}	6.7×10^{-7}	0.17	2.5×10^{1}
Ti	IV	3×10^{-3}	$<9.6 \times 10^{-4}$	0.32	$<1.6 \times 10^{2}$
V	IV	9×10^{-4}	1.2×10^{-3}	1.3	7.9×10^{3}
Cr	IV	1×10^{-3}	2×10^{-4}	0.2	1.6×10^{3}
Mn	III	7×10^{-3}	3×10^{-4}	0.04	3.2×10^{1}
Fe	IV	4×10^{-2}	6×10^{-5}	0.0015	6.9×10^{-1}
Co	III	1×10^{-4}	2×10^{-6}	0.02	2.0×10^{1}
Ni	III	3×10^{-4}	5×10^{-4}	1.7	1.6×10^{3}
Cu	III	7×10^{-3}	3×10^{-4}	0.04	1.0×10^{3}
Zn	III	2×10^{-2}	4×10^{-4}	0.02	1.3×10^{3}
Ga	IV	9×10^{-5}	2×10^{-5}	0.2	7.9×10^{2}
Ge	IV	5×10^{-6}	5×10^{-6}	1	2.0×10^{3}
As	IV	2×10^{-3}	1.7×10^{-3}	0.85	1.0×10^{5}
Se	III	6×10^{-5}	1.3×10^{-4}	2.2	6.3×10^{5}
Br	I	2×10^{-2}	6.7×10^{1}	3350	7.9×10^{8}
Rb	I	1×10^{-3}	1.2×10^{-1}	120	7.9×10^{5}
Sr	I	7×10^{-2}	7.6	109	5.0×10^{6}
Y	III	4×10^{-5}	7×10^{-6}	0.18	1.3×10^{2}
Zr	IV	—	3×10^{-5}	—	1.6×10^{2}
Nb	IV	—	$<5 \times 10^{-6}$	—	$<2.5 \times 10^{2}$
Mo	I	6×10^{-4}	1.1×10^{-2}	18.3	3.2×10^{5}
Ag	III	3×10^{-4}	2.7×10^{-6}	0.009	2.0×10^{4}
Cd	IV	1×10^{-5}	8×10^{-5}	8	7.9×10^{4}
In	IV	—	1×10^{-7}	—	1.0×10^{3}
Sn	IV	4×10^{-5}	5×10^{-7}	0.013	1.3×10^{2}
Sb	III	7×10^{-5}	1.5×10^{-4}	2.1	1.3×10^{5}
I	I	7×10^{-3}	5.6×10^{-2}	8	$<4.0 \times 10^{6}$
Cs	I	2×10^{-5}	2.9×10^{-4}	14.5	4.0×10^{4}
Ba	III	2×10^{-2}	1.4×10^{-2}	0.7	5.0×10^{3}
La	III	4.8×10^{-5}	4.5×10^{-6}	0.094	7.9×10^{1}
Cl	III	7.9×10^{-5}	3.5×10^{-6}	0.044	3.2×10^{1}

Table A.12 (continued)

Element	Classification[a]	Stream water	Seawater	Seawater enrichment	MORT[b]
Pr	III	7.3×10^{-6}	1.0×10^{-6}	0.14	7.9×10^{1}
Nd	III	3.8×10^{-5}	4.2×10^{-6}	0.11	7.9×10^{1}
Sm	III	7.8×10^{-6}	8.0×10^{-7}	0.10	7.9×10^{1}
Eu	III	1.5×10^{-6}	1.5×10^{-7}	0.10	6.3×10^{1}
Gd	III	8.5×10^{-6}	1.0×10^{-6}	0.11	1.0×10^{2}
Tb	III	1.2×10^{-6}	1.7×10^{-7}	0.14	1.0×10^{2}
Dy	III	7.2×10^{-6}	1.1×10^{-6}	0.15	1.0×10^{2}
Ho	III	1.4×10^{-6}	2.8×10^{-7}	0.20	1.3×10^{2}
Er	III	4.2×10^{-6}	9.2×10^{-7}	0.22	1.6×10^{2}
Tm	III	6.1×10^{-7}	1.3×10^{-7}	0.21	1.6×10^{2}
Yb	IV	3.6×10^{-6}	9.0×10^{-7}	0.25	2.0×10^{2}
Lu	IV	6.4×10^{-7}	1.4×10^{-7}	0.22	2.0×10^{2}
Hf	IV	—	$<7 \times 10^{-6}$	—	$<1.3 \times 10^{3}$
Ta	IV	—	$<2.5 \times 10^{-6}$	—	$<2.0 \times 10^{3}$
W	IV	3×10^{-5}	1×10^{-4}	3.3	7.9×10^{4}
Re	IV	—	4×10^{-6}	—	3.2×10^{6}
Au	III	2×10^{-6}	4.9×10^{-6}	2.5	1.6×10^{6}
Hg	III	7×10^{-5}	1×10^{-6}	0.14	7.9×10^{3}
Tl	IV	—	1×10^{-5}	—	6.3×10^{3}
Pb	III	1×10^{-3}	2×10^{-6}	0.002	5.0×10^{1}
Bi	IV	—	2×10^{-5}	—	2.5×10^{4}
Th	IV	$<1 \times 10^{-4}$	6×10^{-8}	~0.0006	3.4
U	I	4×10^{-5}	3.1×10^{-3}	2.7	1×10^{6}

[a]I. Conservative element whose concentration is directly related to the salinity of seawater.

II. Nonconservative element whose concentration varies with depth or regionally within the oceans, or both, generally because of involvement in biological activity.

III. Nonconservative element whose concentration varies irregularly and is not related to salinity, depth, or geographic factors.

IV. Unclassified, but probably nonconservative.

[b]Mean oceanic residence time in years.

SOURCE: Taylor and McLennan (1985).

References

BROWNLOW, A. H., 1979. *Geochemistry*. Prentice-Hall, NJ, 498 pp.

CARR, M. H., R. S. SAUNDERS, R. G. STROM, and D. E. WILHELMS, 1984. *The Geology of the Terrestrial Planets.* NASA SP-469, 317 pp.

GANAPATHY, R., and E. ANDERS, 1974. Bulk compositions of the Moon and Earth, estimated from meteorites. *Proc. 5th Lunar Sci. Conf.,* 1181–1206.

GLASS, B. P., 1982. *Introduction to Planetary Geology.* Cambridge University Press, Cambridge, England, 469 pp.

HENDERSON, P., 1982. *Inorganic Geochemistry.* Pergamon Press, Oxford, England, 353 pp.

HERRMANN, A. G., 1970. Ytttrium and the lanthanides. In K. H. Wedepohl (Ed.), *Handbook of Geochemistry,* vol. 2, part 5, chs. 39, sec. E, 57-71. Springer-Verlag, Berlin.

LUNAR SAMPLE PRELIMINARY EXAMINATION TEAM, 1972. The Apollo 15 lunar samples: a preliminary description. *Science,* **175**:363–375.

OWEN, T., K. BIEMANN, D. R. RUSHNECK, J. E. BILLER, D. W. HOWARTH, and A. L. LAFLEUR, 1977. The composition of the atmosphere at the surface of Mars. *J. Geophys. Res.,* **82**:4635–4639.

OYAMA, V. I., G. C. CARLE, F. WOELLER, and J. B. POLLACK, 1979. Venus lower atmosphere composition: analysis by gas chromatography. *Science,* **203**:802–805.

PHILPOTTS, J. A., C. C. SCHNETZLER, D. F. NAVA, M. L. BOTTINO, P. D. FULLAGAR, H. H. THOMAS, S. SCHUHMANN, and C. W. KOUNS, 1972. Apollo 14: some geochemical aspects. *Proc. Third Lunar Sci. Conf., Geochim. Cosmochim. Acta, Suppl.,* 3, vol. 2, 1293–1305.

ROSS J. E., and L. H. ALLER, 1976. The chemical composition of the Sun. *Science,* **191**:1223–1229.

TAYLOR, S. R., 1975. *Lunar Science: A Post-Apollo View.* Pergamon Press, 372 pp.

TAYLOR, S. R., and S. M. McLENNAN, 1985. *The Continental Crust: Its Compositon and Evolution.* Blackwell Scientific Publ., Oxford, England, 312 pp.

TOULMIN, P. III, A. K. BAIRD, B. C. CLARK, K. KEIL, H. J. ROSE, Jr., R. P. CHRISTIAN, P. H. EVANS, and W. C. KELLIHER, 1977. Geochemical and mineralogical interpretations of the Viking inorganic chemical results. *J. Geophys. Res.,* **82**:4625–4634.

TUREKIAN, K. K., and K. H. WEDEPOHL, 1961. Distribution of the elements in some major units of the Earth's crust. *Geol. Soc. Amer. Bull.,* **72**:175–192.

VINOGRADOV, A. P., 1962. Average contents of chemical elements in the principal types of igneous rocks of the Earth's crust. *Geochemistry,* **1962**(7):641–664.

Appendix B

STANDARD GIBBS FREE ENERGIES (G_f°) AND STANDARD ENTHALPIES OF FORMATION (H_f°)

The elements are listed in alphabetical order by their chemical symbols, except the REEs that follow La and the transuranium elements that follow U. All charged ionic species are considered to be in aqueous solution. The superscript 0 means neutral molecule in aqueous solution, (g) = gas, (l) = liquid, (c) = crystalline, (amorph.) = amorphous. Compounds identified by their mineral names are crystalline. The conversion from kilicalories to kilojoules is 1 kcal = 4.184 kJ. The standard state is $T = 25\,^\circ\text{C}$, P = 1 atm, activity of ions and molecules = 1.

The data are from compilations of the literature by (1) Woods and Garrels (1987); (2) Weast et al., (1986); (3) Drever (1982); (4) Brookins, (1988); (5) Lindsay (1979); (6) Krauskopf (1979); (7) Garrels and Christ (1965); and (8) Tardy and Garrels (1974) and are expressed in kcal/mol. In cases where several sources are indicated, the stated value is the *average* of the data given by those sources. The calculations in the text are based on data from 2, 3, 4, 5, 6, and 7. Brookins (1988) primarily cited data from Wagman et. al. (1982). The values compiled by Woods and Garrels (1987) were averaged when the range of values is less than about 1% but stated as ranges

when the data vary beyond reasonable estimates of analytical errors. These values are stated here to confirm the older data compilations where possible, to detect significant discrepancies in the values of the thermodynamic parameters, and to suggest the magnitude of these discrepancies by the range of values that have been reported in the literature.

The range of thermodynamic parameters compiled from the literature by Woods and Garrels (1987) reminds us that the values of these parameters are *measurements* that have both random and systematic errors. Therefore, the correct values must be *selected* by a review of the analytical data for other solids and ions based on chemical reactions. Such a critical evaluation of the thermodynamic properties of the rock-forming minerals was not attempted in this book but was made by Helgeson et al. (1978). However, many new measurements have since been reported. When calculations are made for research purposes, the original sources should be consulted to assure the accuracy of the data. The values listed by Weast et al. (1986) are from *Technical Notes* 270-3, 270-4, 270-5, 270-6, 270-7 and 270-8 of the National Bureau of Standards.

Contents

(The elements are listed in alphabetical order by name)

Compound or species	Mineral name	G_f°, kcal/mol	H_f°, kcal/mol
Ag (silver)			
Ag(c)		0.00	0.00
Ag^+		+18.433[2,4–7]	+25.234[2,6,7]
Ag^{2+}		+64.20[5,7]	+64.20[1]
		+64.29[4]	
AgO(c)		+2.61[1]	−6.0[1]
Ag_2O(c)		−2.68[2,4]	−7.42[2]
Ag_2O_2(c)		+6.60[1,4]	−5.81[1]
Ag_2O_3(c)		+29.01[2,4]	+8.1[2]
$AgCl^0$		−12.939[2]	−14.718[2]
		−17.43[1]	−17.47[1]
AgCl	chlorargyrite	−26.244[2,5–7]	−30.37[2,6,7]
$AgCl_2^-$		−51.48[1,4]	−58.56[1]
$AgCl_3^{2-}$		−82.67[1]	−58.56[1]
AgF^0		−48.21[2,5]	−54.27[2]
AgF	fluorargyrite	−50.03[5]	−48.9[2,6]
		−44.2[6]	
AgBr	bromargyrite	−23.1[1]	−24.0[1]
AgI^0		−2.94[5]	
AgI(c)	iodargyrite	−15.87[5,7]	−14.91[7]
			−14.77[1]
$AgNO_3^0$		−8.18[2]	−24.33[2]
$AgNO_3$(c)		−8.00[2]	−29.73[2]
$Ag(NH_3)_2^+$		−4.12[1]	−26.66[1]
$Ag(CN)_2^-$		+72.53[1]	+64.55[1]
$Ag(OH)^0$		−21.84[5]	
$Ag(OH)_2^-$		−62.20[4,5]	
$Ag(OH)_3^{2-}$		−85.91[5]	
$Ag_3(OH)_4^-$		−120.78[5]	
AgOH(c)		−29.69[5]	
		−21.98[1]	
$Ag(OH)_2$(c)		−45.86[5]	
$AgMoO_4$(c)		−196.4[1]	
Ag_2MoO_4(c)		−179.2[1]	−200.5[1]
Ag_2CO_3(c)		−104.47[5,7]	−120.97[7]
$AgHS^0$		+2.19[5]	
$Ag(HS)_2^-$		−3.08[5]	
α-Ag_2S	acanthite	−9.63[2,5,6,7]	−7.73[2,6,7]
		−9.72[4]	
β-Ag_2S(c)	argentite	−9.41[2,5,7]	−7.02[2,7]
Ag_2SO_4(c)		−147.82[2]	−171.10[2]
		−147.53[1]	−170.8[1]
$AgSO_4^-$		−161.30[5]	
$Ag_2SO_4^0$		−141.10[2,5]	−166.85[2]

Compound or species	Mineral name	G_f°, kcal/mol	H_f°, kcal/mol
$Ag(SO_3)_2^{3-}$		-225.4[1]	
$Ag(S_2O_3)_2^{3-}$		-246.5[1]	-295.5[1]
Al (aluminum) (additional silicates are listed with Ca, Mg, Fe)			
$Al(c)$		0.00	0.00
Al^{+3}		-117.33[5]	-127.0[2]
		-116.9[1]	-126.5[1]
		-115.92[4]	
AlO_2^-		-196.7[2]	-219.6[2]
		-198.59[4]	
$Al(OH)^{2+}$		-167.17[5]	-182.8[1]
		-165.89[4]	
		-166.3[1]	
$Al(OH)_2^+$		-218.02[5]	
		-216.9[1]	
$Al(OH)_3^0$		-266.94[5]	
$Al(OH)_4^-$		-312.25[5]	-356.2[6]
		-312.0[1]	-357.0[1]
$Al(OH)_5^{2-}$		-354.06[5]	
$Al_2(OH)_2^{4+}$		-337.54[5]	
$Al(OH)_3(amorph.)$		-271.9[6,7]	-305.0[2]
		-272.4[1]	
α-$Al(OH)_3$	bayerite	-275.78[5]	308.0[2]
		-274.9[1]	
γ-$Al(OH)_3$	gibbsite	-276.1[3,4,5]	-306.25[2]
		-275.9[1]	-308.9[1]
$Al(OH)_3$	nordstrandite	-276.3[5]	
α-$AlOOH$	diaspore	-219.2[2,3,5,7]	-238.4[2,3]
		-219.5[1]	-238.6[1]
γ-$AlOOH$	boehmite	-218.36[2,3,5,6,7]	-236.0[2,3,6,7]
		-218.5[1]	-236.2[1]
α-Al_2O_3	corundum	-378.2[2,3,5,6]	-398.6[2,3,6,7]
		-377.9[1]	-399.3[1]
γ-$Al_2O_3(c)$		-373.38[5]	
$AlPO_4$	berlinite	-387.6[2,5]	-414.4[2]
		-386.1[1]	
$AlPO_4 \cdot 2H_2O$	variscite	-505.97[5]	-562.4[1]
		-501.7[1]	
$AlPO_4(OH)_3$	augelite	-660.9[1]	
$Al_3(PO_4)_2(OH)_3(H_2O)_4 \cdot H_2O$	wavelite	-1330.0[1]	
$H_6K_3Al_5(PO_4)_8 \cdot 18H_2O$	K-taranakite	-4014.76[5]	-4520.5[1]
		-4005.9 to 4162.8[1]	
$H_6(NH_4)_3Al_5(PO_4)_8 \cdot 18H_2O$	NH_4-taranakite	-3864.84[5]	
$AlSO_4^+$		-299.64[5]	

Compound or species	Mineral name	G_f°, kcal/mol	H_f°, kcal/mol
$Al(SO_4)_2^-$		$-475.82^{(5)}$	
$Al_2(SO_4)_3^0$		$-765.94^{(5)}$	
$Al_2(SO_4)_3(c)$		$-740.51^{(2,5)}$	$-822.38^{(2)}$
		$-740.4^{(1)}$	$-822.0^{(1)}$
$Al_2(SO_4)_3 \cdot 6H_2O(c)$		$-1104.38^{(2,5)}$	$-1269.53^{(2)}$
$KAl_3(SO_4)_2(OH)_6$	alunite	$-1111.38^{(5)}$	$-1235.6^{(1)}$
		$-1113.6^{(1)}$	
$NH_4Al(SO_4)_2(c)$		$-487.2^{(2)}$	$-562.2^{(2)}$
$NH_4Al(SO_4)_2 \cdot 12H_2O(c)$		$-1180.21^{(2)}$	$-1420.26^{(2)}$
$KAl(SO_4)_2 \cdot 12H_2O(c)$		$-1228.9^{(2)}$	$-1448.8^{(2)}$
$KAl(SO_4)_2 \cdot 3H_2O(c)$		$-711.0^{(2)}$	$-808.1^{(2)}$
$KAl(SO_4)_2(c)$		$-535.4^{(1,2)}$	$-590.4^{(1,2)}$
$AlF_3(c)$		$-341.6^{(1)}$	$-360.5^{(1)}$
$Al_2SiO_4(OH)_2$	hydroxyl-topaz	$-643.7^{(1)}$	
$Al_2SiO_4F_2$	fluor-topaz	$-695.7^{(1)}$	$-737.2^{(1)}$
Al_2SiO_5	andalusite	$-584.25^{(5)}$	$-562.2^{(2)}$
		$-583.5^{(1)}$	$-618.7^{(1)}$
Al_2SiO_5	kyanite	$-583.38^{(5)}$	$-619.5^{(1)}$
		$-583.8^{(1)}$	
Al_2SiO_5	sillimanite	$-582.81^{(5)}$	$-618.0^{(1)}$
		$-582.9^{(1)}$	
$Al_6Si_2O_{13}$	mullite	$-1541.2^{(2)}$	$-1632.8^{(2)}$
		$-1537.7^{(1)}$	$-1628.9^{(1)}$
$CaAl_2SiO_6$	pyroxene	$-745.13^{(5)}$	$-787.0^{(1)}$
		$-746.1^{(1)}$	
$CaAl_2Si_2O_8$	anorthite	$-960.4^{(3,5)}$	$-1010.1^{(2,3,6)}$
	(triclinic)	-954.3 to $961.5^{(1)}$	-1007.8 to $1014.1^{(1)}$
$CaAl_2Si_2O_8$	Ca-glass	$-946.24^{(5)}$	-993.1 to $-997.0^{(1)}$
		-941.9 to $945.7^{(1)}$	
$CaAl_2Si_4O_{12} \cdot 2H_2O$	wairakite	$-1447.65^{(3)}$	$-1579.55^{(3)}$
		-1477.7 to $1483.0^{(1)}$	-1579.5 to $-1586.4^{(1)}$
$CaAl_2Si_4O_{12} \cdot 4H_2O$	laumontite	$-1597.04^{(3)}$	$-1728.88^{(3)}$
		-1597.0 to $1603.0^{(1)}$	-1729.8 to $1733.0^{(1)}$
$CaAl_2Si_2O_8 \cdot 2H_2O$	lawsonite	$-1081.95^{(5)}$	$-1162.6^{(1)}$
		-1073.6 to $-1082.3^{(1)}$	
$Ca_2Al_4Si_8O_{24} \cdot 7H_2O$	leonhardite	$-3155.37^{(5)}$	(see also Ca)
$CaAl_2Si_3O_{10} \cdot 3H_2O$	scolecite	$-1337.9^{(1)}$	$-1445.7^{(1)}$
$KAlSiO_4$	kaliophilite	$-479.70^{(5)}$	$-507.1^{(1)}$
		$-479.4^{(1)}$	
$KAlSiO_4$	kalsilite	-471.3 to $481.7^{(1)}$	-498.9 to $509.5^{(1)}$
$KAlSi_2O_6$	leucite	$-681.6^{(6)}$	$-721.7^{(6)}$
		-681.8 to $-687.4^{(1)}$	-721.0 to $-726.3^{(1)}$
$KAlSi_3O_8$	K-glass	$-885.34^{(5)}$	
$KAlSi_3O_8$	microcline	$-892.8^{(6)}$	$-946.3^{(6)}$
		$893.6^{(1)}$	$-947.3^{(1)}$

Compound or species	Mineral name	G_f°, kcal/mol	H_f°, kcal/mol
$KAlSi_3O_8$	sanidine	$-892.9^{(1)}$	$-945.2^{(1)}$
$KAlSi_3O_8$	high sanidine	$-894.16^{(5)}$	$-946.5^{(1)}$
		$-893.8^{(1)}$	
$NaAlSiO_4$	nepheline	$-472.8^{(2)}$	$-500.2^{(2)}$
		-463.6 to $-477.0^{(1)}$	-491.1 to $-504.3^{(1)}$
$NaAlSi_3O_8$	Na-glass	$-876.32^{(5)}$	-922.6 to $-926.3^{(1)}$
$NaAlSi_3O_8$	albite	$-884.8^{(1)}$	$-938.1^{(1)}$
$NaAlSi_3O_8$	high albite	$-886.5^{(3,5)}$	$-940.1^{(3)}$
		-882.0 to $-884.5^{(1)}$	-933.9 to $-937.0^{(1)}$
$NaAlSi_3O_8$	low albite	$-844.0^{(6)}$	$-937.1^{(6)}$
		-883.3 to $-887.1^{(1)}$	-936.5 to $940.5^{(1)}$
$NaAlSi_2O_6$	jadeite	$-681.7^{(2)}$	$-724.4^{(2)}$
		-676.6 to $-681.7^{(1)}$	-719.3 to $-724.4^{(1)}$
$NaAlSi_2O_6 \cdot H_2O$	analcime	$-738.16^{(3,5)}$	$-790.63^{(3)}$
		-734.4 to $-738.9^{(1)}$	-786.6 to $-791.1^{(1)}$
$NaAlSi_2O_6$	dehydrated analcime	-671.3 to $-675^{(1)}$	-711.0 to $-714.7^{(1)}$
$NaAlSi_2O_6 \cdot 2H_2O$	analcime	$-733.3^{(1)}$	$-785.9^{(1)}$
$Na_2Al_3Si_3O_{10} \cdot 2H_2O$	natrolite	$-1270.7^{(1)}$	$-1366.7^{(1)}$
$4NaAlSi_3O_8 \cdot CaAl_2Si_2O_8$	oligoclase	$-897.3^{(1)}$	$-950.6^{(1)}$
$3NaAlSi_3O_8 \cdot 2CaAl_2Si_2O_8$	andesine	$-912.6^{(1)}$	$-966.0^{(1)}$
$2.5NaAlSi_3O_8 \cdot 2.5CaAl_2Si_2O_8$	labradorite	$-919.9^{(1)}$	$-973.3^{(1)}$
$Na_2Mg_3Al_2Si_8O_{22}(OH)_2$	glaucophane	$-2710.1^{(1)}$	$-2885.4^{(1)}$
$NaCa_2Mg_4Al_3Si_6O_{22}(OH)_2$	pargasite	$-2848.8^{(1)}$	$-3021.7^{(1)}$
$NaCa_2Fe_4^{2+}Al_3Si_6O_{22}(OH)_2$	ferropargasite	$-2518.8^{(1)}$	$-2685.8^{(1)}$
$Al_2Si_2O_5(OH)_4$	dickite	$-908.55^{(5)}$	$-984.1^{(1)}$
		$-906.2^{(1)}$	
$Al_2Si_2O_5(OH)_4$	halloysite	$-904.2^{(3,5)}$	$-980.28^{(3)}$
		$-902.3^{(1)}$	$-979.9^{(1)}$
$Al_2Si_2O_5(OH)_4$	kaolinite	$-906.84^{(3)}$	$-983.5^{(3)}$
		$-905.7^{(1)}$	$-983.5^{(1)}$
$Al_2Si_4O_{10}(OH)_2$	pyrophyllite	$-1256.00^{(3)}$	$-1345.31^{(3)}$
		$-1258.3^{(1)}$	$-1347.4^{(1)}$
$KAl_2(AlSi_3O_{10})(OH)_2$	muscovite	$-1337.45^{(3)}$	$-1427.9^{(3)}$
		-1330.0 to $-1340.6^{(1)}$	-1421.4 to $-1430.3^{(1)}$
$CaAl_4Si_2O_{10}(OH)_2$	margarite	$-1398.0^{(1)}$	$-1489.8^{(1)}$
$KMg_{0.5}Al_2Si_{3.5}O_{10}(OH)_2$	phengite	$-1321.5^{(1)}$	$-1412.6^{(1)}$
$K_{0.6}Mg_{0.25}Al_{2.3}Si_{3.5}O_{10}(OH)_2$	illite	$-1307.49^{(8)}$	
$KMg_3AlSi_3O_{10}(OH)_2$	phlogopite	$-1396.19^{(3)}$	$-1488.07^{(3)}$
		-1389.5 to $-1410.9^{(1)}$	-1412.6 to $-1568.5^{(1)}$
$KMg_3(AlSi_3O_{10})F_2$	fluorphlogopite	$-1461.63^{(5)}$	-1522.8 to $-1527.9^{(1)}$
		-1440.3 to $-1446.7^{(1)}$	
$KFe_3AlSi_3O_{10}(OH)_2$	annite	$-1147.16^{(3)}$	$-1232.20^{(3)}$
		-1145.8 to $1159.2^{(1)}$	-1232.2 to $-1301.1^{(1)}$

Compound or species	Mineral name	G_f°, kcal/mol	H_f°, kcal/mol
$NaAl_3Si_3O_{10}(OH)_2$	paragonite	$-1327.42^{(3,5)}$	$-1416.96^{(3)}$
$NaSi_7O_{13}(OH)\cdot 3H_2O$	magadiite	$-1762.2^{(1)}$	
$NaSi_7O_{13}(OH)_3$	magadiite	$-1589.85^{(3)}$	
$K_{0.33}Al_{2.33}Si_{3.67}O_{10}(OH)_2$	K-montmorillonite	$-1279.9^{(1)}$	$-1369.3^{(1)}$
$Na_{0.33}Al_{2.33}Si_{3.67}O_{10}(OH)_2$	beidellite	$-1277.76^{(3)}$	
$Ca_{0.167}Al_{2.33}Si_{3.67}O_{10}(OH)_2$	Ca-montmorillonite	$-1279.24^{(1,2)}$	$-1368.4^{(1)}$
$Mg_2Al_4Si_5O_{18}$	Mg-cordierite	$-2075.26^{(5)}$	
$Mg_{0.2}(Si_{3.81}Al_{1.71}Fe_{0.22}^{3+}Mg_{0.29})$ $O_{10}(OH)_2$	Mg-montmorillonite	$-1258.84^{(5)}$	
$(Al_{1.29}Mg_{0.445}Fe_{0.335}^{3+})$ $(Si_{3.82}Al_{0.18})O_{10}(OH)_2$	Aberdeen montmorillonite (charges unbalanced)	$-1227.7^{(1)}$	
$(Al_{1.515}Mg_{0.29}Fe_{0.225}^{3+})$ $(Si_{3.935}Al_{0.065})O_{10}(OH)_2$	Spur montmorillonite (charges unbalanced)	$-1237.1^{(1)}$	
$(Mg_{2.71}Fe_{0.02}^{2+}Fe_{0.46}^{3+}Ca_{0.06}K_{0.1})$ $(Si_{2.91}Al_{1.14})O_{10}(OH)_2$	vermiculite	$-1324.38^{(5)}$	
$Mg_5Al_2Si_3O_{10}(OH)_8$	chlorite (clinochlore)	$-1961.70^{(3)}$ $-1975.56^{(5)}$	
As (arsenic) (see also Fe)			
α-As (grey, metallic)		0.00	0.00
As(g)		$+50.7$ to $+62.4^{(1)}$	$+60.6$ to $+72.3^{(1)}$
$As_2(g)$		$+17.5^{(7)}$	$+29.6^{(7)}$
		$+17.5$ to $+41.1^{(1)}$	$+29.6$ to $+53.1^{(1)}$
$As_4(g)$		$+20.9$ to $+25.2^{(1)}$	$+34.3$ to $+35.1^{(1)}$
AsO^+		$-39.15^{(2)}$	
		$-38.3^{(1)}$	
AsO_2^-		$-83.66^{(1,2)}$	$-102.54^{(1,2)}$
AsO_4^{3-}		$-155.0^{(2,6)}$	$-212.27^{(2,6)}$
		$-154.97^{(4)}$	$-211.7^{(1)}$
		$-154.7^{(1)}$	
$HAsO_4^{2-}$		$-170.82^{(2)}$	$-216.62^{(2)}$
		$-170.69^{(4)}$	$-216.7^{(1)}$
		$-170.7^{(1)}$	
$H_2AsO_4^-$		$-180.04^{(2)}$	$-217.39^{(2)}$
		$-180.01^{(4)}$	$-217.6^{(1)}$
		$-180.2^{(1)}$	
$H_3AsO_4^0$		$-183.1^{(6)}$	$-216.2^{(2)}$
		$-183.06^{(4)}$	$-216.0^{(1)}$
		$-183.7^{(1)}$	
$H_3AsO_3^0$		$-152.9^{(4,6,7)}$	$-177.4^{(6,7)}$
		$-153.6^{(1)}$	$-178.1^{(1)}$
$HAsO_3^{2-}$		$-125.31^{(4)}$	
$H_2AsO_3^-$		$-140.35^{(2,4)}$	$-170.84^{(2)}$
		-125.3 to $141.8^{(1)}$	-164.7 to $-172.3^{(1)}$

Compound or species	Mineral name	G_f°, kcal/mol	H_f°, kcal/mol
AsO_3^{3-}		$-107.00^{(4)}$	
$HAsO_2^0$		$-96.25^{(2)}$	$-109.1^{(2)}$
As_2O_3, α	arsenolite, (octahedral)	$-138.4^{(1)}$	$-157.6^{(1)}$
As_2O_3, β	claudetite, (monoclinic)	$-137.9^{(6)}$	$-156.5^{(6)}$
		$-137.66^{(4)}$	
		$-138.7^{(1)}$	$-157.2^{(1)}$
$As_2O_5(c)$		$-187.0^{(2)}$	$-221.05^{(2)}$
		$-186.2^{(1)}$	$-220.2^{(1)}$
$As_2O_5 \cdot 4H_2O$		$-411.1^{(7)}$	$-503.0^{(2)}$
		$-411.1^{(1)}$	$-501.6^{(1)}$
AsS	realgar	$-16.80^{(4)}$	
As_2S_2		$-32.15^{(7)}$	$-34.1^{(2)}$
		$-33.1^{(1)}$	$-33.6^{(1)}$
As_2S_3	orpiment	$-40.3^{(2,4)}$	$-40.4^{(2)}$
		-22.8 to $-40.3^{(1)}$	-23.0 to $-40.4^{(1)}$
Au (gold)			
$Au(c)$		0.00	0.00
Au^+		$+39.0^{(6,7)}$	$+53.1^{(1)}$
		$+41.4^{(1)}$	
Au^{3+}		$+103.6^{(1,6,7)}$	
$AuCl_2^-$		$-36.13^{(2,4)}$	$-41.7^{(1)}$
		$-36.1^{(1)}$	
$AuCl_4^-$		$-56.22^{(2)}$	$-77.0^{(2)}$
		$-56.2^{(1,4)}$	$-77.4^{(1)}$
$HAuCl_4^0$		$-56.22^{(2,4)}$	$-77.0^{(2)}$
$AuCl(c)$		$-3.6^{(6)}$	$-8.3^{(2)}$
$AuCl_3(c)$		$-10.8^{(6)}$	$-28.1^{(2)}$
$AuO_2(c)$		$+48.0^{(1,7)}$	
$Au(CN)_2^-$		$-68.31^{(4)}$	
$Au(SCN)_2^-$		$-60.21^{(4)}$	
$Au_2O_3(c)$		$+39.0^{(1,7)}$	$+19.3^{(1,7)}$
$H_3AuO_3^0$		$-61.8^{(1,7)}$	
$H_2AuO_3^-$		$-52.2^{(2,4)}$	
		-45.8 to $-61.8^{(1)}$	
$HAuO_3^{2-}$		$-34.0^{(2,4)}$	
		-27.6 to $-34.0^{(1)}$	
AuO_3^{3-}		$-12.4^{(2,4)}$	
		-5.8 to $-12.4^{(1)}$	
$Au(OH)_3(c)$		$-75.77^{(2,4)}$	$-101.5^{(2)}$
		-69.3 to $83.4^{(1)}$	-100.0 to $-114.3^{(1)}$
$Au(OH)_3^0$		-67.7 to $75.9^{(1)}$	

Compound or species	Mineral name	G_f°, kcal/mol	H_f°, kcal/mol
B (boron) (see also Na)			
B(c), β		0.00	0.00
B_2O_3 (glass)		$-282.6^{(2)}$	$-299.84^{(2)}$
		$-281.8^{(1)}$	$-299.1^{(1)}$
B_2O_3(c)		$-285.30^{(2,4)}$	$-304.20^{(2)}$
		$-284.7^{(1)}$	$-303.7^{(1)}$
H_3BO_3(c)		$-231.60^{(2)}$	$-261.55^{(2)}$
		$-230.9^{(1)}$	$-260.9^{(1)}$
$H_3BO_3^0$		$-231.56^{(2,4)}$	$-256.29^{(2)}$
		$-231.2^{(1)}$	$-256.1^{(1)}$
$H_2BO_3^-$		$-271.60^{(1,4,7)}$	$-251.8^{(1,7)}$
HBO_3^{2-}		$-200.2^{(7)}$	
		$-200.3^{(1,4)}$	
BO_3^{3-}		$-181.48^{(7,4)}$	
		$-181.5^{(1)}$	
$B(OH)_4^-$		$-275.65^{(2)}$	$-321.23^{(2)}$
		$-275.61^{(4)}$	$-321.1^{(1)}$
		$-275.7^{(1)}$	
BO_2^-		$-162.27^{(2,4)}$	$-184.60^{(2)}$
		-162.3 to $-181.5^{(1)}$	$-184.1^{(1)}$
$HB_4O_7^-$		$-628.3^{(1)}$	
$B_4O_7^{2-}$		$-622.56^{(4)}$	
		-615.0 to $-622.6^{(1)}$	
BF_4^-		$-352.8^{(1)}$	$-372.5^{(1)}$
BH_4^-		$+27.33^{(4)}$	
Ba (barium)			
Ba(c)		0.00	0.00
Ba^{2+}		$-134.02^{(2,4)}$	$-128.50^{(2)}$
		$-132.7^{(1)}$	$-127.2^{(1)}$
BaO(c)		$-132.0^{(6)}$	$-139.1^{(6)}$
		$-124.5^{(1)}$	$-131.3^{(1)}$
		$-125.50^{(4)}$	
BaO_2(c)		$-135.8^{(1,7)}$	$-151.6^{(2)}$
			$-151.0^{(1)}$
$BaO_2 \cdot H_2O$(c)		$-195.0^{(1,7)}$	$-222.3^{(2)}$
			$-222.9^{(1)}$
BaF_2(c)		$-276.5^{(2)}$	$-288.5^{(2)}$
		$-273.9^{(1)}$	$-286.5^{(1)}$
BaF_2^0		$-267.30^{(2)}$	$-287.50^{(2)}$
BaS(c)		$-109^{(2)}$	$-110^{(2)}$
		-104.5 to $-109.0^{(1)}$	-106.0 to $-109.9^{(1)}$
$BaSO_4$	barite	$-325.6^{(2,4)}$	$-352.1^{(2)}$
		$-324.2^{(1)}$	$-351.0^{(1)}$

Compound or species	Mineral name	G_f°, kcal/mol	H_f°, kcal/mol
$BaSO_4^0$		$-311.99^{(2)}$	$-345.82^{(2)}$
$BaCO_3$	witherite	$-271.9^{(2,4)}$	$290.7^{(2)}$
		$-272.3^{(1)}$	$291.2^{(1)}$
$BaCO_3^0$		$-260.19^{(2)}$	$-290.34^{(2)}$
$BaSiO_3(c)$		$-368.1^{(6)}$	$-388.7^{(6)}$
		-338.7 to $-369.1^{(1)}$	-359.5 to $-389.2^{(1)}$
$Ba_2SiO_4(c)$		$-519.8^{(2)}$	$-546.8^{(2)}$
		$-518.1^{(1)}$	$-544.8^{(1)}$
$Ba(OH)_2(c)$		$-204.7^{(7)}$	$-225.8^{(2)}$
		$-204.8^{(1)}$	$-226.2^{(1)}$
$Ba(OH)_2 \cdot 8H_2O(c)$		$-667.6^{(2)}$	$-798.8^{(2)}$
		$-667.1^{(1)}$	$-799.1^{(1)}$
$Ba_3(PO_4)_2(c)$		$-944.4^{(1,7)}$	$-998.0^{(7)}$
			$-998.1^{(1)}$
$BaCl_2(c)$		$-193.7^{(2)}$	$-205.2^{(2)}$
		$-192.7^{(1)}$	$-204.2^{(1)}$
$BaCl_2^0$		$-196.76^{(2)}$	$-208.40^{(2)}$
$Ba(NO_3)_2$	nitrobarite	$-190.42^{(2)}$	$-237.11^{(2)}$
		$-189.5^{(1)}$	$-236.3^{(1)}$
$Ba(NO_3)_2^0$		$-187.24^{(2)}$	$-227.62^{(2)}$
Be (beryllium)			
$Be(c)$		0.00	0.00
Be^{2+}		$-90.75^{(2,4)}$	$-91.5^{(2)}$
		-85.2 to $-91.1^{(1)}$	-91.5 to $-96.5^{(1)}$
$BeO(c)$	bromellite	$-138.7^{(2,4)}$	$-145.7^{(2)}$
		-136.1 to -139.01	-143.1 to $-146.0^{(1)}$
BeO_2^{2-}		$-153.0^{(2,4)}$	$-189.0^{(2)}$
		-153.0 to $-155.3^{(1)}$	-187.8 to $-189.0^{(1)}$
$Be_2O_3^{2-}$		$-298.0^{(1)}$	
$BeCl_2, \alpha(c)$		-106.5 to $-118.9^{(1)}$	-117.2 to $-122.3^{(1)}$
$BeCl_2, \beta(c)$		$-107.3^{(1)}$	-115.7 to $-118.5^{(1)}$
$BeAl_2O_4$	chrysoberyl	$-520.7^{(2)}$	$-549.9^{(2)}$
		$-520.6^{(1)}$	$-549.8^{(1)}$
Be_2SiO_4	phenacite	-485.3 to $-488.6^{(1)}$	-512.6 to $-515.9^{(1)}$
$BeSiO_3(c)$		$-347.5^{(1)}$	$-368.0^{(1)}$
$Be(OH)_2, \alpha(c)$		$-194.8^{(2,4)}$	$-215.7^{(2)}$
		-194.8 to $-196.2^{(1)}$	-215.7 to $-216.8^{(1)}$
$Be(OH)_2, \beta(c)$		$-195.4^{(2)}$	$-216.5^{(2)}$
		$-195.5^{(1)}$	$-216.4^{(1)}$
$Be_3 \cdot (OH)_3^{3+}$		$-430.6^{(2)}$	
		$-431.0^{(1)}$	
$BeS(c)$		$-55.9^{(1,7)}$	$-56.0^{(1,2)}$
$BeSO_4, \alpha(c)$		$-261.44^{(2)}$	$-288.05^{(2)}$
		$-260.6^{(1)}$	-286.0 to $-288.1^{(1)}$

Compound or species	Mineral name	G_f^o, kcal/mol	H_f^o, kcal/mol
$BeSO_4^0$		$-268.72^{(2)}$	$-308.8^{(2)}$
$BeSO_4 \cdot 2H_2O(c)$		$-381.99^{(2)}$	$-435.74^{(2)}$
$BeSO_4 \cdot 4H_2O(c)$		$-497.29^{(2)}$	$-579.29^{(2)}$
Bi (bismuth)			
$Bi(c)$		0.00	0.00
$Bi_2(g)$		$+48.0^{(1,7)}$	$+52.5^{(2)}$
			$+52.5$ to $+59.4^{(1)}$
Bi^{3+}		$+19.8^{(2,4)}$	$+19.3^{(1)}$
		$+14.8$ to $+21.9^{(1)}$	
BiO^+		$-34.8^{(1)}$	
		$-34.99^{(4)}$	
$Bi(OH)^{2+}$		$-39.13^{(7)}$	
		$-34.99^{(4)}$	
		-32.6 to $-39.13^{(1)}$	
$BiO(c)$		$-43.5^{(1,7)}$	$-49.85^{(1,7)}$
$Bi_2O_3(c)$	bismite	$-118.0^{(2)}$	$-137.16^{(2)}$
		$-118.2^{(1)}$	$-137.4^{(1)}$
$Bi_2O_4(c)$		$-109.0^{(1,7)}$	
$Bi_2O_5(c)$		$-91.57^{(1,7)}$	
$Bi_4O_7(c)$		$-232.75^{(1,4)}$	
$Bi(OH)_3(amorph.)$		$-137.0^{(1,7)}$	$-169.6^{(1,7)}$
$Bi_6O_6^{6+}$		$-221.80^{(4)}$	
$Bi(OH)_3(c)$		$-139.3^{(1)}$	$-170.8^{(1)}$
$BiCl_3(c)$		$-76.23^{(7)}$	$-90.61^{(7,1)}$
		-74.8 to $-76.2^{(1)}$	
$BiOCl$	bismoclite	$-77.6^{(1)}$	$-87.8^{(1)}$
Bi_2S_3	bismuthinite	$-33.6^{(2,4)}$	$-34.2^{(2)}$
		-33.6 to $-39.4^{(1)}$	-34.2 to $-43.8^{(1)}$
Br (bromine)			
$Br_2(l)$		0.00	0.00
Br^-		$-24.85^{(2)}$	$-25.05^{(2)}$
		$-24.8^{(1)}$	$-29.0^{(1)}$
$HBr(g)$		$-12.77^{(2)}$	$-8.70^{(1,2)}$
		$-12.7^{(1)}$	
HBr^0		$-24.85^{(2)}$	$-29.05^{(2)}$
$Br_2(g)$		$+0.751^{(5,7)}$	$+7.34^{(7)}$
		$+0.74^{(1)}$	$+7.37^{(1)}$
Br_2^0		$+0.977^{(7)}$	$-1.1^{(7)}$
		$+0.96^{(1)}$	-0.19 to $-1.10^{(1)}$
C (carbon)			
C	graphite	0.00	0.00
C	diamond	$+0.693^{(1,2)}$	$+0.454^{(1,2)}$
$CO(g)$	carbon monoxide	$-32.780^{(1,2)}$	$-26.416^{(1,2)}$

Compound or species	Mineral name	G_f°, kcal/mol	H_f°, kcal/mol
CO^0		$-28.66^{(2)}$	$-28.91^{(2)}$
$CO_2(g)$	carbon dioxide	$-94.254^{(2)}$	$-94.051^{(2)}$
		$-92.27^{(1)}$	$-98.85^{(1)}$
CO_2^0		$-92.26^{(2)}$	$-98.90^{(2)}$
HCN^0		$+28.6^{(1)}$	$+25.6^{(1)}$
CN^-		$-41.2^{(1)}$	$+36.0^{(1)}$
CH_2O^0		$-31.00^{(7)}$	
CO_3^{2-}		$-126.17^{(1,2,4)}$	$-161.8^{(1,2)}$
HCO_3^-		$-140.26^{(1,2,4)}$	$-165.39^{(1,2)}$
$H_2CO_3^0$	carbonic acid	$-148.94^{(1,2)}$	$-167.22^{(1,2)}$
		$-149.00^{(4)}$	
$CH_4(g)$	methane	$-12.13^{(1,2)}$	$-17.88^{(1,2)}$
CH_4^0		$-8.22^{(1,2)}$	$-21.28^{(1,2)}$
		$-8.28^{(4)}$	
CH_3COO^-		$-88.29^{(1,2)}$	$-116.16^{(1,2)}$
$CH_3COOH(l)$		$-93.2^{(2)}$	$-115.8^{(2)}$
CH_3COOH^0	acetic acid	$-94.8^{(2)}$	$-116.10^{(2)}$
$C_2O_4^{2-}$		$-161.1^{(5)}$	$-195.7^{(7)}$
		$-159.5^{(1)}$	-193.2 to $-195.6^{(1)}$
$HC_2O_4^-$		$-166.93^{(5)}$	$-195.7^{(1,7)}$
		$-165.1^{(1)}$	
$H_2C_2O_4^0$	oxalic acid	$-166.8^{(1)}$	$-197.2^{(2)}$
			$-195.6^{(1)}$
Ca (calcium)			
$Ca(c)$		0.00	0.00
Ca^{2+}		$-132.30^{(2,4)}$	$-129.74^{(1,2)}$
		$-132.2^{(1)}$	
$CaHCO_3^+$		$-274.33^{(3,5)}$	
		$-273.7^{(1)}$	
$CaCO_3^0$		$-258.47^{(2)}$	$-291.58^{(2)}$
		$-262.7^{(1)}$	
$CaCO_3$	aragonite	$-269.55^{(2,4)}$	$-288.51^{(2)}$
		$-269.7^{(1)}$	$-288.6^{(1)}$
$CaCO_3$	calcite	$-269.80^{(2,4)}$	$-288.46^{(2)}$
		$-269.9^{(1)}$	$-288.6^{(1)}$
$CaCO_3$	vaterite (hexag.)	$-269.0^{(1)}$	
$CaCO_3 \cdot H_2O(c)$		$-325.4^{(1)}$	$-258.1^{(1)}$
$CaMg(CO_3)_2$	dolomite	$-517.1^{(2)}$	$-556.0^{(2)}$
		-514.3 to $-520.5^{(1)}$	-553.2 to $-557.5^{(1)}$
$CaMg_3(CO_3)_4$	huntite	$-1007.7^{(7)}$	$-1082.6^{(1)}$
		-1004.6 to $-1007.7^{(1)}$	
$CaBa(CO_3)_2$	alstonite	$-543.0^{(1,7)}$	
$CaBa(CO_3)_2$	barytocalcite	$-542.9^{(1,7)}$	

Compound or species	Mineral name	G_f°, kcal/mol	H_f°, kcal/mol
$CaMn(CO_3)_2$	kutnahorite	$-466.2^{(1,7)}$	
$CaCl^+$		$-162.53^{(5)}$	
$CaCl_2^0$		$-195.04^{(2)}$	$-209.64^{(2)}$
$CaCl_2$	hydrophilite	$-178.8^{(2)}$	$-190.2^{(2)}$
		$-179.0^{(1)}$	$-190.0^{(1)}$
CaF_2	fluorite	$-279.0^{(2)}$	$-291.5^{(2)}$
		-277.7 to $-281.3^{(1)}$	-290.3 to $-293.8^{(1)}$
CaF_2^0		$-265.58^{(2)}$	$-288.74^{(2)}$
$CaMoO_4$	powellite	$-342.9^{(2)}$	$-368.4^{(2)}$
		-342.9 to $-346.6^{(1)}$	-368.4 to $-372.3^{(1)}$
$CaMoO_4^0$		$-332.2^{(2)}$	$-368.2^{(2)}$
$CaWO_4$	scheelite	$-367.1^{(2)}$	$-393.2^{(2)}$
		-357.0 to $-367.7^{(1)}$	-381.1 to $-393.2^{(1)}$
$CaNO_3^+$		$-152.61^{(5)}$	
$Ca(NO_3)_2^0$		$-185.52^{(2)}$	$-228.86^{(2)}$
$Ca(NO_3)_2(c)$		$-177.63^{(1,2)}(c)$	$-224.28^{(1,2)}$
$CaOH^+$		$-171.7^{(2)}$	$-182.7^{(1)}$
		$-171.4^{(1)}$	
$Ca(OH)_2^0$		$-207.49^{(2)}$	$-239.68^{(2)}$
$CaO(c)$	lime	$-144.37^{(1,2,4)}$	$-151.79^{(2)}$
			$-151.8^{(1)}$
$Ca(OH)_2$	portlandite	$-214.76^{(2,4)}$	$-235.68^{(2)}$
		$-214.5^{(1)}$	$-235.6^{(1)}$
$CaPO_4^-$		$-386.51^{(5)}$	
$CaHPO_4^0$		$-398.29^{(5)}$	
$CaH_2PO_4^+$		$-406.28^{(5)}$	
$CaHPO_4^0$		$-392.64^{(2)}$	$-438.57^{(2)}$
$CaHPO_4$	monetite	$-401.5^{(2)}$	$-433.65^{(2)}$
		$-401.6^{(1)}$	-433.4 to $-435.2^{(1)}$
$CaHPO_4 \cdot 2H_2O$	brushite	$-515.00^{(2)}$	$-574.47^{(2)}$
		$-514.8^{(1)}$	-574.3 to $-576.0^{(1)}$
$Ca(H_2PO_4)_2(amorph.)$		$-672^{(1,7)}$	$-742.04^{(2)}$
			$-744.4^{(1)}$
$Ca(H_2PO_4)_2(c)$		$-675.1^{(1)}$	-741.9 to $-744.2^{(1)}$
$Ca(H_2PO_4)_2^0$		$-672.64^{(2)}$	$-749.38^{(2)}$
$Ca(H_2PO_4)_2 \cdot H_2O(c)$		$-730.98^{(2)}$	$-814.93^{(2)}$
$\alpha\text{-}Ca_3(PO_4)_2(c)$	high temp.	$-926.3^{(2)}$	$-982.3^{(2)}$
		-925.0 to $-929.7^{(1)}$	-981.0 to $-986.2^{(1)}$
$\beta\text{-}Ca_3(PO_4)_2(c)$		$-928.5^{(2)}$	$-984.9^{(2)}$
		-928.5 to $-932.0^{(1)}$	-984.9 to $-988.9^{(1)}$
$Ca_3(PO_4)_2^0$		$-883.9^{(2)}$	$-999.8^{(2)}$
$Ca_8H_2(PO_4)_6 \cdot 5H_2O$	octacalcium phosphate	$-2942.62^{(3)}$	

Compound or species	Mineral name	G_f°, kcal/mol	H_f°, kcal/mol
$Ca_{10}F_2(PO_4)_6$	fluorapatite	-3094.73[5]	-3287.4[1]
		-3108.2[1]	
$Ca_{10}(OH)_2(PO_4)_6$	hydroxyapatite	-3030.0[3,5]	-3211.0[3]
		-3030.4[1]	-3218.9[1]
$CaAl_3(PO_4)_2(OH)_5 \cdot H_2O$	crandallite	-1336.7[1]	
CaS	oldhamite	-114.1[2]	-115.3[2]
		-112.3 to -114.1[1]	-113.5 to -115.3[1]
$CaSO_4^0$		-310.27[2]	-347.06[2]
		-313.0[1]	
$CaSO_4$	anhydrite	-315.93[1,2,4]	-342.76[1,2]
$CaSO_4 \cdot \frac{1}{2}H_2O, \alpha$	bassanite	-343.41[2]	-376.85[2]
		-343.0 to -349.3[1]	-376.6[1]
$CaSO_4 \cdot \frac{1}{2}H_2O, \beta(c)$		-342.8 to -349.0[1]	-376.1[1]
$CaSO_4 \cdot 2H_2O$	gypsum	-429.60[2]	-483.42[2]
		-429.56[4]	-483.3[1]
		-429.5[1]	
α-$CaSO_4$(soluble)(c)		-313.93[2]	-340.64[2]
		-313.7[1]	-340.4[1]
β-$CaSO_4$(soluble)(c)		-312.87[2]	-339.58[2]
		-312.7[1]	-339.4[1]
$CaSiO_3$	wollastonite	-358.2 to -370.5[1]	-378.6 to -390.8[1]
$CaSiO_3$	α-wollastonite	-369.7[1]	-389.7[1]
$CaSiO_3$	β-wollastonite	-370.39[2]	-390.76[2]
		-370.6[1]	-390.9[1]
$CaSiO_3$	pseudowollastonite	-369.2[2]	-389.2[2]
		-357.4 to -370.0[1]	-377.4 to -390.0[1]
Ca_2SiO_4	larnite	-524.1[2]	-551.5[2]
		-512.7 to -524.3[1]	-538.0 to -551.7[1]
Ca_2SiO_4	Ca-olivine	-526.1[2]	-554.0[2]
		-525.8[1]	-553.7[1]
Ca_2SiO_4, γ	bredigite	-526.1[1]	-554.0[1]
$Ca_3Si_2O_7$	rankinite	-899.0[2]	-946.7[2]
		-890.8 to -902.3[1]	-938.7 to -952.5[1]
$CaMg(SiO_3)_2$	diopside	-724.00[3]	-765.60[3]
		-724.6[1]	-766.3[1]
$Ca_2Mg_5Si_8O_{22}(OH)_2$	tremolite	-2780[2]	$-2954.$[2]
		-2776.7[1]	-2951.5[1]
$Ca_2Fe_5Si_8O_{22}(OH)_2$	actinolite	-2352.1 to -2625.5[1]	-2515.2[1]
Ca_3SiO_5	hatrurite	-665.6[1]	-700.6[1]
$Ca_3Fe_2Si_3O_{12}$	andradite	-1294.4[1]	-1376.8[1]
$Ca_3Al_2Si_3O_{12}$	grossular	-1500.2[1]	-1583.8[1]
$Fe_3Al_2Si_3O_{12}$	almandine	see Fe	
$Ca_2Al_2SiO_7$	gehlenite	-906.1[1]	-953.4[1]
$Ca_2MgSi_2O_7$	akermanite	-879.1[1]	-926.3[1]

Compound or species	Mineral name	G_f°, kcal/mol	H_f°, kcal/mol
$Ca_2Al_3Si_3O_{12}(OH)$	clinozoisite	$-1549.7^{(1)}$	$-1644.2^{(1)}$
$Ca_2Al_3Si_3O_{12}(OH)$	zoisite	$-1293.7^{(1)}$	$-1644.2^{(1)}$
$Ca_2Fe^{3+}Al_2Si_3O_{12}(OH)$	epidote	$-1451.3^{(1)}$	$-1544.4^{(1)}$
$Ca_2Fe_3^{3+}Si_3O_{12}(OH)$	epidote	$-1244.2^{(1)}$	
$Ca_2Al_2Si_3O_{10}(OH)_2$	prehnite	$-1390.8^{(1)}$	$-1480.2^{(1)}$
$Ca_2Al_2SiO_6(OH)_2$	bicchucite	$-973.0^{(1)}$	$-1036.6^{(1)}$
$Ca_2Al_4Si_8O_{24}\cdot7H_2O$	leonhardite (see Al)	$-3150.1^{(1)}$	$-3400.9^{(1)}$
$CaAl_2Si_7O_{18}\cdot6H_2O$	heulandite	$-2331.4^{(1)}$	$-2527.6^{(1)}$
$CaFeSi_2O_6$	hedenbergite	$-639.6^{(1)}$	$-678.9^{(1)}$
$CaMgSiO_4$	monticellite	$-512.2^{(1)}$	$-540.6^{(1)}$
$Ca_3Mg(SiO_4)_2$	merwinite	$-829.0^{(1)}$	$-1090.8^{(1)}$
$Ca_4MgAl_5(H_2O)Si_6O_{23}(OH)_3$	pumpellyite	$-3194.0^{(1)}$	$-3403.4^{(1)}$
$CaTiO_3$	perovskite	$-376.6^{(1)}$	$-397.0^{(1)}$
$CaTiSiO_5$	sphene	$-586.7^{(1)}$	$-619.0^{(1)}$
$Ca_5(SiO_4)_2(CO_3)$	spurrite	$-1331.0^{(1)}$	$-1410.1^{(1)}$
$Ca_5Si_2O_7(CO_3)_2$	tilleyite	$-1436.4^{(1)}$	$-1522.8^{(1)}$
$Ca_4Al_6Si_6O_{24}(CO_3)$	meiyonite	$-3141.2^{(1)}$	$-3321.6^{(1)}$
Cd (cadmium)			
α-Cd(c)		0.00	0.00
Cd^{2+}		$-18.6^{(1,2)}$	$-18.0^{(1,2)}$
		$-18.55^{(4)}$	
$HCdO_2^-$		$-86.7^{(1)}$	
CdO_2^{2-}		$-67.97^{(4)}$	
$CdCO_3^0$		$-150.38^{(5)}$	
$CdHCO_3^+$		$-161.74^{(5)}$	
$CdCO_3$	otavite	$-160.0^{(2,4)}$	$-179.4^{(1,2)}$
		$-160.3^{(1)}$	
$CdCl^+$		$-52.68^{(5)}$	$-57.5^{(1)}$
		-51.8 to $-53.6^{(1)}$	
$CdBr_2(c)$		$-70.6^{(1)}$	$-75.4^{(1)}$
$CdCl_2(c)$		$-82.21^{(2)}$	$-93.57^{(2)}$
		$-82.1^{(1)}$	$-93.4^{(1)}$
$CdCl_2^0$		$-81.286^{(2)}$	$-98.04^{(2)}$
		-84.3 to $-85.9^{(1)}$	$-96.8^{(1)}$
$CdCl_3^-$		$-116.00^{(1,5,7)}$	$-134.0^{(1)}$
$CdCl_4^{2-}$		$-147.50^{(5)}$	
$CdF_2(c)$		$-154.8^{(2)}$	$-167.4^{(2)}$
		$-155.1^{(1)}$	$-166.6^{(1)}$
CdF_2^0		$-151.82^{(2)}$	$-177.14^{(2)}$
$CdNO_3^+$		$-45.67^{(5)}$	
$Cd(NO_3)_2^0$		$-71.76^{(2)}$	$-117.26^{(2)}$
$Cd(OH)^+$		$-61.52^{(1)}$	
		-62.4 to $-64.9^{(1)}$	

Compound or species	Mineral name	G_f°, kcal/mol	H_f°, kcal/mol
$Cd(OH)_2^0$		$-93.73^{(2)}$	$-128.08^{(2)}$
		-105.8 to $-106.7^{(1)}$	
$Cd(OH)_3^-$		$-143.64^{(5)}$	
		-143.6 to $-144.6^{(1)}$	
$Cd(OH)_4^{2-}$		$-180.85^{(5)}$	
		$-181.5^{(1)}$	
$Cd(OH)_5^{3-}$		$-217.57^{(5)}$	
$Cd(OH)_6^{4-}$		$-253.96^{(5)}$	
Cd_2OH^{3+}		$-85.18^{(5)}$	
$Cd_4(OH)_4^{4+}$		$-263.10^{(5)}$	
CdO	monteponite	$-54.6^{(2,4)}$	$-61.7^{(2)}$
		$-54.4^{(1)}$	$-61.5^{(1)}$
$Cd(OH)_2(c)$		$-113.26^{(2)}$	$-134.06^{(2)}$
		$-113.19^{(4)}$	
		$-113.4^{(1)}$	$-134.7^{(1)}$
$CdHPO_4^0$		$-285.00^{(5)}$	
$Cd_3(PO_4)_2(c)$		$-598.16^{(5)}$	
CdS	greenockite	$-37.4^{(2,4)}$	$-38.7^{(2)}$
		-33.6 to $-37.4^{(1)}$	-34.5 to $-38.7^{(1)}$
$CdSO_4(c)$		$-196.65^{(2)}$	$-223.06^{(2)}$
		$-196.33^{(4)}$	
		$-196.4^{(1)}$	$-222.5^{(1)}$
$CdSO_4^0$		$-196.51^{(2)}$	$-235.46^{(2)}$
$CdSO_4 \cdot H_2O(c)$		$-255.46^{(2)}$	$-296.26^{(2)}$
		$-255.2^{(1)}$	$-295.6^{(1)}$
$CdSO_4 \cdot \frac{8}{3}H_2O(c)$		$-350.224^{(2)}$	$-413.33^{(2)}$
		$-350.0^{(1)}$	$-412.8^{(1)}$
CdTe		$-23.1^{(1)}$	$-23.5^{(1)}$
$CdSO_4 \cdot 2Cd(OH)_2(c)$		$-429.63^{(5)}$	
$2CdSO_4 \cdot Cd(OH)_2(c)$		$-519.93^{(5)}$	
$CdSiO_3(c)$		$-264.0^{(2)}$	$-284.20^{(2)}$
		$-263.9^{(1)}$	$-283.9^{(1)}$
Cl (chlorine)			
$Cl_2(g)$		0.00	0.00
Cl^-		$-31.3^{(1,2)}$	$-39.9^{(1,2)}$
Cl_2^0		$+1.65^{(1,2)}$	$-5.6^{(1,2)}$
$HCl(g)$		$-22.77^{(1,2)}$	$-22.06^{(1,2)}$
HCl^0		$-31.35^{(1,2)}$	$-40.0^{(1,2)}$
$HClO^0$		$-19.1^{(1)}$	$-28.6^{(1)}$
ClO^-		$-8.82^{(1)}$	$-25.6^{(1)}$
$HClO_2^0$		$+0.07$ to $+6.8^{(1)}$	$-12.7^{(1)}$
ClO_2^-		$+2.74$ to $+4.1^{(1)}$	-15.9 to $-17.2^{(1)}$

Compound or species	Mineral name	G_f^o, kcal/mol	H_f^o, kcal/mol
$HClO_3^0$		-0.62[1]	-23.5[1]
ClO_3^-		-0.62 to -2.0[1]	-23.5 to -24.9[1]
$HClO_4^0$		-2.47[1,7]	-31.41[1,7]
ClO_4^-		-2.57[7]	-31.41[7]
		-2.0 to -2.6[1]	-31.0[1]
Co (cobalt)			
Co(c)		0.00	0.00
Co^{2+}		-13.0[2,4]	-13.9[1,2]
		-13.1[1]	
Co^{3+}		$+32.03$[2,4]	$+22$[2]
		$+18.7$ to $+32.0$[1]	$+6.0$ to $+22.0$[1]
CoO(c)		-49.0[7]	-55.2[7]
		-51.20[4]	
		-49.0 to -51.4[1]	-55.2 to -56.9[1]
Co_3O_4	cobalt spinel	-184.9[1,2,4,7]	-213[1,2]
$HCoO_2^-$		-82.97[7]	
		-83.0 to -97.4[1]	
$Co(OH)_2$(c, blue, ppt)		-107.6[1,2]	-129.3[7]
		-107.53[4]	
$Co(OH)_2$	transvaalite	-109.8[1]	
$Co(OH)_2$(c, pink, ppt)		-108.6[1,2]	-129.0[1,2]
$Co(OH)_2^0$		-88.2[2]	-123.8[2]
		-101.0[1]	
$CoOH^+$		-56.7[1]	
$Co(OH)_3$(c)		-142.6[1,2,7]	-171.3[2]
			-174.6[1]
$Co(OH_3^-)$		-140.5[1]	
α-CoS(c)		-19.8[1,7]	-19.8[2]
			-19.3[1]
β-CoS(c)		-20.2[1]	-19.8[1]
$CoSO_4$(c)		-187.0[2]	-212.3[2]
		-180.1 to -187.3[1]	-205.5 to -212.6[1]
$CoSO_4^0$		191.0[2]	231.2[2]
$CoCO_3$	sphaerocobaltite	-155.57[7]	-170.8[1]
		-154.8[1]	
$Co(NO_3)_2$(c)		-55.1[1,7]	-100.5[1,2]
$Co(NO_3)_2^0$		-66.2[2]	-113.0[2]
$CoSO_4 \cdot 6H_2O$(c)		-534.35[2]	-641.4[2]
$CoSO_4 \cdot 7H_2O$(c)		-591.26[2]	-712.22[2]
$Co_3(AsO_4)_2$		-387.4[2]	
CoS_2	cattierite	-32.8[1]	-32.1[1]
Co_3S_3	linnaeite	-77.3[1]	-73.4[1]

Compound or species	Mineral name	G_f°, kcal/mol	H_f°, kcal/mol
Cr (chromium)			
Cr(c)		0.00	0.00
Cr^{2+}		$-42.1^{(7)}$	$-34.3^{(2)}$
		-39.3 to $-42.1^{(1)}$	-33.2 to $-34.4^{(1)}$
Cr^{3+}		$-51.5^{(4,7)}$	-50.6 to $-56.4^{(1)}$
		-48.7 to $-53.3^{(1)}$	
$Cr(OH)^{2+}$		$-103.0^{(7,4)}$	$-113.5^{(7)}$
		$-100.0^{(1)}$	$-106.6^{(1)}$
$Cr(OH)_2^+$		$-151.2^{(7)}$	
		-149.1 to $-151.2^{(1)}$	
$Cr(OH)_3(c)$		$-215.3^{(7)}$	$-254.3^{(2)}$
		-202.4 to $-215.3^{(1)}$	-233.2 to $-247.1^{(1)}$
$Cr(OH)_2(c)$		$-140.5^{(1,7)}$	
$Cr(OH)_4(c)$		$-242.4^{(1,7)}$	
CrO_2^-		$-128.00^{(7)}$	
		-123.9 to $-128.1^{(1)}$	
CrO_3^{3-}		$-144.2^{(1)}$	
CrO_4^{2-}		$-173.94^{(4)}$	
		-172.3 to $-176.1^{(1)}$	-209.2 to $-213.7^{(1)}$
$Cr_2O_7^{2-}$		$-311.0^{(2,4)}$	$-356.2^{(2)}$
		-307.7 to $-315.4^{(1)}$	-353.4 to $-364.0^{(1)}$
$CrO_2(c)$			$-142.9^{(1)}$
$CrO_3(c)$			$-140.9^{(1)}$
Cr_2O_3	eskolaite	$-259.9^{(2)}$	$-272.4^{(2)}$
		$-252.89^{(4)}$	
		$-252.0^{(1)}$	$-271.5^{(1)}$
$H_2CrO_4^0$		$-185.92^{(7)}$	$-201.7^{(1)}$
		-179.8 to $-185.9^{(1)}$	
$HCrO_4^-$		$-182.8^{(2,4)}$	$-209.9^{(2)}$
		-181.2 to $-184.9^{(1)}$	-208.5 to $-220.2^{(1)}$
$FeCr_2O_4$	chromite	$-321.2^{(2)}$	$-345.3^{(2)}$
		$-324.5^{(1)}$	$-348.6^{(1)}$
Cs (cesium)			
Cs(c)		0.00	0.00
Cs^+		$-69.79^{(2)}$	$-61.73^{(2)}$
		-67.4 to $-69.8^{(1)}$	-59.2 to $-61.7^{(1)}$
CsOH(c)		$-84.9^{(7)}$	$-97.72^{(2)}$
		-84.9 to $-88.6^{(1)}$	-97.2 to $-99.7^{(1)}$
$Cs_2O(c)$		$-65.6^{(7)}$	$-68.4^{(2)}$
		-65.6 to $-73.6^{(1)}$	-79.9 to $-82.6^{(1)}$
CsF(c)		$-125.6^{(2)}$	$-132.3^{(2)}$
CsCl(c)		$-99.08^{(2)}$	$-105.89^{(2)}$
$Cs_2SO_4(c)$		$-316.36^{(2)}$	$-344.89^{(2)}$
$Cs_2SO_4^0$		$-317.55^{(2)}$	$-340.78^{(2)}$

Compound or species	Mineral name	G_f°, kcal/mol	H_f°, kcal/mol
$CsNO_3(c)$		97.18[2]	−120.93[2]
$Cs_2UO_4(c)$		−429.6[1]	−458.9[1]
Cu (copper)			
$Cu(c)$		0.00	0.00
Cu^+	cuprous ion	+11.95[2,4]	+17.13[2]
		+12.0[1]	+12.4 to +17.2[1]
Cu^{2+}	cupric ion	+15.65[1,2,4]	+15.48[2]
			+15.6[1]
$CuCO_3^0$		−119.69[5]	
		−120.0[1]	
$Cu(CO_3)_2^{2-}$		−250.10[5]	
		−250.6[1]	
$CuHCO_3^+$		−127.46[5]	
$CuCO_3(c)$		−123.65[5]	−142.2[1,7]
		−123.8[1]	
$CuCO_3 \cdot Cu(OH)_2$	malachite	−216.44[7]	−251.3[2]
		−213.58[4]	
		−213.6 to −216.4[1]	−250.6 to −251.9[1]
$2CuCO_3 \cdot Cu(OH)_2$	azurite	−345.8[7]	−390.1[2]
		−314.29[4]	
		−314.4 to −343.7[1]	−389.0 to −390.1[1]
$CuCl$	nantokite	−28.65[2]	−32.8[2]
		−28.5[1]	−32.2 to −32.8[1]
$CuCO_3(OH)_2^{2-}$		−206.0[1]	
$CuCl^0$		−22.95[5]	
$CuCl^+$		−16.3[2]	
		−15.8 to −17.1[1]	
$CuCl_2^0$		−57.4[2]	
		−45.2 to −47.3[1]	
$CuCl_2(c)$		−42.0[2]	−52.6[2]
		−41.1 to −42.1[1]	−51.6 to −52.6[1]
$CuCl_2 \cdot 2H_2O(c)$		−156.8[2]	−196.3[2]
$CuCl_2^-$		−58.16[5,6]	−66.3[6]
		−57.4 to −58.4[1]	
$CuCl_3^-$		−76.30[5]	
$CuCl_3^{2-}$		−89.79[5]	
		−85.5 to −90.0[1]	
$CuCl_4^{2-}$		−103.6[1]	
$Cu_2Cl_4^{2-}$		−119.15[5]	
$\alpha\text{-}CuFe_2O_4$	cupric ferrite	−205.3[5]	
$\alpha\text{-}Cu_2Fe_2O_4$	cuprous ferrite	−229.04[5]	
$CuMoO_4(c)$		−193.01[5]	
$CuNO_3^+$		−11.65[5]	

Compound or species	Mineral name	G_f°, kcal/mol	H_f°, kcal/mol
$Cu(NO_3)_2^0$		$-37.06^{(5)}$	
$CuOH^+$		$-30.52^{(5)}$	
		$-31.1^{(1)}$	
$Cu(OH)_2^0$		$-59.53^{(2)}$	$-94.46^{(2)}$
$Cu(OH)_2(c)$		$-85.3^{(1,6,7)}$	$-106.8^{(2,6)}$
			-106.1 to $-107.5^{(1)}$
$Cu(OH)_3^-$		$-117.90^{(5)}$	
$Cu(OH)_4^{2-}$		$-157.08^{(5)}$	
$Cu_2(OH)_2^{2+}$		$-67.47^{(5)}$	
CuO_2^{2-}		$-43.3^{(7)}$	
		$-43.88^{(4)}$	
		$-43.6^{(1)}$	
$HCuO_2^-$		$-61.42^{(7)}$	
		$-61.8^{(1)}$	
CuO	tenorite	$-30.7^{(2,5,7)}$	$-37.4^{(2,6,7)}$
		$-31.00^{(4)}$	
		$-30.4^{(1)}$	$-37.3^{(1)}$
$CuOH(c)$		$-45.54^{(5)}$	
		$-43.5^{(1)}$	
$Cu_4Cl_2(OH)_6$	atacamite	$-320.5^{(1)}$	$-396.2^{(1)}$
$Cu_4Cl_2(OH)_6$	paratacamite	$-320.3^{(1)}$	
Cu_2O	cuprite	$-35.1^{(1,2,5,6,7)}$	$-40.3^{(2,6,7)}$
			$-40.4^{(1)}$
$CuHPO_4^0$		$-250.72^{(5)}$	
$CuH_2PO_4^+$		$-258.35^{(5)}$	
$Cu_3(PO_4)_2(c)$		$-493.63^{(5)}$	$-533.3^{(1)}$
		-490.3 to $-493.9^{(1)}$	
$Cu_3(PO_4)_2 \cdot 2H_2O(c)$		$-609.60^{(5)}$	
$CuSe$	klockmannite	-7.9 to $-9.9^{(1)}$	-6.6 to $-9.4^{(1)}$
CuS	covellite	$-12.81^{(2,4,6)}$	$-12.7^{(2,6)}$
		-11.7 to $-12.9^{(1)}$	-11.6 to $-12.7^{(1)}$
Cu_2S	chalcocite	$-20.6^{(1,2,4,6,7)}$	$-19.0^{(1,2,6,7)}$
$CuFeS_2$	chalcopyrite	-42.7 to $-45.6^{(1)}$	-42.3 to $-45.5^{(1)}$
Cu_5FeS_4	bornite	-86.7 to $-93.9^{(1)}$	-79.9 to $-90.9^{(1)}$
$CuSO_4$	chalcocyanite	$-158.2^{(1,2)}$	$-184.36^{(2)}$
			$-184.2^{(1)}$
$CuSO_4^0$		$-162.31^{(2)}$	$-201.84^{(2)}$
		$-165.3^{(1)}$	$-200.4^{(1)}$
$Cu_2SO_4(c)$		$-156.41^{(5,7)}$	$-179.6^{(2)}$
		$-156.0^{(1)}$	$-179.4^{(1)}$
$CuO \cdot CuSO_4(c)$		$-187.61^{(5)}$	$-223.8^{(2)}$
$CuSO_4 \cdot 3Cu(OH)_2$	brochantite	$-434.5^{(1,2)}$	$-525.4^{(1)}$
$CuSO_4 \cdot 3Cu(OH)_2 \cdot H_2O$	langite	$-488.6^{(2)}$	$-594^{(1,2)}$
		$-488.5^{(1)}$	

Compound or species	Mineral name	G_f°, kcal/mol	H_f°, kcal/mol
$CuSO_4 \cdot H_2O(c)$		$-219.46^{(2)}$	$-259.52^{(2)}$
		$-219.2^{(1)}$	$-259.2^{(1)}$
$CuSO_4 \cdot 3H_2O(c)$		$-334.6^{(1,2)}$	$-402.56^{(2)}$
			$-402.4^{(1)}$
$CuSO_4 \cdot 5H_2O(c)$	chalcanthite	$-449.3^{(1,2)}$	$-544.85^{(2)}$
			$-544.7^{(1)}$
$CuSO_4 \cdot 2Cu(OH)_2$	antlerite	$-345.8^{(2)}$	
		$-345.6^{(1)}$	
$CuSiO_3 \cdot H_2O$	dioptase	$-288.6^{(1)}$	$-324.8^{(1)}$
$CuSiO_3 \cdot 2H_2O$	chrysocolla (amorph.)	$-345.1^{(1)}$	

F (fluorine) (see also Ag, Ba, Ca, etc.)

$F_2(g)$		0.00	0.00
F^-		$-66.4^{(2)}$	$-79.5^{(2)}$
		$-66.8^{(1)}$	$-79.6^{(1)}$
$HF(g)$		$-65.3^{(2)}$	$-64.8^{(2)}$
		-64.7 to $-65.8^{(1)}$	-64.2 to $-65.3^{(1)}$
HF^0		$-66.64^{(2)}$	$-79.50^{(2)}$
		$-71.0^{(1)}$	$-77.1^{(1)}$

Fe (iron) (see also Ca)

$Fe(c)$		0.00	0.00
Fe^{2+}		$-18.85^{(2,3,4,6)}$	$-21.3^{(2,3,6)}$
		-18.9 to $-22.0^{(1)}$	-21.0 to $-22.1^{(1)}$
Fe^{3+}	ferric ion	$-1.12^{(2,3,4,6)}$	$-11.6^{(2,6)}$
		-1.1 to $-4.3^{(1)}$	-11.1 to $-12.1^{(1)}$
FeO_2^{2-}		$-70.58^{(4)}$	
Fe_3C	cohenite	$+4.3$ to $+4.7^{(1)}$	$+5.9^{(1)}$
$FeCO_3$	siderite	$-160.3^{(2,3,5,6,7)}$	$-177.3^{(2,3,6,7)}$
		$-159.34^{(4)}$	
		-159.3 to $-162.6^{(1)}$	-176.1 to $-180.0^{(1)}$
$FeCl_2$	lawrencite	$-72.3^{(1,2)}$	$-81.7^{(1,2)}$
$FeCl_2^0$		$-81.59^{(2)}$	$-101.2^{(2)}$
$FeCl_3$	molysite	$-79.8^{(1,2)}$	$-95.48^{(2)}$
			$-95.8^{(1)}$
$FeCl_3^0$		$-95.2^{(2)}$	$-131.5^{(2)}$
$FeCl^{2+}$		-34.4 to $-37.7^{(1)}$	-42.9 to $-46.5^{(1)}$
$Fe_{0.95}O$	wüstite	$-58.7^{(2,5,6,7)}$	$-63.6^{(1,2,6,7)}$
		$-58.6^{(1)}$	
$FeO(c)$		$-60.1^{(1,5)}$	$-65.0^{(1,2)}$
$\alpha\text{-}Fe_2O_3$	hematite	$-177.6^{(3,5,6,7)}$	$-197.3^{(2,3,6)}$
		$-177.39^{(4)}$	
		$-177.5^{(1)}$	$-197.0^{(1)}$

Compound or species	Mineral name	G_f°, kcal/mol	H_f°, kcal/mol
$\gamma\text{-}Fe_2O_3$	maghemite	$-173.75^{[5]}$	
Fe_3O_4	magnetite	$-242.69^{[2,3,4,5,6,7]}$	$-267.1^{[2,3,6,7]}$
		-241.6 to $-243.1^{[1]}$	-265.9 to $-267.3^{[1]}$
$\alpha\text{-}FeOOH$	goethite	$-116.86^{[3,5,6]}$	$-133.6^{[2,3,6]}$
		$-116.77^{[4]}$	
		$-117.0^{[1]}$	$-133.7^{[1]}$
$\gamma\text{-}FeOOH$	lepidocrocite	$-115.50^{[5]}$	
		$-112.7^{[1]}$	
$FePO_4(c)$	heterosite	$-283.19^{[5]}$	$-299.6^{[7]}$
		-272.0 to $-283.2^{[1]}$	-299.6 to $-310.1^{[1]}$
$FePO_4 \cdot 2H_2O$	strengite	$-396.2^{[2]}$	$-451.3^{[1,2]}$
		-396.2 to $-397.4^{[1]}$	
$Fe_3(PO_4)_2 \cdot 8H_2O$	vivianite	$-1058.36^{[5]}$	
		$-1046.2^{[1]}$	
$FeSO_4(c)$		$-196.2^{[2]}$	$-221.9^{[2]}$
		-196.2 to $-198.3^{[1]}$	-220.5 to $-221.9^{[1]}$
$FeSO_4^0$		$-196.82^{[2]}$	$-238.6^{[2]}$
$FeSO_4^+$		$-187.63^{[5]}$	
$Fe_2(SO_4)_3(c)$		$-537.4^{[1]}$	$-616.7^{[1]}$
$Fe(SO_4)_2^-$		$-367.26^{[5]}$	
$FeS(amorph.)$		$-21.3^{[3]}$	
FeS	Fe-rich pyrrhotite	$-24.0^{[1,2]}$	$-23.9^{[1,2]}$
$Fe_{0.877}S$	pyrrhotite	$-25.6^{[1]}$	$-25.2^{[1]}$
FeS	mackinawite(synthet.)	$-22.3^{[1,3]}$	
FeS	troilite, α	$-23.87^{[3,5,6]}$	$-24.0^{[3,6]}$
		$-24.0^{[1,4]}$	$-23.9^{[1]}$
FeS_2	pyrite	$-39.0^{[2,3,5]}$	$-41.8^{[2,3,6]}$
		$-39.89^{[4]}$	
		-36.0 to $-39.9^{[1]}$	-41.0 to $-42.6^{[1]}$
FeS_2	marcasite	$-37.4^{[2,5]}$	
		$-37.9^{[1]}$	
$FeOH^+$		$-66.3^{[3]}$	$-77.6^{[3]}$
		-64.2 to $-69.6^{[1]}$	-77.6 to $-78.4^{[1]}$
$Fe(OH)_2^0$		$-113.29^{[5]}$	
		$-109.5^{[1]}$	
$Fe(OH)_2$	amakinite	$-116.3^{[2,4]}$	$-136.0^{[2]}$
		-115.6 to $-117.8^{[1]}$	-135.8 to $-137.0^{[1]}$
$Fe(OH)_3^-$		$-147.0^{[3]}$	
		-147.9 to $-148.5^{[1]}$	
$Fe(OH)_4^{2-}$		$-185.27^{[5]}$	
		-184.0 to $-185.5^{[1]}$	
$Fe_3(OH)_4^{2-}$		$-230.23^{[5]}$	
$Fe(OH)^{2+}$		$-54.83^{[3]}$	$-69.5^{[3]}$
		-54.8 to $-58.0^{[1]}$	$-69.7^{[1]}$

Compound or species	Mineral name	G_f°, kcal/mol	H_f°, kcal/mol
$Fe(OH)_2^+$		$-104.7^{(3)}$	$-131.8^{(1)}$
		-104.7 to $-109.3^{(1)}$	
$Fe(OH)_3(c)$		$-166.5^{(2,4)}$	$-196.7^{(2)}$
		-166.0 to $-171.7^{(1)}$	-197.0 to $-201.8^{(1)}$
$Fe(OH)_3^0$		$-156.6^{(2)}$	
		-157.6 to $-161.9^{(1)}$	
$Fe(OH)_4^-$		$-201.32^{(5)}$	
		$-201.4^{(1)}$	
$Fe_2(OH)_2^{4+}$		$-117.46^{(5)}$	$-146.3^{(1)}$
		$-111.7^{(1)}$	
$FeSO_4\cdot7H_2O$	melanterite	$-599.9^{(1,2)}$	$-720.5^{(1,2)}$
$FeSO_4\cdot H_2O$	szomolnokite		$-297.3^{(1)}$
$KFe_3(SO_4)_2(OH)_6$	jarosite	$-792.66^{(5)}$	
$FeSiO_3$	ferrosilite	$-267.16^{(3,4)}$	$-285.63^{(3)}$
		-256.9 to $-268.5^{(1)}$	-276.1 to $-288.2^{(1)}$
$FeS_3O_3(OH)_8^0$		$-898.00^{(4)}$	
$FeSiO_3$	clinoferrosilite	$-267.5^{(1)}$	$-286.0^{(1)}$
$FeSiO_3$	orthoferrosilite	$-266.4^{(1)}$	$-285.0^{(1)}$
Fe_2SiO_4	fayalite	$-329.6^{(2)}$	$-353.7^{(2)}$
		-319.8 to $-330.7^{(1)}$	-343.7 to $-354.8^{(1)}$
$FeMgSi_2O_6$	hypersthene	$-619.7^{(1)}$	$-658.6^{(1)}$
$FeMoO_4(c)$		$-232.14^{(5)}$	$-257.5^{(7)}$
		-232.9 to $-234.8^{(1)}$	$-257.1^{(1)}$
$FeWO_4(c)$	ferberite	$-250.4^{(7)}$	$-274.1^{(7)}$
		-250.4 to $-259.8^{(1)}$	-274.1 to $-283.9^{(1)}$
$FeAsS$	arsenopyrite	-12.0 to $-26.2^{(1)}$	-10.0 to $-25.2^{(1)}$
$FeAl_2O_4$	hercynite	-442.3 to $-449.3^{(1)}$	-470.0 to $-476.9^{(1)}$
Fe_2SiO_4	fayalitic spinel	$-327.1^{(1)}$	$-351.9^{(1)}$
$FeTiO_3$	ilmenite	$-277.9^{(1)}$	$-295.5^{(1)}$
$Fe_3Si_2O_5(OH)_4$	greenalite	$-720.0^{(8)}$	
		$-719.9^{(1)}$	
$Fe_2^{3+}Si_4O_{10}(OH)_2$	minnesotaite	$-1055.0^{(8)}$	$-1152.6^{(1)}$
		$-1069.9^{(1)}$	
$Fe_7Si_8O_{22}(OH)_2$	grunerite	$-2141.0^{(1)}$	$-2294.0^{(1)}$
$Fe_3Al_2Si_3O_{12}$	almandine	$-1187.8^{(1)}$	$-1267.2^{(1)}$
	pyrope	see Mg	
	andradite	see Ca	
	grossular	see Ca	
$Fe_2Al_4Si_5O_{18}$	ferrocordierite	$-1902.7^{(1)}$	$-2019.7^{(1)}$
	cordierite	see Mg	
$FeAl_2SiO_5(OH)_2$	chloritoid	$-709.8^{(1)}$	$-762.7^{(1)}$
$Fe_5Al_2Si_3O_{10}(OH)_8$	Fe-clinochlore	$-1801.2^{(1)}$	
$Fe_4Al_4Si_2O_{10}(OH)_8$	daphnite	$-1678.4^{(1)}$	$-1823.7^{(1)}$
$Fe_2Al_9Si_4O_{23}(OH)$	staurolite	$-2674.8^{(1)}$	$-2882.1^{(1)}$

Compound or species	Mineral name	G°_f, kcal/mol	H°_f, kcal/mol
Ga (gallium)			
Ga(c)		0.00	0.00
Ga^{3+}		-38.00[4]	
GaO_3^{3-}		-147.94[4]	
$Ga(OH)^{2+}$		-90.89[4]	
$Ga(OH)_2^+$		-142.78[4]	
$Ga(OH)_3(c)$		-198.69[4]	
$Ga_2O_3(c)$		-238.52[4]	
Ge (germanium)			
Ge(c)		0.00	0.00
Ge^{2+}		-8.8[1]	
Ge^{4+}		-6.6[1]	
GeO_3^{2-}		-157.9[1,4]	
$HGeO_3^-$		-175.2[1,4]	
$H_2GeO_3^0$		-186.8[1,4]	
$HGeO_2^-$		-92.1[1]	
GeO(brown, c)		-57.7[2,4]	-62.6[2]
		-50.7 to -56.7[1]	-56.7 to -62.6[1]
GeO(yellow, c)		-49.5[1,2]	
GeS(c)		-17.09[4]	
GeO(hydrated, a,c)		-69.1[1]	
GeO(hydrated, b,c)		-62.7[1]	
GeO_2(hexagonal)		-119.7[1]	-132.6[1]
GeO_2(quartz type)		-118.8[1]	-131.7[1]
		-118.79[4]	
$GeO_2(c)$		-124.7[1]	-138.7[1]
GeO_2(tetragonal)		-136.1[1]	
GeO_2(amorph.)		-113.6[1]	-125.8 to -129.1[1]
H (hydrogen)			
$H_2(g)$		0.00	0.00
H_2^0		$+4.2$[1,2]	-1.0[1,2]
H^+		0.00	0.00
OH^-		-37.594[2]	-54.970[2]
		-37.595[1]	-54.971[1]
$H_2O(l)$		-56.687[2]	-68.315[2]
		-56.684[1]	-68.307[1]
$H_2O(g)$		-54.636[1,2]	-57.791[1,2]
$H_2O_2^0$		-31.5 to -32.0[1]	-45.7[1]
$H_2O_2(l)$		-28.8[1]	-44.9[1]
Hf (hafnium)			
Hf(c)		0.00	0.00
$HfO_2(c)$		-260.09[4]	
Hf^{4+}		-156.80[4]	

Compound or species	Mineral name	G_f°, kcal/mol	H_f°, kcal/mol
Hg (mercury)			
Hg(l)		0.00	0.00
Hg(g)		+7.6[1,2]	+14.6[1,2]
Hg^{2+}		+39.3[1,2,4]	+40.9[2]
			+41.0[1]
Hg_2^{2+}		+36.70[2,4]	+41.2[1,2]
		+36.6[1]	
$Hg(NO_3)^+$		−12.72[5]	
$Hg(NO_3)_2^0$		−12.06[5]	
$HgOH^+$		−12.69[5]	−20.2[1]
		−12.5[1]	
$Hg(OH)_2^0$		−65.7[1,2]	−84.9[1,2]
$Hg(OH)^+$		−12.50[4]	
$Hg(OH)_2(c)$		−70.47[5]	
$Hg(OH)_3^-$		−101.92[5]	
HgO(red)	montroydite	−13.995[2]	−21.71[2]
		−14.0[1]	−21.7[1]
HgO(red, hexagonal)		−13.92[2]	−21.4[1,2]
		−13.9[1]	
HgO(c)		−58.56[4]	
HgO(yellow)		−13.964[1,7]	−21.62[2]
		−14.0[1]	−21.6[1]
$HHgO_2^-$		−45.48[4]	
$Hg_2(OH)_2(c)$		−69.49[5]	
$HgCO_3(c)$		−117.62[5]	
$Hg_2CO_3(c)$		−111.9[2]	−132.3[1,2]
		−105.8 to −111.9[1]	
$HgCl_2(c)$		−42.7[2]	−53.6[2]
		−42.7 to −44.4[1]	−53.6 to −55.0[1]
Hg_2Cl_2	calomel	−50.37[1,2,4]	−63.39[2]
			−63.37[1]
$HgCl^+$		−1.3[1]	−4.5[1]
$HgCl_2^0$		−41.4[1]	−51.7[1]
$HgCl_3^-$		−73.9[1]	−92.9[1]
$HgCl_4^{2-}$		−106.79[4]	
		−107.2[1]	−132.4[1]
HgCl(g)		+13.9 to +15.0[1]	+18.9 to +20.1[1]
$HgBr_2(c)$		−35.2 to −36.6[1]	−40.6[1]
$Hg_2Br_2(c)$		−43.1[1]	−49.4[1]
$HgBr_4^{2-}$		−88.3[1]	−99.9 to −103.0[1]
$HgI_2(red)$	coccinite	−24.4[1]	−25.2[1]
$Hg_2I_2(yellow)$		−26.6[1]	−29.0[1]
$Hg_2HPO_4(c)$		−242.23[5]	
HgS(red)	cinnabar	−12.1[2]	−13.9[2]

Compound or species	Mineral name	G°_f, kcal/mol	H°_f, kcal/mol
		-10.9 to $-12.1^{(1)}$	-12.7 to $-13.9^{(1)}$
HgS(black)	metacinnabar	$-11.4^{(2)}$	$-12.8^{(2)}$
		-10.3 to $-11.8^{(1)}$	-10.4 to $-12.9^{(1)}$
Hg_2S(c)		$-17.49^{(5)}$	
HgS_2^{2-}		$+45.90^{(4)}$	
		$+10.0$ to $+11.6^{(1)}$	
$HgSO_4$(c)		$-143.14^{(5)}$	$-168.3^{(7)}$
		$-141.0^{(1)}$	-168.3 to $-169.1^{(1)}$
$HgSO_4^0$		$-140.51^{(2)}$	
Hg_2SO_4(c)		$-149.589^{(2)}$	$-177.61^{(2)}$
		$-149.4^{(1)}$	$-177.5^{(1)}$
Hg_2CrO_4		-147.6 to $-155.7^{(1)}$	
I (iodine)			
I_2(c)		0.00	0.00
I^-		$-12.33^{(2)}$	$-13.19^{(2)}$
		$-12.4^{(1)}$	$-13.5^{(1)}$
HI(g)		$+0.41^{(2)}$	$+6.33^{(2)}$
HI^0		$-12.33^{(2)}$	$-13.19^{(2)}$
I_2^0		$+3.92^{(1,2)}$	$+5.4^{(2)}$
			$+5.2^{(1)}$
In (indium)			
In(c)		0.00	0.00
In^{2+}		$-12.12^{(4)}$	
In^{3+}		$-23.42^{(4)}$	
In_2O_3(c)		$-198.54^{(4)}$	
$InOH^{2+}$		$-74.81^{(4)}$	
$In(OH)_2^+$		$-125.48^{(4)}$	
InS(c)		$-31.50^{(4)}$	
In_2S_3(c)		$-98.59^{(4)}$	
InS(c)		$-31.50^{(4)}$	
Ir (iridium)			
Ir(c)		0.00	0.00
Ir_2S_3(c)		$-49.8^{(4)}$	
IrS_2(c)		$-30.4^{(4)}$	
IrO_2(c)		$-28.0^{(4)}$	
IrO_4^-		$-47.0^{(4)}$	
K (potassium) (see also Al, Fe)			
K(c)		0.00	0.00
K^+		$-67.70^{(2)}$	$-60.32^{(2)}$
		$-67.6^{(1)}$	$-60.2^{(1)}$
KCl	sylvite	$-97.79^{(2)}$	$-104.385^{(2)}$

Compound or species	Mineral name	G_f°, kcal/mol	H_f°, kcal/mol
		$-97.7^{(1)}$	$-104.3^{(1)}$
KCl^0		$-99.07^{(1)}$	$-100.27^{(1)}$
$K_2CO_3(c)$		$-254.2^{(1,2)}$	$-275.1^{(2)}$
			$-274.4^{(1)}$
$K_2CO_3^0$		$-261.18^{(5)}$	
$KOH(c)$		$-90.61^{(2)}$	$-101.521^{(2)}$
		$-90.4^{(1)}$	$-101.6^{(1)}$
KOH^0		$-105.29^{(2)}$	$-115.29^{(2)}$
$KH_2PO_4(c)$		$-339.65^{(5)}$	
$K_2HPO_4(c)$		$-392.14^{(5)}$	
KSO_4^-		$-246.4^{(1,3,5,7)}$	$-276.6^{(1)}$
K_2SO_4	arcanite	$-315.83^{(2)}$	$-343.64^{(1,2)}$
		$-315.6^{(1)}$	
$K_2SiO_3(c)$		$-343.3^{(6)}$	$-365.9^{(6)}$
		$-350.1^{(1)}$	-368.5 to $-372.5^{(1)}$
$K_2O(c)$		$-76.2^{(7)}$	$-86.4^{(7)}$
		$-76.6^{(1)}$	$-86.5^{(1)}$
$K_2S(c)$		$-87.0^{(2)}$	$-91.0^{(2)}$
		-87.0 to $-96.6^{(1)}$	-91.0 to $-99.9^{(1)}$
K_2S^0		$-114.9^{(2)}$	$-112.7^{(2)}$
$KBr(c)$		$-90.9^{(1)}$	$-94.1^{(1)}$
$KJ(c)$		$-77.4^{(1)}$	$-78.5^{(1)}$
$K_2SO_4^0$		$313.37^{(2)}$	$-337.96^{(2)}$
$KHSO_4$	mercallite	$-246.5^{(2)}$	$-277.4^{(1,2)}$
		-246.5 to $-247.4^{(1)}$	
KNO_3	niter	$-94.39^{(2)}$	$-118.22^{(2)}$
		$-94.2^{(1)}$	$-118.1^{(1)}$
$KHSO_4^0$		$-248.39^{(2)}$	$-272.40^{(2)}$
KNO_3^0		$-94.31^{(2)}$	$-109.88^{(2)}$

RARE EARTH ELEMENTS

La (lanthanum)

α-La(c)		0.00	0.00
La^{3+}		$-163.4^{(2,4)}$	$-169.0^{(2)}$
		-163.4 to $-174.5^{(1)}$	-169.0 to $-176.2^{(1)}$
$La_2O_3(c)$		$-407.7^{(2,4)}$	$-428.7^{(2)}$
		-407.6 to $-426.9^{(1)}$	-428.7 to $-458.0^{(1)}$
$La(OH)_3(c)$		$-313.2^{(1,7)}$	$-337.0^{(2)}$
		$-305.8^{(4)}$	-337.0 to $-345.0^{(1)}$
$La_2(CO_3)_3(c)$		$-750.9^{(2,4)}$	
$La_2S_3(c)$		$-301.2^{(1)}$	-288.9 to $-306.8^{(1)}$

Ce (cerium)

Ce(c)		0.00	0.00

Compound or species	Mineral name	G_f°, kcal/mol	H_f°, kcal/mol
Ce^{3+}		$-160.6^{(2,4)}$	$-166.4^{(2)}$
		-160.6 to $-170.5^{(1)}$	-166.4 to $-173.8^{(1)}$
Ce^{4+}		$-120.9^{(1)}$	-128.4 to $-137.9^{(1)}$
		$-120.44^{(4)}$	
CeO_2	cerianite	$-244.9^{(2)}$	$-258.80^{(2)}$
		$-244.4^{(4)}$	
		-219.0 to $-245.1^{(1)}$	-233.0 to $-260.2^{(1)}$
$Ce_2O_3(c)$		$-407.8^{(2,4)}$	$-429.3^{(1,2)}$
		$-408.1^{(1)}$	
$Ce(OH)^{3+}$		$-187.7^{(7)}$	$-194.1^{(1)}$
		-179.1 to $-187.7^{(1)}$	
$Ce(OH)_2^{2+}$		$-244.0^{(1,7)}$	
$Ce_2(CO_3)_3(c)$		$-744.1^{(4)}$	
$Ce(OH)_3(c)$		$-311.63^{(7)}$	$-336.2^{(1)}$
		-303.2 to $-311.6^{(1)}$	
		$-303.6^{(4)}$	
$CePO_4$	monazite	$-434.0^{(1)}$	$-464.6^{(1)}$
$CeSO_4^+$		$-343.3^{(2)}$	$-380.2^{(2)}$
$Ce(SO_4)_2^-$		$-523.6^{(2)}$	$-595.9^{(2)}$
$Ce_2(SO_4)_2 \cdot H_2O(c)$		$-1320.6^{(2)}$	
$CeS(c)$		$-107.9^{(1)}$	$-109.8^{(1)}$
$CeS_2(c)$		$-151.5^{(1)}$	-146.3 to $-153.9^{(1)}$
$Ce_2S_3(c)$		$-293.1^{(1)}$	$-298.7^{(1)}$
Pr (praseodymium)			
$Pr(c)$		0.00	0.00
Pr^{3+}		$-162.3^{(4)}$	
$Pr(OH)_3(c)$		$-307.1^{(4)}$	
$Pr_2(CO_3)_3(c)$		$-747.1^{(4)}$	
Nd (neodymium)			
$Nd(c)$		0.00	0.00
Nd^{3+}		$-160.5^{(4)}$	
$Nd(OH)_3(c)$		$-305.2^{(4)}$	
$Nd_2O_3(c)$		$-411.3^{(4)}$	
$Nd_2(CO_3)_3$		$-741.4^{(4)}$	
$Nd(OH)_4^-$		$-336.7^{(4)}$	
Sm (samarium)			
$Sm(c)$		0.00	0.00
Sm^{3+}		$-159.3^{(4)}$	
$Sm(OH)_3(c)$		$-306.9^{(4)}$	
$Sm_2O_3(c)$		$-414.6^{(4)}$	
$Sm_2(CO_3)_3$		$-741.4^{(4)}$	

Compound or species	Mineral name	G_f°, kcal/mol	H_f°, kcal/mol
Eu (europium)			
Eu(c)		0.00	0.00
Eu^{3+}		−137.2[4]	
Eu^{2+}		−129.10[4]	
$Eu(OH)_3(c)$		−285.5[4]	
$Eu_2O_3(c)$		−372.2[4]	
$Eu_2(CO_3)_3(c)$		−697.3[4]	
Gd (gadolinium)			
Gd(c)		0.00	0.00
Gd^{3+}		−158.0[4]	
$Gd(OH)_3(c)$		−306.9[4]	
$Gd_2(CO_3)_2(c)$		−738.9[4]	
$Gd(OH)_4^-$		−338.0[4]	
Tb (terbium)			
Tb(c)		0.00	0.00
Tb^{3+}		−155.8[4]	
$Tb(OH)_3(c)$		−303.4[4]	
$Tb_2(CO_3)_3(c)$		−734.5[4]	
Dy (dysprosium)			
Dy(c)		0.00	0.00
Dy^{3+}		−159.0[4]	
$Dy(OH)_3(c)$		−307.2[4]	
$Dy_2O_3(c)$		−423.4[4]	
$Dy_2(CO_3)_3(c)$		−739.3[4]	
$Dy(OH)_4^-$		−339.9[4]	
Ho (holmium)			
Ho(c)		0.00	0.00
Ho^{3+}		−161.0[4]	
$Ho(OH)_3(c)$		−310.1[4]	
$Ho_2O_3(c)$		−428.1[4]	
$Ho_2(CO_3)_3(c)$		−744.9[4]	
Er (erbium)			
Er(c)		0.00	0.00
Er^{3+}		−159.9[4]	
$Er(OH)_3(c)$		−309.5[4]	
$Er_2O_3(c)$		−432.3[4]	
$Er_2(CO_3)_3(c)$		−742.7[4]	
$Er(OH)_4^-$		−342.2[4]	

Compound or species	Mineral name	$G_f°$, kcal/mol	$H_f°$, kcal/mol
Tm (thullium)			
Tm(c)		0.00	0.00
Tm^{3+}		-158.2[4]	
$Tm(OH)_3(c)$		-307.8[4]	
$Tm_2O_3(c)$		-428.9[4]	
$Tm_2(CO_3)_3(c)$		-739.3[4]	
Yb (ytterbium)			
Yb(c)		0.00	0.00
$Yb(OH)_3(c)$		-301.7[7]	
		-303.9[4]	
		-301.7 to -310.3[1]	
$Yb(OH)_4^-$		-336.0[4]	
Yb^{2+}		-126[2]	
		-129.0[1]	
		-153.9[2,4]	-161.2[2]
Yb^{3+}		-153.9 to -156.8[1]	-160.9[1]
$Yb_2O_3(c)$		-412.7[1,2,4]	-433.7[1,2]
$Yb_2(CO_3)_3(c)$		-729.1[4]	
Lu (lutetium)			
Lu(c)		0.00	0.00
Lu^{3+}		-150.0[4]	
$Lu(OH)_3(c)$		-300.3[4]	
$Lu_2O_3(c)$		-427.6[4]	
$Lu_2(CO_3)_3(c)$		-722.9[4]	

End of Rare Earths

Compound or species	Mineral name	$G_f°$, kcal/mol	$H_f°$, kcal/mol
Li (lithium)			
Li(c)		0.00	0.00
Li^+		-70.10[2]	-66.56[1,2]
		-70.0[1]	
LiOH(c)		-104.92[2]	-115.90[2]
		-105.4[1]	-116.1[1]
$LiOH^0$		-107.82[7]	-121.511[7]
		-104.9[1]	-121.5[1]
$Li_2O(c)$		-134.13[2]	-142.91[2]
		-134.2[1]	-142.8[1]
LiCl(c)		-91.87[1,2]	-97.66[2]
			-97.7[1]
$LiCl^0$		-101.48[2]	-106.51[2]
LiF(c)		-140.47[2]	-147.22[2]
$Li_2SO_4(\beta, c)$		-315.91[2]	-343.33[2]
		-316.1[1]	-343.1[1]

Compound or species	Mineral name	G_f°, kcal/mol	H_f°, kcal/mol
$Li_2SO_4^0$		$-318.18^{(2)}$	$-350.44^{(2)}$
		$-317.8^{(1)}$	$-350.0^{(1)}$
$LiNO_3(c)$		$-91.1^{(2)}$	$-115.47^{(2)}$
		-91.1 to $-93.1^{(1)}$	$-115.4^{(1)}$
$LiNO_3^0$		$-96.7^{(2)}$	$-116.12^{(2)}$
		$-96.6^{(1)}$	$-115.9^{(1)}$
$Li_2CO_3(c)$		$-270.58^{(2)}$	$-290.6^{(2)}$
		$-270.3^{(1)}$	$-290.1^{(1)}$
$Li_2CO_3^0$		$-266.66^{(7)}$	$-294.74^{(7)}$
$LiAlO_2(c)$		$-269.7^{(1)}$	$-284.5^{(1)}$
$Li_2SiO_3(c)$		$-372.2^{(2)}$	$-393.9^{(2)}$
		-368.7 to $-372.5^{(1)}$	-390.0 to $-393.9^{(1)}$
$\alpha\text{-}LiAlSi_2O_6$	spodumene	$-688.71^{(2)}$	$-730.1^{(2)}$
		$-688.0^{(1)}$	$-729.4^{(1)}$
$\beta\text{-}LiAlSi_2O_6$	spodumene	$-683.79^{(2)}$	$-723.4^{(2)}$
		$-683.0^{(1)}$	$-722.7^{(1)}$
$LiAlSiO_4$	eucryptite	$-479.7^{(1)}$	$-507.0^{(1)}$
Mg (magnesium) (see also Ca)			
$Mg(c)$		0.00	0.00
Mg^{2+}		$-108.7^{(2,4)}$	$-111.58^{(2)}$
		$-108.8^{(1)}$	$-110.9^{(1)}$
$MgCl_2^0$		$-171.71^{(5)}$	$-191.48^{(2)}$
$MgCl_2$	chloromagnesite	$-141.45^{(7)}$	$-153.28^{(2)}$
		$-141.5^{(1)}$	$-153.3^{(1)}$
$MgHCO_3^+$		$-250.74^{(3,5,7)}$	
		$-250.8^{(1)}$	
$MgCO_3^0$		$-239.73^{(3,5,7)}$	
		$-239.5^{(1)}$	
$MgCO_3$	magnesite	$-246.1^{(3,6,7)}$	$-266.1^{(3,6,7)}$
		$-241.90^{(4)}$	
		-241.9 to $-246.1^{(1)}$	-261.9 to $-266.1^{(1)}$
$MgCO_3 \cdot 3H_2O$	nesquehonite	$-412.6^{(2)}$	-472.6 to $-474.6^{(1)}$
		$-412.4^{(1)}$	
$MgCO_3 \cdot 5H_2O$	lansfordite	$-525.7^{(2)}$	
		$-525.8^{(1)}$	
$Mg_5(CO_3)_4(OH)_2 \cdot 4H_2O$	hydromagnesite	$-1401.7^{(1)}$	$-1557.1^{(1)}$
$Mg_4(CO_3)_3(OH)_2 \cdot 3H_2O(c)$		-1100.2 to $-1108.3^{(1)}$	
$Mg_2(OH)_2(CO_3) \cdot 3H_2O$	artinite	$-613.9^{(1)}$	$-698.0^{(1)}$
$MgCa(CO_3)_2$	dolomite	$-517.9^{(3,5,6)}$	$-556.7^{(3,6)}$
$Mg(NO_3)_2(c)$		$-140.9^{(2)}$	$-188.97^{(2)}$
		$-140.8^{(1)}$	$-188.9^{(1)}$
$Mg(NO_3)_2^0$		$-161.9^{(2)}$	$-210.70^{(2)}$
$MgOH^+$		$-149.8^{(3,5,7)}$	
		-149.3 to $-152.5^{(1)}$	

Compound or species	Mineral name	G_f°, kcal/mol	H_f°, kcal/mol
$Mg(OH)_2^0$		$-183.91^{(2)}$	$-221.52^{(2)}$
MgO	periclase	$-136.04^{(3,5,6)}$	$-143.84^{(3,5,7)}$
		$-136.1^{(1)}$	$-143.8^{(1)}$
$Mg(OH)_2$	brucite	$-199.39^{(2,3,5,6,7)}$	$-221.14^{(2,3,6,7)}$
		$-199.21^{(4)}$	
		$-199.4^{(1)}$	$-221.2^{(1)}$
MgOHCl(c)		$-174.9^{(1)}$	$-191.2^{(1)}$
MgF_2	sallaite	$-255.8^{(2)}$	$-268.5^{(2)}$
		$-255.9^{(1)}$	$-268.6^{(1)}$
$MgFe_2O_4$	magnesioferrite	-314.8 to $-322.9^{(1)}$	-341.4 to $-349.9^{(1)}$
$MgAl_2O_4$	spinel	-518.9 to $-523.5^{(1)}$	-544.8 to $-553.3^{(1)}$
$MgCr_2O_4$	picrochromite	$-398.9^{(1)}$	-418.5 to $-426.3^{(1)}$
$Mg_3(AsO_4)_2(c)$		$-679.3^{(1)}$	-731.3 to $-739.2^{(1)}$
$MgTiO_3$	geikelite	$-354.6^{(1)}$	$-375.8^{(1)}$
$MgHPO_4^0$		$-375.01^{(5)}$	
		$-373.7^{(7)}$	
$MgHPO_4 \cdot 3H_2O$	newberryite	$-549.04^{(5)}$	
$MgKPO_4 \cdot 6H_2O(c)$		$-776.31^{(5)}$	
$MgNH_4PO_4(c)$		$-390^{(7)}$	
		-388.5 to $-390.0^{(1)}$	
$MgNH_4PO_4 \cdot 6H_2O$	struvite	$-731.24^{(5)}$	$-880.0^{(1)}$
		$-729.9^{(1)}$	
$Mg_3(PO_4)_2$	farringtonite	$-845.8^{(2)}$	$-903.6^{(2)}$
		-845.8 to $-904.0^{(1)}$	-903.6 to $-961.5^{(1)}$
$Mg_3(PO_4)_2 \cdot 8H_2O$	boberrite	$-1304.99^{(5)}$	
$Mg_3(PO_4)_2 \cdot 22H_2O(c)$		$-2096.01^{(5)}$	
$MgSO_4^0$		$-289.6^{(2,7)}$	$-324.1^{(2)}$
		-289.6 to $-298.6^{(1)}$	-324.1 to $-327.2^{(1)}$
MgS(c)		$-81.7^{(2)}$	$-82.7^{(2)}$
		-81.7 to $-83.6^{(1)}$	-82.7 to $-84.4^{(1)}$
$MgSO_4(c)$		$-279.8^{(2)}$	$-307.1^{(2)}$
		-278.6 to $-280.5^{(1)}$	-305.5 to $-307.1^{(1)}$
$MgSO_4 \cdot 6H_2O(c)$		$-629.1^{(2)}$	$-737.8^{(2)}$
$MgSO_4 \cdot 7H_2O$	epsomite	$-681.1^{(1)}$	$-809.8^{(1)}$
$MgSiO_3$	enstatite	$-348.7^{(1)}$	$-364.9^{(1)}$
$MgSiO_3$	clinoenstatite	$-349.46^{(2)}$	$-370.22^{(2)}$
		$-349.4^{(1)}$	$-370.3^{(1)}$
$Mg_{1.6}Fe_{0.4}SiO_4$	olivine	$-460.08^{(5)}$	
Mg_2SiO_4	forsterite	$-491.5^{(2,3,5,6)}$	$-519.6^{(2)}$
		$-491.3^{(1)}$	$-519.8^{(1)}$
$Mg_2Si_3O_6(OH)_4$	sepiolite	$-1020.95^{(5)}$	
		$-1020.5^{(1)}$	
$Mg_4Si_6O_{15}(OH)_2 \cdot 6H_2O$	sepiolite	$-2211.19^{(3)}$	$-2418.00^{(1,3)}$
		$-2211.2^{(1)}$	

Compound or species	Mineral name	G_f°, kcal/mol	H_f°, kcal/mol
$Mg_3Si_2O_5(OH)_4$	chrysotile	$-965.1^{(2)}$	$-1043.4^{(2)}$
		-962.0 to $-965.1^{(1)}$	-1042.5 to $-1071.2^{(1)}$
$Mg_3Si_4O_{10}(OH)_2$	talc	$-1320.38^{(3,5)}$	$-1410.92^{(3)}$
		-1318.1 to $-1324.9^{(1)}$	-1410.9 to $1481.9^{(1)}$
$Mg_3Si_4O_{10}(OH)_2 \cdot 2H_2O$	vermiculite	$-1422.85^{(5)}$	
$Mg_6Si_4O_{10}(OH)_8$	serpentine	$-1933.85^{(5)}$	
$Mg_3Si_2O_5(OH)_4$	serpentine	$-964.87^{(3)}$	$-1043.12^{(3)}$
		$-966.0^{(1)}$	$-1044.0^{(1)}$
$Mg_2Al_4Si_5O_{18}$	cordierite	$-2055^{(2)}$	$-21.77^{(2)}$
		$-2075.26^{(5)}$	-2177.0 to $-2204.5^{(1)}$
		-2055.0 to $-2088.9^{(1)}$	
$Mg_7Si_8O_{22}(OH)_2$	anthophyllite	$-2715.4^{(3)}$	$-2888.75^{(3)}$
		-2708.6 to $-2723.7^{(1)}$	-2882.0 to $-2898.1^{(1)}$
Mg_2SiO_4	forsteritic spinel	$-483.4^{(1)}$	$-513.3^{(1)}$
$Mg_2Al_4SiO_{10}$	sapphirine	$-1190.6^{(1)}$	$-1261.5^{(1)}$
$Mg_3Al_2Si_3O_{12}$	pyrope	-1415.1 to $-1428.8^{(1)}$	-1512.1 to $-1520.6^{(1)}$
$Mg_4Al_4Si_2O_{10}(OH)_8$	amesite	$-2013.0^{(1)}$	$-2167.1^{(1)}$
$Mg_5Al_2Si_3O_{10}(OH)_8$	7-Å clinochlore	$-1957.1^{(1)}$	$-2113.2^{(1)}$
$Mg_5Al_2Si_3O_{10}(OH)_8$	14-Å clinochlore	$-1961.70^{(3)}$	$-2116.96^{(3)}$
		-1961.7 to $-1968.9^{(1)}$	-2117.0 to $-2123.1^{(1)}$
$Mg_5Al_2Si_3O_{10}(OH)_8$	clinochlore	$-1974.3^{(1)}$	
$Mg_5Al_2Si_3O_{10}(OH)_8$	chlorite	$-1973.5^{(1)}$	$-2127.5^{(1)}$
$Mg_4Al_{10}Si_7O_{31}(OH)_4$	yoderite	$-4016.5^{(1)}$	$-4271.2^{(1)}$
$Mg_{0.167}Al_{2.33}Si_{3.67}O_{10}(OH)_2$	montmorillonite	$-1274.9^{(1)}$	$-1364.4^{(1)}$
$Mg_{0.49}Fe_{0.22}^{3+}Al_{1.71}Si_{3.81}O_{10}(OH)_2$	montmorillonite	$-1255.8^{(1)}$	

Mn (manganese)

Compound or species	Mineral name	G_f°, kcal/mol	H_f°, kcal/mol
α-Mn(c)		0.00	0.00
Mn^{2+}		$-54.52^{(2,4)}$	$-52.76^{(2)}$
		$-54.6^{(1)}$	$-52.9^{(1)}$
Mn^{3+}		$-20.26^{(5)}$	$-27.0^{(7)}$
		-19.6 to $-20.4^{(1)}$	$-24.0^{(1)}$
MnO_4^-		-105.2 to $-107.4^{(1)}$	-127.4 to $-129.7^{(1)}$
MnO_4^{2-}		-118.1 to $-120.4^{(1)}$	$-156.1^{(1)}$
$HMnO_2^-$		$-120.9^{(1)}$	
$MnHCO_3^+$		$-197.84^{(5)}$	
		$-195.0^{(1)}$	
$MnCO_3$	rhodochrosite	$-195.20^{(2,4)}$	$-213.7^{(2)}$
		$-195.1^{(1)}$	$-212.9^{(1)}$
$MnCO_3^0$		$-179.96^{(5,7)}$	$-213.9^{(1,7)}$
		$-179.6^{(1)}$	
$Mn(OH)^+$		$-97.8^{(3,5)}$	$-107.7^{(3)}$
		$-96.80^{(4)}$	
		$-97.1^{(1)}$	$-107.0^{(1)}$

Compound or species	Mineral name	G_f°, kcal/mol	H_f°, kcal/mol
$Mn(OH)_2$	pyrochroite	$-147.76^{(5)}$	$-167.3^{(1)}$
		$-146.99^{(4)}$	
		$-147.2^{(1)}$	
$Mn(OH)_2(amorph.)$		$-147.0^{(1,2)}$	$-166.2^{(1,2)}$
$MnOH^{2+}$		$-77.49^{(5)}$	
$Mn(OH)_3(c)$		$-181^{(7)}$	$-212^{(1,7)}$
		$-180.9^{(1)}$	
$Mn(OH)_3^-$		$-178.35^{(3,5)}$	
		$-177.87^{(4)}$	
$Mn(OH)_4^{2-}$		$-215.99^{(5)}$	
Mn_2OH^{3+}		$-152.45^{(5)}$	
$Mn_2(OH)_3^+$		$-247.69^{(5)}$	
MnO	manganosite	$-86.74^{(2,4)}$	
		$-86.7^{(1)}$	$-92.0^{(1)}$
MnO_2, γ	pyrolusite	$-111.18^{(2,4)}$	$-124.29^{(2)}$
		$-111.2^{(1)}$	$-124.3^{(1)}$
$\delta\text{-}MnO_{1.8}$	birnessite	$-108.59^{(3,5,7)}$	
		$-108.3^{(1)}$	
$\gamma\text{-}MnO_{1.9}$	nsutite	$-109.4^{(5,7)}$	
		$-109.1^{(1)}$	
Mn_2O_3	bixbyite	$-210.6^{(2,4)}$	$-229.2^{(2)}$
		$-211.2^{(1)}$	-228.9 to $-232.1^{(1)}$
Mn_3O_4	hausmannite	$-306.7^{(2,4)}$	$-331.7^{(2)}$
		$-306.3^{(1)}$	$331.5^{(1)}$
$\gamma\text{-}MnOOH$	manganite	$-133.2^{(1,3,5,7)}$	
$MnHPO_4(c)$		$-332.5^{(2)}$	
$Mn_3(PO_4)_2(precipitated)$		$-683.1^{(1)}$	-744.7 to $-771.0^{(1)}$
$Mn_3(PO_4)_2(c)$		$-692.96^{(5)}$	$-744.9^{(2)}$
$MnCl_2$	scacchite	$-105.3^{(1)}$	-113.0 to $-115.0^{(1)}$
$MnS(green)$		$-52.2^{(2)}$	$-51.2^{(2)}$
MnS	alabandite	$-52.16^{(3,5,6)}$	$-51.15^{(3,6)}$
		$-52.20^{(4)}$	
		-49.9 to $-52.2^{(1)}$	-48.8 to $-51.2^{(1)}$
MnS_2	hauerite	$-55.53^{(5)}$	$-58.5^{(1)}$
		$-55.5^{(1)}$	
$MnSO_4(c)$		$-228.83^{(2)}$	$-254.60^{(2)}$
		$-228.8^{(1)}$	$-254.5^{(1)}$
$MnSO_4^0$		$-232.5^{(2)}$	$-270.1^{(2)}$
$\alpha\text{-}MnSO_4 \cdot H_2O(c)$		$-289.11^{(5)}$	$-329.0^{(2)}$
$Mn_2(SO_4)_3(c)$		$-590.14^{(5)}$	
$MnSiO_3$	rhodonite	$-296.5^{(2)}$	$-315.7^{(2)}$
		$-296.7^{(1)}$	$-315.6^{(1)}$
Mn_2SiO_4	tephroite	$-390.1^{(2)}$	$-413.6^{(2)}$
		$-389.8^{(1)}$	$-413.3^{(1)}$
$MnWO_4$	huebnerite	$-287.8^{(1)}$	$-312.0^{(1)}$

Compound or species	Mineral name	G_f°, kcal/mol	H_f°, kcal/mol
Mo (molybdenum) (see also Ag, Ca, Fe, Pb)			
Mo(c)		0.00	0.00
Mo^{3+}		$-13.8^{(1,7)}$	
MoO_2^+		$-122.80^{(4)}$	
$H_2MoO_4^0$		$-227^{(7)}$	
MoO_4^{2-}		$-199.90^{(2)}$	$-238.5^{(1,2)}$
		-199.9 to $-205.4^{(1)}$	
$HMoO_4^-$		$-213.6^{(7)}$	
		$-205.62^{(5)}$	
		-207.1 to $-213.6^{(1)}$	
$H_2MoO_4^0$		$-211.08^{(5)}$	$-240.8^{(1)}$
		-209.6 to $-227.1^{(1)}$	
$H_2MoO_4(c)$		$-218.08^{(5)}$	$-250.0^{(1,2)}$
		$-283.7^{(1)}$	
$MoO_2(c)$		$-127.4^{(2,4)}$	$-140.76^{(2)}$
		-120.0 to $-127.4^{(1)}$	$-140.7^{(1)}$
MoO_3	molybdite	$-159.66^{(2,4)}$	$-178.08^{(2)}$
		-159.6 to $-162.0^{(1)}$	-178.1 to $-180.3^{(1)}$
MoS_2	molybdenite	$-54.0^{(2,4)}$	$-56.2^{(2)}$
		$-53.9^{(1)}$	$-55.9^{(1)}$
$Mo_3O_8(c)$		$-480.00^{(4)}$	
$MoO_3 \cdot H_2O(c)$		$-283.7^{(1)}$	
N (nitrogen)			
$N_2(g)$		0.00	0.00
N_2^0		$+2.994^{(7)}$	$-2.51^{(1)}$
		$+2.98$ to $+4.35^{(1)}$	
$NH_3(g)$	ammonia	$-3.94^{(1,2)}$	$-11.02^{(1,2)}$
		$-3.98^{(4)}$	
NO_2^-		-0.88 to $-8.2^{(1)}$	$-25.1^{(1)}$
NHO_2^0		$-12.1^{(1)}$	$-28.4^{(1)}$
NH_3^0	aqueous ammonia	$-6.35^{(1,2)}$	$-19.19^{(2)}$
			$-19.2^{(1)}$
NH_4^+		$-18.97^{(2,4)}$	$-31.67^{(2)}$
		$-18.9^{(1)}$	$-31.7^{(1)}$
NH_4OH^0		$-63.04^{(2)}$	$-87.05^{(2)}$
		$-63.0^{(1)}$	$-87.5^{(1)}$
$NH_4OH(l)$		$-60.74^{(2)}$	$-86.33^{(2)}$
$NO(g)$		$+20.69^{(2)}$	$+21.57^{(2)}$
		$+20.7^{(1)}$	$+21.6^{(1)}$
NO^0		$+24.41^{(5)}$	
$NO_2(g)$		$+12.26^{(2)}$	$+7.93^{(2)}$
		$+12.3^{(1)}$	$+7.99^{(1)}$
$NO_3(g)$		$+27.75^{(5)}$	

Compound or species	Mineral name	G_f°, kcal/mol	H_f°, kcal/mol
$N_2O_3(g)$		$+33.32^{(2)}$	$+20.01^{(2)}$
$N_2O_4(g)$		$+23.38^{(2)}$	$+2.19^{(2)}$
		$+23.4^{(1)}$	$+2.26^{(1)}$
$N_2O_5(g)$		$+27.5^{(2)}$	$+2.7^{(2)}$
$HNO_3(g)$		$-17.87^{(2)}$	$-32.28^{(2)}$
$HNO_3(l)$	nitric acid	$-19.31^{(2)}$	$-41.61^{(2)}$
		$-19.2^{(1)}$	$-41.5^{(1)}$
NO_3^-		$-26.61^{(2)}$	$-49.56^{(2)}$
		$-26.5^{(1)}$	$-49.4^{(1)}$
		$-25.99^{(4)}$	
HNO_3^0	nitric acid	$-26.46^{(2)}$	$-49.56^{(2)}$
		$-26.5^{(1)}$	$-49.5^{(1)}$
$NH_4NO_3(c)$		$-43.98^{(2)}$	$-87.37^{(2)}$
		$-43.9^{(1)}$	$-87.4^{(1)}$
$NH_4NO_3^0$		$-45.58^{(2)}$	$-81.23^{(2)}$
N_2O^0		$+24.90^{(2)}$	$+19.61^{(2)}$
NH_4Cl	sal ammoniac	$-48.51^{(2)}$	$-75.15^{(2)}$
		$-48.6^{(1)}$	$-75.2^{(1)}$
NH_4Cl^0		$-50.34^{(2)}$	$-71.62^{(2)}$
$(NH_4)_2SO_4$	mascagnite	$-215.56^{(2)}$	$-282.23^{(2)}$
		$-215.5^{(1)}$	$-282.2^{(1)}$
$NH_4VO_3(c)$		-212.2 to $-221.8^{(1)}$	$-251.7^{(1)}$
Na (sodium) (see also Al)			
$Na(c)$		0.00	0.00
Na^+		$-62.593^{(2)}$	$-57.39^{(2)}$
		$-62.6^{(1)}$	$-57.4^{(1)}$
$NaCl^0$		$-93.96^{(5,7)}$	$-97.302^{(5,7)}$
$NaCl$	halite	$-91.815^{(2)}$	$-98.268^{(2)}$
		$-91.8^{(2)}$	$-98.3^{(1)}$
$NaCO_3^-$		$-190.50^{(3,5,7)}$	$-223.7^{(1)}$
		-189.5 to $-190.5^{(1)}$	
$Na_2CO_3^0$		$-251.36^{(2)}$	$-276.62^{(2)}$
$Na_2CO_3(c)$		$-249.64^{(2)}$	$-270.24^{(2)}$
		$-250.2^{(1)}$	$-270.3^{(1)}$
$NaHCO_3^0$		$-202.88^{(3,5,7)}$	$-222.5^{(7)}$
		$-202.8^{(1)}$	-222.5 to $-225.6^{(1)}$
$NaHCO_3$	nahcolite	$-203.4^{(2)}$	$-227.25^{(2)}$
		-195.0 to $-203.6^{(1)}$	-218.3 to $-227.2^{(1)}$
$NaOH^0$		$-99.91^{(5)}$	$-112.36^{(2)}$
$NaOH(c)$		$-90.709^{(2)}$	$-101.723^{(2)}$
		$-90.6^{(1)}$	$-101.8^{(1)}$
$NaOH \cdot H_2O(c)$		$-149.0^{(7)}$	$-175.17^{(7)}$
		-149.0 to $-150.4^{(1)}$	$-175.4^{(1)}$

Compound or species	Mineral name	G_f°, kcal/mol	H_f°, kcal/mol
$NaSO_4^-$		$-241.26^{[3,5,7]}$	$-273.6^{[1]}$
		$-241.2^{[1]}$	
$Na_2SO_4(c)$	thenardite	$-303.59^{[2]}$	$-331.52^{[2]}$
		$-303.4^{[1]}$	$-331.5^{[1]}$
$NaSO_4^0$		$-303.16^{[2]}$	$-332.10^{[2]}$
$Na_2SO_4 \cdot 10H_2O(c)$		$-871.75^{[2]}$	$-1034.24^{[2]}$
$NaHSO_4(c)$		$-237.3^{[2]}$	$-269.01^{[2]}$
$Na_2SiO_3(c)$		$-349.19^{[2]}$	$-371.63^{[2]}$
		-341.1 to $-353.2^{[1]}$	-363.0 to $-375.2^{[1]}$
$NaNO_3^0$		$-89.20^{[2]}$	$-106.95^{[2]}$
$NaNO_3$	soda niter	$-87.73^{[2]}$	$-111.82^{[2]}$
		$-87.7^{[1]}$	$-111.8^{[1]}$
$Na_2Si_2O_5(c)$		$-555.0^{[2]}$	$-589.8^{[2]}$
$NaSi_{11}O_{20.5}(OH)_4 \cdot 3H_2O$	kenyaite	$-2603.5^{[1]}$	
Na_2S^0		$-104.7^{[2]}$	$-106.9^{[2]}$
$Na_2S(c)$		$-83.6^{[2]}$	$-87.2^{[2]}$
		-83.6 to $-86.6^{[1]}$	-87.2 to $-89.2^{[1]}$
$Na_2O(c)$		$-89.74^{[2]}$	$-99.00^{[2]}$
		$-90.0^{[1]}$	$-99.5^{[1]}$
NaF	villiaumite	$-129.902^{[2]}$	$-137.105^{[2]}$
		-129.3 to $-130.6^{[1]}$	-136.0 to $-137.8^{[1]}$
NaF^0		$-129.23^{[2]}$	$-136.89^{[2]}$
$NaBr(c)$		$-83.1^{[7]}$	$-86.030^{[7]}$
		$-83.3^{[1]}$	$-86.2^{[1]}$
$NaI(c)$		$-56.7^{[7]}$	$-68.84^{[7]}$
		-56.7 to $-68.7^{[1]}$	$-69.0^{[1]}$
$Na_2CO_3 \cdot H_2O$	thermonatrite	$-307.35^{[2,7]}$	$-342.08^{[2]}$
		$-307.5^{[1]}$	$-342.2^{[1]}$
Na_3AlF_6	cryolite	$-751.6^{[1]}$	$-790.9^{[1]}$
$NaAlCO_3(OH)_2$	dawsonite	$-426.9^{[1]}$	$-469.4^{[1]}$
$Na_2B_4O_7 \cdot 4H_2O$	kernite		$1077.3^{[1]}$
$Na_2B_4O_7 \cdot 10H_2O$	borax	$-1318.4^{[1]}$	$-1501.9^{[1]}$
$Na_2CO_3 \cdot 10H_2O$	natron	$-819.36^{[2,7]}$	$-975.46^{[2]}$
		$-819.6^{[1]}$	$-975.5^{[1]}$
$Na_2CO_3 \cdot 10H_2O$	mirabilite	$-871.5^{[3]}$	$-1034.2^{[3]}$
		$-871.6^{[1]}$	$-1034.3^{[1]}$
$Na_2CO_3 \cdot 7H_2O(c)$		$-648.8^{[2]}$	$-764.81^{[2]}$
$NaHCO_3 \cdot Na_2CO_3 \cdot 2H_2O$	trona	$-570.40^{[3,7]}$	$-641.7^{[1]}$
		$-569.7^{[1]}$	
$Na_3PO_4(c)$		$-427.55^{[2]}$	$-458.27^{[2]}$
$Na_3PO_4^0$		$-431.3^{[2]}$	$-477.5^{[2]}$
$Na_2UO_4(c)$			$-500.9^{[1]}$
$Na_2UO_4(\alpha,c)$		-422.7 to $-424.9^{[1]}$	-451.0 to $-452.5^{[1]}$
$Na_3UO_4(c)$		$-453.9^{[1]}$	$-483.6^{[1]}$

Compound or species	Mineral name	G_f°, kcal/mol	H_f°, kcal/mol
Nb (niobium)			
Nb(c)		0.00	0.00
Nb^{3+}		$-76.0^{(1,4,7)}$	
NbO_3^-		$-222.78^{(4)}$	
NbO(c)		$-90.5^{(2,4)}$	$-97.0^{(2)}$
		-90.5 to $-93.7^{(1)}$	-97.0 to $-100.3^{(1)}$
$NbO_2(c)$		$-177.0^{(2)}$	$-190.3^{(2)}$
		$-190.30^{(4)}$	
		$-176.5^{(1)}$	$-190.2^{(1)}$
$Nb_2O_4(c)$		$-362.4^{(1,7)}$	$-387.8^{(1,7)}$
$Nb_2O_5(c)$		$-422.1^{(2)}$	$-454.0^{(2)}$
		$-453.99^{(4)}$	
		$-421.9^{(1)}$	$-453.7^{(1)}$
Ni (nickel)			
Ni(c)		0.00	0.00
Ni^{2+}		$-10.9^{(2,4)}$	$-12.9^{(2)}$
		$-11.0^{(1)}$	-12.8 to $-15.3^{(1)}$
NiO	bunsenite	$-50.6^{(1,2,4)}$	$-57.3^{(1,2)}$
$NiO_2(c)$		$-47.5^{(1,7)}$	
$Ni_3O_4(c)$		$-170.1^{(1)}$	
$Ni(OH)_2(c)$		$-106.9^{(2,4)}$	$-126.6^{(2)}$
		-106.9 to $-109.7^{(1)}$	-126.6 to $-128.6^{(1)}$
$Ni(OH)_2^0$		$-86.1^{(2)}$	$-122.8^{(2)}$
$NiOH^+$		$-54.40^{(4)}$	
		-52.6 to $-55.9^{(1)}$	$-68.8^{(1)}$
$HNiO_2^-$		$-83.46^{(7)}$	
		$-83.5^{(1)}$	
$Ni(OH)_3(c)$		$-129.5^{(1,7)}$	$-160^{(2)}$
γ-NiS(c)		$-27.3^{(1)}$	$-162.1^{(1)}$
α-NiS(c)	millerite	$-19.0^{(2,4)}$	$-19.6^{(2)}$
		-17.7 to $-20.6^{(1)}$	-19.6 to $-20.3^{(1)}$
$Ni_3S_2(c)$	heazlewoodite	$-47.1^{(1,2)}$	$-48.5^{(1,2)}$
$NiCO_3(c)$		$-146.4^{(2,4)}$	$-158.7^{(7)}$
		-146.3 to $-147.0^{(1)}$	-158.7 to $-164.7^{(1)}$
$Ni_3O_4 \cdot 2H_2O(c)$		$-283.5^{(1)}$	
$NiO_2 \cdot 2H_2O(c)$		$-164.8^{(1)}$	
$Ni_2O_3 \cdot H_2O(c)$		$-168.9^{(1)}$	
$NiSO_4(c)$		$-181.6^{(2,4)}$	$-208.63^{(2)}$
		-181.6 to $-184.9^{(1)}$	-208.6 to $-213.0^{(1)}$
$NiSO_4^0$		$-188.9^{(2)}$	$-230.2^{(2)}$
$NiSO_4 \cdot 6H_2O(green, \alpha)$	retgersite (tetrag.)	$-531.7^{(1,2)}$	$-641.21^{(2)}$
			-641.2 to $-645.0^{(1)}$
$NiSO_4 \cdot 6H_2O(blue, \beta,c)$	(monocl.)	$-531.0^{(1,7)}$	$-638.7^{(2)}$

Compound or species	Mineral name	G_f°, kcal/mol	H_f°, kcal/mol
			-638.7 to $-642.5^{(1)}$
$NiCl_2(c)$		$-61.9^{(1)}$	$-73.0^{(1)}$
$NiSO_4 \cdot 7H_2O$	morenosite	$-588.4^{(1)}$	$-711.3^{(1)}$
$NiFe_2O_4$	trevorite	$-232.2^{(1)}$	$-258.1^{(1)}$
$NiAl_2O_4(c)$		$-434.8^{(1)}$	-457.9 to $-463.6^{(1)}$
$NiSO_3(c)$		$-269.6^{(1)}$	
Ni_2SiO_4	olivine	$-308.1^{(1)}$	$-333.8^{(1)}$
Ni_2SiO_4	spinel	$-306.2^{(1)}$	
$Ni_2SiO_4(c)$		-307.4 to $-314.8^{(1)}$	-334.3 to $-341.7^{(1)}$

O (oxygen)

$O_2(g)$		0.00	0.00
$O_2^0(aqueous)$		$+3.91^{(1)}$	$-2.80^{(1)}$
O_2^-		$+12.9^{(1)}$	
$O_3(g)$	ozone	$+39.0^{(5)}$	

Os (osmium)

$Os(c)$		0.00	0.00
OsS_2		$-27.49^{(4)}$	
$Os(OH)_4(c)$		$-160.97^{(4)}$	
$H_2OsO_5^0$		$-128.81^{(4)}$	
$HOsO_5^-$		$-112.38^{(4)}$	
OsO_4^{2-}		$-89.25^{(4)}$	

P (phosphorus) (see also Na, Al, Ca, Mg, K, Mn, Fe, etc.)

α-P(white)		0.00	0.00
$P_2(g)$		$+24.7^{(1)}$	$+34.2^{(1)}$
HPO_3^0		$-215.8^{(5)}$	$-234.8^{(5)}$
HPO_3^{2-}		$-191.07^{(5)}$	$-233.8^{(7)}$
		$-194.0^{(1)}$	-231.6 to $-233.8^{(1)}$
$H_2PO_3^-$		$-200.33^{(5)}$	$-231.7^{(1)}$
		$-202.3^{(1)}$	
$H_3PO_3^0$		$-202.38^{(5)}$	$-232.2^{(7)}$
		$-204.8^{(1)}$	-230.6 to $-232.2^{(1)}$
PO_4^{3-}		$-245.18^{(5,7)}$	$-306.9^{(7)}$
		$-243.5^{(3,4)}$	$-305.3^{(3)}$
		$-243.8^{(1)}$	$-305.6^{(1)}$
HPO_4^{2-}		$-261.33^{(3,5,7)}$	$-308.83^{(3)}$
		$-260.31^{(4)}$	
		$-260.6^{(1)}$	$-309.2^{(1)}$
$H_2PO_4^-$		$-271.57^{(3,5,7)}$	$-310.6^{(3,7)}$
		$-270.14^{(4)}$	
		$-270.4^{(1)}$	$-310.2^{(1)}$
$H_3PO_4^0$	phosphoric acid	$-274.0^{(3,5,7)}$	$-308.1^{(3,7)}$
		$-273.07^{(4)}$	

Compound or species	Mineral name	G_f°, kcal/mol	H_f°, kcal/mol
$H_3PO_4(c)$		-273.1 to -274.2[1]	-307.9 to -308.2[1]
		-267.5[2]	-305.7[2]
		-266.6[1]	-304.7[1]
$(H_2PO_4)_2^{2-}$		-543.22[5]	
PH_3^0		$+6.06$[4]	
$P_2O_5(\text{orthorhomb.})$		-328.1[1]	-359.7 to -363.5[1]
$P_2O_5(\text{hexagonal})$		-322.4[1]	-356.6[1]
Pb (lead)			
$Pb(c)$		0.00	0.00
Pb^{2+}		-5.85[2,4,5,6,7]	-0.4[2,6]
		-5.83[1]	-0.38[1]
PbO_3^{2-}		-66.3[1]	
PbO_4^{4-}		-67.4[1]	
$HPbO_2^-$		-80.9[1,4]	
Pb^{4+}		$+72.3$[1,7]	
$PbCO_3$	cerussite	-149.9[1,2,6,7]	-167.2[2,6,7]
		-149.5[4]	-167.5[1]
$Pb_2CO_3Cl_2$	phosgenite	-227.94[5]	
$Pb_3(CO_3)_2(OH)_2$	hydrocerussite	-409.08[5]	-457.5[1]
		-406.1 to -409.7[1]	
$PbCl^+$		-39.45[5]	
		-39.4[1]	
$PbCl_2^0$		-68.57[2]	-80.3[2]
		-71.0 to -71.9[1]	
$PbCl_3^-$		-102.31[5]	
		-101.69[4]	
		-101.9[1]	
$PbCl_4^{2-}$		-133.27[5]	
$PbCl_2$	cotunnite	-75.98[2]	-85.90[1,2]
		-75.1[1]	
$PbClOH$	laurionite	-93.5[1]	-110.1[1]
PbF^+		-75.86[5]	
PbF_2^0		-139.11[2]	-159.4[2]
		-144.86[5]	
PbF_3^-		-214.35[5]	
PbF_4^{2-}		-281.85[5]	
$PbF_2(c)$		-147.5[2]	-158.7[2]
		-147.5 to -150.0[1]	-158.5 to -162.2[1]
$Pb(OH)^+$		-52.09[4,5]	
		-52.6 to -54.1[1]	
$Pb(OH)_2^0$		-95.06[5]	
		-95.8 to -97.6[1]	
$Pb(OH)_3^-$		-137.64[5,6]	

Compound or species	Mineral name	G_f°, kcal/mol	H_f°, kcal/mol
$Pb(OH)_4^{2-}$		$-137.6^{(1)}$	
		$-178.78^{(5)}$	
$Pb_2(OH)^{3+}$		$-59.76^{(5)}$	
$Pb_3(OH)_4^{2+}$		$-211.86^{(5)}$	
$Pb_4(OH)_4^{4+}$		$-211.85^{(5)}$	
$Pb_6(OH)_8^{4+}$		$-429.45^{(5)}$	
$Pb(OH)_2(c)$		$-108.1^{(2,5)}$	$-123.3^{(2)}$
		-100.6 to $-108.1^{(1)}$	$-123.0^{(1)}$
$PbWO_4$	stolzite	$-243.9^{(1)}$	$-268.1^{(1)}$
$PbCrO_4$	crocoite	-195.2 to $-203.6^{(1)}$	-218.7 to $-225.2^{(1)}$
$PbMoO_4$	wulfenite	$-227.62^{(5)}$	$-265.8^{(7)}$
		-226.7 to $-231.7^{(1)}$	-250.7 to $-265.8^{(1)}$
$Pb_2O_3(c)$		$-98.42^{(1,7)}$	
PbO(yellow)	massicot	$-45.00^{(1,2,5,7)}$	$-51.77^{(2,7)}$
PbO(red)	litharge	$-45.20^{(1,2,5,6,7)}$	$-52.37^{(1,2,6,7)}$
		$-45.16^{(4)}$	
PbO_2	plattnerite	$-51.95^{(2,4,5,6,7)}$	$-66.21^{(2,6,7)}$
		-51.5 to $-52.3^{(1)}$	$-66.0^{(1)}$
Pb_3O_4	minium	$-143.75^{(2,5)}$	$-171.7^{(2)}$
		$-143.69^{(4)}$	
		-143.6 to $-147.6^{(1)}$	-171.7 to $-175.6^{(1)}$
$PbO \cdot PbCO_3(c)$		$-195.72^{(2,7)}$	$-220.0^{(7)}$
		$-195.4^{(1)}$	$-219.7^{(1)}$
$PbHPO_4^0$		$-272.16^{(5)}$	
$PbH_2PO_4^+$		$-279.80^{(5)}$	
$PbHPO_4(c)$		$-283.55^{(5)}$	$-234.5^{(7)}$
$Pb(H_2PO_4)_2(c)$		$-563.04^{(5)}$	
$Pb_3(PO_4)_2(c)$		$-568.57^{(5)}$	$-620.3^{(1,7)}$
		-565.0 to $-581.4^{(1)}$	
$Pb_5(PO_4)_3Cl$	chloropyromorphite	$-910.59^{(5)}$	
		-838.1 to $-906.2^{(1)}$	
$Pb_5(PO_4)_3F$	fluoropyromorphite	$-930.68^{(5)}$	
$Pb_5(PO_4)_3OH$	hydroxypyromorphite	$-907.39^{(5)}$	
$PbSO_4^0$		$-187.42^{(5)}$	
$Pb(SO_4)_2^{2-}$		$-366.53^{(5)}$	
PbS	galena	$-22.9^{(2,5,6,7)}$	$-24.0^{(2,6)}$
		$-23.59^{(4)}$	
		-22.2 to $-23.1^{(1)}$	-22.5 to $-24.0^{(1)}$
$PbSe$	clausthalite	$-24.3^{(1)}$	$-24.6^{(1)}$
$PbTe$	altaite	$-16.6^{(1)}$	$-16.7^{(1)}$
$PbSO_4$	anglesite	$-194.42^{(2,5,6)}$	$-219.75^{(1,2,6,7)}$
		$-194.35^{(4)}$	
		$-194.2^{(1)}$	
$PbSO_4 \cdot PbO$	lanarkite	$-246.70^{(1,5)}$	$-282.5^{(7)}$

Compound or species	Mineral name	G_f°, kcal/mol	H_f°, kcal/mol
			$-280.0^{(1)}$
$PbSO_4 \cdot 2PbO(c)$		$294.00^{(5)}$	
$PbSO_4 \cdot 3PbO(c)$		$-341.2^{(5)}$	
$PbSeO_4(c)$		$-120.7^{(1)}$	-145.6 to $-147.9^{(1)}$
$PbSiO_3$	alamosite	$-253.86^{(2,5,6)}$	$-273.83^{(2,6)}$
		$-253.5^{(1)}$	$-273.5^{(1)}$
$Pb_2SiO_4(c)$		$-299.4^{(1,2,5)}$	$-325.8^{(2)}$
			$-326.0^{(1)}$
Pd (palladium)			
$Pd(c)$		0.00	0.00
Pd^{2+}		$+42.2^{(1,2)}$	$+35.6^{(1,2)}$
		$+42.49^{(4)}$	
$PdO(c)$		$-14.4^{(1,7)}$	$-20.4^{(1,2)}$
		$-15.31^{(4)}$	
$PdO_3(c)$		$+24.1^{(1,7)}$	
$Pd(OH)_2(c)$		$-72.0^{(1,7)}$	$-94.4^{(4)}$
			-92.1 to $-94.4^{(1)}$
$Pd(OH)_4(c)$		$-126.2^{(1,7)}$	$-171.1^{(2)}$
			$-169.4^{(1)}$
$PdS(c)$		$-16.01^{(2,4)}$	$-18.^{(2)}$
$PdS_2(c)$		$-17.81^{(2,4)}$	$-19.4^{(2)}$
$Pd_4S(c)$		$-16.^{(2)}$	$-16.^{(2)}$
Po (polonium)			
$Po(c)$		0.00	0.00
Po^{2+}		$+16.97^{(4)}$	
Po^{4+}		$+70.00^{(4)}$	
$Po(OH)_4(c)$		$-130.00^{(4)}$	
$Po(OH)_2^{4+}$		$-113.05^{(4)}$	
$PoS(c)$		$-0.96^{(4)}$	
$PoO_2(c)$		$-46.60^{(4)}$	
Pt (platinum)			
$Pt(c)$		0.00	0.00
Pt^+		$+60.9^{(2)}$	
Pt^{2+}		$+54.8^{(4,7)}$	
		$+44.4$ to $+60.9^{(1)}$	
$Pt(OH)_2(c)$		$-68.2^{(1,4,7)}$	$-84.1^{(2)}$
			-84.1 to $-87.2^{(1)}$
PtS	cooperite	$-18.2^{(2)}$	$-19.5^{(2)}$
		$-18.3^{(1)}$	$-19.6^{(1)}$
		$-21.6^{(4)}$	
$PtS_2(c)$		$-23.8^{(1,2)}$	$-26.0^{(1,2)}$
		$-25.6^{(4)}$	
$PtO_2(c)$		$-20.0^{(4)}$	

Compound or species	Mineral name	G_f°, kcal/mol	H_f°, kcal/mol
Ra (radium)			
Ra(c)		0.00	0.00
Ra^{2+}		−134.20[4]	
$RaSO_4$(c)		−326.39[4]	
$RaCO_3$(c)		−271.69[4]	
Rb (rubidium)			
Rb(c)		0.00	0.00
Rb^+		−67.87[2]	−60.03[1,2]
		−67.4 to −69.7[1]	
RbCl(c)		−97.47[2]	−104.05[2]
Rb_2SO_4(c)		−314.76[2]	−343.12[2]
$Rb_2SO_4^0$		−313.71[2]	−337.38[2]
$RbNO_3$(c)		−94.61[2]	−118.32[2]
$RbNO_4^0$		−94.48[2]	−109.59[2]
Rb_2CO_3(c)		−251.2[2]	−271.5[2]
		−250.0[1]	−270.2[1]
$Rb_2CO_3^0$		−261.9[2]	−281.90[2]
RbOH(c)		−87.1[1,7]	−99.95[1,2]
$Rb(OH)^0$		−105.05[7]	−113.9[7]
Rb_2O(c)		−69.5[7]	−81.[2]
			−78.9[1]
Rb_2S(c)		−80.6[1]	−83.2 to −86.2[1]
$RbHCO_3$(c)		−206.4[2]	−230.2[2]
$RbHCO_3^0$		−208.13[2]	−225.42[2]
Re (rhenium)			
Re(c)		0.00	0.00
Re_2O_3(c)		−138.64[4]	
ReO_2(c)		−87.95[4]	
ReO_3(c)		−127.96[4]	
ReO_4^-		−165.99[4]	
ReS_2(c)		−38.10[4]	
Rh (rhodium)			
Rh(c)		0.00	0.00
Rh_2O(c)		−20.0[4]	
Rh_2O_3(c)		−50.0[4]	
Rh^{3+}		+55.3[4]	
Ru (ruthenium)			
Ru(c)		0.00	0.00
RuS_2(c)		−45.03[4]	
RuO_2(c)		−60.49[4]	

Compound or species	Mineral name	G_f°, kcal/mol	H_f°, kcal/mol
$Ru(OH)_2^{2+}$		$-53.01^{[4]}$	
RuO_4^{2-}		$-73.28^{[4]}$	
RuO_4^-		$-59.78^{[4]}$	
S (sulfur) (see also Ca, Fe, Mn, Ni, Co, etc.)			
S(c, rhombic)		0.00	0.00
$S_2(g)$		$+18.96^{[2,6]}$	$+30.68^{[2,6]}$
		$+19.0^{[1]}$	$+30.7^{[1]}$
S^{2-}		$+20.51^{[3,4,5,6]}$	$+7.9^{[3,6]}$
		$+20.6^{[1]}$	$+7.81^{[1]}$
HS^-		$+2.93^{[3,5,6,7]}$	$-4.2^{[3,6,7]}$
		$+2.89^{[1,4]}$	$-4.19^{[1]}$
$H_2S(g)$		$-8.02^{[2,5,6]}$	$-4.93^{[2,6]}$
		$-8.00^{[1]}$	$-4.90^{[1]}$
H_2S^0		$-6.66^{[2,3,5,6]}$	$-9.5^{[2,3,6]}$
		$-6.65^{[1,4]}$	$-9.57^{[1]}$
SO(g)		$-4.741^{[2]}$	$+1.496^{[2]}$
		-4.73 to $+12.8^{[1]}$	-1.48 to $+19.0^{[1]}$
$SO_2(g)$		$-71.66^{[2,3,5,6]}$	$-70.94^{[1,2,6,7]}$
		$-71.8^{[1]}$	
SO_2^0		$-71.871^{[2,5]}$	$-77.194^{[2]}$
SO_3^{2-}		$-116.24^{[1,2,3,4,5,7]}$	$-151.9^{[1,2,3,7]}$
HSO_3^-		$-126.05^{[3,5,7]}$	$-149.88^{[3,7]}$
		$-126.1^{[1,4]}$	$-149.8^{[1]}$
$H_2SO_3^0$		$-128.56^{[2,5,7]}$	$-145.51^{[2,7]}$
		$-128.6^{[1]}$	$-145.5^{[1]}$
SO_4^{2-}		$-177.75^{[2,3,5,6,7]}$	$-217.32^{[2,3,6]}$
		$-177.8^{[1]}$	$-217.3^{[1]}$
		$-177.95^{[4]}$	
HSO_4^-		$-180.69^{[2,3,4,5,6]}$	$-212.08^{[2,3,6]}$
		$-180.4^{[1]}$	$-212.2^{[1]}$
$H_2SO_4^0$		$-177.75^{[2,5,7]}$	$-217.32^{[2]}$
$H_2SO_4(l)$		$-164.938^{[2]}$	$-194.548^{[2]}$
		$-164.9^{[1]}$	$-194.5^{[1]}$
Sb (antimony)			
Sb(c)		0.00	0.00
$Sb_2(g)$		$+40.^{[7]}$	$+52.^{[7]}$
		$+39.9$ to $+44.7^{[1]}$	$+52.1$ to $56.3^{[1]}$
SbO^+		$-42.33^{[2,4]}$	
		$-42.2^{[1]}$	
SbO_2^+		$-65.5^{[1,7]}$	
SbO_3^-		$-122.9^{[1]}$	
SbO_2^-		$-81.32^{[2,4]}$	

Compound or species	Mineral name	G_f°, kcal/mol	H_f°, kcal/mol
		-81.3 to $-82.5^{(1)}$	
$HSbO_2^0$		$-97.4^{(1,2,4)}$	$-116.6^{(1,2)}$
$Sb_2O_4(c)$		$-190.2^{(2,4)}$	$-216.9^{(2)}$
		-165.9 to $-190.2^{(1)}$	-193.3 to $-216.9^{(1)}$
$Sb_2O_5(c)$		$-198.2^{(2,4)}$	$-232.3^{(2)}$
		-198.2 to $-206.7^{(1)}$	-232.3 to $-240.8^{(1)}$
Sb_4O_6	cubic senarmontite	$-298.0^{(7)}$	$-336.8^{(7)}$
		-298.0 to $-306.4^{(1)}$	-336.8 to $-334.4^{(1)}$
Sb_4O_6	ortho. valentinite	$-294.0^{(7)}$	$-338.7^{(1)}$
		-294.0 to $-302.0^{(1)}$	
$Sb(OH)_3(c)$		$-163.8^{(2,4)}$	
$Sb(OH)_3^0$		$-154.1^{(2)}$	$-184.9^{(2)}$
$SbCl_3(c)$		$-77.5^{(1)}$	$-91.3^{(1)}$
$SbF_3(c)$		$-199.8^{(1)}$	-217.2 to $-218.8^{(1)}$
SbS_3^{2-}		$-32.0^{(1,7)}$	
Sb_2S_3	stibnite	$-41.5^{(1,2,4)}$	$-41.8^{(1,2)}$
SbS_2^-		$-13.0^{(1,7)}$	
$Sb_2S_4^{2-}$		$-11.9^{(1)}$	$-26.2^{(1)}$
		$-23.78^{(4)}$	
Sc (scandium)			
$Sc(c)$		0.00	0.00
Sc^{3+}		$-140.2^{(2,4)}$	$-146.8^{(2)}$
		$-143.7^{(1)}$	-146.8 to $-151.0^{(1)}$
$Sc_2O_3(c)$		$-434.85^{(2,4)}$	$-456.22^{(1,2)}$
		$-434.8^{(1)}$	
$Sc(OH)^{2+}$		$-193.7^{(7)}$	$-205.9^{(1)}$
		$-191.49^{(4)}$	
		$-193.9^{(1)}$	
$Sc(OH)_3(c)$		$-294.8^{(2,4)}$	$-325.9^{(2)}$
		-293.5 to $-297.0^{(1)}$	-325.9 to $-328.7^{(1)}$
$Sc(OH)_4^-$		$-327.60^{(4)}$	
$ScF_3(c)$		$-371.8^{(2)}$	$-389.4^{(2)}$
Se (selenium)			
$Se(c, hexagonal, black)$		0.00	0.00
$Se_2(g)$		$+23.0^{(2)}$	$+34.9^{(1,2)}$
		$+21.2$ to $+23.0^{(1)}$	
Se^{2-}		$+30.9^{(1,2,4)}$	$+31.6^{(1,7)}$
			$+15.3^{(1)}$
$SeO_2(c)$		$-41.5^{(7)}$	$-53.86^{(1,2)}$
		$-41.2^{(1)}$	
$H_2SeO_3^0$		$-101.8^{(1,4,7)}$	$-121.29^{(1,2)}$
$HSeO_3^-$		$-98.34^{(1,4,7)}$	$-123.5^{(7)}$

Compound or species	Mineral name	G_f°, kcal/mol	H_f°, kcal/mol
			$-123.1^{(1)}$
SeO_3^{2-}		$-89.33^{(7)}$	$-122.39^{(7)}$
		-87.0 to $-89.3^{(1)}$	$-121.7^{(1)}$
		$-88.38^{(4)}$	
$H_2SeO_4^0$		$-105.42^{(1,7)}$	$-145.3^{(1,7)}$
$HSeO_4^-$		$-108.2^{(7)}$	$-143.1^{(7)}$
		$-107.9^{(1)}$	$-138.1^{(1)}$
		$-108.08^{(4)}$	
SeO_4^{2-}		$-105.42^{(7)}$	$-145.3^{(7)}$
		$-105.5^{(1)}$	$-143.2^{(1)}$
		$-105.47^{(4)}$	
$H_2Se(g)$		$+3.80^{(1)}$	$+7.10^{(1)}$
H_2Se^-		$+5.31^{(1)}$	$+4.59^{(1)}$
		$+3.80^{(4)}$	
HSe^-		$+10.5^{(1,4)}$	$+3.80^{(1)}$
Si (silicon) (see also Al, Ca, Fe, Mg, etc.)			
$Si(c)$		0.00	0.00
$SiO_2 \cdot nH_2O(amorph.)$		$-202.9^{(1)}$	$-215.6^{(1)}$
$SiO_2(amorph.)$		$-203.33^{(2,4)}$	$-215.94^{(2)}$
		$-203.1^{(1)}$	$-215.7^{(1)}$
SiO_2	chalcedony	$-204.3^{(1)}$	$-217.3^{(1)}$
SiO_2	coesite	$-203.44^{(5)}$	$-216.4^{(1)}$
		$-203.3^{(1)}$	
SiO_2	cristobalite	$-204.46^{(2)}$	$-217.37^{(2)}$
		$-204.2^{(1)}$	$-217.0^{(1)}$
SiO_2	quartz	$-204.75^{(2)}$	$-217.72^{(2)}$
		$-204.7^{(1,4)}$	$-217.7^{(1)}$
SiO_2	tridymite	$-204.42^{(2)}$	$-217.27^{(2)}$
		$-204.2^{(1)}$	$-217.1^{(1)}$
SiO_2	stishovite	$-191.8^{(1)}$	$-205.8^{(1)}$
SiO_4^{4-}		$-249.99^{(5)}$	
$HSiO_4^{3-}$		$-267.86^{(5)}$	
$H_2SiO_4^{2-}$		$-283.1^{(3)}$	
$H_3SiO_4^-$		$-299.1^{(3)}$	$-342.1^{(3)}$
		$-295.53^{(4)}$	
		$-299.5^{(1)}$	$-340.7^{(1)}$
$H_4SiO_4^0$		$-312.66^{(1,3,5)}$	$-348.3^{(6)}$
		$-314.53^{(4)}$	$-349.0^{(1)}$
$SiF_4(g)$		$-375.9^{(1,6)}$	$-386.0^{(1,6)}$
SiF_6^{2-}		$-367.6^{(1)}$	$-572.0^{(1)}$
Sn (tin)			
$Sn(white, c)$		0.00	0.00

Compound or species	Mineral name	$G_f^°$, kcal/mol	$H_f^°$, kcal/mol
Sn^{2+}		$-6.5^{(2,4)}$	$-2.1^{(2)}$
		$-6.49^{(1)}$	$-2.12^{(1)}$
Sn^{4+}		$+0.6^{(4,6,7)}$	$+7.3^{(6)}$
		$+0.14^{(1)}$	$+7.29^{(1)}$
SnO	romarchite	$-61.4^{(1,2,4)}$	$-68.3^{(2)}$
			$-68.4^{(1)}$
SnO_2	cassiterite	$-124.2^{(1,2,4)}$	$-138.8^{(1,2)}$
$HSnO_2^-$		$-98.0^{(1)}$	
		$-48.0^{(4)}$	
SnO_3^{2-}		$-137.4^{(1,4)}$	
$Sn_2O_3^{2-}$		$-141.1^{(1)}$	
$SnO \cdot OH^+$		$-113.29^{(4)}$	
$Sn(OH)^+$		$-60.6^{(7)}$	$-68.4^{(1)}$
		$-60.90^{(4)}$	
		$-60.5^{(1)}$	
$Sn(OH)_2(c)$		$-117.5^{(2)}$	$-134.1^{(2)}$
		$-117.6^{(1)}$	$-138.3^{(1)}$
$Sn(OH)_4(c)$		$-227.5^{(1,7)}$	$-265.3^{(2)}$
			$-270.5^{(1)}$
$SnCl_4(g)$		$-105.2^{(2)}$	$-112.2^{(2)}$
$SnCl_4^0$		$-124.9^{(2)}$	$-152.5^{(2)}$
$SnCl_2(c)$		$-72.2^{(1,7)}$	$-77.7^{(1,2)}$
$SnCl_2^0$		$-71.6^{(2)}$	$-78.8^{(2)}$
SnS	herzenbergite	$-23.5^{(2)}$	$-24.^{(2)}$
		$-25.02^{(4)}$	
		-23.5 to $-25.0^{(1)}$	-23.9 to $-25.5^{(1)}$
SnS_2	berndtite	$-38.0^{(6,4)}$	$-40.0^{(6)}$
		$-34.7^{(1)}$	$-36.7^{(1)}$
$Sn(OH)_6^{2-}$		$-310.5^{(1,6,7)}$	
$Sn(SO_4)_2(c)$		$-346.8^{(7)}$	$-393.4^{(7)}$
		-344.6 to $-346.8^{(1)}$	-389.4 to $-393.4^{(1)}$
SnF_6^{2-}		$-419.9^{(1)}$	$-474.7^{(1)}$
Sr (strontium)			
$Sr(c)$		0.00	0.00
Sr^{2+}		$-133.71^{(2,4)}$	$-130.45^{(2)}$
		$-133.5^{(1)}$	$-130.4^{(1)}$
$Sr(NO_3)_2(c)$		$-186.46^{(2)}$	$-233.80^{(1,2)}$
		$-186.3^{(1)}$	
$Sr(NO_3)_2^0$		$-186.93^{(2)}$	$-229.57^{(2)}$
$SrO(c)$		$-134.3^{(2,4)}$	$-141.5^{(2)}$
		$-134.0^{(1)}$	$-141.2^{(1)}$
$SrSO_4$	celestite	$-320.5^{(1,2,4)}$	$-347.3^{(1,2)}$

Compound or species	Mineral name	G_f°, kcal/mol	H_f°, kcal/mol
$SrSO_4^0$		$-311.68^{(2)}$	$-347.77^{(2)}$
$SrCO_3$	strontianite	$-272.5^{(2,4)}$	$-291.6^{(2)}$
		$-272.1^{(1)}$	$-291.4^{(1)}$
$SrCO_3^0$		$-259.88^{(2)}$	$-292.29^{(2)}$
$Sr(OH)_2(c)$		$-207.8^{(1,4,7)}$	$-229.2^{(2)}$
			$-229.3^{(1)}$
$Sr(OH)^+$		$-172.39^{(4)}$	
		$-175.3^{(1)}$	$-186.8^{(1)}$
$SrCl_2(c)$		$-186.7^{(1,2)}$	$-198.1^{(1,2)}$
$SrF_2(c)$		$-278.4^{(2)}$	$-290.7^{(2)}$
		$-278.1^{(1)}$	$-290.5^{(1)}$
$Sr_3(PO_4)_2(c)$		$-932.1^{(1,7)}$	$-985.4^{(2)}$
			-985.4 to $-987.3^{(1)}$
$SrHPO_4(c)$		$-403.6^{(2)}$	$-435.4^{(2)}$
		-399.7 to $-403.6^{(1)}$	-431.3 to $-435.4^{(1)}$
$SrBr_2(c)$		$-166.5^{(1)}$	$-177.5^{(1)}$
$SrS(c)$		-97.4 to $-111.8^{(1)}$	-108.1 to $-112.9^{(1)}$
$Sr_3SiO_5(c)$		$-690.0^{(1)}$	-709.9 to $-724.4^{(1)}$
$SrSiO_3(c)$		$-370.4^{(2)}$	$-390.5^{(2)}$
		$-372.9^{(1)}$	$-390.6^{(1)}$
$Sr_2SiO_4(c)$		$-523.7^{(2)}$	$-550.8^{(2)}$
		-523.7 to $-532.4^{(1)}$	-550.0 to $-556.6^{(1)}$
$SrWO_4(c)$		$-366.^{(2)}$	$-391.9^{(2)}$
		$-366.2^{(1)}$	-391.9 to $-398.3^{(1)}$
Ta (tantalum)			
$Ta(c)$		0.00	0.00
TaO_2^+		$-201.39^{(4)}$	
$Ta_2O_5(c)$		$-456.8^{(2,4)}$	$-489.0^{(2)}$
		$-456.9^{(1)}$	$-489.1^{(1)}$
Te (tellurium)			
$Te(c)$		0.00	0.00
Te^{2-}		$+42.6^{(1)}$	$+24.2^{(1)}$
Te_2^{2-}		$+38.9^{(1)}$	
Te^{4+}		$+52.38^{(1,7)}$	
$Te_2(g)$		$+29.0^{(7)}$	$+41.0^{(7)}$
		$+28.1^{(1)}$	$+40.2^{(1)}$
HTe^-		$+37.7^{(7)}$	
$H_2Te(g)$		$+33.1^{(7)}$	$+23.8^{(2)}$
H_2Te^0		$+21.4^{(1)}$	$+18.6^{(1)}$
		$+34.1^{(4,7)}$	
TeO_2	tellurite	$-64.6^{(2,4)}$	$-77.1^{(2)}$
		$-64.9^{(1)}$	$-77.0^{(1)}$

Compound or species	Mineral name	G_f°, kcal/mol	H_f°, kcal/mol
$H_6TeO_6(c)$		$-245.04^{(1,7)}$	$-310.4^{(1,2)}$
TeO_3^{2-}		$-93.79^{(7)}$	$-127.3^{(1)}$
		$-93.6^{(1)}$	
$H_2TeO_3(c)$		$-114.36^{(1,7)}$	$-146.5^{(1)}$
		$-113.84^{(4)}$	
$H_2TeO_3^0$		$-76.2^{(2)}$	
$HTeO_3^-$		$-104.34^{(7)}$	$-133.0^{(1)}$
		$-108.1^{(1)}$	
$HTeO_4^-$		$-123.27^{(1,7)}$	
TeO_4^{2-}		$-109.1^{(1,7)}$	
$H_2TeO_4^0$		$-131.66^{(1,7)}$	
$TeOOH^+$		$-61.6^{(1)}$	
$HTeO_2^+$		$-62.5^{(1)}$	
$TeCl_6^{2-}$		$-137.4^{(1)}$	
$Te(OH)_3^+$		$-118.57^{(4)}$	
Th (thorium)			
$Th(c)$		0.00	0.00
Th^{4+}		$-168.5^{(2,5)}$	$-183.8^{(2)}$
		-168.6 to $-175.2^{(1)}$	-181.6 to $-183.8^{(1)}$
ThO_2	thorianite	$-279.35^{(1,2,4)}$	$-293.12^{(2)}$
			$-293.2^{(1)}$
$Th(OH)^{3+}$		$-220.00^{(4)}$	
$Th(OH)_4(c)$	"soluble"	$-379.1^{(1,7)}$	$-421.5^{(1,7)}$
$Th(OH)_2^{2+}$		$-272.68^{(4)}$	
$Th(OH)_4(c)$		$-382.2^{(1)}$	$-423.6^{(1)}$
$Th(OH)_6^{2-}$		$-445.0^{(4)}$	
$ThS_2(c)$		$-148.3^{(2)}$	$-149.7^{(2)}$
$Th_2S_3(c)$		$-257.4^{(2)}$	$-259.0^{(2)}$
		$-257.6^{(1)}$	-259.1 to $262.0^{(1)}$
$Th(SO_4)^{2+}$		$-353.90^{(4)}$	
$ThF_4(c)$		$-477.30^{(2)}$	$-499.90^{(2)}$
$ThCl_4(c)$		$-261.6^{(2)}$	$-283.6^{(2)}$
Ti (titanium) (see also Ca, Mg, Fe)			
$Ti(c)$		0.00	0.00
TiO^{2+}		$-138.00^{(6,7,4)}$	
		$-137.9^{(1)}$	
TiO_2^{2+}		$-111.7^{(1,7)}$	
$TiO(c)$		$-117.0^{(1)}$	$-124.1^{(1)}$
$\alpha\text{-}TiO(c)$		$-118.3^{(2)}$	$-124.2^{(2)}$
		-116.9 to $-118.3^{(1)}$	$-124.1^{(1)}$
TiO_2	rutile	$-212.6^{(2)}$	$-225.8^{(2)}$
		$-212.5^{(1)}$	$-225.7^{(1)}$

Compound or species	Mineral name	G_f°, kcal/mol	H_f°, kcal/mol
TiO_2	anatase	$-211.4^{(2,4)}$	$-224.6^{(2)}$
		$-211.1^{(1)}$	$-224.4^{(1)}$
TiO_2	brookite		$-225.1^{(1)}$
TiO_2(c, hydrated)		$-196.3^{(1)}$	$-207.0^{(1)}$
TiS_2(c)		$-78.9^{(6)}$	$-80.0^{(6)}$
Ti^{2+}		$-75.1^{(1,7)}$	
Ti^{3+}		$-83.6^{(1,7)}$	
$HTiO_3^-$		$-228.5^{(1)}$	
$Ti(OH)_3$(c)		$-250.9^{(1,7)}$	
Ti_2O_3(c)		$-342.8^{(2,4)}$	$-363.5^{(2)}$
		$-342.6^{(1)}$	$-363.3^{(1)}$
Ti_3O_5(c)		$-553.87^{(4)}$	
		$-553.7^{(1)}$	$-587.7^{(1)}$
Ti_4O_7(c)		$-768.0^{(1)}$	$-813.7^{(1)}$
$TiO(OH)_2$(c)		$-253.0^{(1,4)}$	
$TiCl_3$(c)		$-156.3^{(1)}$	$-172.3^{(1)}$
$FeTiO_3$	ilmenite	$-268.9^{(7)}$	$-288.5^{(7)}$
		$-277.0^{(1)}$	$-295.5^{(1)}$
Tl (thallium)			
Tl(c)		0.00	0.00
Tl^+		$-7.74^{(4)}$	
$Tl(OH)_2^+$		$-58.48^{(4)}$	
$Tl(OH)^{2+}$		$-3.80^{(4)}$	
Tl_2O(c)		$-35.20^{(4)}$	
Tl_2O_3(c)		$-74.50^{(4)}$	
Tl_2O_4(c)		$-82.98^{(4)}$	
$Tl(OH)_3$(c)		$-121.18^{(4)}$	
Tl_2S(c)		$-22.39^{(4)}$	
U (uranium) (see also Cs, Na)			
U(c)		0.00	0.00
U^{3+}		$-124.4^{(7)}$	$-123.0^{(7)}$
		-113.6 to $-124.4^{(1)}$	-116.9 to $-123.0^{(1)}$
U^{4+}		$-126.9^{(2)}$	$-141.3^{(2)}$
		-126.9 to $-139.5^{(1)}$	-141.1 to $-146.7^{(1)}$
UO_2^+		$-237.6^{(2)}$	$-247.4^{(7)}$
		$-231.0^{(1)}$	$-247.1^{(1)}$
UO_2^{2+}		$-227.9^{(2)}$	$-243.7^{(2)}$
		$-228.8^{(1)}$	$-244.2^{(1)}$
		$-227.66^{(4)}$	
UO_2^0		$-230.1^{(2)}$	
$(UO_2)_3(OH)_5^+$		$-945.2^{(1)}$	
$UO_2CO_3^0$		$-367.4^{(1)}$	$-406.4^{(1)}$
$UO_2(CO_3)_2^{2-}$		-502.9 to $-506.8^{(1)}$	

Compound or species	Mineral name	G_f°, kcal/mol	H_f°, kcal/mol
$UO_2(CO_3)_3^{4-}$		$-502.97^{[4]}$	
		$-640.0^{[7]}$	
		$-638.4^{[1]}$	
		$-635.57^{[4]}$	
$UO_2(CO_3)_2(H_2O)_2^{2-}$		$-622.0^{[1,7]}$	
$UO_2SO_4^0$		$-413.7^{[7]}$	$-467.3^{[7]}$
		$-411.4^{[1]}$	$-456.6^{[1]}$
UO_2	uraninite	$-246.6^{[1,2,4]}$	$-259.3^{[1,2]}$
$UO_3(c)$		$-273.0^{[1]}$	
$UO_3(\alpha,c)$			$-291.4^{[1]}$
$UO_3(\beta)$	orthorhomb. orange-red	$-273.0^{[1]}$	$-291.8^{[1]}$
$UO_3(\gamma)$	orthorhomb.	$-274.1^{[1]}$	$-292.6^{[1]}$
$UO_3 \cdot H_2O(\alpha,c)$			$-365.2^{[1]}$
$UO_3 \cdot H_2O(\beta,c)$		$-333.4^{[1]}$	$-366.6^{[1]}$
$UO_3 \cdot H_2O(\epsilon,c)$			$-366.0^{[1]}$
$UO_3 \cdot 2H_2O$		$-389.8^{[1]}$	$-436.6^{[1]}$
$U(OH)^{3+}$		$-193.5^{[7]}$	$-204.1^{[7]}$
		-182.7 to $-193.5^{[1]}$	-197.7 to $-204.1^{[1]}$
$U(OH)_3(c)$		$-263.2^{[1,7]}$	
$U(OH)_4(c)$		$-351.6^{[1,7]}$	
$U(OH)_5^-$		$-389.77^{[4]}$	
$UO_2(OH)_2(c)$		-322.6 to $-334.7^{[1]}$	$-366.8^{[1]}$
$U_3O_8(c)$		$-805.35^{[4]}$	
$UO_2(OH)_2 \cdot H_2O$	schoepite	$-390.6^{[1]}$	$-436.7^{[1]}$
$UCl_3(c)$		-191.0 to $-197.0^{[1]}$	-207.1 to $-213.0^{[1]}$
$UCl_4(c)$		$-222.1^{[1]}$	$-234.5^{[1]}$
$UF_3(c)$		$-344.2^{[2]}$	$-360.6^{[2]}$
UF_4	monoclinic	$-437.4^{[2]}$	$-459.1^{[2]}$
		-435.8 to $-438.3^{[1]}$	-457.5 to $-450.0^{[1]}$
$UF_6(c)$		$-494.4^{[2]}$	$-525.1^{[2]}$
$UOF_2(c)$		$-341.5^{[2]}$	$-358.3^{[2]}$
$UO_2(NO_3)_2(c)$		$-264.1^{[2]}$	$-322.5^{[2]}$
β-$UO_2SO_4(c)$		$-402.4^{[2]}$	$-441.0^{[2]}$
UO_2CO_3	rutherfordine	$-373.5^{[2,4]}$	$-404.2^{[2]}$
		-373.4 to $-337.0^{[1]}$	-404.2 to $-405.4^{[1]}$
$USiO_4$	coffinite	$-452.0^{[1]}$	$-478.0^{[1]}$
		$-445.0^{[4]}$	
$MgUO_4(c)$		$-418.2^{[2]}$	$-443.9^{[2]}$

TRANSURANIUM ELEMENTS

Np (neptunium)

Np(c)		0.00	0.00

Compound or species	Mineral name	G_f°, kcal/mol	H_f°, kcal/mol
Np^{3+}		-123.59[4]	
Np^{4+}		-120.20[4]	
$Np(OH)_5^-$		-384.08[4]	
$NpO_2(c)$		-244.22[4]	
NpO_2^{+1}		-218.69[4]	
NpO_2^{2+}		-190.20[4]	
$NpO_2(OH)(amorph.)$		-297.57[4]	
$Np(CO_3)_2(c)$		-389.63[4]	
$NpO_2(OH)_2CO_3^{2-}$		-424.05[4]	
$Np_2O_3(c)$		-346.08[4]	
$Np_2O_5(c)$		-481.12[4]	
$NpO_2(OH)_2(c)$		-294.57[4]	
Pu (plutonium)			
$Pu(c)$		0.00	0.00
Pu^{3+}		-138.15[4]	
Pu^{4+}		-114.96[4]	
$Pu_2S_3(c)$		-233.99[4]	
$PuO_2(c)$		-238.53[4]	
$Pu(OH)_5^-$		-378.10[4]	
PuO_2^-		-203.10[4]	
$PuO_2(OH)_2CO_3^{2-}$		-413.99[4]	
$Pu(OH)_4(c)$		-340.82[4]	
$Pu_2(CO_3)_3(c)$		-697.44[4]	
Am (americium)			
$Am(c)$		0.00	0.00
Am^{3+}		-143.19[4]	
$Am_2(CO_3)_3(c)$		-716.11[4]	
$Am(OH)_3(c)$		-279.16[4]	
$Am(OH)_3(amorph.)$		-289.38[4]	
$AmO_2(c)$		-210.43[4]	
Am^{4+}		-89.20[4]	
$Am_2O_3(c)$		-385.97[4]	
$Am(OH)_5^-$		-356.58[4]	
$Am(OH)^{2+}$		-189.65[4]	

End of Transuranium Elements

V (vanadium) (see also N)			
$V(c)$		0.00	0.00
V^{2+}		-51.9[1]	-54.3[1]
V^{3+}		-57.8[1]	-62.9[1]
VO^{2+}		-106.7[1,2,4]	-116.3[2]
			-116.6[1]

Compound or species	Mineral name	G_f°, kcal/mol	H_f°, kcal/mol
VO_2^+		$-142.6^{(7)}$	$-155.4^{(1)}$
		$-140.3^{(1,4)}$	
VO_3^-		$-137.28^{(4)}$	
$V_4O_9^{2-}$		$-665.3^{(1)}$	
$V_2O_7^{4-}$		$-410.85^{(4)}$	
$V_{10}O_{28}^{6-}$		$-1862.4^{(1)}$	
$HV_{10}O_{28}^{5-}$		$-1840.8^{(4)}$	
		-1840.8 to $-1875.3^{(1)}$	$-2077.9^{(1)}$
$H_2V_{10}O_{28}^{4-}$		$-1845.84^{(4)}$	
		-1845.8 to $-1875.2^{(1)}$	
VO_4^{3-}		$-214.87^{(4)}$	
HVO_4^{2-}		$-233.0^{(2,4)}$	$-277.0^{(2)}$
$H_2VO_4^-$		$-244.0^{(2,4)}$	$-280.6^{(2)}$
$VO(c)$		$-96.6^{(1,2)}$	$-103.2^{(1,2)}$
$V_2O_2(c)$		$-189.0^{(1)}$	
V_2O_3	karelianite	$-272.3^{(1,2,4)}$	$-291.3^{(1,2)}$
$V_2O_4(c)$		$-315.1^{(1,2,4)}$	$-341.1^{(1,2)}$
$V_2O_5(c)$		$-339.3^{(1,2,4)}$	$-370.6^{(1,2)}$
$V_3O_5(c)$		$-434.^{(2)}$	$-462.^{(2)}$
$V(OH)_3(c)$		$-218.0^{(1,7)}$	
$V(OH)^{2+}$		$-112.8^{(7)}$	
		-110.5 to $-112.8^{(1)}$	
$VO(OH)_2(c)$		$-213.6^{(7)}$	
		$-212.8^{(1)}$	
$NH_4VO_3(c)$		$-221.8^{(7)}$	
$VCl_2(c)$		$-97.0^{(1)}$	$-108.0^{(1)}$
$VCl_3(c)$		$-122.2^{(1)}$	$-138.8^{(1)}$
W (tungsten) (see also Ca, Fe, Mn, Pb)			
$W(c)$		0.00	0.00
$WO_2(c)$		$-127.61^{(1,2,4)}$	$-140.94^{(1,2)}$
WO_3(yellow, c)		$-182.67^{(2)}$	$-201.45^{(2)}$
		$-182.5^{(1)}$	$-200.8^{(1)}$
$WO_3(\alpha, c)$		$-182.6^{(1,4)}$	$-201.5^{(1)}$
$W_2O_5(c)$		$-360.9^{(1,7)}$	$-337.9^{(1,7)}$
WO_4^{2-}		$-220.00^{(7,4)}$	$-257.1^{(2)}$
		$-222.6^{(1)}$	$-256.8^{(1)}$
WS_2	tungstenite	$-46.2^{(7)}$	$-50.^{(2)}$
		$-71.20^{(4)}$	
		-46.2 to $-71.2^{(1)}$	-46.3 to $-71.3^{(1)}$
Y (yttrium)			
$Y(c)$		0.00	0.00
Y^{3+}		$-165.8^{(2,4)}$	$-172.9^{(2)}$

Compound or species	Mineral name	G_f°, kcal/mol	H_f°, kcal/mol
		$-164.1^{(1)}$	$-167.8^{(1)}$
$Y(OH)_3(c)$		$-307.1^{(7)}$	$-339.5^{(7)}$
		$-308.6^{(4)}$	
		-307.1 to $-310.3^{(1)}$	-339.5 to $-342.0^{(1)}$
$Y(OH)_4^-$		$-341.2^{(4)}$	
$Y_2O_3(c)$		$-434.19^{(2,4)}$	$-455.38^{(2)}$
		$-434.2^{(1)}$	$-455.4^{(1)}$
$Y_2O_3(cubic)$		$-434.2^{(1)}$	$-455.4^{(1)}$
$Y_2(SO_4)_3(c)$		$-866.8^{(2)}$	
$Y_2(CO_3)_3(c)$		$-752.4^{(2,4)}$	
Zn (zinc)			
$Zn(c)$		0.00	0.00
Zn^{2+}		$-35.14^{(2,4)}$	$-36.78^{(2)}$
		$-35.2^{(1)}$	$-36.7^{(1)}$
$ZnCO_3$	smithsonite	$-174.85^{(2,4)}$	$-194.26^{(2)}$
		$-174.9^{(1)}$	$-194.3^{(1)}$
$ZnCl^+$		$-67.14^{(5)}$	
		$-65.8^{(1)}$	
$ZnCl_2^0$		$-97.88^{(2)}$	$-116.68^{(2)}$
		$-96.5^{(1)}$	
$ZnCl_3^-$		$-129.98^{(5)}$	
		$-129.2^{(1)}$	
$ZnCl_4^{2-}$		$-160.94^{(5)}$	
		$-159.2^{(1)}$	
$ZnCl_2(c)$		$-88.296^{(2)}$	$-99.20^{(2)}$
		$-88.3^{(1)}$	$-99.3^{(1)}$
$ZnB_2(c)$		$-74.5^{(1)}$	$-78.5^{(1)}$
$ZnF_2(c)$		$-170.5^{(2)}$	$-182.7^{(2)}$
$ZnFe_2O_4$	franklinite	$-256.54^{(5)}$	
$ZnMoO_4$	zinc molybdate	$-241.15^{(5)}$	
$ZnNO_3^+$		$-62.37^{(5)}$	
$Zn(NO_3)_2^0$		$-88.36^{(2)}$	$-135.90^{(2)}$
$Zn(OH)$		$-81.39^{(5)}$	
		$-78.90^{(4)}$	
		$-79.2^{(1)}$	
$HZnO_2^-$		$-109.26^{(2)}$	
		$-109.42^{(4)}$	
		-109.2 to $-110.9^{(1)}$	
ZnO_2^{2-}		$-91.85^{(2,4)}$	
		-91.8 to $-93.0^{(1)}$	
$Zn(OH)_2^0$		$-110.33^{(2)}$	$-146.72^{(2)}$
		-124.9 to $-128.0^{(1)}$	
$Zn(NH_3)_4^{2+}$		-72.2 to $-73.5^{(1)}$	$-127.5^{(1)}$

Compound or species	Mineral name	G_f°, kcal/mol	H_f°, kcal/mol
$Zn(OH)_3^-$		$-167.49^{(5)}$	
		-165.9 to $-168.4^{(1)}$	
$Zn(OH)_4^{2-}$		$-205.23^{(2)}$	
		-205.2 to $-208.2^{(1)}$	
ZnO	zincite	$-76.08^{(2,4)}$	$-83.24^{(2)}$
		$-76.6^{(1)}$	$-83.7^{(1)}$
$Zn(OH)_2$(amorph.)		$-131.54^{(5)}$	$-153.5^{(2)}$
		$-131.7^{(1)}$	
α-$Zn(OH)_2$(c)		$-131.93^{(1,5,7)}$	
β-$Zn(OH)_2$(c)		$-132.31^{(2)}$	$-153.42^{(1,2)}$
		$-132.36^{(4)}$	
		$-132.6^{(1)}$	
γ-$Zn(OH)_2$(c)		$-132.38^{(2)}$	
		$-132.8^{(1)}$	
ε-$Zn(OH)_2$(c)		$-132.68^{(2)}$	$-153.74^{(1,2)}$
		$-133.0^{(1)}$	
$Zn_5(OH)_6(CO_3)_2$	hydrozincite	$-750.8^{(1)}$	$-828.3^{(1)}$
$ZnHPO_4^0$		$-301.72^{(5)}$	
$ZnH_2PO_4^+$		$-309.22^{(5)}$	
$Zn_3(PO_4)_2 \cdot 4H_2O$	hopeite	$-870.82^{(5)}$	
$ZnSe$	stilleite	$-34.7^{(7)}$	$-34.^{(7)}$
		$-39.6^{(1)}$	$-39.1^{(1)}$
α-ZnS	sphalerite	$-48.11^{(2,4)}$	$-49.23^{(2)}$
		$-48.3^{(1)}$	$-49.4^{(1)}$
β-ZnS	wurtzite	$-45.37^{(5)}$	$-46.04^{(2)}$
		-44.2 to $-46.2^{(1)}$	-45.3 to $-46.6^{(1)}$
$ZnSO_4$	zinkosite	$-209.0^{(2)}$	$-234.9^{(2)}$
		$-208.1^{(1)}$	$-234.5^{(1)}$
$ZnSO_4^0$		$-213.11^{(2)}$	$-254.10^{(2)}$
$ZnSO_4 \cdot H_2O$(c)		$-270.58^{(2)}$	$-311.78^{(2)}$
		$-270.2^{(1)}$	$-311.2^{(1)}$
$ZnSO_4 \cdot 6H_2O$	bianchite	$-555.64^{(2)}$	$-663.83^{(2)}$
		$-555.1^{(1)}$	$-663.7^{(1)}$
$ZnSO_4 \cdot 7H_2O$	goslarite	$-612.59^{(2)}$	$-735.6^{(2)}$
		$-612.3^{(1)}$	$-735.4^{(1)}$
$ZnO \cdot 2ZnSO_4$(c)		$-492.07^{(5)}$	
$Zn(OH)_2 \cdot ZnSO_4$(c)		$-351.47^{(5)}$	
$ZnSiO_3$(c)		$-274.8^{(7)}$	$-301.2^{(2)}$
		$-287.1^{(1)}$	$-306.9^{(1)}$
Zn_2SiO_4	willemite	$-364.06^{(1,2)}$	$-391.19^{(1,2)}$
$ZnWO_4$	sanmartinite	$-268.6^{(1)}$	$-294.6^{(1)}$
$ZnAl_2O_4$	gahnite		$-493.6^{(1)}$
$ZnTiO_4$(c)		$-366.7^{(1)}$	$-393.8^{(1)}$

Compound or species	Mineral name	G_f°, kcal/mol	H_f°, kcal/mol
Zr (zirconium)			
Zr(c)		0.00	0.00
Zr^{4+}		-142.0[7]	
		-141.00[4]	
		-125.4 to -142.0[1]	
ZrO_2	baddeleyite	-249.24[2,4]	-263.04[2]
		-248.6[1]	-262.7[1]
$ZrO(OH)_2(c)$		-311.5[1,7]	-338.0[1,7]
$Zr(OH)_4(c)$		-370.0[1,4,7]	-411.2[1,7]
$ZrSiO_4$	zircon	-458.7[2]	-486.0[2]
		-456.7 to -458.7[1]	-483.3 to -486.0[1]
ZrO^{2+}		-201.5[1]	
		-200.90[4]	
$HZrO_3^-$		-287.7[1,4]	

References

BROOKINS, D. G., 1988. *Eh–pH Diagrams for Geochemistry.* Springer-Verlag, Berlin, 176 pp.

DREVER, J. I., 1982. *The Geochemistry of Natural Waters.* Prentice-Hall, Upper Saddle River, NJ, 388 pp.

GARRELS, R. M., and C. L. CHRIST, 1965. *Solutions, Minerals, and Equilibria.* Harper & Row, New York, 450 pp.

HELGESON, H. R., J. M. DELANEY, H. W. NESBITT, and D. K. BIRD, 1978. Summary and critique of the thermodynamic properties of rock-forming minerals. *Amer. J. Sci.,* 278A, 229 pp.

KRAUSKOPF, K. B., 1979. *Introduction of Geochemistry.* McGraw-Hill, New York, 617 pp.

LINDSAY, W. L., 1979. *Chemical Equilibria in Soils.* Wiley, New York, 449 pp.

TARDY, Y., and R. M. GARRELS, 1974. A method of estimating the Gibbs energies of formation of layer silicates. *Geochim. Cosmochim. Acta,* **38**:1101–1116.

WAGMAN, D. D., W. H. EVANS, V. B. PARKER, R. H. SCHUMM, I. HALOW, S. M. BAILEY, K. L. CHURNEY, and R. L. NUTTALL, 1982. The NBS tables of chemical thermodynamic properties. Selected values for inorganic and C1 and C2 organic substances in SI units. *J. Phys. Chem. Ref. Data,* **11** (suppl. 2), 392 pp.

WEAST, R. C., W. J. ASTLE, and W. H. BEYER (Eds.), 1986. *CRC Handbook of Chemistry and Physics.* CRC Press, Boca Raton, Fl.

WOODS, T. L., and R. M. GARRELS, 1987. *Thermodynamic Values at Low Temperature for Natural Inorganic Materials: An Uncritical Summary.* Oxford University Press, New York, 242 pp.

Author Index

Subject Index

UNITS AND CONVERSIONS

1 Definition of International System of Units (SI)

Length: meter (m)
The meter is the length of the path traveled by light in vacuum during a time interval of 1/299 792 458 of a second.

Mass: kilogram (kg)
The kilogram is the unit of mass; it is the mass of the international prototype of the kilogram.

Time: second (s)
The second is the duration of 9 192 631 770 periods of the radiation corresponding to the transition between the two hyperfine levels of the ground state of the cesium-133 atom.

Electric current: ampere (A)
The ampere is the constant current that, if maintained in two straight parallel conductors of infinite length, of negligible circular cross section, and placed 1 meter apart in a vacuum, would produce between these conductors a force equal to 2×10^{-7} newton per meter of length.

Thermodynamic temperature: kelvin (K)
The kelvin, unit of thermodynamic temperature, is the fraction 1/273.15 of the thermodynamic temperature of the triple point of water.

Amount of substance: mole (mol)
The mole is the amount of substance of a system that contains as many elementary entities as there are atoms in 0.012 kilogram of carbon-12. When the mole is used, the elementary entities must be specified and may be atoms, molecules, ions, electrons, other particles, or specified groups of such particles.

2 SI Prefixes

deci (d): 10^{-1}	nano (n): 10^{-9}	deca (da): 10	giga (G): 10^9
centi (c): 10^{-2}	pico (p): 10^{-12}	hecto (h): 10^2	tera (T): 10^{12}
milli (m): 10^{-3}	femto (f): 10^{-15}	kilo (k): 10^3	peta (P): 10^{15}
micro (μ): 10^{-6}	atto (a): 10^{-18}	mega (M): 10^6	exa (E): 10^{18}

3 Physical Units and Conversions

A. Distance
1 kilometer (km)	$= 10^3$ meter (m)
1 centimeter (cm)	$= 10^{-2}$ m
1 millimeter (mm)	$= 10^{-3}$ m
1 micrometer (μm)	$= 10^{-6}$ m
1 statue mile	$= 5280$ ft
1 statute mile	$= 1609.344$ m
1 inch	$= 25.4$ mm (defined)
1 light year	$= 9.46055 \times 10^{12}$ km

B. Weight
1 kilogram (kg)	$= 10^3$ gram (g)
1 milligram (mg)	$= 10^{-3}$ g
1 microgram (μg)	$= 10^{-6}$ g
1 nanogram (ng)	$= 10^{-9}$ g
1 picogram (pg)	$= 10^{-12}$ g
1 tonne (t)	$= 10^3$ kg
1 short ton	$= 907.184$ kg
1 long ton	$= 1016.0469$ kg
1 pound (lb) (avoirdupois)	$= 0.453 592 37$ kg
1 ounce (oz) (avoirdupois)	$= 28.349 527$ g
1 ounce (oz) (troy)	$= 31.103 486$ g